14th Textile Bioengineering and Informatics Symposium Proceedings (TBIS 2021)

Advanced Material, Smart Wearables, Industry 4.0

Online
6-9 July 2021

Editors:

Yi Li
Xianyi Zeng
Timjun Li

ISBN: 978-1-7138-3559-2

Printed from e-media with permission by:

Curran Associates, Inc.
57 Morehouse Lane
Red Hook, NY 12571

Some format issues inherent in the e-media version may also appear in this print version.

Printed with permission by Curran Associates, Inc. (2022)

For permission requests, please contact Textile Bioengineering and Informatics Society (TBIS) at the address below.

Textile Bioengineering and Informatics Society (TBIS)
Unit 503, 5/F. Silvercord
Tower 2, 30 Canton Roa
Tsim Sha Tsui
Hong Kong

Phone: (852) 66382781

tbis@tbisociety.org

Additional copies of this publication are available from:

Curran Associates, Inc.
57 Morehouse Lane
Red Hook, NY 12571 USA
Phone: 845-758-0400
Fax: 845-758-2633
Email: curran@proceedings.com
Web: www.proceedings.com

PROCEEDINGS OF TBIS 2021

TABLE OF CONTENT

Track 1 E-textiles and Wearables Technologies

Track 2 Textile Biomedical Engineering

Track 3 Sustainable Textile Bioengineering

Track 4 Digitalization and Bioinformatics

Track 5 Textile Material Bionegineering

Track 6 Clothing Design and Aesthetics

Track 7 Textile Bioenegineering Technologies

Track 8 Clothing Bioengineering

TBIS

Textile Bioengineering and Informatics Symposium Proceedings

2021

ISSN:19423438

Edited by
Yi Li
Xianyi Zeng
Timjun Li

The 14th Textile Bioengineering and Informatics Symposium

July 6-9, 2021

TBIS

TBIS Secretariat
Email: tbis@tbisociety.org
www.tbisociety.org

Published by:
Textile Bioengineering and informatics Society

Welcoming Message

Academician Mu Yao
Honorary Chairman, ISC TBIS 2021
Chinese Academy of Engineering
Xi'an Polytechnic University
China

Hello! Hosts, Experts and Everyone!

Warmly celebrate the successful opening of TBIS 2021 Conference!

The COVID-19 pandemic is an unprecedented tragedy that has caused substantial impact on humanity and economy in a global scale with over 177 million confirmed cases and over 3.8 million deaths around the world as reported by WHO on 18 June 2021. In combating this pandemic, great efforts have been put into development of COVID-19 Vaccines with significant progress, resulting 287 candidate vaccines, 103 in clinical phase and 184 in pre-clinical phase. (WHO: https://www.who.int/publications/m/item/draft-landscape-of-covid-19-candidate-vaccines).

A number of vaccines have been used globally for vaccination with evidence of significant reduction of infection rates and death rates. Research is still ongoing into how much vaccines protect not only against the disease but also against infection and transmission. Concerns have been raised on the safety, side effects, effectiveness of different types of vaccines, particularly against SARS-CoV-2 variants. World economy and social life have been somewhat effected with very slow recovery.

The pandemic monitoring and control practices vary significantly among nations/countries in different regions of the world, resulting in substantially different impact on the safety, social life and economy in individual countries. The results demonstrated that non pharmaceutical technology innovations and measurements are as important and effective as the development and use of vaccines. This includes the effective execution of disease prevention strategies such as wearing masks, washing hands, and the implementation of social distancing and epidemic contact tracing etc.

Development of smart e-textile technologies enables wearable techniques for the real time and on-spots detection of viruses, blood pressure, breathing, heart rate, blood oxygen and other health-based benchmarks for more infection control, monitoring and accurate patient diagnosis.

This development is crucial in improving the process of detecting COVID-19 in asymptomatic patients. Therefore, systematic approach to pandemic management is essential, which need to integrate both pharmaceutical and non-pharmaceutical technology innovations. COVID-19 pandemic has caused a catastrophic effect on most countries in the world, which continues up-to-now. International cooperation under the leadership of the International Health Organization is urgently needed to overcome the difficulties and restore the world economy and normal life of humanity.

M. Yao

Welcoming Message

Academician Professor Meifang Zhu
TBIS International Executive Committee Chair, TBIS 2021
Member, Chinese Academy of Sciences
Director, State Key Laboratory for Modification of Chemical Fibers and Polymer Materials
Dean, College of Materials Science and Engineering of Donghua University
Director, Innovation Base of Advanced Fabrication Technology of Fiber Materials

As the director of the State Key Laboratory for Chemical Fiber and Polymer Materials (SKLFPM, Donghua University, China), it is our great honor to support the 14th International Conference of TBIS with the International Society of Textile Bioengineering and Information. TBIS has been successfully held for 13 consecutive sessions since 2008, which is an international grand forum in textile and clothing technology research. And the theme of each symposium is closely related to the global hot topics.

In light of the critical challenges arising from the global COVID-19 pandemic, textile industry plays an extremely important role in providing personal protective equipment for combatting coronavirus infections. Responding to these urgent global epidemic prevention needs, the conference aims to explore sustainable "healthy and green" solutions for combatting the global COVID-19 pandemic and future potential pandemics with science and technological innovations. By applying and integrating the latest progress in functional materials, personal protective equipment (PPEs), biosensors, smart materials and e-fibres, e-yarns and e-fabrics, nano-microelectronics with clinical medicine, epidemiology, wearable electronics, information technology and big data, we will work together to develop advanced cost-effective and practical technical solutions, as well as establish technology platforms for monitoring and controlling pandemics. **We are living in the same world; we should fight the pandemic together and manage this global crisis!**

As the only key state-level scientific research institution of fibers and textiles in China, the State Key Laboratory for Chemical Fiber and Polymer Materials (SKLFPM, Donghua University, China) has made great contributions to the development of chemical fiber industry in China. The SKLFPM was founded under the approval of State Development Planning Commission in 1992. In 2018, SKLFPM was rated as "Excellent State Key Laboratory". Last year, we worked together to establish the International Digital Health and Intelligent Materials Alliance (IDHIMIA) in TBIS 2020. This year, we work together again to organize the 10th International Conference on Advanced Fibers and Polymer Materials, which will be held in October 17-20, 2021, Shanghai, China (https://icafpm2021.scimeeting.cn). Here, we sincerely invite you to join and actively contribute your expertise and networking resources for the benefit of the world and humanity. At the same time. We are waves of the same sea, leaves of the same tree, flowers of the same garden! Finally, I want to say, SKY we share, together we stand. Many thanks for your participation and contributions!

Meifang Zhu

Welcoming Message

Professor Junwang Tang
Member of Academia Europaea,
Fellow of European Academy of Sciences,
Fellow of Royal Society of Chemistry,
Professor of Materials Chemistry & Engineering
Chemical Engineering
University College London

I am very pleased and honoured to welcome you all to attend the 14th Textile Bioengineering and Informatics Symposium, co-organised by the TBIS, Ensait and University of Manchester with the central theme of Advanced Materials, Smart Wearables and Industry 4.0, which reflects the urgent need for a sustainable and healthy society.

This year, the symposium has to be held virtually due to the outbreak of the COVID-19 pandemic. This pandemic presents us from the normal lives, including face to face meetings. This also reflect the importance and urgency of the symposium which targets to develop innovative materials, fibers, devices, and apparel products for protection, prevention, diagnosis and treatment of diseases.

I have been in the related research areas for two decades and fully agree that timely information exchange and brainstorming meetings are crucial to achieve breakthroughs and disruptive technologies. Furthermore, the interdisciplinary knowledge and collaboration is a catalyst for such development. I believe that the symposium would be one of the best sites to accelerate such multidisciplinary information exchange, to stimulate new ideas and to promote multidisciplinary science and technology collaborations.

Finally I sincerely invite you to actively engage in the discussions of the topics of the symposium online and wish you all a successful and enjoyable conference.

Welcoming Message

Professor Guangwei Fu
Chairman of China Textile Engineering Society
Deputy Secretary of the Party Committee of China Textile Information Center
Director of CNTAC Testing Center
Chairman and General Manager of the National Textile Fabric Museum,
Director of CTES Textile Industry Research Institute
Deputy Director of Standardization Committee of CNTAC
China

Dear honorary speakers and delegates.

Please allow me to extend my gratitude to TBIS 2021 on behalf of the China Textile Engineering Society.

As COVID-19 pandemic continues worldwide and has generated sustainable negative impact on economy and business activities and social stability in many countries on a global scale. It has been demonstrated that science and technology innovations textile industry has played a vital role in combating the SARS-Cov-2 viruses since the beginning of the pandemic. China has been the major manufacturing base to produce personal protective equipment such as facemasks and protective clothing as the supply to provide protections for people and healthcare workers in the world. The pandemic has changed the working environment, lifestyles, economies, and business practices in the world. Digitalization has been accelerated and sustainability against environmental changes and public health crisis have become the major concerns and focus of science and technology innovation.

I am glad to see that TBIS 2021 continues to focus on the key challenges from the pandemic with extension to industry 4.0, suitability and digital health to address these important issues. The workshops on textile sustainability with international collaboration of training PhD students and the international collaborative project on fashion business big data business model are timely for discussing and debating the science and key technical problems and deriving technical solutions by close international collaborations. The EU Horizon Project Application International Cooperation Forum is excellent initiative to promote collaborations between countries and continents to work together to end the pandemic and develop a sustainable and better world for all human beings. Thank you for your attention. I sincerely wish you have successful and fruitful conference.

Albert Fu.

Welcoming Message

Professor Xianyi Zeng
Conference Chair, OC TBIS 2021
Ecole Nationale Supérieure des Arts et
Industries Textiles (ENSAIT)
Roubaix, France

Welcome to TBIS 2021.

TBIS, an acronym for Textile Bioengineering and Informatics Symposium, is a well-established international research forum to exchange scientific views and new technology developments as well as promote cross-disciplinary research from material sciences, nanotechnology, fiber sciences, textile engineering, eco technology, textile and clothing CAD, with more emphasis on interactions of human/material/environment.

TBIS 2021 is the fourteenth in this conference series. It was planned to be held in Roubaix. Unfortunately, in the current severe health situation, we have to cancel travels and face-to-face activities and perform a webinar conference for all sessions. We sincerely hope that all participants will have opportunities to visit the city of Lille/Roubaix after the pandemic and enjoy our local fashion history and culture.

Founded in 1881, ENSAIT is a fashion and technology-oriented French higher education and research institution training engineers at master level. Its research activities cover MTC (Mechanics and Textile Composites), MTP (Multifunctional Textiles and Processes) and HCD (Human Centred Design). ENSAIT has been working through various European and national projects on textile sustainable development (bio-sourced materials, recycling, LCA, traceability, etc.), smart textile and wearable system development for various applications (health, security, sport, etc.), textile product and supply chain digitalization and intelligentalization. Most of the themes proposed by TBIS2021 are consistent with the research activities of ENSAIT. With the experiences and competences acquired through the past research activities, we can effectively contribute to technological innovations in textile industry.

Zeng

Welcoming Message

Professor Yi Li
Chairman, TBIS 2021 International Scientific Committee
Professor and Chair of Textile Science and Engineering
Department of Materials, School of Natural Sciences
The University of Manchester
Manchester, M13 9PL, UK

I am delighted to welcome you all to TBIS 2021. TBIS (Textile Bioengineering and Informatics Symposium) is a well-known international academic platform exploring the cross-disciplinary interfaces of materials, biology, textile science and engineering, computer science and informatics, and life sciences. This year, TBIS has a focused theme on "Advanced Materials • Smart Wearables • Industry 4.0".

The global COVID-19 pandemic has changed the world and global textile and fashion industry substantially. Over one year, we are still in the process of combating COVID-19 pandemic. The textile industry is continuously playing a pivotal role in providing personal protective equipment with functional textile materials and devices and remote diagnosis and contact tracing with smart e-textile wearables for combatting coronavirus infections. TBIS 2021 will review the status of the pandemic and rethinking the critical issues for combating a global pandemic. Thank you all for support and contributions. In TBIS 2021, we aim to explore sustainable "healthy and sustainable" solutions for combatting the global pandemic and future supply chain management with science and technological innovations. By applying and integrating the latest advances in advanced materials, digital technologies, smart materials, e-textiles with information technology and big data, we can work together to develop advanced cost-effective and practical technical solutions, as well as establish digital technology platforms for sustainable development. I sincerely invite you to join and actively contribute your expertise and networking resources for the benefit of the world and humanity.

With continuous success and enthusiastic support from authors and experts around the world, TBIS has been recognized internationally as a brand characterized by high-quality conferences with vision, creativity, excitement, and global networks. TBIS proceedings have been indexed by renowned worldwide scientific and academic databases such as EI Compendex, Scopus and CPCI/ISI Web of Knowledge. TBIS has established an effective platform for academic institutions and individual scholars, researchers and students to publish their research and communicate with peers and experts in the field to receive recognition for their work and increase its impact. TBIS is a peer-reviewed conference with the aim to provide a unique global platform for researchers and industrialists working in the whole supply chain from materials, fibers, textiles and apparel and technical textiles to services such as the medicine, healthcare, fashion, sports, firefighting and security, who have the common interests in providing high-quality products and services for human populations living in various environments and exposed to various pandemics. The cross-disciplinary feature of TBIS has created a unique platform to discuss, debate and promote globalized sustainable supply chains ecosystems for textile industry from advanced materials, smart manufacturing to digitalized services in big data economy. On behalf of the organizing committee and international scientific committee, we cordially welcome you to TBIS 2021 on -line conference and look forward to your active participation.

Yi Li

Welcoming Message

信州大学
SHINSHU UNIVERSITY

Dr. Kanji Kajiwara
Vice Chair, ISC TBIS 2021
Research Fellow
Faculty of Textile Science and Technology
Shinshu University
Ueda, Japan

Under the cloud of COVIT-19 pandemic, we are obliged to have TBIS 2021 through web again. On the occasion of TBIS 2020, many expected to have TBIS 2021 face-to-face, but we were too optimistic. COVIT-19 has taught us that our society is not safe, and we need more collaboration rather than competition in order to have a sustainable and safe society. TBIS provides an open platform to discuss our concerns with sustainable and safe society, as TBIS 2021 focuses on "Advanced Materials ▪ Smart Wearables ▪ Industry 4.0", which will play a key role for restructuring a sustainable and safe society in the next generation.

I should repeat the last-year's message again to emphasize the role of textile materials in the next generation. I recall one occasion when the comfort of the face mask was discussed at TBIS 2019. At that time, nobody thought the significance of face masks, and now the face mask is regarded as a simple yet vital PPE against coronavirus. We may further investigate to make face masks smarter to monitor our health by measuring vital signs of breathing/temperature/smell and even infections. Who imagined the face mask could be a vital tool for us to protect from the coronavirus infection? There are many other examples at TBIS symposia where we could not notice the significance of the presentations at that time but the topics inspired the scientist from other fields.

We should look around what other colleagues are investigating. A philosopher once told me that philosophy is a simple science to combine two unrelated subjects together and create a new concept. We all know the paradigm shift is a key for innovation. Don't you think that the innovative technology is a kind of philosophy?

Kanji Kajiwara

Welcoming Message

Professor Wei Lin Xu
Vice Chairmen, IEC TBIS 2021
President of Wuhan Textile University
Vice President of China Textile Engineering Society
Changjiang Scholar
China

Welcome, my fellow honorable experts and scholars.

First, please allow me to express my sincere gratitude to the TBIS society, the organization holding this conference, for providing such a precious opportunity for us to interact with each other in this delicate time. Since January 2020 when the COVID pandemic started, we have all walked through a long journey of uncertainty and uneasiness to hope and cooperation. Along the journey, many people and things have moved us, and many unforgettable experiences have been engraved in our hearts.

One thing I would like to share today is that in-light-of this pandemic, we can see science and technology innovations in textile industry have played critical roles in our battle against the virus with wide applications, including the disposable surgical masks, medical personal protective equipment, alcohol prep pad, nonwoven geotextile for blocking the sharp particles in the solum and protect the geomembrane, as well as hollow fiber membrane for making the extra-corporeal membrane oxygenation. The power of the modern textile technology innovations and its strategic significance have fully demonstrated.

Further, as a participant and witness of this challenging test of pandemic, I have experienced the warmth of the people in offering their hands to each other and together, combatting the virus. Up to now, the whole world is still struggling to fight the pandemic, we have no choice but to continue to work together to establish a global defense force against the pandemic and rebuilding world economy with sustainability.

Welcome to TBIS 2021. I would like to offer my best wishes to the great success of this conference and good health and happiness to you all.

TBIS 2021 Organizing and International Scientific Committees

Organizing Committee

Conference Chair:	Prof. Xianyi Zeng (France)		
Vice Chairmen:	Prof. Ludovic Koehl (France)	Prof. Vladan Koncar (France)	Prof. Francois Boussu (France)
Secretariat:	Ms Marion Houyvet (France)	Ms Mai Li (UK)	
Members:	Prof. Francois Boussu (France)	Asso Prof. Aurelie Cayla (France)	Asso Prof. Cedric Cochrane (France)
	Kuo Cao (UK)	Johanna Ferreira Da Silva (France)	Jin Duan (China)
	Prof. Xuemei Ding (China)	Asso Prof. Hayriye Gidik (France)	Dr. Junyan Hu (HongKong, China)
	Dr. Haibo Jia (UK)	Dr. Xiao Liao (China)	Zhangchi Liu (China)
	Prof. Xiaonan Luo (China)	Tim Li (HongKong, China)	Dorothee Mercier (France)
	Dr. Senthilkannan S. Muthu (Hong Kong, China)	Sandrine Pesse (France)	Weiguang Qi (China)
	Prof. Fabien Salaun (France)	Asso Prof. Xuyuan Tao (France)	Prof. Ruomei Wang (China)
	Prof. Xiongying Wu (China)	Dr. Zhong Wang (China)	Jiangpeng Xiang (China)
	Shan Xue (China)	Dr. Boan Ying (China)	Dr. Lei Yao (Hong Kong, China)
	Prof. Xin Zhang (China)	Prof. Yaolin Zhu (China)	

International Scientific Committee

Honorary Chairman:	Academician Mu Yao	China		
Chairman:	Prof. Yi Li	UK		
Vice Chair:	Prof. Kanji Kajiwara	Japan		
Members:	Prof. Seung Kook An (Korea)	Dr. Simon Annaheim (Switzerland)	Dr. Shaghaeygh Rezaei Arangdad	Prof. Aizheng Chen (China)
	Prof. Yan Chen (China)	Prof. Xuemei Ding (China)	Prof. Zhaoqun Du (China)	Prof. ElSayed Ahmed ElNashar (Egypt)
	Prof.A. Blanton Godfrey (USA)	Prof. Youfan Hu (China)	Aris Huang (China)	Prof. Shigeru Inui (Japan)
	Dr. Liya Ju (France)	Dr. KyoungOk Kim (Japan)	Dr. Dana Kremenakova (Czech Republic)	Prof. Gang Li (China)
	Prof.Budimir Mijovic (Croatia)	Prof. Rajesh Mishra (Czech Republic)	Prof. Xiumei Mo (China)	Prof. Guowen Song (USA)
	Prof. Tetsuya Sato (Japan)	Prof.Runjun Sun (China)	Prof. Masayuki Takatera (Japan)	Prof. David James Tyler (UK)
	Dr. Mohanapriya Venkataraman (Czech Republic)	Prof. Lijing Wang (Australia)	Dr. Xin Wang (Australia)	Prof.Xiaoqin Wang (China)
	Prof. Bingqing Wei (USA)	Dr. Long Wu (China)	Prof. Maobin Xie (China)	Dr. Lei Yao (Hong Kong, China)
	Dr.ChunhongZhu (Japan)	Dr.Malgorzata Zimniewska (Poland)	Prof. Yueqi Zhong (China)	Prof. Zijian Zheng (Hong Kong, China)

TBIS International Executive Committee

Organizing and Supporting Institutions

Co-organized by

ENSAIT

Textile Bioengineering and Informatics Society

The University of Manchester

Supported by

FBD_Model

FBD_Foundation

European Commission

SKLFPM, Donghua University

IDHIMIA

Beijing Institute of Fashion Technology

Technical University of Liberec

Xi'an Polytechnic University

Wuhan Textile University

Soochow University

Shinshu University

Zhejiang Sci-Tech University

Empa

RMIT University

IOWA State University

The Czech University of Life Sciences Prague

Association of British Chinese Professors

XiHua University

BEYOND Fashion Group

Best Pacific International Holdings Limited

Digital Clothing Limited

Golden Data Ltd

Haier IoC Research Institute

Ningbo Kechuang Assistant Technology Service Co., Ltd

Shenzhen Smart Clothing Technology Co, Ltd

Swift Chain

ZYPHR

UK-China Association for Talent, Science and Technology Exchange

Global Science Press

Cite as: Bilqees et al.,
TBIS 10.3993/tbis (2021).

A Flexible Bombyx Mori Silk Fibroin Film with Conductive Screen Printed AgNPs Pattern

Bilqees[a], Mujahid Mehdi[a], Muhammad Yousif[a], Wangkai Jiang[a], Yong Lin[b], Yuan Wei[b], Jianchen Hu[a*], Ke-Qin Zhang[a*]

[a]*National Engineering Laboratory for Modern Silk, College of Textile and Clothing Engineering, Soochow University, Suzhou 215123, China*
[b]*Printable Electronics Research Centre, Suzhou Institute of Nano-Tech and Nano-Bionics, Chinese Academy of Sciences, Suzhou 215123, China*

*Corresponding author's email: hujianchen@suda.edu.cn; kqzhang@suda.edu.cn

Abstract

The patterning of silver nanoparticles (AgNPs) on highly flexible silk fibroin film is subject to critical challenges, which have significantly limited their practical applications in printed electronics. This study describes a simple and efficient method to combine screen printing with vacuum dryer for patterning AgNPs network onto flexible bombyx mori silk fibroin (BSF) film. Conductive AgNPs based patterns were printed over the Si wafer through a screen printing technique. Then, the glycerol (30 wt%) added BSF solution was prepared. This BSF/Glc solution was poured over the AgNPs pattern and kept in a vacuum dryer for 12 hours. Finally, the conductive screen printed BSF/Glc/AgNPs film was peeled off, and the AgNPs based pattern was successfully printed onto the flexible BSF/Glc film. The effect of vacuum dryer, morphological structure, chemical properties of the films was investigated. EDX results confirmed the presence of Ag (silver) in conductive BSF/Glc/AgNPs film. The maximum sheet resistance achieved was 49 (Ω/sq) during 1000 bending cycles. Moreover, conductivity analysis showed that this conductive patterned film could quickly light up 8 LEDs when connected to a 9V battery.

Keywords: Printed Electronics; Bombyx Mori Silk Fibroin Film; Glycerol; Screen Printing; AgNPs.

1. Introduction

Despite its popularity as a luxury fabric in its natural fiber form, silk is now being studied in various material formats for high-tech applications, including films, nanofibers, gels, and sponges [1, 2]. Due to its specific physical, chemical, and biological properties, as well as its possible applications in the fields of tissue engineering, bio-sensing, electronic, optics, and photonics, silk fibroin (SF) film has gained increased interest as the primary form of regenerated silk materials [3, 4]. Furthermore, many materials have been used in the past to increase the flexibility and mechanical strength of silk [5]. Electronic passive components made of silk fibroin include energy storage precursors, dissoluble substrates for conformal bioelectronics, and resorbable substrates for compostable electronics and energy storage prototypes [6]. Silk fibroin has been suggested as an active component in resistive-switching memory devices and organic field-effect transistors by some researchers [7]. Tao et al. demonstrated that previously silk solution was used to make electronic and photonic devices using different techniques, like spin coating, nano-imprinting, inkjet printing [8]. Printed electronics is a form of electronics made using traditional printing technologies that has gotten much attention in recent years [9]. It enables the fabrication of massive electronics and hybrid systems on flexible printed substrates using inorganic, organic, and bio inspired materials, focusing on ease of integration [10]. By printing usable inks comprising soluble or dispersed materials, printed electronics has been investigated to create bendable and elastic electronic devices [11].

Printing electronic patterns over flexible biomaterial substrates have been an area of beneficial interest to the researchers due to their economical fabrication and feasibility of achieving multifunctional applications in large areas. In recent years, multiple printing technologies have been evolved over the years to pattern a wide variety of electronic materials. Also, conductive nanomaterials, namely copper nanoparticles [12], silver nanoparticles

(AgNPs) [13], nanowires (AgNWs) [14], carbon nanotubes [15], and graphene [16], are promising building blocks for printed electronics. Bombyx mori silk fibroin, due to its bio-sustainable and biodegradable nature, configurable mechanical properties, and excellent optical and electronic properties, has recently seen a huge increase in its use for applications in flexible electronic devices [17, 18]. Advances in silk fibroin (SF)-based flexible printed electronics could approach using biomaterials (natural or synthetic biopolymers) in high-performance design and high-biocompatible electronics in the future. Previously, spray-coating and spin-coating methods have been performed to fabricate silk fibroin based conductive films by incorporating different conductive nanomaterials. However, these methods cannot produce highly visual and impactful patterns or circuits, and their areas of applications in electronics are also limited.

Screen printing is being considered as one of the potential solution to this issue. It brings simplicity, affordability, speed, and adaptability to the fabrication process, making it faster and more flexible than other printing tools [19]. Screen printing, unlike many other manufacturing methods, does not necessitate a significant capital investment. Because of the low capital costs, compact designs for patterns, screen printing technology is the most widely utilized traditional mass-printing technique in flexible electronics due to its low waste and compatibility with a wide range of valuable inks and solvents [16]. With a squeegee moving forward in the printing process, ink is pressed via a decoration screen onto various substrates. A porous mesh, such as steel material and transparent fabric, is usually used for the screen.

For applications of printed electronics, flexible and high strength substrates have always played a vital role. In this way, our study is based on bombyx mori silk fibroin (BSF) film, whose flexibility has been firstly increased by the addition of glycerol (Glc). Chemical properties, along with the flexibility of BSF film, were also improved by the addition of Glc. Furthermore, screen printing of AgNPs has been carried out in this work. The as-prepared conductive BSF/Glc/AgNPs film showed the maximum sheet resistance around 49 (Ω/sq) during 1000 bending cycles. Also, conductivity analysis showed that this conductive patterned film could quickly light up 8 LEDs when connected to a 9V battery.

2. Experimental

2.1 Materials

Bombyx mori silkworm cocoons were purchased from Shanghai, China. Anhydrous sodium bicarbonate (Na_2CO_3), Lithium Bromide (LiBr) and Glycerol (Glc) were purchased from Sinopharm Chemical Reagent Co., Ltd. TDSP(cleaning water for the screen) was purchased from Zhongyi Ink & Paint Co., Ltd. Screen Printing Silver Ink Paste (SCR-60) is purchased from Qingdao Nayin New Material Technology Co., Ltd. All other materials, solvents and reagents were obtained from a commercial source and used without further purification.

2.2 Characterizations

SEM samples were coated with gold and examined at an accelerating voltage of 3.0 kV on Scanning Electron Microscope (FE-SEM) (S-4800; Hitachi Ltd. Japan) to analyze surface and cross-sectional morphology of BSF, BSF/Glc and BSF/Glc/AgNPs films. Chemical structural changes in pure BSF and BSF/Glc films were assessed using FT-IR spectroscopy (Thermo Nicolet 5700, Thermo Fisher Scientific Inc. USA). The measurement wavenumber range was 500 cm^{-1} to 4000 cm^{-1}. In order to observe and check the crystalline structure of the samples in the transmittance mode, the crystallinity of the prepared films was checked using X-ray diffraction (X Pert-Pro MPD) (PANalytical, Netherlands). The beam that was received through incidence called the incident beam had a wavelength of around 0.154 nm (Cu Kα radiation), a scanning range of diffraction angle 2θ was 5°-50°. There were corrections made in the intensity to get the desired changes in the background, sample absorption and the incident beam density. The transmission of light in the Pure BSF and BSF/Glc films was measured by a UV-Vis Spectrophotometer (UltraScan PRO-Hunter Lab). The wavelength range was kept from 350 nm to 900 nm. Screen Printing Machine (AT-45PA) from Atma Tung Yuan M/C IND. (KUNSHAN) Co., Ltd., is used for printing conductive patterns. Linear motor (GG10-186-045-S100-I-A) purchased from Dongguan Gaogong Intelligent Transmission Co., Ltd. for bending movement of conductive film. DMM7510-KEITHLEY – Digital multimeter is used to measure the sheet resistance.

2.3 Screen Printing Process

The Screen printer and a (500 mm × 400 mm) precision screen mesh (325 mesh count) were used in experiments. Both the screen and a squeegee were attached to the machine. Parameters of the machine were set accordingly, along with the machine moving speed (110 mm/sec). Firstly, the Si wafer substrate was placed under the screen board. The screen was slowly and gradually lowered down on to the printing board. Then, silver nanoparticles (AgNPs) based ink was placed to the top end of the screen, and simultaneously a squeegee attached at a 45° angle was used to spread the ink along the entire length of the screen. It pushed the AgNPs ink through the meshed open holes of the screen, imprinting the pattern on the Si wafer underneath. Both the patterns were printed in the same way. After the printing process, the screen was washed properly with TDSP (washing water) for better cleaning. The screen was dried and could be used for printing process again. Fig.1 shows the schematic design of screen printing process.

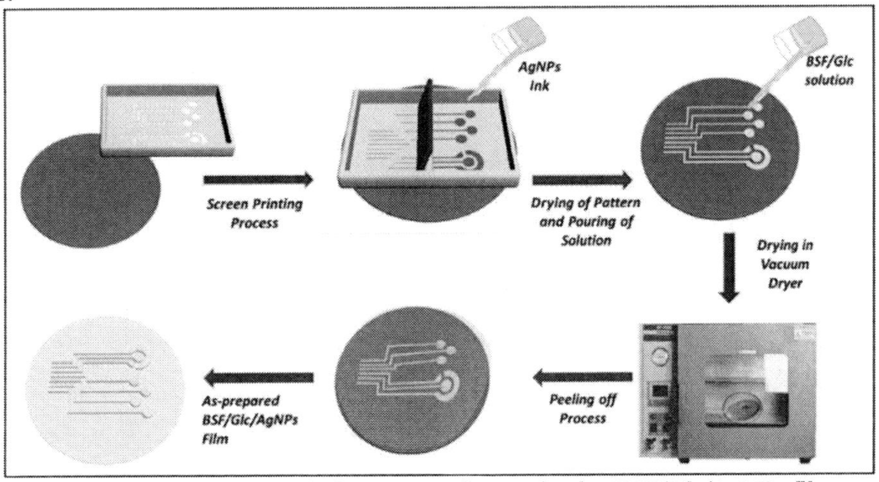

Fig.1 Schematic screen printing process for conductive BSF/Glc/AgNPs film.

2.3.1 Post-Treatment

Sintering is very important process for printing. Just after screen printing of AgNPs, the substrates were kept in oven for 1 to 2 hours. Screen printed AgNPs require almost 150°C sintering temperature [20]. Curing of Ag printed patterns plays a vital role in printing process as the size and conductivity of Ag ink is also dependent on curing. So, the printed pattern on Si wafer was cured/heated at 120°C for 20-25 minutes. Silk film can't withstand with this high degree temperature directly as it becomes more brittle and stiff. So, this work was carried out to successfully get screen printing on bombyx mori silk fibroin film.

2.4 Preparation of Pure BSF Film

To better understand physical and chemical changes that occurred in silk film while the addition of glycerol, pure BSF film with a thickness (70 μm) was also prepared and compared with BSF/Glc film in further characterizations. The pure BSF film without glycerol with 70 μm thickness was prepared by casting solution into a petri dish and drying at room temperature for 38 hours. The film was then peeled off and stored. In order to ensure the silk fibroin film has good performance in characterizations, it is usually placed under constant temperature and humidity (temperature around 20 °C, and relative humidity around 65%).

2.5 Preparation of Conductive BSF/Glc/AgNPs Film

Bombyx mori silkworm cocoons were boiled in an aqueous solution of 0.5 % (w/v) Na_2CO_3 for 1hour and washed with deionized (DI) water at different times to remove sericin, maintaining the temperature at 60°C. The degummed BSF fibers were dissolved in 9.3 mol·L^{-1} LiBr solution at 60 °C for 1hour. The regenerated BSF aqueous solution was obtained after dialysis in DI water using a dialysis tube (MWCO 6000-8000 Da, Spectra/Por, USA.) for 2-3 days to remove the LiBr salts. The final concentration of BSF solution prepared by the above method was generally between 5-8%. Finally, the obtained solution was stored in a refrigerator.

The BSF/Glc solution was prepared by adding (0.225g) glycerol in 15 mL silk solution. After that, the glycerol added silk solution was kept on a stirrer for 15 minutes at 250 rpm to make a homogeneous mixture. While the stirring process, the BSF/Glc solution got bubbles, so in order to overcome this problem, the solution was placed in a fume hood for 10 minutes and printed patterned Si wafer was placed in a petri dish. Then, the solution of BSF/Glc/AgNPs was poured onto the petri dish and kept in a vacuum dryer for 12 hours, and then film was peeled off from Si wafer. Fig.2 shows the actual size measurements of conductive BSF/Glc/AgNPs film and film thickness was 220 μm.

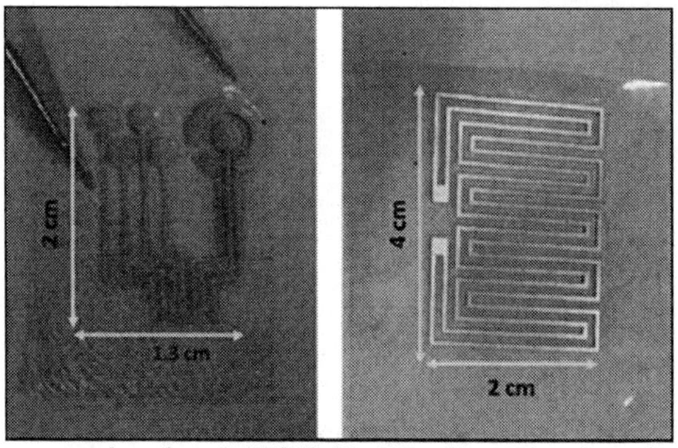

Fig.2 Size of the patterns printed on BSF/Glc/AgNPs film. The size of AgNPs was 250 nm.

3. Results and Discussion

3.1 Surface Morphology

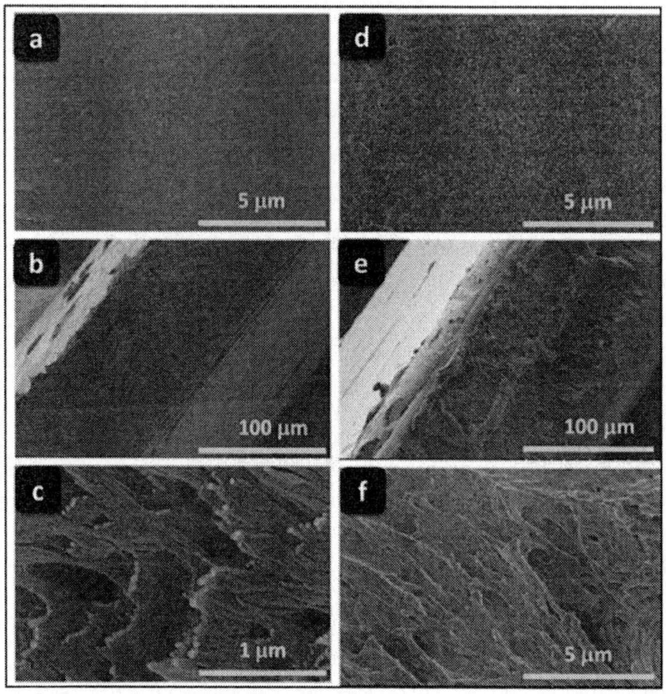

Fig.3 (a) The surface morphology of pure BSF film, (b,c) The cross sectional view of BSF film, (d) The surface morphology of BSF/Glc film, (e,f) The cross sectional view of BSF/Glc film.

As shown in Fig.3, the surface and cross-sectional morphology of the pure BSF and BSF/Glc blend films were observed with a scanning electron microscope (SEM). In Fig.3(a,d), the pure and blended film showed a relatively flat surface as expected, indicating the suitable film-forming property of the fibroin-plasticizer solution. Among the many requirements of a plasticizer, the plasticizer must be compatible with the polymer to achieve the plasticization purpose [21]. However, the microstructures of the films were discovered by close observation in Fig.3(b,c,e, and f). Previously, the unique nanostructure of BSF film was studied [22]. The pure BSF film showed that the nanofibrous structure originated from the silk nano-fibrils. The preserved nano-fibril endowed the pure BSF film with good biodegradable properties. As shown in Fig.3f, the glycerol is blended adequately with silk nano-fibrils, resulting in the higher flexibility of BSF/Glc films. In short, the BSF/Glc blended films showed a smooth surface and specific nanostructure, as well as the compatibility of glycerol plasticizer with BSF.

Another SEM characterization was done to get familiar with the type, shape and size of Ag nanoparticles printed on flexible BSF/Glc film. As shown in Fig.4, a very well-ordered, minimal-defect AgNPs array on BSF/Glc film can be produced. As shown in Fig.4a, it can be seen that Ag particles are completely covered over the surface of BSF/Glc film without any cracks and left spaces. The actual shape and size of Ag nanoparticles has been shown in Fig.4d. It clearly indicates that individual Ag hexagonal nanoparticle has sharp tips. BSF matrix has well distributed AgNPs, along with a particle size of 250 nm. The cross-sectional morphology images of conductive BSF/Glc/AgNPs film were also taken. Fig.5 shows the clear images that Ag printed pattern is present on BSF/Glc film. Here, in Fig.5a, we can see the presence of a pattern line over BSF/Glc film. Also, it can be seen in Fig.5(b, c) that the Ag ink used in the screen printing process showed adhesion property with BSF/Glc film. Ag ink used for our experiments has shown good adhesion ability with BSF/Glc film.

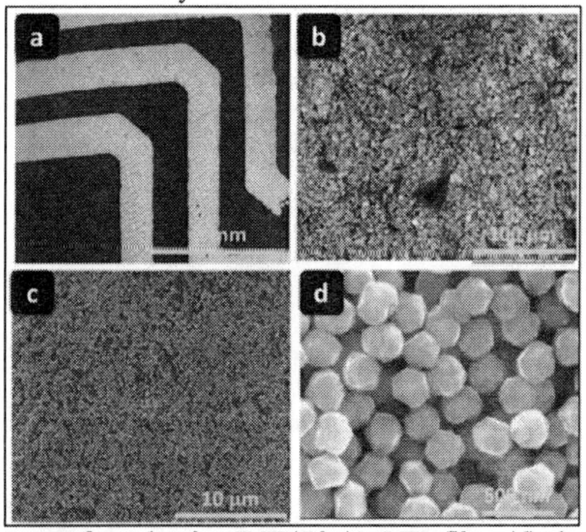

Fig.4 (a-d) Morphological structure of conductive BSF/Glc/AgNPs film. (d) shows the actual shape of AgNPs.

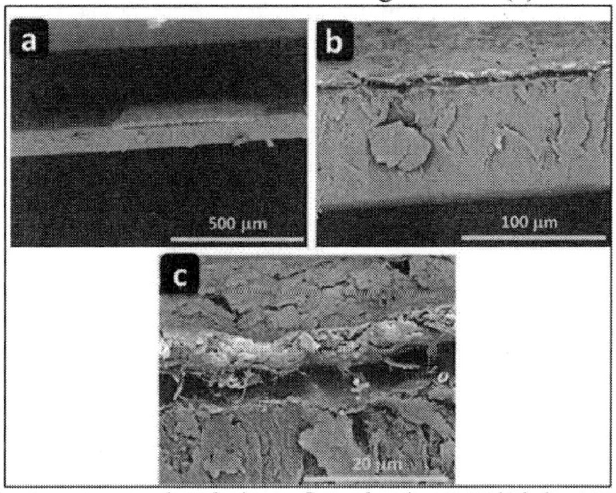

Fig.5 (a-c) Cross-sectional view of conductive BSF/Glc/AgNPs film.

3.2 Chemical Analysis

Fourier transform infrared spectroscopy was used to investigate the transition of the secondary structure of the BSF and BSF/Glc film (shown in Fig.6). Glycerol (30 wt%) as a plasticizer was added in pure BSF films to improve the flexibility of films. As plasticizers change the intermolecular and intra interaction between the chains of polymers, further investigations to observe the secondary structure of the BSF film was carried out following the glycerol content. There is a stretching C=O vibration of the amide groups present in the proteins through which the absorbance between 1000 cm^{-1} and 1650 cm^{-1} was investigated. The amide I structure is represented through this region, and it is also known to provide important information regarding silk film's secondary structure. Also, it was observed that the absorption peak of glycerol was not affected within this region in any way. At 1647 cm^{-1}, observations were made for getting the amide I band's maximum peak for the pure BSF film. The addition of glycerol caused a high absorption peak at 1650 cm^{-1}, which corresponds to the β-sheet structure. Glycerol has been shown to cause protein compression, allowing the protein molecule to be packed more tightly [23]. Therefore, a similar effect was expected to be seen here too. The characteristic peaks assigned to the random coil configuration of the β-sheet conformation for amide I to be 1623 cm^{-1}, amide II to be 1527 cm^{-1}, and amide III be 1265 cm^{-1}, suggested that the predominantly β-sheet but with quite a certain quantity of random coil structure in BSF/Glc film.

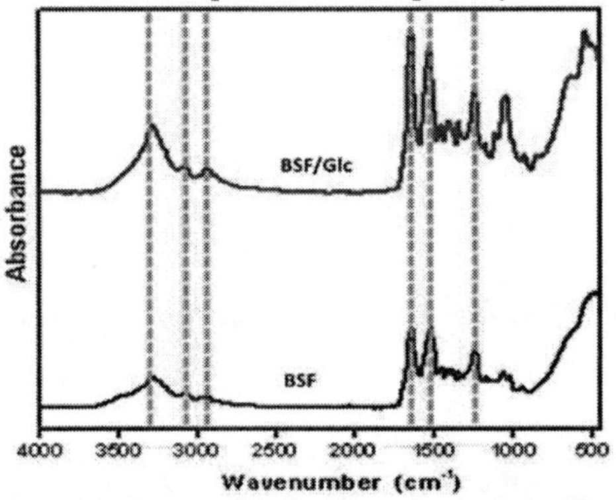

Fig.6 FT-IR spectra of BSF and BSF/Glc film.

In order to study and observe the crystalline structure of the pure BSF/Glc and BSF films, an X-ray diffraction was carried out. In Fig.7, the uncross linked pure BSF film showed arc shaped scattering at around 20° and weak diffraction at around 12.3°, corresponding to the random coil and α-helical structure [24]. And, the BSF/Glc film showed a very broad peak at 20°, which was assigned to the superposition of the two kinds of crystallization peaks. The presence of this broad peak showed the increased regularity or crystallinity in the films made of BSF/Glc. We can't explicitly link the increase in crystallinity to the increase in the structure of the β-sheets. This result showed that the β-sheet conformation from random coil and/or α-helical structure can be induced through the crosslinking reaction and it was also stable with FTIR spectra results [23].

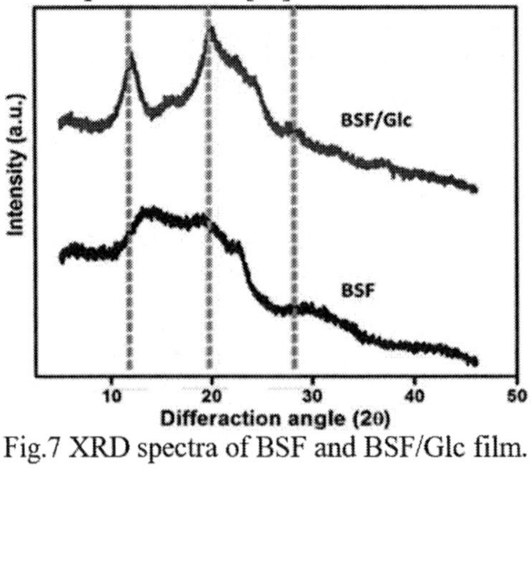

Fig.7 XRD spectra of BSF and BSF/Glc film.

3.3 Optical Transmittance

In order to explain fabricated film's transmittance further, the optical transmittance of the pure BSF and BSF/Glc/AgNPs films were measured, which is conveyed in Fig.8. In the range of the visible spectrum wavelength, which starts from 350 nm and goes to 900 nm, shown as the black line in Fig.8, the maximum optical transmittance of the original pure BSF film is 91.5%. Compared to that, the maximum optical transmittances of the BSF/Glc film is 89.9%, respectively, which indicates the addition of glycerol in pure silk solution leads to a slight reduction in the optical transmittance. This slight difference appeared because presence of glycerol increased the crystallinity of BSF/Glc film (shown in XRD result) and crystalline films show semi-transparent characteristics.

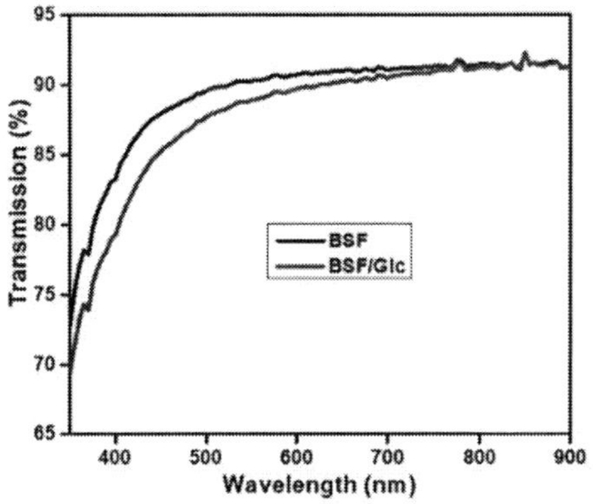

Fig.8 Optical transmittance curves of pure BSF and BSF/Glc film.

3.4 Energy Dispersive X-ray (EDX)

The presence of silver (Ag) nanoparticles on the surface of conductive BSF/Glc/AgNPs film were verified using EDX in this research, showing that silver (Ag) particles are uniformly printed on the BSF/Glc film surface. The EDX spectroscopy results showed the presence of 100% pure silver with no other contaminants. The optical absorption peak of silver at 3 keV is solid and typical of the absorption of metallic silver nanocrystallites due to surface plasmon resonance. Other bands for carbon (C) and oxygen (O) elements peaks appeared due to scattering caused by the compounds attached to the surface of silver (Ag), suggesting that the applied cationic surfactants acted as silver nanoparticle capping agents (AgNPs). In contrast to the bare BSF/Glc film, the silver coating was detected to have a bit darker (white arrow), as shown in the Fig.9. The components became colour coded yellow (silver), green (oxygen), and red (carbon) in these illustrations. In the silver (Ag) part, a different yellow colour was visualized as AgNPs layer on BSF/Glc film. The change in colour in the other sections was evident (C and O). It showed that the silver (Ag) on the BSF/Glc/AgNPs film was well presented.

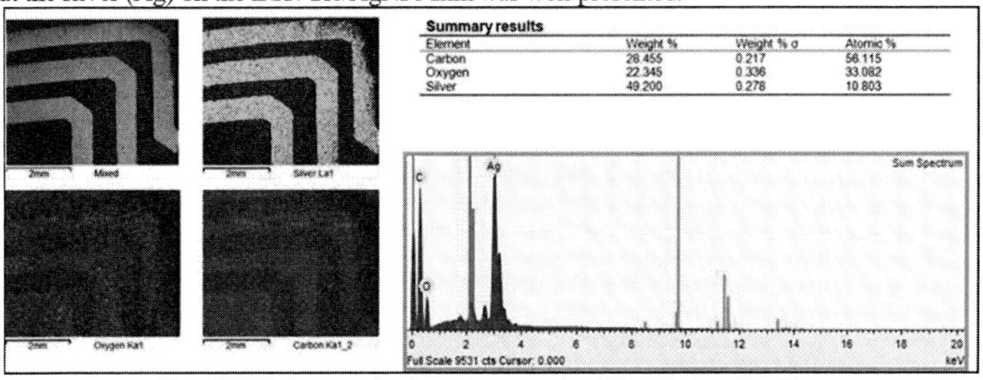

Fig.9 EDX characterization of conductive BSF/Glc/AgNPs film.

3.5 Effect of Pressure and Temperature on AgNPs based Pattern

Fig.10 shows the actual images of defective printed BSF/Glc/AgNPs film. The film was dried at average room temperature by keeping it in a fume hood for 38 hours. Nevertheless, it did not work. The pattern was not adequately printed to the BSF/Glc film. In order to cope with this problem, a vacuum drying technique was used.

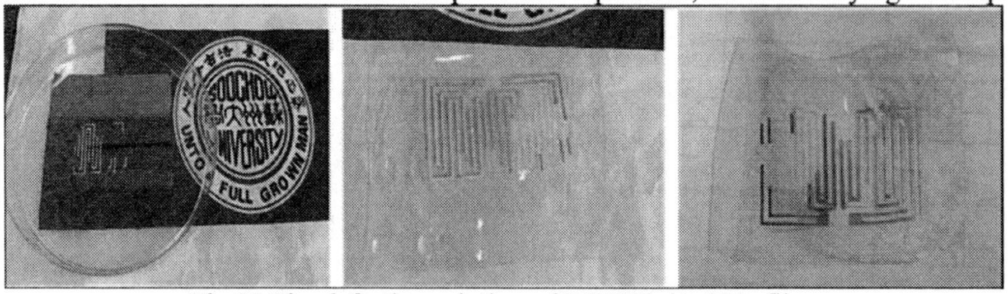

Fig.10 The defective printing of BSF/Glc/AgNPs film.

The Si wafer substrate is first screen printed with AgNPs based pattern and then sintered at 120°C temperature. The BSF/Glc solution was poured over the printed pattern and kept in a vacuum dryer (temperature 40 °C and pressure 0.08 MPa) for 12 hours. After that, the fabricated conductive BSF/Glc/AgNPs film was peeled off from the Si substrate, and the pattern was printed neatly and cleanly on the film, as shown in Fig.11. A vacuum dryer in fact provides pressure and temperature together, which enabled the patterns to be easily detached from the Si wafer and have print onto the film.

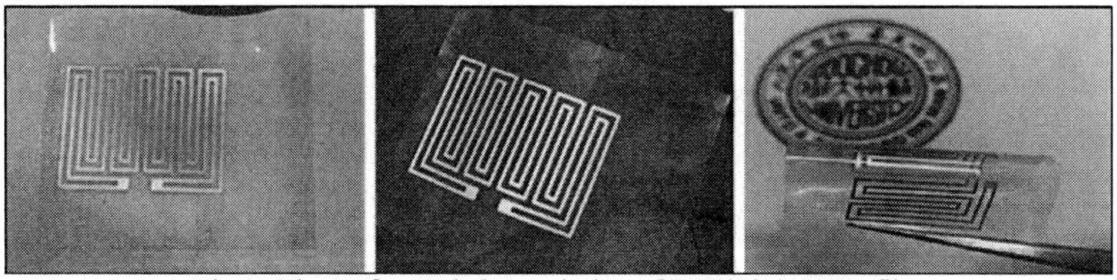

Fig.11 The perfect and clean printing of BSF/Glc/AgNPs film

3.6 Sheet Resistance and Electrical Conductivity

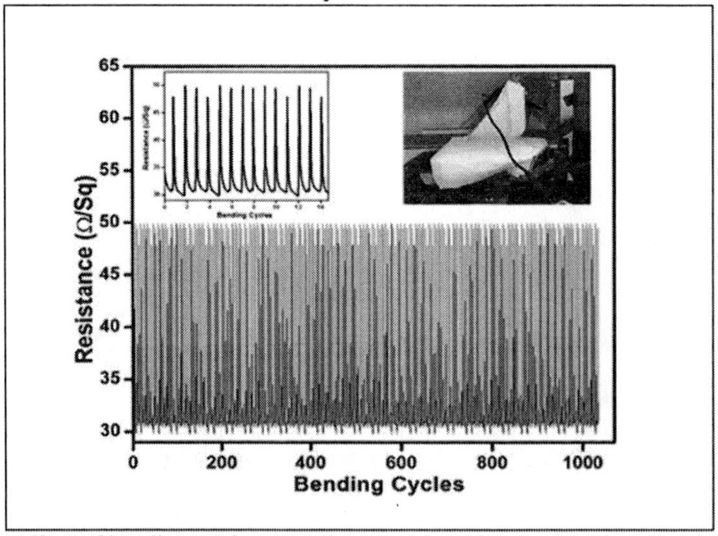

Fig.12 Effect of bending cycles on resistance of conductive BSF/Glc/AgNPs film.

The conductive properties of the film with different bending cycles were calculated. As shown in Fig.12, the maximum sheet resistance during bending cycles was 49 (Ω/sq) and the minimum was 29 (Ω/sq). Resistance increases under normal conditions, but it decreases under bending due to the closing of a conductive path printed on BSF/Glc film. The pattern showed the optimized resistance change up to 1000 bending cycles. This result showed minor changes in sheet resistance under bending stress, which is due the presence of the high flexibility of conductive BSF/Glc/AgNPs film. The most significant advantage of this BSF/Glc/AgNPs film is that it is flexible. It can bend freely and maintain electrical conductivity throughout the bending process, which is due to the characteristics of the soft matter of the film. Conductivity of film is analyzed in Fig.13. The pattern was connected to the 8 LEDs, which were connected to the 9.0 V batteries. The LEDs were working smoothly due to the conductive path of AgNPs on silk film.

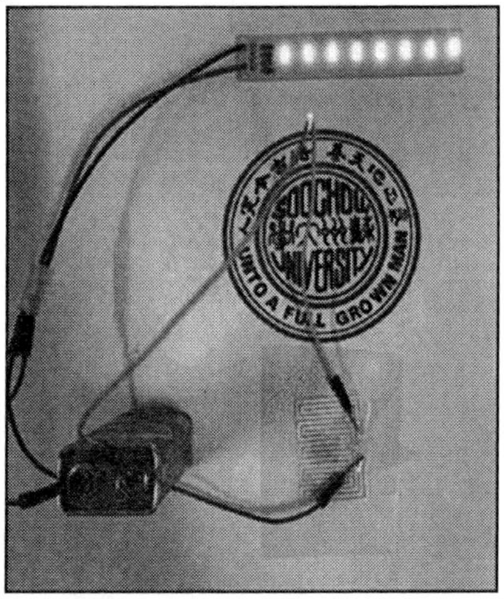

Fig.13 Photograph for the conductive track used for the light up of 8 LEDs with 9.0V battery.

4. Conclusion

In this study, an effort was made to prepare a flexible and conductive patterned BSF/Glc/AgNPs film using screen-printing technology. The flexibility of bombyx mori silk fibroin (BSF) film was enhanced by adding glycerol (Glc) in solution. Hence, the addition of glycerol gave high flexibility to BSF/Glc films. Characterizations like FT-IR, SEM, XRD and optical transmittance were performed on pure BSF and BSF/Glc films to get familiar with the effect of glycerol on BSF film. Also, the vacuum drying process is used for drying BSF/Glc/AgNPs film. The conductive BSF/Glc/AgNPs film was prepared by applying pressure 0.08 MPa and temperature 35-40°C for 12 hours. Luckily, the pattern was completely and cleanly printed onto the film. Further, characterizations like EDX and SEM were performed to study the surface morphology and element mapping of the film. The size of AgNPs was calculated through image J software which was approximately 250 nm. The sheet resistance of conductive BSF/Glc/AgNPs film was a maximum of 49 (Ω/sq) during 1000 bending cycles. Also, conductivity analysis showed that this conductive film could easily light up 8 LEDs when connected to a 9V battery.

5. Acknowledgment

The work was supported by National Engineering Laboratory for Modern Silk, College of Textile and Clothing Engineering, Soochow University, Suzhou 215123, China.

References

[1] Rockwood DN, Preda RC, Yücel T, Wang X, Lovett ML, Kaplan DL. Materials fabrication from Bombyx mori silk fibroin. Nature protocols: 2011; 6 (10): 1612.

[2] Wang Q, Ling S, Liang X, Wang H, Lu H, Zhang Y. Self-healable multifunctional electronic tattoos based on silk and graphene. Advanced Functional Materials: 2019; 29(16): 1808695.
[3] Koh LD, Cheng Y, Teng CP, Khin YW, Loh XJ, Tee SY, Low M, Ye E, Yu HD, Zhang YW, Han MY. Structures, mechanical properties and applications of silk fibroin materials. Progress in Polymer Science: 2015; 46: 86-110.
[4] Liu Y, Qi N, Song T, Jia M, Xia Z, Yuan Z, Yuan W, Zhang KQ, Sun B. Highly flexible and lightweight organic solar cells on biocompatible silk fibroin. ACS applied materials & interfaces: 2014; 6 (23): 20670-5.
[5] Lu S, Wang X, Lu Q, Zhang X, Kluge JA, Uppal N, Omenetto F, Kaplan DL. Insoluble and flexible silk films containing glycerol. Biomacromolecules: 2010; 11 (1): 143-50.
[6] Yun YS, Cho SY, Shim J, Kim BH, Chang SJ, Baek SJ, Huh YS, Tak Y, Park YW, Park S, Jin HJ. Microporous carbon nanoplates from regenerated silk proteins for supercapacitors. Advanced materials: 2013; 25 (14): 1993-8.
[7] Wang H, Meng F, Zhu B, Leow WR, Liu Y, Chen X. Resistive switching memory devices based on proteins. Advanced Materials: 2015; 27 (46): 7670-6.
[8] Tao H, Kaplan DL, Omenetto FG. Silk materials–a road to sustainable high technology. Advanced materials: 2012; 24 (21): 2824-37.
[9] Mohammed MG, Kramer R. All-printed flexible and stretchable electronics. Advanced Materials: 2017; 29 (19): 1604965.
[10] He J, Nuzzo RG, Rogers JA. Inorganic materials and assembly techniques for flexible and stretchable electronics. Proceedings of the IEEE: 2015; 103 (4): 619-32.
[11] Han Y, Dong J. Electrohydrodynamic (EHD) printing of molten metal ink for flexible and stretchable conductor with self‐healing capability. Advanced Materials Technologies: 2018; 3 (3): 1700268.
[12] Tam SK, Ng KM. High-concentration copper nanoparticles synthesis process for screen-printing conductive paste on flexible substrate. Journal of Nanoparticle Research: 2015; 17 (12): 1-2.
[13] Hyun WJ, Lim S, Ahn BY, Lewis JA, Frisbie CD, Francis LF. Screen printing of highly loaded silver inks on plastic substrates using silicon stencils. ACS applied materials & interfaces: 2015; 7 (23): 12619-24.
[14] Liang J, Tong K, Pei Q. A water-based silver-nanowire screen-print ink for the fabrication of stretchable conductors and wearable thin-film transistors. Advanced Materials: 2016; 28 (28): 5986-96.
[15] Cho SY, Yun YS, Kim E, Kim MS, Jin HJ. Stem cell response to multiwalled carbon nanotube-incorporated regenerated silk fibroin films. Journal of nanoscience and nanotechnology: 2011; 11 (1): 801-5.
[16] He P, Cao J, Ding H, Liu C, Neilson J, Li Z, Kinloch IA, Derby B. Screen-printing of a highly conductive graphene ink for flexible printed electronics. ACS applied materials & interfaces: 2019; 11 (35): 32225-34.
[17] Liu LX, Chen W, Zhang HB, Wang QW, Guan F, Yu ZZ. Flexible and multifunctional silk textiles with biomimetic leaf-like MXene/silver nanowire nanostructures for electromagnetic interference shielding, humidity monitoring, and self-derived hydrophobicity. Advanced Functional Materials: 2019; 29 (44): 1905197.
[18] Wang C, Xia K, Zhang Y, Kaplan DL. Silk-based advanced materials for soft electronics. Accounts of chemical research: 2019; 52 (10): 2916-27.
[19] Krebs FC, Alstrup J, Spanggaard H, Larsen K, Kold E. Production of large-area polymer solar cells by industrial silk screen printing, lifetime considerations and lamination with polyethyleneterephthalate. Solar energy materials and solar cells: 2004; 83 (2-3): 293-300.
[20] He P, Cao J, Ding H, Liu C, Neilson J, Li Z, Kinloch IA, Derby B. Screen-printing of a highly conductive graphene ink for flexible printed electronics. ACS applied materials & interfaces: 2019; 11 (35): 32225-34.
[21] Vagenende V, Yap MG, Trout BL. Mechanisms of protein stabilization and prevention of protein aggregation by glycerol. Biochemistry: 2009; 48 (46): 11084-96.
[22] Asakura T, Kuzuhara A, Tabeta R, Saito H. Conformational characterization of Bombyx mori silk fibroin in the solid state by high-frequency carbon-13 cross polarization-magic angle spinning NMR, x-ray diffraction, and infrared spectroscopy. Macromolecules: 1985; 18 (10): 1841-5.
[23] Zhang H, Deng L, Yang M, Min S, Yang L, Zhu L. Enhancing effect of glycerol on the tensile properties of Bombyx mori cocoon sericin films. International journal of molecular sciences: 2011; 12 (5): 3170-81.
[24] Wang Y, Wang X, Shi J, Zhu R, Zhang J, Zhang Z. Flexible silk fibroin films modified by genipin and glycerol. RSC advances: 2015; 5 (123): 101362-9.

Cite as: B. Richardson et al.,
TBIS 10.3993/tbis (2021).

Designing Wearable Electronic Textiles to Detect Early Signs of Neurological Injury and Disease: A Review

Bethany Richardson[1], Prof. Henry Yi Li[1]

[1]University of Manchester, Oxford Road, Manchester, M13 9PL, United Kingdom

*Corresponding author's email: bethany.richardson@manchester.ac.uk

Abstract

Neurological disorders affect the central and peripheral nervous system covering the brain, spinal-chord, cranial nerves, nervous system, and neuromuscular junction. There is an unsatisfied need for a non-invasive ambient measuring system which can record patients' vital body levels autonomously and in real-time. There is also an acute need to detect diseases at prodromal stages in patients that may carry asymptomatic characteristics of underlying disease. The opportunities offered through innovations in wearable electronic textiles can provide a solution by measuring biomarkers for Alzheimer's Disease detection, and other neurological disorders, through a non-invasive biosensor that detects protein levels in saliva. This is an area which has large gaps in research on the application of these technologies for the early diagnosis of injury, and for differentiating between stages of illness to provide more accurate and bespoke neurological healthcare.

Keywords: Brain injury; Alzheimer's Disease; Wearable Electronics; Biosensors

1. Introduction

Neurological injury covers areas of the central and peripheral nervous system and can include temporary or permanent damage to the brain, spinal chord and peripheral nerves. Due to the complex and intricate structure of the brain, the responses triggered by injury are equally multifaceted and complex [1]. Severe brain injuries can result from damaged tissues or cells from high-impact blows to the skull, penetrative injury or acceleration and rotational forces. Brain injuries resulting from traumatic brain injury (TBI) can include subdural or epidural hematoma, subarachnoid hemorrhage (SAH), contusion and diffuse axonal injury. Acquired brain injury (ABI) results from injuries suffered by means of stroke, tumors or degenerative diseases [2]. These injuries include anoxia and hypoxia as a result of restricted or low blood pressure to the brain which can starve the organ of oxygen and cause devastating long-term damage [3].

Severe head injuries also place the patient under high risk of developing neurological diseases such as Alzheimer's disease and dementia. A study produced by the Care Policy and Evaluation Centre at the London School of Economics and Political Science modelled projections of dementia care in England. In 2019, 885,000 people in the UK were diagnosed with dementia [4]. These figures are expected to rise in the future driven by an increasingly ageing population in the United Kingdom. The ONS population projections predicts that between the years 2019-2040 the proportion of people aged 65-74 will increase by 20% and the amount of people above 85 years over will increase by 114% [5]. This leads to an increase in older people suffering from dementia by 80% by the year 2040 affecting 1.6 million people. The costs associated with dementia care are substantial and are expected to rise over subsequent decades. The total cost of dementia care for older people in the UK in 2019 was £34.7 billion and is expected to triple by 2040. The development of a point-of-care, wearable electronic device, which can detect early signs of Alzheimer's disease by measuring Amyloid beta protein in saliva can be a solution. It could assist with the early detection of Alzheimer's disease treatment, make healthcare more accessible to lower income households, and a move towards a neurological point-of-care health system.

2. Literature Review

This review will outline the developments in wearable technologies in the application of monitoring vital health signs through the application of biosensor materials. There has been increased interest in this area due to their accessibility, affordability and progression towards personalised medical care. This review will cover conductive materials, nanobiosensors, biomarkers, protein detection, sensor device technologies and device applications. The first section will cover critical analysis of conductive materials and their desirable properties as a biosensor component. The second section will cover biomarkers used for protein detection with a summary of the knowledge gaps in this areas. The last section will look at sensor device technology and its applications in healthcare monitoring and e-textile developments. This will be concluded with a summary of the knowledge gaps covered in the review followed with an assessment of how these will be developed into future research objectives.

3. Conductive Biosensing Materials

Table 1. Critical analysis of biosensors in detecting Tau protein in biological samples [8-12]

Biosensor	Examples	Limited detection	Advantage	Disadvantage	Biologica l sample	Type of Tau
Optical biosensor	SPR biosensors - Immune-based SPR biosensor - Multimode optical fibers SPR biosensor - Fluorescentbio-sensors - Opticalwave interferometry biosensors - SERSbased biosensors	Femtomol ar Picomolar	Direct, real-time; label-free detection; highly specific and sensitive	Sensor surface functionalization challenging and bulky optical equipment	Serum, aCSF, plasma	Human tau Tau12 Tau 381
Electrochem ical biosensors	- Biocatalytic biosensors - Affinity type biosensors - Fourelectrode electrochemical biosensor - Goldelectrodes Electrochemical biosensors	Picomolar Nonomol ar	Simplicity of using devices with high sensitivity; capable of operating with a very low sample concentration	Narrow or limited temperature range; limited shelf life; cross-sensitivity of other gases	Au surface, serum, equilibriu m mixture	Tau441 Tau381 Tau410
Piezoelectric biosensors	Immunosensing piezoelectric biosensors	Nanomola r	Reliable analytical tools; label-free detection; real-time data; small and robust; and capable of giving a rapid response.	Interference from atmospheric humidity; the difficulty in using them for the determination of material in solution	Human Tau,	aCSF

The key properties that conductive biosensing materials require are to be lightweight, durable, highly sensitive, flexible and hold the ability to be utilised non-invasively. Innovations in novel micro or nanonstructured materials provide great potential for medical applications to monitor physiological changes in the body [6]. The use of graphene or carbon-treated fibres has proven a successful material for monitoring bodily functions in the short-term but remain challenging when worn in the long-term [7]. The promising applications of this technique include treating natural fibres such as cotton, silk and flax with graphene oxide (GO) or novel combinations through reduction (rGO). Further desirable qualities of an effective conductive fibrous medium include breathability for comfort of wear by the user, for example stretchability and moisture wicking capabilities. Within the field of enzyme-based biosensors the area of glucose detection is the most saturated in terms of research. However, the conventional conducting-based glucose sensors will have limitations such as unsatisfactory sensitivity or detection limit and relatively high applied potential. Another limitation is that interference can occur from other oxidisable substances such as ascorbic acid, acetaminophen or uric acid. Therefore developing a hybridised biosensor which incorporates organic and synthetic materials could prove a viable material for this application. The table below compares advantages and disadvantages of biosensors in detecting different biological samples.

3.1 Summary and Knowledge Gap

In summary, the area of enzymatic biosensors is large in the area of glucose detection, due to the wide application in the field of diabetic health monitoring. However, these biosensors are still limited by often not being sensitive enough and having a detection limit which can limit the accuracy or extent of analysis gained from the sample. Moreover, they can react with other substances such ascorbic acid, acetaminophen or uric acid which can present misleading results. Therefore, more selective sensitivity is needed from combinations of different materials [13]. The knowledge gap in this area is the unexplored applications of electrodeposited polymers and development of new polymer-based biosensors. A further development in the future is of the nanostructuration of electrodeposited polymers since the electrosynthesis of polymer nanowires or nanotubes recently led to strong improvements in the sensing properties of conducting polymers [14]. In addition, the landscape for hybrid conducting polymer systems combining polymers and conducting inorganic materials, especially metallic nanoparticles and carbon nanomaterials, is rich in potential with numerous promising materials, each with their own chemical, electrical and physical properties, yet to be explored for biosensing.

4. Biomarkers for Protein Detection

4.1 Protein Detection

Amyloid beta protein analysis is one of the core molecules observed in detection of Alzheimer's disease (AD). This protein is of importance in detecting brain disorders as is it produced in the brain and passes into the blood across the blood brain barrier [15]. Deposits of amyloid β-protein (Aβ) are characterised by amounts of senile plaques detected in the brain parenchyma. Historically these plaque build-ups would be detected in post-mortem investigations and was the primary method of detection until the early 2000's [16]. Following this, the use of magnetic resonance imaging (MRI) was used for detection and increasingly the use of positron emission technology (PET) is used more commonly today. The presence of Aβ in senile plaques contains 4.2 Total Tau (T-Tau) and Phosphorylated Tau (P-Tau) [17]. Currently the procedure for detecting this is through an invasive lumbar puncture to the lower spine to extract Cerebrospinal fluid (CSF). This procedure carries health risks with it such as incorrect insertion and painful discomfort to the patient and adverse reactions such as chronic back pain. This review aims to propose an alternative non-invasive method for amyloid β-protein (Aβ) investigation through the collection of proteins present in saliva.

Table 2. Critical analysis of nanomaterials in detecting biomarkers [18-22]

Nanomaterial	Properties	Application	Advantages	Disadvantages	Biosensor examples	Detection limit
Graphene Quantum Dots (GDQs)	Atom-thin graphitized planes (< 2nm thick) Photoluminescence (PL) Water soluble Biocompatible <10 nm size	Ion detection Biomarkers detection (Asorbic Acid (AA), dopamine, DNA and Amino Acid) Disease detection (Cancer) Tumour imaging	PL can be adjusted by changing dimensions or dopant Good conductivity Time-saving Convenient Cost-effective Highly-sensitive	Limited aqueous solubility without doping	Tumour imaging pH-responsive fluorescent sulfur- nitrogen-doped GQDs probe (pRF-GQDs). Green PL in pH below 6.8 Blue PL in pH overtop 6.8	5.1 ppb Linear range 400.0 nM Detection limit 0.6 nM
Magnetic Nanoparticles (MNPs)	Size range 1-100 nanometers High saturation magnetisation (e.g. Fe, Co, Ni, Fe_3O_4, $CoFe_2O_4$)	Hypothermia treatment Biosensing Drug delivery Early stage cancer diagnosis	viable solutions to low detection limits and non-specific effects ultralow detection of proteins, disease biomarkers and pathogens	Need specific and stable coating to control surface properties	Protein capture for specific analyst synthesized anti-prostate-specific antigen (PSA)-labeled MNPs to selectively capture protein analyte PSA.	The detection limit of this nanopore sensor was 0.8 fM,
Gold Nanoparticles (AuNPs)	Large surface area Unique optical properties - high Localized Surface Plasmon Resonance (LSPR) High electrical conductivity	Optical and electrical biosensors Cancer detection Neurological disease Protein, nucleic and pathogen detection	High electric conductivity makes enhanced electron transfer between redox centre and electrode	Cost possible aggregation	HBsAg - Hepititus B detection. gold nanoparticle-based immunosensor to detect HBsAg with a linear concentration range of 0.12 ~ 30 ng/mL	Detection limit of 14 pg/mL
Silver nanoparticles (AgNPs)	Antifungal and antibacterial More prone to electrochemical oxidation Higher extinction coefficient	Medical research Diagnosis of tumours Cancer treatment	Excellent recognition ability and specific response in tumor microenvironment in cancer detection	Fewer applications due to cytotoxicity concerns	Nanosensor MnO_2-AgNPs for GSH sensing. AgNPs fluorescence quenched by MnO_2 through internal filter effect and static quenching effect recovered by GSH due to MnO_2 decomposition.	The detection limit of GSH was as low as 0.55μM

4.2 Summary and Knowledge Gaps

In summary, the advantages of using electrochemical biosensors are that they are low cost, are simple in operation and can detect analytes at low concentration. However, current methods in use cannot control the material structure at atomic level. The use of graphene coated materials can be twisted but not stretched which would present difficulties in application to textile materials as the recovery of material when stretched is a vital property when used on the body. Another challenge is in maintaining conductive, mechanical and electrical properties over time. This is an area which is necessary to ensure the consistent reliability of data when monitoring patients' vital body levels. Therefore, the knowledge gap to be considered further in this area is how to create a resilient material as a biosensor which can be stretched and recovered when used on the body whilst retaining its conductive properties.

4.3 Applications of Biosensors in Electronic Textiles

Electronic textiles, or "e-textiles", are intelligent textiles which have electronic components integrated into the fabric to give it unique properties. These can be integrated by different techniques such as sewing, weaving, printing, knitting, and embroidery, braiding or chemical treatments [23].

4.3.1 Health Monitoring

Table 3. Applications of biosensors in protein detection and breath sampling [25-31]

	Biosensing			Breath sampling		
	Tooth enamel biosensor	**Enzymatic electrode printed onto mouthguard**	**Mouthguard sensing platform**	**Graphene-oxide-based flexible breath humidity sensor**	**Graphene-oxide-based sensor for respiratory monitoring**	**Wearable humidity sensor**
Device	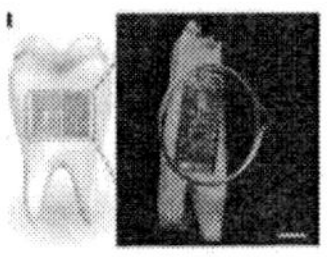		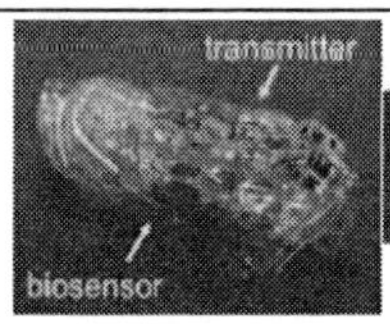			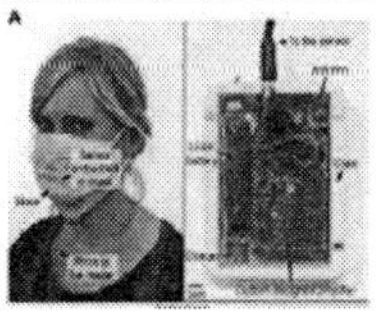
Mechanism	Antimicrobial peptide on graphene binding to saliva based bacteria. Printed on water-soluble thin film substrates Integrated with electrodes patterned with inductive coil antennae	Crosslinking uricase enzyme with electro-polymerizing o-phenylenediamine .Sensor fabricated on flexible PET substrate and integrated with wireless amperometic circuitry and mounted on mouthguard SUA detection	Glucose oxidase entrapped with poly(MPC-co-EHMA) (PMEH) onto working electrode. 3-electrode system combined with wireless transmitter onto a polyethyleneterephthalateglycol (PETG) mouthguard	Flexible breath humidity sensor with graphene oxide measures sensor performance using a controlled humidity generator	Measures graphene oxide resistance change to monitor respiratory activity in real time Flexible antenna was prototyped on Kapton to enable RFID and wireless data transmission Antenna and the graphene oxide sensor were integrated onto a facemask	Measures change in the ionic conductivity with the water content absorbed in the paper, which makes it possible to deploy the device as a portable humidity sensor to track the real-time respiration

Sensitivity	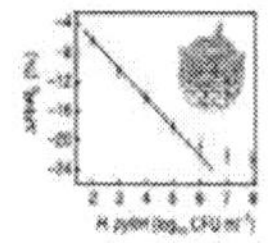	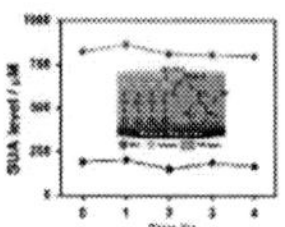	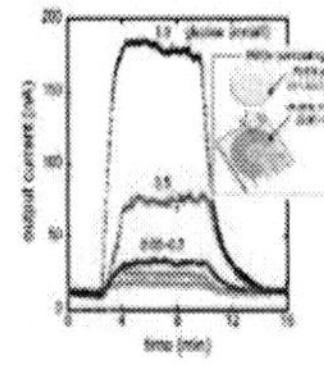		Ability to monitor individual exhalation–inhalation peaks and distinguish normal breath from apnea conditions	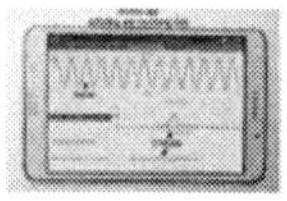
Advantages	Detects highly sensitive individual bacteria. Wireless readout achieved in proximity	Bluetooth wireless data communication achieved	The testing yielded sensitive amperometric outputs over relevant glucose levels in human saliva	Ultrafast response to a modulated humid flow (B30 ms response and recovery time) Device is highly flexible, transparent, wearable size, low cost of production. Distinguishes different patterns of human behaviors (speaking/talking)	Measured respiration in real time. Ability to monitor individual exhalation–inhalation peaks and distinguish normal breath from apnea conditions	Accessible, simple, and wearable breath testing by integrating sampling, collection, and analysis steps into a single device
Disadvantages	Interference from other inputs such as food residue which can secrete protein and other active chemicals can interfere with target analyte	Real-time monitoring was not achieved	Voluntary and in-voluntary mouth muscle movements in daily life, such as talking, which could present mechanical stress on the sensors	Possible interference of dehydration in the body when analysing the sample. Other interference from external factors, such as humidity.	Daily lifestyle habits such as smoking, diet, fitness, water intake	Low biomarker concentrations, contamination from other breath molecules. Expiratory flow rate

Flexible and stretchable sensors have increased in interest in wearable health monitoring fields in recent years due to their non-intrusive monitoring of physiological signals [24]. Ambient biosensors can monitor bodily functions such as motion, breath, heart rate and skin temperature in real-time. The interest in this area is in the materials which present opportunities through their application in new areas to detect early signs of critical and degenerative diseases. Table 3 below lists different applications which could be considered for the utilisation of new designs to combine biosensing elements for detection of proteins in saliva.

4.4 Summary and Knowledge Gap

From this research is it clear that salivary monitoring detection is possible in-situ and it is also possible to communicate data from the vessel to an external system through bluetooth wireless connection. The knowledge gap in the research is to develop a mechanism which can deliver real-time monitoring data communication. The other knowledge gap is in the collection of samples which require innovation to create a design which can collect saliva samples without using a mouthguard which is bulky and obtrusive method of wearing.

Further challenges that lie in creating an in-mouth based biosensor are in its biocompatibility. Interference from other inputs such as food residue which can secrete protein and other active chemicals can interfere with target analyte sensing and requires the sensor to be highly sensitive and accurate. Further risks include sensor damage from mechanical use over time which could corrode and interfere with the sensors' accuracy. Therefore this suggests further evidence that using breath sampling could be a more accurate solution for proteinuous salivary collection. A further limitation in through voluntary and in-voluntary mouth muscle movements in daily life, such as talking. This could present mechanical stress on the sensors and reinforces the priority that the biosensors used in this study should have a priority of recoverability after stretching in order to maintain accuracy of results. A further limitation lies in the interference of dehydration in the body when analysing the sample. This is a gap in research to look further into how to establish a baseline level of measurement which can be analysed without interference from external factors, such as humidity, or internal factors such as dehydration.

5. Conclusion

In conclusion the following knowledge gaps have been highlighted in this review. Firstly the landscape for hybrid conducting polymer systems combining polymers and conducting inorganic materials, especially metallic nanoparticles and carbon nanomaterials is largely unexplored. It is rich in potential with numerous promising materials, each with their own chemical, electrical and physical properties, yet to be developed for the detection of protein biomarker Amyloid Beta. Secondly, another area which research is needed is in a recoverable biosensing material that can retain its conductive properties after being stretched or deformed when worn on the body. Thirdly Existing in-mouth biosensors are still very bulky and invasive to the user. Therefore developing a non-invasive method to collect saliva, for example using breathomic humidity techniques to collect saliva through a mask that holds a high range of sensitivity range need to be developed. Finally, the limitation with current means of communicating data is that it can be achieved in real-time with bluetooth wireless communication. However, more advanced data communication is available, such as using an integrated NFC antenna applied to fabric, and should be explored for future data communication methods.

References

[1] Chodobski A, Zink B, Szmydynger-Chodobska J. Blood–Brain Barrier Pathophysiology in Traumatic Brain Injury, Transl. Stroke Res: 2011; 2: 492-516.
[2] Information on https://www.biausa.org/
[3] Chodobski A, Zink B, Szmydynger-Chodobska J. Blood-Brain Barrier Pathophysiology in Traumatic Brain Injury, Transl. Stroke Res: 2011; 2: 492-516.
[4] Wittenberg R, Hu B, Barraza-Araiza L, Rehill A. Projections of older people with dementia and costs of dementia care in the United Kingdom: 2019-2040, CPEC Working Paper 5, Care Policy and Evaluation Centre, London School of Economics and Political Science.
[5] Office for National Statistics, website database available here: https://www.ons.gov.uk/
[6] Yang Y, Gao W. Wearable and flexible electronics for continuous molecular monitoring, Chemical Society Reviews: 2019; 6.
[7] Xu L, Liu Z, Zhai H, Chen X, Sun R, Lyu S, Fan Y, Yi Y, Chen Z, Jin L, Zhang J, Li Y, T. Ye T. Moisture-Resilient Graphene-Dyed Wool Fabric for Strain Sensing, ACS Appl. Mater. Interfaces: 2020; 12: 13265-13274.
[8] Mehrota P. Biosensors and their applications - A review, journal of oral biology and craniofacial research 6: (2016); 153-159.
[9] Feng L, Zhu A, Shi H. Recent Advances in Optical Biosensors for Environmental Monitoring and Early Warning: (2013).
[10] Mollarasouli, Kurbanoglu, Ozkan. The Role of Electrochemical Immunosensors in Clinical Analysis, Biosensors.
[11] Naresh, Lee. A Review on Biosensors and Recent Development of Nanostructured Materials-Enabled Biosensors: (2021).
[12] Wang Z, Hu T, Liang R, Wei M. Application of Zero-Dimensional Nanomaterials in Biosensing, Frontiers in Chemistry: 2020; 8: 320.
[13] Lakard B. Electrochemical Biosensors Based on Conducting Polymers: A Review, Chemical Society Reviews: 2019; 48 (6): pp. 1423-1954.

[14]Jaworski S, Sawosz E, Grodzik M, Winnicka A, Pra- sek M, Wierzbicki M, Chwalibog A. Int. J. Nanomedicine: 2013; 8: 413.
[15]Yoo K, Kim G, Park D, Kim J, Kim Y, Kim H, Yang S, Lee J, Hwang K. Gold nanoparticles assisted sensitivity improvement of interdigitated microelectrodes biosensor for amyloid-β detection in plasma sample, Sensors and Actuators: B. Chemical: 308 2020; 127710.
[16]Loof A, Schoofs L. Alzheimer's Disease: Is a Dysfunctional Mevalonate Biosynthetic Pathway the Master-Inducer of Deleterious Changes in Cell Physiology? OBM Neurobiology: 2019.
[17]Zou K, Abdullah M, Michikawa M. Current Biomarkers for Alzheimer's Disease: From CSF to Blood, J. Pers. Med: 2020; 10: 85.
[18]Wang Z, Hu T, Liang R, Wei M. Application of Zero-Dimensional Nanomaterials in Biosensing, Frontiers in: April 2020; Volume 8: Article 320.
[19]Siontorou C, Georgopoulos K, Nikoleli G, Nikolelis D, Karapetis S, Bratakou S. Protein-Based Graphene Biosensors: Optimizing Artificial Chemoreception in Bilayer Lipid Membranes, Membranes: 2016; 6: 43.
[20]Ameri M, Shabaninejad Z, Movahedpour A, Sahebkar A, Mohammadi S, Hosseindoost S, Ebrahimi M, Savardashtaki A, Karimipour M, Mirzaei H. Biosensors for detection of Tau protein as an Alzheimer's disease marker, International Journal of Biological Macromolecules: 162; 2020: 1100-1108.
[21]Hansson O, Lehmann S , Otto M , Zetterberg H, Lewczuk P. Advantages and disadvantages of the use of the CSF Amyloid β (Aβ) 42/40 ratio in the diagnosis of Alzheimer's Disease, Alzheimer's Research & Therapy: 2019; 11: 34.
[22]Li S, Lin C, Wei K, Huang C, Hsu P, Liu H, Lu Y, Lin S, Yang H, Ma C, Non-invasive screening for early Alzheimer's disease diagnosis by a sensitively immunomagnetic biosensor: 2016, Scientific Reports; 6: 25155.
[23]Langenhove V, Lieva, ed. Smart textiles for medicine and healthcare: materials, systems and applications. Elsevier: 2007.
[24]Liu Y, Wang H, Zhao W, Zhang M, Qin H, Xie Y. Flexible, Stretchable Sensors for Wearable Health Monitoring: Sensing Mechanisms, Materials, Fabrication Strategies and Features, Sensors: 2018; 18: 645.
[25]Mannoor MS, Tao H, Clayton JD, Sengupta A, Kaplan DL, Naik RR, Verma N, Omenetto FG, McAlpine MC. Nat. Commun.: 2012; 3: ncomms1767.
[26]Kim J, Vald´es- Ram´ırez G, Bandodkar AJ, Jia W, Martinez AG, Ram´ırez J, Mercier P. Wang J. Analyst: 2014; 139: 1632-1636.
[27]Kim J, Imani S, de Chemistry Araujo WR, Warchall J, Ram´ırez G, Paix˜ao TRLC, Mercier PP, Wang J. Biosens. Bioelectron: 2015; 74: 1061-1068.
[28]Borini S, White R, Wei D, Astley M, Haque S, Spigone E, Harris N, Kivioja J, Ryh¨anen T. ACS Nano: 2013; 7: 11166-11173.
[29]Hatice CA, Anna B, Felix VS, Gerald A. Urban, Firat Güder, Arben Merkoçi, Susanna Maria Früh, and Can Dincer*, Integrated Devices for Non-Invasive Diagnostics Adv. Funct. Mater: 2021: 31; 2010388.
[30]Yang YR, Gao W. Wearable and flexible electronics for continuous molecular monitoring Chem. Soc. Rev: 2019; 48: 1465.
[31]Güder F, Ainla A, Redston J, Mosadegh B, Glavan A, Martin TJ, Whitesides GM, Chem A. Int. Ed: 2016; 55: 572.

Endowing Electric Conductivity to Common Fabrics with Common Structures Via Electroless Deposition Technique

Muhammad Yousif[a], Bilqees[a], Sadam Hussain[a], Suibo Zhou[a], Jianchen Hu[a*], Ke-Qin Zhang[a*]

[a] *National Engineering Laboratory for Modern Silk, College of Textile and Clothing Engineering, Soochow University, Suzhou 215123, China*

[*]Corresponding author's email: hujianchen@suda.edu.cn; kqzhang@suda.edu.cn

Abstract

Herein, we successfully synthesized different conductive fabrics by depositing two transition metals, i.e., Cu, Ni via electroless deposition (ELD) technique. The results were compared with different fabrics (cotton, PET, and nylon) and fabric weave designs (plain and mesh). Each successful step for fabrication of conductive fabrics (plasma treatment, salinization, polymerization, and electroless deposition) was directly dependent on water contact angle, and the obtained results were found satisfactory accordingly. The surface morphology validated the successful deposition of metals. Furthermore, the effect of crystallinity was explained using X-ray diffraction technique. It was observed that the PET fabric (i.e., Cu-PET having a resistance of 0.18 Ω sq^{-1} and Ni-PET having resistance of 0.116 Ω sq^{-1}) shows much favourable metal deposition over cotton/nylon fabrics.

Keywords: Textile Fabrics; Weave Design; Transition Metals; Copper, Nickle; Electroless Deposition

1. Introduction

Wearable electronics is one of the emerging areas of electronics due to its wide range of applications in wearable displays, solar cells, actuators, data management devices, biomedical sensors, and real-time health monitoring devices. [1] Wearable electronics rely heavily on metal-coated textiles that require high-performance flexible conductive interconnects, contacts and electrodes. In addition, such electronics provide properties such as electrical conductivity, flexibility, lightweight, electrostatic discharge, and electromagnetic interference protection. [2-5]

Over the past few years, remarkable manufacturing strategies have been achieved in converting various organic and inorganic conductive materials, such as metals, carbon materials, nanoparticles, nanowires, and polymers, into versatile, stretchable, and wearable substrates. Due to these properties, including extremely high electrical conductivity, low cost, simplicity to synthesize, and stability, metals have consistently been recognized as promising materials for fabricating flexible conductors. [6-11]

From the cost perspective, copper, and nickel (Cu, Ni) show significantly higher conductivity to price ratios than noble metals (Au, Ag, Pt, and Pd). However, the spontaneous surface oxidation of Cu in the air poses significant challenges for its practical applications. On the other hand, Ni, another important transition metal, shows a relatively similar conductivity to price ratio but with much better resistance to oxidation than Cu. Therefore, Ni is a great metal for flexible electronics, especially for electrochemical energy storage devices. Hence, the engineering of well-defined metal structures is decorated on the surface of flexible textile via various techniques, i.e., physical vapor deposition, galvanic deposition, chemical vapor deposition, atomic layer deposition, electrochemical plating, and electroless deposition (ELD). [12-18]

Among the techniques mentioned above, the ELD technique is much attractive due to its versatility, site-selectivity, easy solution-process, low-cost, and scalable method, which renders it particularly suitable for fabricating large-scale metal structures on flexible textile surfaces. [19-21] Recently, various polymers, thin films, paper, rubber, foils, and fabrics have been developed as flexible electrode devices by the ELD technique. Among the different available textile fabrics, polyethylene terephthalate (PET), cotton and nylon, attract much attention due to available multipurpose structures, low cost, lighter weight, high strength, and facile processability. [22-24]

Herein, we are trying to deposit two metals (Cu and Ni) on different types of fabrics (cotton, PET, and nylon) and different weave design (plain and mesh) via the ELD technique. Thus, we aim to extend the applications of everyday textiles with typical weave structures in the fast-developing wearable electronic devices. By connotation of this strategy, the pristine weave structures of the original textiles are maintained, making it easy to integrate with fabrics by the techniques such as seaming. Furthermore, systematic guidance is expected to be established for choosing suitable metals and fabrics for the potential applications by comparing the performance of these two metals and analyzing the effects of these metals on fabric weave designs, i.e., deposition time, towards resistance in the ELD technique. We believe that our comparative study will help scientists understand the effects of ELD process factors and help them to choose the best performing material for the ELD technique.

2. Experimental Section

2.1 Materials

The 100% Cotton, PET, and Nylon fabrics with an areal density of 50 g/m^2, with different weave designs (plain and mesh structures), were supplied by Changle HG Textile Fujian, China. In addition, 2-methoxyethanol, ammonium tetrachloropalladate (II) ($(NH_4)_2PdCl_4$), Vinyltrimethoxy silane (VTMS), [2-(methacryloyloxy) ethyl] trimethyl ammonium chloride (METAC) (80 wt% aqueous solutions), Potassium persulfate (KPS) ($K_2S_2O_8$), Nickel sulfate hexahydrate ($NiSO_4 \cdot 6H_2O$), Copper(II) sulfate pentahydrate ($CuSO_4 \cdot 5H_2O$), Potassium sodium tartrate tetrahydrate ($KNaC_4H_4O_6 \cdot 4H_2O$) and Formaldehyde (HCHO) were purchased from Sigma-Aldrich. Chemical reagents, Sodium citrate [$HOC(COONa)(CH_2COONa)_2 \cdot 2H_2O$], Lactic acid ($CH_3CH(OH)COOH$), Dimethylamine borane ($(CH_3)_2NH \cdot BH_3$), Sodium hydroxide (NaOH), were purchased from Sinopharm Chemical Reagent Co., Ltd (China).

2.2 Characterizations

The surface morphology of fabrics before and after ELD were examined at an accelerating voltage of 15 kV on a Scanning electron microscope (FE-SEM) (S-4800; Hitachi Ltd. Japan). The crystallinity of fabrics before and after metal deposition was investigated through X-ray diffraction (XRD) (model D/max-IIB, Rigaku). Plasma treatment was carried out on Harrick Plasma cleaner PDC-002. An OCA15EC Contact angel instrument (Data physics Germany) was applied to determine the water contact angle of fabrics to ensure the successful surface functionalization of the ELD process. Finally, the fabric resistance value was measured using ST-225*C multifunctional Digital 4 Probe meter.

2.3 Fabrication of Conductive Fabrics

Fig.1 indicates the complete fabrication process for the conductive fabrics, consisting of 5 steps (Plasma treatment, salinization, polymerization, ion exchanger, and electroless deposition).

2.3.1 Plasma Treatment

All the fabrics were washed with DI-water and detergents to remove impurities before surface functionalization and the ELD process. The hydrophobic nature of fabrics (PET and nylon) needs plasma treatment before salinization, creating functional hydroxyl groups for further reactions. [25] At constant pressure (30 Pa) and power (45 W), plasma was used for 5 minutes.

2.3.2 Salinization

The Plasma-fabrics and hydrophilic fabrics were immersed into an ethanol solution containing 2% (v/v) VTMS for 15 min. After VTMS treatment, the surface of fabrics became largely hydrophobic due to dense vinyl groups. The VTMS-fabrics were finally rinsed with DI water and immediately dried at 150 °C for 15 min.

2.3.3 Polymerization

The VTMS-Fabrics sample was immersed into 100 mL of 20 % (v/v) METAC solution containing 60 mg KPS and placed at 80°C for 1 hour to carry out the polymerization of PMETAC brushes while stirring. Finally, the polymerized samples were carefully washed with DI water and dried at 45°C in a vacuum dryer for 2 hours.

2.3.4 Ion Exchange

The PMETAC-Fabrics were immersed into a 5 mM $(NH_4)_2PdCl_4$ solution for 15 mins for the ion exchange. During an ion exchange $[PdCl_4]^{2-}$, catalytic ions were immobilized onto the quaternary ammonium groups of PMETAC polymer chains, and samples were gently washed with DI before further reaction.

2.3.5 Electroless Deposition

The deposition of Copper (Cu) and Nickel (Ni) via the ELD technique was followed with previously reported methods. [25-26] Finally, the sample was immersed in the Cu and Ni ELD bath. The Cu ELD bath contained the 1:1 mixture of solution A and B, where solution A contained the 13 g/L copper (II) sulfate pentahydrate, 29 g/L potassium sodium tartrate tetrahydrate, and 12 g/L sodium hydroxide. At the same time, solution B contained 9.5 mL/L formaldehyde in water. The Ni ELD bath contained the 4:1 mixture of solution C and D, where solution C contained the 40 g/L nickel sulfate hexahydrate, 20 g/L sodium citrate, and 10 g/L lactic acid. In contrast, solution D contained 1.0 g/L dimethylamine borane in water. The reaction was performed for 10-50 min and the effect of time was analyzes on the ELD process. After the activity, the samples were washed with DI water and ethanol. Finally, the obtained samples were dried in a vacuum dryer.

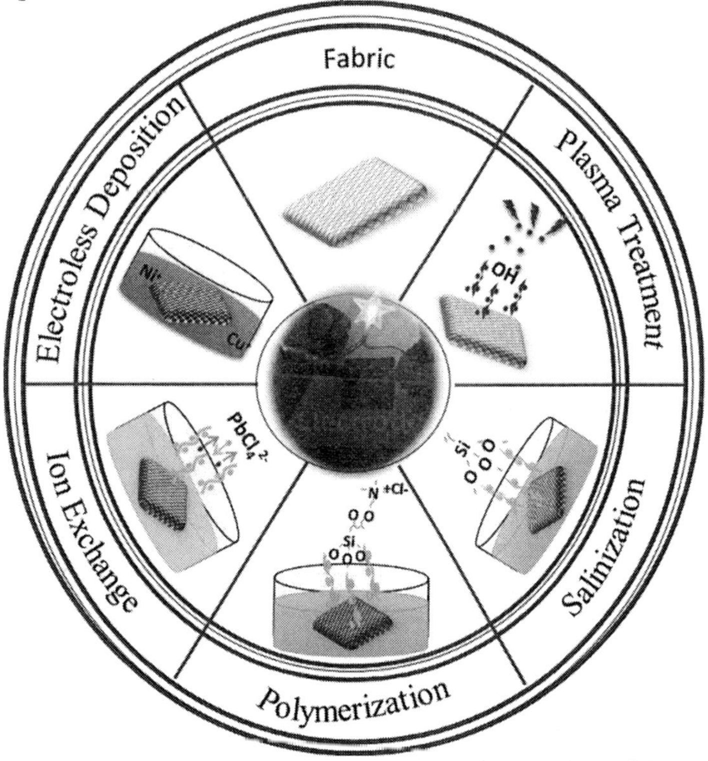

Fig.1 Fabrication steps of electroless deposition of noble metals on different fabrics

3. Results and Discussion

3.1 Water Contact Angle

Water contact angle measurements were examined to evaluate the wettability of neat fabrics (cotton, PET, and nylon) and metal deposited fabrics (Cu and Ni) to confirm the effective surface modification of each stage in the ELD process. It was found that the contact angles differ for neat fabrics, air-plasma treated-fabrics, VTMS-fabrics, PMETAC-fabrics, and ELD-fabrics. Fig.2 indicates the water contact angle of different fabrics after each process of ELD. The neat 100% cotton fabric showed hydrophilic nature with a water contact angle of 1.7°, while 100% PET and nylon fabrics show hydrophobic nature with a water contact angle of 134° and 131°, respectively. It was found in the literature that coating of VTMS needs the hydrophilic nature of substrate. [27] And hydrophilic nature is mainly due to the presence of hydroxyl groups in any material. Plasma treatment is a well-known method of creating hydroxyl groups onto hydrophobic materials and changing the nature of material towards hydrophilic. Our experiment observed that plasma treatment changes the hydrophobic nature of PET and nylon into hydrophilic nature with a 1.5° and 1.2° water contact angle, respectively. This hydrophilicity is due to the successful creation of hydroxyl groups onto PET and nylon fabrics, as reported earlier. [19] These hydrophilic fabrics (cotton, plasma-PET, and plasma-nylon) got hydrophobic nature again after treatment with VTMS in an ethanol solution, having contact angles of 133°, 140°, and 144°, respectively. This change in nature indicates successful chemical attachment of silane molecules with hydroxyl groups of fabrics. [27] Due to the presence of vinyl groups in fabrics, they show hydrophobic nature. [28] Consequently, the VTMS-fabrics (cotton, PET, and nylon) were further treated with METAC monomer and a KPS initiator for a specific time. This specific allowing time helped to bind chemically METAC monomer and KPS onto fabric surface and create PMETAC polymerization on fabrics surface. It was observed that after successful polymerization, PMETAC-fabrics (cotton, PET, and nylon) dramatically change the water contact angle to 59°, 74°, and 72°, respectively. The change towards hydrophilicity suggests successful polymerization onto fabrics as PMETAC contains hydroxyl groups. [29] Finally, surface-functionalized PMETAC polymer brushes onto fabrics are well enough to strongly self-attraction (via ionic attraction) and densely pack the metals (Cu and Ni) in an ELD bath. After the densely packed metals onto PMETAC-fabrics, the Cu-cotton, Cu-PET, Cu-nylon, Ni-cotton, Ni-PET, and Ni-nylon show once again hydrophobic nature having a contact angle of 140°, 144°,145°, 143°, 146°, and 148° respectively. This super hydrophobicity is mainly due to densely packed metal particles onto the surface of fabrics. [30] Overall results of the water contact angle are good evidence of a successful ELD process in each step.

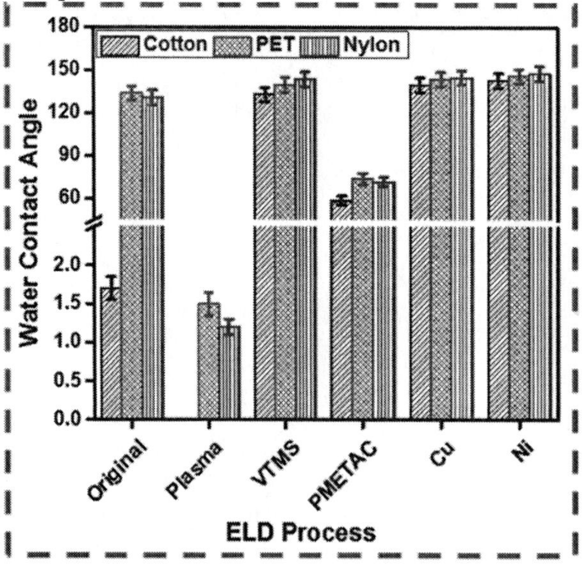

Fig.2 Water contact angle at different steps during electroless deposition.

3.2 Surface Morphology

The surface changes on neat fabrics and metal deposited fabrics were studied via SEM at different magnifications. SEM surface morphology in Fig.3 (a, b, c, j, k, and l) indicates the smooth fibrous morphology of neat cotton, PET, and nylon fabrics, respectively. While SEM surface morphology in Fig.3 (d, e, f, g, h, i, m, n, o, p, q, and r) indicates the metal deposited fabrics which were densely covered with metal particles. Moreover, increased fabric thickness from 11 μm to 23μm suggests the proper deposition of metal particles. [31] In Fig.4, the average fiber diameter of neat fabrics, Cu-fabrics, and Ni-fabrics was calculated from SEM images.

It was observed that fiber diameter increased after metal deposition and overall fiber diameter of nickel deposition is higher than copper deposition. This is due to higher deposition rate of Ni particles than copper particles and higher tendency of surface polymerization of PMETAC to attract nickel particles over copper particles. Moreover, Fig.5 indicates the photographic and physical assessment of neat fabrics over metal deposited fabrics. After Cu deposition leading to noticeable color change of fabrics from white to brown (light to dark). The color variation from light brown to dark brown after copper deposition on different fabrics depends on sheet resistance, as shown in Fig.6A (lowest sheet resistance shows the light brown color. In contrast, the highest sheet resistance shows the dark brown color). Furthermore, after Ni deposition, the color change of fabrics from white to grey (light to dark). The color variation from light grey to dark grey after nickel deposition on different fabrics also depends on sheet resistance, like copper deposition as shown in Fig.6B (lowest sheet resistance shows the light grey color. In contrast, the highest sheet resistance shows the dark grey color).

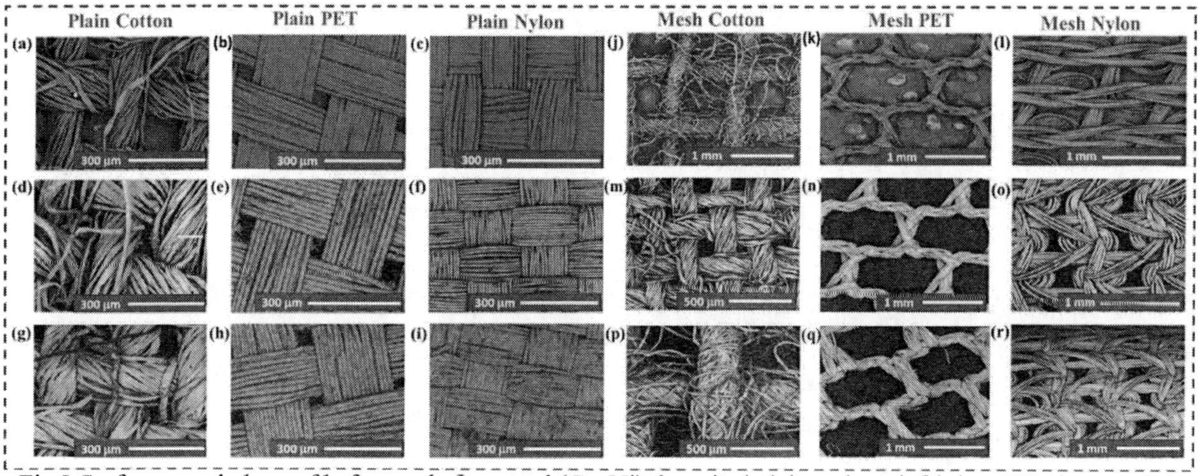

Fig.3 Surface morphology of before and after metal (Cu, Ni) deposited plain and mesh fabrics. SEM image of a) Neat Plain Cotton, b) Neat Plain PET, c) Neat Plain Nylon d) Cu-Plain Cotton, e) Cu-Plain PET, f) Cu-Plain Nylon, g) Ni-Plain Cotton, h) Ni-Plain PET, i) Ni-Plain Nylon, j) Neat Mesh Cotton, k) Neat Mesh PET, l) Neat Mesh Nylon, m) Cu-Mesh Cotton, n) Cu-Mesh PET, o) Cu-Mesh Nylon, p) Ni-Mesh Cotton, q) Ni-Mesh PET and r) Ni-Mesh Nylon.

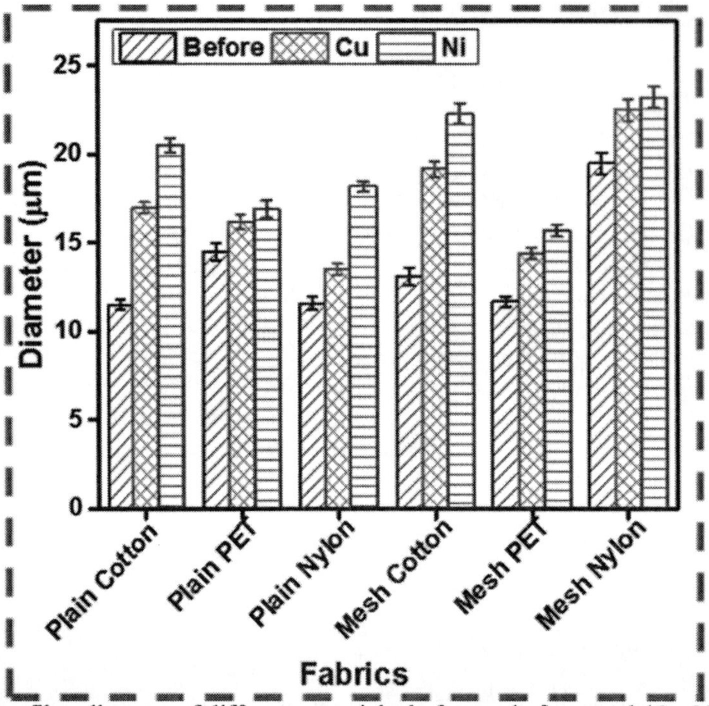

Fig.4 Average fiber diameter of different materials, before and after metal (Cu, Ni) deposition.

Fig.5 Photograph of different fabric types before and after metal deposition.

3.3 Effect of Time on Resistance

The effect of time is one of the most important factors for the proper deposition of metal particles. The effect of time for resistance measurement was examined from 10 to 50 min, and results are disclosed in Fig.6 (A, B). It was observed that change in resistance decreases with increasing time, and nearly after 30 min, the change in resistance gets equilibrium or again increases. Meanwhile, the weight of metal deposited fabrics were measured concerning time, and results disclosed that weight increases with increasing time and obtains equilibrium after 30 min or decreases. [32] The equilibrium or decrease in weight is due to the saturation limit of fabric for further deposition, [33] and this can result in increased resistance at higher deposition time. Moreover, the lowest/optimized resistance on plain Cu-cotton, Cu-PET, Cu-nylon, Ni-cotton, Ni-PET, and Ni-nylon fabric was measured as 4.3 Ω sq^{-1}, 0.18 Ω sq^{-1}, 0.24 Ω sq^{-1}, 0.4 Ω sq^{-1}, 0.11 Ω sq^{-1} and 0.2 Ω sq^{-1} respectively. Simultaneously, the lowest/optimized resistance on mesh Cu-cotton, Cu-PET, Cu-nylon, Ni-cotton, Ni-PET, and Ni-nylon fabric was measured as 4.4 Ω sq^{-1}, 0.21 Ω sq^{-1}, 0.44 Ω sq^{-1}, 0.4 Ω sq^{-1}, 0.12 Ω sq^{-1} and 0.32 Ω sq^{-1} respectively. The results concluded that metal deposited on plain fabrics gets relatively lower resistance than mesh fabrics. Fig.7 shows the resistance measurement and conductivity of Cu-PET and Ni PET after the ELD process.

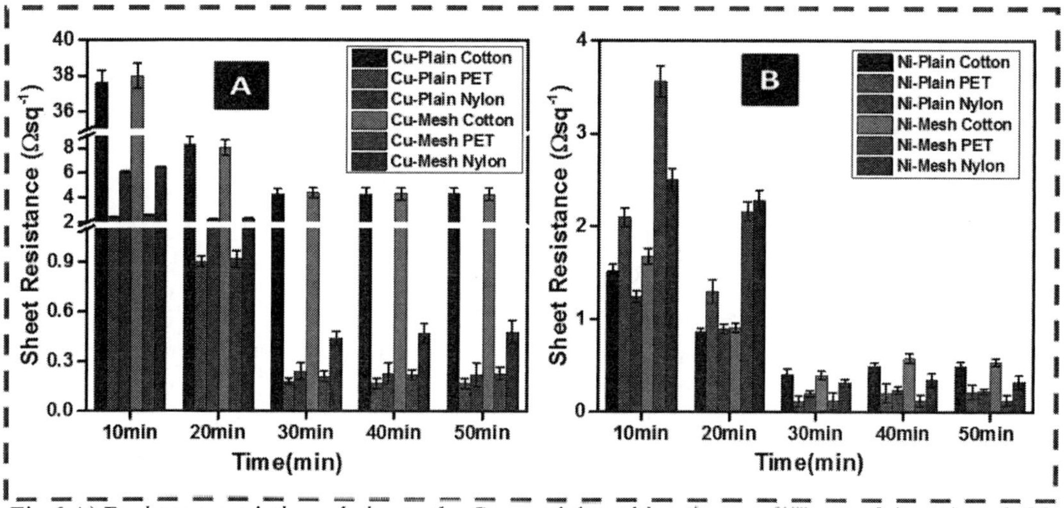

Fig.6 A) Resistance variation relating to the Cu metal deposition time on different plain and mesh fabrics B) Resistance variation relating to the Ni metal deposition time on different plain and mesh fabrics.

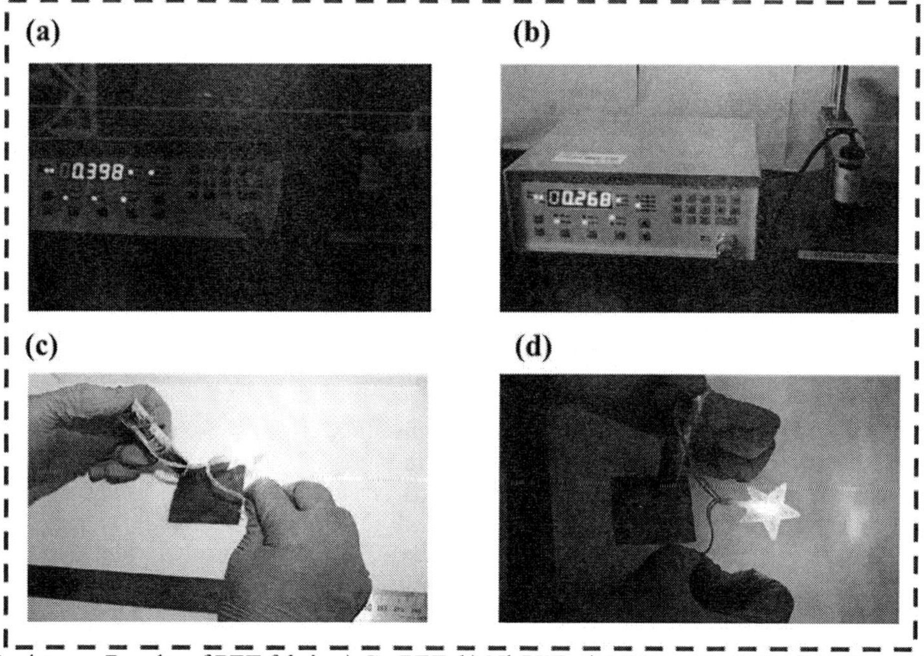

Fig.7 Resistance Results of PET fabric a) Cu-PET, b) Ni-PET, c) LED on Cu-PET, d) LED on Ni-PET.

3.4 X-Ray Diffraction

The XRD analysis was carried out to determine the crystallinity of neat fabrics (cotton, PET, and nylon) and metal-deposited (Cu and Ni) fabrics. Fig.8 (a) shows the XRD patterns of neat cotton, Cu-cotton, and Ni-cotton. The diffraction peaks 2θ of the neat cotton were observed at 14.85°, and 22.7°, corresponding to (101), (002) planes of cotton fabric (cellulose I structure). [34]

Fig.8 (b) shows the XRD pattern of neat PET, Cu-PET, and Ni-PET. The diffraction peaks 2θ of the neat PET was observed at 17.5°, 22.7° and 25.6°, corresponding to planes of polyester fabric. [35] Fig.8 (c) shows the XRD pattern of neat nylon, Cu-nylon, and Ni-nylon. The diffraction peaks 2θ of the neat nylon was observed at 21.7°, corresponding to the plane of nylon fabric. [36] The peaks of Cu-deposited fabrics (cotton, PET, and nylon) located at 2θ = 43.4°, 50.5°, corresponds to (111), (200) crystal faces of copper in all fabrics. Further, two peaks were observed at 36.6°, corresponding to copper. In contrast, the peaks at 36.6° and 74.3° are corresponding to Cu_2O, representing the oxidation of copper. [35, 37] The peaks of Ni-deposited fabrics (cotton, PET, and nylon) located at 2θ = 44.6° corresponding to (111) crystal plane of nickel. [37] In contrast, no nickel oxide peak was observed, which explains the good stability of Ni-deposited fabrics in the air.

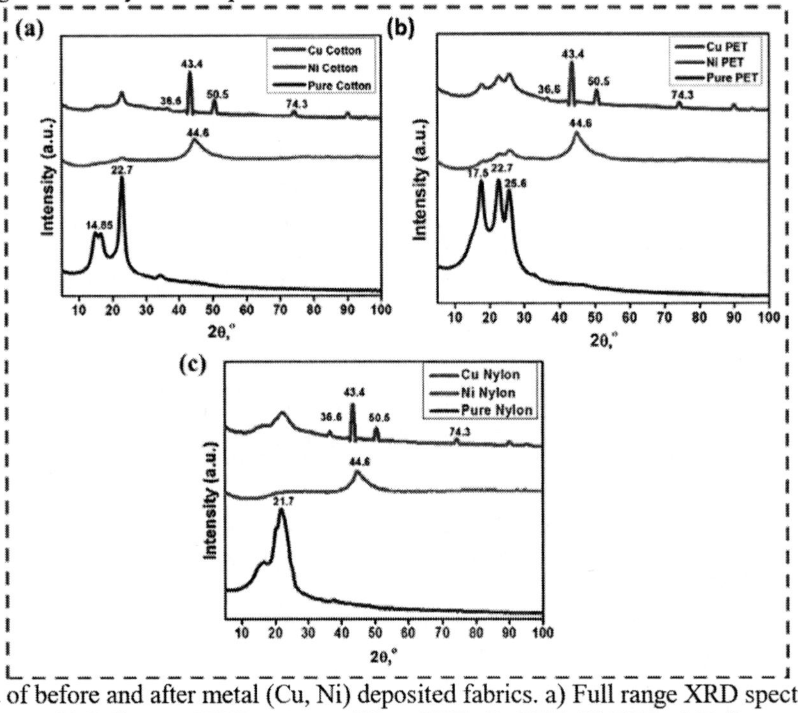

Fig.8 XRD spectra of before and after metal (Cu, Ni) deposited fabrics. a) Full range XRD spectra of Neat-Cotton, Cu-Cotton, Ni-Cotton, b) Full range XRD spectra of Neat-PET, Cu-PET, Ni-PET, c) Full range XRD spectra of Neat-Nylon, Cu-Nylon, Ni-Nylon.

4. Conclusion

The study was conducted to explore the synthesis and application of conductive fabrics. Different conductive fabrics were successfully developed by depositing two transition metals, i.e., Cu, Ni via the electroless deposition (ELD) technique. Both metals effects and performance on different fabrics (cotton, PET, and nylon) with different fabric weave designs (plain and mesh) were compared. The surface morphology validated the successful deposition of metals. It was observed that Ni deposited fabrics have a comparatively lower resistance (i.e., Ni-PET having a resistance of 0.116 $\Omega\ sq^{-1}$) than that of copper deposited fabrics (i.e., Cu-PET having a resistance, of 0.18 $\Omega\ sq^{-1}$) due to higher deposition of Ni. More ever, it was observed that fabric weave design has a significate role in the effectiveness of the ELD process, for Ni/Cu compact weave structures are more stable than loose weave structures. To conclude, the ELD technique is highly recommended for Ni-deposition over Cu-deposition, PET fabrics over cotton/nylon, and plain weave design over mesh weave design. These synthesized, conductive fabrics have great potential in the field of wearable electronics. They can be used in various applications such as flexible and electronic textile sensors, flexible and wearable computing, smart clothing, and energy storage devices.

5. Acknowledgment

The work was supported by National Engineering Laboratory for Modern Silk, College of Textile and Clothing Engineering, Soochow University, Suzhou 215123, China.

References

[1] Bae JH, Oh HJ, Song J, Kim DK, Yeang BJ, Ko JH, Kim SH, Lee W, Lim SJ. Nano-and microfiber-based fully fabric triboelectric nanogenerator for wearable devices. Polymers: 2020; 12 (3): 658.

[2] Ryu S, Lee P, Chou JB, Xu R, Zhao R, Hart AJ, Kim SG. Extremely elastic wearable carbon nanotube fiber strain sensor for monitoring of human motion. ACS nano: 2015; 9 (6): 5929-36.

[3] Li Z, Shen J, Abdalla I, Yu J, Ding B. Nanofibrous membrane constructed wearable triboelectric nanogenerator for high performance biomechanical energy harvesting. Nano Energy: 2017; 36: 341-8.

[4] Hu L, Pasta M, La Mantia F, Cui L, Jeong S, Deshazer HD, Choi JW, Han SM, Cui Y. Stretchable, porous, and conductive energy textiles. Nano letters: 2010; 10 (2): 708-14.

[5] Lee YH, Kim JS, Noh J, Lee I, Kim HJ, Choi S, Seo J, Jeon S, Kim TS, Lee JY, Choi JW. Wearable textile battery rechargeable by solar energy. Nano letters: 2013; 13 (11): 5753-61.

[6] Sekhar SC, Nagaraju G, Yu JS. Conductive silver nanowires-fenced carbon cloth fibers-supported layered double hydroxide nanosheets as a flexible and binder-free electrode for high-performance asymmetric supercapacitors. Nano Energy: 2017; 36: 58-67.

[7] Chen L, Chen L, Ai Q, Li D, Si P, Feng J, Zhang L, Li Y, Lou J, Ci L. Flexible all-solid-state supercapacitors based on freestanding, binder-free carbon nanofibers@ polypyrrole@ graphene film. Chemical Engineering Journal: 2018; 334: 184-90.

[8] Xu T, Zhang Z, Qu L. Graphene-based fibers: recent advances in preparation and application. Advanced Materials: 2020; 32 (5): 1901979.

[9] Du D, Tang Z, Ouyang J. Highly washable e-textile prepared by ultrasonic nanosoldering of carbon nanotubes onto polymer fibers. Journal of Materials Chemistry C: 2018; 6 (4): 883-9.

[10] Wu C, Kim TW, Li F, Guo T. Wearable electricity generators fabricated utilizing transparent electronic textiles based on polyester/Ag nanowires/graphene core–shell nanocomposites. Acs Nano: 2016; 10 (7): 6449-57.

[11] Huang Q, Liu L, Wang D, Liu J, Huang Z, Zheng Z. One-step electrospinning of carbon nanowebs on metallic textiles for high-capacitance supercapacitor fabrics. Journal of materials chemistry A. 2016; 4(18): 6802-8.

[12] Yu Y, Zhang Y, Li K, Yan C, Zheng Z. Bio-inspired chemical fabrication of stretchable transparent electrodes. Small: 2015; 11(28): 3444-9.

[13] Ohno K, Kayama Y, Ladmiral V, Fukuda T, Tsujii Y. A versatile method of initiator fixation for surface-initiated living radical polymerization on polymeric substrates. Macromolecules: 2010; 43 (13): 5569-74.

[14] Kimura M, Yamagiwa H, Asakawa D, Noguchi M, Kurashina T, Fukawa T, Shirai H. Site-selective electroless nickel plating on patterned thin films of macromolecular metal complexes. ACS applied materials & interfaces: 2010; 2 (12): 3714-7.

[15] Pandey PA, Bell GR, Rourke JP, Sanchez AM, Elkin MD, Hickey BJ, Wilson NR. Physical vapor deposition of metal nanoparticles on chemically modified graphene: observations on metal–graphene interactions. Small: 2011; 7 (22): 3202-10.

[16] Grodzicki A, Łakomska I, Piszczek P, Szymańska I, Szłyk E. Copper (I), silver (I) and gold (I) carboxylate complexes as precursors in chemical vapour deposition of thin metallic films. Coordination chemistry reviews: 2005; 249 (21-22): 2232-58.

[17] Little BK, Li Y, Cammarata V, Broughton R, Mills G. Metallization of kevlar fibers with gold. ACS applied materials & interfaces: 2011; 3 (6): 1965-73.

[18] Jur JS, Sweet III WJ, Oldham CJ, Parsons GN. Atomic layer deposition of conductive coatings on cotton, paper, and synthetic fibers: conductivity analysis and functional chemical sensing using "all-fiber" capacitors. Advanced functional materials: 2011; 21 (11): 1993-2002.
[19] Liang S, Li Y, Zhou T, Yang J, Zhou X, Zhu T, Huang J, Zhu J, Zhu D, Liu Y, He C. Microfluidic Patterning of Metal Structures for Flexible Conductors by In Situ Polymer - Assisted Electroless Deposition. Advanced Science: 2017; 4 (2): 1600313.
[20] Yu Y, Yan C, Zheng Z. Polymer-Assisted Metal Deposition (PAMD): A Full-Solution Strategy for Flexible, Stretchable, Compressible, and Wearable Metal Conductors. Advanced Materials: 2014; 26 (31): 5508-16.
[21] Liu L, Yu Y, Yan C, Li K, Zheng Z. Wearable energy-dense and power-dense supercapacitor yarns enabled by scalable graphene–metallic textile composite electrodes. Nature communications: 2015; 6 (1): 1-9.
[22] Muthukumar N, Thilagavathi G, Kannaian T. Polyaniline-coated foam electrodes for electroencephalography (EEG) measurement. The Journal of The Textile Institute: 2016; 107 (3): 283-90.
[23] Yang Y, Huang Q, Niu L, Wang D, Yan C, She Y, Zheng Z. Waterproof, ultrahigh areal - capacitance, wearable supercapacitor fabrics. Advanced Materials: 2017; 29 (19): 1606679.
[24] Huang Q, Wang D, Zheng Z. Textile-based electrochemical energy storage devices. Advanced Energy Materials: 2016; 6 (22): 1600783.
[25] Zhen H, Li K, Chen C, Yu Y, Zheng Z, Ling Q. Water-borne foldable polymer solar cells: one-step transferring free-standing polymer films onto woven fabric electrodes. Journal of Materials Chemistry A: 2017; 5 (2): 782-8.
[26] Yu Y, Xiao X, Zhang Y, Li K, Yan C, Wei X, Chen L, Zhen H, Zhou H, Zhang S, Zheng Z. Photoreactive and Metal-Platable Copolymer Inks for High-Throughput, Room-Temperature Printing of Flexible Metal Electrodes for Thin-Film Electronics. Advanced Materials: 2016; 28 (24): 4926-34.
[27] DEMİR A, Bozaci E, Gülümser TÜ, Sarikanat M. An ecological approach for the surface modification of aramid fibers. Tekstil ve Konfeksiyon: 2016; 26 (3): 256-61.
[28] Wang C, Yang H, Chen F, Peng L, Gao HF, Zhao LP. Influences of VTMS/SiO2 ratios on the contact angle and morphology of modified super-hydrophobic silicon dioxide material by vinyl trimethoxy silane. Results in Physics: 2018; 10: 891-902.
[29] Politakos N, Azinas S, Moya SE. Responsive Copolymer Brushes of Poly [(2-(Methacryloyloxy) Ethyl) Trimethylammonium Chloride](PMETAC) and Poly (1H, 1H, 2H, 2H-Perfluorodecyl acrylate)(PPFDA) to Modulate Surface Wetting Properties. Macromolecular rapid communications: 2016; 37 (7): 662-7.
[30] Liu S, Hu M, Yang J. A facile way of fabricating a flexible and conductive cotton fabric. Journal of Materials Chemistry C: 2016; 4 (6): 1320-5.
[31] Shiue ST, Yang CH, Chu RS, Yang TJ. Effect of the coating thickness and roughness on the mechanical strength and thermally induced stress voids in nickel-coated optical fibers prepared by electroless plating method. Thin Solid Films: 2005; 485 (1-2): 169-75.
[32] Sridhar N, Bhat KU. Effect of deposition time on the morphological features and corrosion resistance of electroless Ni-high P coatings on aluminium. J. Mater: 2013; 2013: 1-7.
[33] Tai Y, Xu C, Chen H. Silver-coated glass fabric composites prepared by electroless plating. Materials Letters: 2016; 180: 144-7.
[34] Zhang H, Shen L, Chang J. Comparative study of electroless Ni-P, Cu, Ag, and Cu-Ag plating on polyamide fabrics. Journal of Industrial Textiles: 2011; 41 (1): 25-40.
[35] Gan X, Wu Y, Liu L, Shen B, Hu W. Electroless copper plating on PET fabrics using hypophosphite as reducing agent. Surface and Coatings Technology: 2007; 201 (16-17): 7018-23.
[36] Muthukumar N, Thilagavathi G, Kannaian T. Analysis of piezoresistive behavior of polyaniline-coated nylon Lycra fabrics for elbow angle measurement. The Journal of The Textile Institute: 2017; 108 (2): 233-8.
[37] Wang C, Yang F, Cao Y, He X, Tang Y, Li Y. Cupric oxide nanowires on three-dimensional copper foam for application in click reaction. RSC advances: 2017; 7 (16): 9567-72.

Cite as: H.H. Liu et al.,
TBIS 10.3993/tbis (2021).

Research on Emotion Recognition of Heart Rate Variability for Smart Clothing

Huan-Huan Liu[1,2], Zhao-Hui Wang[1,2] *

[1] *College of Fashion &Design, Donghua University, 1882West Yan'an Road, Shanghai 200051, China*
[2] *Key Laboratory of Clothing Design&Technology, Donghua University, Ministry of Education, Shanghai, 20051,China*

*Corresponding author's email: wzh_sh2007@dhu.edu.cn

Abstract

For the development of emotion recognition smart clothing, this research first designed a smart luminous clothing model for emotion recognition, and proposed an emotion recognition method based on heart rate variability. Using 24 university students as experimental subjects, the ECG signals of four emotions, namely joy, anger, sadness and happiness, were collected with the help of Mindware multi-channel physiometer, denoised using FIR filter and a median filter. Thirteen heart rate variabilities of time-domain, frequency-domain and non-linear feature parameters were extracted. Feature selection was optimised using principal component analysis. Finally, a support vector machine, the random forest and extreme gradient boosting (XGBoost) algorithm and six other algorithms were used to implement emotion classification. The results showed that the positive and negative emotions based on arousal emotions were better classified, and the extreme gradient boosting (XGBoost) algorithm had a high recognition accuracy of 80%. The design of this study uses information from a heart rate variability feature as an emotion recognition method which can then be applied in smart clothing oriented to human mental health monitoring and providing a favourable tool for emotion recognition and psychological intervention.

Keywords: Emotion Recognition; ECG Signals; Heart Rate Variability; Smart Clothing; Support Vector Machines

1. Introduction

The increasing pace of life today has led to an increase in psychological stress. Physical health problems are closely related to the psychological and emotional state, and psychological health can directly affect one's physical health. [1]. In recent years, there has been a gradual increase in questions about sudden death and mood disorders such as depression [2]. At present, smart clothing is gradually improving its functions in transmitting information, entertainment and communication and physiological health monitoring, but it is still in the preliminary research stage in areas such as emotion recognition and monitoring [3].

Emotion recognition methods are mainly based on facial expressions, voice intonation, and body posture [4], which are easily controlled by subjective will, whereas physiological signals are controlled by the neurological and endocrine systems and have objective reality. Current classification methods based on multimodal physiological signal recognition are widely used, and ECG signals play an important role in emotion recognition. At the same time, considering the lightness of signal collection equipment in intelligent clothing, ECG collection equipment is more portable and collected thanother physiological signal equipment, and this is conducive to the marketability of intelligent clothing for emotion recognition.

Heart rate variability (HRV) refers to the small variability characteristic of the ECG signal between beat-to-beat intervals and is, regulated by sympathetic and parasympathetic nerves, which reflects abnormalities in mental health. Thisis a highly sensitive indicator of autonomic tone [5]. The European Heart Association ESC and the North American Society of Pacing and Electrophysiology NASPE recognizesHRV as a valid indicator of autonomic function [6]. Studies have shown that a lower HRV value means that the sympathetic nervous system is active, indicating that the body is in a state of excitement and tension whereasa higher HRV value means that the parasympathetic nervous system is active and the body is in a state of relaxation and rest. Picard et al. [7]

demonstrated experimentally the feasibility of an approach to emotion recognition using physiological signal signature parameters. P Rainville et al. [8] extracted time-domain features such as median RR interval and variance of heart rate variability under different emotions and combined them with respiratory signal features. This is used to select features using principal component analysis and to construct a decision tree for emotion recognition. Yi Hui et al. [9] constructed a model using the characteristic parameters of the time-domain and frequency-domain indicators of heart rate variability and uses this as the basis for identification. Goshvarpouret et al. [10] found that the shape of the logged Poincare plot in the relaxed state resembled a cigar, and that the shape of the Poincare in joy, sad and fearful emotions were circular. Compared to traditional time and frequency domain analysis, there is little research on the non-linear characteristics of emotional physiological signals, and further exploration is needed. This paper proposes an emotion recognition method based on heart rate variability using time-domain, frequency-domain and non-linear feature values as the basis for emotion recognition classification, and also in constructing an emotion recognition model based on heart rate variability. Smart clothing for human emotion monitoring uses ECG collection devices embedded in clothing to complete the collection and analysis of ECG signals and to correctly grasp changes in emotional states in time for psychological intervention and emotional guidance, thus improving human physical and mental health.

2. Theoretical Model for Emotion-recognising Intelligent Clothing

Emotion recognition smart clothing includes three modules: signal acquisition, signal processing and result display, and each module uses wireless communication to achieve signal transmission, as shown in Fig.1. The model uses a flexible sensor on the smart garment to capture the ECG signal and convert the analogue signal into a digital one through A/D conversion. The signal processing module selects and classifies the collected ECG signals after pre-processing noise filtering and extracts feature values to reflect the emotional state of the human body, thus diagnosing the emotional health of the human body. The recognition results can be fed back to an alert system on the garment and simultaneously transmitted to an intelligent terminal or medical centre for data storage and intelligent analysis, allowing for a more comprehensive evaluation of the wearer's emotional health in the future.

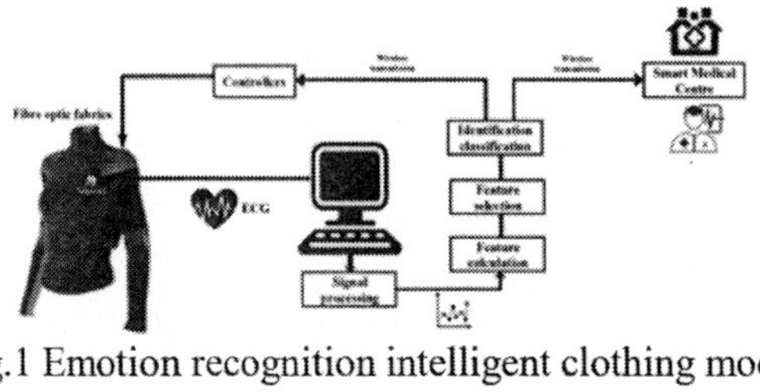

Fig.1 Emotion recognition intelligent clothing model

3. Experimental Work

3.1 Perimental Subjects and Instrumentation

3.1.1 Experimental Subjects

Twenty-four school students without psychological or cardiovascular disease, aged 20-27 years, who had not consumed alcohol, tea, coffee or tobacco for 24 hours prior to the experiment, who were not taking any other drugs that might affect the function of the autonomic nervous system, and who were not exercising strenuously, were selected as subjects.

3.1.2 Experimental Equipment

MindWare ECG Collection Device: The equipment has good flexibility, reliability and ease of operation. Its audio and video stimulus response system records the subject's response and aids in signal localisation acquisition. Monitoring data can be exported in .txt text format or Excel tables for analysis in the accompanying BioLab analysis system.

Other instruments and equipment: electrode patches, alcohol swabs, scissors, forceps and medical tape, etc.

3.2 Experimental Methods

The temperature of the laboratory was 22-24°C and the experiments were conducted in a relatively quiet and closed environment in order to reduce disturbances and to better elicit emotions from the subjects. Subjects entered the laboratory 20 minutes in advance. The laboratory is equipped with two computers. One is used to play the induction material and record changes in the subject's facial expressions, and the other is used to record the collected ECG signals. Before the formal start of the experiment, subjects were helped into a sitting position and was ensuredthat the subject did not move significantly during the collection.

This study used video material for target emotion elicitation. The criteria for selecting emotions were that the episodes were easy to understand and that each video evoked only a single emotion, ensuring that it was similar to everyday emotions and that it was easy to recover in a short period of time.

Emotions were categorized by a two-dimensional model of validity-arousal, and the four emotions selected for this study were joy, anger, sadness and pleasure. The space model for these four emotions is shown in Fig.2.

After several trials, it has been shown that a particular sequence of emotional induction clips;sadness,pleasure, anger, and then happiness, resulted in better ECG data collection. The subject was then asked to fill in a subjective emotion questionnaire immediately after each session to ensure that the data collected corresponded to the subject's emotions. The subjective questionnaire is shown in Table 1. The experimental flow is shown in Fig.3.

To ensure successful elicitation of the corresponding emotions in the subjects, a post-hoc comparative analysis of the elicitation material was conducted, and the subjective questionnaire data of the subjects were selected for t-test. The results of the test showed that all four emotions induced success. Fig.4 shows the ECG signal collected from one subject after sequential emotional induction.

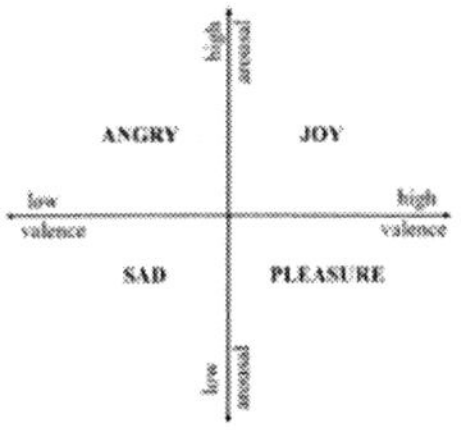

Fig.2 Two-dimensional spatial model of emotions

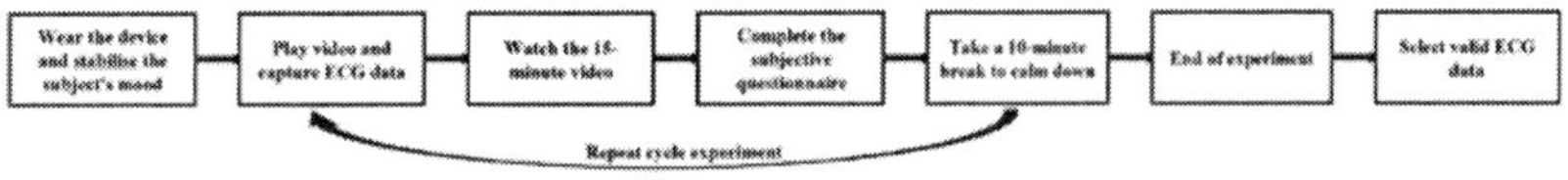

Fig.3 Flow chart of the emotion induction experiment

Table 1. Scale of subjective perception

	No.		Gender		Age	
	Level / Video	Slightly 0 marks	In a small extent 1 mark	Moderately 2 marks	Significantly 3 marks	Fiercely 4 marks
Emotion	Episode 1					
	Episode 2					
	……					

4. Experimental Data and Analysis

4.1 Pre-processing of ECG Signals

The acquired ECG signal usually contains a variety of noises, such as 50Hz industrial frequency interference caused by the circuit system, low frequency baseline drift caused by factors for respiratory movements and poor electrode contact, and high frequency myoelectric interference caused by human activity and muscle tension. Although the acquisition instrumentation unit has filtered out most of the electromagnetic and industrial frequency interference from the ECG signal, the acquired signal is still noisy and therefore it has to be pre-processed. For noise problems, noise reduction is carried out using the FIR filter 50Hz trap filter and the median filtering method. Firstly, the FIR filter 50HZ trap filter was designed using the window function method. The Kaiser window is a window function that is close to the optimal window structure, as it can adjust the indicators of the filter according to different parameters, so the Kaiser window function is used for filter design band trap filters. See Fig.5 or a comparison before and after the removal of industrial frequency interference. Next, the median filtering method was used to process the baseline drift. The comparison before and after treatment of baseline drift is shown in Fig.6

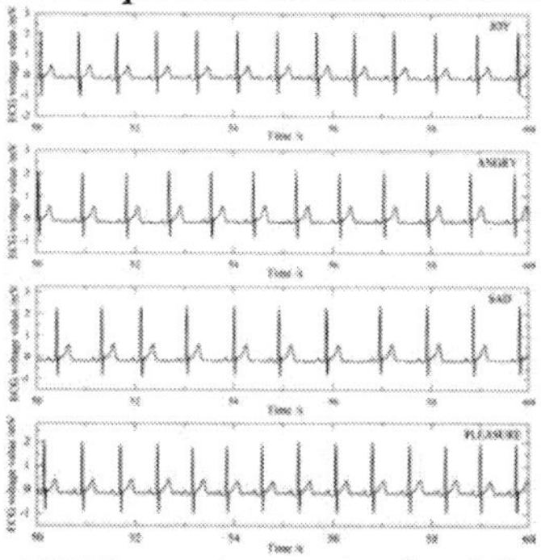

Fig.4 ECG raw time series for different emotions

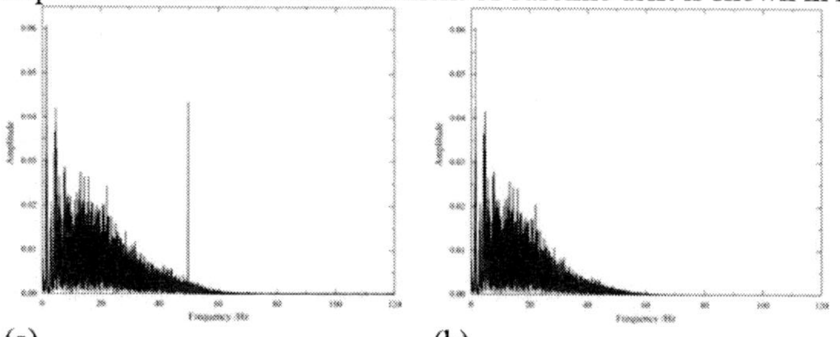

(a) (b)

Fig.5 Comparison chart for removal of industrial frequency interference, (a) Spectrogram of the original presence of industrial frequency interference,(b) Spectrogram after removal of industrial frequency interference

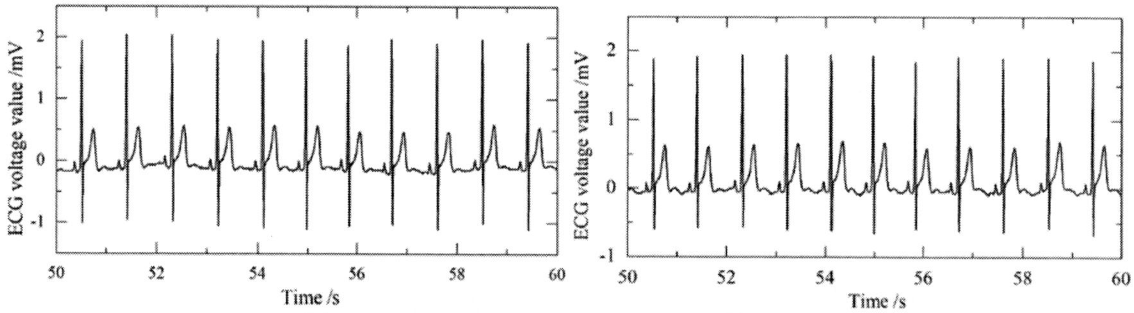

(a) (b)

Fig.6 Timing diagram for removal of EMG interference and baseline drift，(a) Original timing diagram,(b) Timing diagram after filtering and processing baseline drift

4.2 Feature Extraction

To perform heart rate variability RR interval feature extraction, the first step is to find the R-wave peak. The method of the algorithm to find the location of the R-peak is to find a sample point whose voltage value is greater than both the value of the previous sample point and the value of the next sample point, then remove the pseudo-peak and finally obtain the R-peak. The resulting R wave is then examined and the iso-pulse is removed. Finally, the sampling time of one peak is subtracted from the sampling time of the previous peak to obtain the RR interval value.

4.2.1 Time-domain Characteristics

The time domain features are mainly extracted from the relevant parameters of the RR interval, and the corresponding features are extracted from the waveform of the ECG signal from a statistical point of view. The HRV time domain characteristic values and physiological significance are shown in Table 2.

Table 2. HRV time-domain analysis eigenvalues and physiological significance

Eigenvalue	Description	Formula	Physiological significance
NNVGR	Mean value of all RR intervals	$NNVGR = \frac{1}{N}\sum_{i=1}^{N} RR$	Overall level of heart rate variability, intensity of emotional signals
SDNN	Standard deviation of normal RR interval	$SDNN = \sqrt{\frac{1}{N}\sum_{i=1}^{N}\left(RR_i - \overline{RR}\right)^2}$	Level of overall heart rate variability, sympathetic and parasympathetic tone; degree of marked mood change
RNSSD	Root mean square of the difference between adjacent RR intervals	$RNSDD = \sqrt{\frac{1}{N-1}\sum_{i=1}^{N-1}(RR_{i+1} - RR_i)}$	Reflects the magnitude of the fast-varying component of heart rate and the magnitude of parasympathetic tone
SDSD	Standard deviation of the difference between all adjacent RR intervals	$SDSD = \sqrt{\frac{1}{N-1}\sum_{i=1}^{N-1}\left(RR_i - \overline{RR}\right)^2}$	Slow-varying component of heart rate variability, sympathetic tone magnitude
NN50	Number of adjacent RR intervals with a difference > 50ms		The size of the component reflecting rapid change in heart rate
PNN50	RR50 divided by the total number of NN intervals as a percentage		Sensitive nerves that respond to the magnitude of the component of rapid heart rate variability, parasympathetic tone

4.2.2 Frequency-domain Characteristics

The frequency domain features mainly extract the relevant parameters of RR interval energy and quantifies the RR interval time series as a function of frequency versus energy. By describing its energy distribution in different frequency bands, the pattern of heart rate variation is analysed from the perspective of spectral analysis [11]. The HRV frequency-domain eigenvalues and physiological significance are shown in Table 3.

Table 3. Common parameters and physiological significance of HRV frequency-domain analysis

Parameters	Unit	Description (frequency range)	Physiological significance
TP	Hz	Total power (0-0.4Hz)	Reaction HRV size
LF	Hz	Low frequency power (0.04-0.15Hz)	Reflects dual sympathetic and parasympathetic regulation
HF	Hz	High frequency power (0.15-0.4Hz)	Parasympathetic nerve activity, associated with RNSSD and pNN50
LF/HF		HF/TP-VLF *100	Autonomic nervous system homeostasis i.e. sympathetic-vagal interaction, positively correlated with sympathetic activity

4.2.3 Non-linear Characteristics

Non-linear analysis is an analytical method based on the mechanics of flying non-periodic chaos. The Poincare scatter plot is an analysis of heart rate variability using the principles of chaos science [12]. A two-dimensional coordinate system was established using Poincare scatter plots, with the horizontal axis representing the RR interval RRn, and the vertical axis representing RR (n+1). An ellipse fitting method was used to analyse the Poncelet scatter plot, using the short axis SD1 and the long axis SD2 as indicators for the analysis [13]. The HRV non-linear characteristic values and physiological significance are shown in Table 4. A scatterplot of the different moods of the Pongarets is shown in Fig.7. The ratios of SD1 and SD2 for JOY, SAD, ANGRY and PLEASURE were 0.191, 0.756, 0.518 and 0.438 respectively, which shows that the ratio of negative emotions is significantly higher than that of positive emotions, and the higher the SD1/SD2 value is when the individual is more negative. Table 7 lists the eigenvalues of heart rate variability extracted for one subject across the four moods.

Table 4. HRV non-linear analysis parameters

Parameters	Formula	Description
SD1	$SD1 = \frac{\sqrt{2}}{2} SD\left(RR_i - RR_{i+1}\right)$	Transient changes in emotion
SD2	$SD2 = \sqrt{2SD\left(RR_i\right)^2 - \frac{1}{2} SD\left(RR_i - RR_{i+1}\right)^2}$	Differences in overall changes in emotions
SD1/SD2		Complexity of emotional change

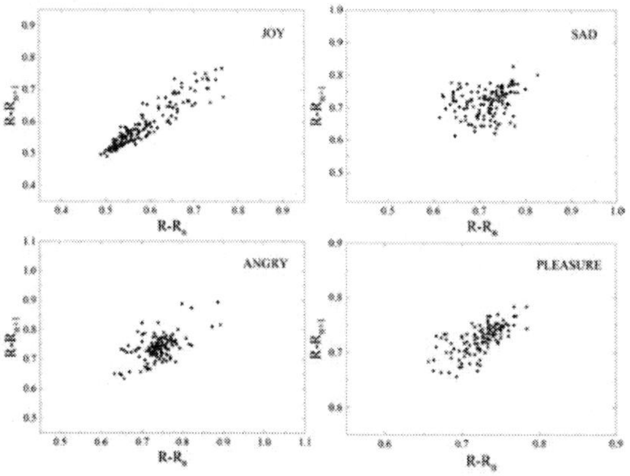

Fig.7 Scatterplot of Pomegranate for different moods

Table 7. Selected characteristic values for different moods

	JOY	ANGRY	SAD	PIEASURE
NNVGR	0.6838	0.8661	0.9282	0.7526
SDNN	0.0449	0.0634	0.0663	0.0302
RNSSD	0.0250	0.0888	0.0716	0.0204
SDSD	0.0160	0.0674	0.0517	0.0112
NN50	7	23	37	2
PNN50	0.0402	0.1805	0.3750	0.0127
TP (*1000)	0.0396	0.1704	0.1028	0.0661
LF (*1000)	1.6311	1.2287	1.7122	1.0190
HF (*1000)	1.930	5.160	4.479	2.055
LF/HF	0.8453	0.2381	0.3823	0.4958

Continued Table 7

	JOY	ANGRY	SAD	PIEASURE
SD1	0.0178	0.0630	0.0508	0.0145
SD2	0.0610	0.0642	0.0787	0.0402
SD1/SD2	0.2909	0.9815	0.6456	0.3594

4.3 Feature Selection

Generally speaking, while each eigenvalue variable provides some information, its importance varies. In most cases, there is correlation between the variables, allowing for some degree of overlap in the information provided by the eigenvalue variables. Principal Component Analysis (PCA) reduces the dimensionality of the features to the required number, ensuring that the features retained after dimensionality reduction are seen to be free of covariance. So, by obtaining the number of solution correlations for the features, the new variables with fewer complementary correlations are used to reflect the absolute majority of information provided by the original variables, thereby improving classifier performance and reducing computational effort. Some of the heart rate variability eigenvalue values are of such a large order of magnitude that the individual features needed to be normalised and then correlations between features had to be observed. Fig.7 shows the violin plot for each eigenvalue after the extreme difference criterion. As can be seen from Fig.8, the individual eigenvalues have all changed to the [0,1] interval, and a two-by-two scatter plot between the eigenvalues is plotted for the normalised data. The results show a strong linear relationship between some features and therefore require dimensionality reduction using principal component analysis. After principal component analysis, the cumulative variance contribution reached 99% when nine principal components were retained.

4.4 Recognition of Classification Algorithms

After selecting the feature values, six different classification algorithms were used for emotion recognition in this study, and finally the recognition results were compared. The six algorithms are SVM (support vector machine), KNN (K-nearest neighbor), decision tree, MLPClassifier, random forest and XGBoost (extreme Gradient Boosting) [14, 15]. Due to the individual differences in the ECG data, the characteristic values are processed. The value obtained by subtracting the characteristic value of various emotions from the characteristic value of the pleasant state is divided by the characteristic value of the pleasant state. The dataset was first subjected to four classifications, of which an arbitrary 80% was selected as the training set and 20% as the test set, with a maximum classification accuracy of 45%. The classification results were poor. This may be related to the insignificant difference in sentiment on the validity dimension, combined with the presence of a small sample size. Positive (happy, happy) and negative (angry, sad) emotions are selected for identification and classification based on arousal. Thirteen eigenvalues were used as the data set, approximately 80% of the subjects with a total of 19 individuals were selected as the training set, and 20% of the subjects with a total of 5 individuals were selected as the test set. The results of the recognition accuracy of the six algorithms are shown in Fig.9. The results show that the average recognition rate of the six classification algorithms is at 72.5%. One of the classifiers with the highest recognition rate is the Extreme Gradient Boost (XGBoost), which achieves 80% recognition rate and is able to identify both positive and negative emotions better. In contrast, the decision tree has a recognition rate of only 60%. This may vary with different classifiers having different requirements for the amount of data.

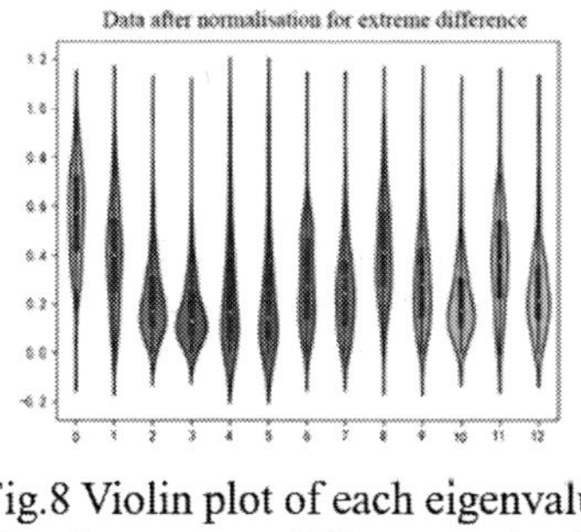

Fig.8 Violin plot of each eigenvalue after the extreme difference criterion

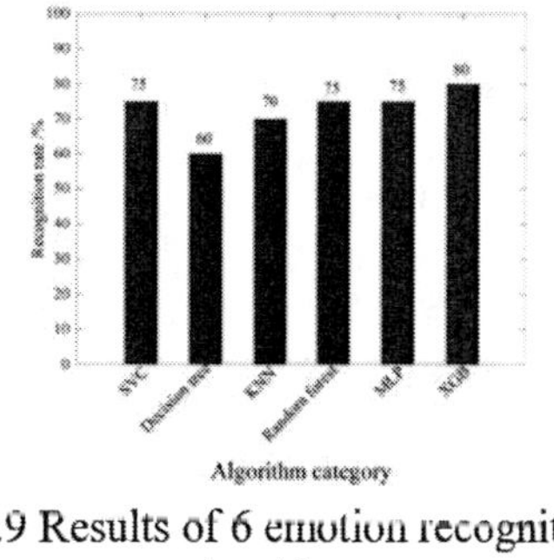

Fig.9 Results of 6 emotion recognition algorithms

5. Conclusion

This study proposes an intelligent clothing system for emotion recognition based on heart rate variability and its framework, which classifies four emotions - happy, angry, sad and pleasure - by collecting ECG signals under different emotions. The XGBT algorithm was the best, with the best results of 80% in the classification of positive (happy, happy) and negative (angry, sad) emotions selected on the basis of arousal. Future research can further explore the following aspects: expanding the sample pool of emotions to improve the recognition accuracy and by combining other modal physiological signals to improve the recognition accuracy in response to the low recognition rate of emotions in the validity dimension. The intelligent clothing for emotion recognition based on HRV feature information proposed in this study has important future applications in mental health monitoring, distance education, traffic driving and other fields.

6. Acknowledgement

The authors would like to acknowledge the financial support from the Fundamental Research Funds for the Central Universities (Grant NO. 2232020G-08).

References

[1] Wang YJ. An analysis of the impact of human mental health on physical health. Science and Technology Innovation Heraid: 2015; 12: p. 248-249.

[2] Jiang GG, LI DY, REN ZH, et al. The status quo and characteristics of Chinese mental health literacy. Acta Psychologica Sinica: 2021; 53: p. 182-198.

[3] Lü XJ, Sun YL, QI XX, et al. Development status and suggestions of intelligent clothing technology. Wool Textile Journal: 2020; 48: p. 96-99.

[4] Barrett LF, Adolphs R, Marsella S, et al. Emotional Expressions Reconsidered: Challenges to Inferring Emotion From Human Facial Movements. Psychological Science in the Public Interest: 2019; 20: p.1-68.

[5] Schmidt P, Reiss A, Durichen R, et al. Wearable-Based Affect Recognition-A Review. Sensors: 2019; 19: p. 42.

[6] Zhu J P, Ji L Z, Liu C Y. Heart rate variability monitoring for emotion and disorders of emotion. Physiological Measurement: 2019; 40: p. 11.

[7] Picard R W, Healey J. Affective wearables. First International Symposium on Wearable Computers - Digest of Papers: 1997; p. 90-97.

[8] Rainville P, Bechara A, Naqvi N, et al. Basic emotions are associated with distinct patterns of cardiorespiratory activity. International Journal of Psychophysiology: 2006; 61: p. 5-18.

[9] Yi H, Chen RJ, Deng GH, et al. Research on emotion recognition based on heart rate variability. Journal of Biomedical Engineerinng Research: 2020; 39: p. 128-132.

[10] Goshvarpour A, Abbasi A. Indices from lagged Poincare plots of heart rate variability: an efficient nonlinear tool for emotion discrimination. Australasian Physical & Engineering Sciences in Medicine: 2017; 40: p. 277-287.

[11] Shi P, Yu HL. Application of Linear and Nonliner Characteristics of Heart Rate Variability in Assessment of Autonomic Nervous System Activity. Journal of Biomedical Engineering: 2014; 31: p. 302-307.

[12] Yan LL, Luo H, Wei J, et al. The Measurement of Mental Stress and New Methodologies. Chinese Journal of Applied Psychology: 2019; 25: p. 32-47.

[13] Chi XL, Zhou JH, Shi P, et al. Analysis Methods of Short-term Non-linear Heart Rate Variability and Their Application in Clinical Medicine. Journal of Biomedical Engineering: 2016; 33: p. 193-200.

[14] Jiang DZ, Wu KC, Chen DC, et al. A probability and integrated learning based classification algorithm for high-level human emotion recognition problems. Measurement: 2020; 150: p. 11.

[15] Wang XW, Nie D, Lu BL. Emotional state classification from EEG data using machine learning approach. Neurocomputing: 2014; 129: p. 94-106.

Modeling of an Electrically Conductive Nonwoven Strip for Electromagnetic Shielding

Sundaramoorthy Palanisamy[1*], Veronika Tunakova[1], Mohanapriya Venkataraman[1], Daniel Karthik[1], Jakub Wiener[1], Jiri Militky[1]

[1]*Department of Material Engineering, Faculty of Textile, Technical University of Liberec, Studentska 1402/2, 461 17 Liberec, Czechia.*

[*]Corresponding author's email: sundaramoorthy.palanisamy@tul.cz; sundar2026@gmail.com

Abstract

Different textile material structures have unique geometrical shapes and designs. In this work, Copper/ nickel-coated ultrathin nonwoven fabric was prepared in a strip form. The 3 mm, 6 mm, and 9 mm thick strips were laid at different gaps, angles and layered to study the effect of factors on EM shielding effectiveness as per ASTM D4935-18 standard. The design of the experiment analyzes the three factors and three levels of the strip properties and its influence on the result of electromagnetic shielding. From the DoE screening design findings, factors such as the thickness of the strips, the gap between the strips, and the strips laid angle have a statistically significant effect on electromagnetic shielding effectiveness.

Keywords: Electromagnetic Shielding; Ultrathin Nonwoven; Copper-nickel Coating; Design of Experiment; Textile Material.

1. Introduction

The universe is filled with an electromagnetic (EM) field naturally as well as man-made. The natural source of the EM fields are the earth's magnetic fields, and lightning. The artificial or man-made sources of EM fields are from all electronic devices and electric power transmission lines[1]. EM ionized radiation can be divided into four categories: static field, extremely low-frequency EM fields, intermediate frequency EM fields, and radiofrequency EM fields [2]. EM radiation of intermediated and radio frequency ranges may be harmful to living beings and electronic devices. It also travels at the speed of light. Living beings exposed to an extreme level of EM radiations may suffer cancer, tissue damage, lymphoma, and leukemia [3], and electronic devices can malfunction when exposed to EM radiations due to electromagnetic interference (EMI) [4]. A recent problem is electromagnetic hypersensitivity due to the increased use of electronic devices [5].

Electrically conductive metals like silver, copper, stainless steel, iron, gold, nickel, and brass are best suited for EM shielding applications as they are good conductors of electricity and has the ability to reflect and absorb most of the EM radiations [6]. Metals have some drawbacks, as they are heavier, susceptible to corrosion, and is flexible. The use of metal-coated textile materials can overcome the drawbacks of the metal properties. Particularly in textile materials, different techniques can be applied to transform conventional textile material into a conductive textile material. The techniques used are coating conductive polymers [7] [8], metal coating of fiber, yarn, and fabric [9] [10], metal fibers blending [11] [12], metalcore wires [13] [14], in-laying of conductive fabric [15], and carbonization of fabric, yarn, and fibers [16].

EM radiation shielding runs on three main mechanisms, with the reflection being primary, absorption as secondary and multiple reflections as tertiary. Beyond these mechanisms, the area covered by the shielding material plays a prominent role in shielding. The gap or space known as aperture in the textile structure has the drawback of radiation penetration. This study has measured and calculated EM shielding of conductive yarn woven fabric. The difference between the measured and calculated EM shielding is between 2-7 dB within ASTM D 4935 with a random error of ±5 dB[17]. Electrical conductivity, fabric pore size, and fabric thickness have a significant effect on EM shielding. The EM shielding influenced by the aperture or pore thickness and maximum pore size as per the model was found [18].

Many studies are available about the effect of EM SE on the gap between the conductive materials in textile structures. However, there are very few studies for the gap between and angle of laying conductive material in textile structures. In this study, the copper-coated nonwoven fabric was cut into strips of different thicknesses and placed at various distances to form a textile structure. A design of the experimental technique chose to study the significant effect on EM SE.

2. Materials and Methods

2.1 Material Specifications

The copper/nickel (Cu/Ni) coated ultrathin nonwoven fabric (Cu/Ni NW) is named 'MEFTEX 20'. Metallization of Cu/Ni on ultrathin nonwoven using the chemical and electrochemical metal deposition method using the 'roll on roll' coating technique. The fabric is finish with anti-corrosion resistance on the metal surface. The 'Cu/Ni NW' fabric parameters are given in table 1.

Table 1. Cu/Ni NW fabric parameters and their values.

Parameters*	Values
GSM [gram per sq. meter]	24.56 ± 2.69
Thickness [mm]	0.087 ± 0.006
Tensile strength in MD [N/50mm]	25[19]
Tensile strength in CD [N/50mm]	16[19]
Surface resistivity [Ω]	2.1 ± 0.28

*GSM, thickness and surface resistivity are measured as per the ASTM method. Note: MD means 'machine direction' and CD means 'cross direction'.

Note: 'Meftex 20' was procured from BOCHEMIE a.s., Bohumin, Czechia. Milife ® is trade mark of the Nippon company, Japan.

2.2 Samples Preparation

In this study, the 'Cu/Ni NW' sample is cut into 3, 6 & 9 mm thick strips and laid with 3, 6, & 9 mm gaps between each other. Each thickness was laid at the above gaps and formed a single sheet layer, as shown in Fig.1. For example, the single-layer 3 mm thick strip laid at 3, 6, & 9 mm gap is in Fig.1(a-c). Likewise, 3 mm thick strip, 6 mm, and 9 mm thick strips were laid at 3, 6, & 9 mm gaps. Two-layer strips were formed by laying two single layer strips at 90°, 45° & 30° angle with the same gap, 9 mm thick strips of 3 mm gap laid at 90°, 45° & 30° angle are shown in Fig.1(d-f). Also, 9 mm thick strips of 6 mm and 9 mm gap are laid at 90°, 45° & 30° angles to form two-layer strip samples. Likewise, the 3 mm and 6 mm thick strips were formed as two-layer strip samples with various gaps and laid angles. The two-layer samples were taken for the EM SE test as per the ASTM method to analyze the gap between and angle between the conductive strips' material. Each thickness of the strips was laid in 9 different ways and tested for EM SE. The overlaying of the strips at various angles is prepared with the same strip thickness and the same gap between the strips, as shown in Fig.1 (d-f).

2.3 Design of Experiment

Table 2. Factors and its levels used for SFD.

Factors	Levels			Response
Strip thickness [mm] (A)	3	6	9	
Gap between the strips [mm] (B)	3	6	9	EMI SE [dB]
Angle of layers [°] (C)	0	45	90	

Screening factorial design (SFD) was chosen to entirely and systematically study the interaction between factors and identify significant factors. In this experiment, there are three variables and three levels used to design an experimental study. The factors and their levels of the design are in table 2. The factorial experimental design was analyzed using Minitab® software.

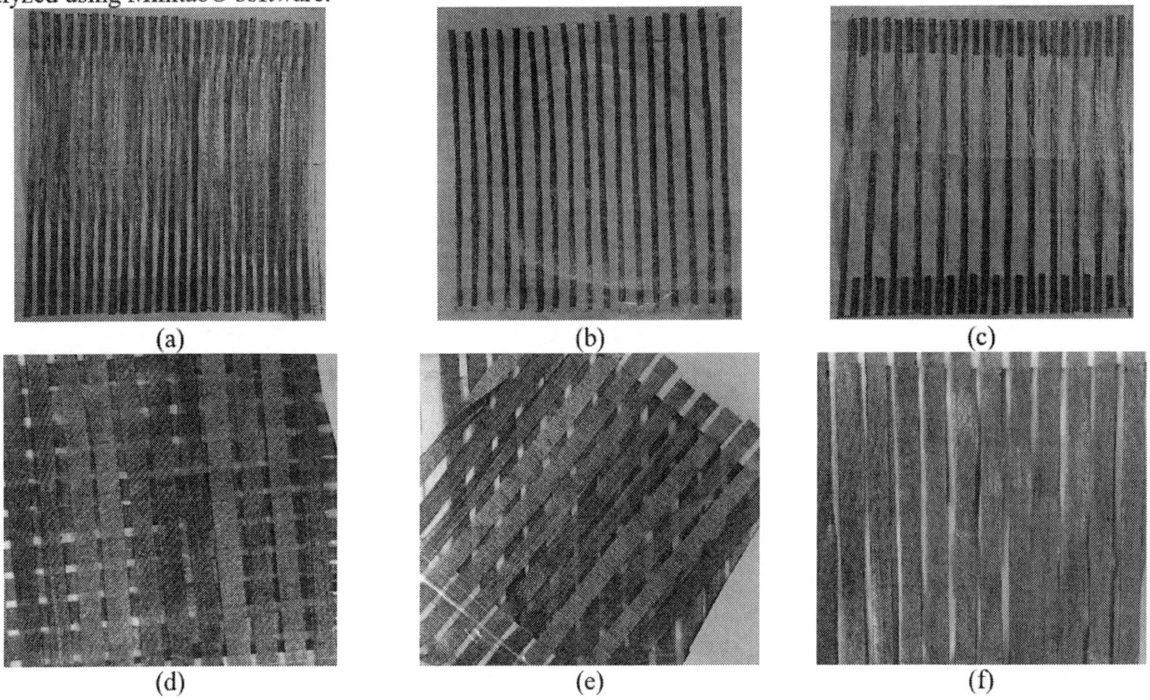

Fig.1 A photographic image of the single-layer Cu/Ni NW strips of 3 mm thickness at a distance (a) 3 mm (b) 6 mm and (c) 9 mm and two-layer Cu/Ni NW strips of 9 mm thickness laid at (d) 90° (e) 45° and (f) 0° angles.

2.4 Electromagnetic Shielding Effectiveness

SE of the sample set measured according to the ASTM D4935-10 [20], for the planar materials using a plane-wave, the far-field EM wave at the temperature T = 21°C, and the relative humidity RH = 54 %. SE of the samples measured over the frequency range of 30 MHz to 1.5 GHz. The set-up consisted of a sample holder with its input and output connected to the network analyzer. A shielding effectiveness test fixture (Electro-Metrics, Inc., model EM-2107A) was used to hold the sample. The design and dimension of the sample holder follow the ASTM method mentioned above. A vector analyzer Rohde & Schwarz ZN3 was used to generate and receive the electromagnetic signals. The standard mentioned above determines the shielding effectiveness of the fabric using the insertion-loss method. A reference measurement for the empty cell is required for the shielding effectiveness assessment. A "through" calibration with the help of the reference sample was made first. A load measurement was performed on a solid disk shape sample subsequently. The reference and load specimens must be of the same material and thickness. Both the reference and load samples geometries are according to the ASTM D 4935-10. The measurements are performed at five different places of the strip samples because of the subsequent statistical analysis.

3. Results and Discussions

In DoE, the SFD model consists of the three factors at three levels, as shown in table 2 is taken for the analysis. In the SFD model, there are 13 base runs, and 39 total runs in the design.

3.1 Analysis of Variance

Analysis of variance (ANOVA) of the SFD model is performed to identify the factors that significantly affect SE at 1.5 GHz frequency. The three main factors, its square and cross effects, are analyzed and calculated for *F*-value and *p*-value. If the *p*-value is less than the significant level (α=0.05), then the factor is statistically significant on SE value and vice versa. ANOVA test shows the factors lesser than significant are A, B, C, C*C, B*C, and A*C, and all other factors are higher than the significant level.

3.1.1 Regression Equation in Uncoded Units

Eq.1 is showing the regression equation in terms of uncoded units. The regression equation helps to predict the SE value, and its R^2 of 97.10% proves that the predictability of SE is highly possible.

$$SE = 10.52 + 1.20\,A - 2.48\,B + 0.5475\,C - 0.092\,A*A + 0.137\,B*B + 0.003025\,C*C + 0.0036\,A*B + 0.01821\,A*C - 0.01768\,B*C \quad 1)$$

3.2 Standardized Effect Charts

The standardized effects of the factors and their combinations are represented in Fig.2 at a 5% significance level (α = 0.05) for SE at 1.5GHz frequency. The normal plot of the standardized effect is used to determine the significance and insignificant of each effect of factors more accurately. In a normal probability plot of effect, the factors A, C, and AC have a significant positive effect on SE, and factors B, BC, and CC have a significant negative effect on SE. Factor 'C' has a higher significant effect on SE value, i.e., SE increases with increasing the laid angle of the strip. The negative effect is saying, 'that the SE is increasing with a decrease in the gap between the strips (B).' The factors AA, BB, and AB, show no significant difference in SE value, which are all very close to the straight line, *i.e.*, normal distribution.

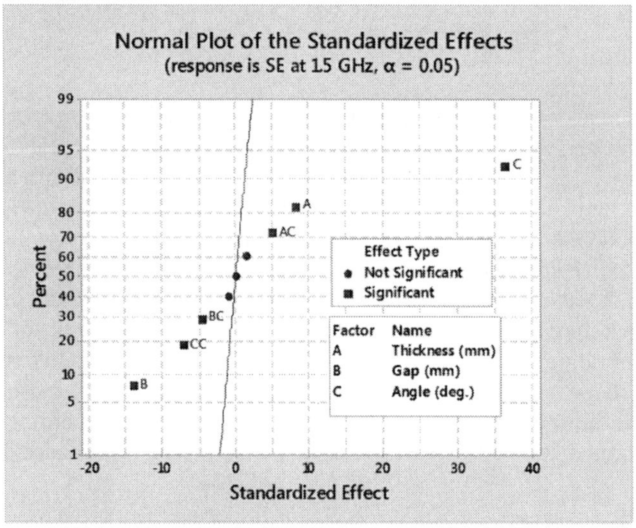

Fig.2 The normal plot of the standardized effects.

3.3 Main Effect Plot

Fig.3 shows the main effects plot for SE [dB] at 1.5 GHz frequency for the factors A, B, and C. The plot is mainly showing the effects of the factors and their levels on SE value. Factors, A (thickness) and C (angle) is showing a positive effect on SE value in the main effect plot, factor B (gap) has a negative effect on SE value. An increase in thickness and angle has increased the SE value.

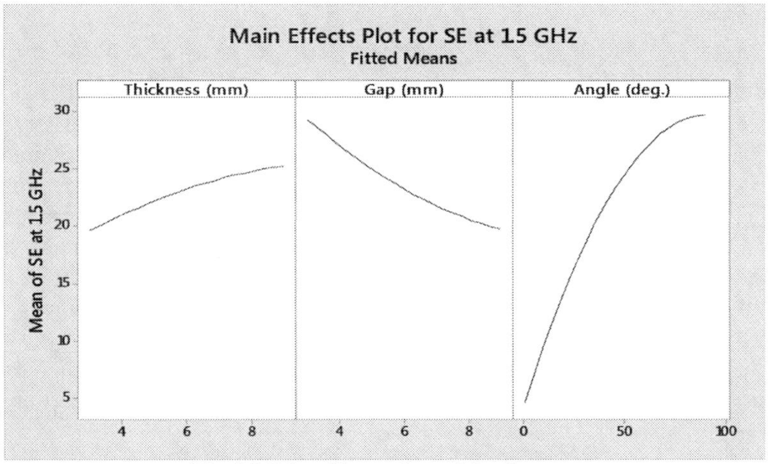

Fig.3 Main effects plot of the factors.

3.4 Interaction Plot

Fig.4 shows the interaction plot for SE [dB] at 1.5 GHz frequency to visualize the factors and their interactions. The interaction plot mainly indicates the interaction between the factors; the parallel line indicates no interaction, out of parallel line indicates interactions, and the cross line indicates more interaction. The highest SE obtain with a higher level of angle (i.e., 90°) and a lower level of gap (i.e., 3 mm), as shown in the gap*angle interaction plot (fig. 3). For interactive effects of gap*thickness and gap*angle, SE decreases with an increasing gap. For thickness*gap and thickness*angle, SE increases with increasing thickness. The angle*thickness and angle*gap interaction show the increasing SE with an increase of angle (from 0° to 90°).

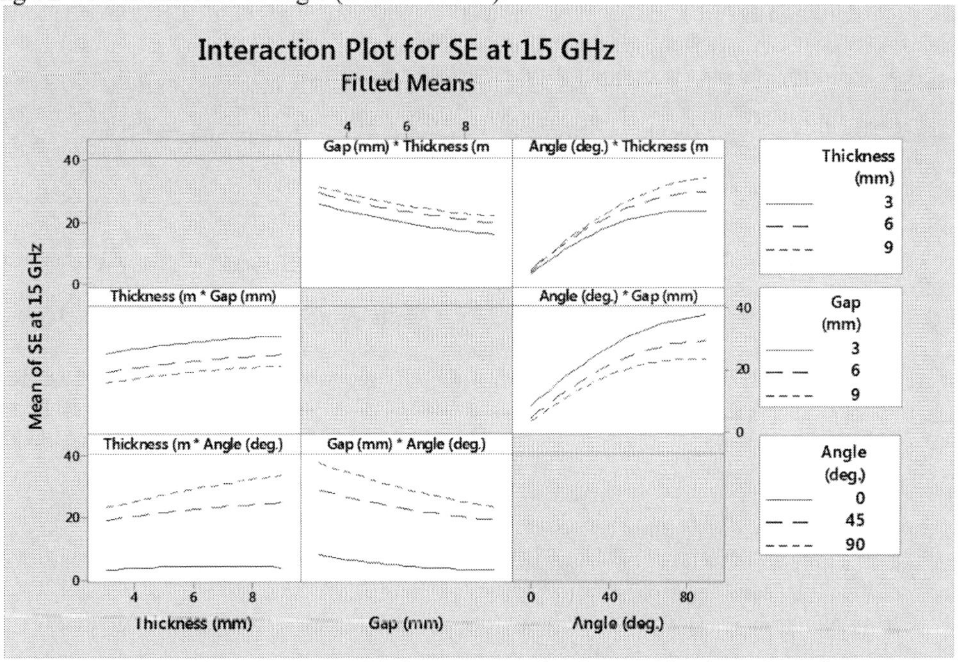

Fig.4 Interaction plot for SE [dB] at 1.5 GHz for the combination of two factors.

4. Conclusions

The Copper/Nickel coated 100% polyester ultrathin nonwoven fabric was studied. The SFD model from the DoE technique is used to analyze factors having a significant effect on EM SE at 1.5 GHz frequency. The ANOVA and coefficient of determination (R^2 = 0.97) predict the significant factors and good EM SE predictably with the regression model. The factors A, B, C, A*C, B*C, and C^2, have a significant effect on the SE value is confirmed in the normal plot. The main effect plot concludes that the SE value was increases with an increase in factor A, and C; decreases with increases in factor B. The highest SE value of 38 dB was found in the interaction plot with the combination of higher C (i.e., 90°) and a lower B (i.e., 3 mm), which is graded as excellent for general use [21].

Hence, the higher percent cover area and lower area per aperture structure have been recommended for achieving higher EM SE. DoE is concluded that the combination of the larger strip thickness, the lower gap between the strips, and the higher laid angle of strips has an excellent EM SE. This model could be helpful to construct the optimal fabric or composite structures based on the required level of shielding for electromagnetic shielding application.

5. Acknowledgement

This work was supported by the Ministry of Education, Youth and Sports of the Czech Republic and the European Union - European Structural and Investment Funds in the frames of Operational Programme Research, Development and Education under project Hybrid Materials for Hierarchical Structures [HyHi, Reg. No. CZ.02.1.01/0.0/0.0/16_019/0000843] and the 'Student Grand Competition' (SGS 2021) under title "Development of conductive textile materials for multi-functional applications" and its internal number is SGS-2021-6008.

References

[1] Barnes FS, Greenebaum B (eds). Bioengineering and Biophysical Aspects of Electromagnetic Fields. Third. Boca Raton, Florida, USA.: Taylor & Francis Group: 2006.

[2] Martin R (ed). Epidemiology of Electromagnetic Fields. Second. Boca Raton, Florida, USA.: Taylor & Francis Group: 2008.

[3] Lin JC (ed). Electromagnetic Fields in Biological Systems. Boca Raton, Florida, USA.: Taylor & Francis Group: 2012.

[4] Ott HW. Electromagnetic Compatibility Engineering. Hoboken, NJ, USA: John Wiley & Sons, Inc. Epub ahead of print 12 August: 2009. DOI: 10.1002/9780470508510.

[5] Stein Y, Udasin IG. Electromagnetic hypersensitivity (EHS, microwave syndrome) – Review of mechanisms. Environ Res: 2020; 186: 109445.

[6] Geetha S, Kumar KKS, Rao CRK, et al. EMI shielding: Methods and materials-A review. J Appl Polym Sci: 2009; 112: 2073-2086.

[7] Kaynak A, Håkansson E. Characterization of conducting polymer coated fabrics at microwave frequencies. Int J Cloth Sci Technol: 2009; 21: 117-126.

[8] Yanılmaz M, Sarac AS. A review: Effect of conductive polymers on the conductivities of electrospun mats. Text Res J: 2014; 84: 1325-1342.

[9] Uzun S, Han M, Strobel CJ, et al. Highly conductive and scalable Ti3C2T -coated fabrics for efficient electromagnetic interference shielding. Carbon N Y: 2021; 174: 382-389.

[10] Wu X, Chen Y, Liang K, et al. Fe2O3 Nanowire Arrays on Ni-Coated Yarns as excellent electrodes for High Performance Wearable Yarn-Supercapacitor. J Alloys Compd: 2020; 158156.

[11] Liang R, Cheng W, Xiao H, et al. A calculating method for the electromagnetic shielding effectiveness of metal fiber blended fabric. Text Res J: 2018; 88: 973-986.

[12] Šafářová V, Militký J. Electromagnetic shielding properties of woven fabrics made from high-performance fibers. Text Res J: 2014; 84: 1255-1267.

[13] Ortlek HG, Alpyildiz T, Kilic G. Determination of electromagnetic shielding performance of hybrid yarn knitted fabrics with anechoic chamber method. Text Res J: 2013; 83: 90-99.

[14] Perumalraj R, Nalankilli G, Balasaravanan TR, et al. Electromagnetic shielding tester for conductive textile materials. Indian J Fibre Text Res: 2010; 35: 361-365.

[15] Li TT, Wang R, Lou CW, et al. Manufacture and effectiveness evaluations of high-modulus electromagnetic interference shielding/puncture resisting composites. Text Res J: 2013; 83: 1796-1807.
[16] Karthik D, Baheti V, Tunakova V, et al. Development of electrically conductive activated carbon fabric from Kevlar fabric for effective EMI shielding applications. In: Textile Bioengineering and Informatics Symposium Proceedings 2017 - 10th Textile Bioengineering and Informatics Symposium, TBIS 2017: 2017.
[17] Neruda M, Vojtech L. Electromagnetic shielding effectiveness of woven fabrics with high electrical conductivity: Complete derivation and verification of analytical model. Materials (Basel): 2018; 11: 1657.
[18] Henn AR, Cribb RM. Modeling the shielding effectiveness of metallized fabrics. In: International Symposium on Electromagnetic Compatibility. Anaheim, CA, USA: IEEE Xplore: pp. 283-286.
[19] MEFTEX 20. BOCHEMIE a.s, Czech Republic., https://www.meftex.cz/en/meftex-20/p-3/ (2020, accessed 1 October 2020).
[20] ASTM D4935-10. Standard Test Method for Measuring the Electromagnetic Shielding Effectiveness of Planar Materials. West Conshohocken, PA. Epub ahead of print: 2010. DOI: 10.1520/D4935-10.
[21] FTTS-FA-003. Specified Requirements of ElectromagneticShielding Textiles. New Taipei City, Taiwan, https://www.ftts.org.tw/images/fa003E.pdf (2005).

Cite as: X.P. Zhou et al.,
TBIS 10.3993/tbis (2021).

Finite Element Model Construction and Simulation Analysis on Shielding Effect of Electromagnetic Radiation Protective Clothing for Pregnant Women

Xiao-Ping Zhou, Bo-An Ying*, Long Wu, Jing Qi

Apparel & Art Design College, Xi'an Polytechnic University, Xi'an, Shaanxi, 710048, P.R. China

*Corresponding author's email: yingba2006@163.com

Abstract

Related studies have shown that pregnant women have a certain sensitivity to electromagnetic radiation. At present, there is a lack of digital evaluation system for the simulation model of pregnant women affected by electromagnetic radiation and shielding effectiveness of pregnant women's anti-electromagnetic radiation clothing. In this paper, the simulation model of pregnant women and fetuses is proposed, and virtual fitting software is used to model the clothing. Finally, the model is combined with three-dimensional electromagnetic simulation software and the real situation is simulated. In this paper, the influence of electromagnetic radiation on pregnant women and fetus before and after wearing anti-electromagnetic radiation clothing was studied by digital evaluation method. The results show that the effects of electromagnetic radiation on pregnant women are concentrated in the neck, chest, abdomen and legs. Under 900MHz electromagnetic irradiation, the electric field value on the surface of human body concentrates between 0 and 2V/m; anti-electromagnetic radiation clothing to a certain extent to reduce the effects of radiation on pregnant women and fetuses. A digital evaluation method is proposed in this paper, providing a novel idea for detecting the shielding effectiveness of electromagnetic radiation in pregnant women.

Keywords: Radiation Proof Clothing; Pregnant Woman; Finite Element Model Construction; Shielding Effectiveness

1. Introduction

At present, the frequency band used in mobile communication is microwave frequency band, closely related to people's daily life. Microwave is a physical occupational hazard factor, and its influence on human body has drawn attention [1]. In the physical world, electromagnetic wave radiation is omnipresent. Electromagnetic radiation is colorless, silent, and is easy to ignore. Studies have shown that pregnant women, are more vulnerable to electromagnetic radiation damage [2]. Therefore, electromagnetic radiation protective maternity clothing are vital.

As a kind of functional clothing, how to evaluate the protective effect of radiation-proof maternity clothing has aroused many experts' research in related fields. Scholars at home and abroad have studied the effectiveness of electromagnetic shielding of clothing and its influencing factors. In terms of physical experiments, considering the harm of electromagnetic radiation to human body, the real dressed human body is generally not used for experiments. Zhang Peng et al. [3] tested the shielding effectiveness of the dummy's chest and abdomen by using the full-wave darkroom method. Zhang Xiaoxia [4] made experimental FRP pregnant woman table and experimental sample, and analyzed the factors affecting the shielding efficiency of clothing based on the structural characteristics of clothing. Chaochan Chen [5] et al. established a performance testing system for darkroom electromagnetic shielding suit. In the aspect of finite element simulation experiments, Zhang Lili [6] et al. used Ansoft HFSS to simulate and analyze the influence of changes in the body type of pregnant women during different pregnancies on the shielding efficiency of clothing. Yao Li [7] used HFSS to simulate the GTEM chamber shielding test system. Sun Ruili [8] established a simulation model of garment pores from the perspective of garment structure. Ma Liang [9] established a simplified model of human body and clothing, and simulated the electromagnetic radiation values at different tissue levels of the human body by using electromagnetic field finite element technology. Once upon a time people, in the research on the shielding effectiveness of the clothing, the use of finite element method greatly simplifies the experimental process, and can accurately reflect the distribution of electromagnetic field on the human body, this method can not only save the economic cost and time cost, also can have certain guidance on the design process of clothing.

In the evaluation of shielding effectiveness of pregnant women's clothing, most studies mainly focus on the influence of clothing structure, clothing porosity, dummy model and other aspects on its shielding effectiveness, while few people study the biological electromagnetic model and clothing model with similar real situation.

The research content of this paper will be based on the finite element simulation, the human body modeling of pregnant women and fetuses and the modeling and simulation of radiation protective clothing. At present, in the field of clothing performance simulation, biomechanical model and clothing mechanical performance simulation are mostly used. Based on previous studies, this study created a more complete biological electromagnetic model of pregnant women and fetuses, and combined with virtual fitting software to make clothing modeling more in line with the real situation. The purpose of this study is to provide a safe and convenient method for testing the shielding effectiveness of clothing and to provide a new idea for the optimization of clothing products.

2. The Establishment of Simulation Model

2.1 Mannequin Building

The simulation model mainly includes two parts, which are pregnant woman model (late pregnancy) and fetus model (late pregnancy).The two parts of the model are built by the method of reverse modeling. The general steps of reverse modeling are as follows: determine the scanning scheme -- entity point cloud scanning -- carry out point cloud data -- establish surface -- entity modeling. In this study, the non-contact 3D body scanning equipment used for the pregnant woman model is Germany's Human Solution laser 3D body scanner. At present, the pregnant woman model has been obtained according to previous studies [10]. The scanner used in the fetal model was Android portable high-precision 3D scanner. The scanner has embedded computing power, is based on "cloud" computing and storage technology, and the scanning process can be fully automated and easy to use. The specific operation is as follows: after scanning the fetal model, the point cloud data of the three-dimensional human body will be obtained, and the software GEOMAGIC will be used to preprocess the point cloud data obtained, and then polygon editing will be carried out, and the final simulation model will be obtained by accurate curved surface.

2.1.1 Fetal Model

After scanning the existing fetal model with Android portable high-precision 3D scanner, the point cloud model of the fetus is obtained, as shown in Fig.1 (a) and (b).The software 3dModelProcess that comes with the flatbed scanner is used for the preliminary processing of the point cloud model. In the scanning process, because all the point cloud data cannot be obtained at one time, it is necessary to scan the fetal model from multiple angles and finally assemble a complete point cloud model. As shown in (c) in Fig.1, the two fetal models scanned were pieced together into a whole model, and then derived into an OBJ model.

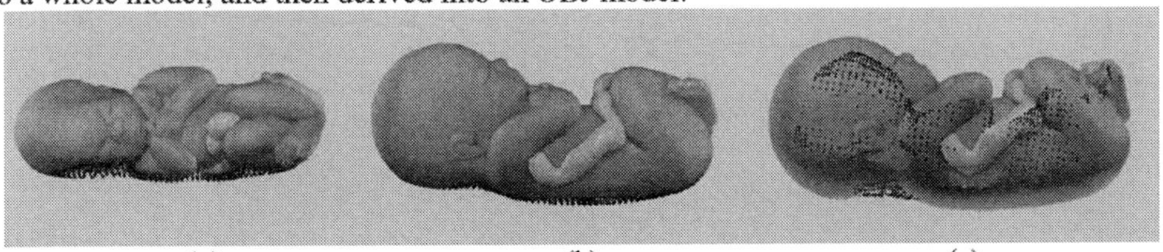

(a) (b) (c)

Fig.1 Fetal model processing

Import the exported OBJ model into GEOMAGIC for point cloud and polygon processing. The principle of GEOMAGIC reverse design is to use many small spatial triangles to approximate and restore the solid CAD model, and use NURBS surface pieces to fit the NURBS surface model. In point cloud processing, we should mainly pay attention to deleting some unnecessary points, as shown in Fig.2-2, to optimize the model. The point cloud processing process includes optimizing the point cloud data through operations such as denoising, sampling and point complement, so as to prepare for the grid of the point cloud in the future. Due to the sparsity of the basic data points and the measurement errors, the point cloud model will have poor surface quality or some model vacancy after the packaging operation enters the polygon stage. Therefore, it is necessary to fill holes and smooth the model data. Fig.2-2 shows the polygon processing process. In the process of polygon processing, according to the different needs of different parts, we need to choose different commands to process it.

(a) Fill holes (b) Relaxation fairing

Fig.2 Point cloud processing and polygon processing

Once the polygons are processed, it's time to use Geomagic's accurate surface functionality to generate the required surface mesh and eventually generate the required NURBS surface model.The specific operation flow chart is shown in Fig.3.

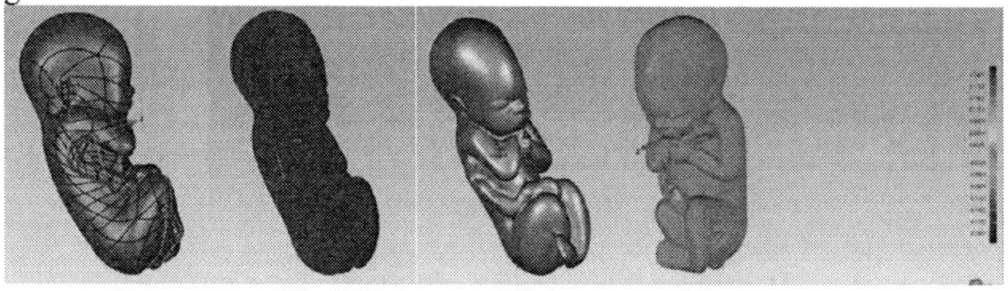

Fig.3 Accurate surface processing

2.1.2 Assembly of Mannequins

The assembly of the model of pregnant woman and fetus refers to the mathematical model of pregnant woman and fetus established by Jing Chen [11] and the model of Chinese reference human pregnant woman established by Tang Xiaobin [12] et al, which mainly includes the establishment of the uterus of pregnant women in the late trimester and the assembly of fetus and uterus. The model of the pregnant woman's uterus is mainly divided into three stages: the first, second and third trimester, as shown in the figure. Among them, the uterine model in the third trimester is mainly surrounded by two hemispheres and a cone in the middle. In order to simplify the model and reduce the amount of calculation, the uterine wall, placenta and amniotic fluid are not considered here.

As shown in Fig.4, the software SimCenter is used to establish the geometric model of the pregnant woman's uterus. The process of establishing the geometric model mainly includes: section selection, sketch drawing, rotation and subtraction, and finally the combination of the fetal model. The combination method of the human body model can be more simple to combine the human body model, so that the final simulation can clearly express the situation of electromagnetic radiation inside and outside the pregnant woman's body.

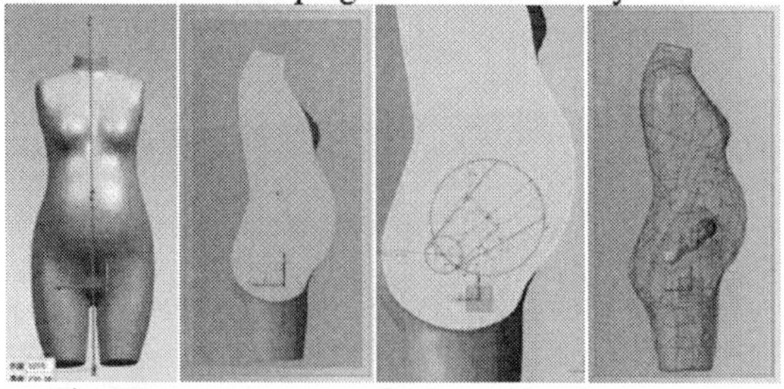

Fig.4 Pregnant woman fetal combination process

2.2 The Antenna Model

Considering the complexity of the antenna model, common antenna structures are selected for simulation. The half-wave dipole antenna is a kind of basic line antenna with simple structure. It is also a kind of more classical and simple antenna. The establishment of antenna model is based on the book HFSS Antenna Design [13]. The antenna is composed of two straight wires with the same diameter and length. The length of the wires is 1/4 wavelength, and the diameter is much smaller than the working wavelength, and the distance between the middle end points is much smaller than the working wavelength.

The antenna model designed here is a half-wave dipole antenna with a central frequency of 900MHz. The antenna material is an ideal conductor, with a total length of 0.48λ and a radius of λ/200. The antenna feed is excited by a centralized port, and the impedance is set at 50Ω.According to the calculation formula in the figure below, when the frequency of the electromagnetic wave is 900MHz, the wavelength of the electromagnetic wave is about 0.33m.The establishment parameters of the antenna model are shown in Table 1.

The propagation velocity of electromagnetic wave in vacuum: $c = 3\times10^{8}\,\mathrm{m/s}$

Frequency of electromagnetic wave: $f = 900MHz = 9\times10^{8}\,Hz$

Wave length of electromagnetic wave: $\lambda = \frac{c}{f} = \frac{3\times10^{8}\,\mathrm{m/s}}{9\times10^{8}\,Hz} \approx 0.33\mathrm{m}$

Table 1. The establishment of antenna model

Variable meaning	A variable's value(unit:mm)
Working wavelength	λ
Total antenna length	0.48 λ
Port distance	0.24
Single pole length	(0.48 λ -0.24) /2
Radius of the antenna	λ /200

2.3 Fashion Model

The clothing model adopts a hybrid modeling method, mainly including forward design (CLO3D software) -- to build the electromagnetic radiation protective clothing model for pregnant women, and reverse design (GEOMAGIC software) -- to process the forward designed clothing model, and finally assemble with the human body model in the electromagnetic simulation software. The main process of modeling is as follows: firstly, the clothing plate is made according to the body shape of the pregnant woman, and then the drawing plate is imported into the virtual fitting software for fitting, and then the clothing model established in the virtual fitting software is exported to the OBJ format. Since the exported garment model cannot be simulated directly in the electromagnetic simulation software, it is necessary to process the garment model in GEOMAGIC before simulation.

As shown in Fig.5(a), the pattern is first made in Clothe3D.The processing process of the garment model in GEOMAGIC is shown in the figure above. Since the imported garment model has many pleated structures, the model needs to be processed and smoothing in and out by removing features(b), simplifying polygons(c), loosening polygons, deleting nails, accurate surface(d) and other steps. The processing sequence can be adjusted according to the situation of the model. Finally, after stretching with SpaceClaim, it is imported into the electromagnetic simulation software for simulation,Fig.5(e).

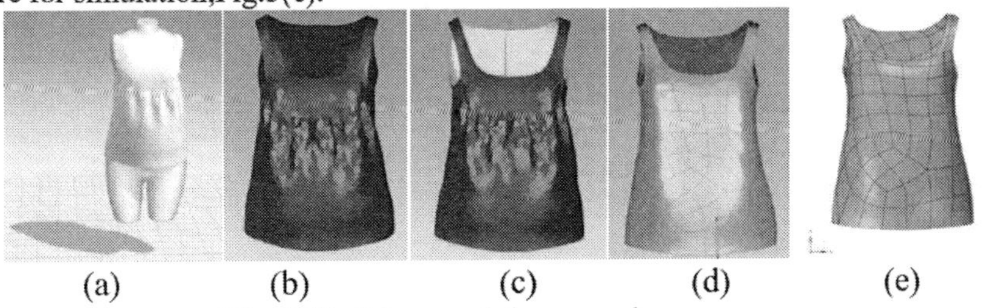

(a) (b) (c) (d) (e)

Fig.5 Clothing model processing process

3. Verification Simulation Experiment of the Model

3.1 Assembly of Simulation Model

HFSS electromagnetic simulation software is a three-dimensional simulation software based on electromagnetic field theory to solve electromagnetic problems. It needs to solve Maxwell's equations under certain boundary conditions and excitation conditions. The simulation model established above needs to be assembled and run in the electromagnetic simulation software, as shown in Figure 6 below. The antenna model, clothing model and human body model need to be assembled together. As described in GB/T2017, the human model faces the transmitting antenna, and the main lobe of the antenna model faces the abdomen of the pregnant woman model, so that the transmitting antenna and the test point are at the same height [14]. Then give the body and clothing corresponding material parameters. The dielectric constant of the mannequin will change with different frequencies. At 900MHz, the dielectric constant of the mannequin is about 41.5.The dielectric constant of the garment model is approximately the dielectric constant of the silver fiber, which is 1.Among them, the material parameters of the human body are set according to GB/T 33615-2017 Clothing Electromagnetic Shielding Effectiveness Test Method.

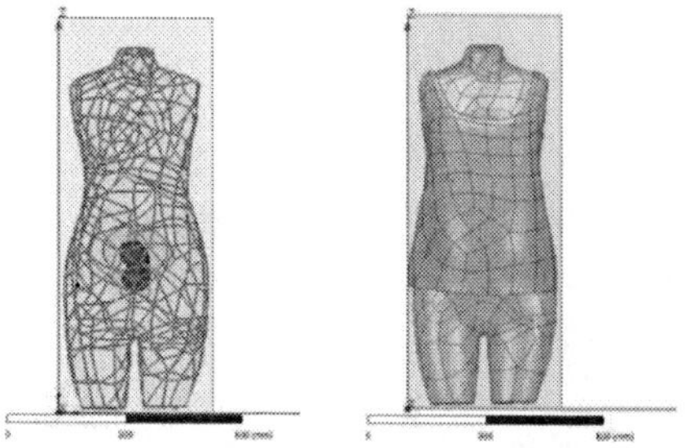

Fig.6 Pregnant woman body assembly (left) and Body clothing assembly (right)

3.2 Simulation Environment Setting

Before the simulation, the radiation boundary conditions should be set to simulate the open free space. In finite element solution, setting the boundary condition at an infinite distance will affect the solving speed of simulation solution. Antenna is a radiation structure, and its radiation boundary is theoretically at an infinite distance. The free space can be effectively simulated by setting the radiation boundary. The radiation boundary is not less than 1/4 wavelength from the human body. The shape of the radiation boundary is shown in Fig.7, which is a rectangular model. As shown in the Fig.7, the cuboid air box needs to cover the antenna and the mannequin. The distance between the mannequin and the antenna is about 1.3m, so the length of the air box is set as 1.8m and the width is set as 0.4m.The radiation boundary of the antenna is the size of the air box. After assembling the models required for the simulation experiment, the two groups of simulation experimental models before and after the human body wearing anti-electromagnetic radiation clothing were respectively run (the difference between the two groups of models only lies in whether there is a clothing model).

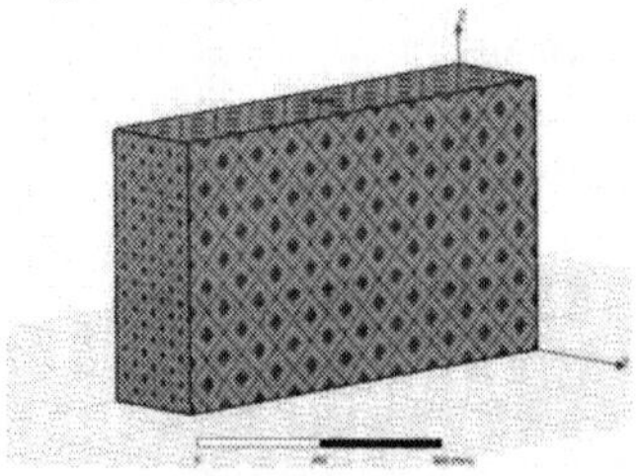

Fig.7 Air Box Settings

3.3 The Running and Verification of Simulation Model

There are various methods to verify the simulation model. This paper selects the method of literature verification to verify the model constructed. Ma Jian [15] pointed out in the paper that when the antenna is irradiated in the abdomen, the neck, upper chest, lower chest and thighs of pregnant women are exposed to the most serious radiation. And when the surface of the human body has a greater curvature, there will be more radiation around that area. This is consistent with the conclusion of the electric field radiation cloud map in this paper. Fig.8 below shows the radiation cloud map of the electric field intensity received by the human body before wearing clothing. Fig.9 shows the radiation cloud of the electric field intensity that the human body is exposed to after wearing clothing.

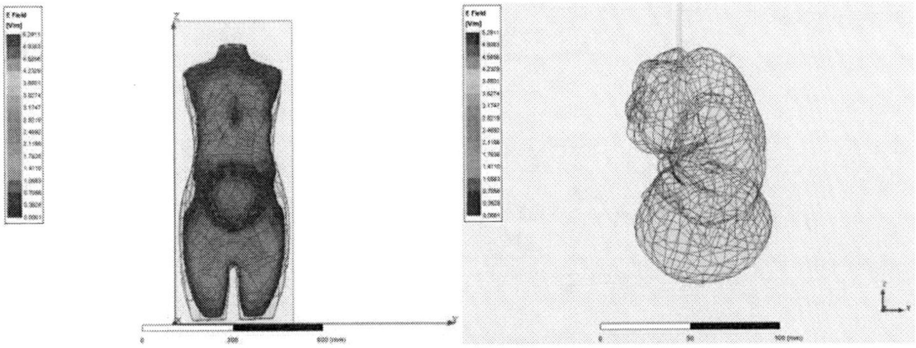

Fig.8 Unclothed model radiation cloud image

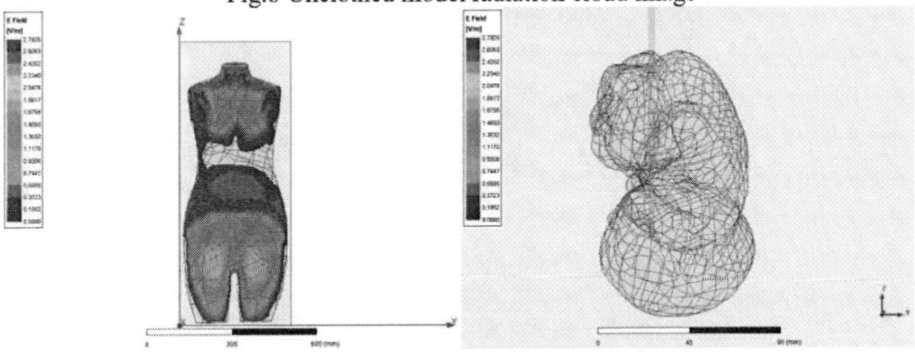

Fig.9 Model radiation cloud image after dressing

4. The Running Results and Analysis of the Simulation Model

From the simulation results in the figure above, we can observe that the radiation received by the front of the human model is significantly greater than that received by the side of the human model, and the electric field radiation received by the pregnant woman and fetus without electromagnetic radiation protection clothing is significantly higher than that received by the wearing clothing. When not wearing anti-electromagnetic clothing, pregnant women's chest, abdomen, upper thighs of these three parts of the radiation degree is higher, the fetus will be exposed to a small dose of electric field radiation; After wearing the clothes, we can obviously observe that the abdomen of the pregnant woman model was almost avoided by the electric field radiation, but the chest and thigh parts were still affected by the electric field radiation, while the fetal model was almost not affected by the electric field radiation. Fig.10 and 11 show the changes in the distribution of electric field values of human body before and after wearing clothes, drawn according to the specific values of the simulation cloud image. The broken line on the left is the electric field intensity of the pregnant woman model. The blue curve is the field intensity before wearing clothes, and the yellow part is the field intensity after wearing clothes. It was clearly observed that after wearing clothing, the pregnant woman's exposure to electric field radiation decreased significantly. The field intensity distribution of the fetal model showed the same trend, and the radiation exposure decreased significantly after dressing.

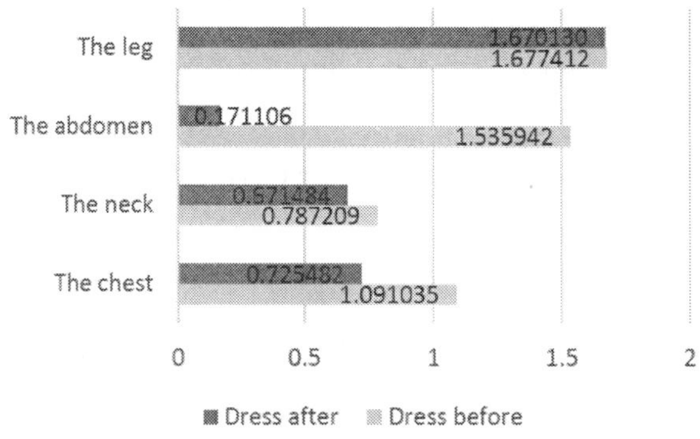

Fig.10 Electric field radiation values in different parts of pregnant women

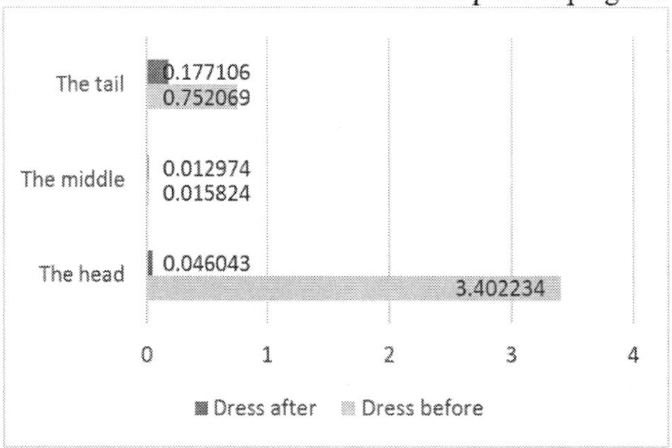

Fig.11 Electric field radiation value of different parts of fetus

Fig.12 and 13 are the cloud images of electromagnetic radiation energy absorbed by various parts of the human body before and after wearing anti-electromagnetic clothing. SAR ratio absorptivity is often used in some research fields related to the biological effects of microwave to represent the degree of radiation received by organisms [14]. In the SAR cloud image, we can also observe similar results: the neck, chest, abdomen and upper thighs of the pregnant woman model absorbed the highest electromagnetic radiation energy when they were not wearing anti-electromagnetic radiation clothing, and the abdomen of the pregnant woman model absorbed the highest electromagnetic radiation energy level. The back of the fetal model also absorbed a smaller dose of electromagnetic radiation energy. After wearing clothing, the neck and upper thighs of the pregnant woman model absorbed more electromagnetic radiation energy, while the abdomen absorbed less electromagnetic radiation energy. The foetal model absorbs almost no electromagnetic radiation energy.

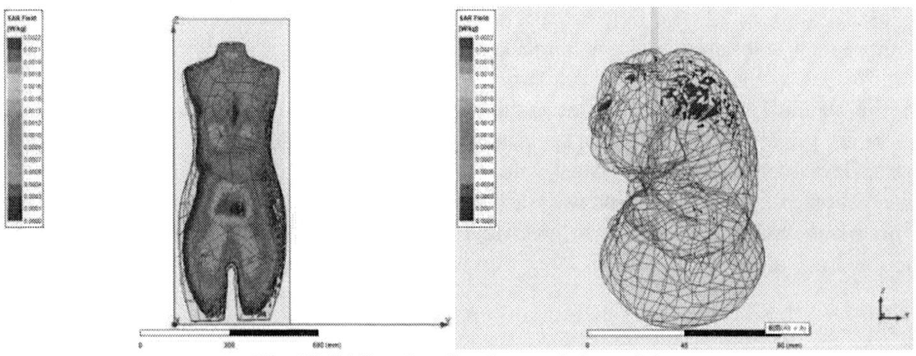

Fig.12 SAR value for the unclad model

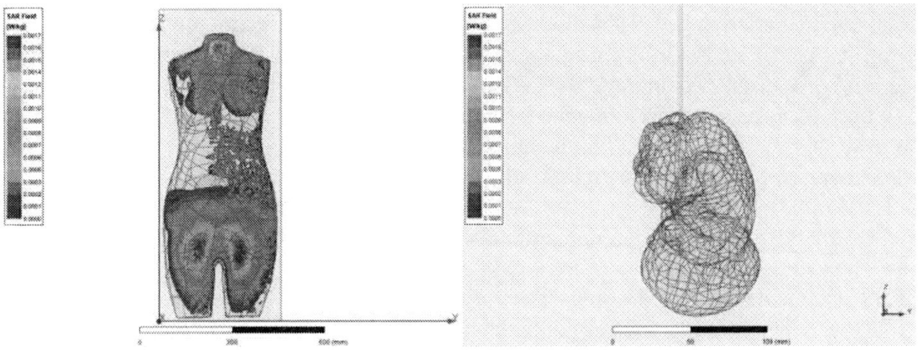

Fig.13 SAR value of the model after dressing

From the above analysis, we can conclude that the radiation exposure of the fetal model is proportional to the radiation exposure of the pregnant model, and the greater the radiation exposure of the pregnant model is, the greater the radiation exposure of the fetal model will be.

5. Conclusion

In this paper, by establishing the model of pregnant women and fetuses and conducting simulation experiments in the three-dimensional electromagnetic simulation software, we studied the electromagnetic radiation exposure of pregnant women and fetuses when wearing anti-electromagnetic clothing and wearing clothing under 900MHz electromagnetic radiation. Through the study found that pregnant women in the neck, chest, abdomen, upper thighs by the influence of electromagnetic radiation is greater, and the fetus by the influence of electromagnetic radiation is smaller. Wearing anti-electromagnetic radiation clothing can effectively reduce the range of electromagnetic radiation, through three-dimensional electromagnetic simulation software, combined with virtual fitting software, we can better design anti-electromagnetic radiation clothing structure, so that it can more effectively protect pregnant women and fetuses.

6. Acknowledgement

This research is supported by Shaanxi Engineering Research Center Of Apparel Intelligent Design And Manufacture platform.

References

[1] Sun ZH, Wang DW, Li YL. Effect of microwave radiation on female reproductive system [J]. Chinese Journal of Labor Health and Occupational Diseases: 2017; 35 (12): 956-960.
[2] Qu YL, Cao ZJ, Wang Q. Effect of electromagnetic radiation on pregnancy and progeny [J]. Journal of Environment and Health: 2010; 27 (4): 367-369.
[3] Zhang P, Gu H, Li WD, ect. Study on shielding effectiveness of different types of radiation protective clothing for pregnant women [J]: 2018; 36 (3): 31-35.
[4] Zhang XX. Study on shielding performance and structure design of radiation-proof maternity clothing [D]. Textile journals. Shang hai: Donghua university: 2009.
[5] Chen C, Yu S, Yi M. The research on shielding effectiveness measurement for electromagnetic shielding garment 2017 [C]. IEEE: 2017.
[6] Zhang LL. Testing and simulation of electromagnetic protection performance of radiation-proof maternity clothing [J].Textile journals: 2011; 32 (10): 108-112.
[7] Yao L. Research on testing method of shielding efficiency of electromagnetic shielding suit [D]. Zhenzhou: Zhongyuan Institute of Technology: 2013.
[8] Sun RL. Research on shielding effectiveness of hole and seam area of electromagnetic shielding garment

based on computer simulation [D]. Zhenzhou: Zhongyuan Institute of Technology: 2015.
[9] Ma L, Zhang X, Ying BA. A bio-electromagnetic modeling approach for electromagnetic radiation resistant clothing design [J]. Journal of clothing: 2018; 3 (6): 497-505.
[10] Kang XL. Study on protective performance of pregnant woman waist support belt based on finite element simulation analysis [D]. Xi'an: Xi 'an Polytechnic University: 2019.
[11] Jing C. Mathematical Models of the embryo and fetus for use inradiological protection [J]. Health Physics: 2004; 86 (3): 285-295.
[12] Tang XB. The establishment of a reference human pregnant woman model and dosimetric study of internal exposure in China [J].Chinese Journal of Medical Physics: 2012; 29 (1): 3088-3095.
[13] Li MY. HFSS antenna design [M]. Beijing: Electronic Industry Press: 2011.
[14] Zhang SY. Research on testing method of shielding effect of shielding service [D]. Shanghai: Southeast university: 2016.
[15] Ma J. Study on shielding mechanism analysis and evaluation method of individual anti-electromagnetic clothing [D]. Xi'an: Xi 'an Polytechnic University: 2020.

The Effects of Different Density Microneedles on Drug Release

Le-Hao Zhang, Zhen-Zhen Qi, Bin Tian, Zu-Qiang Yin, Shen-Zhou Lu*

National Engineering Laboratory for Modern Silk, College of Textile and Clothing Engineering, Soochow University,Suzhou, China

*Corresponding author's email address: lushenzhou@suda.edu.cn

Abstract

The effect of different microneedles densities on drug release was studied by using silk fibroin with good biocompatibility as the substrate of microneedles. The mixture of D-sorbitol and silk fibroin at the ratio of 2:10 was poured into a polydimethylsiloxane mold. After vacuum extraction, constant temperature and humidity drying, the silk fibroin microneedles were obtained. Three kinds of microneedles arrays of different densities were prepared with the area of 1cm^2: 5×5, 10×10 and 20×20. The length of silk fibroin microneedles was about 600μm, and D-sorbitol/silk fibroin blend film was used as the control group. Melatonin was selected as the model drug and three kinds of microneedles and blend film were used to study the drug release in vitro. The aggregation structure of D- sorbitol/silk fibroin microneedles was measured by X-ray diffraction (XRD) and infrared spectroscopy. The anti-compression properties of three kinds of microneedles were measured by texture analyzer (TMS-PRO). The results showed that: (1) D-sorbitol could change the silk fibroin from random coil structure to the crystal structure of Silk Ⅰ, and the dissolution rate of Silk fibroin microneedles was low and the swelling property of silk fibroin was certain. (2) Three kinds of microneedles all have good compression resistance and puncture performance. (3) All three kinds of microneedles have good drug release performance. With the increase of microneedles density, drug release rate and cumulative drug release rate increase. With the increase of microneedles density, the compressive performance of microneedles is improved. The three kinds of silk fibroin microneedles have good penetration effect and can achieve the slow release effect. Compared with silk fibroin film without microneedles, silk fibroin microneedles can significantly improve the speed of drug release by opening skin micropores.

Keywords: Silk Fibroin; Microneedles; Melatonin; Drug Release

1. Introduction

The cocoon is mainly composed of silk fibroin (SF) and sericin (SS). The content of silk fibroin is 72-81%, and sericin is 20-30%. Silk fibroin (SF) has high mechanical strength, good flexibility, light weight and good biocompatibility [1, 2]. There are α-helical conformations, β-sheet conformations, β-angular conformations and random coiled conformations in silk fibroin. The uncrystallized Silk fibroin presents a random coil conformation, and the crystalline structure of silk fibroin is mainly divided into two kinds: Silk Ⅰ and Silk Ⅱ [3, 4].

Transdermal drug delivery refers to the drug delivery mode wherein the drug passes through the epidermis, penetrates the skin, absorbed by the capillaries and enters the circulation of the human body [5]. The advantage is that the drug entering the human body is not altered by the changes in the digestive tract environment, and the drug plays a constant and slow role, avoiding adverse reactions caused by the high blood drug concentration caused by other rapid administration methods, and avoiding the first pass effect of liver [6]. Microneedles have been extensively studied as a tool in the 3rd generation transdermal system (TDDS). In recent years, many drug-carrying microneedles have entered clinical trials, and can be soon used for disease treatment [7]. Most microneedles do not need to rely on the help of instruments, and can carry a drug storehouse for drug delivery [8, 9]. Compared with the traditional transdermal patch, microneedles create many micron-sized holes on the skin surface to increase the penetration of drugs. This transdermal drug delivery method eliminates the barrier of the cuticle layer of the skin and breaks the limitation that the transdermal drug delivery system is only suitable for small molecule lipophile drugs [10, 11].

As a kind of natural polymer with non-toxic, non-irritating, good stability and excellent mechanical properties, silk fibroin is expected to become an excellent microneedles substrate [12]. Zhu P Yin et al. [13, 14] studied a kind of drug delivery system of swelling silk fibroin microneedles. Using silk fibroin with good biocompatibility as the substrate of the microneedles and using mold casting method, a kind of silk fibroin microneedles with good swelling property, low solubility, good compression resistance and good transdermal drug release effect was obtained.

At present, the relationship between the density of microneedles and the efficacy of transdermal drug release is not clear. In order to explore the relationship between microneedles density and drug release rate, this paper used melatonin as the model drug and set up three kinds of silk fibroin microneedles with different array densities. In this paper, the silk fibroin film without microneedles was used as the control group, and D-sorbitol and silk fibroin were mixed in a ratio of 2:10 to prepare microneedles with certain swelling property and low solubility loss [15]. It is expected to obtain relevant experimental results through the study of microneedles patches with different array densities, so as to provide some references for the application of silk fibroin microneedles.

2. Materials and Methods

2.1 Experimental Materials

Home cocoon shell silkworm silk biological technology Co. , Ltd. (Suzhou), anhydrous sodium carbonate, sodium bicarbonate, potassium dihydrogen phosphate, twelve hydration disodium hydrogen phosphate, sodium chloride, potassium chloride (national medicine group chemical reagent Co. , Ltd.), D - sorbitol (Jiangsu argon krypton xenon materials technology Co. , Ltd.), melatonin, lithium bromide (Tiancheng chemical Co. , Ltd.),Dialysis bag (Shanghai Puyi Biotechnology Co. , Ltd. , with intercalated molecular weight of 8 ~ 10kDa), the experimental water was all deionized water.

2.2 Experimental Instruments

DF-101S heat-collecting magnetic constant temperature agitator (Henan Gongyi Yuhua Instrument Co., Ltd.), polydimethylsiloxane (PDMS) mold, laser engraving machine, fluorescent microscope, Nicolet5700 type Fourier transform infrared spectrometer (Nicolet USA),Automatic X 'PERTPRO MPD X-ray diffractometer(Panaco, Panalytical Company), TMS-PRO texture analyzer (Food Technology Corporation, USA),Precision balance, Franz diffusion cell (Tianjin Zhengtong Technology Co., Ltd.), ShimadzauUV-1800 double beam UV-Vis spectrophotometer, etc.

2.3 Preparation of Regenerated Silk Fibroin Solution

Take the fresh cocoon shell, peel off the surface of the miscellany silk, weigh 80g cocoon, cut it into pieces. Using an induction cooker and a degumming pan, add 4 L of deionized water and turn on power to bring the water to a boil. 3g anhydrous $NaHCO_3$ and 1 g anhydrous Na_2CO_3 were weighed and used in the process of three degumming. The cocoon shell was boiled in a pot for 30 min, and then extracted and washed repeatedly with deionized water to remove sericin. The cocoon shell was then placed in a drying box at 60°C to dry silk fibroin. The silk fibroin was dissolved in 9. 3 M LiBr at 3:20 (w/v) and bathed at 65°C for 1 h. The dissolved solution was put into a dialysis bag (14kDa interception of molecules), and dialysis was carried out in ionic water at 4C for three days. The absorbable cotton was filtered to obtain the silk fibroin aqueous solution, which was bottled and sealed and stored in a refrigerator at 4°C for later use.

2.4 Preparation of Melatonin Silk Fibroin Microneedles

By blending the drug solution with silk fibroin solution, as shown in Fig.1, the microneedles can be prepared by casting and vacuum, without external stimulation throughout the whole process. The drugs can be wrapped in the microneedles in a gentle way to preserve the drug activity and improve the storage stability. The drug loading method can determine the dosage, so as to have the ability to regulate the drug loading of microneedles and meet the drug dosage demand in various situations. 2ml of the prepared blend solution of D-sorbitol/SF and melatonin was placed on a PDMS microneedles mold with the size of 10mm×10mm and the microneedles density distribution of 5×5, 10×10 and 20×20, and placed in a vacuum drying oven to remove the bubbles inside (the vacuum degree was 0. 09 MPa). Then placed in a constant temperature and humidity environment drying molding.

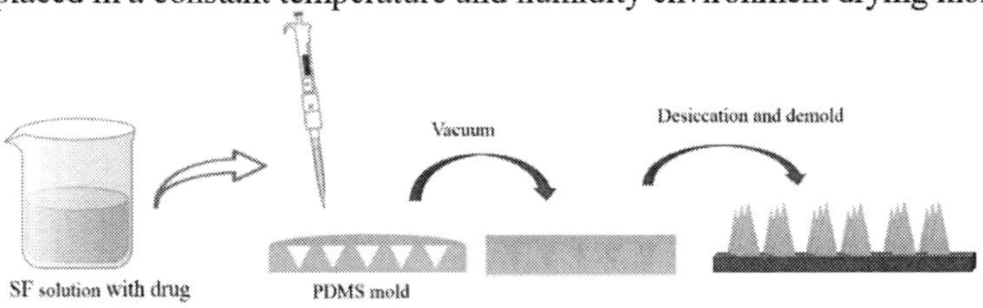

Fig.1 Preparation of silk fibroin microneedles

2.5 Measure the Length of the Microneedles

Fluorescence microscope was used to measure the length of the above three kinds of microneedles. To ensure that the three different densities of microneedles have the same length and can be successfully pierced without causing pain.

2.6 Swelling and Dissolution Loss Property of SF Microneedles was Detected

The dissolution and swelling property of microneedles is an important property that affects the drug delivery efficiency of microneedles. The lower dissolution property maintains the needle type of microneedles, while the higher swelling property opens the drug delivery pathway and improves the delivery efficiency.

0.1g of the prepared microneedles flakes were cut and placed in a circular bottom centrifugal tube. 10ml of deionized water was added at a bath ratio of 1:100 and then vibrated in a 37°C water bath thermostatic oscillator for 24h. The centrifuge tube was removed and the incomplete microneedles flakes were washed with deionized water for three times, and then the water on the surface of the flakes was dried with a filter paper. Then the centrifuge tube was centrifuged for 15 min, the supernatant was taken at 278nm and the absorbance was measured by ultraviolet absorbance. The concentration of silk fibroin in the supernatant was obtained by comparing with the standard curve, and the dissolution rate of microneedles at D-sorbitol /SF=2:10 was calculated to be 7%. The swelling degree and dissolution rate of the prepared film measured by the above method had no difference with that of the microneedles.

2.7 Mechanical Properties of Silk Fibroin Microneedles were Measured

In terms of the mechanical properties of the microneedles, the maximum bearing force of the microneedles during compression and deformation was mainly tested to reflect the mechanical strength of the microneedles.

Three kinds of microneedles of 5×5, 10×10 and 20×20 specifications (the size is 1cm^2) were prepared. The tip of the microneedles is placed upward on the steel plate operating table at the bottom of the TMS-Pro texture instrument (Fig.2). The probe moves downward at a certain speed, and the probe contacts the needle to generate compression force. The induction device can test the compression force until the tip is bent and broken. Detection conditions: the initiation force is 0.02N, the force sensing element is 25N, the compression degree is 80%, and the probe loading rate is 10mm/min. Measure the compressive strength of all microneedles of 1cm^2 size.

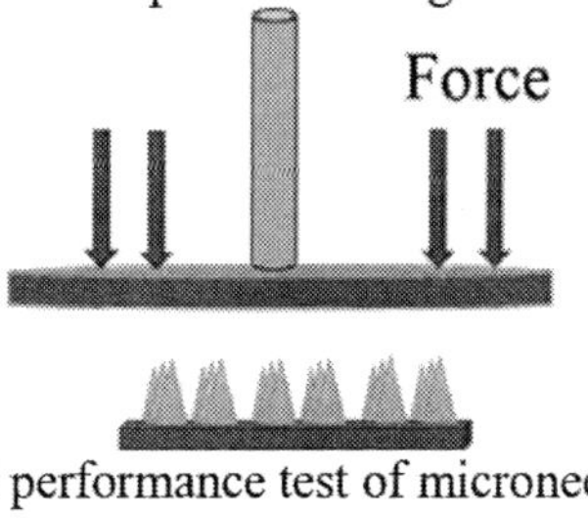

Fig.2 Mechanical performance test of microneedles compression

2.8 X-ray Diffraction Detection of Silk Fibroin Microneedles and Film

The silk fibroin microneedles and film were cut into powder, passed through an 80-mesh sieve, and the screened powder was dispersed in the quartz sample chamber, and the groove was compacted with a glass slide. The sample was tested by X-ray diffraction (XRD) with automatic X 'PERTPROMPD. The diffraction intensity curve of 2θ=5 ~ 45° was recorded under the test conditions of scanning speed 10°/min, tube voltage 40kV and tube current 35mA to determine the aggregate structure of the microneedles and the film.

2.9 Detection of Silk Fibroin Microneedles and Film by FTIR

The three kinds of microneedles and the blend film were ground into powder with appropriate particle size. The sample to be tested was mixed with potassium bromide, and the tablet pressing mechanism was used to prepare the tablet in accordance with the regulations. According to the operating requirements of the experimental instrument, the scanning range of 400~4000cm^{-1} was used to obtain the infrared absorption spectrum on the Nicolet 5700 intelligent Fourier transform infrared spectrometer.

2.10 In Vitro Release of Silk Fibroin Melatonin Microneedles

Three kinds of microneedles were cut into the size of 10mm×10 mm, and another group of D-sorbitol /SF blend film was added as the control group, which was also cut into the size of 10mm×10mm. Two groups of microneedles with uniform length and good looking tip and blend film were selected for weighing, and three kinds of microneedles with three specifications on the same piece were also selected for weighing, and then the three kinds of microneedles were placed in an oven at 110°C for drying for two hours to measure the solid content. The melatonin concentration set in this experiment was 0.5 mg/ml. According to the ratio of silk fibroin, D-sorbitol and melatonin, the drug content M in a single piece of microneedles was obtained.

$$M = m \times \omega \times [\frac{\alpha}{1+\alpha+S}] \quad 1)$$

In the formula, M(μg) is the drug loading of microneedles or film. m(μg) is the mass of the dry microneedles or film, ω is the solid content of the microneedles or film, S is the mass ratio of D-sorbitol to SF, and α is the mass ratio of melatonin to SF.

The full wavelength scanning of melatonin was carried out with ShimadzauUV-1800 double beam UV-Vis spectrophotometer. In the Horiba fluoromax-4 type fluorescence spectrometer with different excitation wavelength to measure the emission spectrum, find the position of the emission wavelength, and then use this emission wavelength to measure the excitation spectrum, find the best excitation wavelength. A series of melatonin PBS solutions with concentration gradient were prepared by dilution with PBS buffer at constant volume. Linear regression analysis was carried out on the concentration of melatonin solution and fluorescence intensity within a certain range, and the standard curve of the melatonin was obtained.

Preparation of skin in vitro: anesthesia in SD rats, with a razor to abdominal hair cut short, dip in with cotton ball for hair removal cream brush coating on the surface of the skin, 2 to 3 minutes, with paper towel first, again will kill the rat cervical dislocation, scissors cut hair removal, skin with blades scrape off the fat layer and the capillary plexus, after wash with physiological saline put-20°C cold storage for later use. Before use, put the refrigerator at 4°C to thaw in normal saline.

In order to prevent melatonin from directly penetrating into the skin, Paraflim sealing film with good ductility was simulated as the stratum corneum, since melatonin is a small molecule drug that is fat soluble. The three kinds of microneedles were inserted into the extended sealing film. After the microneedles were confirmed to be able to penetrate, they were then inserted into the abdominal skin of the fresh SD rats treated above, and the needles were fixed in the middle of the drug release pool (Fig.3) with the tip down. The extended sealing film was also placed between the blend film and the rat skin as the control group. A magnetic agitator and 12ml PBS buffer solution were added to the sampling port at a rotating speed of 300r/min and a water bath temperature of 32°C. Sample 1mL at time intervals and add 1mL fresh PBS buffer solution. The sample was placed in a fluorescence spectrometer with an excitation wavelength of 286 nm and a scanning range of 296-500nm. The fluorescence intensity at 352nm was read, and the corresponding FITC-Ins concentration was calculated as C. The cumulative drug release rate W at each time point was calculated, and the cumulative drug release rate curve of the three microneedles and the film was plotted over time. A total of 14 samples were taken at regular intervals.

$$W = \frac{C_n \times \sum_{i=1}^{n-1} C_i \times V_i}{M} \times 100\% \qquad 2)$$

In the formula, W(%) is the cumulative drug release rate of transdermal release of microneedles or film, V is the volume of the receiving chamber (ml), V_i is the volume of each sampling (ml), C_n and C_i are the drug concentration (μg/ml) in the receiving chamber during the nth and ith sampling respectively, and M is the drug loading amount.

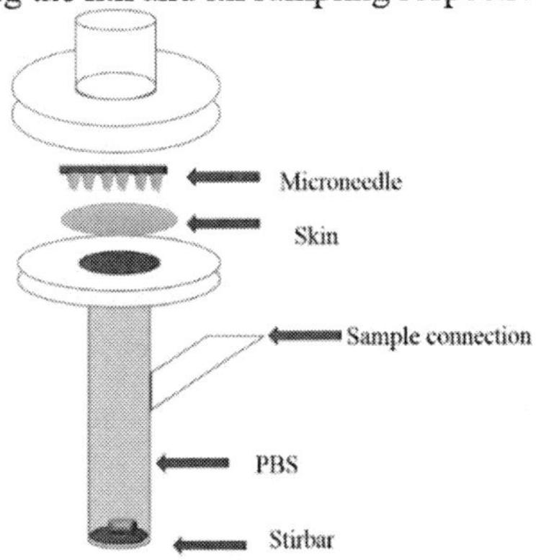

Fig.3 Schematic diagram of Franz transdermal diffusion experiment

3. Analysis of Experimental Results

3.1 Length and Morphology Analysis of Microneedles

The corneous layer of the skin is about 10 ~ 20μm. It is a layer of dead tissue without blood vessels and nerves. It is the main obstacle of drug absorption through the skin. The epidermis is located about 50 ~ 100μm below the skin and contains a few living cells and nerves without blood vessels. Further down, the dermis contains an abundance of living cells, nerves, and blood vessels. The thickness of the cuticle of human skin is less than 40μm, and the average thickness of the epidermal layer is 150μm. The length of the microneedles should be designed to penetrate the skin to achieve therapeutic effect without touching the nerve tissue to avoid pain. As can be seen in Fig.4, the prepared silk fibroin microneedles of the three specifications maintain a high degree of uniformity, and the microneedles have a complete morphology. It shows that the microneedles manufactured by casting method can fit the cavity completely and maintain good morphology. The length of microneedle is about 600μm.The microneedles can be pierced into the stratum corneum without penetrating into the blood vessels and nerves, and the piercing process is painless [16].

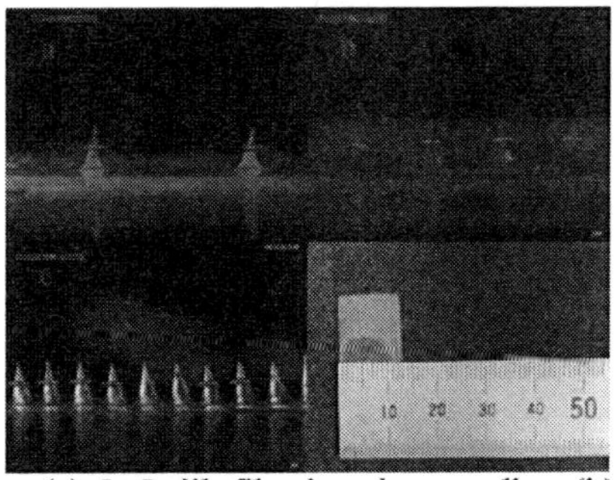

Fig.4 morphologies of microneedles (a) 5×5 silk fibroin microneedles, (b) 10×10 silk fibroin microneedles, (c) 20×20 silk fibroin microneedles, (d) D-sorbitol/silk fibroin blend film

3.2 Analysis of Mechanical Properties of Microneedles Compression

The injection of drugs requires a high degree of expertise, and repeated injection may cause tissue damage and skin infection, which will bring great pain to patients. The small size and simple operation of microneedles enable patients to take in drugs more comfortably and conveniently, so it is necessary to ensure that the microneedles have a certain strength enough to penetrate the skin. The displacements and loads of different microneedles (Fig.5) showed that the three kinds of D-sorbitol/fibroin microneedles all had good mechanical properties: 20×20 microneedles had the best compressive properties, followed by microneedles with densities of 10×10 and 5×5.

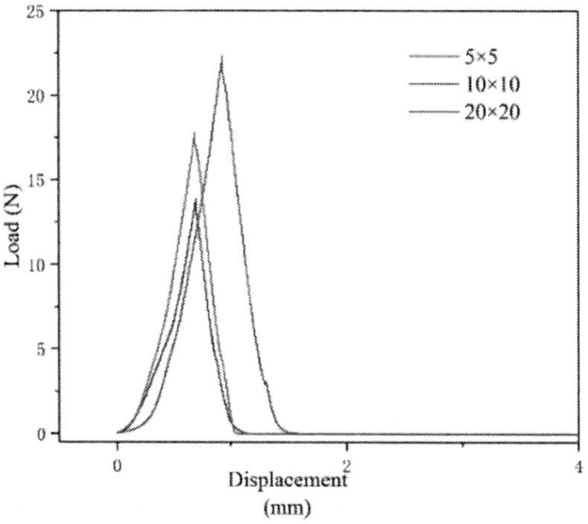

Fig.5 Displacement-load of different microneedles

3.3 X-ray Diffraction and FTIR Analysis

In the X-ray diffraction curve as shown in Fig.6(a), crystal diffraction peaks of Silk I appear at 12.2°, 19.7°, 32.3°, 36.8° and 40.1° [17]. The XRD curve of pure silk fibroin showed a steamed bread peak without obvious absorption peak. The results indicated that pure silk fibroin was in random coil state and did not form crystal structure. Silk I crystal is not very dense, so the microneedles after the modification has a certain swelling property. Silk II crystal is more regularity and dense, which will restrict swelling property and solubility of microneedles. The addition of D-sorbitol transformed the silk fibroin material into Silk I crystal structure. The pores formed by the swelling of silk fibroin microneedles could provide channels for drug molecules, and drug molecules showed different release rates with different swelling properties. Moreover, the XRD curves showed that the structure of the microneedles prepared by the silk fibroin solution with the addition of D-sorbitol was the same as that of the blend film. Based on this rule, a kind of microneedles release and drug delivery system can be built to make the drug release controllable, reduce the risk and improve the safety of use. Silk fibroin microneedles with excellent performance can be obtained by modifying silk fibroin by adding D-sorbitol.

FTIR is an important method specially used to detect and analyze the secondary structure of fibers. The wave number of the secondary structure ranges from 1700cm^{-1} to 1500cm^{-1}. The secondary structure of pure SF belongs to random coil type, and 1650cm^{-1} is the infrared characteristic peak of random coil and α-helical [18]. As shown in the Fig.6(b), the silk fibroin microneedles added with D-sorbitol have the same structure as the film. With the addition of D-sorbitol, the characteristic peak at 1654cm^{-1} in the microneedles and the film disappeared, and the characteristic peak at 1635cm^{-1} was prominent, which was the characteristic peak of β-sheet structure. These results indicate that both α-helix and β-sheet structures exist in the microneedles and the film, and D-sorbitol can transform the silk fibroin from random coil structure to β-sheet structure, which is consistent with the results of X-ray diffraction test above.

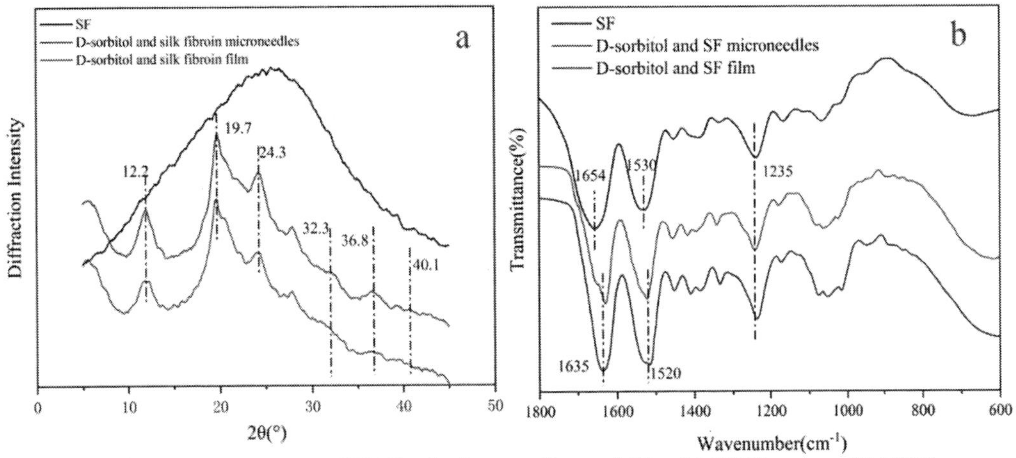

Fig.6 XRD curves and FTIR of microneedles and film (a) XRD curves (b) FTIR

3.4 Analysis of in Vitro Release Results of D-sorbitol /SF Silk Fibroin Melatonin Microneedles

The melatonin full-wavelength scan showed that the peak of the wave was 278nm, the trough of the wave was 248nm, and the maximum absorption wavelength was 226nm. Therefore, the range of an excitation wavelength is determined. The emission spectrum is measured by different excitation wavelengths within the range on the fluorometer to find the position of the emission wavelength, and then the excitation spectrum is measured by this emission wavelength to find the optimal excitation wavelength. Quartz fluorescence dish with four passes of light was used for testing sample solution. The fluorescence detection conditions of melatonin were determined by fluorescence spectrometer. The excitation wavelength was 286 nm, the emission wavelength was 352nm, the slit width was 5nm, and the sample volume was 0.2ml. Try to sample and test on site, and avoid light during the whole process.

According to fluorescence spectrometry, as shown in the Fig.7 the linear regression equation of melatonin concentration and fluorescence intensity was obtained as follows:

$$Y = 420438.9662X + 5964.39364 \qquad 3)$$

R^2=0. 99932, the linear range was 0.1-1.5μg/ mL

Where, Y represents fluorescence intensity (CPS) and X represents melatonin concentration (μg/ mL).

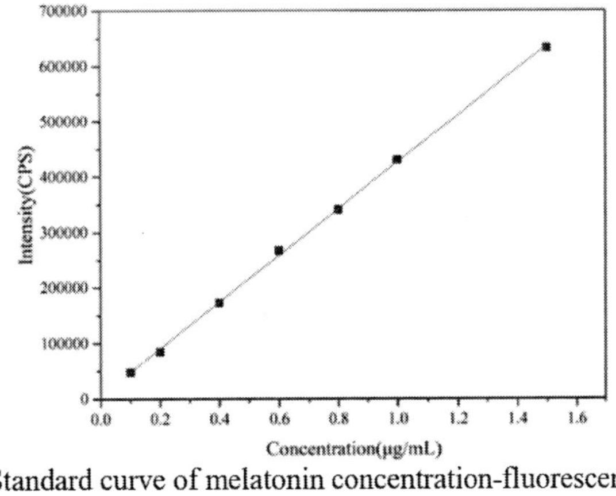

Fig.7 Standard curve of melatonin concentration-fluorescence intensity

As shown in Fig.8, melatonin microneedles cumulative drug release rate analysis can be obtained: the melatonin release rate of the three kinds of microneedles was faster within 60min, and the release rate gradually slowed down after 200min. The cumulative drug release rate of 20×20 microneedles was the highest, reaching 69% at 8h, followed by 10×10 microneedles and 5×5 microneedles. In the control group, the cumulative drug release rate was only 4% after 8 hours, and melatonin was not released. Therefore, the transdermal drug release effect of the microneedles has a certain relationship with the density of the microneedles: with the increase of the density of the microneedles, the speed of the drug release increases, and the rate of cumulative drug release in the same time increases.

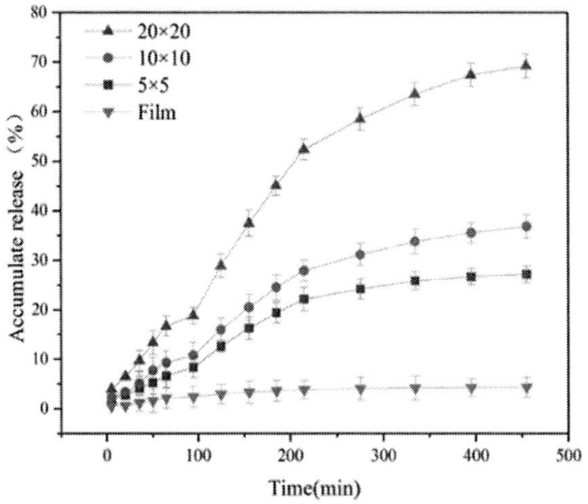

Fig.8 Cumulative release curve of melatonin microneedles in vitro

4. Conclusion

In this paper, by adding D-sorbitol, Silk fibroin can be transformed from random coil structure to crystal structure of Silk Ⅰ, so that the dissolution rate of Silk fibroin microneedles is low and it has certain swelling property. With the increase of microneedles density, drug release rate and cumulative drug release rate increased. With the increase of microneedles density, the compressive performance of microneedles is improved. The three kinds of silk fibroin microneedles have good penetrating effect on skin and slow release effect. Compared with silk fibroin film without microneedles, silk fibroin microneedles significantly improved the rate of drug transdermal release by opening micropores in the skin. Moreover, with the increase of microneedles density, the release rate of melatonin increased significantly in the early stage. Drug release rate could be controlled by controlling the density of microneedles.

5. Acknowledgments

This work was supported by College Nature Science Research Project of Jiangsu Province, China (Grant No. 20KJA540002), PAPD and Six Talent Peaks Project in Jiangsu Province (Grant No. SWYY-038).

References

[1] Tsukada M, Freddi G, Gotoh Y, et al. Physical and chemical properties of tussah silk fibroin films [J]. Journal of Polymer Science Part B: Polymer Physics: 1994; 32 (8): 1407.
[2] Rockwood DN, Preda RC, Yücel T, et al. Materials fabrication from Bombyx mori silk fibroin [J]. Nature Protocols: 2011; 6 (10): 1612.
[3] Tanaka K, Inoue S, Mizuno S. Hydrophobic interaction of P25, containing Asn-linked oligosaccharide chains, with the H-L complex of silk fibroin produced by Bombyx mori [J]. Insect Biochemistry and Molecular Biology: 1999; 29 (3): 269.
[4] Lu Q, Hu X, Wang XQ, et al. Water-insoluble silk films with silk I structure [J]. Acta Biomaterialia: 2010; 6 (4): 1380.
[5] Lee H, Song C, Baik S, et al. Device-assisted transdermal drug delivery [J]. Advanced Drug Delivery Reviews:

2017; 11 (6): 5992.
[6] Waghule T, Singhvi G, Dubey SK, et al. Microneedles: A smart approach and increasing potential for transdermal drug delivery system [J]. Biomedicine & Pharmacotherapy: 2019; 109: 1249.
[7] Münch S, Wohlrab J, Neubert RHH. Dermal and transdermal delivery of pharmaceutically relevant macromolecules [J]. European Journal of Pharmaceutics and Biopharmaceutics: 2017; 119: 235.
[8] Cheung K, Das DB. Microneedles for drug delivery: trends and progress [J]. Drug Delivery: 2016; 2338.
[9] Stinson JA, Raja WK, Lee S, et al. Silk Fibroin Microneedles for Transdermal Vaccine Delivery [J]. ACS Biomaterials Science & Engineering: 2017; 3 (3): 360.
[10] Moniz T, Costa Lima SA, Reis S. Marine polymeric microneedles for transdermal drug delivery [J]. Carbohydrate Polymers: 2021; 266: 118098.
[11] Delgado-Charro MB, Guy RH. Effective use of transdermal drug delivery in children [J]. Advanced Drug Delivery Reviews: 2014; 73: 63.
[12] Stinson JA, Raja WK, Lee S, et al. Silk Fibroin Microneedles for Transdermal Vaccine Delivery [J]. ACS Biomaterials Science and Engineering: 2016; 3 (3): 360.
[13] Yin Z, Kuang D, Wang S, et al. Swellable silk fibroin microneedles for transdermal drug delivery [J]. International Journal of Biological Macromolecules: 2018; 106: 48.
[14] Konstantinos, Tsioris, Waseem, et al. Fabrication of Silk Microneedles for Controlled-Release Drug Delivery [J]. Advanced Functional Materials: 2011; 22 (2): 330.
[15] Zhao L, Wen Z, Jiang F, et al. Silk/polyols/GOD microneedles based electrochemical biosensor for continuous glucose monitoring [J]. RSC Advances: 2020; 10: 6163.
[16] Kelton PLJ. Physiology, Biochemistry, and Molecular Biology of the Skin [J]. Plastic and Reconstructive Surgery: 1993; 92 (4): 761.
[17] Cebe P, Partlow BP, Kaplan DL, et al. Silk I and Silk II studied by fast scanning calorimetry [J]. Acta Biomaterialia: 2017; 55: 323.
[18] Cai S, Singh BR. Identification of β-turn and random coil amide III infrared bands for secondary structure estimation of proteins [J]. Biophysical Chemistry: 1999; 80 (1): 7.

Cite as: W. Long et al.,
TBIS 10.3993/tbis (2021).

Simulation Model of Sports Bra and Breast Contact in Walking State Based on Finite Element Co-simulation

Wu Long*, Chang-Lin Zhu, Bo-An Ying, Jing Qi

Apparel & Art Design College, Xi'an Polytechnic University, Xi'an, Shaanxi, 710048, China

*Corresponding author's email: wulong7398@163.com

Abstract

To study the stress state of women wearing a typical compression sports bra during running, a simulation model was established using the combined simulation finite element method of statics and dynamics to simulate women wearing a sports bra while walking at a speed of 2km/h. This paper uses the Adams Co-Simulation Interface to realize the information transfer between multi-body dynamics and nonlinear finite element software; it establishes the finite element model of the human body and the sports bra contact and the multi-body dynamic human movement model. The information is transmitted between the models through the interaction points GFORCE and RIGID SURFACE. The walking movement displacement of the human body is transmitted to the contact model of the human chest and the sports bra and the dynamic force change of the sports bra on the human body during the movement is obtained. The model is verified through the human motion non-contact optical capture experiment to evaluate the effectiveness of the finite element model. The model established in this paper can accurately predict breast deformation and pressure comfort.

Keywords: Finite Element; Sports Bra; Force State; co-simulation

1. Introduction

With the development of competitive sports, sportswear has attracted more and more attention. Fit and comfortable sportswear is very important for athletes. It not only improve athletes' competitive state, it also reduces athletes' injuries. According to a survey conducted by Hart and Dewsnap [1], nearly 80% of the women participating in the survey have never worn a suitable sports bra. In a survey of 13,000 women by Greenbaum [2], about 90% of women said they felt discomfort, pain, fatigue, inflammation, allergic rash and tightness from wearing a bra. For sports athletes, the sports bra has a good support and restraint effect on the breasts, reducing the vibration of the breasts, thereby improving the athlete's sports state, while improving sports performance [3]. The research on the force state of sports bras mainly includes the analysis of the wearing force in the static state and the force analysis in the motion state. The force analysis of the sports bra on the breast in the motion state mainly includes the pressure of squeezing and the inertial force of the breast movement causes the breast to vibrate. Therefore, it is necessary not only] analyze the pressure distribution of the sports bra in the process of wearing the sports bra when static, it is also important to analyze the interaction between the breast and the sports bra when dynamic, to provide a design basis for the development of the sports bra and develop a sports Bra with enhanced comfort fit.

Earlier research on the dynamic state of the breast force by sports bras was conducted by Haycock [5] in 1978 by marking the breast points of 5 subjects, conducting running tests at speeds of 4.8km/h and 9.7km/h, and taking camera shoots at a frequency of 100HZ when the subject is wearing a normal bra and a sports bra. The movement displacements of the breast points in different states were compared. Chen Xiaona and Ruan Jia [6] used the Labview stress tester to test 8 subjects wearing typical compression sports bras at rest and walking or walking on a treadmill at speeds of 2.5km/h, 5km/h and 7.5km/h, respectively. When running, the static and dynamic pressure on the bra cup, lower circumference, shoulders and back were measured. Li-Hua Chen, Sun-Pui Ng [7] proposed a method of simulating nonlinear chest movement by using a three-dimensional (3D) dynamic human body model (FEM) to construct a chest model with skin layers and breasts. Qiu Jiang yuan [4] constructed a solid chest model and a bra solid model as well as established a static and dynamic sports bra and a finite element contact model of the chest. However, the constructed dynamic sports bra and the human body contact finite element model did not simulate the motion state of the human body. It is to input the displacement curve of the human body movement as the boundary condition.

In this paper, the three-dimensional human body modeling technology is used to realize the information transfer between the multi-body dynamics software Adams and the nonlinear finite element software Marc through using the Adams Co-Simulation Interface to realize the co-simulation and establish the movement of the human body in the running state. The finite element model of the bra in contact with the human body can develop more suitable sports bras for athletes of different sports types.

2. Construction of Co-simulation Model of Sports Bra Under exercise

2.1 Adams-Marc Co-simulation Technology

Adams-to-Marc Co-simulation Interface (ACSI) realizes the information communication between the multi-body dynamics software Adams and the nonlinear finite element software Marc, and organically combines the multi-body dynamics and nonlinear finite element tools. It is an effective method to solve nonlinear structural problems in system dynamics simulation. For example, the problems of self-contact, material nonlinearity, geometric nonlinearity and contact nonlinearity appearing in the process of mechanism movement are effective means for coupling multi-body dynamics with simulation of other disciplines.

Data transmission is carried out between Adams model and Marc model through interaction points. In Adams model, the interaction point is GFORCE point. In the Marc model, the interaction point is RIGID SURFACE point. Adams transfers the displacement to Marc to act on RIGID SURFACE through GFORCE point. At the same time, Marc transmits the force/moment to Adams through the GFORCE point.

The sports bra and human motion dynamic co-simulation system in this article is shown in the Fig.1. The finite element model of the contact between the sports bra and the human body is completed in Mrac, and the multi-body dynamics model of the human walking motion is completed in Adams. Two model information is transferred through the interaction points RIGID SURFACE and GFORCE. Adams transfers the displacement data of the human walking to Marc as a boundary condition to realize the finite element calculation of the contact between the human body and the sports bra in the walking motion state.

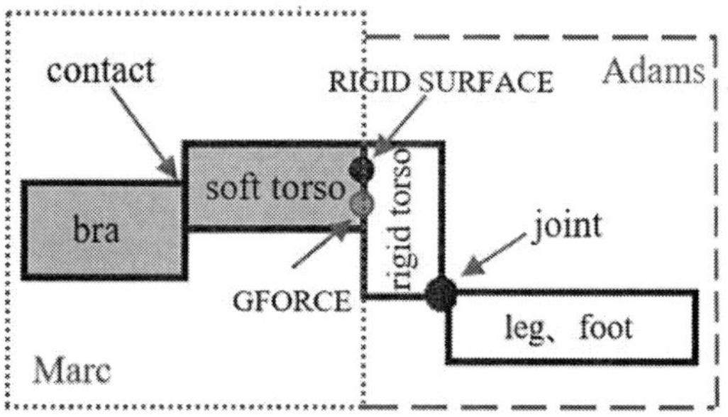

Fig.1 Sports bra and human body motion dynamic co-simulation system

2.2 Construction of the Human Body Model

1) human body model

In this paper, the human body model is established using V-stitcher software. First, the size of the sports test model is obtained, and then the model size parameters are input to V-stitcher software to construct the human body geometric model.

In order to construct a finite element model of the human chest and a multi-body dynamics model of the human body, the data of the human body model parameters is measured and parameterized to generate the same three-dimensional human body model as the test model. The constructed human body model and the required measurement data is shown in Fig.2.

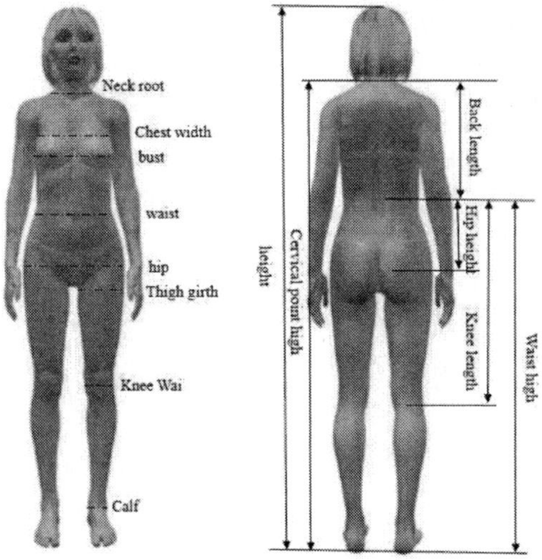

Fig.2 Measurement data and human body model

2) Human tissue model

For the modeling of the breast layer, it is first necessary to determine the boundary of the breast to segment the breast tissue from the human body. The researcher Yip [8] defined the boundary of the breast as a circle, as shown in Fig.4, but did not clearly define it. boundary. Brotherston [9] established human body markers to define breast boundaries. The key landmarks defined are shown in Fig.3. These include the nipple (N), medial (M) and lateral (L) under the breast crease, as well as the point of the under-breast crease (IC) and the base point of the breast (B).

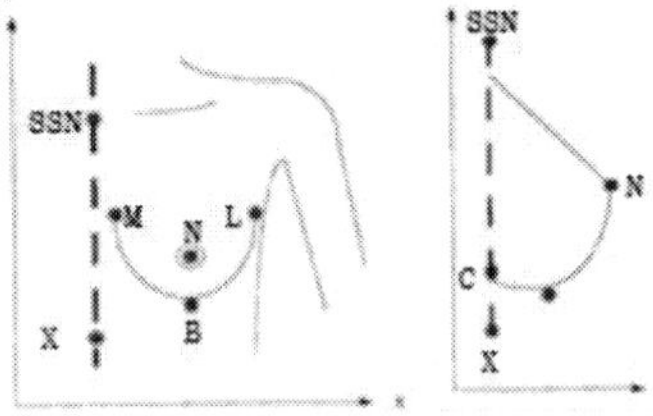

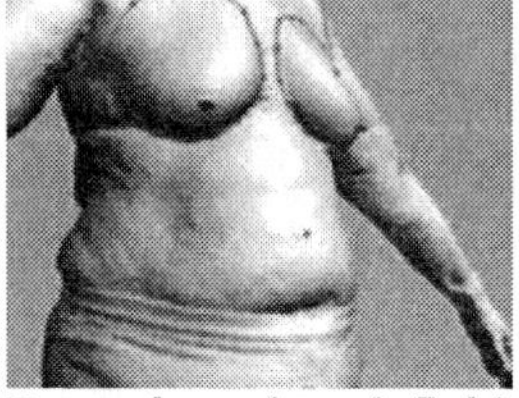
Fig.3 Breast boundary marking points [9] Fig.4 Breast boundary definition [8]

According to the research of Brotherston and YIP, the left and right breasts are cut according to the position of the pectoralis major muscle. The final human body model is shown in Fig.5.

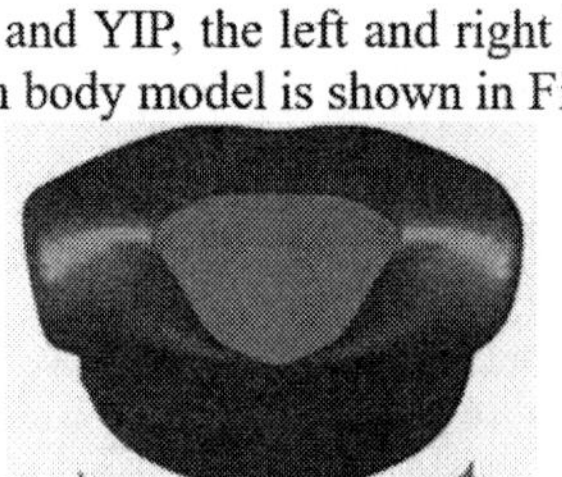
Fig.5 Breast and human body model

2.3 Construction of a Sports Bra Model

1) The Design of the Front and Back Pieces of the Bra Pattern

The size of the front and back pieces of the underwear is designed according to the size of the test model. The sports bra pattern is imported into the virtual fitting software for virtual fitting. The front and back sheet patterns of the sports bra are generated as shown in Fig.6.

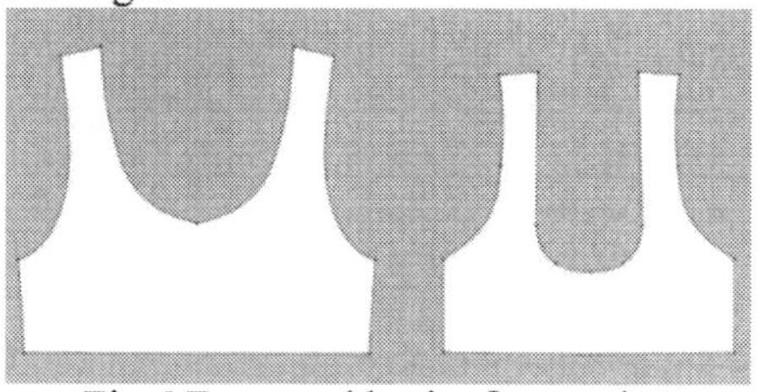

Fig.6 Front and back of sports bra

2) Virtual Try on Sports Bra

The generated sports bra pattern are imported into the three-dimensional fitting software V-stitcher for virtual stitching, and the fabric properties are set to elasticity. The generated sports bra model corresponds to the size of the sports bra that is worn, and fits perfectly with the human body model, such as Shown in Fig.7.

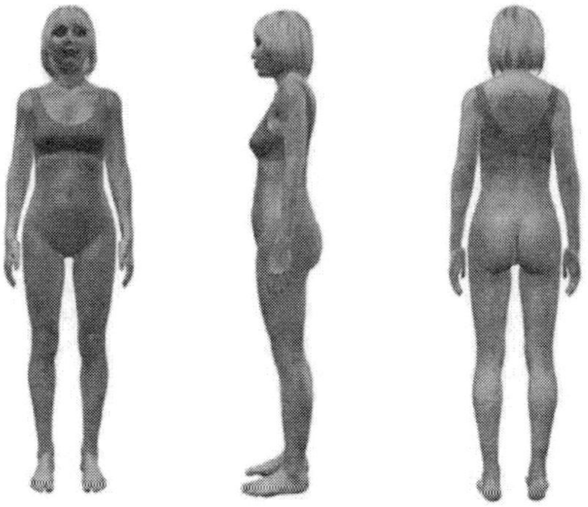

Fig.7 Virtual fitting diagram of sports bra

2.4 Creation of Geometric Model of Human Body Dynamics

The obtained geometric model of the human body is segmented according to the joints to generate the required structural parts of the human body. In the walking movement of the human body, the movement of the human body is mainly the movement of the torso and legs, so this article is divided into torso, thighs, calves, and feet. The four parts are used for modeling and assembly. This article regards the human body as a rigid body based on the following three simplifications:

1) In every part, human soft tissues and bones are regarded as rigid bodies;

2) In this article, Adams is required to calculate the kinematic parameters of the human body during movement, so the deformation of the soft tissue has no major influence on the joint movement;

3) Construct geometric models of various parts of the human body according to the size of the human body, wherein the thigh and calf models are simplified into regular cylinders, and the rest are real human body geometric shapes;

4) During low-speed running, the swing of the arm is small, and the degree of freedom of the arm joints is more complicated. In this article, the arm model is simplified without considering the swing of the arm.

According to the above model simplification, the various components are assembled to generate a complete human body model. The constructed human body model is shown in Fig.8.

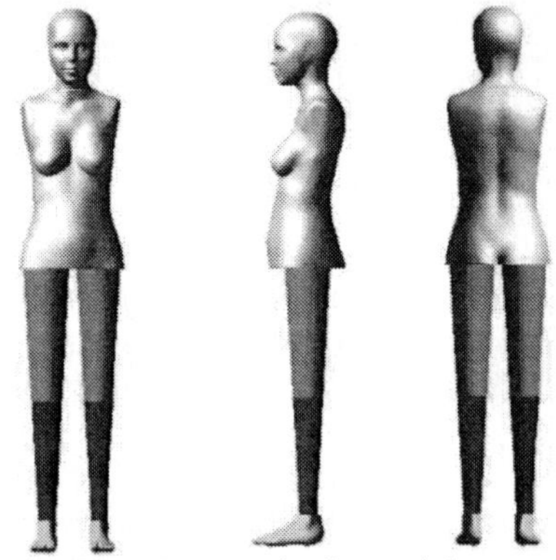

Fig.8 Human body model

2.5 Definition of the Mass and Moment of Inertia of the Human Body Dynamics Model

To analyze the motion of the human body, inertial parameters are essential. The inertial parameters of the human body mainly include human body mass, center of mass position and moment of inertia [10]. DRILLIS, R, CONTINI, R, etc. have a certain proportional relationship between human body weight and various tissues of the human body [11]. The mass of each part of the human body in this article is defined according to the weight of the experimental object, which is consistent with the model attribute of the experimental object. The specific mass of each part is calculated with the 60kg experimental object in this experiment as shown in Table 1.

Table 1. Mass of various parts of the human body

body structure	Head, neck, torso	thigh	Calf	foot
Percentage of body weight/%	53.42	11.18	4.39	1.83
Mass (kg)	27.07	9.77	2.99	0.95

The moment of inertia of each part of the human body is determined according to the 《Measurement of Normal Human Inertial Parameters and Large Sample Statistics in China》 and 《Study on the Center of Mass of the Adult Human Body in China》. There is a correlation between the mass of each part of the human body and the moment of inertia. The regression equation can calculate the moment of inertia of each part of the human body as shown in Table 2.

Table2. Moment of inertia of various parts of the human body

body structure	JX (kg.cm2)	JY (kg.cm2)	JZ (kg.cm2)
Head, neck, torso	581.70	429.70	277.80
thigh	156.30	156.30	32.10
Calf	36.50	35.10	6.00
foot	0.60	2.50	2.70

2.6 Dynamic Model Constraint Addition and Joint Drive Creation

There are two types of kinematics analysis, which are divided into forward dynamics and inverse dynamics [11]. Forward dynamics is the process of solving the trajectory of the object by adding driving force to the part of the object; reverse dynamics is the process of solving the trajectory of the object by adding driving force to the part of the object. The motion trajectory is added to the process of solving the force and moment of the object. In this paper, the method of inverse dynamic analysis is selected.

1) Driving of human lower limb joints

The human knee joints and hip joints are set as rotating joints to add on joint drive. The CGA joint angle data will be collected through a motion capture system. In this paper, the motion analysis uses CGA data from the

International Society of Biomechanics, Hong Kong Polytechnic University, and Dundee Limb Accessories Center [12]. The experiment will simulate the walking condition of the human body at a speed of 2km/h, and the adopted joint angular displacement curve is shown in Fig.9.

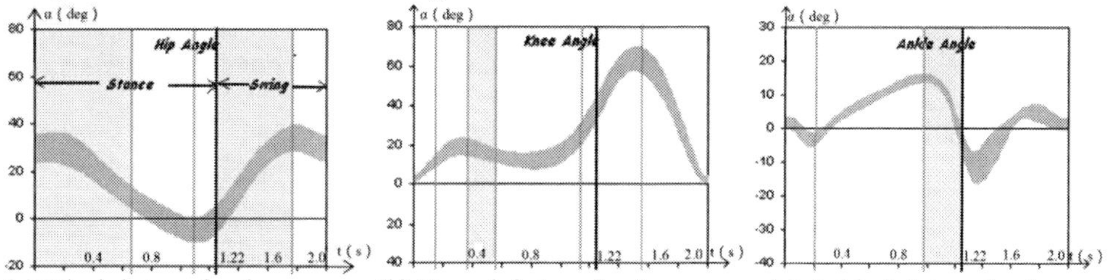

(a) Hip joint angle change (b) Knee joint angle change (c) Ankle joint angle change

Fig.9 Angular displacement curve of each joint of the human body [12]

There are many drive control functions in Adams. This article selects the CUBSPL control function. The joint drive Spline curve is the curve of angular displacement changing with time. The angular displacement curve is imported in Fig.10, and the final generated CUBSPL control function is CUBSPL (time ,0,.MODEL_1.time, 0).

2) Foot-to-ground contact

The human body model walks on the ground. The human body has gravity, and the ground exerts upward supporting force on the human body. At the same time, the human foot will generate friction with the ground. The mechanical analysis is relatively complicated. The contact function is used in Adams to calculate the contact force. The contact force parameters of the human foot and the ground are shown in table 3.

Table 3. Human body and ground contact setting table

Parameter	Parameter value
Stiffness	380.00
Force Exponent	2.20
Damping	0.30
Penetration Depth	1.00
Static coefficient	0.10
Dynamic coefficient	9.0E-02
Stiction transition velocity	1000.00
Friction transition velocity	1.0E+05

3) Create interaction points in Adams

In the co-simulation process, it is necessary to create corresponding interactive nodes in Adams and Marc respectively to facilitate the transmission of force and displacement information. The interaction node in Adams is created by General Force. It is a 6-degree-of-freedom node force and moment input method, including three-direction force and three-direction torque, which can completely transfer the movement and rotation of the human body during movement. Go to the discrete chest model in Marc software. The General Force point is created at the xiphoid point of the human chest. The general force creation method and creation result are shown in Fig.10.

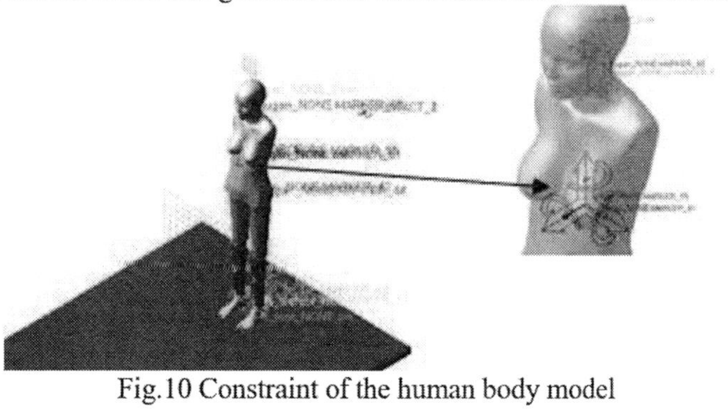

Fig.10 Constraint of the human body model

2.7 Finite Element Model Creation in Marc

2.7.1 Model Meshing

The acquired human chest and sports bra models are meshed, and the unit sizes, unit types, and number of units for the human chest and sports bra models are shown in Table 4.

Table 4. Grid data of human body and sports bra

part	Mesh size (mm)	Mesh type	Element type
torso	10	Tetrahedral entity	134
breast	3	Hexahedral entity	157
Sports bra	4	Quadrilateral shell	75

2.7.2 Model Boundary Conditions

1) Initial boundary conditions

The initial boundary conditions define the initial state of the model. The initial condition of the motion state after simulating the wearing of the sports bra is that the sports bra has a certain pressure distribution on the human body during the wearing process, and the static lower node pressure that has been calculated in my literature is taken as the initial pressure value is input [13], which requires the extraction of the previously calculated model node pressure in the Marc software to ensure that the model calculation reflects the real situation.

2) interaction points

The interaction node established in the finite element model is at the position of the human chest, which corresponds to the Gforce point of the multi-body dynamics model and is numbered 1024. This node connects all the nodes of the human body's torso node together and is connected through the RBE2 node. The node constraints are shown in Fig.11.

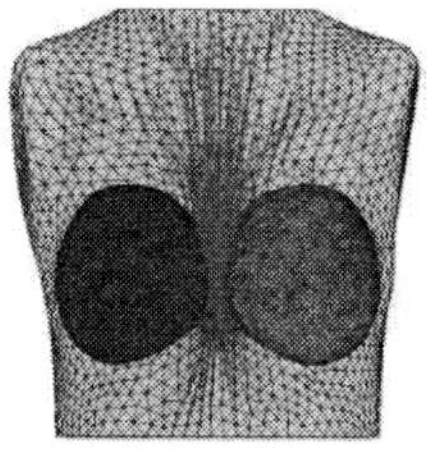

Fig.11 Model node constraints

3) Contact Conditions

There is no displacement and movement between the human body torso and the breast, which are set in a bonding relationship. There is a relative displacement between the human body, the breast and the sports bra, which are in a relatively separated state, and are set in a contact relationship. The specific contact relationship is shown in Table 5.

Table 5. Contact table of human body and sports bra

Model parts	torso	breast	Sports bra
torso		glue	touch
breast	glue		touch
Sports bra	touch	touch	

3. Validation and Analysis of Co-simulation Finite Element Model

3.1 Breast Dynamic Capture Experiment While Walking in a Sports Bra

This experiment is based on the development of a new technical system, based on the working principle of optical non-contact dynamic capture, to build a human chest dynamic tracking measurement system, according to the change characteristics and laws of the chest surface shape of the human body in the state of motion, to obtain the state of motion as well as the changing process of the shape of the human breast.

In this experiment, the test wearer with a size of 75B (chest circumference of 86cm and lower chest circumference of 72cm) is selected to wear a 75B sports bra and run a dynamic test experiment on a treadmill. After the preparations are ready, the treadmill was turned on, and the testers walked on the treadmill at a speed of 2 km/h. The sampling frequency of the XTDIC system is 10 frames per second.

Fig.12 Full field displacement measurement

3.2 Co-simulation Validation of Finite Element Contact Model

According to the above simulation experiment, the experiment simulated the running process of the human body wearing a sports bra with an elastic modulus of E=0.58N/mm2 at a speed of 2km/h, and the movement curve of the human body at a speed of 2km/h and the movement curve of the running experiment were simulated. For comparison, it is verified that the displacement of the running human body in the experiment corresponds to the displacement of the simulated human body. Fig.13 shows the comparison curve between the Y-direction displacement of the human body and the Y-direction displacement of the simulation experiment.

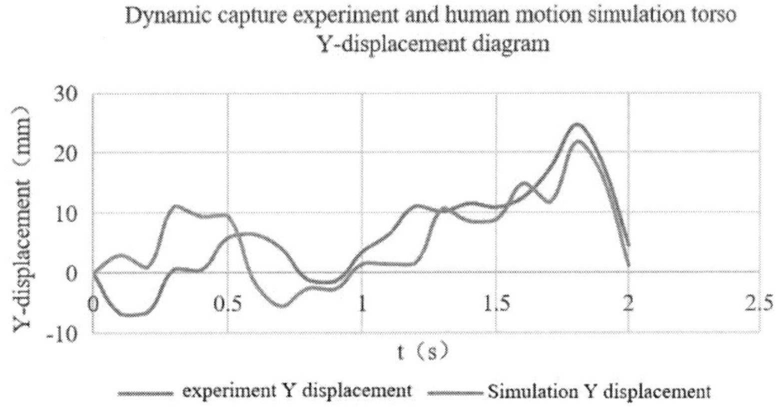

Fig.13 Comparison of Y-direction displacement of human chest in experiment and simulation

Through dynamic full-field displacement measurement, a cloud map of the movement breast displacement change in 0-2s is obtained, and the corresponding simulation time lasts 2s. The dynamic displacement change of the breast is recorded. By selecting the displacement change cloud diagrams of 4 time nodes for comparative analysis, the maximum displacement of the sports bra in the experiment at 0.5s is 31.86mm, the maximum displacement in the simulation is 28.02mm, and the maximum displacement of the sports bra in the experiment at 1.0s is 44.25mm, the maximum displacement in the simulation is 52.60mm, the maximum displacement of the sports bra in the experiment at 1.5s is 28.89mm, the maximum displacement in the simulation is 12.10mm, the maximum displacement of the sports bra in the experiment at 2.0s is 7.66mm,. In the simulation the maximum displacement is 4.36mm. Fig.14 demonstrates that the simulation experiment and the experimental test breast movement displacement are basically the same.

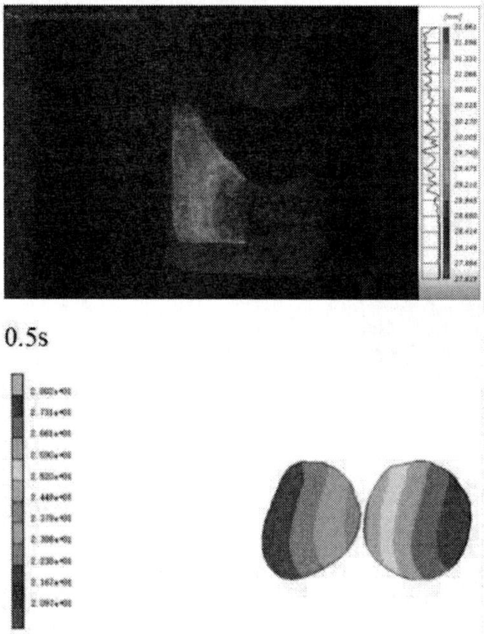

Fig.14-a 0.5s displacement change

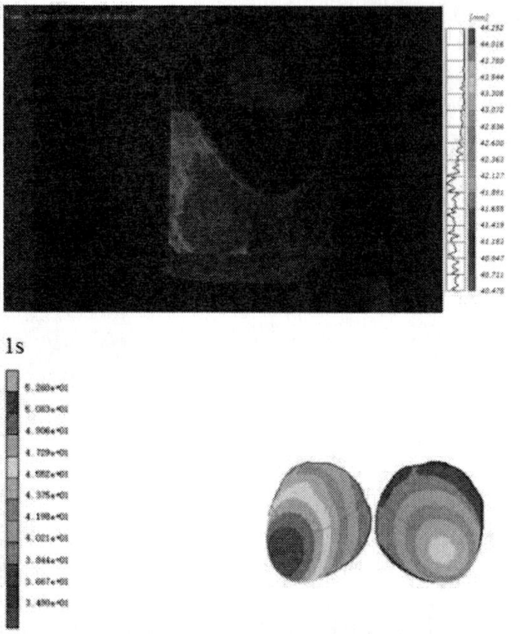

Fig.14-b 1.0s displacement change

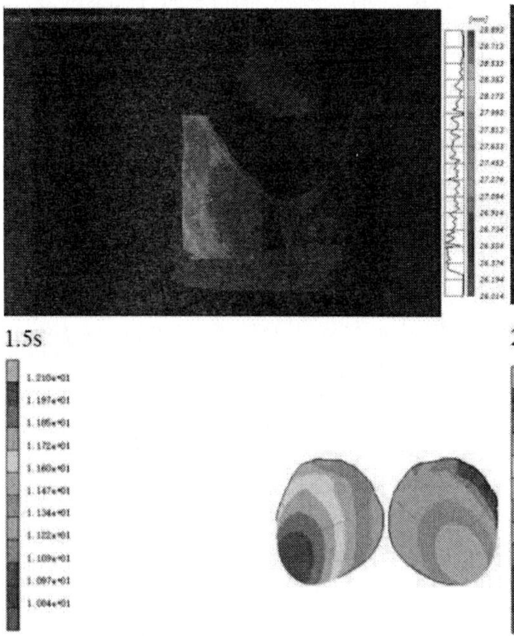

Fig.14-c 1.5s displacement change

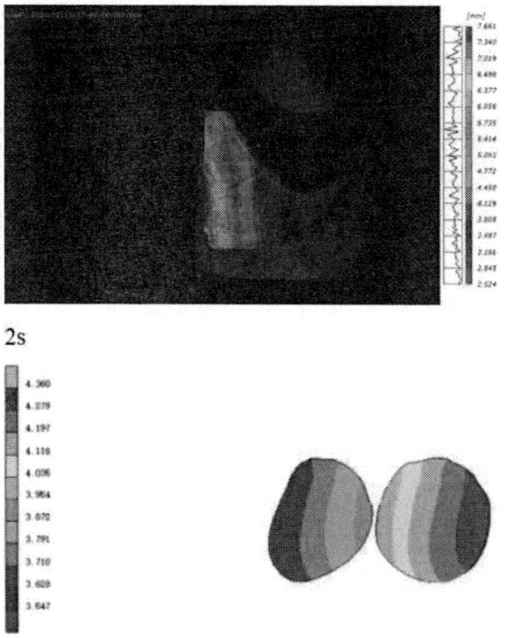

Fig.14-d 2.0s displacement change

Fig.14 The displacement changes of experimental and simulated breasts at different time

3.3 Analysis of the Results of Dynamic Breast Force

1) The contact stress distribution of the breast

When the speed is 2km/h and the elastic modulus of the sports bra is E=0.58N/mm2, the movement of the breast changes under dynamic conditions. The contact force of the sports bra on the breast under dynamic changes is constantly changing. In order to characterize the change of the breasts' contact force, the contact force cloud diagram of the breast at the maximum vibration displacement of 1.5s was obtained and is shown in Fig.15.

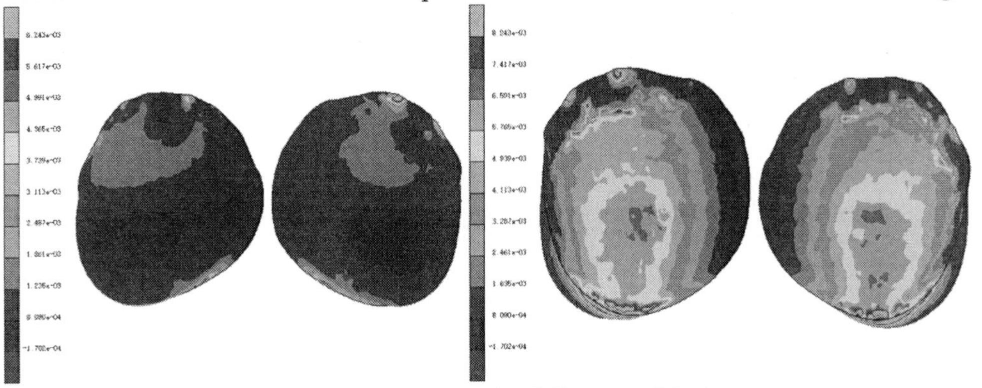

Fig.15 The contact stress cloud diagram of the breast

It can be seen from the cloud diagram that when the movement speed is 2.5km/h and the maximum breast amplitude is 1.5s, the contact force is mainly in the lower and upper areas of the breast, and the force value is between 1.86Mpa-2.49Mpa. The larger contact force value indicates that the up-and-down vibration of the breasts during exercise leads to greater stress on the connection between the breast muscles of the human body and the breast, which causes discomfort during exercise.

2) Contact stress distribution of sports bra

The contact force cloud of the sports bra at 1.5s is also obtained as shown in Fig.16. It can be seen from the figure that the contact force of the sports bra near the breast is mainly distributed on the upper and lower parts of the breast, which is caused by the shaking of the breast during the movement process. The contact force of the sports bra at the shoulder straps is relatively large, indicating that the straps exert a greater pulling force on the sports bra during the exercise, indicating that the shoulder straps have a positive effect on the sports bra. Pay attention to the design of the shoulder straps when designing the bra to prevent excessive pulling force.

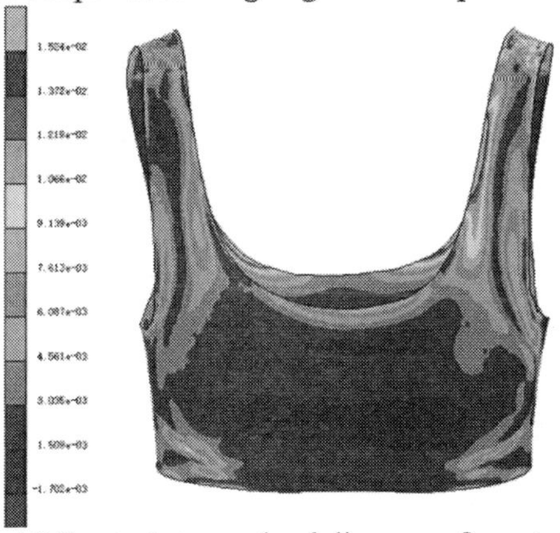

Fig.16 Contact stress cloud diagram of sports bra

4. Conclusion

1) Existing research on the dynamic simulation of sports bras has found that there are fewer studies on the static and dynamic stress state of sports bras, while the combined simulation of statics and dynamics can realize the static and dynamic forces of sports bras.

2) Through statics and dynamics combined simulation technology to simulate the force state of the breast and sports bra in walking state, the contact stress between the sports bra and the human body and the result of the movement displacement of the breasts are obtained.

3) Through the non-contact full-field dynamic capture system, the cloud map of the full-field displacement change of the human chest in the walking state is obtained. Through comparing with the simulation analysis, it is verified that the co-simulation finite element simulation analysis model is effective. Through the establishment of a co-simulation model, an effective method is provided for studying the force state analysis of clothing under dynamic conditions.

5. Acknowledgement

This work was supported by the Shaanxi Engineering Research Center of Apparel Intelligent Design and Manufacture.

References

[1] Hart C, Dewsnap B. (2001). An exploratory study of the consumer decision process for intimate apparel [J]. Journal of Fashion Marketing Management: 5; 108-119.
[2] Greenbaum AR, Heslop T, Morris J, Dunn KW. (2003). An investigation of the suitability of bra fit in women referred for reduction mammaplasty. British Journal of Plastic Surgery: 56 (3); 230-236.
[3] Chang LX, Qi J, Zhang X. Sports bras at home and abroad [J]. Frontiers of Science and Technology: 2008; (4): 168-171.
[4] Qiu JY. Sports bra wearing simulation based on chest mechanics model [D]. Soochow University: 2016.
[5] Haycock CE, Shierman G, Gillette J. The female athlete-does her anatomy pose problems[C].American Medical Association 19th Conference on the Medical Aspects of Sports: 1978; 1- 8.
[6] Chen XN, Ruan J, Zhao MM, Li YM, Xie H, Wang YZ, Xia L. Research on the dynamic pressure distribution of young women wearing typical compression sports bras [J]. Journal of Beijing Institute of Fashion Technology (Natural Science Edition): 2018; 38 (03): 40-46.
[7] Chen LH, Ng SP. 2013.A study of breast motion using non-linear dynamic FEanalysis. Taylor & Francis No. 5: 868-878.20.
[8] Jia MY, Naila M, Rebecca MJ, Daisy EV, Richard JW, Nicola RD. Accurate Assessment of Breast Volume: A Study Comparing the Volumetric Gold Standard (Direct Water Displacement Measurement of Mastectomy Specimen) With a 3D Laser Scanning Technique[J]. Annals of Plastic Surgery: 2012; 68 (2).
[9] La TP, Brown H, Ringrose C, Hyland RE, Cole AA, Brotherston TM. A method of assessing female breast morphometry and its clinical application [J]. British Journal of Plastic Surgery: 1999; 52 (5).
[10] Zhang WQ. Lower limb exoskeleton robot design and simulation research [D]. Shandong University of Technology: 2015.
[11] Jin DW, Zhang JC. Rehabilitation Engineering and Biomechanics [J]. Chinese Journal of Rehabilitation Medicine: 2011; 26 (05): 414.
[12] Ren JP, Yan Y, Zhou XL. The effect of sports bra on breast amplitude in different speeds of walking and running [J]. Journal of Capital Institute of Physical Education: 2015; 27 (02): 172-177.
[13] Chang L, Zhu W, Long BA, et al. Establishment of the Finite Element Model of Human Body Wearing Sports Bra [C]: 2020.

Comfort Assessment of Adaptive Jackets for Wheelchair Users of Different Types

Si-Yu Li, Qi Wang, Jing Guo*, Jin-Hui Ren, Li-Min Shi, Wei-Wei Li, Ji-Hong Zhang

Universal Fashion Lab, Beijing Institute of Fashion Technology, Beijing, 100029, China

*Corresponding author's email: guoj@uflbift.com

Abstract

The study designed and developed two types of adaptive jacket for wheelchair users and carried out try-on experiments, to investigate the the comfort assessment of adaptive jackets on different wheel chair users. First, via a questionnaire survey, the study collected assessment opinions of 83 wheelchair users and classified the wheelchair users based on their gender, age, figure and use of a wheelchair. Then the study conducted overall and difference analysis of size and comfort assessment given by the selected wheelchair users in different groups respectively. The analysis demonstrates that the two types of jacket both received an overall positive assessment; wheelchair users in different gender, age, figure, and use of a wheelchair showed significant difference in their assessment of some size items; wheelchair users of different gender and figure displayed significant difference in assessment of some comfort items. The study provides a reference for adaptive garment design and development tailored for wheelchair users of different types.

Keywords: Adaptive Garment; Wheelchair User; Comfort Assessment; Questionnaire Survey

1. Introduction

According to the estimation of the World Health Organization (WHO) in 2010, the global population of disabled people surpassed one billion, accounting for 15 percent of the total world population [1]. In China, there are about 24.72 million people with physical disability, wherein 45.61 percent are disabled in the lower limbs [2]. Most of them use wheelchairs in their daily life. Disabled people differ in gender, age, and degree of reliance on a wheelchair. In terms of the degree of reliance on a wheelchair, some people can only make an outing with the assistance of a wheelchair, while others often or occasionally use a wheelchair. Hence, it is necessary to differentiate different disabled groups in design and studies of adaptive garment.

In recent years, many scholars have carried out studies in this field. Sara Bragança et al [3] and Marija Nakić et al [4] analyzed and compared the existing studies on anthropometric measuring of wheelchair users and put forward suggestions on the future study direction of the field. Andreja Rudolf et al [5] constructed a paraplegia anthropometric dummy through 3-D scanning and created an apparel prototype. Park J. et al [6] conducted measurement analysis of the upper body shape of wheelchair users, providing reference to the design of suit jackets for wheelchair users. In terms of functionality design, Yunyi Wang et al [7] developed a type of adaptive garment and tested it on wheelchair users in their daily activities with the result showing that the adaptive clothes could effectively reduce the time the wearer takes in dressing and undressing. Liana Babenko et al [8] simulated heat dissipation process of wheelchair users' lower limbs on thermal wear via mathematical modeling, giving suggestions on designing adaptive thermal protection clothing. J. Webber et al [9] tested four types of warning signs on adaptive garment and conducted a comparative analysis on the efficacy of the four warning signs by comparing the time drivers take to respond to the warning signs. Imran A. [10] designed adaptive underwear equipped with electronic devices, with the function of testing a range of physiological indices for the wearer along with the merit of making dressing and undressing easier. Sarita Rani et al. [11] designed a light-sensitive smart clothing for people with visual impairments that can facilitate their daily lives. Yuhui Wei et al. [12] designed nursing clothing that is easy to put on and take off for the handicapped and the experiments show that it is convenient to wear and take off. With regard to the sports element, Sara Bragança et al [13] carried out a questionnaire survey among wheelchair-bound American football players and proposed suggestions on designing football sportswear for wheelchair-bound players. I. Rice et al [14] conducted a performance evaluation of two types of gloves for wheelchair racing players with the result showing that the harder one demonstrated better comprehensive performance. Katy E. et al [15] reviewed the cooling methods used in garment for spinal cord injury sufferers, giving suggestions for cooling design of clothes tailored for professional athletes with spinal cord injury. The above studies have all made their headway in wheelchair users' body measurement and functionality design of various adaptive garment. However, difference analysis of different wheelchair users in their needs for clothing is scarce. Moreover, a market study demonstrates that the current available adaptive garment in the market pays more attention to functionality and convenience, but lack tailored design for different types of wheelchair users.

Based on an investigation, the study designed and produced two types of adaptive jacket, then through a questionnaire survey collected wheelchair users' assessment on size fitness of different clothing parts and overall comfort. With the data analysis software SPSS 25, the study classified the assessment information based on age, gender, figure, and wheelchair use of the subjects, and made a comprehensive analysis and difference analysis of different wheelchair user groups in terms of their assessment of size fitness of different parts of the jackets and overall comfort. Based on the analysis, the study evaluated the performance of the two types of adaptive jacket and analyzed the preference of different wheelchair using groups in adaptive jackets.

2. Methodology

2.1 Apparel Design

In light of anthropometric measurement and investigation, the study designed two types of adaptive jacket targeting the special needs of wheelchair users with consideration of their physical characteristics and behavioral habits. The size-setting was based on wheelchair users' anthropometric measurement data. In terms of the structural design, the two jackets feature three-dimensional sleeves, reduce the front garment length, increase the rear garment length and the depth of the front collar, making it more suitable for sitting, as well as an adjustable zipper along the side seam of the jackets. In terms of fabric selection, the study chose to use a type of elastic water-proof fabric for the main part of the jacket with wear-resisting cloth for the inner side of the sleeve. Based on try-on feedback of a small set of samples on the first type of jacket, the neckline and outside sleeves were added on the second type of jacket without any other change made. The effect chart of garment is shown in Fig.1.

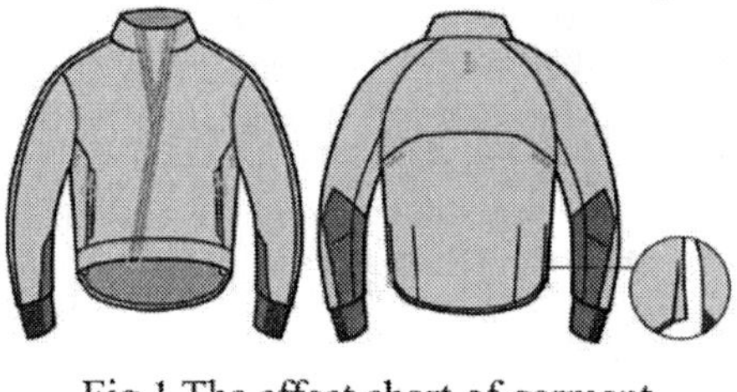

Fig.1 The effect chart of garment

2.2 Questionnaire Design

Table 1. Content of the questionnaire

Basic information	Size assessment	Comfort assessment
Gender, age, height, weight, use of wheelchair (standing position, ease of stand and sitting, or sitting position), type and size of the tried jacket	Neck base girth, collar width, shoulder width, chest girth, waist girth, bottom hem girth, front piece length, front piece width, back piece length, back piece width, sleeve length, cuff girth, pocket height and the position pockets (front or back)	Overall appearance, overall comfort of the fabric, partial wear resistance of the fabric, overall comfort of the jacket when the wearer sits still, overall comfort of the jacket when the wearer uses a wheelchair, and overall comfort of the jacket when the wearer holds arms in a horizontal position

The questionnaire mainly consists of three parts: basic information, size assessment, and comfort assessment, with specific items of each part shown in Table 1. There are three options in the size assessment section. small, suitable, and large. In the comfort assessment section, each item is assessed by Hollies' five level scale with choices of the scores ranging from 1-5, 1 for very uncomfortable and 5 for very comfortable.

2.3 Questionnaire Distribution and Collection

Aided by an organization, the study recruited 83 wheelchair users. Based on their height and weight, they were given the adaptive jacket of the corresponding size for a try-on experiment. The wearers were then requested to complete the questionnaire sheet. Thus, the study collected the wearers' evaluation feedback on size and comfort of the jacket. After sifting, 76 copies of effective evaluation feedback were obtained.

3. Analysis and Discussion

3.1 Data Processing

3.1.1 Descriptive Analysis

Descriptive statistics on the selected wheelchair users' basic information are shown as Table 2.

Table 2. Basic information of wheelchair users

	Minimum value	Maximum value	Mean value	Standard deviation
Age	30.00	68.00	53.66	9.58
Height[cm]	135.00	194.00	163.03	9.58
Weight[kg]	38.00	92.50	62.37	11.74
BMI	18.30	34.30	23.35	3.15

BMI=kg/m^2

3.1.2 Pre-processing of Data

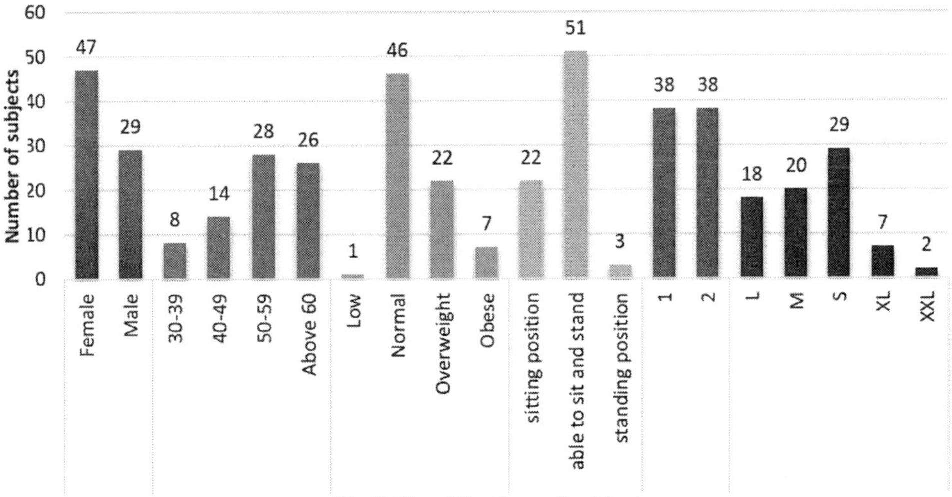

Fig.2 Classification of subjects

In light of the basic information, the study classified the selected wheelchair users in different groups respectively in terms of gender, age, BMI, use of wheelchair, type and size of the tried jacket. To have a difference of analysis, the subjects were put into four age groups of 30-39, 40-49, 50-59, 60 and above. Per their BMI value, they were categorized in four groups, low (below 18.5), normal (18.5-23.9), overweight (24.0-27.0), and obese (28.0-32.0).

The classification of the subjects is shown as Fig.2.

As for the size assessment segment, for the convenience of statistical analysis, set “large” as the score “1”, “suitable” as “0”, and “small” as “-1”. Therefore, in the analysis of size evaluations, the much closer the score is to “0”, the better the fitness of the jacket; the score below “0” indicates that the jacket is small for the subject, and the one above “0” means that the jacketis large for the subject.

3.2 Data Analysis

3.2.1 Overall Analysis of Size and Comfort Assessment

The mean value of the evaluation scores on the size of the adaptive jacket is shown in Fig.3(a). On the whole, the absolute values of the evaluation scores on the size are all within 0.2, which can be interpreted as pretty good performance. Analysis of evaluation scores on neckline circumference and collar width shows that the two parts are large, and analysis of the scores on the position of pockets indicates that the pocket positions are relatively high and front; and analysis of scores on other items shows that sizes of the related parts are small. The mean values of the evaluation scores on front piece length, back piece length, and bottom hem circumference are all below -0.1, indicating that the related parts of the jacket are relatively ill-fitted to subjects. The absolute values of evaluation scores on other parts are all below 0.1, implying the good fitness.

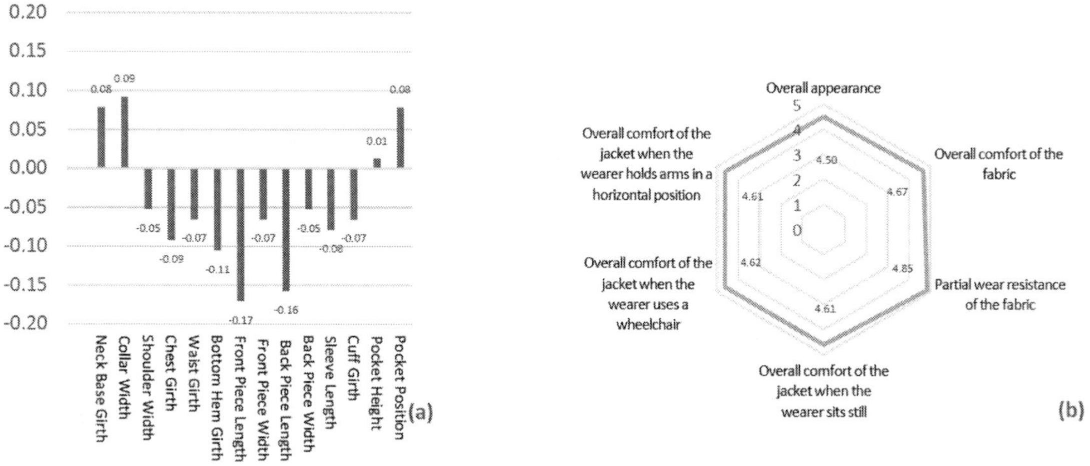

Fig.3(a) Mean value of evaluation scores on size of different parts (b) Mean value of evaluation scores on comfort items

Mean values of the evaluation scores on overall comfort of the jacket is shown in Fig.3(b). With the Hollies’ five-level scale, the mean values of all items regarding comfort are above 4.5, which means that the comfort level of the jacket is pretty good.

3.2.2 Divergence Analysis of Size Assessment

Results of divergence analysis of size evaluations indicate that subjects of different gender in size evaluation of the five parts “shoulder width, chest circumference, front piece length, back piece length, and sleeve length” show significant difference ($P<0.05$) as demonstrated in Table 3. The male subjects’ evaluations show that the size of the five parts are too small. Different age groups show significant difference in size evaluation ($P<0.05$) on shoulder width, front piece length, and sleeve length. In particular, the size evaluation of the 40-49 age group on shoulder width is “large”; the evaluation of the front piece length and sleeve length of the 30-39 group tends to be “small”. Due to the difference in use of wheelchair, subjects show significant difference ($P<0.05$) in their size evaluation on front piece length, back piece length, and sleeve length, as shown in Table 4. Compared with other subjects, size evaluation, ease of standing and sitting, front piece length and back piece length is closer to “0”.Subjects in standing position tend to consider the sleeve length to be big. The size evaluations of subjects with different figures show significant difference (P<0.05) in waist circumference, chest circumference, and front piece width. In their evaluation of the three items, it can also be observed that the bigger the subjects BMI value is the fatter the, they are and have a smaller size evaluation.

While trying on the two types of adaptive jacket, wheelchair users show a significant difference (P<0.05) in the size evaluation of neckline circumference and shoulder width, as shown in Table 5. Subjects evaluate the neckline circumference of type 2 jacket as big, and its shoulder width as small. In terms of size setting, there is no significant difference in subjects' size evaluation.

Table 3. Results of size divergence analysis for different gender groups

	Gender	Mean value	Standard deviation	t	Sig
Shoulder width	female	0.11	0.38	4.04	0.00
	male	-0.31	0.47		
Chest girth	Female	0.00	0.42	2.39	0.02
	Male	-0.24	0.44		
Front piece length	Female	-0.02	0.49	2.87	0.01
	Male	-0.41	0.63		
Back piece length	Female	-0.04	0.29	3.04	0.00
	Male	-0.34	0.48		
Sleeve length	female	0.00	0.36	2.23	0.03
	Male	-0.21	0.41		

Table 4. Results of size divergence analysis for different wheelchair using groups

	Use of wheelchair	Mean value	Standard deviation	Kruskal-Wallis test	Degree of freedom	Sig
Front piece length	Sitting position	-0.41	0.73	10.07	2.00	0.01
	Able to sit and stand	-0.04	0.45			
	Standing position	-0.67	0.58			
Back piece length	Sitting position	-0.36	0.49	9.52	2.00	0.01
	Able to sit and stand	-0.06	0.31			
	Standing position	-0.33	0.58			
Sleeve length	Sitting position	-0.23	0.53	6.99	2.00	0.03
	Able to sit and stand	-0.04	0.28			
	Standing position	0.33	0.58			

Table 5. Results of size divergence analysis for different jacket types

	Type	Mean value	Standard deviation	t	Sig
Neck base girth	1	-0.08	0.36	-3.47	0.00
	2	0.24	0.43		
Shoulder width	1	0.05	0.40	2.04	0.05
	2	-0.16	0.50		

From the above analysis, the followings can be observed: Among wheelchair users of different gender, male subjects tend to prefer bigger shoulder width, chest circumference, sleeve length, front piece length and back piece length. Among different age groups, the 40-49 group tend to prefer smaller shoulder width, and the 30-39 group favors bigger front piece length and sleeve length. In terms of their use of wheelchair, subjects in sitting and standing positions prefer smaller front and back piece length, and subjects in standing position tend to favor smaller sleeve length. In terms of figures, the fatter the subject is, the bigger their preferred size for chest and waist circumference. The difference in the size evaluation analysis of the two types of jacket shows that subjects evaluate the neckline circumference of the type 2 jacket as bigger and its shoulder width to be smaller.

3.2.3 Divergence Analysis of Comfort Assessment

In the divergence analysis, wheelchair users of different gender show significant difference (P<0.05) in their comprehensive comfort evaluation of the adaptive jacket while raising their arms horizontally. The mean value of comfort evaluation by female subjects is higher than the ones by male subjects. Wheelchair users of different figures show significant difference (P<0.05) in their comprehensive comfort evaluation of the adaptive jacket while raising their arms horizontally. In terms of other groupings, wheelchair users' comfort evaluation of the jacket has not displayed significant difference statistically. Based on the mean values of comfort evaluations for different groupings, it is found that in terms of the use of wheelchair, subjects able to both sit and stand have seen the mean values of their comfort evaluations bigger than those of subjects in sitting position or standing position, as shown in Fig.4.

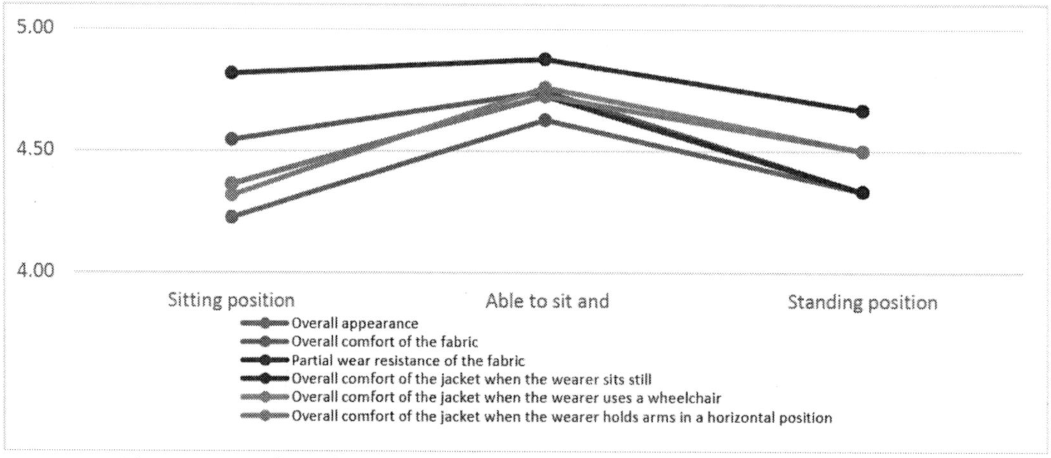

Fig.4 Scores of comfort evaluations by subjects in different use of wheelchair

4. Conclusion

The paper designed two types of adaptive jacket. By recruiting wheelchair users for try-on experiments, the study collected their evaluations on the size and comfort of the jackets. Analysis of their overall evaluations shows that the two types of jackets meet their expectations for fitness and comfort, conforming to their physical characteristics and needs of daily activities.

Through divergence analysis of subjects in different groups, the study drew the following conclusions: In regards to size evaluation, wheelchair users in different gender, figure and use of wheelchair show significant difference on some items. Male subjects tend to prefer bigger circumference and length on some items; subjects who frequently using a wheelchair favor short front and back piece length and long sleeve length; and the fatter the wheelchair user is, the bigger their preferred circumference. Subjects displayed significant difference in their evaluation on some size items of the two types of jacket. In regards to comfort evaluation, wheelchair users of different gender and figure showed significant difference in their evaluation on the overall comfort of the jacket when raising their arms horizontally with the score of female subjects higher than that of male ones. Subjects with a normal BMI is higher than that of subjects in other groups. The study's statistical analysis didn't find significant difference in other groupings.

Based on the above conclusions, the paper suggested that the future market-oriented development of adaptive garment should pay attention to special needs of different wheelchair-using groups. The paper also proposed that in the follow-up designs of adaptive clothes, different designs should be given targeting different wheelchair-using groups in terms of gender, age, figure, and use of the wheelchair, thus providing more garment options for wheelchair users.

5. Acknowledgment

The authors wish to acknowledge the programs below:

1) Beijing Engineering Research Center of Apparel Digitalization.
2) Education and teaching reform project of Beijing Institute of fashion Technology (No. JG-1704)

References

[1] WHO / World Bank. World Disability Report: 2010.

[2] Wu J. Development and evaluation of thermal pants for the limbs disabled. Beijing Institute Of Fashion Technology: 2010.

[3] Braganca S, Castellucci I, Costa E, et al. Anthropometric data for wheelchair users: a systematic literature review [J]. Int J Occup Saf Ergon: 2020; 26 (1): 149-172.

[4] Nakic M, Bogovic S. Computational Design of Functional Clothing for Disabled People [J]. Tekstilec: 2019; 62 (1): 23-33.

[5] Rudolf A, Cupar A, Kozar T, et al. Study regarding the virtual prototyping of garments for paraplegics [J]. Fibers and Polymers: 2015; 16 (5): 1177-1192.

[6] Park J, Park K, Lee B, et al. Classification of Upper Body Shapes Among Korean Male Wheelchair Users to Improve Clothing Fit [J]. Assist Technol: 2019; 31 (1): 34-43.

[7] Wang Y, Wu D, Zhao M, et al. Evaluation on an ergonomic design of functional clothing for wheelchair users [J]. Appl Ergon: 2014; 45 (3): 550-555.

[8] Rudoy D, Babenko L, Molev M, et al. Aspects of applying theory of human ecology to design safe heat-protective clothing for the disabled people [J]. E3S Web of Conferences: 2020; 175: 15004.

[9] Webber J, Wuschke J, Sawatzky B, et al. Evaluating common approaches to improve visibility of wheelchair users [J]. Assist Technol: 2019; 1-5.

[10] Imran A, Dream E, Schacher L, et al. Adaptive bra designs for the individuals with special needs [J]. IOP Conference Series: Materials Science and Engineering: 2017; 254: 072012.

[11] Rani S, Hooda A. Design & Implementation of Smart Clothing for Persons with Disabilitics [J]. Journal of Basic and Applied Engineering Research: 2018; 5 (1).

[12] Wei Y, Su Z, Wang Z, et al. Design and ergonomic evaluation of nursing clothing for disabled people with one side of body [C]. Journal of Physics: Conference Series. IOP Publishing: 2021; 1790 (1): 012013.

[13] Braganca S, Castellucci I, Gill S, et al. Insights on the apparel needs and limitations for athletes with disabilities: The design of wheelchair rugby sports-wear [J]. Appl Ergon: 2018; 67: 9-25.

[14] Rice I, Dysterheft J, Bleakney AW, et al. The Influence of Glove Type on Simulated Wheelchair Racing Propulsion: A Pilot Study [J]. Int J Sports Med: 2016; 37 (1): 30-35.

[15] Griggs KE, Price MJ, Goosey-Tolfrey VL. Cooling athletes with a spinal cord injury [J]. Sports Med: 2015; 45 (1): 9-21.

Antibacterial and UV Protection Properties of TiO_2 Nanoflowers Coated Fabrics

[1]Muhammad Zaman Khan, [1]Jiri Militky, [1]Dana Křemenáková, [2]Michal Petru, [1]Azam Ali

[1]*Department of Material Engineering, Faculty of Textile Engineering, Technical University of Liberec, Studentská 2, Liberec 46117, Czech Republic*
[2]*Department of Machinery Construction, Institute for Nanomaterials, Advanced Technologies and Innovation (CXI), Studentská 1402/2, 461 17 Liberec 1, Technical University of Liberec, 46117 Liberec,*

*Corresponding author's email: zamankhan017@yahoo.com

Abstract

This study aims to develop TiO_2 nanoflowers for the functionalization of polyester fabric to impart antibacterial and UV protection properties. The TiO_2 nanoflowers were grown on polyester fabric by a two-step method. Firstly, the cotton fabric was seeded using sol-gel method and subsequently nanoflowers were grown by hydrothermal method. The morphology and elemental analysis were carried out by using Scanning electron microscopy (SEM), energy dispersive X-ray spectroscopy (EDS) and atomic force microscopy (AFM). The functional properties were examined based on antibacterial activity and UV protection. The TiO_2 nanoflowers coated fabric demonstrated the ability to kill the bacteria and to prevent its growth. The zone of inhibition of 2 mm and 3 mm was found for E. coli and S. aureus respectively. The TiO_2 nanoflowers coated polyester fabric exhibited excellent UV blocking properties. A maximum UPF value of 228 was observed for the TiO_2 nanoflowers coated polyester fabric.

Keywords: Titanium Dioxide; Nanoflowers; Antibacterial; UV Protection

1. Introduction

Nanotechnology is used to impart functional properties to the textile materials. Anti-bacterial, UV-protection, self-cleaning, stain repellent, antistatic, moisture management, water proof, fire retardant and thermoregulation properties can be developed at a molecular level by using nanotechnology without affecting comfort and flexibility of the fabric [1]. The use of nanotechnology allows textiles to become multifunctional and produce fabrics with special functions, including antibacterial, UV protection, easy-clean, water- and stain repellent and anti-odour. Nanotechnology deals with structures of sizes between 1 to 100 nm in at least one dimension and involves developing materials or devices possessing dimension within that size. Nanotechnology creates structures that have excellent properties by controlling atoms and molecules, functional materials, devices and systems on the nanometer scale by involving precise placement of individual atoms [2]. Materials at nanoscale can exhibit different properties as compared to a macro scale level. Properties of material change quickly when it reduces to nanoscale. Nanostructure materials have a grain size of less than 100 nm, which increases the surface area of the material and cause excellent change in physical and surface properties [3].

The UV protection property of a textile fabric can be achieved by dye, pigment, delustrant (TiO_2) or UV absorber finish. These materials can provide protection against UV radiation by absorbing and blocking its penetration through a fabric. To achieve UV protection several nanomaterials or nanoparticles can be applied on textile fabric [4-6]. Titanium dioxide and zinc oxide nanoparticles are mostly used for this purpose because they have good UV protection properties [7-9]. They give protection by reflecting, scattering or absorbing harmful UV radiations of sun. They are more stable as compared to organic UV-blocking agents. Nanoparticles of ZnO can be prepared by the wet chemical method [10-12]. Titanium dioxide nanoparticles can be imparted on textile fabric by treating in TiO_2 nanoparticles aqueous solution using a pad-drying-cure method [4, 13]. Organic UV absorbers are mainly derivatives of O-hydroxy benzophenones, O-hydroxy phenyl triazines, O-hydroxy phenyl hydrazine's and they are less stable than inorganic UV absorbers [14]. The humidity, sweat and moisture provides suitable environment for the growth of many types of microbes (pathogenic or non-pathogenic) on the textile inner or outer wear. These microbes can cause diseases, bad odor for the wearer and reduce strength and color properties of the fabric. To avoid all these microbes, we need anti-microbial finish on the fabric. Many antibacterial finishes and disinfection techniques have been developed for all types of textiles. Many materials are used for the antibacterial properties such as nano-sized copper, silver, titanium dioxide and zinc oxide particles. Anti-bacterial effects can be maximized on textile fabric by using nanoparticles because nanoparticles increases the surface area [15]. Some of important antibacterial agents that are used in textile finishing are quaternary ammonium, triclosan, chitosan and metallic salts. Preparation of nano-sized metals and metal oxides mainly silver, titanium dioxide, zinc oxide and cooper oxide has enabled the development of a new generation of biocides [16].

The present study is focused on development of multifunctional polyester fabric. Here, the growth of TiO_2 nanoflowers on surface of polyester fabrics using two step approaches of sol–gel technology and hydrothermal method is reported. Compared to previous studies, it has number of advantages such as simple, inexpensive, high yield, and effective control over phase and morphology of TiO_2 at low temperatures. In situ growth of TiO_2 nanoflowers on polyester fabric was achieved by low temperature hydrothermal method. The functional properties investigated include UV protection and antibacterial activity.

2. Problem Formulation

2.1 EXPERIMENTAL

2.1.1 Material

Titanium(IV) isopropoxide (97%) and Titanium(IV) butoxide [97%, $Ti(OC_4H_9)_4$], HCl (37%), granulated caustic soda, ethanol (99.9%) and petroleum ether were purchased from Sigma Aldrich. A plain woven polyester fabric (areal density 97 g/m^2) was supplied by a local industry. The pathogenic bacterial strains obtained from the Czech collection of microorganisms of Masaryk University Brno were used to test the antibacterial properties.

2.1.2 Growth of Nanoflowers

At the beginning, the polyester fabric was treated with 20 g/L caustic soda solution at 80 °C for 30 mins. The growth of TiO_2 nanoflowers were carried out on the caustic treated polyester fabric according to our recently published work [17]. In a typical method, a semi-transparent TiO_2 sol was prepared by continuously stirring a mixed solution of Titanium butoxide (6 mL), ethanol (60 mL) and acetic acid (2 mL) for 3 h at 60°C. The as-prepared TiO_2 sol was deposited onto the caustic treated fabric by the pad-dry-cure method. The seeded polyester fabric was placed in the solution containing titanium(IV) isopropoxide (2 mL), hydrochloric acid (10 mL) and deionized water (60 mL) and heated at 125°C for 1 h. Finally, the fabric was removed from the reactor, rinsed with deionized water and dried. TiO_2 nanoflowers array was obtained on the polyester fabric. Fig.1 shows the schematic growth of TiO_2 nanoflowers onto polyester fabric.

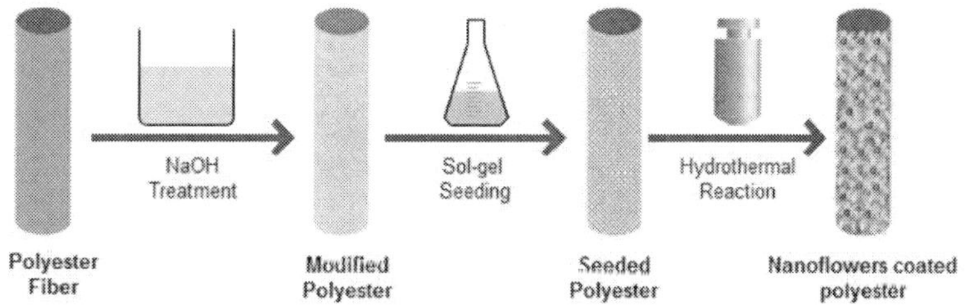

Fig.1 The schematic approach of growing TiO_2 nanoflowers on polyester fabric

2.2 Characterizations

2.2.1 Surface Morphology and Elemental Analysis

The surface morphology of control and coated fabrics were observed by using a Scanning Electron Microscope TS5130 VegaTescan at 30 kV acceleration voltage and UHR-SEM Zeiss Ultra Plus with an accelerating voltage 2 kV equipped with an Energy Dispersive X-ray spectrometer Oxford X-max 20. The elemental analysis of coated polyester fabric was carried out at 10 kV accelerating voltage using Energy-dispersive X-ray spectroscopy. In this study, NanoWizard 3 NanoScience AFM from JPK Instruments was used in non-contact mode to determine the surface roughness and to observe the topography of the coated samples.

2.2.2 UV Protection Properties

UV protection behaviour of the coated fabric was determined by measuring percentage transmission of UV radiations on UV-vis-NIR Spectrophotometer (UV-3101PC) as per AATCC 183-2000 standard. The samples were conditioned for 4 hours at 21 ± 1°C temperature and 65 ± 2 % relative humidity. For each sample, four measurements were performed and the average of scans was taken to calculate the UPF factor from Equation 1.

$$UPF = \frac{\sum_{280nm}^{400nm} E_\lambda S_\lambda \Delta_\lambda}{\sum_{280nm}^{400nm} E_\lambda S_\lambda T_\lambda \Delta_\lambda} \quad 1)$$

Where S_λ is the solar spectral irradiance, E_λ is the relative erythemal spectral response, T_λ is the average spectral transmittance of the sample and Δ_λ is the measured wavelength interval in nanometres.

2.2.3 Evaluation of Antibacterial Activity

The pathogenic bacterial strains from the Czech collection of microorganisms of Masaryk University Brno were used to test the antibacterial properties.

1) Escherichia coli (E.C.) - CCM 2024 (ATCC 9637), gram negative bacteria (G-).
2) Staphylococcus aureus (S.A.) - CCM 2260 (ATCC 1260), gram positive bacteria (G+).

2.2.4 AATCC Test Method 147-2004

This method is qualitative, tentative and it should be done prior to the method AATCC 100. The antibacterial sample is placed on the agar which is standardly applied with the bacterial inoculum in several lines. A modification of this method was used in this case where the bacterial inoculum was applied to the whole surface of the agar. After 24 hours of incubation there was rated grow of the bacteria under the tested sample as well as the inhibitory zone around the sample.

2.2.5 AATCC Test Method 100-2004

This test method provides a quantitative procedure for the evaluation of the degree of antibacterial activity. The result is a number of survivor bacteria colonies (CFU) and from these number there is calculated inhibition degree (in %). It is always necessary to compare the treated sample with an untreated one (control).

3. Problem Solution

3.1 Results and Discussion

SEM analysis was conducted to observe the surface morphology of pristine and nanoflowers coated polyester fabric which can be seen in Fig.2. Uniform flower-like hierarchical micro/ nano structures of TiO_2 with an average size around 500 nm were dispersed on the surface of smooth polyester fiber after hydrothermal reaction (Fig.2b).

From the SEM image, a uniform and continuous layer of nanoflowers can be seen on the surface of the polyester fibers. A good adhesion between the TiO_2 particles and polyester fibers can be attributed to the presence of hydroxyl groups and formation of hydrogen bonds between them.

Fig.2 FESEM images of: (a) pristine polyester; (b) TiO_2 nanoflowers coated polyester. Inset (b) is the image at high resolution.

The chemical composition of TiO_2 nanoflowers coated fabric was characterized by energy dispersive X-ray spectroscopy (EDS) measurement. Fig.3 shows the EDS spectra of TiO_2 nanoflowers coated fabric. The main elements were determined to be O, C and Ti. However, the pristine polyester was mostly consisted of O and C elements. It should be noted that the Au element was caused by sputtering gold on TiO_2 coated sample for the scanning electron microscopy (SEM) characterization. The relative atom ratio of O/C/Ti is about 34.1%/33.5%/32.4%, respectively. The EDS map clearly showed that these uniformly distributed particles were mainly composed of Ti and O elements.

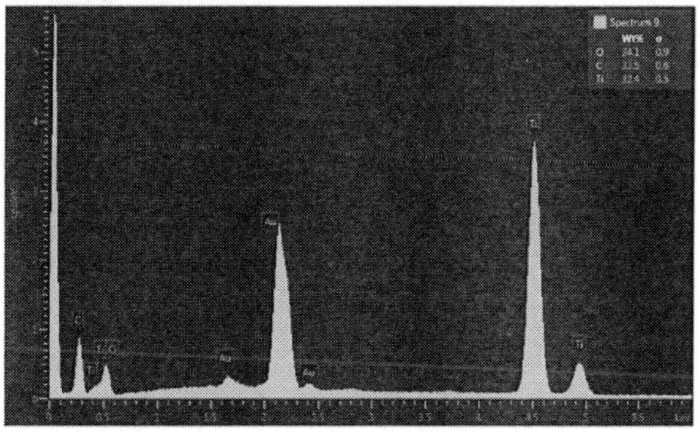

Fig.3 EDS spectrum of TiO_2 nanoflowers coated polyester fabric, the image inset is corresponding element percentages

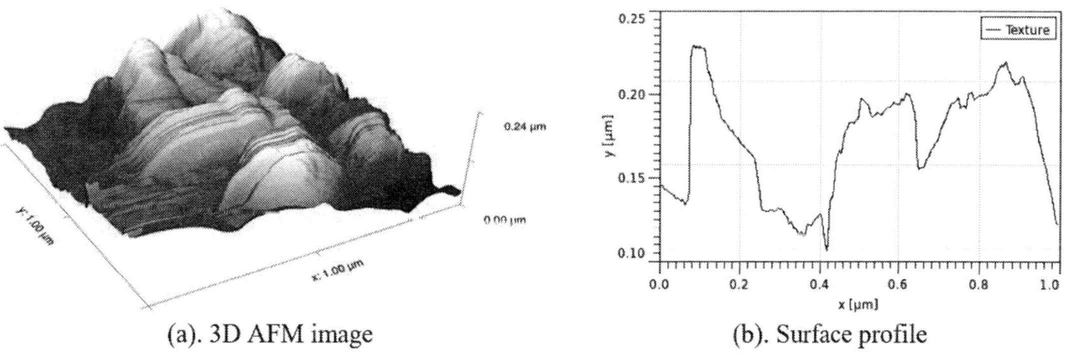

(a). 3D AFM image (b). Surface profile

Fig.4 AFM analysis of the TiO_2 nanoflowers coated polyester fabric

Fig.4 shows the topographical AFM images and surface profiles of the TiO_2 coated polyester fabrics. The increase in surface roughness of the coated polyester fabric can be seen from the topographical image. The average roughness (Ra) of 78.8 ± 9.3 nm was estimated for the TiO_2 nanoflowers coated polyester fabric. The 3D AFM image of the coated fiber shows the topographical structure of the surface. The micro/nano structures enhance the roughness, which is necessary to form air cushion and validation of the Cassie–Baxter equation.

3.2 Antibacterial Activity

The antibacterial property of the nanoflowers coated fabrics was tested against Gram-negative E. coli and Gram-positive S. aureus. Fig.5 shows the zones of inhibition around fabric samples after incubation took place in a thermostat with 37°C for 24 hours. The pristine polyester fabric without TiO_2 coating showed no antibacterial activity. However, the zone of inhibitions was evidenced against both types of bacteria S. aureus and E. coli after the TiO_2 coating. Moreover, S. aureus depicted the highest sensitivity as compared to E. coli. The zone of inhibitions for E. coli was 2 mm, while for S. aureus it was 3 mm for nanoflowers coated fabrics. The mechanism associated with the antibacterial behaviour of TiO_2 is due to the photocatalytic activity. It is well-known that the photocatalytic performance of TiO_2 depends strongly on its size and morphology. Titanium dioxide is mainly used as nanoparticles for any study of its photoactivity. However, other well-defined morphologies (i.e., nanoflowers) have shown outstanding performances for the photocatalytic annihilation of pathogenic microorganisms.

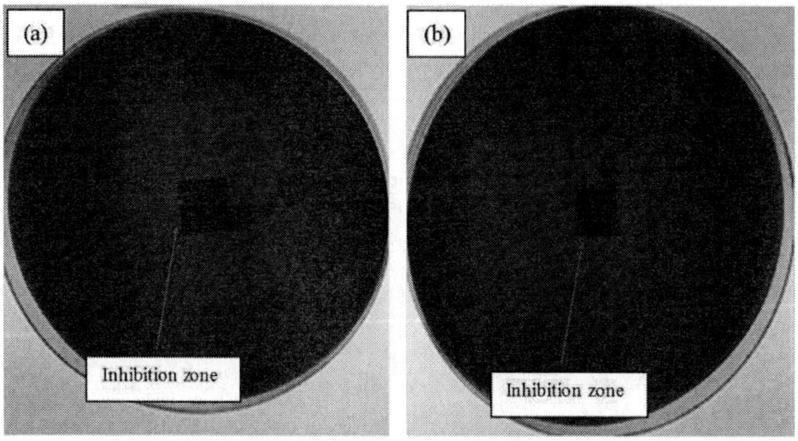

Fig.5 Antibacterial properties of TiO_2 coated polyester fabrics. (a) E. coli and (b) S. aureus

The quantitative assessment of the antibacterial activity of the functionalized polyester fabric was performed on Staphyloccocus aureus and Escherichia coli. This method is quantitative and there is rated a reduction factor which states the reduction (in percent) of the inoculated concentration of the bacteria due to the effect of the coated sample. The result is a number of survivor bacteria colonies (CFU) and from these numbers inhibition degree is calculated (%).It is always necessary to compare the treated sample with an untreated one (Control). Our findings show that the antibacterial activity of functionalized fabric is much higher on S. aureus as compared to E. coli. The quantitative results obtained for E. coli show a decrease of the initial cell count from 6.9 log CFU/mL to 3.5 log CFU/mL after 24 h of incubation. Moreover, results from S. aureus showed a decrease of 2 log CFU/mL from initial cell count of 6.8 log CFU/mL. The 96% and 99% of bacterial reduction was found for E. coli and S. aureus, respectively. From the results of the AATCC 100 assay, it is evident that the coated fabrics exhibited very good inhibition.

3.3 UV Protection

Sunlight is composed of ultraviolet (UV), visible and infrared radiation (IR). UV radiation is composed of UVA (400-315 nm), UVB (315-290 nm) and UVC (290-200 nm). Ultraviolet protection factor (UPF) ratings of less than 15, between 15 and 50, and more than 50 are generally classified as bad, good, and excellent UV-blocking performance for textiles, respectively. A textile having a UPF value of less than 15 is considered inadequate for protection against UV radiation. The absorption of UV radiations is a natural property of the TiO_2. The UPF value is used to evaluate the UV absorption efficiency of the fabricated samples. The high UV absorption intensity leads to higher UPF value. The results of ultraviolet protection factor (UPF) values and the percentages of UV blocking for UVA (320-400 nm) and UVB (290-320 nm) ranges are described in Table 1. The polyester fabric coated with TiO_2 nanoflowers showed excellent UV-blocking properties (higher UPF value and lower UVA/UVB transmittance) as compared to pristine polyester fabric. This clearly confirmed that the application of nanostructured TiO_2 to polyester fabric increased UV light absorbance at all the measured wavelengths. These results ascribed to the high UV absorbance and scattering properties of TiO_2 nanoflowers on the fiber surface that can provide effective shielding of UV radiation. The UPF values of 17 and 228 were found for pristine and TiO_2 nanoflowers coated polyester fabrics, respectively. These results indicate that UPF values are strongly related to amount of TiO_2 particles deposited on substrates.

Table 1. UV-protection properties of the pristine and TiO_2 nanoflowers coated polyester fabrics.

Sample	UPF	UVA Blocking %	UVB Blocking %
Pristine polyester	17	85.22	95.74
TiO_2 nanoflowers polyester	228	97.92	99.66

3.4 Durability of TiO_2 Nanoflowers Coated Fabric

For practical applications, the durability of the functional coating on the fabric is an important factor. The mechanical stability of the TiO_2 nanoflowers coated fabric was qualitatively assessed by adhesive tape pressing. However, no visible particles were seen on the adhesive tape, which confirm the strong bonding of TiO_2 nanoflowers on the PET fabric surface. In another test, the durability of the flower-like TiO_2 coated fabrics were evaluated against washing according to method ISO 105-C01 standard. After washing change in UPF values was measured and it has been observed that there is no significant change in UPF (214) after washings which shows strong attachment of the TiO_2 nanoflowers on the fabric surface.

4. Conclusion

In this study Titanium dioxide (TiO_2) nanoflowers were successfully grown on 100% polyester fabric by hydrothermal process. TiO_2 nanoflowers were grown on polyester fabric by a two-step method. The flower-like structures were grown on the surface of the polyester fibers via sol–gel technology and hydrothermal method. The antibacterial and UV protection properties of the TiO_2 nanoflowers coated fabrics were studied. The coated fabric showed good antibacterial activity against the S. aureus and E. coli after 24 hours incubation due to its photocatalytic nature. Moreover, excellent results of UV-protection properties were achieved. The nanoflowers coated fabric exhibited 97.92% and 99.66% blockage of UVA and UVB, respectively. The high UV blocking ability of the coated fabric is due to complete covering of polyester fibers with TiO_2 nanoflowers.

5. Acknowledgement

This work was supported by the Ministry of Education, Youth and Sports of the Czech Republic and the European Union (European Structural and Investment Funds-Operational Program Research, Development and Education) in the frames of the project Hybrid Materials for Hierarchical Structures (HyHi, Reg. No. CZ.02.1.01/0.0/0.0/16_019/0000843). The authors also acknowledge the support of project "Textile structures combining virus protection and comfort", reg. no. CZ.01.1.02/0.0/0.0/20_321/0024467.

References

[1] Gupta D. Design and engineering of functional clothing. Indian J Fibre Text Res: 2011; 36: 327-335.
[2] Patra JK, Gouda S. Application of nanotechnology in textile engineering. J Eng Technol Res: 2013; 5: 104-

111.
[3] Joshi M, Bhattacharyya a. Nanotechnology - a new route to high-performance functional textiles. Text Prog: 2011; 43: 155-233.
[4] Khan MZ, Baheti V, Ashraf M, et al. Development of UV Protective, Superhydrophobic and Antibacterial Textiles Using ZnO and TiO2 Nanoparticles. Fibers Polym: 2018; 19: 1647-1654.
[5] Khan MZ, Baheti V, Militky J, et al. Superhydrophobicity, UV protection and oil/water separation properties of fly ash/Trimethoxy(octadecyl)silane coated cotton fabrics. Carbohydr Polym: 2018; 202: 571-580.
[6] Khan MZ, Ashraf M, Hussain T, et al. In situ deposition of TiO2 nanoparticles on polyester fabric and study of its functional properties. Fibers Polym: 2015; 16: 1092-1097.
[7] Kang X, Liu S, Dai Z, et al. Titanium dioxide: From engineering to applications. Catalysts; 9. Epub ahead of print: 2019. DOI: 10.3390/catal9020191.
[8] Rodríguez-González V, Obregón S, Patrón-Soberano OA, et al. An approach to the photocatalytic mechanism in the TiO2-nanomaterials microorganism interface for the control of infectious processes. Appl Catal B Environ: 2020; 270: 118853.
[9] Kajbafvala A, Ghorbani H, Paravar A, et al. Effects of morphology on photocatalytic performance of Zinc oxide nanostructures synthesized by rapid microwave irradiation methods. Superlattices Microstruct: 2012; 51: 512-522.
[10] Khan MZ, Militky J, Baheti V, et al. Growth of ZnO nanorods on cotton fabrics via microwave hydrothermal method: effect of size and shape of nanorods on superhydrophobic and UV-blocking properties. Cellulose: 2020; 27: 10519-10539.
[11] Becheri A, Durr M, Lo Nostro P, et al. Synthesis and characterization of zinc oxide nanoparticles: Application to textiles as UV-absorbers. J Nanoparticle Res: 2008; 10: 679-689.
[12] Wang J, Zhao J, Sun lu, et al. A review on the application of photocatalytic materials on textiles. Text Res J: 2015; 85: 1104-1118.
[13] Haydon B. Nanomaterials and their Applications in Textiles - Domestic Standardization for Canadian Manufacturers and Importers and International Standardization Developments: 2012.
[14] Saravanan D. UV protection textile materials. AUTEX Res J: 2007; 7: 11-18.
[15] Wong YWH, Yuen CWM, Leung MYS, et al. Selected applications of nanotechnology in textile water repellence: 2006; 6: 1-8.
[16] Shahidi S, Wiener J. Antibacterial Agents in Textile Industry: 2012.
[17] Khan MZ, Baheti V, Militky J, et al. Self-cleaning properties of polyester fabrics coated with flower-like TiO2 particles and trimethoxy (octadecyl)silane. J Ind Text: 2020; 50: 543-565.

Effect of Antibacterial Yarn on Fabric Antibacterial Property

Tian-Chen Huang, Li-Hua Chen*

School of Material Science and Engineering, Beijing Institute of Fashion Technology, Beijing, 100029, China

*Corresponding author's email: wclh@sina.com

Abstract

In order to explore the effect of antibacterial yarn on fabric antibacterial property, 11 kinds of antibacterial fabrics with different structures were designed from three aspects of type, content and arrangement of antibacterial yarn. These 11 fabrics were tested against Staphylococcus aureus and Escherichia coli. The results show that antibacterial yarn arrangement has no significant effect on fabric antibacterial property, the type and content of antibacterial yarn has a significant effect on fabric antibacterial property. Fabrics with different antibacterial yarn have different performance against different bacteria, and antibacterial yarn content has a non-linear positive correlation with fabric antibacterial rate. In the development of antibacterial fabrics, antibacterial yarns should be reasonably designed to achieve a good antibacterial effect.

Keywords: Antibacterial Yarn; Type; Content; Arrangement; Fabric Antibacterial Rate

1. Introduction

People are in constant contact with bacteria in daily life. Traditional textiles provide a good carrier for the growth and reproduction of microorganisms. Under suitable conditions and environments, pathogenic bacteria multiply and endanger the health of people [1]. The outbreak of novel coronavirus in 2020 has made people aware on the importance of resisting bacteria and viruses and improving immunity in daily life, which has also put forward higher demands on the textile industry. Market demand for green, safe and environmentally friendly textiles have grown, and now antibacterial textiles have become a key product [2]. The properties of fabric are mainly affected by fabric structure, which generally reflects the geometric form of yarn in fabric [3]. The antibacterial property of fabric is also affected by the antibacterial yarn. In this study, analyzing fabric antibacterial property from the perspective of antibacterial yarn's type, content and arrangement, two natural antibacterial yarns, Artemisia multifilament and Tentea blended yarn, were selected to design weft backed fabrics for antibacterial testing against Escherichia coli (E. coli) and Staphylococcus aureus (S. aureus). Variance analysis and curve fitting was conducted on the results, the results have certain reference value for design and development of antibacterial textiles.

2. Experiment

2.1 Experimental Materials and Specifications

2.1.1 Materials Selection and Yarn Specifications

The yarn specifications are shown in Table 1.

Table 1. Yarn specifications

Yarn	Yarn number	Type	Composition and ratio (%)	Fineness (tex)
Weft	A	Artemisia multifilament	Viscose 100	13.3
Weft	B	Tentea blended yarn	Viscose/Cotton 30/70	14.6
Weft	C	Viscose filament	Viscose 100	13.3
Warp	/	DTY HIM filament	Polyester 100	11.1

Artemisia multifilament is composed of Artemisia fibers without twist. Artemisia fiber is a kind of pure natural plant antibacterial fiber, which is organically combined with natural cellulose extracted from the herbaceous plant Artemisia annua and based on viscose [4, 5].

Tentea blended yarn is made by mixing and twisting Tentea fibers and cotton fibers. Tentea fiber is a new type of functional viscose fiber by blending tea polyphenols extracted from mountain tea and natural wood pulp, using nanotechnology to uniformly disperse antibacterial ingredients into the fiber [6, 7].

2.1.2 Microstructure Observation of Antibacterial Yarn

Used JSM-7500F scanning electron microscope to magnify 600 times and 800 times respectively to observe the microstructure of fiber longitudinal surface and cross section in antibacterial yarn, as shown in Fig.1.

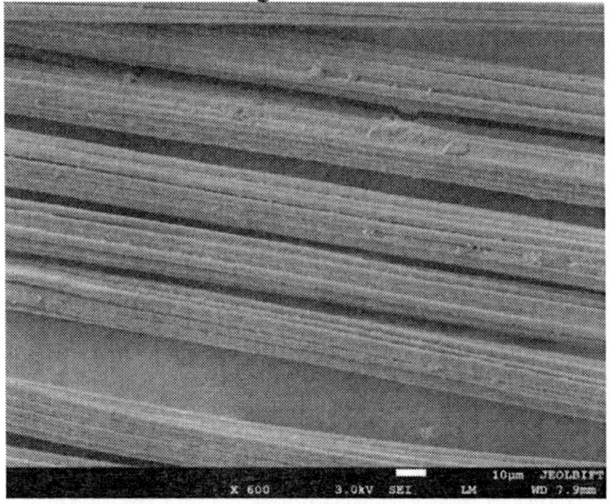

(a1) Longitudinal surface of Artemisia multifilament

(a2) Cross section of Artemisia multifilament

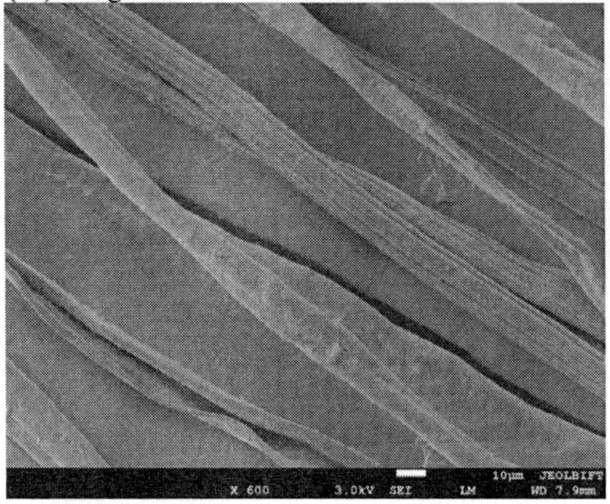

(b1) Longitudinal surface of Tentea blended yarn

(b2) Cross section of Tentea blended yarn

Fig.1 Microstructure of antibacterial yarns

In Fig.1, it can be seen from (a1) and (b1) that Artemisia fiber and Tentea fiber are not as smooth as ordinary viscose fiber longitudinally with some grooves. At the same time, it can be seen that a lot of particles are attached to the fiber surface, which are compound antibacterial agents and nano antibacterial particles added in the fiber spinning process [8]. From (a2) that the cross section of Artemisia fiber is polygonal and has different components, because the fiber is mixed with antibacterial substances; from (b2) that the cross section of cotton fiber is oblateness, and Tentea fiber has an irregular cross section without obvious antibacterial component.

2.1.3 Tensile Property of Antibacterial Yarn

Refer to GB/T 3916-2013 "Determination of Tensile load and Extension at break of Single Yarn of Textile Package Yarn (CRE)", Constant Rate of Extension machine was used to test the tensile fracture of antibacterial yarn in dry and wet state. The clamping distance was set as 2cm, and the pretension was adjusted according to the fineness of tested yarn. Each yarn was tested 20 times and average value was taken. The test results are shown in Table 2.

Table 2. Tensile test data of antibacterial yarns

Yarn number	Tensile load (N)		Tensile strength (N/tex)		Extension at break (%)	
	Dry	Wet	Dry	Wet	Dry	Wet
A	1.88	0.91	0.141	0.068	11.4	21.6
B	1.51	1.25	0.103	0.086	5.25	8.64

Table 2 shows that tensile strength of Artemisia multifilament is higher than Tentea blended yarn in dry state, but lower than Tentea blended yarn in wet state, which will affect the performance of its fabrics in daily wet and hot wearing environment. This is because, the strength of viscose fiber is higher than cotton fiber in dry state, and lower than cotton fiber in the wet state [9]. Compared to dry state, the breaking extension of Artemisia multifilament is nearly doubled in wet state, which will seriously affect the weaving process. Tentea blended yarn is a viscose-cotton blended yarn, which can reduce the influence on yarn's tensile property from humidity.

2.1.4 Antibacterial Fabric Design and Specifications

The fabric specifications are shown in Table 3.

Table 3. Fabric specifications

Fabric number	Weft arrangement in weave repeat unit		Antibacterial yarn content (%)
	Top weft/Inner weft	Antibacterial weft yarn interweaving ratio (%)	
1	AAAA/AAAA	100	66.37
2	AAAC/AAAA	87.5	60.62
3	ACAC/AAAA	75	51.96
4	ACCC/AAAA	62.5	43.3
5	CCCC/AAAA	50	34.64
6	CCCC/AAAC	37.5	25.98
7	CCCC/ACAC	25	17.32
8	CCCC/ACCC	12.5	8.66
9	ACAC/ACAC	50	34.64
10	AAAC/CCCA	50	34.64
11	BBBB/BBBB	100	67.63

Note: The fineness of yarn A and C are the same.

The weft backed weave is 1/3 right twil on the top and 3/1 right twill on the inner, and filling ratio of top and inner wefts is 1:1. Excluding the measurement error, the fabric warp density is 170/10cm, the weft density of fabric 1 to fabric 10 is 320/10cm, and of fabric 11 is 310/10cm.

The content of antibacterial yarn in fabric is calculated by formula (1):

$$X = \frac{N_{texW} \times P_W \times H}{N_{texT} \times P_T + N_{texW} \times P_W} \times 100\% \qquad 1)$$

In the formula: X—Antibacterial yarn content; N_{texW}—Weft yarn fineness; N_{texT}—Warp yarn fineness; P_W—Fabric weft density; P_W—Fabric warp density; H—Antibacterial weft yarn interweaving ratio.

2.2 Antibacterial Testing of Fabrics

2.2.1 Testing Method

Refer to GB/T 20944.3-2008 "Evaluation of Antibacterial Property of Textiles Part 3: Oscillation Method", conducted quantitative antibacterial performance tests on fabric 1 to fabric 11. The bacteria used for the test are E. coli and S. aureus.

2.2.2 Experimental Equipment

Main Instruments: autoclave, spectrophotometer, constant temperature incubator, light shaker, refrigerator, household double tub washing machine.

Main Reagents: beef extract, peptone, AGAR powder, anhydrous disodium hydrogen phosphate, anhydrous sodium dihydrogen phosphate, distilled water.

2.2.3 Sample Preparation

Cut antibacterial fabric into 5cm×5cm, and weigh 0.75g as a set of samples, wrap them with small paper sheets, and make at least 3 sets of samples for testing.

2.2.4 Testing Indicator

Antibacterial property of fabric is expressed by antibacterial rate. The standard stipulates that when antibacterial rate for E. coli or S. aureus is ≥70%, the fabric has an antibacterial effect. Fabric antibacterial rate is calculated by formula (2):

$$Y = \frac{W_t - W_b}{W_t} \times 100\% \qquad 2)$$

In the formula: Y—Average antibacterial rate of fabric sample; W_t—Average value of viable bacteria concentration in blank group; W_b—Average value of viable bacteria concentration in sample group.

2.3 Analysis Method

As shown in Table 3, for antibacterial yarns, fabric 1 and 11 have similar content, same arrangement, but different types; fabric 1 to 8 have the same type, different contents and arrangements; fabric 5, 9 and 10 have same type and content, but different arrangements. Given significance level of 0.05, using repeatable Two-way ANOVA to analysis the effect of antibacterial yarn type, content and arrangement on fabric antibacterial rate. Then using Boltzmann function to fit the relationship between antibacterial yarn content and fabric antibacterial rate, and draw the fitting curve [10].

3. Results and Discussion

3.1 Effect of Antibacterial Yarn Type on Fabric Antibacterial Property

3.1.1 Analysis of Test Results

The antibacterial test data of fabric 1 and 11 against E. coli are shown in Table 4, and the antibacterial test data against S. aureus in Table 5.

Table 4. Antibacterial test data of fabric 1 and 11 against E. coli

Number	Test photos	Dilution times	Average viable bacteria concentration (CFU·mL^{-1})	Average antibacterial rate (%)
Blank group	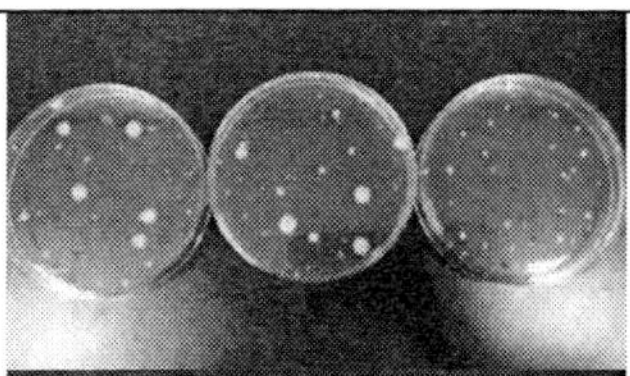	100000	3.7×10^{5}	0
Fabric 1	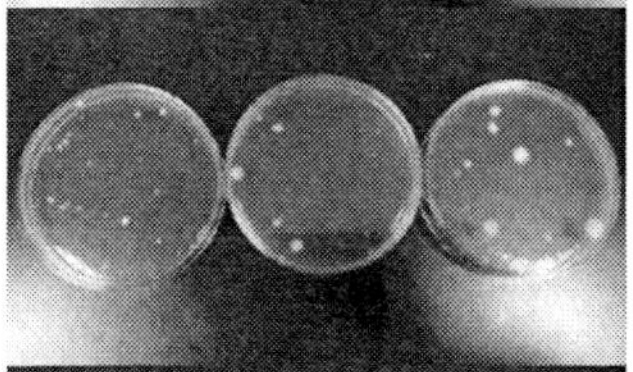	10000	2.1×10^{4}	93.6
Fabric 11	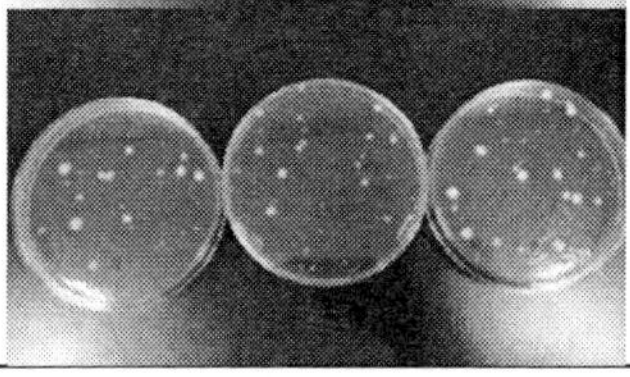	10000	4.8×10^{4}	87.0

It can be seen from Table 4 that fabric 1 has a higher antibacterial rate than fabric 11 against E. coli, so Artemisia multifilament has a better antibacterial property against E. coli than Tentea blended yarn.

Table 5. Antibacterial test data of fabric 1 and 11 against S. aureus

Number	Test photos	Dilution times	Average viable bacteria concentration (CFU·mL^{-1})	Average antibacterial rate (%)
Blank group	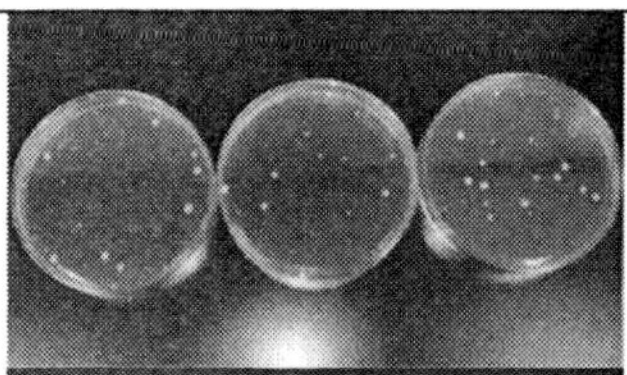	10000	7.1×10^{4}	0
Fabric 1	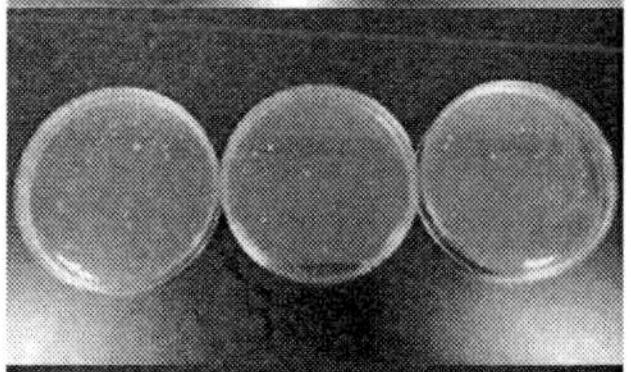	100	1.1×10^{2}	99.8
Fabric 11	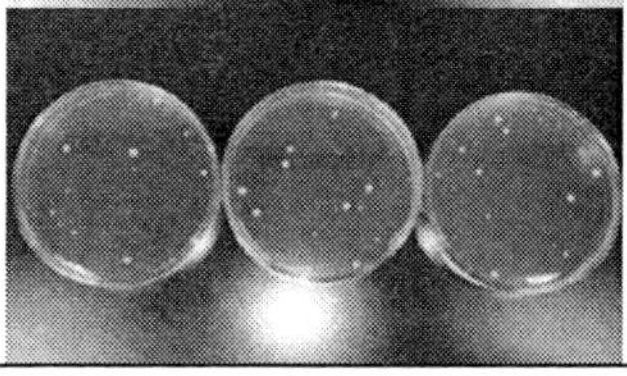	1000	1.2×10^{3}	98.3

In Table 5, fabric 1 has a higher antibacterial rate than fabric 11 against S. aureus, so Artemisia multifilament has better antibacterial property against S. aureus than Tentea blended yarn. Based on the results of Table 4 and Table 5, Artemisia multifilament has stronger antibacterial properties than Tentea blended yarn, and has the best antibacterial performance on S. aureus.

3.1.2 Significance Analysis

The significance analysis of fabric 1 and 11 against E. coli and S. aureus is shown in Table 6.

Table 6. Two-way ANOVA of AT and BT

Source of variance	Sum of squares	DF	Mean square	F value	P value
AT (Antibacterial yarn type)	6.183×10^{-3}	1	6.183×10^{-3}	30.864	5.375×10^{-4}
BT (Bacteria type)	2.037×10^{-2}	1	2.037×10^{-2}	101.670	7.979×10^{-6}
AT×BT	1.208×10^{-3}	1	1.208×10^{-3}	6.030	3.960×10^{-2}
Error	1.603×10^{-3}	8	2.004×10^{-4}		
Total	2.936×10^{-2}	11			

According to Table 6 that P_{AT}, P_{BT} and $P_{AT\times BT}$ are all less than 0.05, so antibacterial yarn type and bacteria type have significant effects on fabric antibacterial rate and have significant interaction. Among them, bacteria type has the greatest effect on fabric antibacterial property.

3.2 Effect of Antibacterial Yarn Arrangement on Fabric Antibacterial Property

3.2.1 Analysis of Test Results

The average antibacterial rates of fabric 5, 9 and 10 against E. coli are 79.3%, 80.4% and 80.2%, and the average antibacterial rates against S. aureus are 99.7%, 99.7% and 99.6%. The overall difference of antibacterial rates on above fabrics against each bacterium is not big, so a significant analysis of its effect is needed.

3.2.2 Significance Analysis

The significance analysis of fabric 5, 9 and 10 against E. coli and S. aureus is shown in Table 7.

Table 7. Two-way ANOVA of AA and BT

Source of variance	Sum of squares	DF	Mean square	F value	P value
AA (Antibacterial yarn arrangement)	1.843×10^{-4}	2	9.217×10^{-5}	3.033	8.6×10^{-2}
BT (Bacteria type)	8.904×10^{-2}	1	8.904×10^{-2}	2930.084	1.041×10^{-15}
AA×BT	1.03×10^{-4}	2	5.15×10^{-5}	1.695	2.248×10^{-1}
Error	3.647×10^{-4}	12	3.039×10^{-5}		
Total	8.969×10^{-4}	17			

According to Table 7 that P_{AA} and $P_{AA\times BT}$ are greater than 0.05, so antibacterial yarn arrangement has no significant effect on fabric antibacterial rate, and has no significant interaction with bacteria type. When testing fabric 1 to 8, the effect of yarn arrangement on results can be ignored.

3.3 Effect of Antibacterial Yarn Content on Fabric Antibacterial Property

3.3.1 Analysis of Test Results

The antibacterial rates of fabric 1 to 8 against E. coli and S. aureus are shown in Fig.2.

It can be seen from Fig.2 that the antibacterial rate against E. coli from fabric 1 to 8 is gradually decreasing, the antibacterial rates of fabric 6, 7 and 8 are all less than 70%, which does not meet the antibacterial standard and does not have an antibacterial property against E. coli. But the antibacterial rates against S. aureus from fabric 1 to 8 all reach 99%. Unable to find a critical value that meets the standard, curve fitting and function analysis were performed in order to show the relation between antibacterial yarn content and fabric antibacterial rate intuitively, seen in Fig.3.

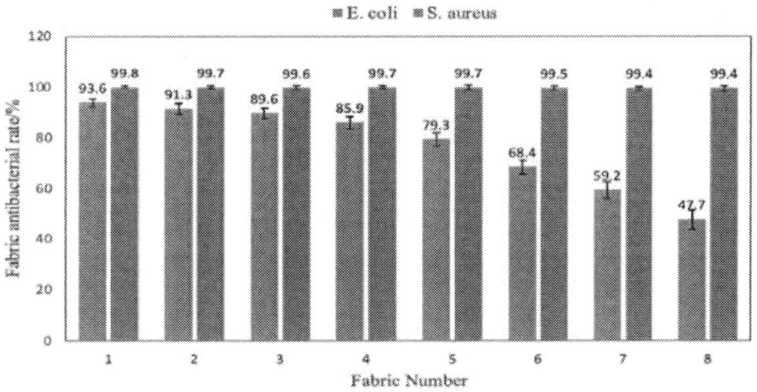

Fig.2 Antibacterial rates of fabric 1 to 8 against E. coli and S. aureus

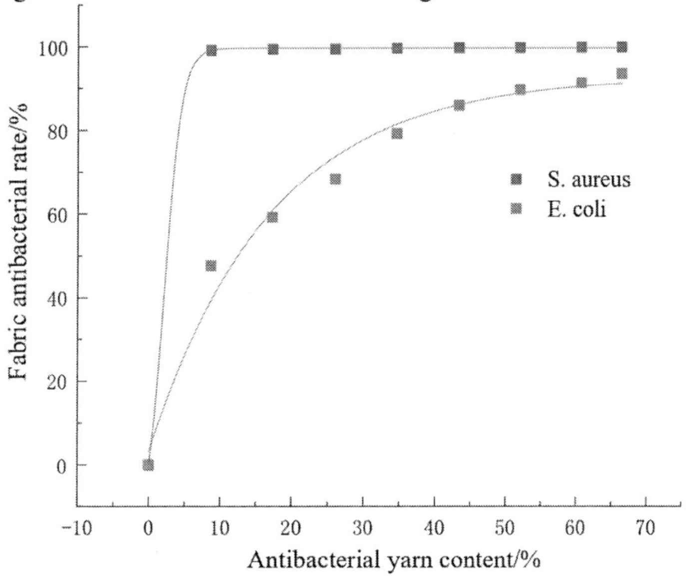

Fig.3 Relation between antibacterial yarn content and fabric antibacterial rate

In Fig.3, the equation of Boltzmann function of fitted curve is $y=A_2+(A_1-A_2)/(1+\exp((x-x_0)/dx))$, parameters of equation seen in Table 8. It can be calculated by the equation that when fabric antibacterial rate against E. coli is 70%, the antibacterial yarn content is about 23.72%, when fabric antibacterial rate against S. aureus is 70%, the antibacterial yarn content is about 4.25%.

Table 8. Parameters of regression equation

Curve name	A_1	A_2	x_0	dx	Adjusted R^2
E. coli	-1.053×10^6	92.951	-157.819	16.843	0.973
S. aureus	-17.313	99.667	2.126	1.215	0.999

The R^2 value is above 0.9, that is, the curve fitting is significantly related, so the calculation result of regression equation is reliable.

3.3.2 Significance Analysis

The significance analysis of fabric 1 to 8 against E. coli and S. aureus is shown in Table 9.

Table 9. Two-way ANOVA of AC and BT

Source of variance	Sum of squares	DF	Mean square	F value	P value
antibacterial yarn content (AC)	2.384×10^{-1}	7	3.405×10^{-2}	710.052	3.088×10^{-33}
Bacteria type (BT)	4.561×10^{-1}	1	4.561×10^{-1}	9510.444	3.585×10^{-41}
AC×BT	2.226×10^{-1}	7	3.225×10^{-2}	672.466	7.324×10^{-33}
Error	1.535×10^{-3}	32	4.796×10^{-5}		
Total	9.218×10^{-1}	47			

According to Table 9 that P_{AC}, P_{BT} and $P_{AC\times BT}$ are all less than 0.05, so antibacterial yarn content and bacteria type have significant effects on fabric antibacterial rate and have significant interaction.

4. Conclusions

By studying the effect of antibacterial yarn on fabric antibacterial property, the following conclusions can be obtained:

1) The antibacterial yarn arrangement has no significant effect on fabric antibacterial rate. The type and content of antibacterial yarn have significant effects on fabric antibacterial rate, and have significant interactions with bacteria type.

2) Among the two tested yarns, Artemisia multifilament has better antibacterial effects against E. coli and S. aureus than Tentea blended yarn, and Artemisia multifilament has the strongest antibacterial performance against S. aureus.

3) Calculated by fitting function that when Artemisia multifilament content reaches about 23.72%, the fabric antibacterial property against E. coli meets antibacterial standard; when Artemisia multifilament content reaches about 4.25%, the fabric antibacterial property against E. coli meets antibacterial standard.

4) The tensile property of Artemisia multifilament in wet state will be affected. When designing its antibacterial fabrics, the conclusion (3) can be combined to appropriately reduce the use of antibacterial yarns to save development costs, and interweave with other yarns to enhance the mechanical property of antibacterial fabrics.

References

[1] Bao YT. Development status of antibacterial textiles [J]. Textile Report: 2020; 39 (08): 12-13.

[2] Zhang HT, Zhang X, Liu MM, Xiao FJ. The development status of natural antibacterial textiles [J]. Textile Science and Technology Progress: 2020 (03); 8-11.

[3] Jing ML. Fabric Structure and Design, 5th Edition [M]. Beijing: China Textile Publishing House: 2014.

[4] Li LL. Analysis of antibacterial activity of different kinds of endophytic bacteria of the medicinal plant Artemisia annua [J/OL]. Guangxi plants: 1-12 [2021-04-20].

[5] Xinxiang Chemical Fiber successfully developed pure natural plant antibacterial fiber [J]. Printing and Dyeing: 2021; 47 (01): 83.

[6] Wang J, Pei YH, Cheng LJ, etc. Research progress in antibacterial effects of tea polyphenols and their development prospects as natural antifungal agents [J]. China Food Additives: 2020; 31 (11): 130-138.

[7] Tian FZ, Wang YL, Zhao GH, Tian GL, Peng Y. Research progress on antibacterial effects of tea polyphenols [J]. Modern Agriculture: 2020 (05); 29-32.

[8] Lu Y. The influence of two antibacterial fiber content on fabric properties [J]. Cotton Textile Technology: 2014; 42 (05): 29-32.

[9] Zong YN, Zhang HX. Textile Materials Science [M]. Shanghai: Donghua University Press: 2019; 309.

[10] Zhang HX, Lu Y, Tian W, Shi LJ, Zhu CY, Bi JW. The influence of silver-plated fiber content on the antibacterial properties of fabrics [J]. Journal of Textile Research: 2011; 32 (09): 38-41.

Adaptive Trousers Structure Design for Wheelchair Curlers

An-An Zhang, Jing Guo*

Universal Fashion Lab, Beijing Institute of Fashion Technology, Beijing, 100029, China

*Corresponding author's email: guoj@uflbift.com

Abstract

This focus of this research is on the waist and legs of wheelchair curlers. By studying the extension of the body surface skin of the curlers during exercise, it provides theoretical support for the optimal design of the curlers' trousers. Currently in China, the research and development of adaptive clothing for the wheelchair curling project is still in its infancy. This paper firstly analyzes the status and shortcomings of China's adaptive wheelchair trousers' market through literature research. Secondly, based on the theoretical support of the literature research, a human body experiment is carried out to measure and analyze the skin extension under common exercise postures. Finally, according to skin extension rules obtained from the experiment, this paper proposes an optimized design plan of the trousers structure for wheelchair curlers.

Keywords: Wheelchair Curling; Skin Extension; Square-Network Measurement; Adaptive Structure Design

1. Introduction

Since the establishment in 2007, the China Wheelchair Curling Team has repeatedly set new records in international competitions such as the World Championships and Paralympic Games. After winning the championship at the Pyeongchang Winter Paralympics, Chinese wheelchair curling achieved the Grand Slam in only one year. The rapid development of wheelchair curling in China has brought attention to disabled groups in society. [1] At present, the development of sportswear for this special group is lacking. The clothing of curling athletes on the market cannot fully meet the needs of wheelchair curling athletes in their career and life.[2] Since 2010, there has been an increasing number of scholars advocating for the design of functional clothing for the disabled, not just aimed to hide their physical disabilities, but to meet their special demand and help them to live better. [3, 4] At the same time, some scholars have conducted research and experiments on the development of trousers for wheelchair users. For example, Sara Bragança et al. analyzed the current satisfaction and actual needs of wheelchair athletes for trousers through a questionnaire, and proposed accordingly adaptive clothing optimization plan for wheelchair athletes. [5] Xuemeng Wu conducted research on clothing that is easy to wear and take off for the disabled through the structural optimization of certain parts [6]; In 2019, Slavica Bogović presented a research on improving trousers design, adjusted to the special needs and demands of an individual by applying 3D scanning technology in. [7] Na Zhu also investigated disabilities of the legs of wheelchair users and conducted detailed skin measurement experiments [8]. However currently there are no specific study focusing on the design of adaptive trousers of wheelchair curlers.

Skin extension experiment belongs to a category of anthropological experiment with strong theoretical support and practical experience. Nowadays, the methods for skin extension measurement are diversified and precise. In 1981, the Service Department of the Japanese Society of Human Engineering, used a legal analysis on the influence of major joint movements on skin extension, and provided data for the study on the influence of human joint movements on clothing. [9] In 2001, Kan Liu et al. used body drawing measurement to analyze the extension rules of skin of the trunk under different exercise states. [10] In 2007, Weiping Wang used the grid network method to dynamically measure the lower body skin to improve the structure of professional women's pants. [11] In 2014, Lan Luo and Jianping Wang applied a gel substance to rub and sample the skin of lined lower extremified to study the rule of skin extension of the lower limbs during cycling. [12] The similar method was also applied by Sunyoon Choi and Susan P Ashdown [13] as well as Xiaofang Guan and Yongmei Liu. In 2018, they used the same method to measure and analyze skin deformation around the knee joint in three human postures. [14] However at present, there are no research on wheelchair curling athletes in the field of skin extension measurement.

2. Methodology

2.1 Experimental Subject

According to the experimental requirements and achievable conditions, the subjects were men whose pre-collected basic physical data were similar to the national wheelchair curling athletes and they were able to simulate important curling postures. Taking the physical data of wheelchair curling national team athletes as a reference, the final four selected subjects had an average age of 21.75 years, an average weight of 60.88kg and an average height of 178.25cm.

2.2 Experimental Method

This experiment was carried out in a laboratory of about 20 square meters. The temperature was set at 24°C±2.5°C, and the humidity did not change significantly.

According to preliminary observations on the changes in body posture of subjects during pitching, their legs undergo varying degrees. Therefore, in this experiment, the body surface drawing measurement is the most suitable method for this subject. By dividing the skin of the legs into small cells of equal size and individually measuring them, the accuracy of the experimental results is ensured.

The main tools used in this experiment in the stage of measurement are skin gel (medium solution soluble in oil), an oily marker, plain weave non-woven fabric (0.5 mm), transparent wide tape and scissors. The main tools used in the data statistics stage are a 2D scanner, Photoshop and Excel.

2.3 Experimental Process

The process of this experiment can be divided into preparation, sampling and analysis shown in Fig.1.

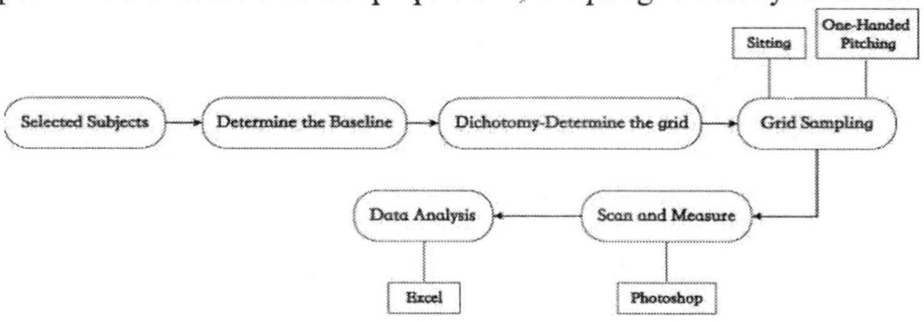

Fig.1 Framework of experimental steps

In the preparation phase, a tester drew a grid network on the subject's waist and legs based on dichotomy. As is shown in Fig.2, the tester takes the main marking lines in the clothing technique as the basic lines and draws a network on the surface of skin on the waist and legs of the subject. Each unit grid is about a square with a side length of 3cm-4cm.

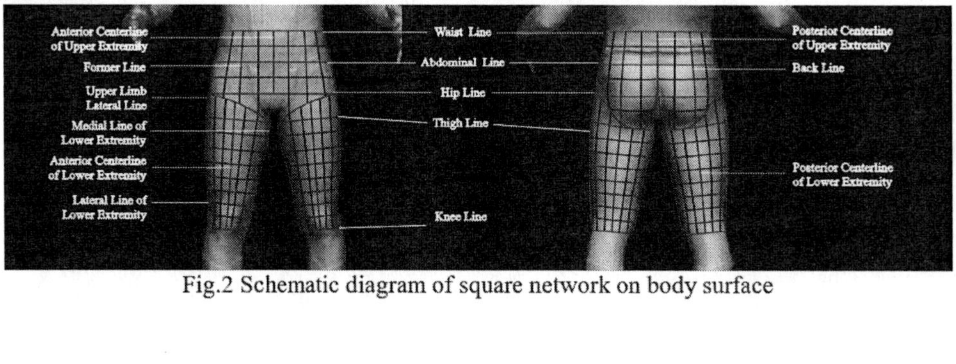

Fig.2 Schematic diagram of square network on body surface

Sampling refers to using a piece of non-woven fabric to sample the body surface mesh of the subjects when they make key movements. The measured movements are divided into sitting movements and one-handed pitching movements.

To standardize accuracy and consistency, before conducting the formal experiments, the tester measures the forward inclination angle and knee joint angle of the Chinese athletes in the sitting posture and the extreme posture of one-handed pitching. Subjects made a similar inclination angle and knee joint angle (The difference from the average angle presented by the athlete is less than five degrees) when they are performing the relevant movement.

In the process of data analysis, the tester scanned the samples and formed a two-dimensional plan, then recorded the pixel number of each grid. After numbering each unit grid according to row and column, the tester imported the numbers into Excel. According to the calculation method in "Skin Deformation of Male Lower-Limb Based on Motion Function for Cycling Shorts," “Pij” is set as the pixel of the grid in the i-th row and the j-th column in the starting position, “Rij” is set as the pixel in the i-th row and j-th column of the grid in the ending position, and “a” is set as the grid in the i-th row and j-th column from the starting position to the ending position. The extension rate of the grid; i = 1, 2, 3,... 15, 16; j = 1, 2, 3,..., 15, 16. Then a=(Pij/ Rij)-1.[15]

3. Analysis and Discussion

3.1 Analysis of Changes in Waist Body Surface

In this experiment, 36 grids are distributed on the front surface of the waist of the human body and 46 grids are distributed on the back. In the standing state, the average area of the unit grid is 16cm^2. A detailed data summary is shown in Table 1.

Table 1. Rate of extension of waist area

		Valid value (number)	Invalid value (number)	Median	Maximum	Minimum	Positive average	Negative average
sitting-down movements	Front	36	0	11.02%	28.36%	-33.4%	17.05%	-12.48%
	Back	46	0	5.52%	9.05%	-5.24%	18.30%	-9.33%
One-handed pitching movement	Front	36	0	-10.5%	14.91%	-31.5%	8.04%	-12.81%
	Back	46	0	0.02%	18.19%	-25.4%	7.17%	-9.15%

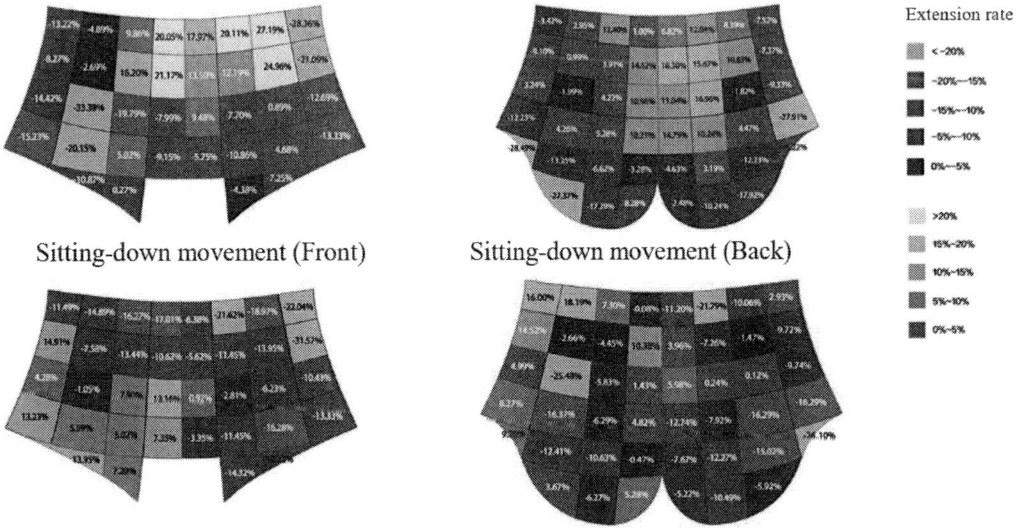

Sitting-down movement (Front) Sitting-down movement (Back)

Single-handed pitching movement (Front) Single-handed pitching movement (Back)

Fig.3 Distribution of extension rate of waist

Generally, when sitting down, the skin on the waist tends to expand, and the expansion trend of the skin on the back of the human body (4.49%) is relatively more obvious than the front (2.29%). The expansion and contraction trends of the skin on the front and back of the human body are symmetrical along the front center line and the back center line. When the one-handed pitching action is performed, the skin on the waist tends to shrink to a certain extent while the skin on the front tends to shrink more obviously (-2.39%). The skin on the front and back of the human body shows no obvious symmetry; one side of the pitching arm shrinks (-8.75%), and the other side expands (6.22%).

According to the data distribution shown in Figure 5, the skin on the side near the former lines shows a general expansion trend (8.63%). The skin on both sides of the anterior and posterior center lines of the upper extremities also shows an overall expansion trend (7.41% and 10.37%). The skin on both sides of the upper limb lateral line of the back of the human body shows an obvious contraction trend (-6.93%) while the skin on both sides of the back lines shows a slight expansion trend (4.86%).

When performing one-handed pitching movements, the skin on the side of the pitching arm of the upper limb lateral line of the front of human body shows an obvious contraction trend (-13.00%). Among them, the contraction trend of the skin near the lateral line is the most obvious (-19.34%). Then the skin on the other side shows a slight expansion trend (2.86%). Among them, the skin near the upper limb lateral line has an obvious expansion trend (10.95%), and the skin on both sides of the back lines shows a slight contraction trend (-4.45%).

3.2 Analysis of Changes in the Body Surface of the Legs

In this experiment, 54 grids are distributed on the front surface of the leg and 51 grids are distributed on the back. In a standing state, the average area of the unit grid is $9cm^2$. A detailed data summary is shown in Table 2.

Table 2. Rate of extension of leg area

		Valid value (number)	Invalid value (number)	Median	Maximum	Minimum	Positive average	Negative average
sitting-down movements	Front	54	0	4.92%	92.34%	-27.5%	20.43%	-6.45%
	Back	51	0	2.43%	57.62%	-91.0%	15.39%	-11.57%
One-handed pitching movement	Front	108	0	3.06%	93.97%	-35.0%	17.10%	-12.39%
	Back	102	0	-0.10%	66.23%	-59.2%	10.89%	-18.37%

Generally, when sitting down, the skin of the legs demonstrates a tendency to expand, and the expansion trend of the skin on the front of the human body (6.99%) is relatively more obvious than the back (1.91%). The skin elasticity of the human legs is the same.

When the one-handed pitching movement is performed, the body surface skin of the leg on the side of the pitching arm tends to expand (17.30%). The expansion trend of the skin on the front of the human body (14.55%) is relatively more obvious than the back (5.50%); the side of the non-pitching arm also has a tendency to expand (3.30%), and the expansion trend of the skin on the back of the human body (4.24%) is relatively more obvious than the expansion trend of the skin on the front (2.37%). The skin on the legs of the human body shows a significantly different expansion and contraction trend, and the extension of the skin on the surface of the leg on the side of the pitching arm is more severe.

3.3 Data Summary and Analysis

Based on the division of human body surface muscles, the extension of the muscles of the key parts of the waist and legs of wheelchair curlers during sitting and one-handed pitching is summarized in Fig.4.

In general, the extension rate of the skin in the same area often shows a certain degree of uniformity when different actions are performed. Skin that is more likely to have a violent expansion trend is prone to have a violent contraction trend, and the skin that does not easily produce a violent expansion trend at the same time is also hard to show a sharp contraction trend.

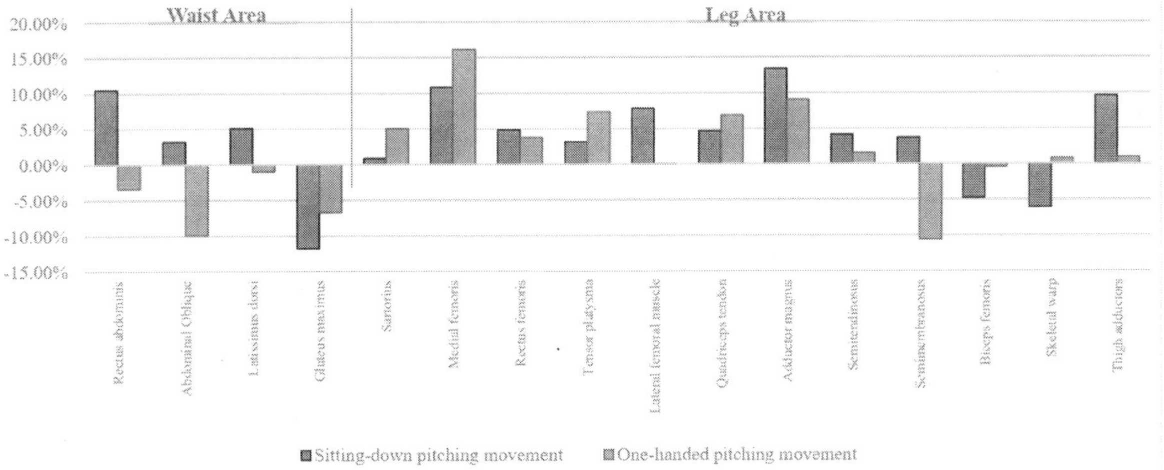

Fig.4 Skin extension of different areas of waist and legs

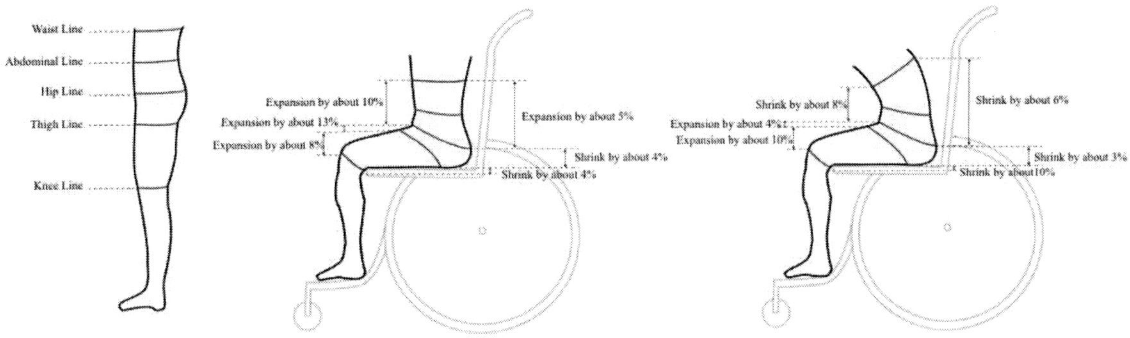

Fig.5 Diagram of skin extension

From the sub-regional analysis of the above data, visually shown in Fig.5, the skin on the front and back of the waist has a certain degree of elasticity, while the skin of the rectus abdominis between the front waist-line and the hip line shows the largest elasticity. As for the crotch, the areas of the front gear are more elastic. When wheelchair curling athletes wear normal trousers, they will feel tightness in the front crotch when they exercise. The knee area is another area where the skin is more flexible. The skin on the knees on the front of the human body usually expands significantly, while the back of the human body usually shrinks. Wearing normal trousers in this case will give people the discomfort of tightness on the front of the knee and many wrinkles on the back of the knee.

4. Adaptive Design

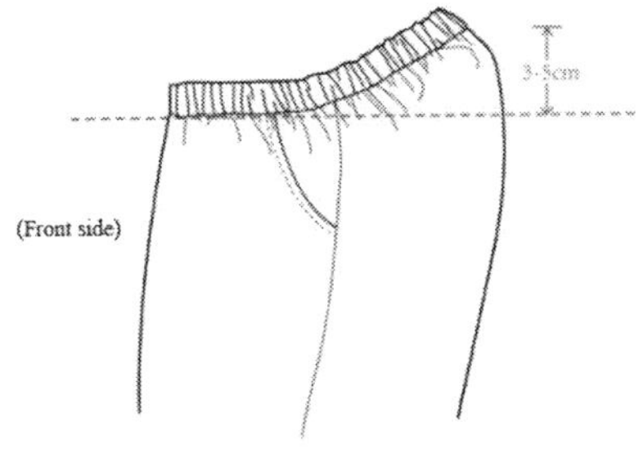

Fig.6 Adaptive waist structure

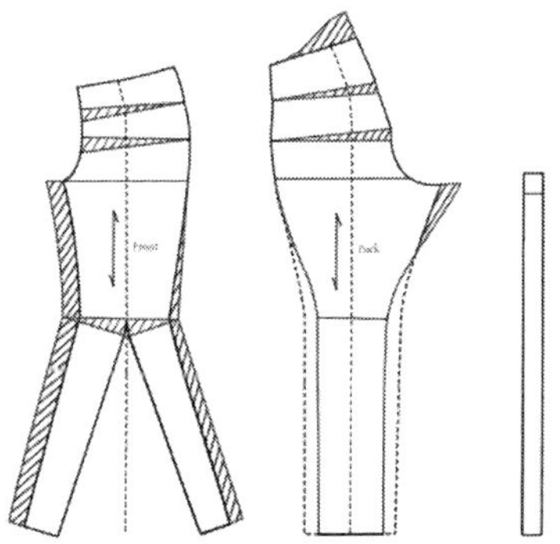

Fig.7 Adaptive pattern design

Based on the above data analysis, the adaptive design of trousers are applied to the skin extension of the waist and lower limbs of wheelchair curlers during sitting and one-handed pitching movements.

Firstly, the ribbed waist is more suitable for wheelchair curlers, as it is better adapted to the intense expansion and contraction of the skin on the athlete's waist and abdomen than the use of zipper. In addition, in order to prevent the waist of the trousers from moving downwards during the pitching action, the ribbed waist of the back piece can be raised to above the waist line, as is shown in Figure 6. Secondly, at the waist area, a certain degree of looseness should be added to the waistline of the trousers to ensure that the waist and abdomen of the curler will not feel tight when sitting and pitching. When calculating the amount of looseness, it is necessary to consider subtracting the original amount of pileup that would otherwise occur in a normal sitting state. Thirdly, for the crotch area, the curvature and ratio of the front and back crotch bends should be adjusted appropriately. The wider front crotch can reduce the tightness of the skin at the base of the thigh when it is expanded. Finally, when designing the pattern of the knee area, the lateral lines can be moved back appropriately, so the looseness of the knee circumference can be transferred to the front piece. In some cases, a lateral dart design can also be carried out at the knees of the back piece to greatly reduce the amount of accumulation of the back piece. The optimized design of trousers' pattern is shown in Figure 7 (the blue lines are the optimized contour line, and the shaded area is the added area).

5. Conclusion

This paper is divided into experimentation and application. Based on anthropometry, the paper combines body sampling and digital calculations to formulate an experimental program that can increase the accuracy of the results. The data is then analyzed, and applied into the adaptive structural design for wheelchair curlers.

Regarding the analysis of the skin extension rate of 4 subjects during the sitting and one-handed pitching movements, the paper draws the following conclusions: in general, the parts that are not easily deformed usually have no obvious stretching and shrinking trends during exercise, and the easily deformed parts are often prone to greater stretching or shrinking. The skin areas with great elasticity, such as the rectus abdominis area girth and the adductor magnus area of the crotch, show an average extension rate of more than 10% when the subjects perform sitting and pitching movements, which are the key areas for adaptive design. According to the experiment outcomes, the paper proposes the adaptive design of wheelchair curlers' trousers, the position that needs to be significantly reduced is the knee areas of the back piece; the ribbed elastic design is suitable for the waist; the design of the crotch can be adopted to increase the width of the back; the knee can be adopted to move the side stitches back and the design of the horizontal cutback near the knee circumference of the back piece.

Based on the above conclusions, the paper suggests that future related studies can continue to carry out pattern design and try-on evaluation, and finally develop the trousers suitable for wheelchair curlers.

6. Acknowledgement

The authors wish to acknowledge National Key R&D Program of China: Construction and demonstration on accessible, convenient and intelligent life service system(No.2019YFF0303300); Subject IV: Universal apparel and accessories research and practice based on body and sporting features of disabled people (2019YFF0303304).

References

[1] Zhuang MH, Li B, et al. Study on the "Three Tune" Training of Chinese Wheelchair Curler [J]. Winter Sports: 2018; 40 (02): 13-16.

[2] Jiang L. The research and development of Chinese wheelchair curling [D]. Harbin Engineering University: 2017.

[3] CURTEZA, Antonela, et al. De- signing functional clothes for persons with loco- motor disabilities [J]. AUTEX Research Journal: 2014; 14 (4): 281-289.

[4] Hong Y, Zeng X, Bruniaux P, Liu K, Chen Y and Zhang X. Collaborative 3D-To-2D Tight-Fitting Garment Pattern Design Process for Scoliotic People [J]. Fibres & Textiles: 2017; 25: 113-117.

[5] BRAGANCA S, CASTELLUCCI I, GILL S, et al. Insights on the apparel needs and limitations for athletes with disabilities: The design of wheelchair rugby sports-wear [J]. Appl Ergon: 2018; 67: 9-25.
[6] Wu XM. The Research on Wear Off for Physically Disabled Clothing [D]. Jiangnan University: 2009.
[7] Nakić, M, Bogović S. Computational Design of Functional Clothing for Disabled People. [J] TEKSTILEC: 2019; 62: 23-33.
[8] Zhu N. Pattern structural research on wheelchair user's trousers [D]. Beijing Institute of Fashion Technology: 2008.
[9] Department of clothing, Institute of human engineering. New quilt and human body [M]. Japan press: 1981: 95-111; 78.
[10] Liu K, Cai X. A study on the relationship between the dynamic parameters of human body and the amount of clothing added [J]. Shanghai Textile Science and technology: 2001; (6): 54-56.
[11] Wang WP. Study on fitted trousers for business women in Shanghai [D]. Donghua University: 2007.
[12] Luo L, Wang JP. Skin deformation of legs in cycling based on human motion capture (HMC) [J]. Journal of Zhejiang textile and clothing vocational and technical college: 2014; 13 (04): 39-43.
[13] Choi S, Ashdown SP. 3D body scan analysis of dimensional change in lower body measurements for active body positions. Textile Research Journal: 2011; 81 (1): 81-93.
[14] Guan XF, Deng YM. Study on the Structure Design of Lower Limb Disabled Pants. [J] Tianjin Textile Technology: 2018 (01); 9-13.
[15] Luo SH. Skin Deformation of Male Lower-Limb Based on Motion Function for Cycling Shorts [J]. Journal of fashion: 2017; 2 (05): 13-18.

Rethinking Critical Processes for Combating a Pandemic

Shaghayegh R. Arangdad[1*], A. Blanton Godfrey[2]

[1,2]*North Carolina State University, 1020 Main Campus Drive, Raleigh, North Carolina, 27695, USA*

[*]Corresponding author's email: srezaei@ncsu.edu

Abstract

Over the past twelve months, we have been intensely engaged in research driven by the COVID-19 pandemic in the United States. Our current research is focused on two related areas: the vaccination rollout process in the United States and the management of COVID-19 in major universities throughout the world. The US Federal Government decided the method of administration of vaccines in each of the 50 states. This research uses modern benchmarking processes across the 50 states to discover best practices and recommends high-performing process designs that can be widely replicated in the US and other countries.

Keywords: COVID-19; Pandemic; Personal Protective Equipment; Vaccination Rollout

1. Introduction

In November 2019, COVID-19 first appeared in China [1]. Soon after, in mid-January 2020 [2, 3], Thailand and Japan became the first and second countries outside of China to confirm COVID-19 cases In quick response, the screening of travellers started at three US international airports: Los Angeles (LAX), San Francisco (SFO), and New York (JFK) on January 17, 2020 [4]. The first COVID-19 case in the United States was detected in Washington State on January 20, 2020 [5]. The virus then started spreading rapidly across the country. The US was hit the hardest out of any country in the [6]. The number of deaths rose quickly, and in just four months, late-May 2020, the number of deaths had reached 100,000 [7]. At the end of March 2021, the global total number of cases was approximately 129 million resulting in 2.82 million deaths. Roughly 20% (0.552 million) of these were in the US [8].

Since January 2020, and driven by the Pandemic in the United States, we have been continuously involved in research.. With the support of over thirty graduate students and faculty colleagues in textiles, industrial and systems engineering, management, and other fields we have focused first, on the major shortcomings in the supply chain management system for Personal Protective Equipment (PPE) and other critical supplies. Our research has provided fundamental insights which were included in recently published papers that recommendedmajor changes in the U.S. Strategic National Stockpile and proposed improvements in data governance. Two other papers that described the quick response by US textile and apparel companies who had pivoted to PPE production as well as compared the US' response with other countries had recently been completed and submitted for publication. These rapid increases in production have been supported by university research which led to breakthroughs in technology for N95 respirators and other critical equipment.

Our current research is focused on two related areas: the vaccination rollout process in the United States and the management of COVID-19 in major universities throughout the world. The US Federal Government chose to delegate how vaccinations would be administered to each of the 50 states. Some states, including North Carolina, further delegated the vaccination process to the counties. Public Health resources in most states are understaffed and have limited experience in major health initiatives. The resources and public health skills in individual counties are even more limited. This situation has created hundreds of different processes for vaccination rollouts with extreme variations in process performance. Our research uses modern benchmarking processes across the 50 states and thousands of counties to discover best practices and recommends high-performing process designs that can be widely replicated in the US and other countries.

The variation in management of COVID-19 on college and university campuses has produced results that are extreme. Some campuses have had zero infections while others have had thousands. Comparisons of the results on different campuses and by different universities does not show that there were any surprising ideas or brilliant strategies. Rather, our research shows how different management processes and strong leadership produces significantly different results.

Shortages in PPE, testing capabilities, and basic supplies for the vaccination process still hinder the efforts to reduce COVID-19 infections in the United States. We believe our research will provide a strong foundation for a more systems-orientated approach to reducing some of these problems.

2. COVID-19 and Critical Challenges in Process

In this section we look at three key challenges: the shortages of critical Personal Protective Equipment, various stumbles in the United States' vaccination rollout and the management of COVID-19 on university campuses. Although these three challenges appear quite different, we found many commonalities. We were hoping to discover some radical new practices that could be shared widely. Instead, we found that the best practices were not great secrets. The wide differences were the results of failures in implementation of these practices.

2.1 PPE Shortages

As the coronavirus pandemic rapidly spread around the world, healthcare systems in many different countries quickly discovered major shortages of critical equipment. During COVID-19, the production and distribution of PPE for healthcare organizations and frontline workers became one of the main public health challenges.

There are a number of factors that contributed to the massive shortages of PPE. One of these factors was just in time delivery. Many healthcare organizations and their purchasing bodies had a tradition of just in time delivery [9]. Products were only manufactured in quantities that covered the predicted normal demand with little capability to increase production if needed [10]. The global nature of the medical supply industry is another factor that contributed to the current shortages. Around 50% of masks used in the world are made in one country: China. As the virus hit China, their production was quickly halted by the lockdowns which caused significant shortages of masks globally [11].

In addition, hospitals and other healthcare providers were struggling with extremely high demand for PPE products. The drastic increase in demand significantly affected the supply chain network causing many shortages of supplies and many problems in distribution. The national stockpile was exhausted quickly, and most of the hospitals only had enough PPE products to last a few days when COVID-19 hit their regions. Most hospitals traditionally relied on third-party organizations for supply chain management. These organizations focused mainly on low-cost supplies that were primarily manufactured outside of the US. Many offshore suppliers diverted their production based on their country's needs. Several US companies who supplied a significant percentage of PPE also realized disruption in some parts of their supply chains [12].

2.2 Vaccination Rollout

We also explored the different distribution practices, methods, and policies across the states (and counties) in the U.S. and around the world to identify best practices that could improve the vaccine distribution process and reduce the numbers of cases and deaths related to COVID-19. There were two extremely important areas we felt were beyond the scope of our research focus. The first was the different detection approaches used to identify the presence of the virus. The importance of biosensors in detecting the presence of the virus and controlling the spread cannot be overemphasized. Aziz, Asif, Ashraf, Farooq, Yang and Wang provide an outstanding review of trends in biosensing platforms for SARS-CoV-2 detection and a critical appraisal of standard detection tools [13]. The second important area beyond the scope of this paper that should be addressed is the efficacy of the vaccines. The effectiveness of the different vaccines administered to various populations throughout the world is still a subject under intense study. However, recent research combining the results of clinical trials and early findings from vaccine rollouts show extremely promising results. An updated report (July 1, 2021) by Olliaaro, Torreele and Vaillant in Lancet summarizes how the difficulties in comparing vaccines are due to disparate study protocols, types of placebos, study populations, and background risks of COVID-19 [14]. Despite these difficulties, these researchers and others at the CDC in the U.S., together with othervarious research centers found that the vaccines currently used are highly effective at preventing hospitalizations and deaths due to COVID-19.

Although the availability of vaccines provided hope to millions of people, the distribution of the vaccines was not done well in many localities. Many countries saw a surge of new cases and more deaths, forcing them to go into lockdowns once again after almost a year since the first wave of the pandemic. There have been over 1.5 million deaths since December 2020, which was when the vaccine distribution began. With more than 1.5 million lives lost due to COVID-19, the world leaders at the UN General Assembly requested urgent action to guarantee reasonable distribution of life-saving vaccines [15]. In the U.S., the CDC recommended a 7-day goal of 100% administration of available vaccine doses for all vaccine sites [16].

In many places in the world the vaccines are in short supply since the manufacturers cannot produce enough to satisfy worldwide demand. In our study to understand the vaccine distribution strategies and therefore, to suggest ways to improve the vaccine distribution, qualitative and quantitative data were collected to create a model that can be referenced in order to refine vaccine rollout.

The initial qualitative analysis was conducted to identify major causes of the vaccine distribution problems. Six different groups were categorized (Method, Facilities, Process, Availability, People and Vaccine) as shown in Fig.1.

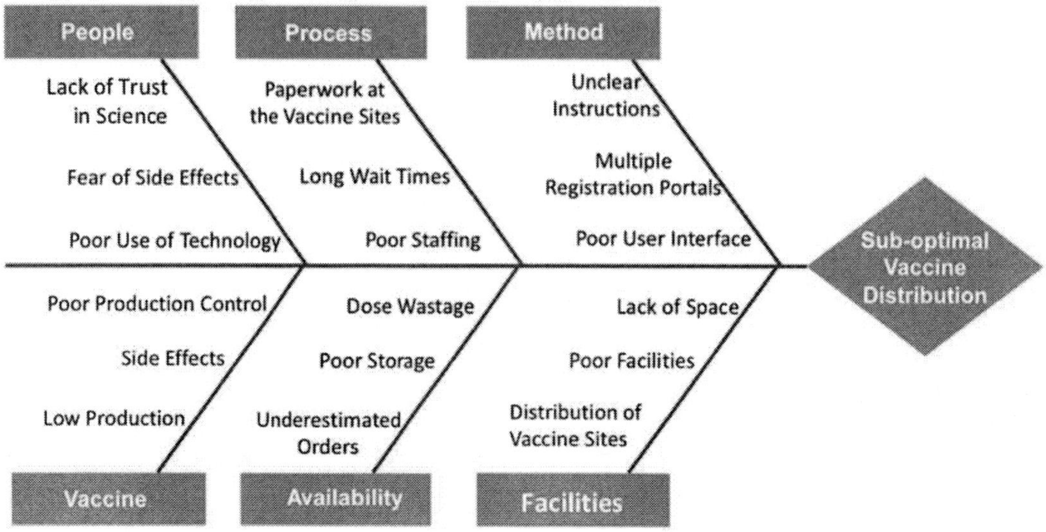

Fig.1 Cause-and-effect diagram regarding sub-optimal vaccine distribution

Due to lack of available data, comparisons of vaccine distribution strategies between countries were not feasible in detail, so the scope of our study was narrowed to a sample of states from the United States. Eighteen states were chosen at random.

Data were obtained from the state government websites and credible news articles. Quantitative analyses were conducted based on data obtained from a GitHub database. This data was compiled on research done at John Hopkins University [17, 18]. The data gathered from this source was collected from January 12, 2021 to April 13, 2021. The database included information such as share distributed, total vaccinations given, and vaccinations/100 people. These data became the basis of our analysis and gave insight into how effectively the states were administering the vaccines they were allocated. Based on our analysis of the eighteen states and areas selected, the District of Columbia, New Mexico and Rhode Island were identified as the best performing states.

States with the best performance have maximized use of their share of doses provided by the Federal government. States with smaller populations like Rhode Island performed much better than states with large populations, e.g., California and Texas. This appeared due to the state authorities having better connections with the local communities and they were able to mobilize the vaccination rollout quickly. Those states which had centralized vaccination distribution systems such as New Mexico were more efficient as they eliminated intra-county conflicts. Federal pharmaceutical partnerships with Walgreens and CVS were effective in some states, e.g., New Mexico and Rhode Island, but fell short in other states, e.g, California and Arizona. Based on our findings a base model was created to improve the vaccination rollout system (Fig.2).

REGISTRATION

- Centralized system
- Easy to access and Use
- Multiple languages
- Auto book appointment for 2nd Dose
- Having alternative methods for booking

AVAILABILITY

- Use all approved vaccines
- Track the doses given
- Consideration of local population
- With the people taking 1st doses also order 2nd doses based on that data

FACILITIES

- Accessible
- Drive-thru or walk-in capabilities
- Capable of meeting storage requirements
- Entering and existing sample/efficient
- Different channels for the elderly people high/risk
- At-home when travel isn't feasible

PROCESS

- Directions easy to follow
- Personnel qualified
- Forms done beforehand
- Fully staffed
- 1st and 2nd appointment done at the same facilities

Fig.2 Baseline improvement model for vaccination rollout

2.3 Management of COVID-19 in Universities

The COVID-19 pandemic has had a profound effect on educational structures across the globe. The pandemic has caused educational institutions to make sudden and drastic changes, shaking up the status quo and spawning an entirely new educational system over the past year. Higher education institutions have had to be especially innovative with engaging students and faculty and pushing forward with the continued delivery of classes. This continuation, despite the numerous and varied efforts by universities to stay as safe as possible given the circumstances, has come at a cost. The COVID-19 pandemic has claimed over 573,000 lives in the United States. College campuses have been hotspots for outbreaks; over 660,000 cases have been reported across American universities to date [19]. Some universities have seen a much higher number of cases than others. Others have been able to successfully control the spread of the virus within their campuses.

The primary research was based on the study done by New York Times on approximately 1,900 universities across the U.S. To narrow it down and selecting the universities to research, four major factors (number of cases, university's size, location, available data) were considered. [19]

While at first glance, it may seem that many universities in the US mimic each other or at least overlap in terms of the type or manner of implementation of safety protocols against the spread of COVID-19, there are several key variables that appeared to be especially significant in distinguishing a school's preparedness both long and short term. Based on our analysis there were a few key tactics/ qualities that distinctly characterize the universities with low COVID-19 spread: multiphase clear strategic procedures, accessible and actively updated online information databases, ultimate transparency in decision making and policy formation, rapid response and subsequent enforcement, and finally, availability of physical resources that allow for continuous testing, transition to remote learning/ instruction, and social distancing.

San Jose State University is an excellent example of a university that, despite having a relatively large student population, was able to keep their COVID-19 spread extremely low [20]. San Jose State University ("SJSU") is not only a prime example of a university offering extensive information to its campus population regarding the spread of COVID-19 among its students and faculty, the information provided is also presented in a designated area on the campus website for optimal accessibility. The SJSU COVID-19 Dashboard offers an extensive yet digestible interface for students and the public to view up-to date information regarding when and where COVID cases have been reported on and off campus [21].

Columbia University is another university that stands apart in its capacity to control the spread of COVID-19. It serves as an example of the complete significance of consolidated leadership providing resources for student/ faculty testing and subsequent publishing of testing results and overall accessible information on appropriate protocols [22].

The University of Nevada is another example demonstrating a distinct ability to control and mitigate the spread of COVID-19. They have offered a unique tracing application to its students. Working in conjunction with the state of Nevada, the university advertises the use of a contact tracing application: COVID Trace [22]. The application tracks an individual's location and, using real time data, will notify them of possible exposure to COVID-19 based on contact tracing. Not only will the application notify the individual without compromising the identity of any party who may have tested positive, it also provides guidance to the user as to the next steps in terms of appropriate quarantine measures and locations for further testing [23].

Based on our study, five key factors were identified that could be indicators of low spread (Fig.3). These factors are the local response from the states and countries within which the universities are located. An early, methodical, and persistent response from states, countries, or the university system to which a college belongs to played a key role in the controlling spread. For example, because the state of New York was one of the first states to see large outbreaks of the virus in the U.S., their preparedness and caution came much earlier than the other states, potentially showing why Columbia University's cases were relatively low. Testing strategies implemented by universities were critical. Contact tracing, especially during the initial outbreaks on campus, was also key in controlling spread. This is a strategy that all universities adopted, but some were much more aggressive and innovative in carrying it out. The National University of Singapore made use of technology to carry out their contact tracing by creating apps that notified nearby cases. A reduced on-campus presence was one of the essential requirements of adapting to the pandemic. Lastly, monitoring gatherings on campus and intervening when they are deemed dangerous is vital. At NC State University, and at many universities in the high and medium categories of cases on campus, initial outbreaks were associated with Greek life and sports activities. Going forward, close monitoring of in-person activities is essential, and prohibiting them is even more critical. The National University of Singapore is relying on technology to monitor their communal spaces and make sure they stay at a safe capacity. They have also been imposing tough penalties for violations.

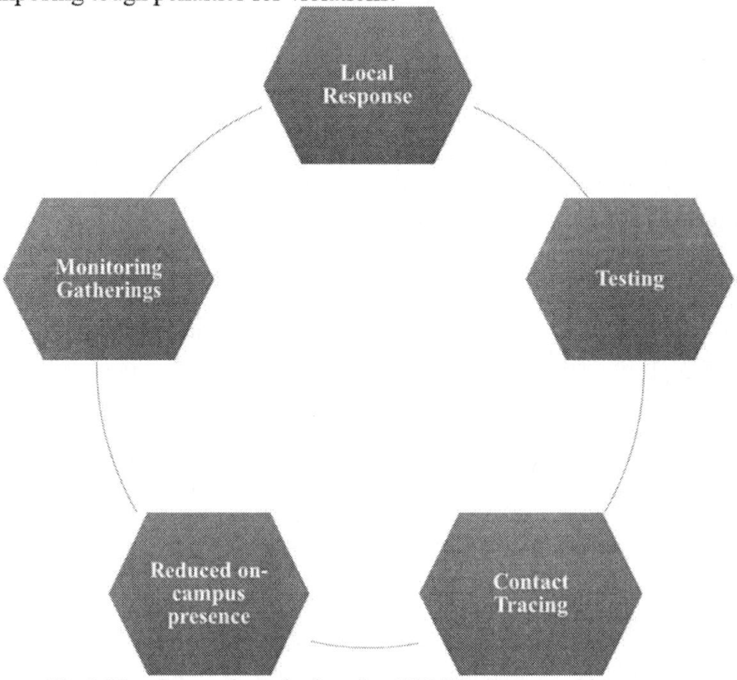

Fig.3 Key factors in reducing the COVID-19 spread in campus

3. Combating the COVID-19 Pandemic

In this study we have focused on research on three critical areas that were especially challenging: providing the critical Personal Protective Equipment to healthcare providers and the public, the methods and procedures used during the vaccination rollouts, and the spread of the virus on university campuses. We have found commonalities of best practices across all three areas. Leadership appears to be critical in controlling the spread of the virus in all three areas. It seems as simple as, "someone needs to be in charge." There are so many interconnected pieces that we must look at all the interactions whether it is the complex supply chains for PPE or for vaccination rollouts. Accurate and timely data are critical. For PPE, we need the right data to understand the demand and to understand the potential for supplies and distribution strategies. For vaccines, we need the data for supplies and actual vaccinations as well as what processes are working. For campuses, we need precise testing data, contact tracing data, and accurate information available to all. It is difficult to make correct decisions without quality data. A systems approach is critical. Just having one or two parts working well does not produce the results needed.

4. Discussion and Conclusion

But perhaps the most obvious result of our study was that the methods and strategies for combating a pandemic are known. Rapid sharing of best practices and rapid adapting and implementing these best practices can dramatically reduce the impact of the pandemic. There was nothing we found in the shortages of PPE that could not be quickly alleviated by using methods created in other locations and governments. Many of the vaccination rollouts improved quickly, often by trial and error, but a systems approach in the beginning based on successful rollouts in other locations could have improved performance considerably. All universities we studied were using similar technologies and strategies and, yet, the performance ranged from almost ten thousand cases on one campus to one or two cases on the best performing campuses. The differences were not in the methods and technologies but in the implementation.

5. Acknowledgement

The authors wish to acknowledge the work of a number of graduate students in our classes whose work informed much of this paper. Much of the original work on the shortages of PPE and the challenges of the PPE supply chains in the U.S. was done by nine graduate students first as projects in a graduate course we teach and who then continued the work throughout the summer of 2020. They have submitted research papers based on their work that are part of our references. Two other groups of graduate students studied the vaccination rollouts in the U.S. and university campuses responses to the pandemic during a graduate course this Spring semester. These students deserve much credit for many of the insights and facts used in this paper. The students working on the vaccination rollout processes were Edgar Garcia Torres, Het Shah, Jazmen Gary, Mohamed Mohamed, Nilu Rajendran and Vaishnavi Yadavalli. The team studying the pandemic on campus were Ananya Suresh, Monika Bhati, Payal Ashtankar, Pravesh Agrawal, Siddhant Patel, and Wesley Brown.

References

[1] Scher I. The first COVID-19 case originated on November 17, according to Chinese officials searching for "patient zero." Business Insider: 2020; March 13. Information on https://www.businessinsider.com/coronavirus-patients-zero-contracted-case-november-2020-3.

[2] Joseph A. Woman with novel pneumonia virus hospitalized in Thailand—The first case outside China. STAT: 2020; January 12, Information on https://www.statnews.com/2020/01/13/woman-with-novel-pneumonia-virus-hospitalized-in-thailand-the-first-case-outside-china/.

[3] Schnirring L. Japan has 1st novel coronavirus case; China reports another death. CIDRAP: 2020; January 16. Information on https://www.cidrap.umn.edu/news-perspective/2020/01/japan-has-1st-novel-coronavirus-case-china-reports-another-death.

[4] CDC. Coronavirus Disease 2019. Centers for Disease Control and Prevention: 2020; February 18. Information on https://www.cdc.gov/media/releases/2020/p0117-coronavirus-screening.html.

[5] Holshue ML, DeBolt C, Lindquist S, Lofy KH, Wiesman J, Bruce H, Spitters C, Ericson K, Wilkerson S, Tural A, Diaz G, Cohn A, Fox L, Patel A, Gerber SI, Kim L, Tong S, Lu X, Lindstrom S, Pillai SK. First Case of 2019 Novel Coronavirus in the United States. New England Journal of Medicine: 2020; 382(10), 929–936 https://doi.org/10.1056/NEJMoa2001191.

[6] Winsor M, Shapiro E, Deliso M. US reaches 100,000 coronavirus deaths. ABC News: 2020; April 27. Information on https://abcnews.go.com/Health/coronavirus-updates-covid-19-cases-us-health-

care/story?id=70897864.
[7] WorldOMeter. (n.d.-a). Coronavirus Update (Live): 100,284,461 Cases and 2,149,460 Deaths from COVID-19 Virus Pandemic - Worldometer. Retrieved January 26: 2021, Information on https://www.worldometers.info/coronavirus/.
[8] Johns Hopkins University & Medicine. (n.d.). COVID-19 Dashboard by the Center for Systems Science and Engineering (CSSE) at Johns Hopkins University (JHU). Johns Hopkins Coronavirus Resource Center. Retrieved January 27: 2021, Information on https://coronavirus.jhu.edu/map.html.
[9] Rondinone D, Godfrey AB, Anil Khamkar A, Padmanabhan Anilkumar A, Jose A, Rucha Girgaonkar R, Arangdad S. The United States Supply Chain Challenges for Personal Protective Equipment (PPE) during the COVID-19 Pandemic. Submitted to Journal of Textile and Apparel, Technology and Management: 2021.
[10] Patel A, D'Alessandro MM, Ireland KJ, Burel WG, Wencil EB, Rasmussen SA. Personal Protective Equipment Supply Chain: Lessons Learned from Recent Public Health Emergency Responses. Health security: 2017; 15(3): 244-252 Information on https://doi.org/10.1089/hs.2016.0129.
[11] Wang X, Zhang X, He J. Challenges to the system of reserve medical supplies for public health emergencies: reflections on the outbreak of the severe acute respiratory syndrome coronavirus 2 (SARS-CoV-2) epidemic in China. Biosci. Trends: 2020; 14 (1): 3-8.
[12] Arangdad S, Godfrey AB, Hoque A, Joshi P, Prasath Manoharan A, Vhanmane A. COVID 19: Situation, action, and tracking of PPE demand in the US. Submitted to Journal of Textile and Apparel, Technology and Management: 2021.
[13] Aziz A, Asif M, Ashraf G, Farooq U, Yang QL, Wang SQ. Trends in biosensing platforms for SARS-CoV-2 detection: A critical appraisal against standard detection tools, Current Opinion in Colloid & Interface Science: 52: 101418: 2021.
[14] Olliaro P, Torreele E, Vaillant M. COVID-19. Vaccine efficacy and effectiveness - the elephant (not) in the room. The Lancet, Comment, Volume 2, Issue 7, E278-E280: July 01; 2021.
[15] The United Nation. 2020, December 3rd. Information on https://www.un.org/press/en/2020/ga12293.doc.htm
[16] Center for Disease Control and Prevention. U.S. COVID-19 Vaccination Program: Vaccine Channel Portfolio by Jurisdiction. Center for Disease Control and Prevention: 2021; April 25; Information on https://www.cdc.gov/coronavirus/2019-ncov/vaccines/distributing/jurisdiction-portfol ios.html.
[17] Git Hub. (n.d.). covid-19-data. Git Hub. Information on https://github.com/owid/covid-19-data/tree/master/public/data/vaccinations.
[18] Git Hub. (n.d.). COVID-19. Git Hub. Information on https://github.com/CSSEGISandData/COVID-19/tree/master/csse_covid_19_data/ csse_covid_19_daily_reports.
[19] Tracking Coronavirus Cases at U.S. Colleges and Universities Information on https://www.nytimes.com/interactive/2021/us/college-covid-tracker.html).
[20] San Jose State University's Quick facts, Information on http://www.iea.sjsu.edu/Students/QuickFacts/.
[21] San Jose State University, COVID-19 Dashboard, Information on https://pdp.sjsu.edu/healthadvisories/covid19-dashboard/index.php.
[22] Columbia University's COVID-19 Testing Program. Information on https://covid19.columbia.edu/content/covid-19-testing-program.)
[23] Contact Tracing. Information on https://news3lv.com/news/local/nevada-launches-covid-trace-contact-tracing-app.

Advances in Intestinal Stents

Yu-Sheng Feng[1], Xu-Sheng Xie[1], Gang Li[1*]

[1]*National Engineering Laboratory for Modern Silk, College of Textile and Clothing Engineering, Soochow University, Suzhou 215123, P.R.China*

*Corresponding author's email: tcligang@suda.edu.cn

Abstract

This paper introduces the incidence, clinical manifestations and limitations of traditional treatment for colorectal cancer (CRC). The necessity, indications, complications and future development trend of stent-assisted therapy, and the performance requirements of intestinal stents were described. The advantages and disadvantages of different materials and processing techniques of intestinal stents were analyzed. In conclusion, the application of new materials, new drug delivery methods and new 3D textile structure design are of great significance to the development of intestinal stents in the future.

Keywords: Intestinal Stents; Colorectal Cancer; Medical Textiles; Medical Devices; Research Status

1. Introduction

1.1 Intestinal Obstruction

Intestinal obstruction is generally caused by adhesions of the small intestine, colorectal cancer (CRC), and paralytic and spastic intestinal wall muscle dysfunction. It is classified into benign obstruction and malignant obstruction. Intestinal obstruction caused by non-tumor causes is generally classified as benign obstruction, including small bowel volvulus, adhesions, diverticulitis, intussusception, and hernia. The disease is usually accompanied by abdominal pain, vomiting, fever and shock caused by severe secondary obstruction, which poses a significant threat to the life of patients. Malignant CRC obstruction (MCO) is a hazard caused by CRC. Patients often need urgent intestinal decompression; otherwise, a series of conditions will follow such as abdominal distension, abdominal pain, loss of body fluids and electrolytes, intestinal circulation disorders as well as necrosis, and secondary infections. This can lead to death in severe cases [1]. According to the Statistics of the World Health Organization in 2018, CRC ranks third and fifth in the incidence and mortality of cancer in the world, respectively [2].

1.2 Invasive Surgical Therapies

1.2.1 Traditional Therapy

In clinical practice, traditional emergency surgical treatments such as small bowel resection, intestinal bypass surgery, endoscopic balloon dilation, and laparoscopic adhesion removal are usually used to alleviate or eradicate intestinal obstruction. However, this may lead to multiple surgeries, with a high probability of various complications, bringing economic and psychological burden to the patients [3]. The introduction of stent therapy has improved this situation.

1.2.2 Stent Therapy

Since the first clinical treatment of self-expanding metal stents (SEMS) [4], CRC stents are now widely used in the treatment of malignant CRC obstruction. Stents implantation not only serves as palliative therapy for non-resectable CRC, but a bridge to resectable CRC surgery, resulting in fewer hospital stays, fewer complications, and lower short-term mortality than emergency surgery. In addition, for CRC patients whose obstruction can be removed by surgical resection, the implantation of a stent allows the intestines to be fully prepared, enabling more time for a thorough stag assessment of cancer and preoperative adjuvant chemotherapy [5].

1.3 Non-invasive Therapies

In addition to invasive surgical therapies, non-invasive therapies such as chemotherapy and immunotherapy can also be used alone or in combination with surgical therapies for intestinal obstruction.

When CRC develops into colorectal obstruction, the use of adjuvant chemotherapy can also reduce the tumor volume as much as possible before surgery, reduce the risk of tumor cell abscission, provide a greater possibility for subsequent complete tumor resection, and reduce the incidence of ostomy and complications [6].

Since the discovery of Programmed cell death protein 1 (PD-1) and cytotoxic T-lymphocyte-associated protein 4 (CTLA-4) associated pathways, significant clinical advances have been made in the treatment of melanoma, renal cell carcinoma and non-small cell lung cancer with associated monoclonal antibodies [7]. By blocking this pathway, CRC has been successfully treated clinically [8]. Immunotherapy provides a new therapy direction for malignant obstruction caused by CRC.

2. Classifications of Intestinal Stents

2.1 Stents of Different Materials

Currently, plastic stents are mostly used for liver, gallbladder and pancreas [9], and less used for intestinal stents. Commercial intestinal stents generally use metal materials [10, 11]. Metal CRC stents, such as SEMS, are safe, reliable, low-risk, and traumatic. It has the advantages of small volume, rapid relief of intestinal obstruction, fewer complications, and a wide range of adjustable design parameters with unique flexibility [12]. Biodegradable metal stents made of magnesium-based, zinc-based and iron-based materials are also appeared [13]. However, metal stents also have some disadvantages, such as the influence of nuclear magnetic resonance results, the long-term implantation of the stent will lead to perforation, improper mechanical properties and other problems. Even worse, they may also induce tumor cells to invade peripheral nerves [10, 14].

The emergence of polymer stents largely makes up for the shortcomings of insufficient biodegradability of metal stents due to good biocompatibility and mechanical properties [15]. Biodegradable materials such as polyglycolic acid (PGA), polylactic (PLA), polylactic-glycolic acid (PLGA), polydioxanone (PDO), polyurethane (PU), and polycaprolactone (PCL) are often used to manufacture stent [16, 17]. As a natural polymer material with good biocompatibility, silk fibroin has been used in the manufacture of fibers, films, microspheres, non-woven fabrics, and other materials and tissue engineering stents in recent years, and has also been applied in the recent research on CRC stents [18, 19]. In general, the application of polymer stents in the treatment of CRC obstruction is very promising, but more improvements are needed in terms of materials.

2.2 Stents of Different Structures

Most commercial stents are reticular structures woven with metal wires, including the Wallflex® stent, Ultraflex® stent, Hanaro® stent, etc. Fig.1 shows intestinal stents with different structures. Bare stents can provide continuous radial support and good visibility to the intestine, facilitating the restoration of intestinal patency. These stents have the advantages of fewer complications after implantation, low slip rate, long opening time, and easy implantation in vivo. However, the implantation of uncoated stents is prone to problems such as the continued growth of tumor tissue into the stent, which may lead to re-obstruction [10]. The coated stent refers to a stent coated with a membrane material on a traditional bare stent. The mulching film acts as a barrier inside the stent, eliminating the potential threat of secondary obstruction caused by tumor cells spreading inward through the mesh. However, the anchoring force of the coated stent is low, and it is easy to slip after implantation [19].

2.3 Stents of Different Functions

2.3.1 Drug-eluting Coated Stents

The outer layer of bare stents is equipped with drug membranes with treatment or anti-intimal hyperplasia. When the stent is implanted, the drug is released to the stenosis by local elution, which makes the embedding drug target close to the tumor tissue, continuously release on the surface of tumor tissue, and achieves sustained-release function. This kind of stent has the ability of intracavitary repair and intracavitary sustained anti-tumor [16, 20].

2.3.2 Radioactive Stents

Traditional bare stents (including metal and plastic stents) generally can only expand the intestines and relieve obstruction, but cannot effectively inhibit, kill, and prevent metastasis of tumor tissue. In recent years, the use of radioactive particles to assist in the treatment of obstruction has received much attention and research, such as 125I, 188Re, 32P, 103Pd, etc [21, 22]. During the past few years, there have been reports of the treatment of CRC by implanting radioactive particles [23], but there are still many problems, including no relevant standards for the dose, method and stage of radioactive particle implantation, lacking relevant research reports and clinical trials.

2.3.3 Photodynamic Therapy Stents

Diamond et al [24] proposed the use of photodynamic therapy for malignant tumors. Since then, photodynamic therapy, which is different from the traditional therapy, has gradually entered into practical application. A recent study [25] manufactured a photodynamic therapy stent by embedding photosensitizer into the stent and found that the photodynamic activity of the stent was maintained for at least 2 months. What's more, the repeated photodynamic treatment was successfully realized, and the apoptosis and degradation of tumor cells were induced, which suggested that the combined photodynamic therapy for CRC was feasible.

3. Property Requirements

3.1 Mechanical Properties

Colorectal stent not only needs the proper supporting force to expand the intestine without causing perforation, but also needs to adapt to the bending stress caused by the continuous peristalsis of the intestine, so that the intestine remains unobstructed without discomfort. The mechanical properties of intestinal stents can generally be divided into radial force and axial force. Radial force can be further subdivided into flexibility that resists expansion from the inside of the intestine to the edge, continuous expansion force, and super-elastic force that can maintain its shape when subjected to external pressure. The axial force is defined as the force to maintain a straight line in the curved intestine, which requires the stent not only to bend properly with the intestine, but also to maintain enough bending adaptability without detachment [20, 26].

3.2 Biocompatibility

The stent material is considered not to cause tissue rejection, including abnormal immune response and inflammation. At the same time, when the stent is implanted in the intestine, it is inevitable that a small amount or a small part of the stent will be degraded, so it is necessary to ensure that the degradation products of stent material can be discharged smoothly from the body and have no adverse effects on the human body [27].

3.3 Multifunctions

The coating stent is essentially a combined device, which is composed of a bare stent and a drug-loaded coating on the surface of the stent. This stent has excellent dual functions, not only can provide good support and expansion performance, but also can release drugs in the location specified, thereby directly contacting the affected area, inhibiting the proliferation of tumor cells, and effectively preventing restenosis in the stent, so it has better safety and therapeutic effect than traditional stents [28]. Due to the particularity of drug loading, the drug-eluting stent has certain requirements on the physical and chemical properties of polymer coating and carrier material system, including the ability to encapsulate drugs efficiently, biocompatibility, controlled drug release performance, etc [10]. Biodegradable synthetic polymers such as PGA, PLA and PLGA have been approved by the US Food and Drug Administration (FDA) for clinical research. The research and development of degradable metals such as magnesium, aluminum and zinc have also been reported [29, 30].

4. Manufacture Technologies

4.1 Laser Cutting Technology

Because of the high degree of automation, low cost and recognized quality, the use of laser cutting technology to manufacture stents is a common method. However, in the traditional process, cleaning, grinding, pickling, passivation, electrochemical polishing and other cumbersome steps are needed after cutting, so how to improve the speed and fineness is the direction of laser cutting technology.

4.2 Braiding Technology

Table 1. Comparison of manufacture technology, structure and feature of stents manufactured by different materials

Material	Manufacture technology	Structure	In vitro/vivo	Feature	References
Nitinol	Braid	Single or double layer braided structure of bare or polyurethane(PU), polytetrafluoroethylene(PTFE) and other materials coated stent	In vitro & vivo	Excellent supporting intestinal performance; Higher flexibility and compliance; The surface of the stent can be functionalized by coating and drug elution; Non-biodegradable	[31, 32, 33]
Mg-Zn-Y-Nd alloy	Extrusion molding after hot processing	Biodegradable metal stent body, bare or outer polymer such as poly-L-lactic acid (PLLA) coated stent	In vitro & vivo	Better corrosion resistance than traditional stent; Drug-eluting targeted therapy; Excellent biodegradability and biocompatibility	[34, 35]
Polydioxanone(PDO)	Weft knitting; Electrostatic spinning	Single layer weft coil structure of bare or natural polymers such as silk fibroin coated stents	In vitro & vivo	Excellent mechanical properties, biocompatibility and biodegradability; Great drug slow - release and continuous intracavitary anti - tumor ability; Effectively prevent cancer tissue from growing into the intestine	[11, 16, 20]
Polylactic acid(PLA)	Weft knitting	Single layer weft knitting coil structure bare stent	In vitro & vivo	Better radial force than metal stent; Excellent biodegradability and biocompatibility; Slower degradation, good mechanical properties can be maintained for a long time	[36, 37]
Polycaprolactone/polydioxanone (PCL/PDO)	3D printing	Flexible length, diameter, bending design, 3D-printed structural stent for personalized dimensions	In vitro & vivo	Outstanding modulus of elasticity; Excellent biodegradability and biocompatibility; Different sizes and shapes can be manufactured according to requirements	[38, 39]

CRC stents are essentially tubular products and therefore can be manufactured using braiding techniques. At present, the new braiding technology can realize the manufacture of ultra-fine monofilament stents that cannot be achieved by laser cutting, and can provide better support after heat setting. Table 1 lists and compares the structure and features of intestinal stents manufactured by different techniques.

4.3 Knitting Technology

Knitting technology used for tissue engineering is divided into warp knitting and weft knitting [15, 16, 40]. The intestinal stent can be manufactured by selecting a cylindrical warp knitting machine with an appropriate caliber, the density of the needle teeth and repeating the loop forming process. The warp-knitted intestinal stent is not easy to be deformed and disassembled, and the structure is stable. The small diameter circular knitting machine can also be used to manufacture the intestinal stent. Due to the particularity of knitted fabric, the stent can be extended in all directions with good elasticity and one-time molding, which can effectively avoid the uneven strength of the stent, and the elastic recovery ability and overall stability are strong [16, 20].

4.4 Electrospinning

Electrospinning technology allows a new manufacturing method for drug-eluting stents. By putting the drug-loading solution into the syringe, the polymer is polarized, stretched, and solidified by a high-voltage electric field; the thin and uniform nanofiber film coating can be formed on the surface of the stent, which has excellent mechanical properties and is easy to be functionalized drug loading [20, 41]. This easy-to-operate and effective stent coating method have a good prospect for industrialization.

5. Key Issues, Development Opportunities and Challenges

5.1 Complications

Intestinal stent implantation has been considered a safe operation, with a mortality rate of less than 1%, but some complications are inevitable, such as failure of stent implantation, bleeding, perforation, stent slippage, and re-obstruction, etc. Tonolini and his colleagues [42] carefully analyzed the various early, middle and late-stage problems caused by the implantation of intestinal stents, elaborated on the causes of various complications from the root and suggested future solutions. With the progress of technology and the deepening of research, complications after intestinal stent implantation can be scientifically reduced, proving the safety of this technology.

5.2 Biomechanical Problems

In order to reduce the perforation, slippage and other complications that occur after the stent is implanted in the body; it is necessary to design according to the actual situation of the patient before implantation and to predict the biomechanical behavior before the implantation of the stent. A finite element simulation method was put forward colon stent, and it can accurately simulate the stent from design to molding of each step, including production process, implantation, release, and changes in bending and radial forces that may be encountered during use [43]. By simulating predictions, specific stents can be manufactured for patients with different needs.

5.3 Material Innovation

Biomedical materials have become a focus and hotspot of future research. They can not only be used for the repair, enhancement or even reconstruction of tissues and organs, but also have excellent biological structures and biological functions. Natural macromolecular biological materials mainly include silk fibroin, cellulose, chitin, and collagen, etc [44]. They can be used to make artificial skin, sutures, heart valves, artificial blood vessel polymer coating, etc. due to their excellent biocompatibility and degradability. Synthetic polymer biomaterials are more widely used, including PLA, PGA and PDO, etc. They have excellent fatigue resistance, high elasticity and biocompatibility, and have been used in a series of medical fields such as medical dressings, stent manufacturing and drug carriers [16, 18]. As for the manufacture of intestinal stents, biomaterials are bound to become the main force in the future due to their good mechanical, biodegradable and auxiliary drug sustained-release properties.

5.4 Structural Design and Manufacture Technology Innovation

A composite stent can be a combination of stent manufacture materials. In recent studies, magnesium particles were combined with PLLA to manufacture a biodegradable composite stent, which combined the excellent toughness and strength of magnesium and the biological characteristics of PLLA. On the other hand, it is a composite of the stent and the coating to form a coated or drug-eluting stent. The polymer-coated stent with excellent biological characteristics can play a more significant role in the treatment of intestinal obstruction in the future [45, 46]. In order to meet the requirements of different patients for stent size and structure, there are more and more reports of using 3D printing technology to make the stent [38]. Adhesive materials such as metal powders and polymers are used to manufacture stents with different parameters through three-dimensional design, layer-by-layer printing and post-processing steps, and the mechanical properties and biocompatibility of the manufactured stents meet the requirements. Individualized stent therapy plans can be customized for patients to meet different groups of people.

5.5 Functionalization

It is reported that directly injecting immune drugs into tumor tissues to achieve the effect of immune infiltration can be more efficient to trigger a greater immune response through local reactions [47]; this also suggests that drug-eluting stents can be used in combination with immunotherapy drugs to directly contact tumor tissues and target to activate more T lymphocytes' immune response to tumor cells, thereby obtaining a better effect of the intestinal stent in the treatment of intestinal obstruction.

A recent study [48] used a polymer-based combination system for gene/chemotherapy therapy of CRC. The good results suggested that the use of gene-loaded nanoparticles is a potential treatment for CRC. Xia et al. [49] studied the targeted therapy of CRC with nanoparticles carrying gene drugs, and induced apoptosis of human CRC cell HT-29 cells by silencing genes and activating related molecules. But the insertion accuracy of gene vectors, the controllability of gene expression in vivo, and the combination ability between the gene and stents need to improve. In order to improve the performance of the stent, a series of physical and chemical modifications can be made to the surface of the stent to enhance the durability, stability, fatigue resistance, and corrosion resistance of the stent itself and improve the biocompatibility of the stent. It can be coated by plasma spraying, low-voltage arc spraying, laser remelting composite coating, vapor phase physical or chemical precipitation coating [50]. The coating material can be inorganic materials such as TiO_2, ZnO, $CaCO_3$, or some biological materials such as PEG, chitosan, silk fibroin. The surface of the stent material can also be modified by mechanical, physical and chemical methods, such as changing the internal crystal structure to obtain the required reaction activity of the stent.

6. Conclusions

Combining stents for targeted therapy and eliminating obstruction is the developing trend. The intestinal stent is a combined system that includes mechanical expansion of the intestine and targeted treatment and inhibition of tumor tissue. The adaptability of the stent itself to the intestines, including physical and biological chemical properties, needs to improve further. The use of interdisciplinary technologies can effectively manufacture the stent using polymers as raw materials. In summary, the future intestinal stent is a comprehensive research topic on mathematics, physics, medicine, materials engineering and medical textiles.

References

[1] Shimura T, Joh T. Evidence-based Clinical Management of Acute Malignant Colorectal Obstruction. Journal of Clinical Gastroenterology: 2016; 50 (4): 273-285.

[2] Bray F, Ferlay J, Soerjomataram I, Siegel RL, Torre LA, Jemal A. Global cancer statistics 2018: GLOBOCAN estimates of incidence and mortality worldwide for 36 cancers in 185 countries. CA Cancer J Clin: 2018; 68 (6): 394-424.

[3] Xu J, Zhang S, Jiang T, Zhao YJ. Transanal drainage tubes vs metallic stents for acute malignant left-sided bowel obstruction: A systematic review and meta-analysis. Medicine (Baltimore): 2020; 99 (2): e18623.
[4] Wright KC, Wallace S, Charnsangavej C, Carrasco CH, Gianturco C. Percutaneous Endovascular Stents - an Experimental Evaluation. Radiology: 1985; 156 (1): 69-72.
[5] Mitra V, Hu M, Majumdar D, Krishnan V, Chaudhury B, Hancock J, Dwarakanath D. Safety and efficacy of self-expandable metal stents for obstructive proximal and distal large bowel cancer. J R Coll Physicians Edinb: 2017; 47 (1): 30-34.
[6] Li ZL, Wang ZJ, Han JG, Yang Y. Successful treatment of obstructing colonic cancer by combining self-expandable stent and neoadjuvant chemotherapy: A case report. World Journal of Clinical Cases: 2019; 7 (3): 335-339.
[7] Siddiqui I, Schaeuble K, Chennupati V, Marraco SAF, Calderon-Copete S, Ferreira DP, Carmona SJ, Scarpellino L, Gfeller D, Pradervand S, Luther SA, Speiser DE, Held W. Intratumoral Tcf1(+)PD-1(+)CD8(+) T Cells with Stem-like Properties Promote Tumor Control in Response to Vaccination and Checkpoint Blockade Immunotherapy. Immunity: 2019; 50 (1): 195-+.
[8] Overman MJ, McDermott R, Leach JL, Lonardi S, Lenz HJ, Morse MA, Desai J, Hill A, Axelson M, Moss RA, Goldberg MV, Cao ZA, Ledeine JM, Maglinte GA, Kopetz S, Andre T. Nivolumab in patients with metastatic DNA mismatch repair-deficient or microsatellite instability-high colorectal cancer (CheckMate 142): an open-label, multicentre, phase 2 study. Lancet Oncology: 2017; 18 (9): 1182-1191.
[9] Costamagna G, Tringali A, Perri V, Familiari P, Boskoski I, Barbaro F, Landi R. Endotherapy of postcholecystectomy biliary strictures with multiple plastic stents: long-term results in a large cohort of patients. Gastrointest Endosc: 2020; 91 (1): 81-89.
[10] Kim EJ. Stents for colorectal obstruction: Past, present, and future. World Journal of Gastroenterology: 2016; 22 (2).
[11] Li G, Chen Y, Hu J, Wu X, Hu J, He X, Li J, Zhao Z, Chen Z, Li Y, Hu H, Li Y, Lan P. A 5-fluorouracil-loaded polydioxanone weft-knitted stent for the treatment of colorectal cancer. Biomaterials: 2013; 34(37): 9451-61.
[12] McKenna CG, Vaughan TJ. An experimental evaluation of the mechanics of bare and polymer-covered self-expanding wire braided stents. Journal of the Mechanical Behavior of Biomedical Materials: 2020; 103.
[13] Wu JY, Zhao DL, Lee B, Roy A, Yao R, Chen SN, Dong ZY, Heineman WR, Kumta PN. Effect of Lithium and Aluminum on the Mechanical Properties, In Vivo and In Vitro Degradation, and Toxicity of Multiphase Ultrahigh Ductility Mg-Li-Al-Zn Quaternary Alloys for Vascular Stent Application. Acs Biomaterials Science & Engineering: 2020; 6 (4): 1950-1964.
[14] Haraguchi N, Ikeda M, Miyake M, Yamada T, Sakakibara Y, Mita E, Doki Y, Mori M, Sekimoto M. Colonic stenting as a bridge to surgery for obstructive colorectal cancer: advantages and disadvantages. Surgery Today: 2016; 46 (11): 1310-1317.
[15] Li G, Li Y, Lan P, He X, Hu H. Polydioxanone weft-knitted intestinal stents: fabrication and mechanics optimization. Textile Research Journal: 2013; 83 (20): 2129-2141.
[16] Li G, Li Y, Lan P, Li J, Zhao Z, He X, Zhang J, Hu H. Biodegradable weft-knitted intestinal stents: fabrication and physical changes investigation in vitro degradation. J Biomed Mater Res A: 2014; 102 (4): 982-90.
[17] Roldan FP, Carro PG, Garcia MCV, Rabih SA, Huidobro MLL, Munoz NSM, Garcia-Escribano OR, Ferrus MY, Martin EB, Carrillo FR. Usefulness of biodegradable polydioxanone stents in the treatment of postsurgical colorectal strictures and fistulas. Endoscopy: 2012; 44 (3): 297-+.
[18] Li G, Li Y, Chen GQ, He JH, Han YF, Wang XQ, Kaplan DL. Silk-Based Biomaterials in Biomedical Textiles and Fiber-Based Implants. Advanced Healthcare Materials: 2015; 4 (8): 1134-1151.
[19] Mashar M, Mashar R, Hajibandeh S. Uncovered versus covered stent in management of large bowel obstruction due to colorectal malignancy: a systematic review and meta-analysis. Int J Colorectal Dis: 2019; 34 (5): 773-785.
[20] Xie X, Zheng X, Han Z, Chen Y, Zheng Z, Zheng B, He X, Wang Y, Kaplan DL, Li Y, Li G, Wang X, Lan P. A Biodegradable Stent with Surface Functionalization of Combined-Therapy Drugs for Colorectal Cancer. Adv Healthc Mater: 2018; 7 (24): e1801213.
[21] Wang Y, Lu J, Guo JH, Zhu GY, Zhu HD, Chen L, Wang C, Teng GJ. A Novel Tracheobronchial Stent Loaded with (125)I Seeds in Patients with Malignant Airway Obstruction Compared to a Conventional Stent: A Prospective Randomized Controlled Study. EBioMedicine: 2018; 33: 269-275.
[22] Bi YH, Li JD, Yi MF, Yu ZP, Han XW, Ren JZ. Self-expanding segmental radioactive metal stents for

palliation of malignant esophageal strictures. Acta Radiologica: 2019.

[23] Kennedy AS, Coldwell D, Nutting C, Murthy R, Wertman DE, Loehr SP, Overton C, Meranze S, Niedzwiecki J, Sailer S. Resin Y-90-microsphere brachytherapy for unresectable colorectal liver metastases: Modern USA experience. International Journal of Radiation Oncology Biology Physics: 2006; 65 (2): 412-425.

[24] Diamond I, Jaenicke R, Wilson CB, Mcdonagh AF, Nielsen S, Granelli SG. Photodynamic Therapy of Malignant Tumors. Lancet: 1972; 2 (7788): 1175-&.

[25] Bae BC, Yang SG, Jeong S, Lee DH, Na K, Kim JM, Costamagna G, Kozarek RA, Isayama H, Deviere J, Seo DW, Reddy DN. Polymeric photosensitizer-embedded self-expanding metal stent for repeatable endoscopic photodynamic therapy of cholangiocarcinoma. Biomaterials: 2014; 35 (30): 8487-8495.

[26] Isayama H, Nakai Y, Toyokawa Y, Togawa O, Gon C, Ito Y, Yashima Y, Yagioka H, Kogure H, Sasaki T, Arizumi T, Matsubara S, Yamamoto N, Sasahira N, Hirano K, Tsujino T, Toda N, Tada M, Kawabe T, Omata M. Measurement of radial and axial forces of biliary self-expandable metallic stents. Gastrointestinal Endoscopy: 2009; 70 (1): 37-44.

[27] Li G, Li Y, He X, Lan P. Mechanical Properties and Degradation of Polydioxanone Stents. Textile Bioengineering and Informatics Society Proceeding: 2012: 184-192.

[28] Arafat M, Fouladian P, Blencowe A, Albrecht H, Song YM, Garg S. Drug-eluting non-vascular stents for localised drug targeting in obstructive gastrointestinal cancers. Journal of Controlled Release: 2019; 308: 209-231.

[29] Venezuela J, Dargusch MS. The influence of alloying and fabrication techniques on the mechanical properties, biodegradability and biocompatibility of zinc: A comprehensive review. Acta Biomaterialia: 2019; 87: 1-40.

[30] Han HS, Loffredo S, Jun I, Edwards J, Kim YC, Seok HK, Witte F, Mantovani D, Glyn-Jones S. Current status and outlook on the clinical translation of biodegradable metals. Materials Today: 2019; 23: 57-71.

[31] Moon CM, Kim TI, Lee MS, Ko BM, Kim HS, Lee KM, Byeon JS, Kim YS. Comparison of a newly designed double-layered combination covered stent and D-weave uncovered stent for decompression of obstructive colorectal cancer: a prospective multicenter study. Dis Colon Rectum: 2010; 53 (8): 1190-6.

[32] Arafat M, Fouladian P, Wignall A, Song Y, Parikh A, Albrecht H, Prestidge CA, Garg S, Blencowe A. Development and In Vitro Evaluation of 5-Fluorouracil-Eluting Stents for the Treatment of Colorectal Cancer and Cancer-Related Obstruction. Pharmaceutics: 2020; 13 (1).

[33] Mauro MA, Koehler RE, Baron TH. Advances in gastrointestinal intervention: The treatment of gastroduodenal and colorectal obstructions with metallic stents. Radiology: 2000; 215 (3): 659-669.

[34] Wang Z, Zheng Q, Guan S, Sun Z, Liu S, Zhang B, Duan T, Xu K. In vitro and in vivo assessment of the biocompatibility of an paclitaxel-eluting poly-l-lactide-coated Mg-Zn-Y-Nd alloy stent in the intestine. Mater Sci Eng C Mater Biol Appl: 2019; 105: 110087.

[35] Wang ZH, Li N, Li R, Li YW, Ruan LQ. Biodegradable intestinal stents: A review. Progress in Natural Science-Materials International: 2014; 24 (5): 423-432.

[36] Tanaka T, Takahashi M, Nitta N, Furukawa A, Andoh A, Saito Y, Fujiyama Y, Murata K. Newly developed biodegradable stents for benign gastrointestinal tract stenoses: a preliminary clinical trial. Digestion: 2006; 74 (3-4): 199-205.

[37] Zilberman M, Nelson KD, Eberhart RC. Mechanical properties and in vitro degradation of bioresorbable fibers and expandable fiber-based stents. J BIOMED MATER RES B: 2005; 74b (2): 792-799.

[38] Fathi P, Capron G, Tripathi I, Misra S, Ostadhossein F, Selmic L, Rowitz B, Pan D. Computed tomography-guided additive manufacturing of Personalized Absorbable Gastrointestinal Stents for intestinal fistulae and perforations. Biomaterials: 2020; 228: 119542.

[39] Elkasabgy NA, Mahmoud AA, Maged A. 3D printing: An appealing route for customized drug delivery systems. International Journal of Pharmaceutics: 2020: 588.

[40] Li G, Zha W, Han Z, Liu L, Chen Y, Lan P, Li Y. Structural Model of a Tubular Stent for the Treatment of Intestinal Cancer. Textile Bioengineering Informatics Society Proceedings: 2015: 642-647.

[41] Li G, Chen Y, Cai Z, Li J, Wu X, He X, Zhao Z, Lan P, Li Y. 5-Fluorouracil-loaded poly-l-lactide fibrous membrane for the prevention of intestinal stent restenosis. Journal of Materials Science: 2013; 48 (18): 6186-6193.

[42] Tonolini M, Bareggi E, Salerno R. Endoscopic stenting of malignant, benign and iatrogenic colorectal disorders: a primer for radiologists. Insights into Imaging: 2019; 10 (1).

[43] Puértolas S, Herrera A, Bajador E, Millastre J, Ibarz E, Gabarre S, Puértolas JA, Gracia L. Chapter 3 - Design,

Simulation, and Experimentation of Colonic Stents. in Advances in Biomechanics and Tissue Regeneration, Doweidar MH, Editor: 2019, Academic Press: 33-61.
[44] Sharip NS, Ariffin H. Cellulose nanofibrils for biomaterial applications. Materials Today: Proceedings: 2019; 16: 1959-1968.
[45] Pant HR, Pokharel P, Joshi MK, Adhikari S, Kim HJ, Park CH, Kim CS. Processing and characterization of electrospun graphene oxide/polyurethane composite nanofibers for stent coating. Chemical Engineering Journal: 2015; 270: 336-342.
[46] Park SB, Lih E, Park KS, Joung YK, Han DK. Biopolymer-based functional composites for medical applications. Progress in Polymer Science: 2017; 68: 77-105.
[47] Sagiv-Barfi I, Czerwinski DK, Levy S, Alam IS, Mayer AT, Gambhir SS, Levy R. Eradication of spontaneous malignancy by local immunotherapy. Science Translational Medicine: 2018; 10 (426).
[48] Chen Y, Li NX, Xu B, Wu M, Yan XY, Zhong LJ, Cai H, Wang T, Wang QJ, Long FY, Jiang G, Xiao HT. Polymer-based nanoparticles for chemo/gene-therapy: Evaluation its therapeutic efficacy and toxicity against colorectal carcinoma. Biomedicine & Pharmacotherapy: 2019; 118.
[49] Xia Y, Tang G, Guo M, Xu T, Chen H, Lin Z, Li Y, Chen Y, Zhu B, Liu H, Cao J. Silencing KLK12 expression via RGDfC-decorated selenium nanoparticles for the treatment of colorectal cancer in vitro and in vivo. Mater Sci Eng C Mater Biol Appl: 2020; 110; 110594.
[50] Xie D, Wan G, Maitz MF, Lei Y, Huang N, Sun H. Characterization and mechanical investigation of $Ti–O_2–x$ film prepared by plasma immersion ion implantation and deposition for cardiovascular stents surface modification. NUCL INSTRUM METH B: 2012; 289: 91-96.

Human Skin Deformation Based on Knee Joint Motions

Ming-Xin Zhou[1*], Wei Cheng[2]

Wuhan Textile University, No.1 Fangzhi Road, Hongshan District, Wuhan, Hubei Province, 430073, China

*Corresponding author's email: Z13657277331@163.com

Abstract

This paper studies human skin deformation of knee joint and its related parts under different movements, to optimize related clothing and improve its dynamic comfort and degree of freedom. Based on the motions of the knee joint, four daily life knee joint motions are selected. Then, tests and analysis was conducted on their characteristics and deformation rate changes in different conditions. The results show that in human physical activities, knee joints are the most significant parts of skin deformation. When the skin around the knee point is moving, it is pulled by the knee joint according to the size of the movement. The skin on the back of the knee joint is in the lateral stretching state and in the longitudinal contracting state. According to the research results and the movement principle of human skin and joints, explanations are given, and relevant suggestions are put forward.

Keywords: Knee joint; Skin Deformation; Analysis; Comfortness

1. Introduction

During human body activities, bones, muscles and tendon movement can cause human skin change which is the direct cause of clothing deformation. Clothing not only needs to conform to the human body shape outline, it also has to meet the needs of the human daily activities. Through surface testing experiment measuring the deformation direction and the change of the human knee joint [1], the research of knee joint skin deformation under different poses and activities brings light to the element of loose quantity in clothing design and function design [2]. Clothing forms and structures adapted to such changes can be obtained and help to design clothing with good fitness along with various functions [3].

1.1 Knee Joint Structure

The human knee movement system is mainly composed of bones, joints and muscles [4]. All kinds of human movements are completed by the contraction of skeletal muscles to produce forces acting on bones and the movement of bones around joints. The knee joint, the most complex joint of the human body structure, is composed of lateral femoral condyle, lateral tibial condyle and patella [2], as shown in the Fig.1.

1.2 Scope of Knee Joint Motions

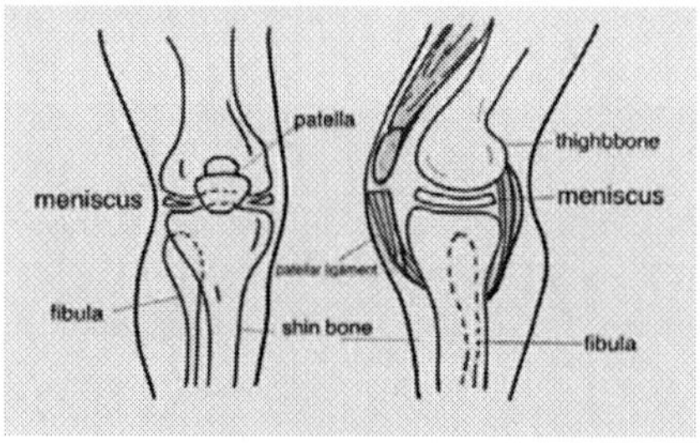

Fig.1 Bones of knee joint

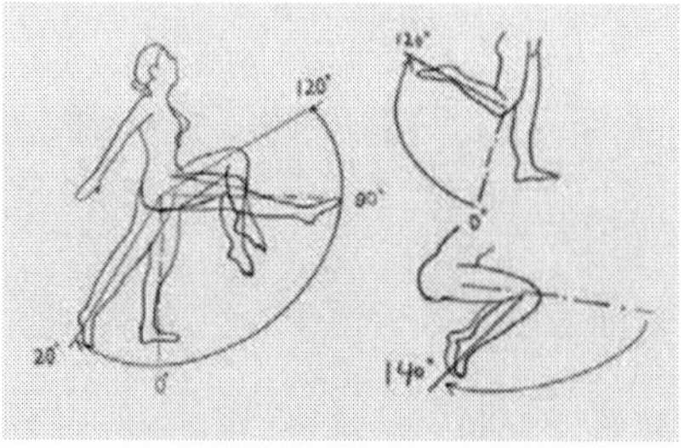

Fig.2 Human motions

The main movement of the knee joint is flexion and extension, and it can also be rotated when the knee is half bended. The knee flexion and extension range is the largest, and the flexion movement is about 0°~130° [5]. The degree of motion in other directions is small and can only be accompanied by movement in other directions during flexion and extension. Rotational movement can reach 30°~50° when the knee is flexed at 90°. as shown in the Fig.2.

2. Skin Deformation Measurement Experiment

2.1 Experimental Conditions, Subjects and Tools

Experimental conditions: room temperature 22-24 °C

Experimental subjects: 10 female college students aged 20-25 with body type A, height 160-165cm, body fat (BMI) 18-23 were selected accordingly. The measured data fluctuated within the permitted error range, and the mean value could be used for basic data analysis [3].

Experimental tools: tape measures, pens, rulers, protractors, recording boards.

2.2 Experimental Steps

The methods of measuring the dynamic deformation of human body include stretching line method, body surface drawing method, gypsum strip method and stamping method [6]. Body surface drawing method was used to establish the measurement site at the corresponding position of the human knee joint and draw lines. To ensure the accuracy of the completion of the experimental posture and correctly describe the skin condition of the knee joint, 6 horizontal lines and 6 vertical lines were drawn at the relevant parts of the knee joint. The intersection points of horizontal and vertical lines were marked as 24 intersection points in total: A1, A2, A3, A4, A5, A6, B1, B2, B3, B4, B5, B6, C1, C2, C3, C4, C5, C6, D1, D2, D3, D4, D5, D6. There are two kinds of measurement methods: total length measurement and segment measurement. First the total length of the line segment was measured with data recorded, and then segment lengths between the intersection points of each segment were measured. The error comparison with the total length and the addition of the segment lengths were made to ensure accuracy. As lines are drawn according to the direction of the muscles, they are not horizontal and vertical lines, as shown in the Fig.3.

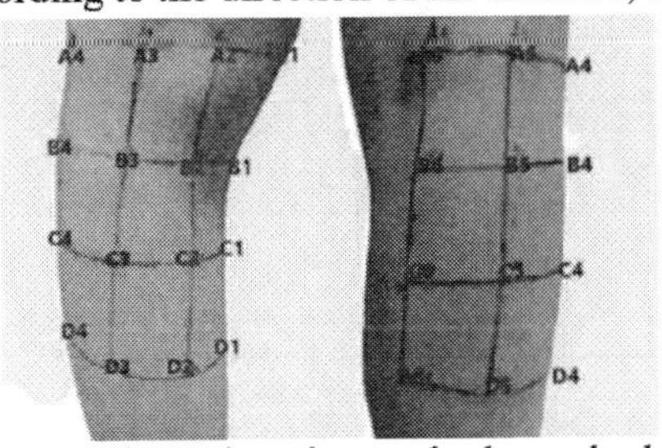

Fig.3 Lineation-on-body method

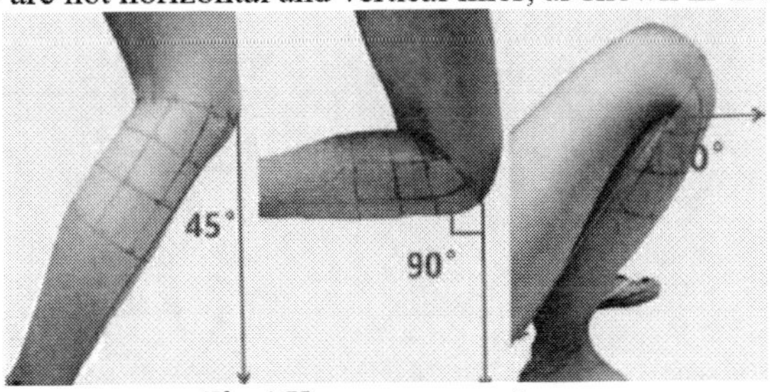

Fig.4 Knee movements

2.3 The Measurement of Postures

The motion of knee joint has a wide and large range, and the strain of clothing is manifested as slippage, fold, extension or restriction of human movement. According to the motion characteristics and range of human knee joint [1], this paper obtains the movements that human body often involved in daily movement. After previous data query [5-6] and pre-experimental test, four movements are selected: natural posture 30°, leg lift 45°, leg lift 90°, and natural squat 130° as is shown in the Fig.4.

2.4 Experimental Process

Throughout the experiment, the subjects were kept upright. When measuring, the ruler was attached to the body surface and was not pressed tightly on the skin [7]. Moreover, the line of sight was vertical when reading data. The surveyors received unified training. In order to control the measurement error, each part was tested 5 times and the average value was taken to ensure that the measurement error was controlled within the permitted range [8]. First, the static value was measured, wherein the measurement range was set as the distance between each intersection point. The static value was measured on the horizontal and vertical axes 5 times and the average value was obtained. Due to the different stretching degrees of the skin, the length of each segment on the grid diagram would change. According to the static measurement method [9] etc, the corresponding data of skin stretching changes under the 4 different states were measured.

2.5 Calculation of Knee Joint Dimensional Deformation Rate

Calculation was made on the static and dynamic length deformation rate of knee skin according to the following formula:

$$\alpha / \% = \frac{\text{Dynamic segment length - static segment length}}{\text{static segment length}} \times 100$$

In the formula, positive value means elongation, and negative value means contraction.

3. Test Results and Analysis

3.1 Data Analysis

The average value for the knee joint skin grid line data and the skin deformation rate of the subjects under four different conditions were calculated from the data measured in the experiment. These include skin grid lines mean value of longitudinal length shown in Table 1, mean value of horizontal length shown in Table 2, skin vertical deformation rate shown in Table 3 and transverse deformation rate shown in Table 4.

Table 1. Mean value of longitudinal length

Position	Static average (cm)	Leg lift 45° Average (cm)	Leg lift 90° Average (cm)	Natural squat 130° Average (cm)
A1-B1	5.4	6.0	7.0	7.7
B1-C1	4.6	5.0	5.5	5.9
C1-D1	5.3	5.0	6.0	5.8
A2-B2	5.0	5.1	6.0	6.5
B2-C2	4.6	5.0	6.0	5.3
C2-D2	5.5	5.8	5.9	5.9
A3-B3	4.5	4.9	4.7	4.7
B3-C3	4.5	4.8	4.0	4.5
C3-D3	5.0	5.3	5.0	5.5
A4-B4	4.0	3.6	3.3	1.9
B4-C4	4.7	4.5	3.6	3.7
C4-D4	5.0	4.9	4.0	4.3
A5-B5	4.0	3.0	1.1	2.5
B5-C5	4.3	3.8	3.3	3.5
C5-D5	5.0	4.0	4.0	3.0
A6-B6	4.0	3.2	0.8	1.5
B6-C6	4.2	3.8	3.0	3.8
C6-D6	4.2	4.0	3.3	3.6

Table 2. Mean value of horizontal length

Position	Static average (cm)	Leg lift 45° average (cm)	Leg lift 90° Average (cm)	Natural squat 130° Average (cm)
A1-A2	2.9	3.2	3.8	3.6
A2-A3	3.1	3.5	3.5	3.5
A3-A4	3.4	4.2	3.9	4.3
B1-B2	2.7	3.7	2.9	3.0
B2-B3	2.9	3.6	2.9	3.5
B3-B4	4.0	4.3	4.6	4.9
C1-C2	2.9	2.7	2.5	2.8
C2-C3	3.8	3.6	3.5	3.7
C3-C4	4.0	4.2	4.2	5.0
D1-D2	3.2	3.3	2.8	3.0
D2-D3	3.7	3.9	4.0	3.8
D3-D4	3.6	4.0	4.5	4.0
A4-A5	3.7	3.8	4.0	4.0
A5-A6	3.5	3.4	4.0	3.0
B4-B5	3.9	4.5	4.9	4.6
B5-B6	3.9	4.0	4.5	5.2
C4-C5	3.9	4.7	4.9	5.2
C5-C6	3.5	4.3	4.9	3.6
D4-D5	3.9	4.6	5.0	5.0
D5-D6	3.8	4.0	4.5	4.5

3.1.1 Leg Lifting 45°

According to the data in the chart, when the subjects' knee joint and body surface skin near the knee joint were raised from static to 45 degrees, the skin movement on the front of the knee joint and the horizontal and vertical directions were stretched, the skin on the longitudinal side and the back side were contracted, and the skin on the transverse back side was stretched, but the deformation rate was not high. The longitudinal change rate of the frontal skin was about 10%, and the deformation rate of A1-B1 was the largest. The transverse skin deformation rate was affected by the knee joint. The closer to the knee joint, the greater the deformation rate, which in turn decreased the contraction state.

3.1.2 Leg Lifting 90°

When the knee joint is bent at 90°, the bending degree is roughly same as when people sit down. According to the chart data, when the subject's knee joint is raised from 45° to 90°, the overall change rate of the subject's body surface skin increased, The closer to the knee, the greater the change value. The frontal skin of the knee joint was stretched in both horizontal and vertical directions. In the part close to the knee, the longitudinal stretching deformation rate of the frontal skin was concentrated between 20% and 30%, and the deformation rate is larger. The longitudinal changing rate of the back skin was negative, and the closer to the knee joint the greater the negative value. The deformation rate was concentrated between -5% and -25%. The lateral deformation rate of the frontal skin was concentrated between 20%.

Table 3. Longitudinal deformation rate of body skin under different knee movements

Position	45° deformation rate (%)	90° deformation rate (%)	130° deformation rate (%)
A1-B1	11	30	43
B1-C1	9	20	28
C1-D1	-6	13	9
A2-B2	2	20	30
B2-C2	9	30	15
C2-D2	5	7	7
A3-B3	9	4	4
B3-C3	7	-1	0
C3-D3	6	0	10
A4-B4	-10	-18	-53
B4-C4	-4	-23	-21
C4-D4	-2	-20	-14
A5-B5	-25	-72	-38
B5-C5	-12	-23	-19
C5-D5	-20	-20	-40
A6-B6	-20	-80	-63
B6-C6	-10	-29	-10
C6-D6	-5	-21	-14

Table 4. Transverse rate of skin deformation in different knee movements

Position	45° deformation rate (%)	90° deformation rate (%)	130° deformation rate (%)
A1-A2	10	31	24
A2-A3	13	13	13
A3-A4	24	15	26
B1-B2	37	7	11
B2-B3	24	0	20
B3-B4	8	15	23
C1-C2	-7	-14	-3
C2-C3	-5	-8	-3
C3-C4	5	5	25
D1-D2	3	-13	-6
D2-D3	5	8	3
D3-D4	11	25	11
A4-A5	3	8	8
A5-A6	-3	14	-14
B4-B5	15	26	18
B5-B6	3	15	33
C4-C5	21	26	33
C5-C6	23	40	3
D4-D5	18	28	28
D5-D6	5	18	18

3.1.3 130° Natural Squat

According to the data in the chart, from static to natural squatting, the longitudinal stretching from the knee joint to the front of the skin below increased. The closer it was to the knee, the greater the stretching, with a concentration of between 30% and 40%, which is the maximum stretching rate of the skin in all experimental active states. On the other hand, the longitudinal tensile deformation rate of the back side decreased. The greater the longitudinal tensile strength of the front side, the smaller the longitudinal tensile strength of the back side. The lateral stretching deformation rate on the front of the knee joint is relatively stable, concentrated between 10% and 25%. As for the skin on the back of the knee, the lateral skin stretching degree varied, the closer to the knee, the greater the stretching.

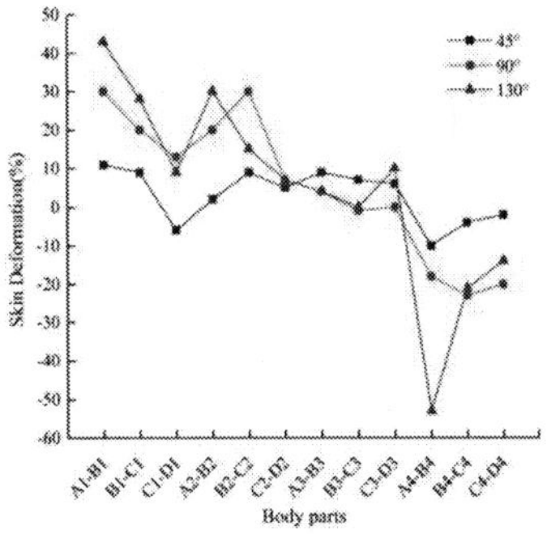

Fig.5 LDF skin of the knee joint

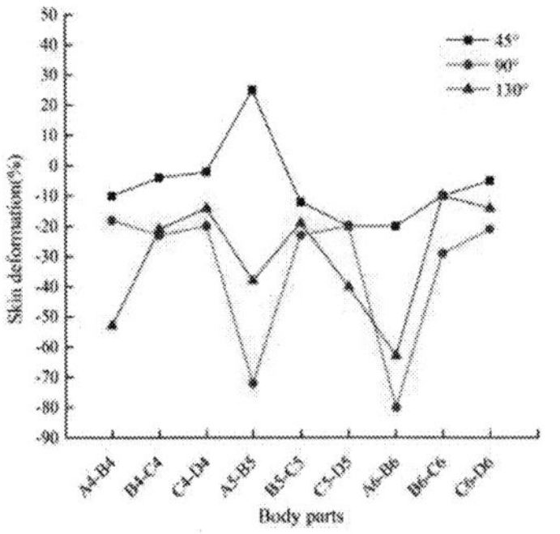

Fig.6 LDB skin of the knee joint

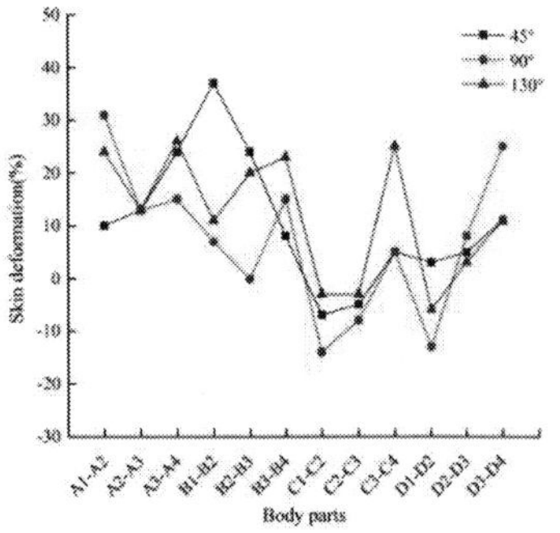

Fig.7 TDF skin of the knee joint

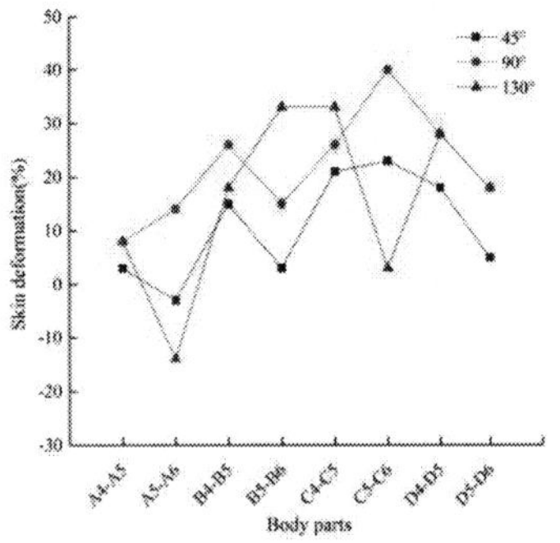

Fig.8 TDB skin of the knee joint

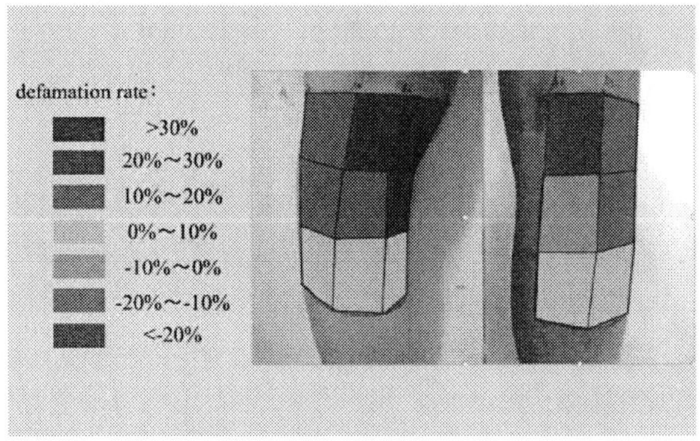

Fig.9 Analysis of deformation rate of body skin

3.2 Chart Analysis

The longitudinal deformation of the frontal (LDF) skin of the knee joint under different activity states are shown in Fig.5. Under these three activity states, the medial side of the A1-D1 knee joint presented an overall stretching state, and the stretching state decreased successively from the knee joint position downward, and the stretching deformation rate of A1-B1 was the largest. When the knee joint was in the natural squatting state of 130°, the stretching degree of A1-B1 was the largest compared with the other two conditions. This part of the mid-line of the A2-D2 knee joint was affected by the medial surface of the A1-D1 knee joint, and the stretching degree fluctuated greatly. The lateral line of the A4-D4 knee joint was affected by the dorsal bending of the knee joint. The contraction state went downward from the knee joint position, and the contraction degree decreased in turn. In summary, when the human body squats at 130° naturally, the longitudinal deformation rate of the skin is the largest, and the stretching degree is the largest.

The longitudinal deformation pattern of the back (LDB) skin of the knee joint is shown in the Fig.6. When the knee joint lifted the leg at 45°, the skin on the back of the knee joint also presented a contraction state compared with the leg lifting at 90° and the natural squatting at 130°, but the difference of the contraction deformation rate was not significant. Under the condition of natural squatting at 130°, the deformation rates of A4-B4,A5-B5 and A6-B6 were the lowest, which were -53%, -38% and -63%, respectively. The skin on the back of the knee joint contracted, and the closer it was to the knee joint, the greater the degree of contraction.

The transverse deformation pattern of the frontal (TDF) skin of the knee joint is shown in the Fig.7. Combined with the measured position of the knee joint and the change pattern figures, the transverse deformation rate of the skin on both sides of the frontal skin of the knee joint is larger, such as A1-A2,A3-A4,B1-B2,B3-B4,C3-C4, and D3-D4. Under the condition of 45° leg lifting, the deformation rate of B1-B2 was the highest with 37%.

The transverse deformation pattern of the back (TDB) skin of the knee joint is shown in Fig.8. Under three different activity states of the knee joint, the skin near the knee joint contracted laterally. Under the condition of natural squatting at 130°, the lowest deformation rate of A5-A6 was -14%, while under the condition of lifting legs at 90°, the highest deformation rate of C5-C6 was 40%.

3.3 Analysis and Summary

Combined with the analysis of the change rate of the skin on the body surface in the Fig.9, it can be seen that:

1) The frontal skin of the knee joint stretched horizontally and vertically, and the closer it was to the knee, the greater the stretching degree. Generally, when people sit down and squat naturally, the knee joint is the place with the largest amount of activity. Therefore, with human activities, the skin deformation at the midpoint of the knee is the most significant.

2) The skin on the back of the knee joint was stretched laterally and contracted longitudinally, in line with the human skin and joint movement principle.

3) The stretching state decreased from the knee joint position downward. In different directions, according to human skin wrinkle mechanism and extension principle [10], the direction indicating large skin extension is perpendicular to the direction of the wrinkle, that is, the skin below the knee has a small deformation rate, but during activities, a certain deformation rate will be generated by the traction of the knee joint according to the size of the extension.

4) In the process of the knee joint movement, the longitudinal change rate of midpoint skin of knee joint is great. When the body naturally squats down at 130 °, the longitudinal deformation rate of the skin is the largest, and the stretching degree is the largest. This is because compared to other body parts, the movement scope of knee point activities is bigger, permitting internal rotation, outward turning, [11], and bending. When the body naturally squats, the bend becomes larger and the natural skin deformation increases.

4. Conclusion

The conclusion drawn from the experimental analysis is as follows:

First, knee skin changes greatly, so it is necessary to combine the dynamic skin deformation patterns to determine the dividing line of clothing and the initial structure design of the parts. According to the deformation pattern, the tensile rate of the appropriate elastic fabric is suggested to be selected, and a certain amount of looseness set aside.

Second, based on the research, it was found that the skin deformation rate at the midpoint of the knee was the largest. Except for the midpoint of the knee, the stretching range of the skin on both sides of the knee was larger. The bending movement in the front and back sides of the knee directly affected the traction and compression of the pants knee, which should be paid attention in the research of functional clothing.

Third, because the range of knee joint position in daily activity movements, such as knee joint lifting leg at 45 °, 90 °, natural squat down at 130 °, is larger than the surrounding parts, its skin deformation rate is also larger. Thus, in the design of pants, designers can include dividing lines and other processes to enhance the structure design by increasing allowance of the knee joint parts.

Fourth, by adopting high elastic fabric in the knee parts design, the range of the knee joint motion can be appropriately increased and the resistance to flexion reduced. Meanwhile, the size of the contraction part on the back of the knee should not be too large, to prevent the fabric stacking during movement.

Finally, in view of the main moving parts, the corresponding parts of the clothing should be adjusted. Human movement is not determined by the amount of clothing relaxation, it is instead affected by the proportion of human body, human activities and other factors. This paper studies the degree of skin deformation in the knee joint and its adjacent areas. It can be found that when we design and study the comfort of trousers related to the knee joint or functional clothing, we need to take full account of the deformation in the knee joint in all directions, and the variation of muscle stretching varied with different postures.

To meet the requirements of the 3D modelling of the human body, and the change of knee bump position, and raised position of the knee, designers can include the dart and pleat designs on the fabric, appropriately increase the amount of loose quantity or the warp and weft elasticity of fabric, change structure lines, splice different fabric, adopt setting technology of shrinking and stretching, etc. In these ways, the relative slip between skin and fabrics, skin friction will be reduced, which enhances the clothing fitness of the knee joint activities and movements harmoniously. To take full consideration of skin deformation of human knees in term of pattern structure design, style design, and fabric design, has very important significance in the design of functional trousers.

References

[1] Huang GF. Elastic clothing [J]Foreign textile technology: 1983 (14); 34-37.

[2] Luo L, Wang JP. Study on skin deformation of riding leg based on human motion capture. Journal of Zhejiang Textile and Clothing Vocational and Technical College: 2014; 13 (4): 39-43.

[3] Wang JL. Study on the customization and development trend of clothing cloud. Textile guide: 2018 (7); 79-80.

[4] Zheng DN, Wang LC, Chen Y. Realization of information interaction in personalized clothing design process. journal of Textile Research: 2014; 35 (11): 157-162.

[5] Zhang HS, Lv L. Structural and Kinematics Characteristics of Knee Joint, Inner Mongolia Medical Journal: 2010; 67-71: 67-71.

[6] Liu RP. Design Principle and Technology of Garment Patterns, Bejing: China Textile Publishing House: 2008; 33-34

[7] Zhang WB, Fang F. Apparel Somatology. Shanghai: Donghua University Press: 2018: 69; 56-68: 120-121.

[8] Suo J. Design and development of material management system for Knitting garment enterprises.Microcomputer information: 2007; 23 (18): 21-23

[9] Wang YJ, Mok PY, Kkok YL. Body measurements of Chinese males in dynamic postures and application. Applied Ergonomics: 2011; 42 (6): 900-912.

[10] Zhong ZY. Human body and clothing, YUAN Guanluo, translate. Bejing:China Textile Publishing House: 2000; 82-84.

[11] Yang CJ, Zhen ZX, Lui H, Yang. Design of flexible knee-joint protection exoskeleton and walking assistance method, Journal of Zhejiang University (Engineering Science): 2021; 55 (2).

Cite as: S.N. Jia et al.,
TBIS 10.3993/tbis (2021).

Product Research on Reducing the Internal Environmental Humidity of Medical Protective Clothing

Song-Ning Jia[1], Ming-Wei Sang[2], Ming-Hai Cui[2*]

Beijing Institute of Fashion Technology, Ying Hua Street, Beijing, 100029, China

*Corresponding author's email: fzycmh@bift.edu.cn

Abstract

Given the poor thermal-humidity comfort of disposable medical protective clothing under high strength exercise, suitable water absorption materials were selected, and the corresponding products were designed to adapt to the use of medical protective clothing. Based on the determination of the total amount of human sweating, the absorbent material (including sodium polyacrylate, wood pulp paper, and water repellent non-woven fabric) were selected to obtain the composite moisture absorption products and verify their properties. The results demonstrated that 1g of composite water-absorbent product can absorb 22.2g of artificial sweat. This product can improve the thermal-humidity comfort of medical protective clothing and provide an experience for humidity treatment in a completely closed clothing environment.

Keywords: Medical Protective Clothing; Thermal and Wet Comfort; Absorbent Material

1. Introduction

At the end of 2019, a new type of coronavirus pneumonia (COVID-19) broke out in China, the situation of prevention and control was difficult, the population was generally susceptible, and the demand for medical disposable protective clothing increased sharply. However, there was a lack of systematic research in the field of medical disposable protective clothing. [1] Medical protective clothing wearing time is generally longer while the human body is not ventilated for a long time. Thus, the skin is stuffy which can cause rash breakouts. Besides, some wearers experience headache, chest tightness, and other symptoms. In the summer of 2020, the outbreak of COVID-19 came back, and the epidemic prevention workers who frequently wore medical protective clothing had a heatstroke phenomenon. It can be discovered that at the present stage, the shortcomings of disposable medical protective clothing such as moisture permeability and poor heat permeability significantly reduce the working efficiency of the staff presenting some health risks. [2] Simultaneously, the comfort of disposable medical protective clothing decreases with the increase in exercise intensity. At high exercise intensity, the amount of sweating rapidly increases to a great extent and even reaches the highest value under the power of metabolic equivalent 8.5. Meanwhile, medical protective clothing moisture permeability quickly decreased, demonstrating that wearing medical protective clothing is not suitable for long term high-intensity activities. [3] Therefore, it is critical to solve the problem of poor thermal-humidity comfort of disposable medical protective clothing.

There are some factors to be considered in this research, including the type of medical protective clothing, the wearing time, the movement state of the wearer, the total amount of human sweating in the process of use, and the selection of material to improve the poor thermal-humidity comfort of medical protective clothing. Hopefully, the total amount of human sweating under the condition of high-intensity exercise in the corresponding population will be obtained by human body experiments. Ideal materials and product design schemes are determined by analyzing the materials acquired by the experiment, the state of deionized water and artificial sweat, the material properties of water-absorbent materials of different magnifications, the liquid absorption speed, the equilibrium moisture content, and the water retention rate. Then, the composite moisture absorption product is designed, and its performance is verified. This study presents a product to handle the environmental humidity in medical protective clothing, provides an experience for humidity treatment in a completely closed clothing environment, and enriches the way of using water-absorbent materials in clothing. It lays a foundation for the improvement of medical protection moisture absorption ability in the future.

2. Literature Review

2.1 Types and Status of use of Medical Protective Clothing

Medical protective clothing, its primary function is to shield and block external infections, effectively isolate external viruses and pollution. Medical protective clothing can be divided into disposable medical protective clothes, isolation clothing and operation clothing. According to the length of service, it can be divided into one-time and repeated use. The fabric can be divided into woven, inorganic silk and composite medical protective clothing. [5-7] At present, the Chinese medical protective clothing standard mainly implements GB19082-2009 medical disposable protective clothing technical request. [8] The relevant evaluation indexes include air permeability (moisture permeability B19082-/(m^2 (m relevant evaluation indexistance (water retention grade ≥3), filtration (filtration efficiency ≥70%), etc. The moisture permeability of most clothing is more than 2 500 g/ (m2 • 24 h), which can maintain the comfort of wearing. Nevertheless, because the disposable medical protective clothing is entirely closed, can not complete the gas exchange with the outside world, our country's current moisture permeability index is low; as early as during SARS, the quarantine area wearing medical staff appeared chest tightness, shortness of breath, sweating and other phenomena, resulting in collapse and fainting. [4] According to the questionnaire survey, the first-line medical staff wears disposable protective clothing more than four hours accounting for more than 70% of the total number of people. When the protective clothing is less than four hours, it is considered that the general and poor air permeability and moisture permeability is more than 60%; when the clothing is worn at 4~8 h, it is believed that the air permeability and moisture permeability is 97.9% when the clothing is more than 8. 52.8% of medical staff suffered from collapse. Besides, from the survey of medical staff, the protective clothing breathable moisture of the parts of the trunk, crotch, head, and neck, legs and arms, [9] Compliance the main distribution of sweat glands in the human body. Trunk and crotch discomfort rates were up to 88.7% and 70.3%. During the new coronavirus pneumonia outbreak, 71.7% of medical staff wore protective clothing longer than four h, so it was necessary to improve humidity comfort.

2.2 Types and Properties of Absorbent Materials

Water absorbing materials generally refer to dry materials that can absorb and maintain water by chemical or physical means. Most of them are porous materials. At present, super absorbent polymers have better water absorbing capacity. At present, a kind of super absorbent polymer has good water absorption ability. Before the invention of super absorbent polymers, the task of absorbent materials had been undertaken by natural products and inorganic substances, such as polysaccharides, cellulose, silica gel, calcium oxide, phosphoric acid, paper, gauze, sponges, dehydrated cotton, and later foam plastics. In the late 1960s, the United States successfully developed a super absorbent polymer, a three-dimensional network of polymers and electrolytes with many hydrophilic groups and moderately cross-linked network structures, insoluble in water and organic solvents. The water absorption capacity can reach hundreds or even thousands of times of its mass, and the water absorption capacity can be restored after drying. According to the source of materials, it can be divided into artificial synthesis; polyacrylate is the most important, starch, cellulose, and other natural products. [10] At present, super absorbent polymers are mainly used in chemistry, health, medicine, and other fields. Because super absorbent polymers are non-toxic, non-irritating, and highly biocompatible, they have been widely used in the area of medical and health supplies. People use super absorbent polymers as absorbent materials to absorb urine, blood, and drugs to make sanitary napkins, adult incontinence supplies, and so on. [10] However, the research and production of super absorbent polymer as an absorbent material to absorb sweat are less present. [11] For the treatment of environmental humidity in medical protective clothing, it is necessary to consider the total amount of sweat absorbed and the rate of absorbing environmental humidity in the clothing. Water absorbent materials with faster moisture absorption are often used as a desiccant. The principle is to adsorb water molecules in their structures physically or to absorb the molecules chemically and change their chemical structures. Commonly used acid desiccant: concentrated sulfuric acid, phosphorus pentoxide; commonly used alkaline desiccant: alkali lime, calcium oxide, solid sodium hydroxide, sodium metal; commonly used neutral desiccant: calcium chloride, silica gel, anhydrous magnesium nitrate. [12]

2.3 Sweat Ingredients and Artificial Sweat Configuration

The study found that the main factors causing human sweating are heat and non-thermal factors, including temperature, humidity, and non-thermal factors for exercise. The order of sweating in different parts of the human body is different. The amount of sweating in the torso cadres accounted for more than half of the total amount of sweating. Different people will have some differences in sweating. The sweat comparison method can be divided into the majority of the population into four kinds of uniform sweating; upper limb sweating insufficient, lower limb sweating insufficient, torso sweating insufficient. Zhang Weiyuan summarized the six factors affecting human sweating: ambient temperature, atmospheric pressure, metabolic rate, environmental humidity, clothing, wind speed. Usually, sweating is positively correlated with ambient temperature and humidity and is greatly affected by clothing permeability and insulation performance. [16] In the static state, each adult sweats 700~1 200 g per day without feeling. When handling goods outside 32~35° I, 1 percent is discharged per day 200~2000 g of work; this feels like sweating. [4] Human sweat is composed of inorganic and organic parts. The various components in sweat show a skewed distribution. Moisture in labor was 99-99.4%, 5996.87 mg/L, of which 82.46% were inorganic, 8.12 % amino acids; Other organic components accounted for 9.42. The leading members of sweat are chlorine root, sodium, calcium, lactic acid, urea, amino acid, glucose, etc. And the pH of sweat PH is generally between 6.5-8.1, i.e., neutral PH7 up and down. The chlorine, sodium, potassium, and calcium contents were the highest, 2.71 g/L,2.04 g/L,0.24 g/L,19.65 mg/L. Chlorine accounted for 54.8 percent of the inorganic component, 70-80% of total anions; Sodium accounts for more than 40% of inorganic elements-the total of the two accounts for about 95% of inorganic components. Amino acids were 486.66 mg/L in sweat (median). The minimum range is 122.07 mg/L, up to 1787.5 mg/l. [13] Exercise or operation in a hot environment to expel electrolyte components, such as K, through sweat pathways, K^+, Na^+, Cl^-, Ca^{2+}, CU^{2+}, Zn^{2+}, Fe^{2+} are more significant. [14] The sweat PH=4-6.8 at rest, and the PH value of work secreted by the sweat gland was 6.5. Therefore, the ratio of reference acid solution is as follows:L- a group of ammonium salt 0.5 g/L, sodium chloride 5.0 g/L, sodium dihydrogen phosphate 5.0, and urea 1.1. After entirely dissolving the above chemical reagents, the mixed solution was adjusted to 6.5 using 0.1 mol/L glacial acetic acids. [15]

3. Problem Solution

3.1 Human Experiment

Subject and Equipment: Li Qian [9]. The questionnaire survey suggested that the height of medical staff was 161~170 cm, the age was 20~50 years old, the majority of them is girls. Besides, 9 women with a height of 165~170, age 22~35 years old, and weight of 50~60 kg were selected to be close to the actual situation. The experimental equipment is: K2 on a treadmill, equipped with a Bluetooth humidity meter, stopwatch.
Measurement method: nude weight difference before and after the exercise method.

The experimental environment and motion state: in the constant temperature and humidity chamber, the temperature is set to (28±2)°C and the relative humidity is (40m10) %. The participants wore colorless experimental clothing, with 220g of cotton. Due to the high intensity of exercise, the amount of sweating rapidly increased to a large amount and even reached the highest value under the intensity of metabolic equivalent 8.5. The medical protective clothing could not penetrate moisture. [3] Moderate and high-intensity exercise with metabolic equivalent greater than 4 and less than 10 was selected to mimic this state. Specifically, the participants exercised on a treadmill at 3.8 km/h for 5 minutes, 6.3 km/h for 25 minutes, and 3.8 km/h for 10 minutes. The participants measured moved for a total of 40 minutes.

Results:

1Table 1. Measurement results of total human sweat

Volunteer Number	1	2	3	4	5	6	7	8	9
Total sweating (unit: kg)	0.43	0.69	0.62	0.66	0.4	0.77	0.69	0.7	0.52

3.2 Water Absorption Materials Experiment

3.2.1 Experimental Environment

Within the constant temperature and humidity chamber, the temperature is set to (2625)°C and the relative humidity is (30nd h

3.2.2 Experimental Materials

Configuration of artificial sweat: refer to the change of human sweat composition and acid solution ratio in high humidity environment, configure artificial sweat. The selection of experimental materials is based on the types of absorbent materials.

2Table 2. Experimental materials

Material Number	Material name	Remarks	Material Number	Material name	Remarks
1	Pectin	99% purity	10	Magnesium chloride	Analytical Reagent
2	Soy protein isolate	99% purity	11	Kaolin	Analytical Reagent
3	Carboxymethylcellulose sodium	99% purity	12	Bentonite	Analytical Reagent
4	Chitosan	99% purity	13	Attapulgite clay	Analytical Reagent
5	Agar powder	99% purity	14	Silicone	Analytical Reagent
6	Sodium alginate	99% purity	15	Polyvinyl alcohol	99% purity
7	Humic acid	99% purity	16	Sodium polyacrylate	99% purity
8	Calcium chloride	Analytical Reagent	17	Degreasing cotton	99% purity
9	Calcium oxide	Analytical Reagent	18	Xylem fiber	99% purity

3.2.3 Liquid Absorbency Rate

Absorption rate refers to the amount of liquid absorbed by a 1 g sample when swelling equilibrium is reached in a liquid solvent.liquid absorbency rate: $Q=(m_2-m_1)/m_1$

Method of measurement: sample of absorbent material of m1(m_1=1 g) weight, soak in sufficient deionized water or artificial sweat for enough time (30 min) to balance swelling, remove the remaining aqueous solution with a mesh of 10 cm, 100 diameters at atmospheric pressure, Test the weight m of the material after suction. The rate of liquid absorption was calculated.

3.2.4 Absorbency under Load

When absorbent materials are used in sanitary or textile fields, the ability to absorbency under load is the most important performance indicator because products are used under pressure. The amount of pressurized absorption specifies the suction capacity of 1 h per gram of absorbent material under a pressure of 2 kPa or 5 kPa. Pressurized Absorption Expression: $AUL=(m_2-m_1)/m_0$(AUL is the absorbency under load, the m2 is the cylinder mass after absorption, the m_1 is the mass before absorption, the m_0 is the sample mass of the absorbent material, the mass before water absorption, and the m_2 is the mass of the material after the absorption).

Measurement method: weighing a certain amount of absorbent material at the bottom with a 300 mesh stainless steel screen, 100 mm cylinder 35 mm high, lay the sample over the screen, 250 g of absorbent material. At 1500 mL of liquid, 250 mm diameter One h, in a disk 30 mm high Determination of the mass of cylinder and sample before and after absorption. The amount of liquid absorbed by the sample is calculated according to the formula.

3.2.5 liquid Absorption Speed

Liquid absorption speed refers to the volume or mass of liquid absorbed by a sample per unit mass in a unit time. Liquid absorption speed : $V=(Q_2-Q_1)/t$

Measurement method: natural filtration method. Take 20 samples of the same kind, soak them in deionized water and artificial sweat several times, soak them for t (T_1 = 1, T_2 = 2... T_{10} = 10) minutes respectively, and then filter them weighing to obtain different time of liquid absorption rate.

3.2.6 Moisture Equilibrium Content

The moisture equilibrium content is the moisture absorption of the water-absorbing material to reach the solid-gas moisture absorption balance. Expression of moisture equilibrium content: $Q_h=(m_2-m_1)/m_1$

Methods: Manufacture of constant humidity environment: Preparation of 100 ml supersaturated $CaCl_2$ Solution, put it into the bottom of a 22 L sealed container balanced for two hours under normal pressure, and the container interior reaches the same relative humidity environment. The experimental temperature is the laboratory temperature (°C). The absorbent material (dry) and humidity instruments are placed in the numbered sealed containers, respectively, and the control variables are the same. The humidity change is detected until humidity in the glass container reaches the balance (about 30min), and the balance of moisture absorption is calculated by taking it out and weighing it.

3.2.7 Moisture Absorption Rate

The moisture absorption rate refers to the amount of vapor absorbed per unit time in moisture absorption.

Measurement method: the manufacture of a constant humidity environment is the same as the equilibrium moisture absorption experiment. Moisture absorption: the absorbent material (dry) is placed in the numbered constant temperature and humidity glass. The control variables are the same, respectively in t (t_1=5, t_2=10...). After 30 minutes, the same material is weighed 6 times until humidity in the glass container is balanced. Calculate the moisture absorption rate.

3.2.8 Mutative Humidity in a Closed Environment

The closed environmental humidity curve refers to the curve automatically generated by the environmental humidity drop temperature meter during the moisture absorption process of the closed environment.

The measurement method is the same as the equilibrium moisture absorption. The humidity meter is put into it at the same time. By detecting the change in humidity, data is recorded once for one second. The experiment takes 30 minutes, data is saved, and the change in humidity is calculated.

3.2.9 Conclusion of the Measurement Experiment of Water Absorbent Materials

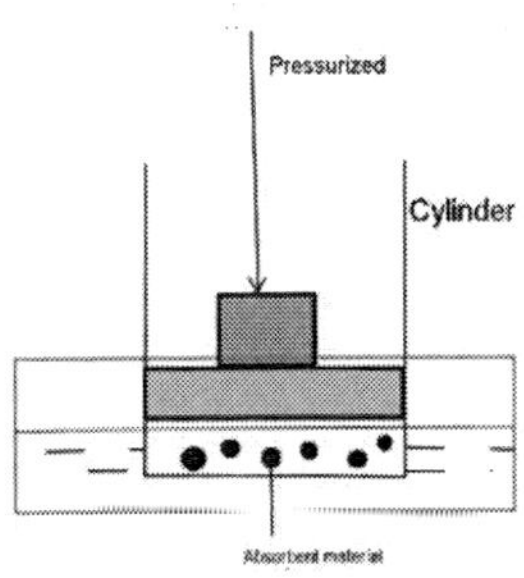

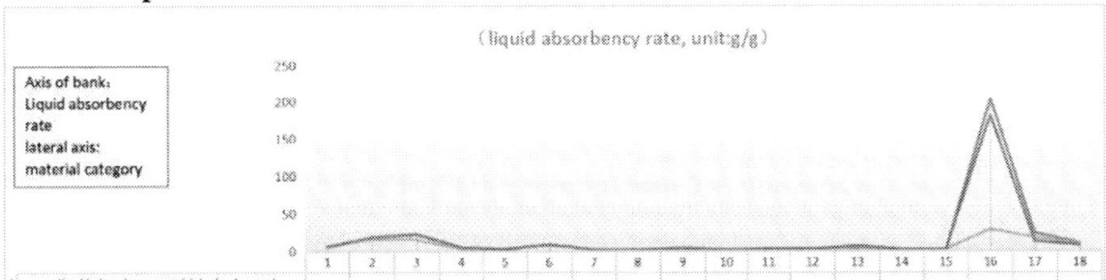

	1	2	3	4	5	6	7	8	9	10	11	12	13	14	15	16	17	18
Liquid absorbency rate(deionized water)	5.26	17.51	21.98	3.84	1.61	7.9	1	0.83	2.51	0.64	1.81	1.25	5.42	0.87	1.79	208.77	23.44	10.23
Liquid absorbency rate(artifical perspiration)	4.99	15.46	14.7	2.44	2.22	7.6	1	0.85	1.45	0.59	1.24	1.31	2.01	0.79	1.89	28.4	17.33	8.84
Absorbency under load	5.43	18.11	22.71	4.53	2.51	8.39	1.96	1.83	3.56	1.65	2.72	2.27	6.07	1.76	2.74	181.24	12.11	7.28

3Fig.1 Experimental instrument of pressure absorption and ratio curve of suction fluid

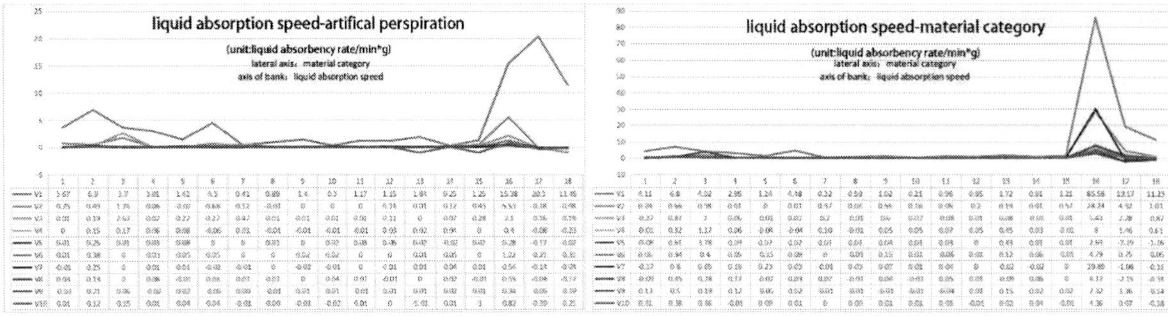

4Fig.2 Liquid absorption speed —— artificial sweat/deionized water

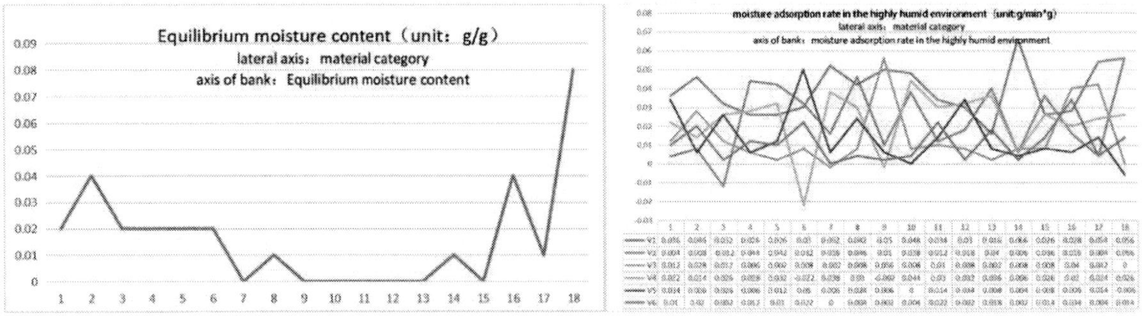

5Fig.3 Equilibrium Moisture content curve and Moisture absorption rate curve

6

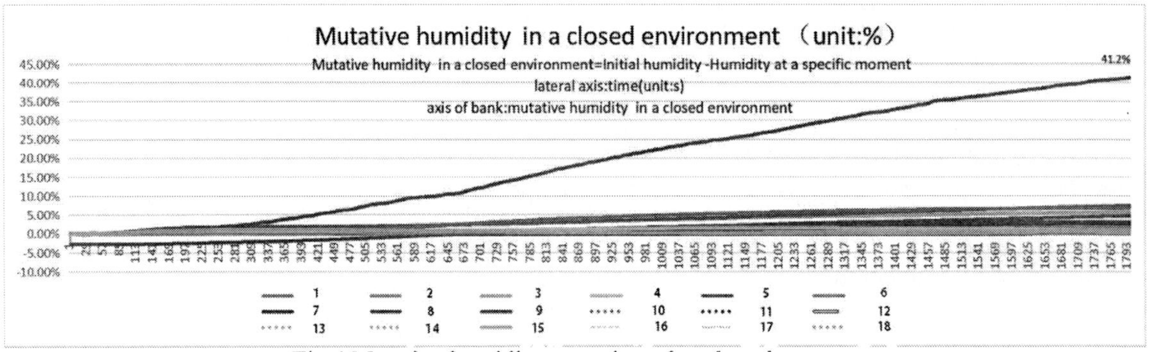

Fig.4 Mutative humidity curve in a closed environment

Conclusion: In the experiment, the humidity variation of most water-absorbent materials is in the range of 0-10%, and the humidity variation of calcium chloride is far more than that of other water-absorbent materials, demonstrating the strong moisture absorption ability of calcium chloride. The equilibrium moisture absorption of wood fiber was the largest, followed by sodium polyacrylate. The reason is that the surface area of wood fiber and sodium polyacrylate of the same quality is much larger than those of sodium polyacrylate, and the contact with air is more sufficient. The moisture absorption rate of most water-absorbent materials first increases and then decreases with time. The largest average moisture absorption rate is calcium chloride, followed by wood pulp fiber and sodium polyacrylate. Additionally, the fastest moisture absorption rate in the first ten minutes was wood pulp fiber. The liquid absorption rate of sodium polyacrylate is several times more than that of ordinary water-absorbent materials. Most of the water-absorbent materials have less absorbency under load than saturated absorption rate, a similar ability to absorb deionized water and artificial sweat, and a much stranger ability of sodium polyacrylate to absorb deionized water than artificial sweat. However, sodium polyacrylate still absorbs more artificial sweat than ordinary absorbent materials. The absorption rate of most materials exhibits a trend of first increasing and then decreasing. With the increase in absorbent time, the total amount of absorbent liquid gradually decreases. Meanwhile, the absorbent ability of sodium polyacrylate is much stronger than that of other absorbent materials.

3.3 Product Output and Validation Tests

The composite water-absorbent material includes the lower layer material-water repellent non-woven fabric, the upper layer material-wood pulp paper, and the intermediate layer sodium polyacrylate. The lower layer material is bonded with the upper layer material, the sodium polyacrylate particle is embedded in the middle, and the thickness of the upper layer material and the lower layer material can be equal.

Processing technology: 1) The lower non-woven fabric is made, coated pulp on the lower non-woven blanket, forming wet surfaces. 2) Spread a layer of sodium polyacrylate particles on the wet surface made in step 1. 3) Composite paper --- spray hot glue on top paper, and then the upper layer paper wet paper made on-line is combined into the lower layer non-woven fabric with one side of sodium polyacrylate particle. 4) Pressing --- after prepressing and pressing rollers, press the upper paper and the lower non-woven fabric to form. The thickness of the paper after preloading is 1.5 times that of the upper paper, and the thickness of the pressed paper is 1.2 times that of the lower non-woven fabric. 5) Drying --- the control inlet temperature is 100°C-150°C, the outlet temperature is 120°C-200, and the feeding rate is 150 m/min-600m/min. Thus, the paper towel after pressing and forming is dried through the drying equipment. 6) Manual testing --- measure the material properties after drying in step 5. 7) Split --- cut packaging, and test the qualified materials with the preset die cutting by cutting equipment; after cutting, separate and save them.

Product performance determination: experimental method, experimental environment reference materials experiment.

7Table 3. Performance of composite hygroscopic products

Product performance (weight :1 g)	Composite hygroscopic products
Liquid absorbency rate- Deionized water (unit: g/g)	136.45
Liquid absorbency rate - Artificial sweat (unit: g/g)	22.2
Absourbency under load (unit: g/g)	136.01
Liquid absorption speed - Deionized water - Maximum value (unit:Liquid absorbency rate/ min*g)	51.14
Liquid absorption speed - Deionized water - Minimum value (unit:liquid absorbency rate/ min*g)	0.5
Liquid absorption speed-Artificial sweat- Maximum value (unit: liquid absorbency rate/ min*g)	13.23
Liquid absorption speed-Artificial sweat-Minimum value (unit: liquid absorbency rate / min*g)	0.02
Moisture adsorption rate - Maximum value (unit: g/min*g)	0.062
Moisture adsorption rate - Minimum value (unit: g/min*g)	0.008
Lumidity drop in enclosed environment (unit:%)	23%
Moisture equilibrium content (unit: g/g)	0.16

4. Conclusion

The material properties of composite moisture absorption products are based on human body experiments and material experiment data. The total amount of sweating in the human body is mainly distributed in the range of 0.4-0.77. However, the total amount of sweating obtained from the experiment can only be used as a reference value due to personal habits and regional differences. This lays a foundation for measuring the basic performance of composite moisture absorption products. The composite moisture absorption products should be employed in medical protective clothing. To improve the wearing comfort, the composite moisture absorption products should have the properties of fast moisture absorption rate, strong water absorption ability, high compatibility with sweat, safe use, light texture, and no significant foreign body sensation. With this intention, the material experiment was conducted. It was revealed that most of the water-absorbent materials are more capable of absorbing water. Besides, sodium polyacrylic acid has stronger absorption rate, absorption rate, and water retention rate compared with other materials, and its equilibrium moisture absorption is second only to wood fiber. Additionally, the largest moisture absorption rate is calcium chloride, followed by wood pulp fiber and sodium polyacrylate. The humidity of most water-absorbent materials is in the range of 0-10%. Therefore, sodium polyacrylate is used as water-absorbent material to compound in water-repellent non-woven fabric and wood fiber. Wood fiber, which can quickly absorb sweat in medical protective clothing and human surface, is adopted as a water-absorbent material to contact with the human body. These three materials are moisture-absorption, not exothermic, safe to use, and suitable for human use.

The composite hygroscopic product presents strong hygroscopic capacity, high hygroscopic rate, light texture, safe use, and no significant foreign body feeling. It is expected to be suitable for people wearing medical protective clothing to work for a long time. In the process of use, it can improve the thermal-humidity comfort of medical protective clothing and reduce the environmental humidity of medical protective clothing. Subsequently, personalized composite moisture absorption products can be developed according to different user needs.

5. Acknowledgement

My deepest gratitude goes first and foremost to professor Ming-Hai Cui for her constant encouragement and guidance. Without her consistent and illuminating instruction, this thesis could not appear.

We would like to thank the support of the Shipei program of high-level talents in Beijing colleges and universities (KJCX21213-30299/006) from the Beijing Municipal Education Commission. The same thanks to the support of 21213 Capacity-building of science and technology innovation services - high-grade discipline construction (KJCX21213-30299/006) from Beijing Institute of Fashion Technology.

References

[1] Luo BJ, Wu XD, Huang WX, Yong X, Chen SD, Liu J, Huang J. Comparative study and research on medical disposable protective clothing standards under the COVID-19 epidemic [J]. China Medical Equipment: 2020; 35 (08); 111-114.

[2] Wang G, Xu X, Wang JY, Wei MF, Chen ZJ. Southern Medical Protective Thermal and Moisture Comfort Technology Research Situation Suit [J]. Journal of Tianjin University of Applied Sciences: 2021 (01); 54-59.

[3] Wang J, Dai HQ, Wang LL. Evaluation of thermal comfort performance of medical protective clothing [J]. Modern Silk Science and Technology: 2016; 31 (03): 92-94+100.

[4] Yang JZ, Wang XY. The structure and moisture permeability of medical protective clothing fabrics [J]. Journal of Textile Research: 2006 (03); 11-15.

[5] Shen JJ, Xu XY, Liu Y, Zhao Y, Jin XY. Research progress of medical protective clothing [J]. Cotton Textile Technology: 2020; 48 (07): 79-84.

[6] Han L, Hao DL, Ma YB, Hu MY. Nonwoven medical protective clothing standards and fabric selection [J]. Basic Science Journal of Textile Universities: 2020; 33 (01): 1-6+14.

[7] Chen TB, Zhou SQ. Development status and trend prospects of medical protective clothing [J]. New Material Industry: 2020 (02); 16-20.

[8] State F, Drug A. Technical requirements for medical disposable protective clothing: GB19082-2009 [S]. China Standard Press (Beijing): 2010.

[9] Li Q, Yang PY, Xu XM, Li L. Analysis of the comfort of disposable medical protective clothing [J]. Journal of Xi'an Institute of Technology: 2020; 34 (05): 8-13.

[10] Lin SB. Super absorbent polymer [M]. Beijing. Beijing Chemical Industry Press: 2013; 7: 1-233.

[11] Zhu ZF, Zhou YY, Zhang WG. Super absorbent material and its application equipment in textile industry [J]. Journal of China Textile University: 1994; 80-86.

[12] Li HS. The application of desiccant in daily life [J]. Science and Technology Wind: 2009 (21); 248-249.

[13] Appendix [J]. Criminal Technology: 1978 (S1); 56-75.

[14] Wu N, Lin D, Qu P, Bao YC, Meng SP. The effect of urea, coolness and sports in sweat under humid and hot environment [J]. Armed Police Medicine: 2012; 23 (01): 8-10.

[15] Huang W. Study on the performance of bluegrass blue dyed viscose fabric under artificial sweat infiltration state [J]. Cellulose Science and Technology: 2018; 26 (01): 55-60.

[16] Wang BJ. Research on the upper body sweating route of young men [D]. Donghua University: 2013.

Cite as: Y. Zhang et al.,
TBIS 10.3993/tbis (2021).

The Study of Leggings' Pressure Based on Restraint Index

Yi Zhang[1], Xue-Wei Jiang[1*], Lu Zheng[1], Run-Yu Tian[1]

[1] *Wuhan Textile and Apparel Digital Engineering Technology Research Centre, School of Fashion, Wuhan Textile University, No.1 Fangzhi Road, Wuhan, Hubei, 430073, China*

*Corresponding author's email: xwjiang@wtu.edu.cn

Abstract

To give a reasonable legging designing scheme, the comfort of leggings was studied based on the restraint index in this paper. The evaluation of comfort adopts a method of combination between the objective and perceptual evaluation. The objective was described by a pressure parameter named restraint index, the objective evaluation was acquired from the three-dimension body scanning technology and physical model. The perceptual evaluation was characterized by the sensation of constriction and tickling. The evaluation system was designed based on the Five-level interval scale. The results suggest that the pressure of legging affects the comfort of wearing and that the sensation of restraint is greater than tickling. Through further analysis, it is found that the larger the constraint the stronger the affection. This provides theoretical guidance for the design of leggings.

Keywords: Pressure of Leggings; Constriction and Tickling Feeling; Restraint Index; Objective Evaluation; Perceptual Evaluation

1. Introduction

Leggings originated in the 1960s. At the time, the miniskirt was designed by the British designer Mary Quinn, it suddenly became popular among women. With the length of the skirt becoming shorter, the trend of stockings and suspenders socks which were previously popular gradually faded. Leggings that combined the style of underwear and socks were born and were well received by women [1].

Leggings refer to clothes worn under dress or skirts to prevent exposure of the private parts. It is a kind of garment that wraps the body from the waist to the feet. Combining knitting technology, physiology, garment ergonomics and professional tights design concept, they are elastic and can exert certain pressure on the legs, so the legs can be lined up. Meanwhile, they can also correct the skin color of the legs, keep legs warm against cold and prevent varicose veins. Leggings meets the current consumers' pursuit of clothing aesthetics and comfort [2-3].

There are many types of leggings and they are classified in different ways according to the intended usage. Some are classified according to the types of materials, while others are classified according to their functions. Leggings can be divided into the types of Lycra, cotton, velvet, wool blends and special materials, depending on the types of material. These materials have good moisture absorption and perspiration function. The blending technology has been used to design new material in the past years. For example, the nano-material of titanium dioxide, Zinc oxide, Silicon dioxide and Silver ion have been added to polyester fiber and polypropylene fiber. By using these modification techniques, the new materials have antibacterial and deodorant functions [4]. According to the function and comfort of the product, leggings can be divided into medical stretch leggings with therapeutic function and pressure leggings with thinning leg function.

Many previous works have studied the physical condition of women who are forcibly tightened by corsets, and the results showed that the pressure of clothing has a great effect on human health [5]. The restraint of clothing pressure on human body can reduce muscle vibrations and improve wear-comfort during exercise [6-7]. However, wearing high pressure clothing for a long time can lead to physiological disorders of the human body. This can affect breathing and blood circulation. It can also affect the health of the wearer [8-9].

Most of the current research on leggings are aimed at the function, and there is little research on the daily style that people wear every day. Most of the research are from a medical point of view, focusing on how to treat and prevent varicose veins, thrombosis, diabetes and improving athletic performance [10-12].

In recent years, there are various excellent brands expanding the leggings industry and products developed from Taobao platform to occupy a larger market share. A large number of leggings with thin leg function were designed without considering the relationship between pressure and body health and comfort. Nowadays, many young women also pursue the so-called thin legged socks for beauty. However, they often feel uncomfortable after wearing unsuitable leggings for daily activities for a day, for example, when they sit, if there is too much pressure at the waist, they may feel constrained. After taking the leggings off, they may even have a tickling feeling or even a red trace of strangulation on their skin. In order to solve this problem, improve the comfort of leggings, this paper studies the objective and perceptual evaluation of leggings.

2. Methodology

Leggings with different functions are made of different materials. For example, leggings with the function of maintaining warmth in the winter are mainly made from wool. Slimming everyday wear leggings are mainly made from velvet, a fabric mix of polyamide and spandex. Leggings with this material can be easily obtained and its parameters can be easily controlled, therefore, three velvet leggings were selected as the experimental objects in this study. They are labeled No. 1, No. 2 and No. 3 respectively. These three leggings have different thicknesses. The thickness of leggings is characterized by denier, it refers to the mass gram of 9000 m long fibre at a public moisture regain, its unit is D. In terms of fabric content, as shown in the Table 1 below, the contents of the three leggings are very close, so that the three different types of pattern were chosen.

Table 1. The thickness and fiber content of three leggings

NO	Thicknesses	Content of polyamide (%)	Content of spandex (%)
1	360D	92	8
2	80D	92.2	7.8
3	680D	92.3	7.7

In our experiment, 36 women were invited to take part in human body measurements. Their ages ranged from 18 to 25 years old. These participants were asked to wear light-colored underwear and their anthropometric data were scanned by 3D human body scanner VITUS SmartXXL. The pulling height of the leggings varies with individual preferences and the surface area of the waistline is measured based on the waistline of the leggings. The exception is the waistline, the rest of the reference lines are based on body. After this scan, each participant was scanned again with No. 1, No. 2 and No. 3 leggings. During the scanning process, the participants remained in a preset position. They were kept upright and stood on the position markings with open arms. After each scan, each participant rated the wear comfort of No. 1, No. 2 and No. 3 leggings.

The perceptual evaluation of pressure comfort has always been a research focus in garment engineering. It not only involves ergonomics, it also covers multiple disciplines such as physiology and psychology [13]. When large pressure are exerted on human projection points, the wearer will have a significant feeling. It should meet the requirements of clothing or human body on the human function point. According to the definition of the restraint index, body surface area is the essential parameter. Therefore, the human bumps, the sensitive point, and the human function point should be considered in the process of measurement and analysis. Furthermore, the larger the circumference of the human bumps the greater the pressure and the corresponding feelings [14]. Applying a certain pressure at these points can meet the requirements of clothing or human body, such as applying pressure on the calf belly can effectively prevent varicose veins [15]. According to the above description, six body parts were selected. These parts include waist, hip, thighs, knees, calves and ankles. Many research on clothing pressure and comfort have been done in the past and many methods have been developed to measure the pressure of garment. The most used devices are the sensors of resistance, fluid and airbag. The sensors of fluid and airbag are too large to cause great leggings deformation and can lead to a errors in the measurement results. The sensor of resistance is too thin and it is also unsuitable for measuring the leggings pressure. This work requires a combination of perceptual and objective evaluations, so it cannot be measured by dummy model [16]. To measure the pressure of leggings more accurately, the restraint index was used as an objective indicator. The restraint index is estimated by the formula (1).

$$k = (B-A)/A. \quad 1)$$

Variable A is the surface area of clothing corresponding to six parts before dressing, B is the surface area of clothing corresponding to six parts after dressing. The units are cm2.

The restraint index of the six parts can be estimated based on the differential method. These positions are 5 cm below the waistline, 2.5 cm above and below the hipline, knee, calf and ankle circumference respectively, with a total length of 5 cm. In this experiment, the finite differential method was used to calculate the surface area of each part, it was regarded as a whole, composed of five regular cylinders. The height of each cylinder was set as 1 cm, and the lateral areas of the five small cylinders are measured at each part, the total surface area is the sum of lateral areas of five cylinders. The height of the lateral area is 1cm, and the upper circumferences of the five cylinders, that are the girth of the body parts, need to be measured.

For the leggings, there is a small difference in circumference from the thigh to the ankle, the surface area of these parts can be easily measured according to their circumference. The surface area of the leggings before wearing could be seen as the side area of the cylinder. The height of the cylinder is calculated by the proportional method. The length of the leggings is h1 and the distance h2 is the height from the waist to the heel after wearing. The h1/h2 is taken as the stretching ratio of the leggings. This ratio is multiplied by the height after dressing (5cm) which is equal to the height before dressing. Therefore the surface area of clothing corresponding to six parts before dressing (Variable A) can be measured.

The measurement methods of girth can be roughly divided into contact measurement methods: Martin measurement method, sliding gauge measurement method and plaster replica method. These methods do not directly obtain the body circumference required by the experiment. The non-contact measurement methods include photography, Mohr diagram, clothing deformation measurement and 3D human scanning system. The former three methods have the same disadvantages as the contact measurement method, which is unable to directly acquire the circumference data.

The 3D human scanning system can not only measure the length, height, depth and width, it can also measure the girth and angle of any part. It is suitable for measuring and conducting computation of the restraint index [17]. Therefore, a 3D human body scanning system was used to measure all the data needed. It is shown in Fig.1.

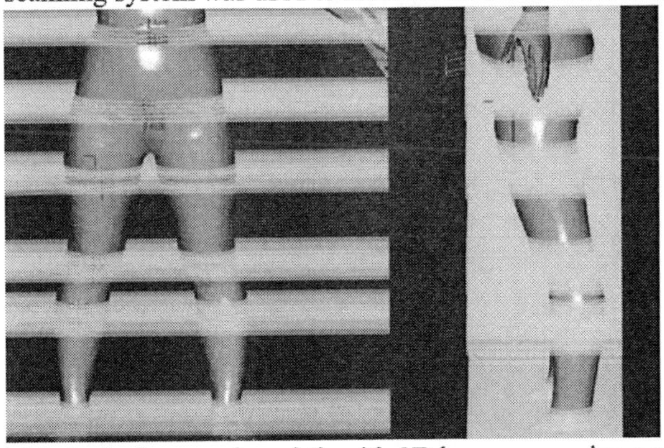

Fig.1 Method of measuring girth with 3D human scanning system

The above is the measuring method for the clothing area corresponding to six parts before dressing (Variable B).

The pressure comfort of leggings is mainly described by perceptual evaluation. It is the organic unity of the physiological state, psychological state and various functions of the garment [18]. The multi-equivalent symmetrical evaluation scales are used as an indicator by many researchers. These scales include the five-point method, the nine-point method and the eleven-point method. These methods assign "complete comfort" as the highest level and "complete discomfort" as the lowest level, and the middle sections are divided by equidistance. This method is fit for the perceptual evaluation of leggings pressure comfort [19].

The perceptual evaluation index describes the garment comfort with restraint, tickling, weight, softness and smoothness sensations [20]. Since the material of the leggings was selected, the feeling of softness and smoothness were not included in this work. The object of this work was the pressure comfort of women's leggings, so the weight value was not included. The quantitative evaluation method was used to evaluate the perceptual feeling. The Hollies' 5-level interval scale [21], the perceptual evaluation index of the itchiness and the restraint constitute the evaluation system.

In this experiment, the objective pressure and perceptual evaluation were carried out at the same time. The participants wore leggings to carry out three-dimensional body scanning, then the perceptual evaluation under five states is carried out, the perceptual evaluation consists of the constriction sensation and the tickling feeling under five states. Several states were selected from our basic lives to test the feelings of constriction and tickling. After a series of pre-experiments, five states were selected. The five states are standing, sitting, squatting, bending and walking. When they finish one process, they can relax for few seconds and then try on the next sample of leggings. The above steps will repeat to ensure that the experimental steps of each sample leggings are correct.

The comfort evaluation of leggings is combined with both objective and perceptual evaluation. The objective evaluation is more accurate, and through the perceptual evaluation, the real feeling of the wearer when they are wearing leggings can be understood. The objective and perceptual evaluation are the same batch of experimental objects, and they complement each other. The results obtained in this way can help businesses design more high-quality products.

3. Results and Discussion

To study the pressure comfort of leggings, with consideration of rigorousness, both the objective and perceptual evaluation are analyzed, the restraint index, constriction feeling, tickling feeling and the relationship among them were discussed in the following sections.

Fig.2 shows the average restraint index of each part. The average restraint index of each part of the No.2 leggings is smallest, and the average restraint index of each part of the No.3 leggings is the largest. The average restraint index values of different parts are also different.

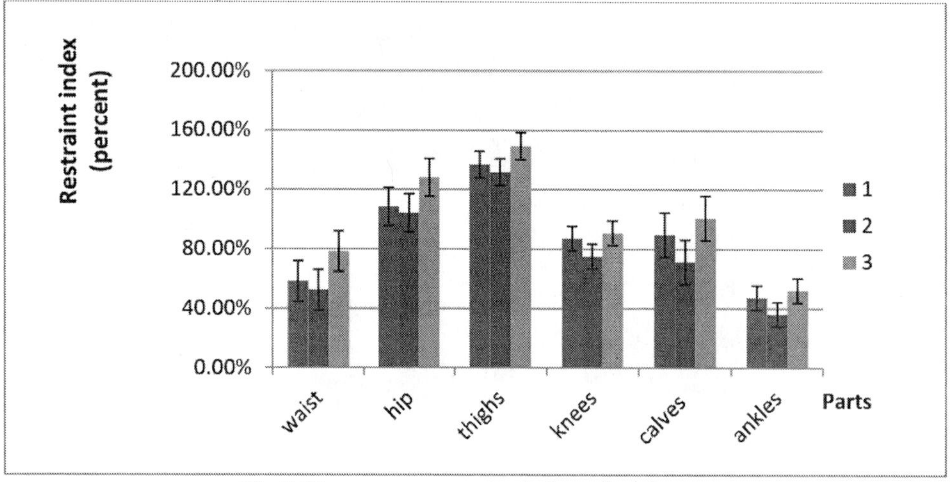

Fig.2 The average restraint index of each part

The average value of the constriction feeling of each part is shown in Fig.3. The constriction feeling of the No.2 leggings is the smallest and the constriction feeling of the No.3 leggings is the largest. There is some difference between the constriction feeling of the three types. No.3 leggings have the best sense of restraint, and the No.2 leggings have the worst. These results indicate that the pattern has an effect on the sense of restraint. Similarly, the average value of tickling sensation for each part is shown in Fig.4. The tickling feeling of No.2 leggings is the largest and the tickling feeling of No.3 leggings is smaller. This result also indicates that the pattern has an effect on the sense of tickling feeling. All the standard deviation of perceptual evaluation are similar, the perceptual feeling of different participants on the same leggings was not significant between these three types.

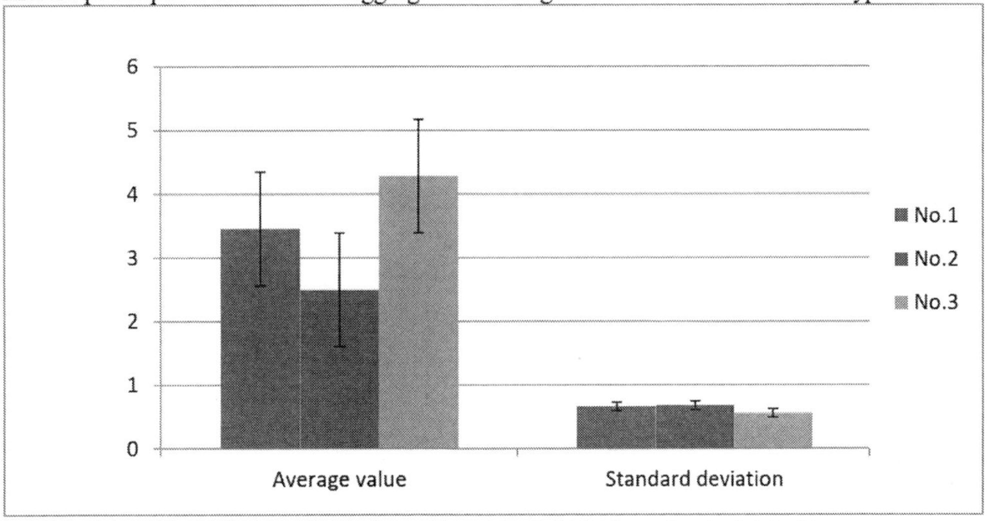

Fig.3 The average value and standard deviation of restraint feeling

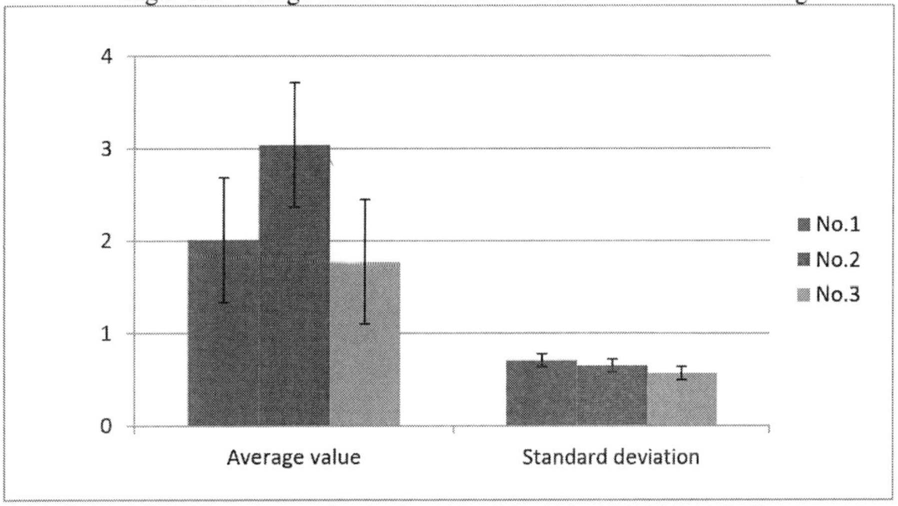

Fig.4 The average value and standard deviation of tickling feeling

The Pearson correlation coefficient has been often used to analyze linear correlations between multiple variables. The correlation analysis with Pearson correlation coefficient was performed between the perceptual evaluation and restraint index. The results show that the correlation coefficients between the restraint sensation and the restraint were positive, and the correlation coefficients between the tickling feeling and restraint were negative. According to the rules of correlation determination, the restraint index and restraint sensation and tickling sensation have some correlation. To analyze the relationship between perceptual evaluation and objective measurement accurately, the details of these correlations are discussed in the following.

As shown in Table 2, when the participants standing with No.1 leggings, the sense of restraint is positively correlated with the restraint indexes of each part, especially with the part of waist and thighs. They are significantly correlated at the 0.05 level with double sides. As mentioned above, the restraint index of the No.2 leggings is smallest, and the No.3 leggings is largest. As the perceptual feeling is weak when the leggings are too tight or loose, the correlation between the restraint index and the sense of restraint is weak.

Table 2. Correlation analysis table of restraint index and perceptual assessment

	Waist	Hip	Thighs	Knees	Calves	Ankles
Standing restraint NO.1	0.34*	0.32	0.40*	0.31	0.33*	0.32
Walking tickling No.2	-0.37*	-0.29	-0.40*	-0.50**	-0.46**	0.05
Bending Tickling No.2	-0.34*	-0.40*	-0.42*	-0.46**	-0.43**	-0.15

When the participants walk with No. 2 leggings, the sense of tickling negatively correlates with the restraint index. There are significant correlations with the waist and thigh parts at the 0.05 level with double sides. There are significant correlations with the parts of knees and calves at the 0.01 level with double sides.

When the participants bend over with No. 2 leggings, the sense of tickling negatively correlates with the restraint index. There are significant correlations with the waist, hip and thigh parts at the 0.05 level with double sides. There are significant correlations with the parts of knees and calves at the 0.01 level with double sides.

As the restraint index of the No.2 leggings is the smallest, the participants can feel the tickling sensation. The correlations between the restraint index and the sense of tickling are significant. When the pressure of the leggings is moderate or tight, the tickling feeling of participants is weak, so there are no correlation between the restraint index and the sense of tickling feeling.

4. Conclusion

In this paper, the objective and perceptual evaluation of pressure comfort of leggings were studied. Businesses cannot just reduce the pressure of leggings for constriction comfort, because the lower the pressure, the stronger the tickling feeling. Therefore, the pressure should not only be moderate, it should also take the constriction and tickling feeling together into consideration during when designing the pressure of the leggings. There are still room for improvement in the experimental design. As the elasticity of leggings may change after repeated try-ons, which is inevitable, in order to save the cost of the experiment, the leggings in this experiment have not been replaced. This may have a slight impact on the experimental results. In addition, different types of material and knitting method of the leggings were not considered in this work. They may be effect the pressure comfort of the leggings. In the future, these factors will be considered, and the details will be discussed. The pressure comfort of leggings is mainly studied in this paper. In the future, further research will be carried out on other aspect, like the effect of leg slimming and warmth retention.

References

[1] Gu T. Elementary Introduction on the Status Quo and Development Trend of Silk Stockings. J Shandong Textile Technology: 2009; 4: 50-51.

[2] Gao L, Geng RY, Hao RM. Design and functional research of female pantyhose. J China National Exhibition: 2020; 02: 181-182.

[3] Zhang M. Prediction Research on Wearing Tactile Comfort of Tight Pantyhose. D Xi'an Polytechnic University, Xi'an: 2016.

[4] Xu ML. Study on comfort of pantyhose. D Donghua University, Shang Hai: 2012.

[5] Jia XR, Xue WL, Wei JB. Study on the clothing pressure and sports comfort. J Melliand China: 2020; 48 (09): 48-53.

[6] Ueda, Atsuhiro. Taping socks prevent ankle sprain. J Advances in Textiles Technology: 2006; 1: 10-11.

[7] Liu T, Sun YC. Research on pressure comfort of leisure sportswear based on clothing looseness. J Modern Silk Science & Technology: 2020; 35 (03): 18-21.

[8] Wang Q, Chen DS, Wei QF. Situation and prospects of the researches on clothing pressure effects on body. J Textile Journal: 2009; 30 (04): 139-144.

[9] Kim NY, Lee HJ. Influence of Clothing Pressure on Blood Flow and Subjective Sensibility of Commercial Sports Compression Wear. J Fashion & Textile Research: 2019; 21 (4): 459-467.
[10] Rui D, Zhen Si. Finite element simulation on the relationship between pressure and displacement for the waist of elastic pantyhose. J International Journal of Clothing Science and Technology: 2020; 33 (2).
[11] Lattimer CR, Kalodiki E, Azzam M, Geroulakos G. Haemodynamic Performance of Low Strength Below Knee Graduated Elastic Compression Stockings in Health, Venous Disease, and Lymphoedema. J Journal of Vascular Surgery: 2016; 64 (2).
[12] Vercruyssen, Easthope, Bernard, etc. The influence of wearing compression stockings on performance indicators and physiological responses following a prolonged trail running exercise. J European Journal of Sport Science: 2014; 14 (2).
[13] Xu DF, Sun Y, Yang YD. Testing and Evaluation of Female Sports Underwear's Wearing Comfort. J Knitting Industries: 2021; 01: 62-65.
[14] Chai Y, Zhao YM, Gong LS, et al. Wearing State Based Pressure Distribution Study of Varicose Veins Care Compressive Stockings. J Knitting industry: 2018; 08: 10-13.
[15] Circaid. Therapeutic compression garment. J Medical Textiles: 2003; 3: 5-6.
[16] Luo SL, Wang YR, Liao YL, et al. Measurement and Evaluation of Compression Garment Performance on Smart Manikin System. J Journal of Donghua University (Natural Science): 2020; 46 (06): 896-901+914.
[17] Cheng NB, Wu ZM. Research method and development tendency of garment pressure comfort. J Silk: 2019; 56 (03): 38-44.
[18] Ju F, Li YM, Gu QT. Design scheme of the shaping pants based on the clothing pressure test and evaluation. J Shanghai Textile Science & Technology: 2014; 42 (12): 41-43.
[19] Zhang QY, Lu X. Analysis of the research status of clothing pressure comfort. J Liaoning silk: 2018; 01: 29-30.
[20] Zhang WB, Fang F. Ergonomics of clothing. Donghua University Press, Shang Hai: 2013.
[21] Zhang ZH, Ying SY, Guo YX, et al. The application of 3D scanning technology on garment ergonomics. J Shanghai Textile Technology: 2015; 43 (08): 40-44.

Network Pharmacology-based Analysis of Cangzhu for the Treatment of Covid-19

Ying-Chen Zhang[1*], Zhen-Zhong Qiu[1], Xia-Nan Zhang[2*], Hong-Yan Wu[1*], Qing-Song Zhang[1], Zhi-Ru Zhang[1], Jia-Hao Li[1], Jia-Qing Zhao[1], Meng-Yao Pan[1]

[1]*The School of Textiles, Zhongyuan University of Technology, Zhongyuan, ZhengZhou, HeNan, 450007, China*
[2]*Fachbereich Chemie und Technologie, Fachhochschule Aachen, Juelich, Nordrhein-Westfalen, 52428, Germany*

[*]Corresponding author's email: yczhang2002@163.com, xianan.zhang@alumni.fh-aachen.de, 416101810@qq.com

Abstract

Cangzhu is a Herbal medicine used to treat symptoms of respiratory diseases in Traditional Medicine system for hundreds of years and can be a new herbal solution for the current world pandemic. By using network pharmacology and molecular docking, this study explores the possible mechanisms of Cangzhu. One target is identified in the results. Further analysis identified 1 key target and 1 key compound. Moreover, 1052 biological processes, 61 cell compositions, and 124 molecular function items are obtained ($P<0.05$). One hundred and thirty pathways are obtained ($P<0.05$). The result of molecular docking shows a stable binding between the active compounds and the target.

Keywords: Cangzhu; Covid-19; Network Pharmacology; Molecular Docking; Pulmonary Fibrosis

1. Introduction

At the end of 2019, an acute respiratory disease, Covid-19 broke out in city of Wuhan and its surrounding cities in the Province of Hubei in China. Major clinical diagnostic feature is Ground-Glass-Opacity (GGO) in lung CT scans, accompanied by more common symptoms of cough, fever and fatigue, and more rarer symptoms as body pain, diarrhoea, loss of taste and smell, body discoloration as well as more severe symptoms such as chest pain, breathing difficulties and loss of mobility [1].

In TCMSP Covid-19 is classified as a Pandemic with a sub-classification of wet toxic pandemic. The basic Pathogenesis is explained as invasion of foreign toxicity, negative influence on lungs and related respiratory system, lack of positive physical inertia, with pathological properties of humid, heat, toxicity, weakness, and inadequate blood flow [2].

Cangzhu, also known as Black Atractylodes rhizome or Rhizoma Atractylodes, is the dried rhizome of Atractylodes Lancea (Thunb.) DC., Atractylodes Chinensis (DC.) Koidz, or certain other local species including Atractylodes Japonica koidz [3]. Compounds isolated from a hexane extract of A. Lancea included atractylochromene, a potent inhibitor of 5-lipoxygenase (IC50 3.3 μ mol L-1) and cyclooxygenase (IC50 0.6 μ mol L-1); 2-[(2E)-3,7-dimethyl-2,6-octadienyl]-6-methyl-2,5-cyclohexadiene-1,4-dione, a selective inhibitor of lipoxygenase (IC50 0.2 μ mol L-1); atractylon and osthol, weak inhibitors of lipoxygenase, and atractylenolides I, II, and III [4]. A prenylated dihydrobenzofuran derivative isolated from A. Lancea, trans-2-hydroxyisoxypropyl-3-hydroxy-7-isopentene-2,3-dihydrobenzofuran-5-carboxylic acid, was found to be cytotoxic to two cancer cell lines tested [5]. In ancient application in TCMSP, A. Lancea is recorded by Compendium of Mate-ria Medica for its ability to counter respiratory inflammation and effectively increase the stability of the blood circulatory system. It is also recorded by the same source, that A. Lancea is widely burned in open air during pandemic to decrease the viral transmission) [6]. In modern TCMSP, it is proven chemically that A. Lancea and the ß-eudesmol, hinesol containing volatile oil extract can significantly improve the human blood circulation and bloody oxygen level [7]. It has been widely used for epidemic prevention and control in China. However, there has been little evidence for the mechanism of action.

Network pharmacology is an approach to drug design that encompasses systems biology, network analysis, connectivity, redundancy, and pleiotropy. Network pharmacology offers a way of thinking about drug discovery that simultaneously embraces efforts to improve clinical efficacy and understand side effects and toxicity—two of the most important reasons for failure [8]. For TCMSP, Network pharmacology is a great tool to modernize the application of proved medicinal herbal compounds. It not only improves efficacy of experience based medical system by more developed modern pharmaceutical database, by cross-linking the modern database with ancient practical summaries, it is possible to find the next Artemisinin, for the future unpredictable pandemic.

Cangzhu, through its long-term application in Traditional Chinese Medicine, has proven its effectiveness against airborne pneumonia like pandemic. The herb itself is inherently complex and is comprised of numerous functional components. Due to characteristics as such, it may affect multiple targets and interact with other herbal pharmaceutical elements. As TCMSP is a summary of functional practices, Cangzhu itself has been mixed with other herbal elements in various functioning recipes. As the applying of the herb is based on the weighing of the original herb instead of the weight of the function group, the effective may vary from season to season even plant to plant. A qualitative and quantitative research on its mechanism is therefore hard to proceed. The present study by performing the cutting-edge method of network pharmacology and through focusing on the main functional herb of Cangzhu, tries to explain the effectiveness of the plant shown during current Covid 19 pandemic. Through analysis with GO, KEGG, and molecular docking, a correlation between the herb and the virous can be constructed, a possible conclusion on its functional mechanism can be raised to explain the therapeutic result at the cell and molecular level.

2. Materials and Methods

2.1 Collection of Molecular Information and Screening of Active Compounds in Cangzhu

To screen the bio-active compounds with anti- COVID-19 functions, the TCMSP and text mining tools were used. The ADME parameter-based virtual screening method is utilized, which with help of two determination factors, could be used to identify anti-covid19 function groups. After controlled with Oral Bioavailability (OB, determination threshold > 30%) and Drug Likeness (DL, determination threshold ≥0.18), common compounds and unique compounds are identified for the next analysis.

2.2 Prediction of Chemical Component Target of Cangzhu

TCMSP is applied in search for the association between potential targets and active compounds. The compound-target network of the compounds and the related targets can be constructed by using Cytoscape 3.7.2.

2.3 Determination of the Disease Related Targets

After active cross referencing and comparison to the recorded in the GeneCards database (http://www.genecards.org/) and the OMIM database (https://omim.org/), a total of 341 targets related to the pneumonia caused by the novel corona-virus (Covid-19) can be identified.

2.4 Prediction of the Targets of Cangzhu for the Treatment of COVID-19

With help of R Programming language, the effective Targets of Cangzhu and the Pathogenic targets on Covid-19-Virus can be cross-linked and analyzed. The intercepted target can be then analyzed by the online STRING database (https://string-db.org/) for the protein-protein interactions (PPI). Cytoscape 3.7.2 is then used to visualize the result of the analysis in form of a protein-protein interaction network.

2.5 Screening of Core Targets and key Compounds

The protein-protein interaction will then be analyzed with R. The first 30 core targets will be visualized, and the corresponding key compounds will be filtered.

2.6 Gene Ontology (GO) and Kyoto Encyclopedia of Genes and Genome (KEGG) Pathway Analysis

The R based Bio-conductor data package is used to perform the GO enrichment analysis and the KEGG pathway enrichment analysis. The intercepted 66 targets are selected under a threshold of P-values (<0.05). The first 20 results will then be visualized with R.

2.7 Molecular Docking Verification

The chemical structures of active compounds can be obtained on PubChem (https://pubchem.ncbi.nlm.nih.gov/), the structures of protein crystals can be obtained on RCSB PDB (https://www.rcsb.org/). The structures will be exported to PDB format which can be used to conduct the molecular docking. The binding energies are used to evaluate the docking stability.

3. Results

3.1 Screening of Active Compounds

When OB 30% and DL 0.18 were selected as filter standards, the compounds which could not meet the filter standards but were proved to be the main effective compounds were retained. After eliminating the repeated compounds, 9 kinds of effective compounds were selected as candidate compounds (Table1). Using a Perl script, a compound target correlation is constructed using Cytoscape 3.7.2 (Fig.1). The analysis of the compound-target network showed 4 nodes (1 drug nodes, 1 gene node, 1 common compound node and 1 target nodes) and three edges in total. The hexagons in the network represents drugs, ellipses represent genes, triangle represents molecular (mol), and diamonds represented diseases. The edges represent the interaction between drugs and compounds or the interaction between compounds and targets.

3.2 Prediction of the Targets on Cangzhu for COVID-19 Treatments and the Analysis of Target Interactions

The 48 targets from Cangzhu are cross matched with 341 targets from the novel Corona-virus, 1 common target is identified. The result can be concluded into Fig.2. The result will then be imported into STRING with the genetic setting of Homo Sapiens. The protein-protein interaction can be obtained after setting the medium confidence to 0.400 and hiding all the disconnected nodes in the network. The protein-protein interactions will be visualized as Fig.3. The network comprised 11 nodes and 51 edges. Further network topology analysis shows that the average node degree is 9.27, with local clustering coefficient to be 0.941.

Table 1. Basic information of the active compounds in Cangzhu.

Mol ID	Molecule Name	OB	DL
MOL000173	wogonin	30.68	0.23
MOL000179	2-Hydroxyisoxypropyl-3-hydroxy-7-isopentene-2,3-dihydrobenzofuran-5-carboxylic	45.2	0.2
MOL000184	NSC63551x`	39.25	0.76
MOL000186	Stigmasterol 3-O-beta-D-glucopyranoside_qt	43.83	0.76
MOL000188	3β-acetoxyatractylone	40.57	0.22
MOL000085	beta-daucosterol_qt	36.91	0.75
MOL000088	beta-sitosterol 3-O-glucoside_qt	36.91	0.75
MOL000092	daucosterin_qt	36.91	0.76
MOL000094	daucosterol_qt	36.91	0.76

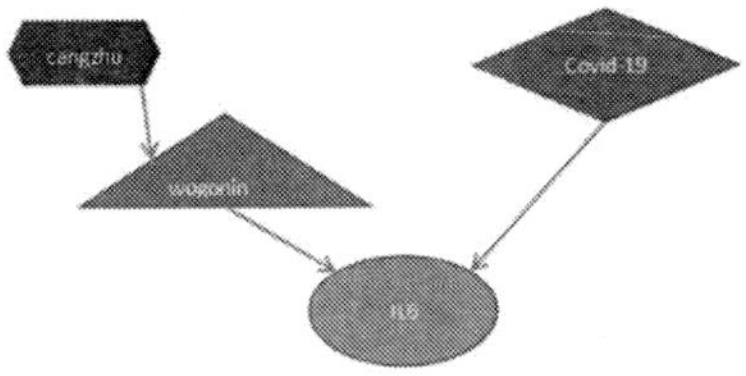

Fig.1 The compound - target interaction network.

Note: The hexagons in the network represented drug, circles represented genes, triangle represents the main function group, and diamond represents the targeted disease. All the edges represented the interaction between drugs and compounds or the interaction between compounds and targets.

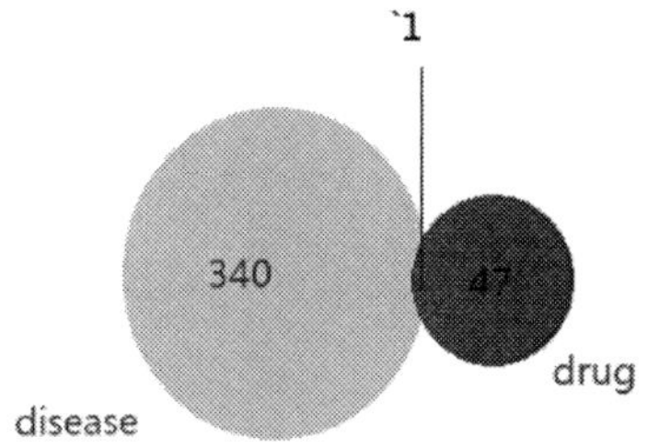

Fig.2 Venn diagram study of the targets.
Note: The intersection part represents the common target.

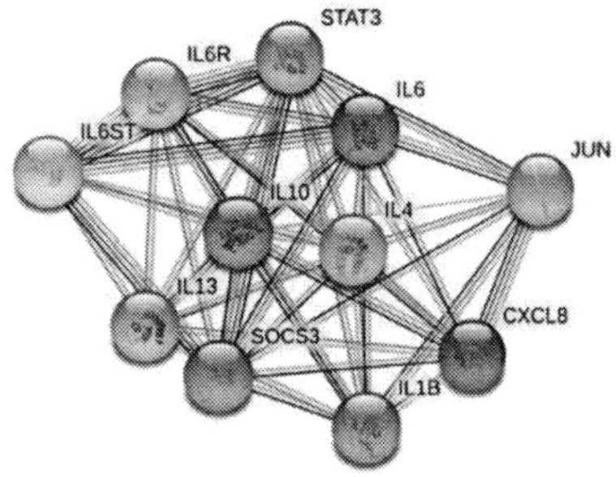

Fig.3 PPI network of the 11 intercepted targets.
Note: The node size and the colour brightness of the node reflects the degree value of the node. In this case the brighter and the bigger nodes represents a great degree value of the node.

3.3 PPI core Targets and Key Compounds

After analysis of protein-protein interactions by cross referencing the related target functioning protein targets on Cangzhu and target disease Covid-19. A total of 30 core targets can be identified. The numbers of top 10 targets are IL10, IL13, IL4, IL6, SOCS3, STAT3, CXCL8, IL1B, IL6R, and JUN. The result is visualized as Fig.4. The functioning compound with these functions group is wogonin.

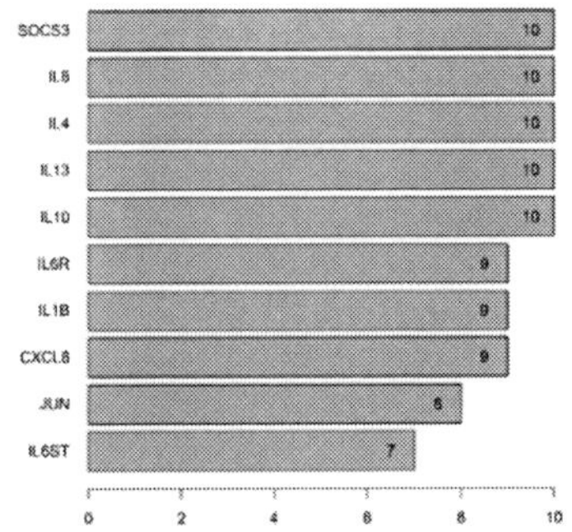

Fig.4 Core targets. Note: A total of 11 core targets were identified. The horizontal axis represented the number of connected nodes.

3.4 Analysis of GO Enrichment Analysis and KEGG Pathway Enrichment.

By performing the GO enrichment analysis, the following items can be obtained, thirteen molecular function (MF), four cell composition (CC), and two hundred sixty-nine biological process (BP). After , top 6 molecular functions (signaling receptor activator activity , receptor ligand activity, cytokine receptor binding, cytokine activity, growth factor activity and growth factor receptor binding), top 2 cell composition (endoplasmic reticulum lumen, plasma membrane signaling receptor complex), and top ten biological process (regulation of apoptotic DNA fragmentation, T-helper 17 cell lineage commitment , response to peptidoglycan , regulation of T-helper 2 cell cytokine production , positive regulation of production of miRNAs involved in gene silencing by miRNA, positive regulation of immunoglobulin secretion, regulation of interleukin-1-mediated signaling pathway, type B pancreatic cell apoptotic process, regulation of glucagon secretion and glucagon secretion) are selected and further processed into Fig.5. to show a broad view of the result.

In the KEGG pathway enrichment analysis, a total of 60 results are identified, mainly involving Antifolate resistance, African trypanosomiasis, Graft-versus-host disease, intestinal immune network for IgA production, Malaria and Legionellosis etc. The top 20 results according to the KEGG enrichment analysis is summarized and shown in Fig.6.

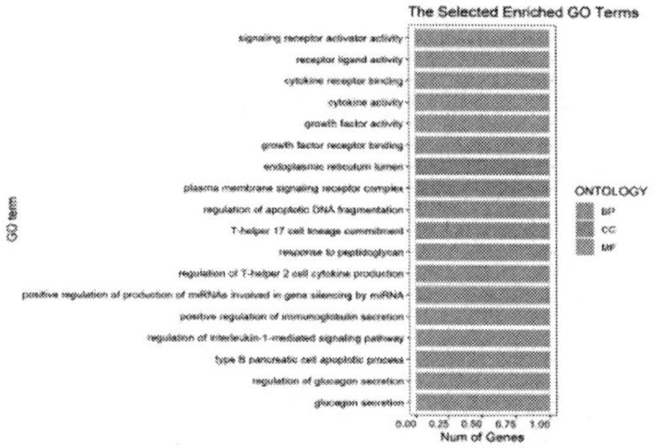

Fig. The results of GO enrichment analysis.

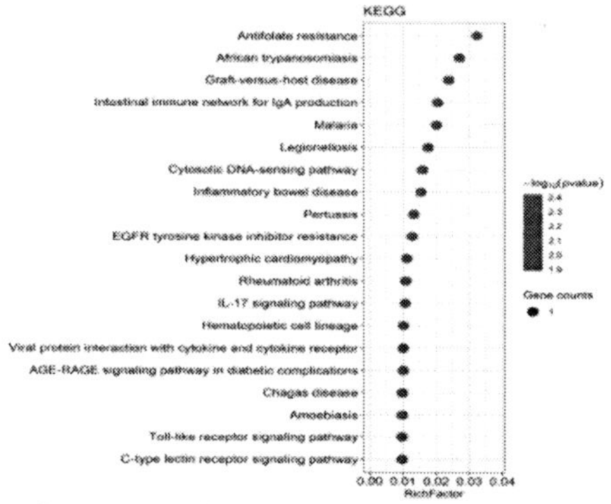

Fig.6 The results of KEGG pathway enrichment analysis (showing the top 20).

Note: The color is scaled from blue to red to represent the magnitude of the adjusted P value, which red represents the smaller end of the scale and the blue presents the larger end of the scale.

3.5 Analysis of Molecular Docking

Binding energy is a good indicator to show the tendency of bonding for ligands. The lower the binding energy, the more stable the binding conformation is. When the value of bonding energy is negative, this means in this case that the ligands and the receptor can bind spontaneously. In this study, Wognin was selected as the representative ligand, and IL6 was selected as the receptor, the threshold is set to be -5.0 kcal/mol(The AutoDock system is evaluated by using semi-empirical free energy evaluation function, and the docking result is stable when the binding energy is lower than-5kcal/Mol, so the energy of molecular docking is set to-5kcal/Mol.), any value under this threshold will be our result (Table 2). The research showed that model is in this case is within our threshold, which indicates a stable and spontaneous bonding between the compound and the target. The result was shown in Fig.7.

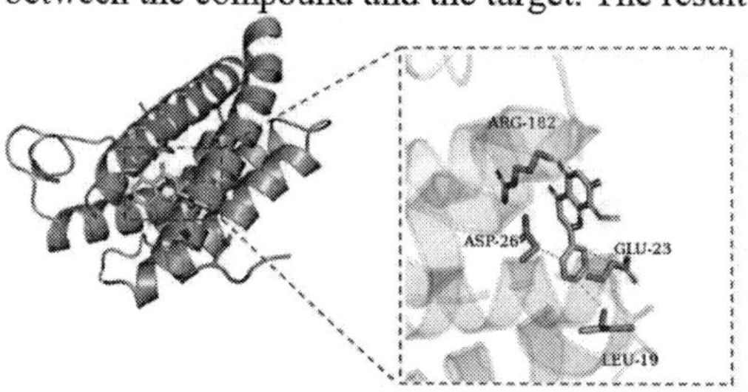

Fig.7 Molecular docking

Table 2. Binding energy values between the active compound and the target.

mode	Affinity (kcal/mol)	dist from rmsd l.b.	best mode rmsd u.b.
1	-5.2	0.000	0.000
2	-4.7	0.951	1.050
3	-4.7	2.634	6.645
4	-4.6	2.380	5.974
5	-4.5	1.913	5.728
6	-4.2	2.163	5.975
7	-4.0	2.269	3.005
8	-4.0	1.758	3.601
9	-4.0	2.129	3.554

4. Discussion

Severe acute respiratory syndrome corona-virus 2 (SARS-CoV-2), the pathogen responsible for corona-virus disease 2019 (COVID-19), has caused morbidity and mortality at an unprecedented scale globally [9]. Scientific and clinical evidence is evolving on the sub-acute and long-term effects of COVID-19, which can affect multiple organ systems [10]. Novel corona-virus disease (COVID-19) has attracted much attention around the world due to its rapid transmission among humans and relatively high mortality rate [11].Current treatments on covid-19 worldwide is based on the symptoms shown and the severity, the symptomatic and mild symptoms carriers are normally issued home isolation, moderate and severe cases are hospitalized. The hospitalized treatments are standard applications of known practices according to vital signs of the patients, from oxygen therapy to mechanical ventilation, from single application of antibiotics antiviral drugs to cocktail therapy, and according to patient's personal situation the application of plasma exchange [12]. The prevention of the disease is mainly by mass vaccination, but with the patent blocks from large pharmaceutical companies and lack of raw materials of the vaccines. Mass vaccination to all population is a dream too far and too good to be true. The research of Cangzhu provides another method which has been applied throughout China's recorded Chinese medical history for respiratory pandemics also for the current covid-19 pandemics as prevention of the spread of the airborne Pneumonia like disease.

No studies so far have examined the mechanism of Cangzhu's antiviral actions though practices have already proven the disinfection effectiveness of the herb. Therefore, a research into the disinfection mechanism of the plant is of great significance while the majority undeveloped but mass populated world is still lacking the opportunity of getting the protection from the expensive and scarce vaccine. With this aim in mind, the Cangzhu related target and covid-19 related targets are analysed, and protein-protein interaction network is constructed. The most important core targets are SOCS3, IL6, IL-10, IL-13, IL-4, CXCL-8, IL-1B, IL-6R, JUN and IL6ST.

SOCS3 is a novel class of immunosuppressive molecules that inhibit a variety of cytokines and regulate the intensity and duration of the effects of cytokines, hormones, and growth factors by inhibiting the JAK/STAT pathway and signaling [13]. IL6 is the core target in PPI network, which indicates that it plays a key role in PPI network. When COVID-19 infects the upper and lower respiratory tract, it can cause a mild or highly acute respiratory syndrome with consequent release of pro-inflammatory cytokines, including interleukin (IL)–6 [14]. It is reported that IL-6 can act on fibro blasts, induce their activation and migration, and promote the occurrence of pulmonary fibrosis. Suppression of IL-6 has been shown to have a therapeutic effect in many inflammatory diseases [15].IL-10 is a multi-cellular and multifunctional cytokine, which regulates cell growth and differentiation, participates in inflammatory and immune responses, and is recognized as an inflammatory and immunosuppressive factor [16, 17]. IL-13 can induce the differentiation of monocyte and enhance the expression of MHC II molecules. Inhibit the secretion of mono-nuclear factor induced by LPS and control the inflammatory response [18, 19]. IL-4 is a growth factor secreted by T cells themselves, and has immunomodulatory effects on B cells, T cells, mast cells, macrophages and building cells [20-22].IL-6R plays a role in multiple processes of the immune response. It can promote the occurrence of adaptive immune response, inhibit the differentiation of Th1 and Treg [23], promote the differentiation of Th2 and Th17 cells, and play an important role in the proliferation and differentiation of B lymphocytes into antibody-producing plasma cells and the conversion of Ig category [24]. IL-1B mainly plays an immunomodulatory role, promoting the growth and differentiation of B cells, and also promoting the formation of antibodies[25, 26].CXCL-8 is a kind of cytokines in the chemo kine family, and its role in participating in and regulating the physiological and pathological processes of human reproduction has been confirmed. One of the mechanisms of action is that it binds to its specific receptor [27, 28]. JUN is a transformation gene of avian sarcoma virus 17, which encodes a protein that has highly similarity to the viral protein and directly interacts with specific target DNA sequence to regulate gene expression [29, 30].IL6ST might play an important role in TNBC progression and could serve as a tumour suppressor gene for diagnosis and treatment [31]. IL6ST is closely related to the incidence of gastric cancer, hepatocellular carcinoma, osteosarcoma, and breast cancer, and may be used as a biomarker for some solid tumours [32].

From the analysis above, Cangzhu may play a great role in the prevention and treatment of covid-19 by regulating the expression levels of these core targets. To further clarify the mechanisms of action, we carried out enrichment analysis of GO and KEGG. GO enrichment analysis shows that the effective compounds of Cangzhu were mainly involved in the regulation of regulation of apoptotic DNA fragmentation, T-helper 17 cell lineage commitment, response to peptidoglycan, regulation of T-helper 2 cell cytokines production, and regulation of production of miRNAs involved in gene silencing by miRNA. KEGG pathway enrichment analysis shows that the effective compounds were mainly involved in the Antifolate resistance, inflammatory response, viral infection, Cytosolic DNA-sensing, and IL-17 related signaling pathways. Previous studies have shown that the development of COVID-19 and its sequelae (pulmonary fibrosis) is closely related to inflammation, apoptosis, and angiogenesis, and this is consistent with the result of our research. The results of molecular docking showed that the binding energy values between effective compounds and targets were less than -5.0 kcal/mol, indicating that there shows an affinity for the compounds and receptors. Based on all the above evidence, we can see that the core effective compounds of Cangzhu may have the intervention effects on the COVID-19 through anti-inflammatory, increase blood oxygen level, regulation of apoptosis, and inhibition of pulmonary fibrosis.

5. Limitations

In this study, we used network pharmacology to research the mechanism behind Cangzhu's antiviral properties. During our study, the active compounds, and targets of Cangzhu is identified. With help of protein-protein interaction network, the possible mechanism theoretically exists. But further, more accurate experimental and clinical verification of the finding is still needed.

6. Conclusion

The overall goal of this study is to explore the mechanism behind the anti-viral actions of Cangzhu for possible prevention and treatment method for the current covid-19 pandemic. Basing on the volk and modern practices of TCMSP, we propose that network pharmacology combining with molecular docking is a feasible method to achieve our goal of research. After systematic analysis, we believe that Cangzhu has the intervention effect on Covid-19 through anti-inflammatory, increase blood oxygen level, regulation of apoptosis, and inhibition of pulmonary fibrosis.

References

[1] Information on https://www.who.int/emergencies/diseases/novel-coronavirus-2019/question-and-answers-hub/q-a-detail/coronavirus-disease-covid-19.
[2] Information on http://www.satcm.gov.cn/xinxifabu/meitibaodao/2020-04-01/14418.html.
[3] Joseph P, et al. The healing power of Chinese herbs and medicinal recipes. Routledge, London: 2008.
[4] Marion R, et al. 5-Lipoxygenase and cyclooxygenase-1 inhibitory active compounds from Atractylodes Lancea. J. Nat. Prod: 2020; 3: 347-350.
[5] Duan JA, et al. A new cytotoxic prenylated dihydrobenzofuran derivative and other chemical constituents from the rhizomes of Atractylodes lancea DC. Archives of Pharmacal Research: 2008; 8: 965-969.
[6] Li SZ. Compendium of Materia Medica: 1590.

[7] Wang JH, et al. Effects of eudesmol, an active constituent from rhizoma atractylodiss on small intestine moveemet in rats. Chinese Pharmaceutical Journal: 2002; 4: 266-269.
[8] Hopkins A. Network pharmacology: the next paradigm in drug discovery. Nat Chem Biol: 2008; 4: 682-690.
[9] Dong E, et al. An interactive web-based dashboard to track COVID-19 in real time. Lancet Infect: 2020; 20: 533-534.
[10] Gupta A, et al. Extrapulmonary manifestations of COVID-19. Nat. Med: 2020; 26: 1017-1032.
[11] Basiri A, Pazhouhnia Z, Beheshtizadeh N, et al. Regenerative Medicine in COVID-19 Treatment: Real Opportunities and Range of Promises. Stem Cell Rev and Rep: 2021; 17: 163-175.
[12] Parasher. ACOVID-19: Current understanding of its Pathophysiology. Clinical presentation and Treatment Postgraduate Medical Journal: 2021; 97: 312-320.
[13] Johnson HM, Lewin AS, Ahmed CMSOCS. Intrinsic Virulence Factors, and Treatment of COVID-19. Frontiers in immunology: 2020; 11: 582102-582102.
[14] Conti P, et al. Induc-tion of pro-inflammatory cytokines (IL-1 and IL-6) and lung inflammation by Coro-navirus-19: anti-inflammatory strategies, J Biol Regul. Homeost Agents: 2020; 32: 327-331.
[15] Zeng YJ, et al. Network pharmacology-based analysis of Zukamu granules for the treatment of COVID-19.European journal of integrative medicine: 2021; 42: 101282-101282.
[16] Chen QG, et al. Immunological and inflammatory profiles during acute and convalescent phases of severe/ critically ill COVID-19 patients. International Immunopharmacology: 2021; 10: 76-85.
[17] Le BN, et al. Highly functional virus-specific cellular immune response in asymptomatic SARS-CoV-2 infection. The Journal of experimental medicine: 2021; 5: 218.
[18] Xun QF, et al. Glucocorticoid induced transcript 1 represses airway remodeling of asthmatic mouse via inhibiting IL-13, periostin,TGF-β1 signaling. International Immunopharmacology: 2021; 97: 107637.
[19] Liao Y, et al. Mechanism of Wumei-Fangfeng herb pair in treatment of allergic asthma based on network pharmacology. Central South Pharmacy: 2020; 1: 138-143.
[20] Wang Q. Changes of IL-4, D-dimer and γ-Interferon in children with mycoplasma pneumonia and different pathological conditions. Clinical Research: 20 21; 5: 145-146.
[21] Panda SK, et al. IL-4 controls activated neutrophil FcγR2b expression and migration into inflamed joints. Proceedings of the National Academy of Sciences of the United States of America: 2020; 6: 3103-3113.
[22] Ueno M, et al. Genetically modified il-4 over-expressing mscs enhance bone healing. Cytotherapy: 2020; 5: 178-179.
[23] Hawkins G. A. et al. IL6R Polymorphisms Predict Measures Of Lung Function In Caucasian Asthmatics: Data From The Severe Asthma Research Program. AMERICAN JOURNAL OF RESPIRATORY AND CRITICAL CARE MEDICINE:2010;2:181-187.
[24] Yu ZJ, et al. Correlations of IL6R gene polymorphism with pediatric asthma susceptibility. INTERNATIONAL JOURNAL OF CLINICAL AND EXPERIMENTAL MEDICINE: 2018; 2: 1331-1336.
[25] Cirelli T, et al. Genetic polymorphisms in the Interleukins IL1B, IL4, and IL6 are associated with concomitant periodontitis and type 2 diabetes mellitus in Brazilian patients. Journal of periodontal research: 2020; 6: 55.
[26] de AJB, et al. Influence of inflammasome NLRP3, and IL1B and IL2 gene polymorphisms in periodontitis susceptibility. PloS one: 2020; 1: 15.
[27] Ma YZ. The Role of CXCL8/FAK Signaling Pathway in the Proliferation, Migration and EMT of Gastric Cancer Cells. Inner Mongolia University: 2020.
[28] Xing ZW, et al. Clinical applicationg of CXCL8 and its receptor in the diagnosis of pulmonary tuberculosis. Journal of Hebei Medical Universty: 2020; 10:1176-1180.
[29] Hartl M, et al. Molecular targets of the oncogenic transcription factor jun. Current Cancer Drug Targets: 2003; 1: 41-55.
[30] Martín G, et al. Jun expression is found in neurons located in the vicinity of subacute plaques in patients with multiple sclerosis. Neuroscience Letters: 1996; 2: 95-98.
[31] Jia R, et al. Bioinformatics Analysis Identifies IL6ST as a Potential Tumor Suppressor Gene for Triple-Negative Breast Cancer. Reproductive sciences: 2021.
[32] Yue YY. Expression of IL-6ST in Colorectal Cancer and Its Effects on Proliferation, invasion and Migration of Colorectal Cancer Cells. China Medical University: 2020.

Cite as: M.W. Sang et al.,
TBIS 10.3993/tbis (2021).

Study on the Structure of Internal-using Hygroscopic Workwear in Medical Protective Clothing

Ming-Wei Sang, Song-Ning Jia, Ming-Hai Cui*

Beijing Institute of Fashion Technology, Ying Hua Street, Beijing, 100029, China

*Corresponding author's email: fzycmh@bift.edu.cn
1st author's email address: 2854938114@qq.com

Abstract

To solve the current lack of thermal and humidity comfort in medical protective clothing for workers in high-risk environments, this paper seeks to adopt moisture-absorbing materials in the design of work clothes for medical staff, thus presenting an innovative design framework for hygroscopic workwear. This paper uses hospital staff in high-risk environment, such as nurses who operate within hospital hot zones and community service personnel, as research subjects. Their activities were recorded and subsequently analyzed Then, the MET values of these actions were discovered and listed. After calculation, the average medium-high-intensity work MET value was ≈ 5.8, and the medium-low-intensity MET value was ≈2.8, with the corresponding speed of 6.3km/h and 2.8km/h, respectively. These two speeds were the references for human motion experiments.

In the human motion experiment, the objects wore medical protective clothing, the ambient temperature was set to 26-28° n, and the motion status was determined to be 2.8km/h:5min-6.3km/h: 25min -2.8km/h:10min. The filtering paper method was employed to measure the regional body sweat. Then, the regional body sweat map was obtained. According to the results, the structural scheme for hygroscopic workwear was proposed to further promote the research process of the hygroscopic workwear.

Keywords: Thermal and Humidity Comfort; Clothing Structure; Medical Protective Clothing; Work Intensity; Sweating Amounts

1. Introduction

In 2019 -2020, because of COVID-19 pandemic, the medical staff, safety patrol personnel and volunteers, who were required by occupation to interact with potential patients constantly, were at high risk of infection. At the same time, due to the particularity of COVID-19, its infectivity was strong and its propagation speed was fast, which seriously endangered their safety. Medical protective clothing can protect the wearers from germs and viruses, by effectively isolating the wears from dangerous environments and protect their health.

However, although the medical protective clothes protect the safety of staff, they are poor in comfort, difficult to wear, poor recognition, and present low physiological adaptability. As the protective clothing functions as a barrier against liquid and gas, the heat and water vapor inside the protective clothing cannot be discharged, causing the wearers to become deeply uncomfortable. After several hours of high-intensity work, the whole body would have been soaked insweat, rendering it impossible to do high-intensity works for a long time when wearing medical protective clothing [1]. Li Qian [2] and others used surveyed frontline medical staff as the research subjects, and analyzed how their comfort relates to their temperature and humidity within the clothing. The results showed that when the protective clothes were worn for 4-8h, 55.8% of the users thought that the air and humidity permeability was average, and 43.5% of the users thought that the air and humidity permeability was poor. Medical staff think that the main problem of low work efficiency is due to the poor moisture control. Therefore, for the sake of comfort, it is necessary to develop medical protective clothing with improvements in thermal and humidity control.

At the same time, as the types of medical protective clothing users and their work sites increases, it produces separate factor due to different working hours and safetyrisks. This context has thus exacerbated the inherent design defects of medical protective clothing with a lack of standardized design that can effectively accommodate everyone's needs. Hou Yu [3] tries to improve the existing disposable medical protective clothing. by by considering the general requirements of medical staff based on factors of: comfort, convenience and protective functions. Yu proposes that the new type of medical protective clothing can be designed with two sets of fabrics: which are internal-using cloths and external-using cloths.

This paper aims to solve the problem of poor thermal and humidity comfort of medical protective clothing under high intensity labor. With the combination of hygroscopic materials and the internal-using workwear, the innovative design of the special hygroscopic work clothes was presented for the protective clothing. Specifically, the outer layer was designed as the main protective isolation layer used to insulate the wearer from the external environment and prevent infection; the inner layer is the hygroscopic workwear. The effect of circulating sweat absorption is achieved through the design of the internal-using workwear, and the replaceable moisture absorption materials are added. The specific design points are described as follows.

1) The workwear is made of special hygroscopic working clothes which are worn inside the medical protective clothes.

2) The internal-using hygroscopic workwear is equipped with patch pockets which are used for the disposable moisture absorbing material (flakes).

3) The internal-using hygroscopic workwear's structure is designed according to the sweat distribution, and the position of patch pockets which were used for moisture absorption material (flakes) were mapped out accordingly.

4) The external workwear is reusable, and only disposable hygroscopic materials (flakes) need to be replaced each time.

In this paper, the investigation results of the intensity and amount of workload of the research subjects can be used as references for motion-based experiments. Besides, the localized sweat distribution while wearing medical protective clothing can be measured through the human motion experiment. We need to use the filtering paper method to measure the distribution of sweat throughout the bodyto obtain a generalized map of sweat distribution. The structural design of the hygroscopic working clothes would then be based on this regional body sweat map, alongside the accurate placement of the disposable moisture absorbing material (flakes) and the idea of the structure design of hygroscopic workwear is provided.

In summary, the methods in this paper include analytical the research, experiments conducted with human subjects and verification experiments.

2. Investigation of the Work Intensity and Workload of the High-intensity Workers

2.1 Investigation on High Intensity Work of Nurses in Red Area

During the beginning and midst of the COVID outbreak, the inpatient nursing department and the inspection department were the departments whose staff had high probability of direct contact with the virus. According to the risk level, nurses in charge of the hospitalization of new patients were generally divided into three categories [4]: green area nurses, yellow nurses and red area nurses. The red area nurses were mainly in charge of thearrangement of discharges, ward checks, treatments, basic nursing, vital signs measurements and other work. Among the nurses, red area nurses are mostly likely to be in direct frequent contacs with new crown patient, and are the most likely to be infected by virus spray. Generally, they require the most protection, but are also required to take on the most intense workloads and require very strong physical and psychological qualities [5].

According to the New Coronavirus pneumonia diagnosis and treatment plan (trial version 8) [6], doctors and nurses need to give different degrees of treatment for COVID patients, including antiviral therapy, immunotherapy, hormone therapy, etc. most of them are clinical treatment methods: injection and drip. ICU emergency rescue takes a long time, high intensity nursing work, which is generally completed by doctors. The working cycle is 4 ~ 6h, and one round of rest.

2.2 Investigation on the High Intensity Work of Community Prevention and Control Workers and Volunteers

Liu Suzhen [7] pointed out that the community prevention and control personnel (such as community medical personnel, neighborhood committees and village committee staff, police, property management personnel), who are infected with the pneumonia prevention and control network of New Coronavirus, undertake community fever clinic screening, follow-up, referral, traffic road point screening, door-to-door registration, and investigation.

Community volunteers also play an important role in community service. From the "community volunteer service action guide" [8], we can see that they have certain physical labor services: help to buy materials, transport support materials, and deliver goods needed for daily life.

2.3 Investigation on High Intensity Work of Security Personnel (Airport as an 3xample)

In the article "Analysis of joint prevention and control mechanism in the capital metropolitan area" [9], it pointed out that in the early stage of the outbreak, nucleic acid samples were collected on a large scale in high-risk areas with communities as the unit.

Novel coronavirus pneumonia joint control mechanisms at Beijing airport port [10] pointed out that the initial working mode of joint prevention and control mechanism has been formed in the inspection and quarantine of entry passengers. Other methods that were used include entry management, personnel diversion, information notification, identity registration, health monitoring, and emergency treatment, etc. Four flights within 24 hours, subject to flight schedule allocation. According to the time and shift, each person on duty for a single time will split their attention between 6-12 flights.

2.4 Work Intensity (MET) of Research Objects with Protection Needs

The daily work content of high-intensity workers with protection needs is compartmentalized into simple repetitive work or action. Wei Yang [11] showed that the comprehensive heating degree and activity intensity of human body can be roughly measured by monitoring their cardiopulmonary function indexes, and the perspiration rate per unit area can be predicted, by drawing a positive correlation between activity intensity and perspiration volume of the human body.

The following table is the compartmentalization of working action taken by subsection of research subkects, and the corresponding MET values in the physical activity compendium [12] are listed one by one.

Table 1. Working action decomposition and MET table

Content	Specific activities	Major heading	Code	METS
Buy supplies	Walking: Round-trip outhouse, purposeful	Volunteers activity	17280	2.5
Delivery of supplies	Stand and walk back and forth in a multi-item integrated motion	miscellaneous	09071	2.5
	Walking: 5.6 km/h, fast, carrying objects less than 11.3 kg	occupation	11810	4.8
	Weight 0.5-6.8 kg: up the stairs, down the stairs, shuttle back and forth	walking	17026	5.0
	Truck driving: loading and unloading trucks, bundling loads, standing, walking and carrying heavy loads	occupation	11766	6.5

We selected 23 representative activities, such as purchasing materials, transporting materials, guiding diversion, station temperature measurement, etc., and divided them into 144 activities, including 39 kinds of activities through the activity category table of the body activity compilation. According to the principle of metabolism counting, 4.5-9 is medium and high intensity work in professional activities or sports. From the Fig.1 below, in the medium and high intensity areas, MET $\approx$ 5 is the most frequent, and after weighted average statistical calculation, the maximum met value is about 5.8, i.e., running: 6.3km / h (105.3m / min). It can be seen from the Fig.1 that in the low intensity area, met is the maximum strength of 2.5, and the maximum met value is about 2.8, namely walking (fast walking): 3.8km / h.

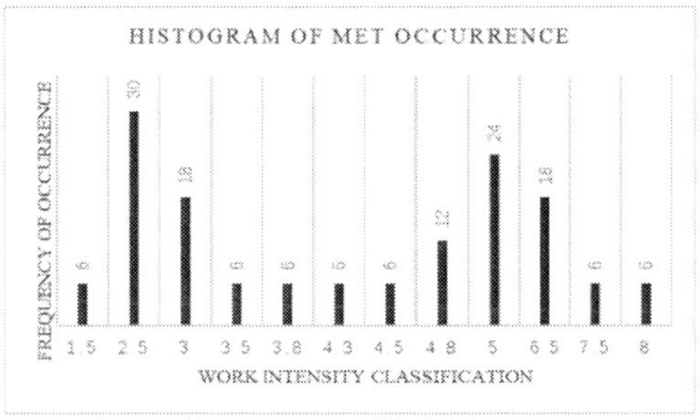

Fig.1 Histogram of MET occurrence

3. Experiment on Measurement of the Regional Body Sweat Amount While Wearing Medical Protective Clothing

3.1 Purpose

Under regulated temperature and humidity under laboratory conditions, the experimenters worn medical protective clothing. By simulating activities of different intensity on treadmills, we used filter paper to collect the sweat from 13 parts of the subject's body. The analysis was efficiently carried out by Atlas and provides a reference for the structure and materials of the internal-using hygroscopic workwear.

3.2 Objects and Equipment

Through the questionnaire survey, Li Qian found that the average height of medical staff was (165±5) cm, the age was 20 ~ 50, and most of them were female. In order to get close to practical conditions, 25 women with height of (165±5) cm, age of 22 ~ 35 and weight of 50 ~ 60kg were selected in the experiment. In the constant temperature and humidity laboratory, the temperature is set at (28 ± 2)° m, and the relative humidity is (40 ± 10)%. The main experimental equipment used are: Keep treadmill K2, stopwatch. According to the results of the work intensity (MET), the speed and time were set as 3.8km/h-5 minutes, 6.3km/h-25 minutes and 3.8km/h-10 minutes respectively. The total exercise time was 40 minutes.

3.3 Process

Table 2. positions

CODE	MAME
A	Inner thigh
B	Lateral thigh
C	Under the leg socket
D	On the chest
E	In the belly
F	Intraabdominal
G	On the back
H	In the back
I	Under the back
J	Under the armpit
K	On the spine
L	Under the spine
M	Under the left arm

3.3.1 Determine Measurement Positions

Verde t [13] determined that the perspiratory positions of various parts of the human body are symmetrical, that is, the s perspiration rate of the right hand and the left hand, the right thigh and the left thigh are symmetrical. In order to simplify the test process and make it operable, the left part of the body was selected for the test.

In this experiment, we marked the neckline, front midline, chest line, waist line, armpit line and back midline by black marking line on the bodies of objects, so as to divide and determine the accurate position of the tested parts. 13 different parts of the left side of the body were tested. The position of the filter paper patch is shown in the Fig.2 and Table 2.

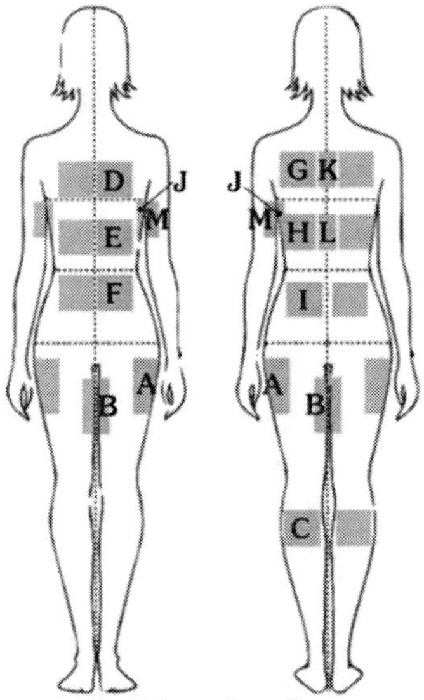

Fig.2 Location diagram

3.3.2 Motion Experiment Process

The self-sealing bag was pre marked with the corresponding movement speed and the position of filter paper to differentiate against each other. Finally, put each self-sealing bag with filter paper on the electronic scale and weigh it twice. After recording the data, the average value was recorded ias according to the weight M of filter paper corresponding to their respective body part before the experiment. According to the position determined in 3.2.1, the filter paper is pasted in the corresponding self-sealing bag, and then quickly put on the pure cotton T-shirt and medical protective clothing. After 40 minutes of exercise, the filter paper attached to each part was torn off from the skin surface one by one and put them into the corresponding self-sealing bags.

After sealing, the bag is weighed immediately. The weighing method is the same as that of filter paper weighing before exercise. The time is controlled within 3 minutes. Finally, the average value of two weighing data is obtained and recorded as N.

3.4 Data Analysis

HAVENIT G FOGARTY A [14] pointed out the perspiration rate of the human body is larger in the front and back midline, while the perspiration rate of the area on the side of the human body is smaller. Considering that there must be a certain margin in the moisture absorption ability, in the area division of sweat, the higher perspiration in the area was taken as the main test subject. And the sweating area was divided according to the perspiration of the main test subject. The weight gain of filter paper before and after exercise was the sweating amount Z in this period.

$$Z = N - M \qquad 1)$$

And The amount of sweating per unit area Zs (g × 10-2 / cm2) was

$$Zs = Z / S \tag{2}$$

Fig.3 is the final data of sweating per unit area of 25 people, and the histogram of perspiration per unit area of A~M is calculated by the average method. Fig.4 is the perspiration per unit area of divided areas by color according to the value. "Red" refers to areas with more sweating, "yellow" refers to areas with less perspiration.

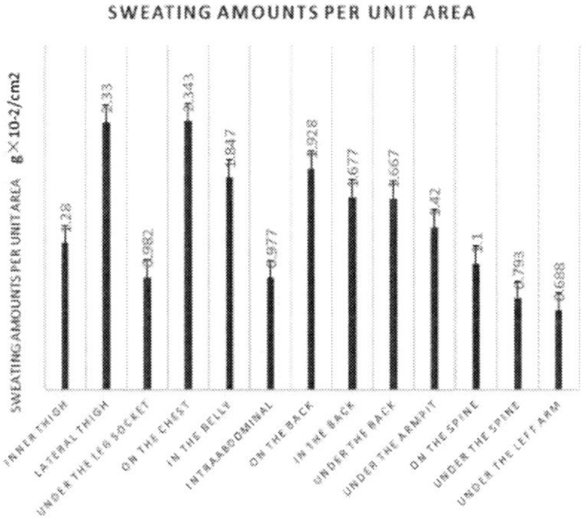

Fig.3 Sweating amounts per unit area

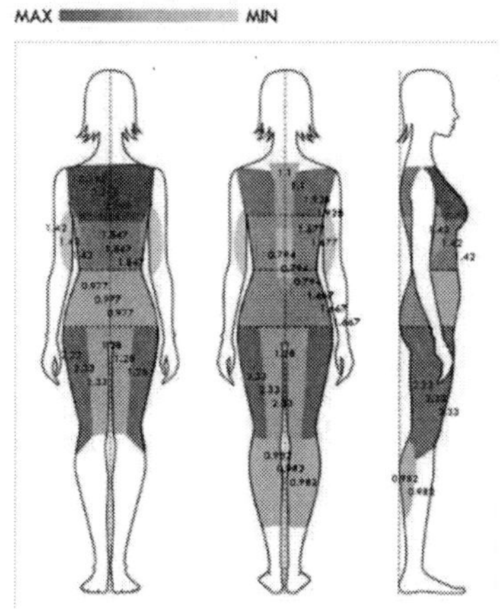

Fig.4 regional body sweat map

4. Design of Hygroscopic Work Clothes

4.1 Structural Design Principles

According to the experimental data and the regional perspiratory patterns of wearing medical protective clothing, the structural design principles of the internal-using hygroscopic workwear are detailed as follows.

1) Positions of the patch pockets; more sweat absorbing material patch pockets were set in the area with more sweat (red areas); the place with large expansion and contraction of human skin surface were avoided during exercise.

2) Pattern design. Through reasonable segmentation, the work clothes were made to be as form-fitting as possible. Since the form-fitting tracksuit is convenient for wearers to work, it can also make sweat-absorbing materials play a full role simultaneously.

3) Patch pockets' shape. According to the regional sweating data, moisture absorption materials were selected, and the amount of each patch bag was scientifically controlled, to achieve efficient moisture absorption. For this purpose, the patch bags were designed as a uniform rectangles, convenient for inserting moisture absorption materials (flakes) and subsequent production.

4) Patch pockets directions. Axillary hygroscopic material (flakes) insertion direction was changed to side oblique insertion to avoid falling when performing demanding movements.

The Fig.5 illustrates the front and back structure of the hygroscopic work clothes. The direction of the arrow indicates the direction of the material inserted into the patch pockets.

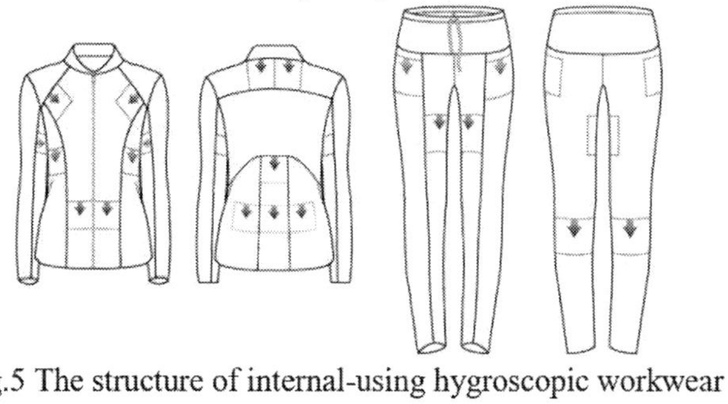

Fig.5 The structure of internal-using hygroscopic workwear

4.2 Fabric Design Principles

In this paper, eight kinds of knitted fabrics suitable for work clothes are selected, including plain knitted fabrics, pick hole knitted fabrics and knitted mesh fabrics.

The purpose is to select the fabrics with excellent performance as the main fabric of moisture absorbing work clothes by testing their moisture permeability. The pictures below show the final designs.

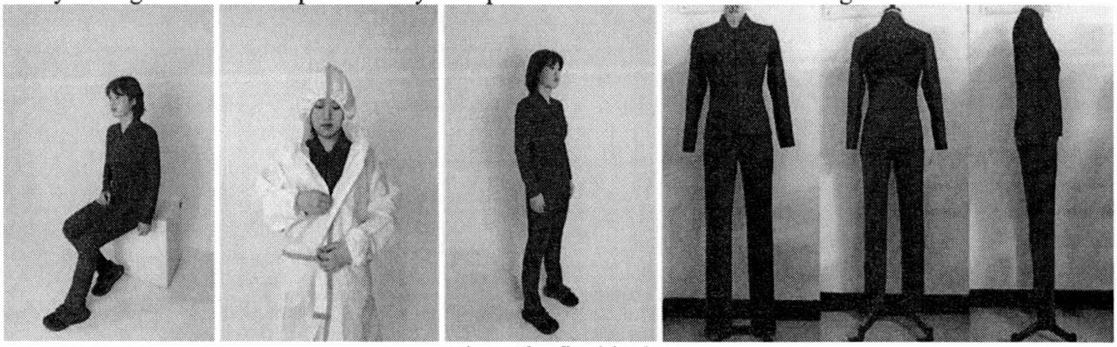

Fig.6 The final looks

5. Verification Experiment

The same test subjets were wearing hygroscopic work clothes with pure cotton upper and lower clothes, and they were wearing medical protective clothing under the same experimental conditions as the perspiration experiment. The efficiency of these hygroscopic work clothes were verified by comparing the initial humidity, the turning point of humidity increase, the humidity growth rate and the highest humidity of the internal environment humidity curve, and through the analysis and comparison of the subjective comfort of the wearer (stuffy, wet, hot, cold, itchy), benign modification suggestions to the design scheme. [15]

After the specified movement state, the maximum humidity of the internal environment is quite different. The maximum humidity of the group wearing hygroscopic work clothes is about 68.2%, while that of the control group is 86.8%. Therefore, the hygroscopic tablets can effectively reduce humidity.

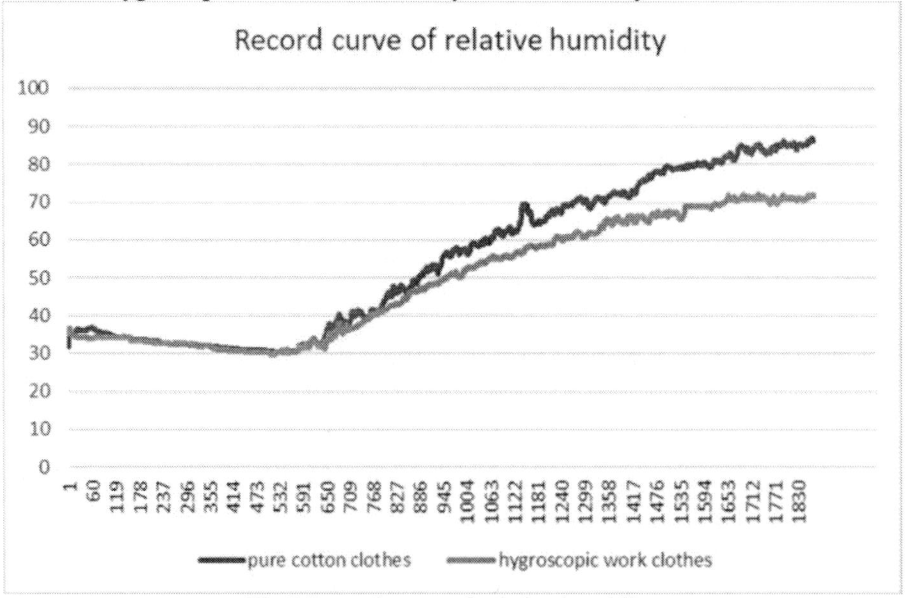

Fig.7 Record curve of relative humidity

In the follow-up development and production stage, different humidity regulatory levels of hygroscopic work clothes can be analyzed. Based on the field investigation of work intensity and the analysis of low, medium and high work intensity as well as the extent of actions taken, personalized hygroscopic work clothes for different types of work can be developed.

6. Conclusion

1) Workers operating in high-risk enviornments, such as nurses and community service personnel in the red area of the hospital, that require protective wear were selected as the research subject. Their work movements were compartmentalized and subsequently analyzed, corresponding to the types of activities in the physical experiments. The MET values were listed and analyzed. The average value for medium and high work intensity (MET) was $\approx$ 5.8, the medium and low intensity (MET) was $\approx$ 2.8, and the corresponding running speeds were 6.3km/h and 2.8 km/h. These results can provide a reference for the motion status in the experiment.

2) In the experiment that measures the regional perspiration of the human body while wearing medical protective clothing, the amount of perspiration per unit area of 13 parts (including the inner thigh, the outer thigh, the lower leg socket, the upper chest, the middle abdomen, the lower abdomen, the upper back, the middle back, the lower back, the lower armpit, the upper spine, the lower spine, and the lower left arm) were calculated based on the experiment data. The perspiration map demonstrated that the chest, abdomen, thigh lateral, and back presented larger concentrations of perspiration.

3) In this study, the hygroscopic materials and work clothes were combined into a complete protective wear, and the innovative design of hygroscopic workwear in protective clothing was proposed. The research results provided experiences for humidity treatment in the fully enclosed clothing environment, as well as new ideas for the improvement of thermal and moisture comfort of medical protective clothing.

7. Acknowledge

First of all, I would like to thank my mentor, Professor Cui Minghai, for giving me suggestions in the formation of the topic selection and framework of the paper, which helped perfect this paper. At the same time, a clear direction is given in the selection of experimental methods to make the experiment results more accurate. Thanks to the volunteers and partner Jia Songning who participated in this experiment. Your active cooperation made the experiment complete smoothly.

Thanks to the financial support provided by the scientific research project of Beijing Institute of fashion for the publication, as well as the financial support provided by the practical training project (Project No.NHFZ20210022/026) in human experiment and material development.

References

[1] Wang J, Dai HQ, Wang LL. Thermal comfort evaluation of medical protective clothing [J]. Modern silk science and technology: 2016; 31 (03): 92-94 + 100.

[2] Li Q, Yang PY, Xu XM, Li L. Comfort analysis of disposable one-piece medical protective clothing [J]. Journal of Xi'an University of technology: 2020; 34 (05): 8-13.

[3] Hou Y. Design of medical protective clothing based on wearer's needs [J]. Acta apparel Sinica: 2020; 5 (03): 238-243.

[4] Peng X, Peng Y, Xiao L, Jin CX, Liu L. Exploration and practice of step-by-step nursing scheduling mode for medical teams in Hubei Province [J]. Chinese PLA Journal of nursing: 2020;37 (02): 10-12.

[5] Bai XL, Li RL, Liu A, Han LY, Liu J, Yan B, Liu D, Zhang YX. Impact of New Coronavirus pneumonia on first-line clinical nurses [J]. China Journal of mental health: 2021;35 (01): 85-87.

[6] New Coronavirus pneumonia diagnosis and treatment plan (Trial Eighth Edition) [J]. Chinese Journal of clinical infectious diseases: 2020;13 (05): 321-328.

[7] Liu SZ, Li JP, Liu CQ, et al. Community prevention and control of New Coronavirus infection [J]. China Clinical Journal of thoracic and cardiovascular surgery: 2020.

[8] Zhang S, Lin JY. New practice and experience of community voluntary service in epidemic prevention and control [J]. Housing and real estate: 2020; No. 574 (15): 10-11 + 28.
[9] Liu XY, Ou YP. Analysis of joint prevention and control mechanism in capital metropolitan area [J]. Frontline: 2020 (3): 65-67.
[10] Novel coronavirus pneumonia joint control mechanism at Beijing airport port [J], Wei Lirong, Wang Linghang, Wu Guoan, Li. China's frontier health and quarantine journal: 2021; 44 (01): 60-62.
[11] Wei Y. Determination of sweating rate of whole body and upper body [D]. Donghua University: 2012.
[12] Ainsworth BE, Haskell WL, Herrmann SD, et al. 2011 Compendium of Physical Activities: a second update of codes and MET values [J]. Medicine & science in sports & exercise: 2011; 43 (8): 1575-1581.
[13] Verde T, Shephard RJ, Corey P, et a1. Sweat composition in exercise and in heat [J]. Human Kinetics Publishers: 1983.
[14] Havenit G, Fogarty A, et a1. Male and female upper body sweat dist-bution during ruruling measured with technical absorbents [J]. Eur J Appl PhySiol: 2008; 104: 245. 255.
[15] Zhang ZH, Li LY, An RP. Evaluation of thermal and wet comfort of head and trunk of duct ventilated clothing [J]. Journal of textiles: 2020; 41 (08): 88-94.

Effectiveness of Copper and Emerging Technologies against Novel Coronavirus

Azam Ali[1], Mehrukh Zehravi[2], Hira khaleeq[3], Muhammad Zaman Khan[1], Dana Kremenakova[1], Jiri Militky[1]

[1]*Department of material engineering, Technical University of liberec, Czech republic*
[2]*Department of clinical pharmacy prince sattam bin abdul aziz university, Kingdom of Saudi Arabia*
[3]*Department of Chemistry Government college university, Lahore*

*Corresponding author's email: mehr_azam91@yahoo.com

Abstract

Coronavirus belongs to the novel virulent strains of respiratory viruses. It is an invisible enemy that is a significant threat to human health. The major problem in the spread of coronavirus is that transmission is not limited to human to human (by contact, fomites, and droplets), it can also be transmitted from contaminated surfaces to humans. These infectious viruses can survive in different non-biocidal materials for a long time. Copper holds a significant position in different biological and biochemical processes because its ions Cu^{+2} and Cu^{+} can carry out oxidation, dioxygen transportation, and electron transference. It is a redox-active metal. It can convert into Cu^{+2} or Cu^{+} state by accepting or donating electrons. Reactive Oxygen Species (ROS) are generated on alloy surfaces. The redox reaction of copper ($Cu^{+2} \leftrightarrow Cu^{+}$) along with the generation of ROS results in enhanced inactivation of the virus. In this review, the effectiveness of copper against coronaviruses is explained. The denaturing of specific proteins of coronavirus by the interaction of copper and its ions has also been reported. Hence, copper coated surfaces can be used in public areas to inhibit the spread of the virus. Furthermore, the review represents the different techniques used for the coating of copper on conductive and non-conductive surfaces.

Keywords: SARS-CoV-2, Copper Binding with Peptide, COVID-19 Respiratory Disease

1. Introduction

Viruses are small intracellular infectious agents or microorganisms. The viral genome consists of DNA or RNA strands coated by a protective, virus-coded protein [1]. Treatment of COVID 19, is facing the crisis point [2]. Coronavirus belongs to an animal virus family. There are different types of human coronaviruses which are generally categorized according to their severity and spreading range. According to researchers, there are about seven different kinds of the coronavirus which can infect the humans. The most common of these kinds include HKU1 (beta coronavirus), OC43 (beta coronavirus), NL63 (alpha coronavirus) and 229E (alpha coronavirus). Some rare types that can cause severe infections include MERS-CoV (Middle East Respiratory Syndrome) and SARS-CoV (Severe Acute Respiratory Syndrome). SARS-CoV is not new as it broke out first in South China and later in Hong Kong [3][4]. It turned into pandemic between November 2002 and July 2003 with 8,273 cases and 775 human loses worldwide. Within weeks, SARS spread from Hong Kong to 37 countries [5]. After 10 years, MERS-CoV appeared in the Kingdom of Saudi Arabia (KSA) in 2012. On 7 February 2014, there were 182 total confirmed cases, out of which 79 (43%) were in critical condition. MERS-CoV belongs to the family of SARS-CoV having a comparable positive sense and a single-stranded RNA genome. The genome structures of single-stranded SARS and MERS-CoV and their transmission routes has been shown in Fig.1 [6]. In the year 2019, a new virus called SARS-CoV-2 broke out, triggering the spread of coronavirus disease 2019 or COVID-19. The curse of this identified virus spread too rapidly worldwide. This virus has not only received wide media coverage but also the attention of scientists. In December 2019, the occurrence of COVID-19 was first recognized in Wuhan, Hubei,

China. The World Health Organization (WHO) on 11 March 2020 identified the COVID-19 as a pandemic [7]. Its transmission rate is very high as compared to that of the previous coronavirus. The main spread is due to close contact with others, respiratory droplets discharge, and by touching a contaminated surface. The risk increased when someone has close contact with animals. The calamity is that the virus does not show its symptoms in the first few days. The incubation period of the virus inside the human body is about 4-7 days. The infected person (during the time of incubation of the virus) keep on visiting different places, touching the surfaces, and remain in contact with people. It gets very late when the person develops the symptoms of infection. Till that time, the virus is delivered to many places, communities, or even to family members. When other people touched these contaminated places, the infectious virus transfers to the facial mucosa. Also, if we are using futile cleaning agents, they may leave behind the residual virus particles that again cause infection [8].

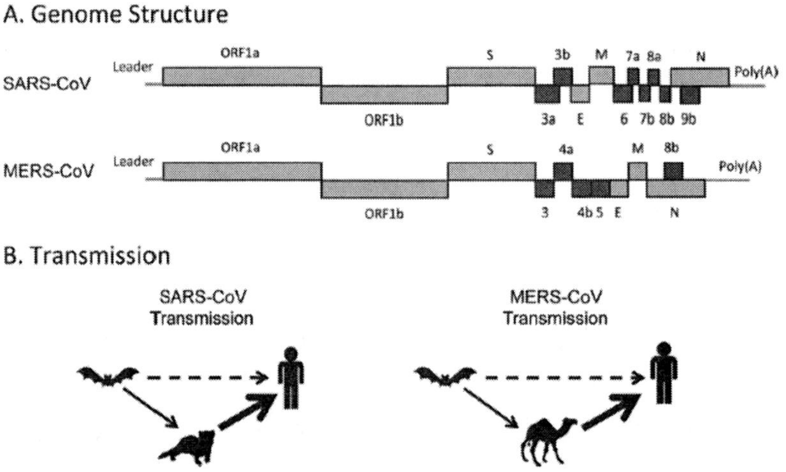

Fig.1 Genome structures of single-stranded SARS and MERS-CoV and their transmission routes [6]

There are number of detection techniques are developed to identify the COVID virus. The CT scan and nucleic acid test are the most commonly used detection techniques for the COVID disease. Instead of syndromic and CT scan the molecular detection approach is most suitable. The molecular detection approach depends on the analysing of proteomic and genomic structure of the microbes. These technique also having some limitations, (1) unavailability in countryside hospitals, (2) denaturing of sample kits during transportation, (3) sometime possibility of false-negative results because the appropriate sampling time for nasopharyngeal swab is still vague for the patients on ventilators. No doubt, PCR testing is the most suitable and easy method and is in use worldwide, but we need to consider and modify the drawbacks associated with this technique. Incase of nonavailability of PCR test an alternative technique of CT scan test is also used for the detection of COVID disease. The major drawback of CT scan method in COVID-19 diagnosis is the lack of specificity because of the overlap of lungs images with other pneumonia. Moreover, the CT scan diagnoses the patients at advanced stage of infection [9].

Copper is essential dietary element for human body. It rapidly absorbed into small intestine and contributes in plasma pool and red cell pool. Copper plays a vital role in the functionality of natural killer cells, B cells, T helper cells, neutrophils, and macrophages etc. These cells are helpful and effective in the control of infectious diseases, cell-mediated immunity and production of specific antibodies. The copper excess level or its deficiency creates abnormality in cellular function or damages that given its central role in host-pathogen interaction. The molecular interplay between the virus and the cellular machinery manages Cu^{2+} flux. Subtle alterations of Cu homeostasis can occur in infectious diseases and results in toxic Cu accumulation to eliminate pathogen. In fact, copper has the capability to kill several infectious viruses such as poliovirus, bronchitis virus, human immunodeficiency virus type 1(HIV-1), single- or double-stranded DNA and RNA viruses, enveloped or nonenveloped viruses [9]. Copper persuaded viral killing may be attributed via ROS, and in this regard, Copper oxide, cuprous oxide and most of the bleaching agents like hydrogen peroxide play the essential roles. By making or coating the surfaces with biocidal materials may help to minimize the danger of spread of coronavirus by touching contaminated surfaces [10].

2. Coronavirus and Structure

Coronavirus is a family of viruses originated from animal source. There are almost 7 types of coronaviruses detected by scientists and severity of disease is different for each of them. The SARS-CoV-2 (Covid-19) consists of about four structural proteins (N, S, E, and M) as shown in Fig.2. The N-protein contains the RNA genome, while the M, E, and S proteins collectively form the coat of the virus body. The S-protein attaches the virus to the membrane of the host cell. Chemically, these are glycoproteins (containing oligosaccharides) covalently attached to the side groups of the polypeptide chains [11].

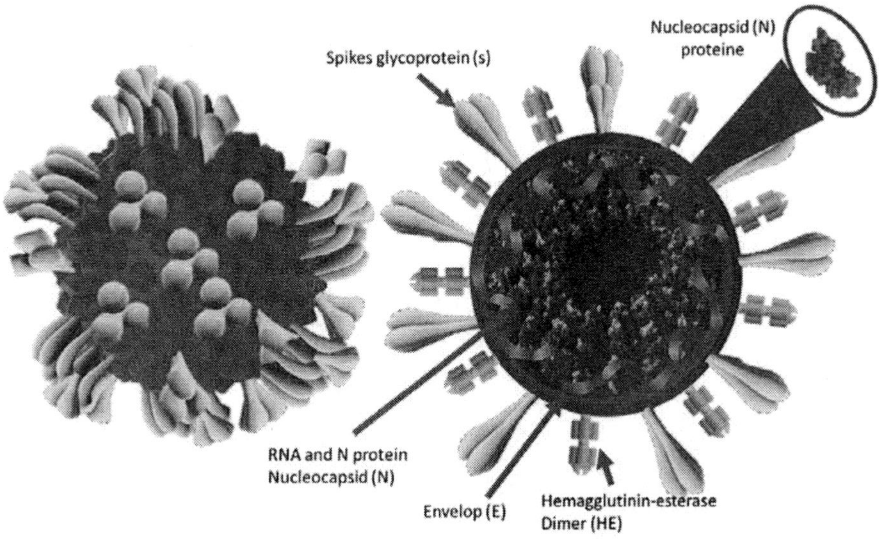

Fig.2 Structure of Coronavirus [11]

2.1 Transmission of Coronavirus

The transmission of coronavirus is similar to as many other respiratory viruses (through tiny droplets). These droplets are released from the mouth and nose of the infected person as he/she coughs or sneezes without covering his/her mouth. Nearly 3,000 droplets are produced as a result of a single cough. The coughed droplets have the average size distribution of 0.62-15.9 microns. These droplets can travel and contaminate the air that is 2 to 6 meters away. Even a single single droplet is enough to infect the individual. Although enveloped respiratory viruses are more susceptible to environmental stresses than non-enveloped viruses, they can still survive on non-biocidal surfaces and contaminate more than 50% of the commonly used surfaces [12]. These particles can transmit to other people, clothing, and surfaces around them during contact. Some of these smaller particles can also stay in the air and by touching or shaking hands with a person who is infected by this virus can pass it to the other individuals. By getting in contact with an object or surface that is contaminated with the virus and then touching the face, nose, eyes, or mouth also causes infection. There are some evidences that the virus can survive for a longer time in faecal matter. So, a person who does not wash his/her hands after visiting the toilet can contaminate anything he/she touches. According to the Centres for Disease Control and Prevention the main spread is due to the touching a contaminated object and then touching one's face. The Centres for Disease Control and Prevention and World Health Organization are still emphasizing to Clean or disinfect the common touched surfaces on regular basis. Both of these activities are very helpful to prevent the spread of Covid-19. So, although it is not known the number of cases which are caused directly by touching the contaminated objects or surfaces, experts advise us to exercise caution [13].

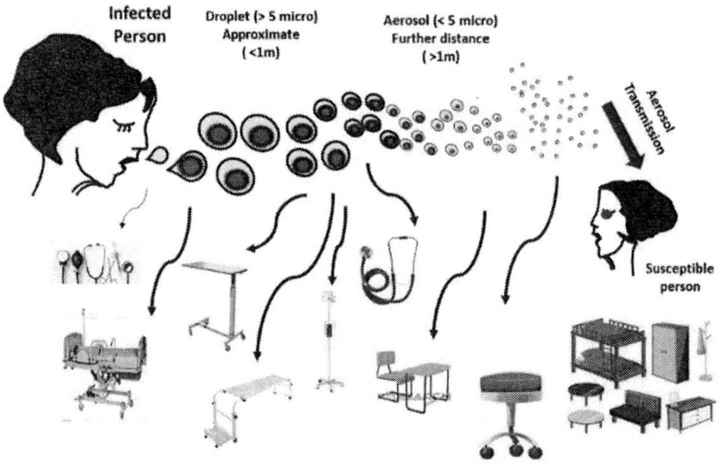

Fig.3 Transmission of Coronavirus [13]

2.2 Survival of Coronavirus Outside of Human Body

Many new researchers have found that SARS-CoV-2 can live from hours to days on range of artificial objects like metal, glass or plastic. Kampf et al. [14] quotes that human coronaviruses can survive till 9 days at room temperature but this range can prolonged for mature coronaviruses for up to 28 days. It is important to highlight that increasing temperatures to 30 °C or more can shorten coronavirus life. Although, there is no built correlation between the fluctuation in temperature and the escalation of coronavirus. The survival of virus in exterior environment rely on couple of factors. The most common factors are temperature, humidity, ventilation of the room and the type of object [14]. The load of SARS-CoV-2 on lifeless objects is still ambiguous and probably changeable, but the sanitizing of the surrounding surfaces seems to diminish the viral load, specifically the places been in contact by the infected person, where the maximum viral load is present [2]. An individual can be affected with the Covid by contacting a surface that has viral particles on it and afterward by touching mouth, nose, or eyes [15]. Van Doremalen et al. [16] analyzed the survival of SARS-CoV-2 noticeable all around and on surfaces. They also described that how infection could be spread by contaminated individuals onto regular surfaces by hacking, wheezing or contacting/dealing with objects. They additionally investigated how long the infection stayed irresistible on these surfaces. They tried viral imperativeness on plastic, tempered steel, copper, and cardboard. They additionally reproduced an airborne suspension having a fog of small beads to check the feasibility that how long the infection could stay in the air. Under these exploratory conditions, SARS-CoV-2 demonstrated to be dynamic and irresistible on plastic and treated steel surfaces from 2 to 3 days, on cardboard for up to 24 h, and on copper for 4 h. SARS-CoV-2 infection was distinguishable in vaporizers for up to 3 h. These occasions ought to shift in genuine conditions, because of the impact of components like temperature, mugginess, ventilation, dust, fingerprints, natural trash, and the measure of infection saved. They further elaborated about the texture of surface. As smooth surfaces were easily disinfected as compared to rough surfaces. These outcomes clearly suggest that individuals can get the SARS-CoV-2 through the air and subsequent to contacting polluted items. For these reasons, incessant sterilization of contacted items and surfaces assume a central part in lessening viral cross tainting [17].

2.3 Pathogenesis of COVID-19

The COVID 19 disease is seriously devastating for health. Due to the limitation of knowledge about the cellular response to the viruses, course of events can be postulated based on past studies with SARS-CoV. Based on the extensive studies the COVID-19 can be divided into three phases that correspond to different clinical stages of the disease. Stage one, is concerned with asymptomatic state with duration of two days. The inhailed virus goes through the nosal cavity and primary airways, where it binds to the epithelial cells membrane (the main receptor for both SARS-CoV2 and SARS-CoV). The virus statts the local propagation with a limited innate immune response.

At this stage the virus can be detected by nasal swabs. In stage second, the virus remain on propagation through upper airways for next few days. At this stage robust immune response is triggered. Almost, upto 80% of the infected patients go through this response. After that the disease will be mild and mostly restricted to the upper and conducting airways. The third stage is very hectic and unfortunately, 20% patients will progress to this stage. The fatality rates develop due to the pulmonary infiltration, which inturn will develop very severe disease. Still there is a huge gap and need to get more knowledge to have better understanding about the pathogenesis of COVID-19 [18].

2.4 The Detection Techniques for SARSCoV-2

The CT scan and nucleic acid test are the most commonly used detection techniques for the COVID disease. Instead of syndromic and CT scan the molecular detection approach is most suitable. The molecular detection approach depends on the analysing of proteomic and genomic structure of the microbes. The Qpcr test is a most reliable version of RT-qPCR technique use for the detection of RNA. This procedure is so easy and give the results within few hours. This technique is much reliable to detect even a small viral genetic material. It works on the principle of detection of conversion of virus RNA to small complementary DNA sequence (cDNA). The DNA produced in the first step is further multiplied through thermal treatment cycling and finally the virus is detected by qPCR testing machine. The technique also having some limitations, (1) unavailability in countryside hospitals, (2) denaturing of sample kits during transportation, (3) sometime possibility of false-negative results because the appropriate sampling time for nasopharyngeal swab is still vague for the patients on ventilators. No doubt, PCR testing is the most suitable and easy method and is in use worldwide, but we need to consider and modify the drawbacks associated with this technique [19]. In case of nonavailability of PCR test an alternative technique of CT scan test is also used for the detection of COVID disease. These images are analysed and compare to detect the difference between the subpleural area of one or both lungs of patients after the appearance of symptoms. The major drawback of CT scan method in COVID-19 diagnosis is the lack of specificity because of the overlap of lungs images with other pneumonia. Moreover, the CT scan diagnoses the patients at advanced stage of infection [20].

2.5 Biosensor Important in Pandemic Outbreaks and their Drawbacks

The analytical technique for the detection of biological molecules is most effective. The biosensors are devices used widely for the detection of particular enzymes and antibodies. These techniques are the analytical methods to detect the pathogens in various specimens with high sensitivity and selectivity without any pre-preparation of sample. The paper based biosensors are easy to fabricate, biodegradable economical and fast method to detect the infectious transmission of diseases. The paper based sensors avid the use of high technology pumps, power supply and gauges that are necessary components of PCR test. Additionally, paper based sensors reduce the possibility of further spread of disease because of their incineration after being used [21]. The ELISA assay method have been used widely to detect antibiotics but it takes long time to complete the assay. The surface plasmon resonance (SPR) sensing approach is a most sensitive detection method to detect specifically large biological molecules. Therefore, surface plasmon resonance detect quantitatively antibiotics of SARS-CoV-2 in human serum samples without any dilution. Furthermore, the portability of this sensor make it more reliable and affective on-site measurements. Overall, SPR based optical sensors are good enough to detect even very low concentration of infectious agents with superb specificity as well as are also able in detecting variants of viral strains [22]. The feld-effect transistor (FET) based biosensors are powerful diagnostic tools to detect and modification of bio-receptor structures as well. High sensitivity to diverse biological specimens of electrochemical sensors makes it convenient in detection of target analytes with benefits of rapid detection using small volume of sample. Another type of biosensor is in-house built biosensor device, which works on the procedure for detection of Covid-19 spiked protein antigen in saliva sample [23]. No doubt, biosensors are most effective and reliable for the detection of covid disease but there are number of drawbacks associated with them. They are highly sensitive, there is need of special instrumentation to perform testing, selective with automated quick response, highly expensive, some of them have low limit of range, lower shelf life and stability. All these challenges need more attention and modifications to enhance the work performance of biosensors [24].

2.6 Strategies Implemented for the Control of the COVID-19 Pandemic

There are number of preventive strategies which are used for the control of COVID-19. The asymptomatic individuals with no fever and no or delicate other indications are the huge source to fabricate the risk of sending the disease if no early logical measures are executed. The speedy and reliable screening of suspected and asymptomatic patients is the so to speak way to deal with fight with COVID-19 at starting stage. It is at this point testing to screen asymptomatic patients from particular families which as well require long time and massive work. This procedure has been used for the revelation of microorganisms and unlawful prescriptions since long time. This procedure the possible intensifying transmission of SARS-CoV-2 through wastewater [25]. The extraordinary neatness and suffcient water quality can altogether address the possible waterborne sicknesses that can be begun from wastewater. Specifcally, the vulnerable countries should forestall the unpredicted flare-ups transmission of SARS-CoV-2 through wastewater. Moreover, there should be genuine cleansing of water and wastewater to control further transmission of disease. The larger part in countries with improper sanitization or open crap systems are at higher threats to get polluted with SARS-CoV-2. Models upgrades are critical to look out for disease spread in the ecological components. The Covids, for instance, SARS-CoV and MERS-CoV are prominent in causing gastrointestinal and respiratory ailments. Surely, even a lot of considers show that the replication of these contaminations occurs in gastrointestinal track. Lately, it has been seen that 2-10% people spoiled with SARS-CoV-2 furthermore had gastrointestinal issues like free insides. Angiotensin-changing over protein 2 (ACE2) works as cell receptor for SARS-CoV-2 and is immensely communicated in little stomach related lot, lungs and oral mucosa. These are the affirmations that help the probability of SARS-CoV-2 to be copied in gastrointestinal track. In past assessments, the viral RNA has been adequately recognized in compost tests. The SARS-CoV-2 viral RNA has moreover been found in poo models. The RT-qPCR area system has been used in the area of SARS-CoV-2 RNA in poop of individuals who were debilitated and had free guts. Another report showed that around 81% people with negative throat swab results can even be recognized positive with SARSCoV-2 RNA in fertilizer tests Currently, it has been taken note that huge degree of COVID-19 patients is asymptomatic, and SARSCoV-2 was come about sure in feces of asymptomatic patients Two late examines moreover uncovered the presence of culturable SARSCoV-2 in the poop models [26].

3. The use of Copper to Prevent the Transmission of SARS-coronavirus

Some common methods used to eliminate viruses include humidity, heating, light irradiation and chemical treatments. The speedier option is the treatment of contaminated surface with some disinfectant. The research has shown that coronaviruses can be inactivated within a minute by disinfecting surfaces with 62-71% alcohol, or 0.5% hydrogen peroxide bleach or household bleach containing 0.1% sodium hypochlorite. Higher temperatures and humidity also tend to result in other coronaviruses dying quicker, although research has shown that a related coronavirus that causes SARS could be killed by temperatures above 56°C or 132°F (hotter than even a bath scalding enough to cause injury) at a rate of about 10,000 viral particles every 15 minutes. Copper is a member of the first row transition series of elements, which consists of Sc, Ti, V, Cr, Mn, Fe, Co, Ni, Cu and Zn, and belongs to group 11 of the periodic table, along with Ag and Au. The element has an atomic number of 29, an atomic mass of 63, two main oxidation states (+1 and +2) and two naturally occurring isotopes (63Cu and 65Cu), with abundances of 69.17% and 30.83% respectively. In spite of a similarity in electronic structure, there are few resemblances between the chemistry of the three elements in group 11, although certain complexes of Cu2+ and Ag^+ are isomorphous (Cotton and Wilkinson 1972). There are 29 isotopes of copper. ^{63}Cu and ^{65}Cu are stable, with ^{63}Cu comprising approximately 69% of naturally occurring copper [27].

Numerous studies evident the antibacterial and antiviral effect of copper. Zerbib et al., 2020 conducted a study in a nursing home for elderly. Antibiotic activity of copper was studied by comparison of surfaces furnished with copper and without copper. In 2 out of 4 zones of nursing home, the door knobs, handles, railings were randomly coated with copper alloy by the researchers. All zones contain the patients in equal proportion. Different outbreaks occur during the course of study including influenza, gastroenteritis caused by norovirus as well as unidentified pathogen and keratoconjuctivitis caused by unknown pathogen probably adenovirus. The copper furnished zones experience notably fewer cases of gastroenteritis and keratoconjuctivitis as compared to non-copper zones. Similarly the risk of transmission of pathogens through contact is relatively low in copper furnished zones. This shows that the copper is effective in decreasing the prevalence of infections that are contagious and can transmit through contact, by air and also in effect to some degree in decreasing infections transmitted via surfaces, in health care systems. A research was performed a research in order to study the effect of copper on influenza virus. Two metals, steel and copper were used and inoculated with 20 microliter of virus contains 2×10^6 particles. Both samples were incubated on 22°C. Fluorescent microscopy analysis showed that 500000 particles were still infectious on stainless steel after day incubation while only 500 particles remain active on copper after incubating for 6 hours. In similar to the study by Zerbib et al. (2020), replacement of steel fixtures with copper may not be able to preclude influenza but it may lessen the risk of transmission [28]. An experiment model by Miyamoto et al. (1998) concluded that copper chelates are effective against apoptosis prompted by influenza virus independent of its type. And also cause inhibition of influenza virus replication and release of virus from infected cells during apoptosis. Minoshima et al. (2016) evaluated insoluble solid state and ionic soluble copper and silver compounds for their antimicrobial activity against enveloped influenza a virus and non-enveloped bacteriophage Qβ. A higher activity was showed by cuprous oxide CuO_2 against influenza A and bacteriophage Qβ as compared to cupric oxide CuO and silver sulphide Ag_2S. Two glycol proteins hemagglutinin (HA) and neuraminidase (NA) are protruded from influenza and are significant in process of causing infection by influenza virus. Cuprous oxide has showed its efficacy in declining the titre of HA beneath the detection value within thirty minutes and of NA in ten minutes of exposure. Silver compounds demonstrated less antiviral activity as compared to cuprous oxide, hence making it a cost effective and accessible option to use in social spaces and medical centers [29].

3.1 Mechanism of Copper Action Against Coronavirus

The exact mechanism for the interaction of copper interferes with proteins in bacterial cells is not fully understood yet. Most of the researchers relay on the hypothesis of mis-metalation. The copper is a stable inactivating metal for coronavirus. The basic mechanism involves the fragmentation of the viral genome on metal surface and ensuring that inactivation is irreversible. Coronavirus exposed to metal surfaces and the positive-stranded viral RNA genome started to reduce the number of copies of replicas by fragmenting. The procedure is more intensive with increasing contact times. Copper ions directly inhibit proteases by reacting with surface cysteine and to inflict damage to the viral genome [29]. The mechanism of bacterial and viral death on copper surfaces is complex, involving not only direct action of copper ion on multiple targets but also the generation of destructive oxygen radicals, resulting in "metabolic suicide. However, it appears that superoxide and hydroxyl radical generation may be important in the inactivation of coronaviruses on copper alloys but that inactivation on 100% copper surfaces is primarily due to the direct effect of copper ions. Following application of a wet droplet to a copper surface, the predominant ionic species to dissolute from the metal surface is Cu(II), but reduction to Cu(I) and the Fenton reaction with oxidative intermediates from cell debris, molecular oxygen, or viral envelope could produce the highly toxic hydroxyl radical. ROS are generated in the natural course of coronavirus infection. Exposure to copper surfaces resulted in significant morphological changes to nonenveloped virus, where possible disassociation of the capsid sub-units exposed the viral genome to copper inactivation. In this study rapid damage, including clumping, breakage, membrane damage, and loss of surface spikes, to the coronavirus particles following exposure to copper, and some particles appeared smaller and seemed to have lost rigidity, folding up on themselves. These changes could not observed with virus recovered from other metals. Analysis of coronavirus genomic RNA from viruses exposed to copper and copper alloys revealed a nonspecific fragmentation of the entire genome that can also be observed at the gene level by the reduction in copy number of a small fragment of proteins, and the extent of damage increased with contact time. It is observed that the reduction in the capsid integrity of norovirus allows access of copper ions to the genome inactivating the virus. For coronavirus, the envelope and nucleoprotein are likewise compromised, and the process occurs more rapidly than with nonenveloped norovirus, which has a resistant capsid, to allow copper ion and/or ROS to destroy the genome. The Cu^+ ions competitively bind to the cysteine thiol versus other electron-rich system such as the guanidino group of arginine. The potential for interactions of cysteine thiol with Cu^+ ions showed that the C-terminus was blocked by addition of an aldehyde group to eliminate the possible interaction of a carboxyl group with Cu^+ ions. However, the guanidino group of arginine is an important ligand for Cu^+ ion but the cysteine side-chain are more preferred in the coordination of Cu^+ ions. Proton from the initial thiol or the sulfonic acid group is transferred to the basic guanidine, resulting in a protonated arginine side chain, and the Cu^+ ion is charge-solvated by the resulting thiolate or sulfonate group and the backbone carbonyls [30].

3.2 Efficacy of Vaccine

COVID-19 vaccines are being progressed by the use of numerous specific platforms. A few of these are conventional approaches, including live attenuated viruses or inactivated virus which have been utilized for measles vaccine and inactivated influenza vaccines respectively. Some approaches use newer platform e.g. vector and recombinant protein. However, DNA and RNA vaccines, have never been utilized in an authorized immunization. Distinctive platforms utilized for COVID 19 vaccines are mentioned here.

Inactivated vaccines SARS-CoV-2 inactivated vaccines are made by generating the virus in cell culture and then chemically inactivating it. To elicit an immunological response, the inactivated virus is frequently mixed with adjuvant or alum in the vaccination. Route of administration for inactivated vaccines is intramuscularly.

Live attenuated vaccines The development of genetically weakened variants of the wild-type virus is used to create live attenuated vaccines; those weakened viruses multiply in the recipient, eliciting an immunological response but not causing infection. Attenuation can be performed by genetically altering the virus or spreading it in un-favourable conditions, resulting in a loss of virulence but maintaining immunogenicity. These vaccines would ideally elicit cellular and humoral immunity to several components of the attenuated virus as a whole.

Recombinant protein vaccines These vaccines are made up of viral proteins which have been communicated in a variety of systems such as mammalian cells, insect cells, plant cells and yeast cells. Typically, these vaccinations are given intramuscularly. Live virus replication does not require, although production yields are dependent on the spike protein's capacity to be expressed, which is changeable.

Replication-incompetent vector vaccines These vaccines utilize the diverse vector virus which has been genetically modified to not duplicate in vivo and to show viral protein that is Immunological target. Adenovirus vectors, Adeno-associated virus (AAV), human parainfluenza virus, Sendai virus are used as candidates in many replication-incompetent vector vaccines one disadvantage of vector vaccinations is that pre-existent immunity to the vector can reduce vaccine immunogenicity [31].

Replication-competent vector vaccines Attenuated or vaccine strains of viruses are used to create vector vaccines. By using replication-competent vectors, the immune response is generally stronger than replication-incompetent vectors, although they proliferate within the vaccinated person and activate the innate immune system. Replication-competent vectors have been developed to express the spike protein in measles vaccine strain vectors, vesicular stomatitis virus-based vectors and influenza virus-based vectors.

Inactivated virus vector vaccines Inactivated virus vectors have been developed to expose the target protein but remain inactive, making them safer because they cannot reproduce indeed within immune-compromised hosts.

DNA vaccines DNA vaccines are made up of plasmid DNA that contains mammalian expression target gene and promotors, allowing the vaccine recipient to express the target protein. DNA vaccines have a poor immunogenicity and require particular delivery systems, i.e. Electroporators which restricting their application. The spike protein gene is the target of SARS-CoV-2 DNA vaccines.

RNA vaccines RNA vaccines were the first SARS-CoV-2 vaccines to be developed, and they represent a completely new vaccine technique. The RNA is translated into the target protein, which is designed to induce an immune response, once it is given. mRNA remains in cytoplasm and does not pass into nucleus; these vaccines do not communicate with recipient's DNA. These vaccines are made entirely in vitro, which makes production easier. Some vaccines must be kept at extremely low temperatures [32].

3.3 Future Challenging

In order to accelerate the treatment of infected persons, the significant diagnostic methodologies like ultrasensitive and some other can be useful in detection of SARS-CoV-2 in very initial stage. Moreover, these techniques are considerable to enable health professionals with direct resources which will lead to combat with COVID-19 in a stronger way. In current scenario, the techniques used for Covid 19 diagnostic propose are CT scan, RT-PCR, and lateral flow immunoassays. The techniques like nucleic acid testing and RT-PCR kits due to their high sensitivity and specificity are favored for the detection of SARS-CoV-2 but the accuracy in their results is mandatory for reliable use of RT-PCR test.

Aiming to control the bottleneck of diagnostic methods, the techniques which can be helpful in early detection and handling of Covid 19 should be inexpensive, easy available, fast response, and with reliable results. Presently, we have described the bio-sensing techniques for the detection of SARS-CoV-2. Biomarker-based diagnosis potential of biosensors will be the captivating approach in combination with clinical observations and risk factors to treat the patients according to the severity of their disease. In recent times, biosensors have detected SARS-CoV-2 from biological samples with high output; therefore the use of bio sensing techniques are preferred in order to diagnose the virus at initial stage. In this time of nanotechnology and quantum computing, the further improvements for bio sensing devices are possible with respect to less detection limits, specificity and reproducibility, leading the devices towards more reliably in both invitro and in-vivo diagnosis. In order to achieve the accuracy, various biomarkers in combination with multiplex bio sensing devices can be effective sensitive technique. For guaranteed speedy measurements early stage, the low cost and simple biosensing devices should be employed. More work can be done to link biosensors with computing system for health professionals to bring in decentralization healthcare which ultimately can be significant tool to diagnose contagious diseases and detect the herd-immunity areas. In addition to this, bio sensing instruments based on colorimetric strips and smartphone would target antibodies or antigens with great potential and can be good measuring devices for home-used testing. Furthermore, fast translation of lab experimental research on bio sensing devices into commercial practicable prototype through industry is demanding in order to fill the gap between lab activities and Industry requirements. Conducive to measure trace level precise and rapid detection of SARS-CoV- epidemiology based on wastewater can be employed to grow effective bio analytical devices. The society demands the enormous protection to contain the huge SARS-CoV-2 such as virus attacks and the better way to overcome in such situations is, the innovation of the super sensitive, reliable, simply accessible point-of-care testing kits and easy to handle for a common man.

4. Conclusion

SARS-CoV-2 a new strain of coronavirus is circulating and causing the coronavirus disease COVID-19. The spread and transmission rate is very fast in almost all countries. Coronavirus persists in an infectious state on common surfaces. The copper based materials (copper and ions) are very effective for inactivation of coronaviruses within time duration of four hours. The rate of inactivation is directly proportional to the percentage of copper and its different ionic forms. Copper ion release and generate the reactive oxygen species (ROS) which are involved in inactivation of coronavirus on copper surfaces. Inactivation of coronavirus on copper and copper alloy surfaces results in fragmentation of the viral genome. These studies clearly showed that cysteine and cysteic acid side chains are important Cu^+ ion ligands, and these metal ion-ligand interactions have a strong influence on the fragmentation chemistry of ions. Analysis of coronavirus genomic RNA from viruses exposed to copper and copper alloys revealed a nonspecific fragmentation of the entire genome that can also be observed at the gene level by the reduction in copy number of a small fragment of proteins, and the extent of damage increased with contact time. The C-terminal carboxyl group is an important Cu ligand, especially for multi-copper peptide ions, whereas the C-terminal amide or the methyl ester group has less influence on the Cu coordination. Cu ions preferentially bind to electron rich systems of amino group (i.e., Nterminal amino group, the side chain of lysine, histidine, and arginine, and the deprotonated C-terminal carboxyl group). As a result, the corona virus will inactive. So, the transmission of virus can be controlled by using copper-based materials or by replacing of different items form daily life with copper.

5. Acknowledgement

This work was supported by the Ministry of Education, Youth and Sports of the Czech Republic and the European Union-European Structural and Investment Funds in the frames of Operational Programme Research, Development and Education-project Hybrid Materials for Hierarchical Structures (HyHi, Reg. No. CZ.02.1.01/0.0/0.0/16_019/0000843).

The research was also supported by 'Textile structures combining virus protection and comfort' reg.c.:cz.01.1.02/ 0.0/0.0//20_321/0024467.

References

[1] Gelderblom HR. Structure and Classification of Viruses. In: Medical Microbiology: 1996; 1-14.
[2] Warnes SL, Little ZR, Keevil CW. Human coronavirus 229E remains infectious on common touch surface materials. MBio: 2015; 6: 6.
[3] "WHO Director-General's opening remarks at the media briefing on COVID-19 - 11 March 2020". World Health Organization: 11 March 2020; Retrieved 11 March 2020.
[4] Van Der Hoek L, Pyrc K, Jebbink MF, Vermeulen-Oost W, Berkhout RJM, Wolthers KC. Identification of a new human coronavirus. Nat Med: 2004; 10: 368-73.
[5] Bean B, Moore BM, Sterner B, Peterson LR, Gerding DN BH. Survival of influenza viruses on environmental surfaces. J Infect Dis: 1982; 146: 47-51.
[6] Ali A, Baheti V, Militky J, Khan Z, Tunakova V, Naeem S. Copper coated multifunctional cotton fabrics. J Ind Text: 2017; 48: 448-64.
[7] Ali A, Baheti V, Vik M, Militky J. Copper electroless plating of cotton fabrics after surface activation with deposition of silver and copper nanoparticles. J Phys Chem Solids: 2020; 13.
[8] Weaver L, Michels HT KC. Potential for preventing spread of fungi in air-conditioning systems constructed using copper instead of aluminium. Lett Appl Microbiol: 2010; 50: 18-23.
[9] Audi G, Bersillon O, Blachot JW. The NUBASE evaluation of nuclear and decay properties. Nucl Phys A: 2003; 729: 3-128.
[10] Villafranca JJ. Copper Proteins: A Review of the Different Classes of Copper Proteins and an In-Depth Investigation of the Active Copper Sites in Dopamine -Hydroxylase. Hydroxylase Biol Oxidat Syst: 1990; 1: 297-327.
[11] Harrison MD1, Jones CE DC. Copper chaperones: function, structure and copper-binding properties. J Biol Inorg Chem: 1999; 4: 145-53.
[12] Lim J, Vachet RW. Development of a Methodology Based on Metal-Catalyzed Oxidation Reactions and Mass Spectrometry to Determine the Metal Binding Sites in Copper Metalloproteins. Anal Chem: 2003; 75: 1164-1172.
[13] Men LJ, Wang YS. Fragmentation of the Deprotonated Ions of Peptides Containing Cysteine, Cysteine Sulfinic Acid, Cysteine Sulfonic Acid, Aspartic Acid, and Glutamic Acid. Rapid Commun. Mass Spectrom: 2006; 20: 777-784.
[14] Cerda BA, Wesdemiotis C. The Relative Copper(I) Ion Affinities of Amino-Acids in the Gas Phase. J Am Chem Soc: 117AD; 9734-9739.
[15] Wu Z, Fernandez-Lima FA, Perez LM, Russell DH. A New Copper Containing MALDI Matrix that Yields High Abundances of [Peptide _ Cu]_ Ions. J Am Soc Mass Spectrom: 2009; 20: 1263-1271.
[16] Van Doremalen N, Bushmaker T, Munster VJ. Stability of Middle East respiratory syndrome coronavirus (MERS-CoV) under different environmental conditions. Eurosurveillance: 2013; 18: 20590, Erratum in Eurosurveillance: 2013; 18: 20599.
[17] Fiorillo L, Cervino G, Matarese M, D'Amico C, Surace G, Paduano V, Fiorillo MT, Moschella A, La Bruna A, Romano GL, et al. COVID-19 Surface Persistence: A Recent Data Summary and Its Importance for Medical and Dental Settings. Int. J. Environ. Res. Public Health: 2020; 17: 3132.
[18] Pathogenesis of COVID-19 from a cell biology perspective Robert J. Mason Affiliation: National Jewish Health, Dept of Medicine, Denver, CO, USA.
[19] Singh AK, Singh A, Shaikh A, Singh R, Chloroquine and hydroxychloroquine in the treatment of COVID-19 with or without diabetes: a systematic search and a narrative review with a special reference to India and other developing countries. Clin Res Rev: 2020; 14: 241-246.
[20] Jiao J, Duan C, Xue L, Liu Y, Sun W, DNA nanoscaffold-based SARS-CoV-2 detection for COVID-19 diagnosis, Biosens Bioelectron: 2020; 167: 112479.
[21] Singh AK, Singh A, Shaikh A, Singh R, Chloroquine and hydroxychloroquine in the treatment of COVID-19 with or without diabetes: a systematic search and a narrative review with a special reference to India and other developing countries, Clin Res Rev: 2020; 14: 241-246.
[22] Jiao J, Duan C, Xue L, Liu Y, Sun W, DNA nanoscaffold-based SARS-CoV-2 detection for COVID-19 diagnosis, Biosens Bioelectron: 2020; 167: 112479.

[23] Du TWZ, Zhu F, Guo F, Yang B. Detection of antibodies against SARSCoV-2 in patients with COVID-19, J. Med;Virol: 2020; 1: 1.

[24] Aziz A, Asif M, Ashraf G, Farooq U, Yang Q, Wang S, ScienceDirect Trends in biosensing platforms for SARS-CoV-2 detection : A critical appraisal against standard detection tools, Curr. Opin. Colloid Interface Scil: 2021; 52: 101418. doi:10.1016/j.cocis.2021.101418.

[25] Lee H, Choi M, Jung Y, Lee SK, Lee CS, Kim J, Kim NH, Kim JKT. A novel rapid detection for SARS-CoV-2 spike 1 antigens using human angiotensin converting enzyme 2 (ACE2), Biosens; Bioelectron: 2021; 171: 112715.

[26] Shushan F, Min W, Shengjun M, Shouguo Z. A Preventive and Control Strategy for COVID-19 Infection: An Experience From a Third-Tier Chinese City: 2020.

[27] Tian ZX., Pawlow A, Poutsma JC, Kass SR. Are Carboxyl Groups the Most Acidic Sites in Amino Acids? Gas-Phase Acidity, H/D Exchange Experiments, and Computations on Cysteine and Its Conjugate Base. J Am Chem Soc: 2007; 129: 5403-5407.

[28] Van Doremalen N, Bushmaker T, Munster VJ. Stability of Middle East respiratory syndrome coronavirus (MERS-CoV) under different environmental conditions. Eurosurveillance: 2013; 18: 20590.

[29] Zerbib S, Vallet L, Muggeo A, de Champs C, Lefebvre A, Jolly D, et al. Copper for the prevention of outbreaks of health care-associated infections in a long-term care facility for older adults. J Am Med Dir Assoc: 2020; 21: 68-71.

[30] Minoshima M, Lu Y, Kimura Nakano R, Ishiguro H, Kubota Y, et al. Comparison of the antiviral effect of solid-state copper and silver compounds. J Hazard Mater: 2016; 312: 1-7.

[31] Julia S, Natalie E, Zachary J, Yang Y, Elizabeth H and Ira L. Efficacy Estimates for Various COVID-19 Vaccines: Reports; doi: https://doi.org/10.1101/2020.10.02.20205906.

[32] Haoyue C, Zhicheng P, Wenliang L, Shuting S, Minjia M, Haibo Z, Xing X, Hui L and Yunxian . Efficacy and Safety of COVID-19 Vaccines in Phase III Trials: A Meta-Analysis: Vaccines: 2021; 9: 582.

Cite as: A. Sheta et al.,
TBIS 10.3993/tbis (2021).

Performance Prediction of Back Support Belt Fabrics

Aida Sheta[1], Sherwet ElGholmy[2], Abeer Mohamed[3]

[1, 2, 3] *Textile Department, Faculty of Engineering, Alexandria University, Lotfy El-Sied st. off Gamal Abd El-Naser - Alexandria, Alexandria Governorate 11432, Alexandria, Egypt*

*Corresponding author's email: abeer.medany@alexu.edu.eg

Abstract

One of the most rapidly expanding sectors in the technical textile market is medical textiles. Lumbar back support belt is an application of medical textiles. Lumbar belt is used to provide additional back support during injuries, weight-lifting, and back pain. Using this belt several times results in fatigue. At the same time, constant exposure to tension causes creep in the belt.. A periodic loading-unloading in the belt affects the mechanical properties and the belt loses its subjected efficiency.
There are many types of lumbar support belts with different materials, structures, and matrices. This paoer studies the physical and mechanical properties of back support belt fabrics to predict its behavior in use. Different back support belt were selected for different cases of medical use. It was found the percent of elastic rubber is the most significant factor that affected the performance of the studied back belts.

Keywords: Medical Textiles; Back Support Belt; Abdominal Belt; Fabric Performance

1. Introduction

Medical textiles are the products and constructions used for medical and biological applications. They are primarily used for first aid, clinical and hygienic purposes. There are two types of medical textiles, implantable and non-implantable fibres [1-3]. Implantable fibres are the textile structures that can be used inside the human body for various purposes, such as replacement surgery. Non-implantable fibres are used for external applications on the body and may or may not contact the skin. The main requirement of medical textile material is bio receptivity and biocompatibility at the application site in human beings. Some examples are protective, healthcare textiles, dressings, bandages, pressure garments, prosthetics, hygiene products, antiseptic wound dressings, and lumbar support belts [4-7]. There are many types of lumbar support belts; lumbo-sacral orthopedic support, lifting belt, clavicle brace, back support belt, back and shoulders support brace, chest binder, and natural leather hernia truss. Each has a specific function designed [8, 9]. Nonwovens are extensively used in lumbar support belts with interlaced yarns to improve their performance and give synthetic filaments cohesion, which could only be obtained by twisting [10, 11]. Others used knitted fabric with rubber for more extensibility and comfort in use. Knitted fabrics and knitting technology play an important role in the fields of technical and medical textiles and their importance is even greater [12]. The durability and comfort of the lumbar back belt must be tested as the air permeability is an important factor in the performance of such textile materials as clothing, fabrics for airbags, etc [13, 14]. This paper studies how to predict the durability and comfort of back support belt fabrics experimentally.

2. Materials and Methods

2.1 Samples' Specifications

In this research, five different types of commercially available back support belts were studied. Table (1) shows the physical and mechanical properties of the studied belts.

Table 1. Physical and Mechanical Properties of Five Different Back Support Belts

Belt type	1	2	3	4	5
Structure	Warp knitted	Interlock	Non-woven	Non-woven	Non-woven
Rubber %	40	33.5	10.6	32	17.6
Ground yarn%	45	21.9	77	19.5	65.6
interlaced yarn%	15	44.6	12.4	48.5	16.8
Weight gm/m2	381	593	775	1030	1068
Max. breaking load (N)	138.18	950.3	176.13	831.1	418.35
Max. breaking elongation %	395.6	319.6	231.2	456.8	390.8

2.2 Testing Methods

2.2.1 Designed Apparatus for Measuring Creep and Fatigue of Belt During use

During the use of back support belts by patients, some dimensional changes occur due to creep and fatigue phenomena. We designed an apparatus to measure the creep and fatigue of the belts. The design consists of a wooden board and a pair of clamps, the upper clamp is fixed to the board and the lower clamp moves freely. The specimen is attached to the two clamps as shown in Fig.(1). The specimen is subjected to a constant load of 488 grams (the weight of the lower clamp). A measuring scale is attached to the board for measuring the changes that occurred in the length. The reading of elongation is taken regularly for 24 hours. The weight is removed and the specimen is set to relax for 24 hours then the elongation is measured again. In this way, the creep can be measured in the samples. By repeating the previous cycle (loading and unloading), fatigue is measured.

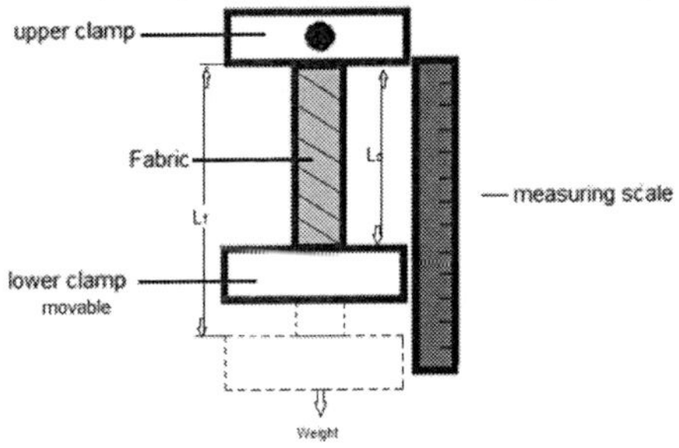

Fig.1 The Designed Apparatus for The Fabric Creep and Fatigue

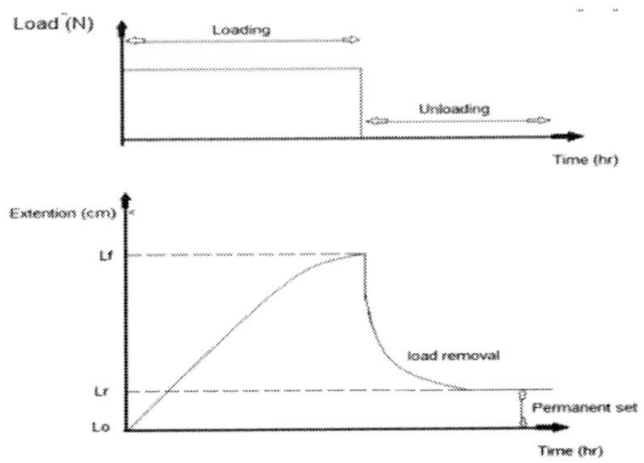

Fig.2 Fabric Creep

Fig.2 illustrates the calculation of the fabric creep. The fabric creep or total extension is composed of two elements; the instantaneous extension and time-dependent extension. Initially, the specimen rapidly extends as the instantaneous effect. By removal of the load, the specimen returns some of its extension rapidly at first and then moves slowly and residual extension which is called permanent set occurs. The instantaneous extension consists of two extensions, the elastic one which is recoverable after load removal, and the plastic extension or permanent extension which is not recoverable [15].

Fig.3 illustrates the fatigue on the fabric. The process of loading and unloading is repeated for several cycles. With every cycle, one can notice that the curves get nearer to each other. The steady state occurs when the curves of consecutive cycles coincide on each other.

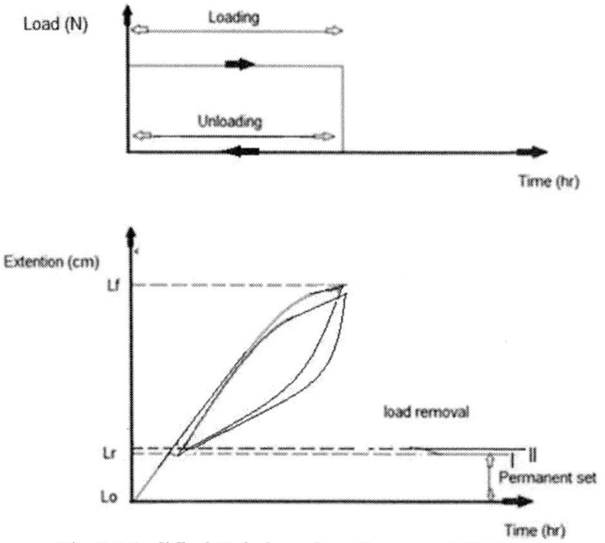

Fig.3 Modified Fabrics after Creep and Fatigue

From Fig.3 the following definitions and calculations can be obtained:

Final extension (E%)= (total extension/initial length) %= $(L_f-L_o)/ L_o$ ....1)

Extension ratio =(final length/initial length) = L_f/L_o................... ....2)

Permanent set (P%)= (relaxed length – initial length)/ initial length = $(L_r-L_o)/ L_o$..................... 3)

Elastic recovery (R%)= elastic extension/total extension= $(L_f-L_r)/ (L_f-L_o)$.........................4)

Creep initial hysteresis= area I – area I'... 5)

Creep steady state hysteresis= area II-areaII'.. 6)

Crush factor (F)= (area II- area I)/ area all.. 7)

Where,
L_o: initial length of the fabric specimen,
L_f: final length of the fabric specimen,
L_r: relax length of the fabric specimen.

Area I, I' II, and II are measured by weighing a paper equivalent to this. Assuming that the density of the paper is the same, the area is directly proportion to the weight.

2.2.2 Experimental Procedure

The designed apparatus shown in Fig.1 is used to measure the creep for the five samples. The dimensions of the samples were 2.5 cm width * 20 cm length. For measuring the fatigue, five cycles were obtained as discussed in Fig.3. The loading time was 24 hours , and the relaxation time was24 hours.

2.2.3 Measuring Mechanical Properties

We used ASTM D 5035 standard test method for measuring breaking force and elongation of textile fabrics for different samples, with dimensions 2.5 cm width * 25 cm length.

2.2.4Measuring Air Permeability

We used ASTM D 737 standard test method for measuring air permeability of textile fabrics for different samples.

3. Results and Discussion

3.1 Creep and Fatigue Results

From equations (1-7), the result for Final extension (E%), Extension ratio, Permanent set (P%), Elastic recovery (R%), and Crush factor(F) were calculated as shown in table (2).

Table 2. calculated creep and fatigue results

Belt number	1	2	3	4	5
Final extension (E%)	25.9	55.4	82.7	59.4	65
Extension ratio	1.25	1.55	1.83	1.59	1.65
Permanent set (P%)	0.5	1.08	3.33	0.52	2.77
Elastic recovery (R%)	97.9	98.1	91	99	96
Crush factor(F)	0.159	0.16	0.39	0.079	0.183

3.3 Effect of Rubber Percentage on the Back-belt Characteristics

The rubber percentage has a significant effect on the final extension (E), extension ratio, permanent set (P%), elastic recovery, and air permeability as illustrated in Table 3. As the rubber percentage increases it makes the belt more supported and its dimensional change by load decreases. Thus the rubber percentage is the most affecting factor on the performance of the back belts as illustrated in Fig.4-6.

3.3.1 The Effect of Rubber Percentage on Final Extension

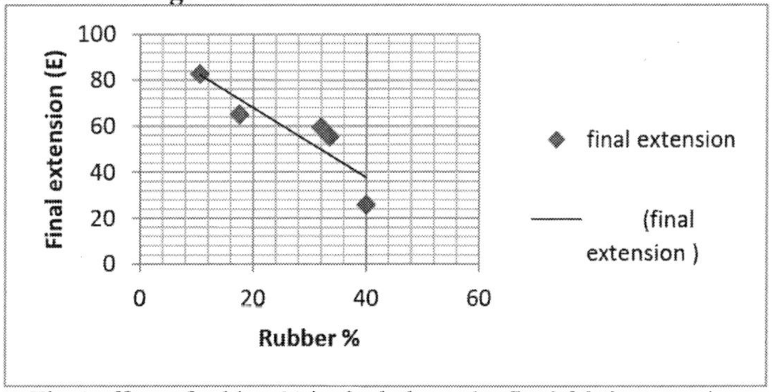

Fig.4 effect of rubber % in the belt on the final fabric extension

3.3.2 The Effect of Rubber % on Extension Ratio

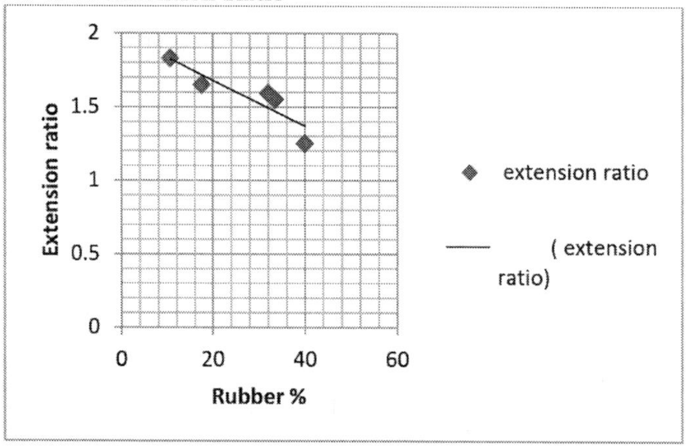

Fig.5 The effect of rubber % on extension ratio of the fabric

3.3.3 The Effect of Rubber % on Permanent set (P%)

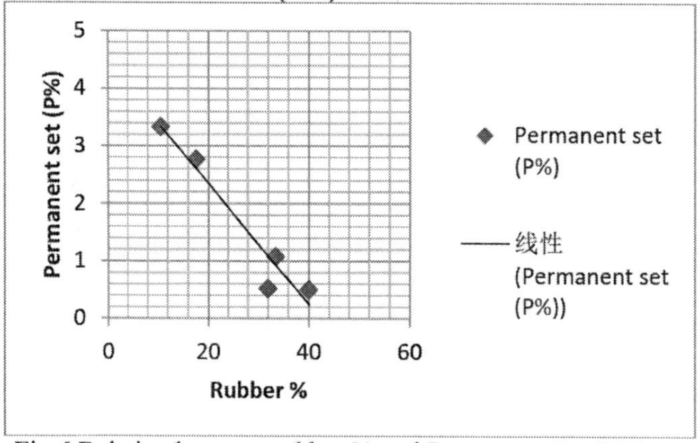

Fig.6 Relation between rubber % and Permanent Set (P%)

3.3.4 The Effect of Rubber % on Elastic Recovery (R%)

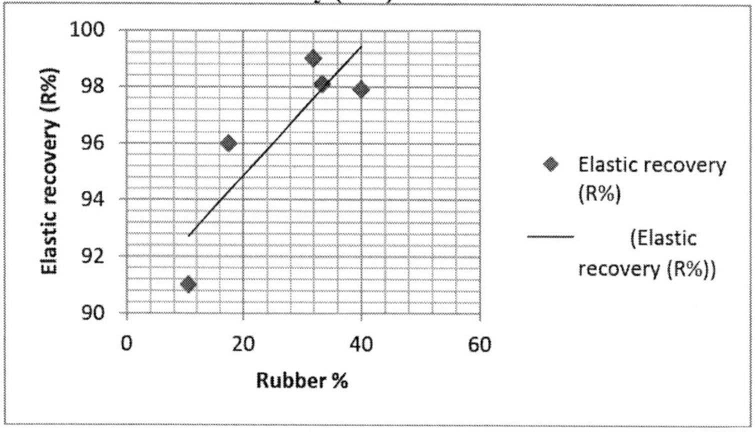

Fig.7 Relation between rubber% and elastic recovery(R %)

The rubber percentage has a positive significant effect on elastic recovery (R %) as shown in Fig.7. This is due to the high recoverability of the rubber, thus increasing the rubber percentage in the back belts increased its elastic recovery.

3.3.5 The Effect of Rubber Percentage on Air Permeability

The rubber percentage also, has a positive effect on air permeability as shown in Fig.8. This may be due to the high rubber percent, which, decreases the yarns intersection or cause the straightening of yarns. Thus, increases the open area and causes an increase in the amount of airflow to give more comfort in use.

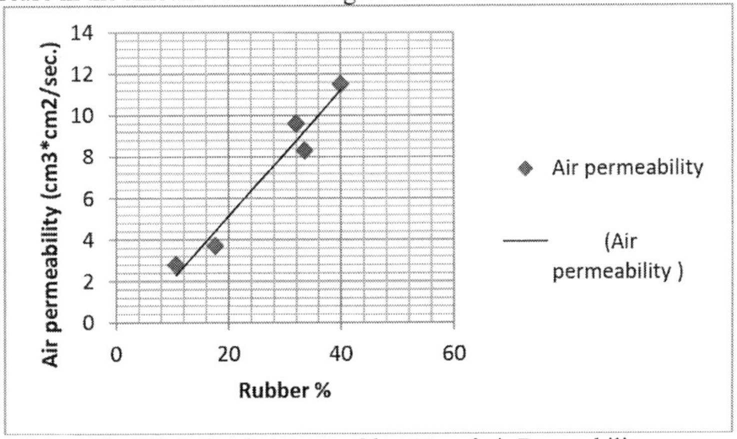

Fig.8 Relation between rubber % and air Permeability

As it can be seen from the Fig.4-7, belt number 4 has 32% elastic rubber percentage gave the highest elastic recovery percentage. On the other hand, this belt gave the least permanent set (P %). Also, belt number 4 with rubber percent 32% gave the optimum required air permeability and gives more comfort during wear. This is a good measure to predict the behavior of the belt during the end-use.

3.4 Effect of Ground Yarn Percent on the Back Belt Fabric Characteristics

The ground yarn percent has a positive significant effect on permanent set and elastic recovery, this is may be due to as the ground yarn percent increases the crimp in this direction decreases which in term decrease the elastic extension region in the fabric and increase the permanent set.

3.5 Effect of Max. Extension on the Crush Factor

Max. Breaking extension percent has a negative significant effect on the crush factor as shown in Fig.9. This is because the crush factor depends on the work done from the initial to the steady-state cycle of creep expressed as a percentage of that of the steady-state. As the belt elongation increases the work done decreases too.

From the previous discussion, it is clear that belt number 4, which was nonwoven in construction as it gave the minimum permanent set, the maximum elastic recovery, and maximum air permeability, gave the optimum required performance.

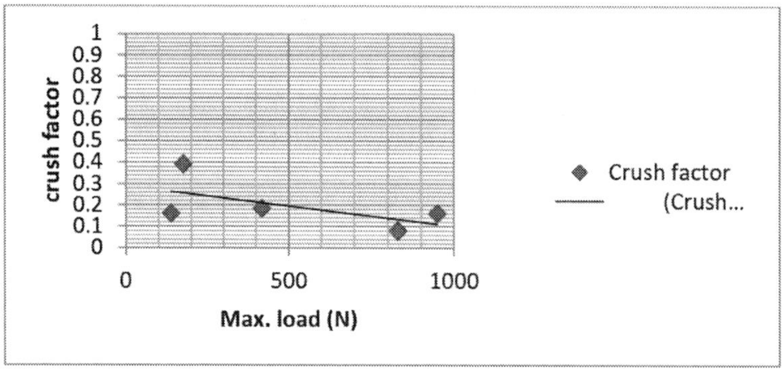

Fig.9 Correlation between max. load and Crush factor

4. Conclusion

We presented a technique for measuring the different characteristics of medical back belt support. Mainly the creep and fatigue properties where found useful to predict the lifetime and durability of the back belt in use and compare between the different types available in the market. From the results, it is clear that the percentage of rubber is the main factor affecting the properties required for measuring the durability of the back belt. Thus, these properties can be improved by increasing the rubber percent in the available range.

Reference

[1] King MW. Overview of opportunities in medical textiles. Canadian Textile Journal: Aug 2001; Vol.118 (4).
[2] Wen Z, An introduction to healthcare and medical textiles, Ph.D, University of Manitoba: 2013.
[3] Chellamani KP, Vignesh Balaji RS, Veerasubramanian D. Medical Textiles: The spun lace process and its application possibilities for hygiene textiles. The South India Textile Research Association (SITRA), Coimbatore-641014, TN, India, J. Acad. Indus Res: May 2013; Vol. 1 (12).
[4] Scarlet R, Deliu R, Manea LR. Implantable Medical Textiles: characterization and applications. Gheorghe Asachi" Technical University, Faculty of Textiles, Leather Engineering and Industrial Management, Iasi, Romania, 7th International Conference -TEXSCI: March 2010.
[5] Chinta SK, Veena KV. Impact of Textiles in Medical Field. D.K.T.E's Textile & Eng. Institute, Ichalkaranji, International Journal of Latest Trends in Engineering and Technology (IJLTET): January 2013; Vol. 2 Issue 1.
[6] Czajka R. Development of Medical Textile Market. Technical University of Łódź Faculty of Textile Engineering and Marketing Institute of World Economy and Textile Marketing ul. Żeromskiego 116, 90-543 Łódź, Poland, FIBRES & TEXTILES in Eastern Europe: March 2005; Vol. 13 January / March 2005.
[7] Liu SR, Wu TY, Dai XQ. Effects of straps and movements on the pressure distribution induced by a backpack, JFBI: 2019; 12 (4): 205-211.
[8] Horrocks AR, Anand SC. handbook of Technical textiles. the textile institute: Jun 2000.
[9] Breteler MR, Nierstrasz VA, Warmoeskerken MMCG. Textile Slow-Release Systems with Medical Applications. Textile TechnologyDepartment of Chemical Technology University of Twente Enschede The Netherlands AUTEX Research Journal: December 2002; Vol. 2.
[10] Miao M, Soong MCC. Air Interlaced Yarn Structure and Properties. Institute of Textiles & Clothing, The Hong Kong Polytechnic University, Hung Hom, Hong Kong, Textile Research Journal: 1995; Vol.65 (8).
[11] Information on http://textilelibrary.weebly.com/tangled--interlace-yarn.html: 2018.
[12] Lazar K. Application of Knitted Fabrics in Technical and Medical Textiles. Hungarian Society of Textile Technology and Science, 45th International Congress IFKT, Ljubljana, Slovenia: May 2010.
[13] Information on https://www.astm.org/Standards/D737.htm: 2018.
[14] Yosuke H, Haruka U, Satoshi T, ShigeruInui. Relationship between Stretchability of Garment Lining and Ease of Body Movement, JFBI: 2020; 13 (4): 169-179.
[15] Booth JE. Principle of Textile Testing. Newes-Butter worths, London: January 1961.

Cite as: Z.Q. Chu et al.,
TBIS 10.3993/tbis (2021).

Life Cycle Modularization for Silk Products and Their Carbon Footprint Quantification Model

Zhao-Quan Chu[1], Dong-Ping Wu[2], Xing-Er Bian[3], Jin-Ping Guan[1], Jian-Mei Xu[1,4*]

[1]*Soochow University, 178# Ganjiang East Rd., Suzhou, 215021, China*
[2]*Zhejiang Silk Technology Co., Ltd, 159# Dayuan Rd., Hangzhou, 311305, China*
[3]*Zhejiang Cathaya International Co., Ltd, 105# Stadium Rd., Hangzhou, 310004, China*
[4]*Jiangsu Engineering Research Center of Textile Dyeing and Printing for Energy Conservation, Discharge Reduction and Cleaner Production (ERC), Soochow University, 199# Renai Rd., Suzhou, 215123, China*

Corresponding author's email: xujianmei@suda.edu.cn

Abstract

When determining system boundary and life cycle inventory (LCI) of silk products, many problems such as accommodating for various kinds of silk products, and their long-life cycle chain are encountered. In order to establish a uniform and universal method in the life cycle assessment (LCA) of silk products, this paper constructed a modularization method in determining the system boundary of the silk products, and analyzed the LCA framework of each module. On this basis, the calculation model of carbon footprint (CFP) for silk products was established, and the CFP conversion factors of energy sources and materials involved in the whole life cycle were collected. The modular construction for system boundary of the silk products makes each module universal. Thus, by carrying out different combinations of the modules, the system boundary of different silk products can be constructed on a universal basis, and the CFP can be calculated on a standardized method.

Keywords: Carbon Footprint; Life Cycle Assessment; System Boundary; Quantification; Conversion Factors

1. Introduction

In recent years, global warming has become a serious threat to the natural environment and human economic development, and the main reason for global warming is that human production activities lead to the high emission of greenhouse gases (GHG). The textile industry has always been a high energy consumption and high carbon emission industry, and its GHG emissions in the production and use stage account for 3% of the global GHG emissions [1]. With the economic development and population growth, the global demand for textiles is still growing every year. In particular, nonrenewable resources such as oil consumption are increasing, which will cause a great burden on the environment. Therefore, it is imperative for the textile industry to realize low-carbon production, establish carbon footprint (CFP) evaluation system, and carry out green ecological production design. CFP is used to measure the emission of GHG in the life cycle of an activity or product. To achieve green production with low energy consumption, pollution and emission, it is necessary to calculate the GHG emissions throughout the whole life cycle of textiles, including raw material acquisition, industrial production, distribution, use, waste or reuse. And then optimize the production process according to the CFP of each stage by using raw materials with low carbon emission, improving energy utilization or using clean energy, and eliminating inefficient equipment. At the same time, the research of textile CFP can help to mark up the CFP information of textile products for consumers, which is helpful in encouraging low-carbon consumption.

Based on the urgent needs of environmental and ecological sustainable development, the CFP research on textiles has become a hot spot in recent years. There are many CFP research on cotton textiles [2-4], while few on silk products. Silk, is considered the "Fiber Queen", its fabrics and garments are gorgeous, noble and luxurious, which can be used as high-end apparel textiles with high added value. Compared with chemical fiber, its production process greatly reduces the consumption of fossil energy, and the carbon sequestration effect of mulberry planting is obvious. Compared with cotton, hemp and other natural fibers, its economic value is higher. The study of the CFP for silk products can reveal the facts that silk textiles is environmentally friendly with low carbon emission, which can enhance the ecological persuasion of their export to the silk textile consumer countries. On the other hand, it can further guide the silk industry to carry out low carbon production. Therefore, as the largest silk producer in the world, the research on CFP of silk products in China is extremely urgent. Astudillo et al. [5] studied the environmental impact from Indian raw silk production stage using life cycle assessment (LCA) method. These impacts included CFP, ecotoxicity, freshwater eutrophication, and water footprint. They also established the life cycle inventory (LCI) of raw silk production in South India, and found that the environmental impact of raw silk production was higher than that of other natural fibers, and the impact on the environment mainly came from the cocoon production, especially fertilization. The result can only reflect the environmental effect of India silk, in stead of silk from other countries. Jiang et al. [6] calculated the carbon emission of whole life cycle of gambiered canton silk based on LCA method, and concluded that the CFP of whole life cycle of 1kg gambiered canton silk was 1.88kg CO_2e, and the majority of carbon emission mainly came from silkworm breeding and silk reeling. Han [7] analyzed the whole raw silk production based on LCA, and listed LCI of raw silk. They also conducted ecological design and low-carbon production of silk products. However, these studies did not consider the carbon sequestration of mulberry trees. Giacomin et al. [8] studied the economic and environmental problems of silk production in Brazil, and pointed out that improving the utilization rate of co-products in silk production can not only create employment and increase income, but also reduce the impact on the environment. At the same time, he also mentioned that the carbon sequestration of mulberry trees can eliminate part of the environmental impact [9], and estimated that the carbon sequestration capacity of mulberry is 735kg CO_2 per kg silk production, but this result is only estimated based on macro statistical data. So far there is no research on CFP quantification of the overall life cycle for silk products.

The biggest problem in the CFP quantification for textile products is that the results are not comparable even for one kind of products. For example, in the CFP quantification of the garment production stage for pure cotton jeans, one study is 9.84 kgCO_2e/pcs[10], while another is 21.513 kgCO_2e/pcs[11]. The huge difference may caused by the different weight, design, processing efficiency of the products. However, there are also other reasons such as different system boundaries, different values of the CFP conversion factor for one material. And these reasons make the results of the same kind of products incomparable. Especially there are many kinds of silk products, and the life cycle chain of each silk product is long and complex, involving the management of mulberry garden, silkworm breeding, industrial production, use, sales, waste or reuse. The subtle changes of the system boundary, the choice of different CFP conversion factors will always get different results when calculating the CFP even for one product. Therefore, this paper proposes a modular division of the whole life cycle of all kinds of silk products. Each silk product can be constructed by combination of these modules, and thus the LCA evaluation framework of silk products including system boundary, material input and output can be established. Based on this LCA evaluation framework, the CFP conversion factors are collected, and the CFP calculating formula is established.

2. Modular Construction of Silk Product System Boundary

The CFP of silk products is generally calculated using LCA method based on PAS2050. PAS2050 is the first standard used to evaluate the CFP of products, and LCA is a method to summarize and evaluate the potential impact of all inputs and outputs of a product on the environment in the whole life cycle, which can be referred to in ISO14040/44. LCA means CFP calculation covers all stages of a product, "from cradle to grave". Firstly we need to define the calculation scope of CFP by determining its system boundary, and all the direct or indirect carbon emissions of products in the life cycle should be covered. Therefore, the definite, complete and clear system boundary are the premise of CFP calculation.

In the CFP calculation, repeated calculation of some shared stages for different silk products often happens. While modularization method can divide the complex and long life cycle chain of silk products into many small modules. Each module has its own simple structure, clear system boundary and LCI, In addition each module will be independent and modifiable, and will not be affected by other modules. Once there is an error in the calculation of a CFP unit, only the problematic module needs amending, while other modules will not be affected, which can improve the working efficiency [12]. Moreover, for all silk products, the life cycle modules are universal. Different modules can be combined to form the system boundary framework of a new product, which can avoid repeated calculation.

The life cycle of silk products is hereby divided into five stages: moriculture and sericulture stage, industrial production stage, sales stage, use stage and waste stage. As shown in Table 1, the moriculture and sericulture stage is regarded as module A; the industrial production stage of silk products is subdivided into module B to M (module I to M is the manufacturing stage of five conventional silk products); the sales stage is module N; the use stage is module O; and the waste or reuse stage is module P. Therefore, different kinds of silk products can be combined by some of the modules to realize the construction of the corresponding system boundary and the corresponding CFP calculation. Taking gray silk fabric using fresh cocoon as an example, the product can be expressed by module combination of ABDFN.

Table 1. Modular construction of silk textile system boundary

Module number	The name of the module	Combination of modules
A	Moriculture and sericulture	
B	Fresh cocoon reeling	Fresh (dry) cocoon raw silk: AB(C)N
C	Dry cocoon reeling	Fresh (dry) cocoon thrown silk: AB(C)DN
D	Yarn twisting	Fresh (dry) cocoon colored silk thread: AB(C)DEN
E	Yarn dyeing	Fresh (dry) cocoon gray silk fabric: AB(C)DFN
F	Weaving	Fresh (dry) cocoon dyed silk fabric: AB(C)DFGN
G	Fabric dyeing	Fresh (dry) cocoon printed silk fabric: AB(C)DFHN
H	Fabric printing	Fresh (dry) cocoon brocade: AB(C)DEFNOP
I	Scarf processing	Fresh (dry) cocoon printed silk scarf: AB(C)DFHINOP
J	Garment processing	Fresh (dry) cocoon printed silk clothing: AB(C)DFHJNOP
K	Tie processing	Fresh (dry) cocoon silk tie: AB(C)DEFKNOP
L	Silk quilt processing	Silk quilt: ALNOP
M	Silk bedding processing	Brocade bedding: AB(C)DEFMNOP
N	Sale	Dyed silk bedding: AB(C)DFGMNOP
O	Usage	
P	Waste or reuse	

2.1 LCA Framework of Moriculture and Sericulture

It is chronicled that silk originated from China, and Chinese silk industry has experienced 5000 years of development. As a big country in the production and export of silk products, China should speed up the promotion and application of energy saving and emission reduction technology in the silk industry, and improve the green production level of the industry.

During the moriculture and sericulture phrase (module A), there are also majority of GHG emission due to the use of fertilizer and energy. Fig.1 is the LCA framework of module A, it defines the system boundary, input and output of this stage. The module A can be roughly divided into three detailed stage. The first stage is the moriculture . Every year mulberry branches should be pruned and fertilized at the end of winter and the beginning of spring, and herbicides should be applied before mulberry leaves grow, so as to reduce the competition between weeds and mulberry trees for nutrients. In the early stage of mulberry leaf growth, pesticides should be used to reduce pests, and the soil moisture of mulberry garden should be adjusted in time. Therefore, the type and amount of chemical fertilizer and pesticide, irrigation method (irrigation, drip irrigation) and water consumption all directly determine the amount of carbon emissions [13]. The second stage is sericulture. Before sericulture, the silkworm rearing room and tools should be disinfected. Then breeding, warming and incubation should be started. After the fifth instar rearing, the silkworm and sericulture tools should still be disinfected in time during this period to prevent the spread of virus or disease such as sepsis, and muscardine in a large scale. After this, it is mounting, cocooning, and harvesting. In the second stage, the use of disinfectant water and silkworm medicine, the oil consumption of mulberry leaves transportation and the consumption of labour are the main reasons for carbon emission. The third stage is cocoon transportation and sale. The carbon emission of this stage is generated by gasoline and electricity consumed in the process of cocoon transportation and storage.

2.2 LCA Framework for Industrial Production, Sales, Use, and Waste Phase of Silk Products

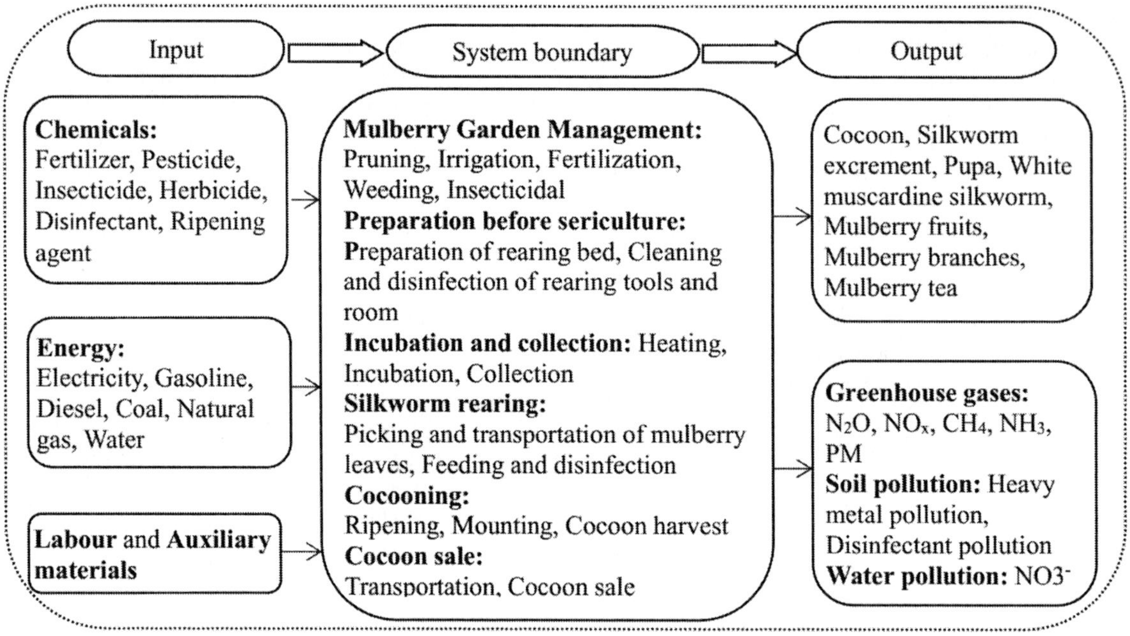

Fig.1 LCA framework of moriculture and sericulture phase

In this study, silk products are divided into two categories: one is business-to-business silk products, such as gray silk fabric, dyed silk fabric, and printed silk fabric; the other is business-to-customer products, such as silk clothing, ties, scarves, and bedding. The two kinds of products have some shared product modules, such as silk reeling module and weaving module. And each product has its own specific processing phase or module. Fig.2 lists the specific process of different modules defined as forementioned, as well as the system of these modules. By selecting and recombining of these modules, it's easy to establish the system boundary of a new silk product.

It can be seen from the figure that the difference between fresh cocoon raw silk and dry cocoon raw silk mainly lies in the difference of cocoon storage and pre-processing. Fresh cocoon raw silk needs to be refrigerated, which consumes more power, while dry cocoon raw silk needs to be dried and cooked at a high temperature, which consumes more energy such as coal and natural gas. Dyeing, printing and finishing processes will use dyes and auxiliaries, and a lot of water resources; the printing and dyeing process will release formaldehyde, nitrogen dioxide, aniline, chlorinated compounds and other gases, which are the main sources of GHG emission. And some consumption such as electricity consumption contributed by the machine running throughout the industrial production, and gasoline consumption contributed transportation throughout the product life cycle, are also the main GHG emission sources. The use stage of silk products mainly includes washing, drying and ironing. The way of washing (machine washing or hand washing) and the times of washing, drying and ironing determine the carbon emission in the use stage, so the carbon emission in this stage is closely related to personal use habits. The waste stage of textiles mainly includes landfill, incineration and reuse. The reuse of textiles can effectively reduce carbon emissions, and incineration causes the largest carbon emission among the three waste treat methods.

3. Calculation of CFP of Silk Products based on Modularization

3.1 Calculation Method of CFP

GHG mainly comes from the use or consumption of chemicals (fertilizers, pesticides, herbicides, dyes, additives, etc.), energy (natural gas, coal, gasoline, electricity, steam, etc.), auxiliary materials (packaging materials, A4 paper, etc.) and labour. The photosynthesis of mulberry leaves can produce carbon sequestration effect, which converts CO_2 in the atmosphere into carbohydrates and sequesters it in plants in the form of organic carbon, and effectively reduces the content of CO_2 in the atmosphere. However, the carbon sequestration effect of plants is

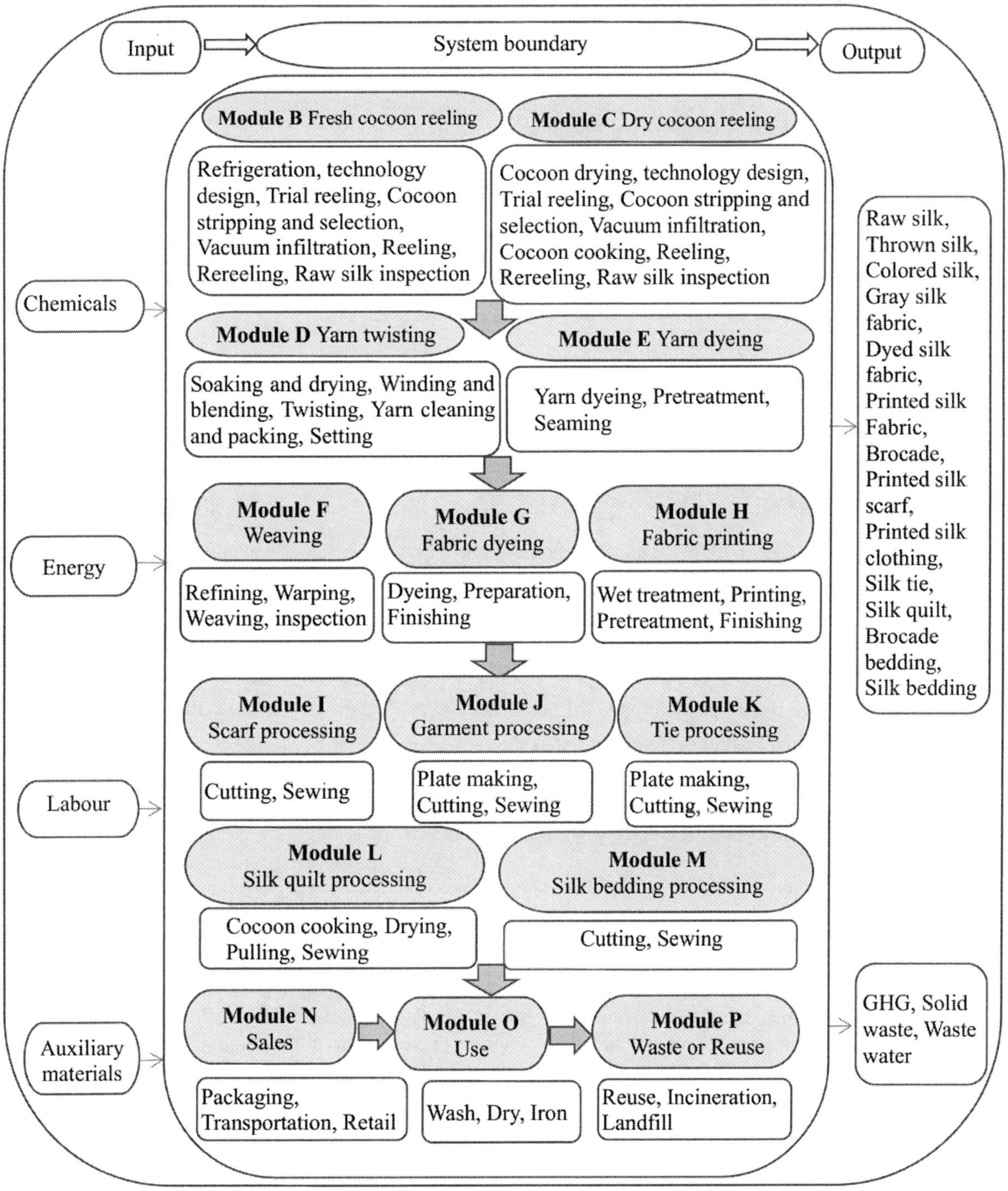

Fig.2 LCA framework of modules B-P expressing different stages of the silk products

seldom considered when calculating CFP in the current literature. In the whole life cycle of silk products, there might be co-products such as silkworm excrement, pupa, white muscardine silkworm, mulberry fruits, mulberry branches, mulberry tea, etc. Therefore the CFP of the silk products can be partially distributed to the co-products according to the physical relationship between them, or their economic values. Therefore, with the increase of the utilization rate and the economic value of co-products, the CFP distributed to co-products will also increase, while the remaining CFP of the silk products will decrease. The CFP accounting of silk products should include six parts: the first part is the carbon emissions generated by chemical input; the second part is the carbon emissions generated by energy use; the third part is the carbon emissions generated by material usage; the fourth part is generated by human consumption; the fifth part is the carbon sequestration generated in the process of plant growth and the sixth part is the carbon emissions allocated by co-products. Among them, the part of carbon emission caused by human consumption in the production process was generally not considered, because it's believed that CO_2 produced by human respiration was generated all the time, and has nothing to do with the product processing. However, for the same product, the consumption of labour caused by different production efficiency is often quite different, and this part of the difference should have different impact on the environment. Especially for labor-intensive products, labour consumption should be within the scope of CFP accounting. The specific calculation formula of CFP for a silk product having m modules can be computed as follows.

$$CFP = \sum_{i=1}^{n_1}\left(f_{CH_i}\sum_{j=1}^{m} CH_{ij}\right) + \sum_{i=1}^{n_2}\left(f_{EN_i}\sum_{j=1}^{m} EN_{ij}\right) + \sum_{i=1}^{n_3}\left(f_{MT_i}\sum_{j=1}^{m} MT_{ij}\right) + f_{PE}\cdot\sum_{i=1}^{m} n_i - \frac{A\cdot Y}{H} - CPF_{co-pro} \quad 1)$$

where CFP is the total carbon emission, n_1, n_2, n_3 are the number of chemical, energy, and material kinds respectively, m is the number of modules, $f_{CH_i}\left(i=1,2,\cdots,n_1\right)$ is the CFP conversion factors of the ith chemical, $CH_{ij}\left(i=1,2,\cdots,n_1;\ j=1,2,\cdots,m\right)$ is the consumption of the ith chemical in the jth module, $f_{EN_i}\left(i=1,2,\cdots,n_2\right)$ is the CFP conversion factors of the ith energy, $EN_{ij}(i=1,2,\cdots,n_2; j=1,2,\cdots,m)$ is the consumption of the ith energy in the jth module, $f_{MT_i}\left(i=1,2,\cdots,n_3\right)$ is the CFP conversion factors of the ith material, $MT_{ij}\left(i=1,2,\cdots,n_3;\ j=1,2,\cdots,m\right)$ is the consumption of the ith chemical in the jth module, f_{PE} is the daily carbon dioxide exhalation of each person, $n_i\left(i=1,2,\cdots,m\right)$ is the number of labor required in module i, A is the carbon absorption rate of mulberry, Y is the economic output of mulberry, H is the economic coefficient of mulberry, and CFP_{co-pro} is the amount of CFP allocated by co-products.

3.2 Carbon Footprint Conversion Factors

The CFP conversion factors (CFP-CF) of various energy sources and materials are obtained by calculating the indirect carbon emission in the production stage and the direct carbon emission in the use stage. For the same material, especially energy, different countries and regions have different CFP conversion factors, we should choose the most suitable CFP conversion factors for silk products. And the suitable CFP conversion factors of the main materials and energy involved in the whole life cycle of silk products are collected in this study, as shown in Table 2. Generally speaking, the CFP conversion factors of fossil energy has three levels, i.e. regional factors, national factors and international factors, and their priority decreases in turn. Usually, we choose the CFP conversion factors of the region nearest to the product processing site. When it is difficult to calculate the CFP conversion factors of a material, the CFP conversion factors of its substitute can be directly used instead. For example the conversion factor of electricity is determined according to the specific region, in China, the electricity network are divided into six sub networks, the data can be referred to the standard T/CNTAC 11-2018.

4. Conclusion

Low-carbon and green silk products have been imperative, meanwhile carbon label has been paid more and more attention. This paper has constructed a modular method in determining the system boundary of a silk product using LCA method. The whole life cycles of the silk products are divided into different modules according to different stages of the products. Based on this method, different modules can be freely combined to form the system boundary of a specific silk product, through which the CFP calculation of the silk product can be quite efficient and accurate. The standardized construction of the LCA framework can facilitate the standardized accounting of the CFP of all kinds of silk products, and make the quantification results of different products comparable.

Table 2. CFP conversion factors of various inputs of silk products throughout their life cycles

Material type	Material inputs	CFP-CF $kgCO_2e/kg$	Involved Modules	Material type	Material inputs	CFP-CF $kgCO_2e/kg$	Involved Modules
Energy	Gasoline	157.48	ABCNP	Pesticides	Dimethoate	24.2	A
[2,14]	Aviation gasoline	0.83[a]	N	[16]	Dichlorvos	28.59	A
	Diesel oil	141.75	ABCNP		Bisultap	48.40	A
	Kerosene	19.05	BC		Glyphosate	70.75	A
	Fuel oil	34.91	ABCNP		Paraquat	71.48	A
	Natural gas	4.55	BCGHL		Nitrofen	42.16	A
	Lubricating oil	7.92	ABCDEFGH IJKLMN		Chlorothalonil	43.99	A
	Standard coal	0.59	ABCGH		Carbendazim	57.46	A
	Steam	0.05	ABCGH		Triadimefon	57.10	A
	Water	0.12[b]	ABCD EGHO	Chemicals	Lime[17]	0.15[a]	AEGH
	Electricity	T/CNTAC 11-2018	All modules		Caustic soda[17]	0.82	BCEGH
Materials	Carton packaging [11]	1.04	All modules		Soda ash[17]	1.31	BCEGH
	Printing paper[13]	0.01[c]	All modules		Concentrated sulfuric acid[18]	0.12	EG
Fertilizers [15]	N-fertilizer	7.76	A		Formic acid[17]	3.14	EGH
	P-fertilizer	2.33	A		Baking soda[17]	1.35	BCEGH
	K-fertilizer	0.66	A		Industrial salt[18]	0.18	BCEGH

Note: a: $kgCO_2e/MJ$; b: $kgCO_2e/t$; c: $kgCO_2e$/ sheet

This study collects the CFP conversion factors of some materials, energy, chemicals involved in the life cycle, whether these factors are the most suitable one still needs further study. The study also establishes a CFP calculation model in which the allocation of CFP and the carbon sequestration of mulberry trees are taken into account.

5. Acknowledgement

This work was supported by Technology Guidance Project of China National Textile And Apparel Council (2021015), Opening Project of Jiangsu Engineering Research Center of Textile Dyeing and Printing for Energy Conservation, Discharge Reduction and Cleaner Production (Q811580621), and the Priority Academic Program Development of Jiangsu Higher Education Institutions.

References

[1] Nørup N, Pihl K, Damgaard A, Scheutz C. Evaluation of a European textile sorting centre: Material flow analysis and life cycle inventory. Resources, Conservation and Recycling: 2019; 143: 310-319.

[2] Wang CX, Wang LH, Liu XL, Du C, Ding D, Jia J, Yan Y, Wu G. Carbon footprint of textile throughout its life cycle: a case study of Chinese cotton shirts. Journal of Cleaner Production: 2015; 108: 464-475.
[3] Sandin G, Peters GM. Environmental impact of textile reuse and recyclin-A review. Journal of Clean Production: 2018; 184: 128-141.
[4] Baydar G, Ciliz N, Mammadov A. Life cycle assessment of cotton textile products in Turkey. Resources Conservation Recycling: 2015; 104: 213-223.
[5] Astudillo MF, Thalwitz G, Vollrath F, Vollrath F. Life cycle assessment of Indian silk. Journal of Cleaner Production: 2014; 81: 158-167.
[6] Jiang T, Chen ZY, Yao TT, Chen H, Yang ZM. Product carbon footprint (PCF) assessment of gambiered canton silk. Dyeing and Finishing: 2012; 38 (08): 39-41.
[7] Han XY. Clean production and ecological design of real silk products. China Textile Leader: 2006; (04): 34-36; 38: 41; 102.
[8] Giacomin AM, Garcia JRJB, Zonatti WF, Silva-santos MC, Laktim MC, Baruque-Romos J. Brazilian silk production: economic and sustainability aspects. Procedia Engineering: 2017; 200: 89-95.
[9] Giacomin AM, Garcia JRJB, Zonatti WF, Silva-santos MC, Laktim MC, Baruque-ramous J. Silk industry and carbon footprint mitigation. IOP Conference Series: Mat Sci and Eng: 2017; 254; (19): 123-156.
[10] Li Y, Wang JT, Wang LL. Demonstration of carbon footprint assessment of jeans industry. Advanced Textile Technology: 2017; 25(06): 58-51.
[11] Tu LH. The carbon footprint research of jeans production on the basis of PAS 2050 standard. Shanghai: Donghua University: 2012; 10-14.
[12] Wang SS. Research on modular calculating method of industrial carbon footprint of textile product and its software development. Shanghai: Donghua University: 2015; 14-52.
[13] Astudillo MF, Thalwitz G, Vollrath F. Life cycle assessment of silk production - a case study from India. Handbook of Life Cycle Assessment (LCA) of Textiles and Clothing: 2015; 255-274.
[14] Yan Y, Wang CX, Ding D, Zhang YN, Wu G, Wang LH, Liu XL, Du C, Zhang YJ, Zhao CL. Industrial carbon footprint of several typical Chinese textile fabrics. Acta Ecologica Sinica: 2016; 36 (3): 119-125.
[15] Chen S, Lu F, Wang XK. Estimation of greenhouse gases emission factors for China's nitrogen, phosphate, and potash fertilizers. Acta Ecologica Sinica: 2015; 35 (19): 6371-6383.
[16] Zhang G, Lu F, Huang ZG, Chen X, Wang XK. Estimations of application dosage and greenhouse gas emission of chemical pesticides in staple crops in China. Chinese Journal of Applied Ecology: 2016; 27 (09): 2875-2883.
[17] Wang LL. Research and Demonstration of carbon footprint and Water Footprint of Textiles and Clothing. Shanghai: Donghua University: 2013: 20-40.
[18] Yang ZP, Zhang JC, Zhang H, Zhang XX, Gao ZQ. Assessing of carbon footprint of hemp product according to PAS2050. Journal of Textile Research: 2012; 33 (08): 140-144.

Green Creativity: A Bioplastic Experimental Study Applied to Fashion and Textile Design

Yu-Tan He[1a], Ming-Ke Wang[2*], Rong Zheng[1b*]

[1a] *School of Fashion and Art Design, Donghua University, 1882 West Yan'an Road, Changning District, Shanghai, 200051, China*
[1b] *Shanghai International Fashion Innovation Centre, Donghua University, 1882 West Yan'an Road, Changning District, Shanghai, 200051, China*
[2] *London College of Fashion, University of the Arts London, 20 John Prince's Street, London, W1G 0BJ, the United Kingdom*

*Corresponding author's email: m.wang0220201@arts.ac.uk, rzheng@dhu.edu.cn

Abstract

In the past decades, fast fashion has sped up the practice of weaving non-degradable petrochemical textiles into their chic and cheap garments. This has caused environmental damages from production, manufacture, and landfilling. Bio-based textile with enhanced performance of degradability, lower carbon footprint and multi-functions is gradually drawing attention from the industry. However, most academic research are theoretical and are based on the perspective of material science. There are few research that discuss it from the angle of design and possible future application scenarios. To break the impediment between designers and accessible technology, this article introduces the classification of bioplastic, analyses its practical value and limitation to fashion and textile design by demonstrating a series of innovative bioplastic experiments. The aim of this study is to provide guidelines to designers for generating "green creativity" in the era where biotechnology has a great potential to create a better future for fashion sustainability.

Keywords: Bioplastic; Sustainability; Fashion; Textile Innovation

1. Introduction

In the past decades, the fast fashion sector has ramped up production of plastic-infused clothes with quick turnaround of new styles, increased number of collections, and lower prices [1]. Oil-based plastic fiber creates pollution along its entire supply chain, from the production, use and end-of-life phases. This has polluted the quality of land, water and air, as well as destroying ecosystems and endangering human health. From the ocean to the land, the equator to the poles, micro polyesters are everywhere [2]. All of the above are attributed to the linear mode of the fashion and textile system: non-renewable and non-degradable resources are extracted to produce garments for short-term usage, whereby the materials are mostly sent to landfills or incinerated.

To reduce the negative impact of petroleum-based textiles on the environment and enhance the sustainable capacity of the fashion and textile industry, biotechnology can now make degradable, low carbon, and safe textiles. A long list of recent collaborations between fashion houses and biotechnology companies, such as Hermès × MycoWorks, indicates that bio-textile is no longer a fantasy but a megatrend worthy of attention. Bioplastic is a general term for biomaterials commonly adopted in the fashion industry with a variety bio-based, bio-sourced and plant-based classifications [3]. Crops, bacteria, fungi, spider silk, seaweed, agar and gelatin have been tested to be the next generation of sources for eco-friendly, degradable and innovative materials.

However, limited by the traditional fashion education, most designers lack interdisciplinary knowledge. There is an invisible wall between designers and accessible biotechnologies, delaying the sustainable progress of the fashion and textile industry. Reaching out beyond mainstream academic research, this article discusses how biotechnology can be applied with design to nurture innovative sustainable ideas. Both qualitative and quantitative approaches are adopted. This article illustrates the classification, characterization and application cases of bioplastic in the context of contemporary fashion and textile industry, seeking potential application of biomaterial through a series of bioplastic experiments, and finally provides a broad reference for designers on how to use biotechnology to generate sustainable design ideas with more creativity, diversity and impact.

2. Literature Review

2.1 Environmental Impact of Petroleum Based Textile

According to the report on Pulse of the Fashion Industry proposed by the Global Fashion Agenda 2017 [4], more than 62 million tons of global apparel are consumed per year, and the figure is projected to see a rise from 63% to 103 million by 2030. More than 26 million tons of old clothes are thrown away per year in China and less than 1% are reused or recycled [5]. As one of the most polluted industries globally, second to the petrochemical industry, the prosperity of the fashion industry has led to serious environmental damages stemming from the stages of production, manufacture and disposal [6]. Many researches have mentioned that fast fashion is mainly responsible for this situation because fast fashion has overwhelmingly produced products for consumers with various styles, excess choices, and induced price with cheap plastic-based textiles [7]. There is increasing evidence showing that the fashion industry relies heavily on cheap petroleum-based textile, particularly polyester, whose production has grown ninefold during the past 50 years [8].

First, the production of petroleum-based textiles are energy intensive. Sourced from oil, gas, and coal, it directly links to carbon emissions and can be attributed to oil spills, methane emission, water pollution and biodiversity loss. The carbon footprint of a single polyester shirt is 5.5kg compared to 2.1kg for a cotton shirt [9]. According to the Ellen MacArthur Foundation (EMF), CO2 emissions for synthetic clothing are six times higher than cotton (530 million tons of CO2 for plastic-based fibres in comparison to 86 million tons for cotton) [10]. Secondly, chlorine, ammonia, sulfur dioxide, carbon oxide, hydrogen sulfide, aniline, nitrobenzene, benzene and nitric acid are emitted during the production, which would inevitably affect the living environment near the manufacturing plant and even damages residents' health. Thirdly, fossil fibers are non-biodegradable and contributes to microplastic pollution. There are some studies that indicate clothing containing synthetic fiber releases filaments into the air, water cycle, and even human body from daily wearing, washing and pulling [11]. From the beginning to the end of usage, synthetic fibers can break into invisible smaller particles, resulting in a total number of 500,000 tons of plastic in the ocean [12]. Plastic textiles could survive in the landfill for more than 200 years, leaching chemicals, shedding microfibers and releasing methane.

2.2 Classification, Characterization and Application of Bio-plastic

The worldwide crisis of non-biodegradable petroleum textiles has drawn attention from governments, who are constantly searching for alternatives and solutions [10]. One of the proposed solutions to this issue is inventing a new plastic that is biodegradable, safe and with enhanced functions called bioplastic. Bioplastic is an assemble of various materials. According to European Bioplastics [13], bioplastic is a bio-polymers combined material that is either biobased, biodegradable, or features both properties. Derived from or synthesized by microbes such as bacteria or by genetically modified plants, bioplastic is mainly regenerated from renewable resources and recycled materials and widely applied to automobile, medical, electronic and textile industries [14].

Similar to plastic, the essence of bioplastics is the "aggregation" of biopolymers, which could be generated from natural and synthetic origins. The natural original bioplastic is the main focus discussed in this article. The properties of plastic are defined by the monomers in the chain and the number of cross-links in its structure. Polysaccharides, such as proteins, lipids and organist or plant-based monomers, are the smaller unit of biopolymer. Table 1 [15] classifies the basic components of natural origins, which are based on agricultural or animal sources. The first known bioplastic, polyhydroxybutyrate (PHB), was discovered in 1926 by a French researcher, Maurice Lemoigne. The significance of the invention was ignored for a long time de to the abundant petroleum supply until the petroleum crisis in the mid-1970s. With the improvement of molecular genetics and recombinant DNA technology, PHB and polyhydroxyalkanoate (PHA) are emerging bioplastics put into practice at the beginning of the 21st century, both of which are synthesized by specialized microbes, as well as polylactic acid (PLA) [16], polymerized from lactic acid monomers produced by microbial fermentation of plant-derived sugars and starches.

Table 1. Natural Origins of Biopolymers

Ppolysaccharides	Starch, cellulose, lignin and chitin
Proteins	Gelatine, casein, silk and wool
Lipids	Plate oil, castor oil, and animal fats
Polyester 1-micro-organism or plants	1-polyhydroxyalcanoates, poly-3-hydroxybutyrate
2-bio-derived monomers	2-polylactic acid

Two major bioplastic features and advantage is that firstly it can save fossil resources by using biomass, secondly it has the potential to achieve carbon neutrality. Even though not all the bioplastic has the property of biodegradability, some of them offer additional methods of recovery at the end of a product's usage. Degradation of the chemical links between the monomers in these plastics is brought about by microorganisms or water, making bioplastics highly desirable materials for fabrication into biodegradable bottles and packaging film [17]. It seems that bioplastics can effectively solve the problem of petroleum-based materials, but in fact, bioplastics are far from being a solution adopted worldwide. One criticism is that to grow the biomass, a considerable amount of food or other agricultural products is needed. Although there are still many shortcomings in using bioplastics, increasing demand for sustainable plastic alternatives is promoted through technological innovations.

2.3 Green Creativity: The Power of Bio-Fashion

In the early 1960s, the concept of sustainable fashion was introduced, and it gained traction in the 1970s with the rise of environmental movements. A growing number of designers started to pay attention to ecological fashion in the 1980s. By 2013, nearly one-third of designers adopted ecological characteristics to their work. As an emerging concept, design is described as the "incorporation of living organisms or ecosystems as essential components." [18], illustrating a completely different image of fashion from what used to be, where innovators began moving into the integration, reconstruction of nature into viable concepts and applications. There is a long list of brands and companies labeled as "bio-fashion." It should not only be regarded as an interdisciplinary achievement but a strategy, enabling designers and enterprises to be much more competitive in the post epidemic market, where the strategic positioning of sustainability and technology is growing in attention. AlgiKnit, for example, is a biomaterial research institute in New York City with the mission to build a cleaner future fashion industry. Alginate biopolymer is produced from seaweed to fuse into yarn and transformed into various textiles and products. AlgiKnit is known to collaborate with Patagonia and Stella McCartney. Bolt Threads used a DNA sample that mimicked spider silk protein and converted the DNA into yeast to produce textiles that is durable and soft as spider silk in a low carbon way. Companies, such as Piñatex, MycoWorks and Modern Meadow, are trying to produce sustainable materials, using a wide range of biopolymers sourced from starch, cellulose, lignin and chitin.

In an era where sustainability plays a vital role in manufacture, marketing and design, fashion is not only about dressmaking but more about the accelerated technological evolution, politics, environment, demography and economic concerns. Green creativity is a mindset to realize and achieve sustainable development through innovative ways, and to market green ideas in an attractive way. Building scenarios and raising green creativity through the power of the designer paves the a way for a sustainable future and can even enhance the soft power of a country. Bio-fashion, as emerging green creativity, transcends its traditional boundaries and aims straight at the core of the moral sphere, shaping the future with enhanced functions, reduced carbon footprint and a higher standard of sustainability and wider applications [18]. Through the analysis of the literature review, it is obvious that petroleum-based plastic has posed a great threat to the environment; the fashion industry is now looking for interdisciplinary cooperation to cultivate bio-textiles for improving the sustainability of industrial development.

3. Bio-textile Experiment and Analysis

3.1 Aim, Methods and Establishment of Experiment

To test the performance of bioplastic materials, a bioplastics experiment including three tests was designed based on a variety of easy-to-access formulas from the Bioplastic Cookbook (2018) [19] published by FAB textile lab.

Combined with qualitative and quantitative methods, this experimental study compares the characteristics of bioplastics under different formulas, materials, and manufacturing processes through the collected performance data from tested samples. The experiment aims to create a guideline for designers to apply bioplastic to fashion and textile design from the perspectives of both production skill and design thinking, thus inspiring designers with wider design methods, media and technology to generate sustainable creativity.

Quantitative and qualitative research work together to ensure the objectivity of the experimental results since this study describes both material characteristics and design performance. The quantitative method is adopted to recode statistic data, such as different proportions of each formula for making bioplastic pieces, to indicate relationships between variables in the analysis stage. Unlike a regular material science experiment, qualitative methods also play a vital role in describing the design outcome in this study since it is difficult to measure the texture, color, and style in an aesthetic way. This experiment is divided into three steps: 1. Basic test: formula preparation, 2. Formula upgrade test and 3. Textile design development. From material selection, formula adjustment to design innovation, a clear step-by-step methodology is created.

3.2 Bio-plastic Textile Experiments

3.2.1 Basic Test: Formula Preparation

According to the Bioplastic Cookbook [20], agar (plant-based), gelatine (animal-based) and sodium alginate (plant-sourced) are selected in this experimental study as the three raw materials. They are referred to as the three typical biopolymers. They are mixed with glycerin (working as a plasticizer) and water in 1:1:30 proportion, and the heated sticky mixture is poured into standard Petri dishes for further observation. Plasticizers are added to bioplastic during the extrusion process to decrease the intermolecular hydrogen bonds, limit microbial growth, and stabilize product properties [21]. As a solvent, water plays the least relevant role in the experiments, which is why the proportion of water remains constant. The three different raw materials (agar agar, gelatine and sodium alginate) are the only independent variables while the other of experimental procedures are kept the same, in order to test the different properties of the textile brought by the material themselves.

Table 2. Texture comparisons of test 1

Drying Time	Bioplastic Materials		
	Agar Agar	Gelatine	Sodium Alginate
0 Hour	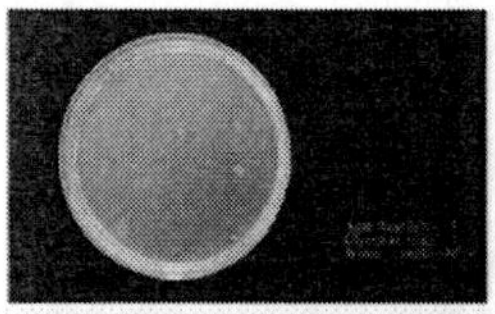	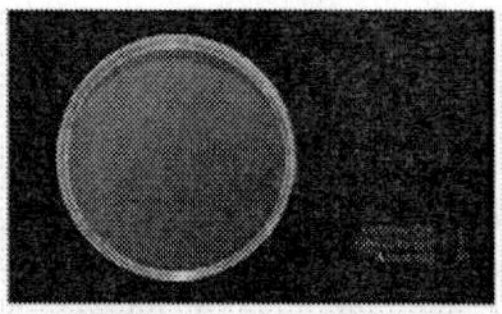	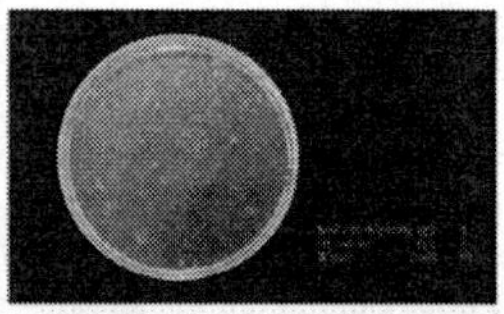
72 Hour	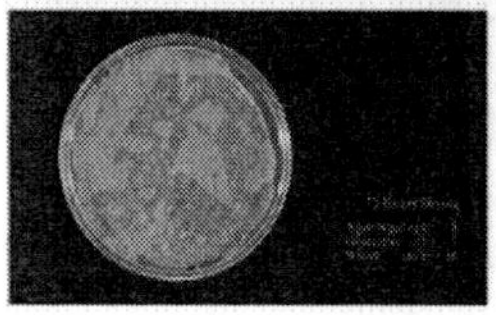	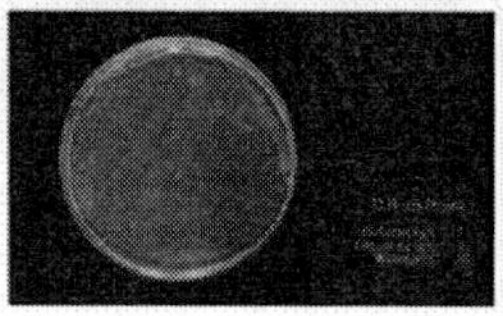	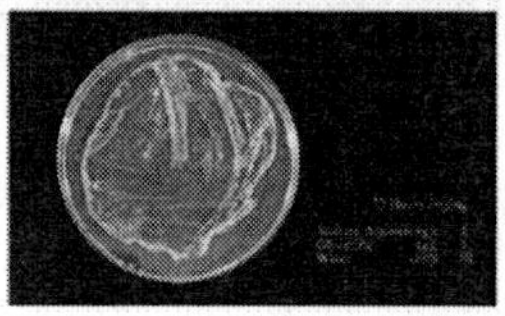

The changes of samples are recorded in the immediate time for 72 hours to explore the feature differences of the three biological components. In Table 2, three samples underwent apparent changes during the drying stage, when the water had mostly evaporated. All the samples were deformed more or less due to water loss, but the deformation of sodium alginate-based sample is the largest, while the gelatine kept its original shape. In terms of texture, the agar-based sample is dry and matte, unlike the original smooth jelly mixture. The gelatine-based sample almost retains its original transparent appearance, but the hardness slightly increased. The original mixture of sodium alginate is filled with tiny bubbles, which disappeared during the drying process. A smooth, delicate, transparent and elastic film is left. With rough evaluation, sodium alginate-based bioplastic is the thinnest and stickiest; it can be easily teared up into pieces and is therefore not suitable to be conduct in further practices. In contrast, the other two materials have moderate toughness and strength with stronger durability.

3.2.2 Formula Upgrade Test

Based on test 1, test 2 expands the proportion of raw materials from 1:1:30 to 5 groups (1:1:30, 1:3:30, 1:5:30, 3:1:30, 5:1:30 respectively) (see Table 3) to explore the role each material plays and how it influences the elasticity, strength, thickness and shape of the outcome. It then undergoes a 72-hour drying process to fully dehydrate the samples. The three groups of different materials show some common changes, among which the agar and gelatine-based samples are still more apparent than the sodium alginate. Intuitively, there is no obvious change in color, transparency and texture of the samples even though they are in five different formulas. The other two groups indicate that the higher the proportion of parameter (m), the thicker the color and the lower the transparency.

Table 3. Formula Comparisons of test 2

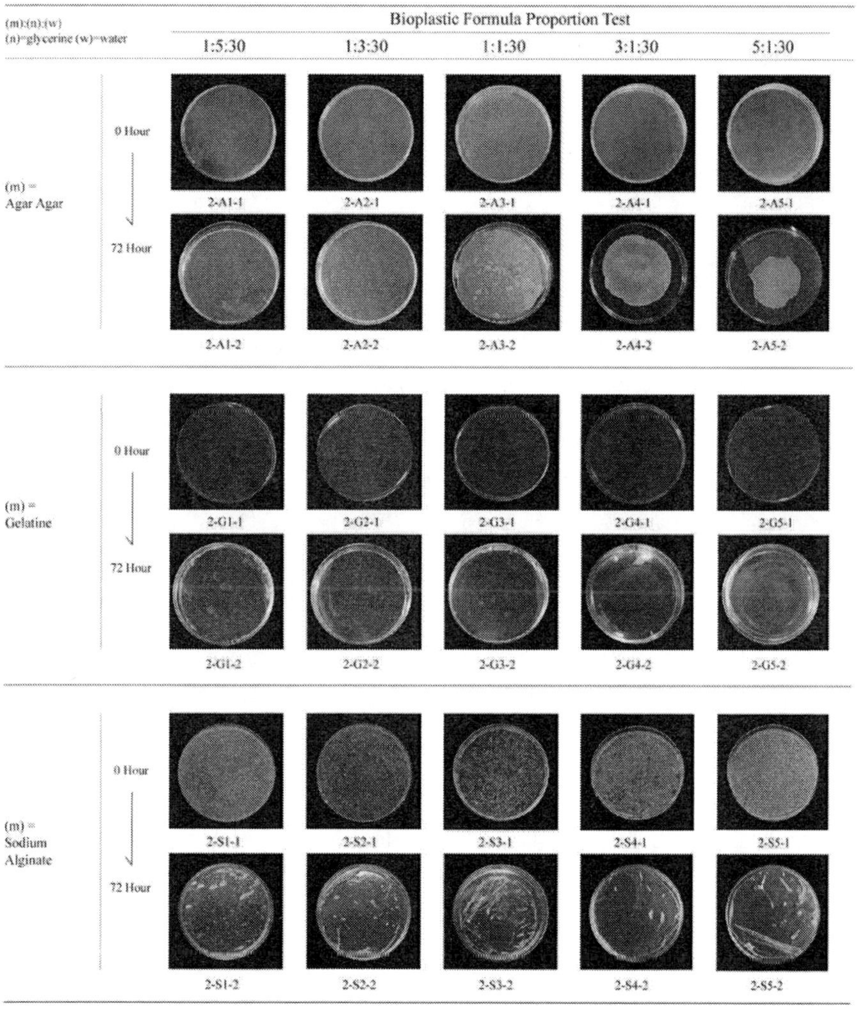

As a practice-based project, first-hand observation and description are used to describe the designed outcome. Considering some properties, such as transparency, smell, hardness, flexibility and elasticity, it also play vital roles in the use of materials. A first-hand test directly works through the senses of touching, stretching and observing. According to a number of biomaterial research, a chart is commonly used to make comparisons, that illustrate the different properties among samples. The shadow of the grey color in the first line represents the transparency of samples: the darker the color, the lower the transparency. The plus sign and the minus sign, respectively, indicate positive and negative results. The number of plus signs varies from 1 to 3, expressing different levels of positivity, from normal, very to the extreme. Table 4 presents comparisons of agar-based samples 2-A1-2 ~ 2-A5-2.

Table 4. Property comparisons (data sourced from agar-based samples: 2-A1-2 ~ 2-A5-2)

Property	2-A1-2	2-A2-2	2-A3-2	2-A4-2	2-A5-2
Transparency					
Smell	-	-	-	-	-
Hardness	+	+	++	+++	+++
Flexibility	+++	+++	++	+	+
Elasticity	+	++	+++	++	+

Agar-based test data (sample numbers from 2-A1-2 to 2-A5-2) are collected to make the comparison because the test results are sharper than the other groups. Even though the comparison table of the Gelatine and Sodium Alginate group is omitted in this chapter, they have similarities in their results. First of all, the proportion of parameter (m), in biomaterial, shows a positive correlation to the hardness. In contrast, the proportion of parameter (n), in glycerine, shows a positive correlation to its flexibility. The property of elasticity reaches its maximum when the parameters (m) and (n) are approaching a balanced ratio. When putting the formula into design practice, the requirement for the material varies due to the different application scenarios. This regular pattern of the ratio provides a reference for further design development.

3.2.3 Textile Design Development

In test 3, a variety of fibres, plants and some other food waste are added to a bioplastic base with relatively stable performance tested through stages 1 & 2 to observe the color, texture, smell, style and other properties of the final samples. At this stage, the analysis of materials was expanded from physical properties to the perspective of aesthetics and style, providing design diversity with bioplastic and paves the way for potential future applications.

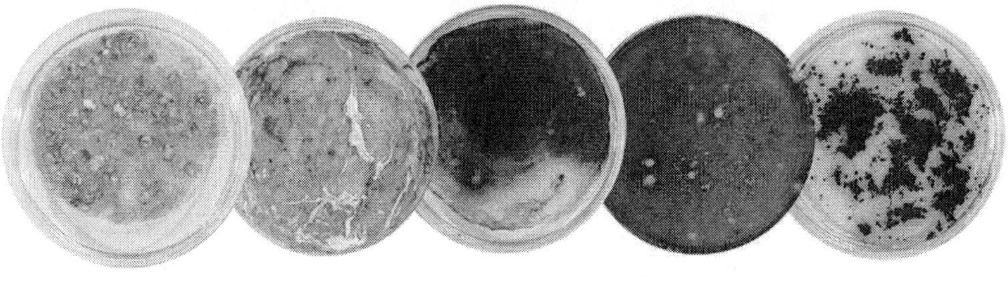

Fig.1 Textile design development

In Fig.1, biomaterials, such as grains, vegetable fibers, colored clay, spirulina and tea bags, are added into agar-based bioplastics, and they are presented differently in appearances, not only in colors but also in textures. When cereal is mixed with bioplastics (left 1), a large amount of moisture is absorbed, resulting in a hard texture and a shrunken shape. In contrast, plant-sourced materials, spirulina (right 2), shows good hydrophilicity, wherein the tiny particles perfectly mix with the agar-based material showing a uniform, natural green color. Food wastes are also considered as part of the experiment. Vegetable residue and tea bags are mixed with bioplastic. Their plant cellulose is insoluble, resulting in gel-like materials with the suspension of plant texture, which speaks a different design language from the others. There is no standard measurement to evaluate whether the materials are successful or lost. This test aims to not define a design result but to provide a practical route for reference.

4. Conclusions

Through the analysis of bioplastic experiments, the results demonstrate that different biomaterial bases has varying influences on the texture and the physical properties, such as hardness, elasticity and flexibility, that can be controlled by changing the proportion of the formulas. To meet the diverse design demands from the perspectives of aesthetics, the bioplastic can be determined by the materials filling to provide more diversity and possibilities to the fashion and textile design. However, the experimental study is still at a starting point, and furthermore, research still needs to be done to refine the bio-approaches design. In this article, only single variables are discussed through the tests, while the interactions between multiple variables are not discussed yet. Due to the limitation of the test equipment, more scientific equipment and methods will be used to measure more accurate material properties to further explore its function and practicability. Sewing and cutting will also be included in test procedures to ensure the durability of bioplastic in long-term design usage.

Bioplastics are being widely tested for their use in fashion and textile design. Environment-friendly bags, biodegradable garments and accessories made with different types of bioplastics have gained much attention in the industry. This study discusses its strategic position, that has been regarded as an essential mean to achieve the demand of sustainable fashion in line with the future circular economy. However, according to the previous experimental study on bioplastic materials, it is apparent that the formulation of bioplastics affects the performance of materials, determining the feasibility of its large-scale commercial use. To help designers better use biotechnology to generate green creativity and expand its future application, three constructive suggestions are provided in the end. First of all, expanding new design methods, techniques and skills are the key strengths in the future fashion industry, where more competitiveness is obtained in an efficient and low-carbon way. Secondly, actively seeking communication with people from various backgrounds accelerates the transformation from accessible technologies into design outcomes. The design generated through interdisciplinary cooperations is deemed to be the main force of innovation and creativity. Lastly, the progress of technology changes the design procedure and leads to the evolution of design thinking. Designers are supposed to be bravely embracing new design thinking that fit with the rapidly changing fashion industry in the future.

References

[1] Information on https://chinadialogueocean.net/14514-how-plastic-pollution-woven-into-fast-fashion.

[2] Palacios- Mateo C, Meer Y, Seide G. Analysis of the polyester clothing value open access chain to identify key intervention points for sustainability. Mateo et al. Environ Sci Eur: 2021; 33 (2): 1-25.

[3] Younes B. Classification, characterization, and the production processes of biopolymers used in the textiles industry. Journal of the Textile Institute: 2016; 108 (5): 674-679.

[4] Information on https://www.globalfashionagenda.com/publications-and-policy/pulse-of-the-industry.

[5] Information on https://www.bloomberg.com/news/features/2020-10-18/china-s-big-fashion-problem-is-recycling-26-million-tons-of-used-clothes.

[6] Silva CJG, de Medeiros ADM, Amorim JDP, et al. Bacterial cellulose biotextiles for the future of sustainable fashion: a review. Environ Chem Lett: 2021; DIO: 10.1007/s10311-021-01214-x.

[7] Friedrich D. Consumer and expert behaviour towards biobased wood-polymer building products: a comparative multi-factorial study according to theory of planned behaviour. Architectural Engineering and Design Management: 2021; DIO: 10.1080/17452007.2020.1865867.

[8] Fossil fashion: the hidden reliance of fast fashion on fossil fuels. Changing Markets Foundation: 2021.

[9] Fixing fashion: clothing consumption and sustainability. House of Commons Environmental Audit Committee. HC 152: 2019.

[10] A new textiles economy: redesigning fashion's future. Ellen MacArthur Foundationn: 2017.

[11] Boucher J, Damien F. Primary microplastics in the oceans. IUCN 2017; DOI: 10.2305/IUCN.CH.2017.01. en

[12] Information on https://www.condorferries.co.uk/plastic-in-the-ocean-statistics.

[13] Information on https://www.european-bioplastics.org/bioplastics.

[14] Vroman I, Tighzert L. Biodegradable Polymers. Materials Basel: 2009; DOI:10.3390/ma2020307.

[15] Image drawing reference: Younes B. Classification, characterization, and the production processes of biopolymers used in the textiles industry. The Journal of The Textile Institute: 2017; DOI: 10.1080/00405000.2016.1180731
[16] Crank M, Patel M, Marscheider-Weidemann F, Schleich J, et all. Techno-economic feasibility of large-scale production of bio-based polymers in Europe - Final report. Spain: Utrecht University & Fraunhofer ISI: 2004.
[17] Chandra R, Rustgi R. Biodegradable biopolymers. Progress in Polymer Science: 1998; 23: 273-1335.
[18] Myers W. Biodesign: Nature, Science, Creativity. London: Thames and Hudson: 2018; p. 6-8.
[19] Information on https://www.bbc.com/future/article/20200310-sustainable-fashion-how-to-buy-clothes-good-for-the-climate.
[20] Information on https://issuu.com/nat_arc/docs/bioplastic_cook_book_3.
[21] Younes B. Investigating the co-effect of twist and hot drawing processes on the bio-based yarns properties. Journal of the Textile Institute: 2020; 111 (2): 202-213.

The Exploration of the Innovative Practice of Plastic Waste in Sustainable Fashion Design

Xiao-Yu Chen[1a], Rong Zheng[1b*], Li Zhou [2*], Xin Yang[1b], Jing Lin[1b]

[1a]*School of Fashion and Art Design, Donghua University, 1882 West Yan'an Road, Changning District, Shanghai 200051, China*
[1b]*Shanghai International Fashion Innovation Centre, Donghua University, 1882 West Yan'an Road, Changning District, Shanghai 200051, China*
[2]*College of Sericulture, Textile and Biomass Science, Southwest University, No. 2, Tiansheng Road, Beibei District, Chongqing 400700, China*

[*] Corresponding authors' email: rzheng@dhu.edu.cn, mydtcazz@126.com

Abstract

In order to solve the contradiction between economic development and environmental problems, and the trend of its recycling strategy is to increase the value of textiles and other wastes in the supply chain. Fashion companies have been trying to develop sustainable clothing made from post-consumer plastic bottles, but few studies have explored the practice and innovation of plastic waste in clothing design from the fashion perspective. Aiming at the complexity of plastic waste recycling, this article explored the possibility of upgrading and recycling plastic waste in clothing fabrics through the analysis and performance analysis of plastic waste materials, further proposed new solutions for the sustainable design of plastics, and demonstrated the effect of plastic waste on clothing visualization via experimental operations, so as to provide more possibilities for sustainable fashion design practices.

Keywords: Plastic Waste; Clothing Innovation Practice; Sustainable Fashion Design

In the important survey report "Our Common Future" drafted by the United Nations World Commission on Environment and Development in 1987, sustainability was defined as "a development that can meet our current needs without affecting the needs of future generations [1]". As a result, many designers focused on sustainable design, and started to guide consumers to aware sustainability and pay close attention to fashion brands with environmental protection certifications. Recycling has become a common approach in sustainable design. In the recycling of plastic waste, the production and disposal methods of plastics are mainly accomplished through the resource consumption of fossil fuels and large amounts of carbon emissions [2, 3]. This process not only cause a serious burden on land resources, but also generate microplastic particles which have unfavorable effects on the marine ecosystem [4]. In addition, plastic waste is very heterogeneous as it contains many different polymers and product types, and it is often contaminated, much of the separated waste is not recycled into new products [5].

In fashion industry, plastic is usually used as the outer packaging of fashion textile clothing, which virtually causes waste. Therefore, many companies and brands have realized the need to reduce waste and have begun to reintroduce the recycling of textile waste, and innovate recycling methods for plastic waste.

1. Introduction

1.1 The Status of Plastics Waste Recycling

In China, the annual the actual consumption of plastic products is large, waste plastic production is equally large in order to become the world's largest recycled plastic market, but the recycling rate for plastics has hovered around 25%. Compared with foreign countries, the recycling rate of plastic waste in the United States is about 50%, and the recycling rate of plastic waste in Germany is about 50%, Germany's waste plastic recycling rate reached 60%, Japan's waste plastic recycling rate as high as 60% [6].

Waste generated during plastic production leads to complex supply chains and difficulties in recycling and sorting. The excessive production of waste emphasized the need for innovative methods to solve unsustainable practices by returning waste to the supply chain system [7]. In the existing research, Reina Pilz proposed the term "up-cycling", that is, by changing the normal use phase of the product life cycle, based on the recycling of waste, the original function is maximized to minimize the process, repeat, and recycle the product life cycle without going through the entire process of product planning, production and consumption [8].

In terms of sustainable fashion design in recent years, the most common way to recycle textile waste is to extend the closed loop of the clothing cycle through by reuse of discarded materials, and to have an impact on the clothing supply chain through recycling programs, and reuse waste clothing [7]. For example, the clothing bank founded by designer Na-yu Zhang re-examined from the perspective of "redesign", using outdated textile and plastic materials to transform and regenerate with the power of design, giving new life to old materials, and creating new works to present the future of sustainable fashion [9].

In the research of the clothing industry, the recycling of plastic fibers is usually used to carry out the sustainable design of clothing. For example, in 2019, ANTA released its environmentally friendly clothing product-"ANTA Training" series. The raw material is recycled plastic waste bottles, using independent research and development. The physical and chemical method of polyester textiles is used to produce clothing fabrics [10].

In order to further understand the recycle methods of waste plastic in sustainable fashion design, the author searched 34 related literatures and reports with "fashion" or "waste plastic" as the key words, and listed the proportion of different recycle methods about waste plastic, as show in Fig.1 and Table 1. As show in the chart 1, most of sustainable fashion design adopts the recycling method of waste plastic products processing and spinning, Only 8.82% of the literature and reports mentioned that waste plastic products are directly applied to fashion design. Therefore, the practical exploration method for the sustainable development of waste plastics is not ample.

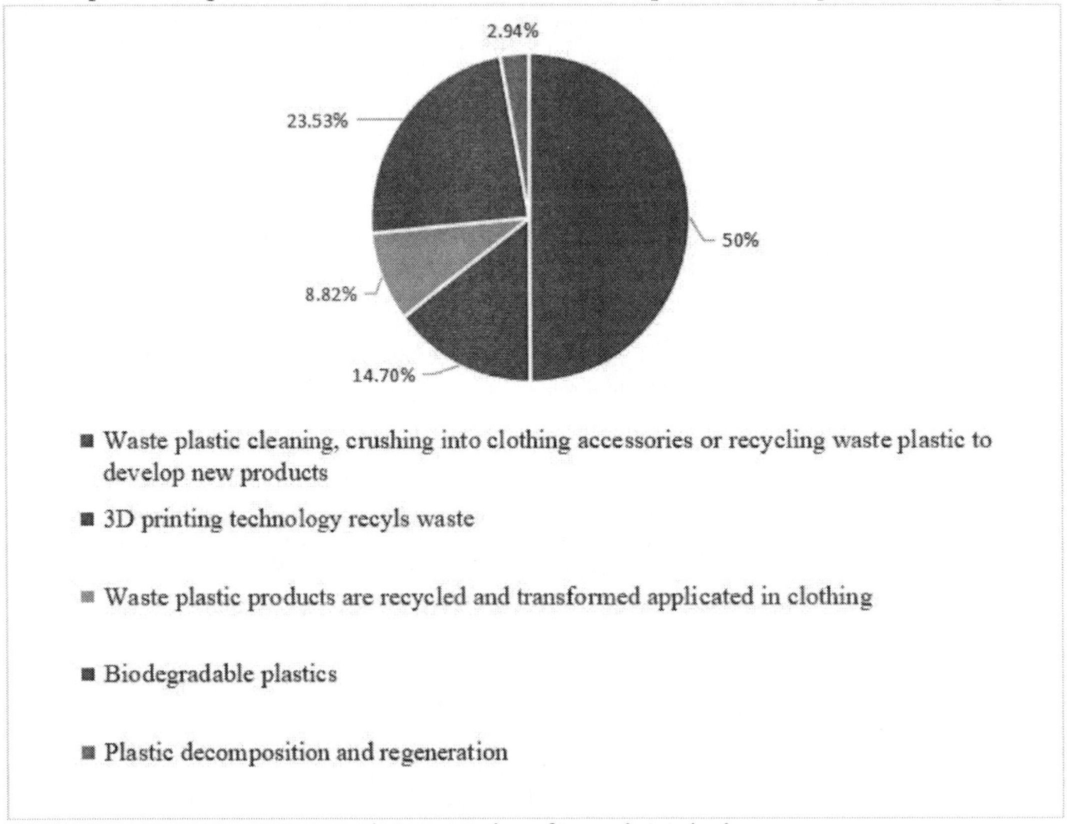

Fig.1 Proportion of Recycle Methods

Table 1. The Analysis of Literature and Report Collection

No.	Recycle method	Number	Accounted	Main Reference
1	Waste plastic cleaning, crushing into clothing accessories or recycling waste plastic to develop new products	17	50.00%	ANTA, Adiads, Nike, Xiake environmental protection, G-Star RAW, Musselblomma, Meyer, Waste2Wear, Coca Cola The (Re)Collection, boots, patagonia, Prada, Repreve, H&M etal.
2	3D printing technology recyls waste	5	14.70%	Polymaker, The "Dolphin Board of Awesome" project, the "Million Waves Project"in 2018 etal.
3	Waste plastic products are recycled and transformed applicated in clothing directly	3	8.82%	Hungarian Designer Katalin huszár, Dutch artist Suzanne Jongmans, Conilie M etal.
4	Biodegradable plastics	8	23.53%	PLA, PHAs, PGA, Muhammad Reza Cordova, "Living Seawall"project、Keron、Roya Aghighi, Carolyn Raff, Piero D'Angelo etal
5	Plastic decomposition and regeneration	1	2.94%	Eastman (China)

1.2 Reasearch Direction

In the entire field of sustainable fashion design, eco-design, and life cycle assessment (LCA) are used to test and verify the redesign process [11]. It is common to extend the product life cycle and reduce the use of waste through redesign. However, in most literature, the research on textile waste and plastic waste is mainly through recycling and fiber reprocessing, and there is still a lack of fashion practical solutions based on material properties. Therefore, this research sorted out the basic material properties of common plastics in daily life and the recycling rate of plastic waste and discussed the basic methods of its design on clothing, which in order to try to determine the operability of plastic waste in clothing fashion design and list three experimental schemes for the innovative application of plastic waste in clothing. The design methods using plastic waste to transform made clothing visually displayed with both fashion and sustainability. At the same time, this research provided more design possibilities and methods for sustainable research on plastic waste used in clothing.

2. Definition and Recycling Rate of Plastic Products

The plastic we use is normally not a single component. It is a polymer compound that is polymerized through addition polymerization or condensation polymerization with monomers as raw materials, commonly known as plastics or resins [12]. Among them, high molecular polymer is the main component of plastic. In addition, in order to change its performance and produce different kinds of plastic products, auxiliary materials must be added to the high molecular compound to finally become various kinds of plastics with good characteristics. The addition of various auxiliary materials makes it difficult for plastics to be recycled and sorted.

In the recycling of plastic products, due to its low processing cost and strong plasticity under certain temperature and pressure conditions, there are many types of plastics and large amounts of them, making it difficult to collect, sort and recycle. Common plastic materials in daily life include PET, HDPE, PVC, PP, etc. After production and use, plastic wastes are generated through being mixed with other materials. These wastes mainly come from four aspects: industry, agriculture, society, and domestic waste [13]. In terms of its recycling rate, there are certain differences according to the production methods and processing methods. Therefore, the sources and recycling rates of plastic waste are sorted out, as shown in Table 2.

It can be seen from the table that plastic products are used in various fields of our lives and have become one of the important materials that is indispensable for human beings in daily life. In recent years, due to the rapid development of express delivery, catering, packaging and other industries, the proportion of plastic packaging products in urban domestic waste has gradually increased, but the recycling rate is very low, and the problem of plastic recycling in domestic waste has become increasingly prominent. In consideration of the overall design materials of the practical plan, the plastic waste collected and sorted from the urban solid waste have the characteristics of light volume, corrosion resistance, easy processing, etc. The materials are relatively easy to obtain and reuse, and it is still a kind of recyclable renewable resources. Practical plastic waste can be obtained from material packaging, disposable plastics, plastic products, and other materials in urban domestic garbage. In this paper, the criterion of waste plastics is summarized according to the national standard compilation of classification, Grade and code for waste plastic, which is published in China [13-15].

Table 2. The Recycling Rate of Common Plastic waste

The source of plastic waste	Generation Channels	Main plastic materials	Recycling Rate
Industrially produced plastics	1) Residues from resin production and by-products 2) Leftovers, inferior products, etc. produced by the production and processing of plastic products	PET (polyethylene terephthalate), PC (polycarbonate), ABS (resin), PVC (polyvinyl chloride)	Generally, the plastics can be recycled in the factory, no excessive waste is generated, and the recycling rate is high
Agriculturally produced plastics	Agricultural equipment such as agricultural mulch film, greenhouse film, agricultural water pipe, rope, etc. are produced after long-term use	HDPE (high density polyethylene), PVC (polyvinyl chloride), PA (polyamide)	For the waste plastics produced, agricultural plastic is generally cultivated equipment, such as this kind of material use life is longer, so compared with packaging plastic recycling rate is higher.
Socially produced plastics	Commodity distribution channel: disposable packaging materials; commodity consumption channel: food boxes, beverage bottles and other plastic sundries sold in hotels, restaurants, and trains	LDPE (low density polyethylene), PS (polystyrene), PET (polyethylene terephthalate), PVC (polyvinyl chloride)	This kind of plastic waste itself is basically non-polluting, and it can be processed through regeneration treatment by one time recycle and classification
Urban household waste produced plastic	Household plastic products, disposable plastics: mainly for life packaging materials, foam materials, woven bags, etc.; non-disposable plastics: various plastic containers, stationery, cosmetics, etc.	PET (polyethylene terephthalate), LDPE (low density polyethylene), PP (polypropylene)	This type of plastic waste accounts for a relatively large amount, most of which are stored in cities, accounting for about 4%-10%[13] of the entire urban domestic waste and showing an upward trend. It is difficult to deal with in the early stage, high cost, and low recycling rate.

3. Practical Exploration of Plastic Waste in Sustainable Clothing Design

In the fashion industry, due to the extremely high chemical stability and strong moldability of plastic materials, many companies in China and abroad used this feature to change the material form of plastics through recycling technology and turn plastic waste into fibers [16] and then spin them into clothing fabrics. To improve the recycling rate of plastic waste, the main steps are as follows, as shown in Fig.2.

Fig.2 The recycling process of plastic waste in making fibers

In the process of spinning plastic products, the regeneration of plastic waste is mainly performed through fiber reprocessing technology and woven into fabrics. For the fashion and aesthetics of fabrics, prototypes of discarded plastic materials are rarely seen on actual clothing products. In the stage of clothing sales, it is difficult for consumers to have psychological recognition of the sustainable design of discarded products and affect their attitudes towards the products and purchase intentions. Therefore, the visual display of plastic waste in sustainable fashion design is essential, and the innovative application of plastic waste in clothing design has become a new breakthrough.

Based on the visual effects of plastic waste in sustainable clothing design, this research developed three design practices based on the characteristics of plastic materials. In order to maximize the performance of plastic waste in clothing design, different types of domestic plastic waste were used for recycling and redesign in this research practice. Through three different processes and fabric recycling attempts, it was able to demonstrate the possibility of combining plastic materials and clothing.

Firstly, display clothing patterns in two-dimensional space. Plastic waste pattern printing as one of the means of fabric design can directly and clearly convey the theme and ideas of waste recycling. Therefore, the first design scheme selected plastic waste accumulation patterns with higher picture pixels and clear resolution to display the patterns on the clothing fabrics through digital printing. In the production process, in order to compare the visual effects of different fabrics and different temperature environments so as to select the most suitable method of operation, a printing experiment was adopted, as shown in Table 3.

Next, in order to highlight the three-dimensional and texture effects of plastic waste materials, in the experiment of the second scheme, plastic packaging bags of different colors and textures, plastic straws and other disposable household daily necessities were selected from domestic waste, the main components of which are LDPE (low-density polyethylene) mixture. Then crushing, cutting, pulling, stacking, burning, accumulating and other destructive methods were applied to conduct multiple fabric sample in experiments. The analysis and comparation of the experimental effects of various plastic waste materials are as shown in Table 4.

It can be concluded from the experiment that through the visual effect experiment of different plastic products, when the scheme is successfully established in the clothing design, it can effectively recycle low-quality scrap plastics, thereby reducing the complexity of recycling and sorting plastic waste and cost. The solution still needs to be improved. Considering the needs of the clothing market and the comfort of wearing, it is not suitable for large-scale applications in the innovative design of clothing. It is costly to use broken plastic products to combine and fix large areas on clothing. What is more, it takes certain periods of time and there is a certain randomness in the process of cutting, stacking and burning plastic blocks.In the exploration attempt of experiment 1 and 2, I have tried to make a preliminary experiment by combining the two experimental techniques. The results of the evaluation and display were not good according to the description of the wearer, so experiments 1 and 2 were not implemented into the fashion design. Therefore, in terms of innovative practices, the sustainable design of waste plastics will focus on the part of experiment 3 and be transformed into ready-to-wear.And the experiment 1 and 2 will final display of the fabric transformation effect.

Table 3. Plastic Printing Pattern Experiment

Fabric Number	Temperature (°C)	Fabric Texture	Effect Evaluation	Results Image
01	180	Pure cotton	Slightly light fabric color	
01	200	Pure cotton	Fabric printing effect better than 180°C	
02	180	Nylon	Printing colors are bright and clear	
03	180	Nylon	The printing effect is average, the overall pattern is light, and the color fastness is not enough	
04	200	Plastic coating Laminated fabric	The temperature is high, the coating melts, and the cloth sticks	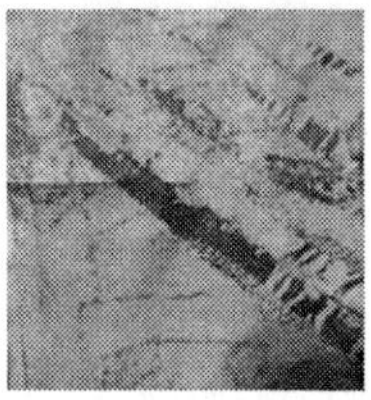
04	100	Plastic coating Laminated fabric	Printing color is light	

Table 4. Different types of plastic accumulation experiments

No.	Material composition	Process Method	Experimental Effect Evaluation	Experimental Process
01	Straw	Overlaid	The accumulation pattern has strong three-dimensional effect and appealing effect	
02	Plastic bags	Laminating, stacking, deforming	Good accumulation effect, random patterns, not easy to manage	
03	Straw+ Plastic bags	Laminating, stacking, deforming	The accumulated pattern is in disorder and has no aesthetic feeling	
04	Plastic rope + Latex	Laminating, stacking	Plastics are highly reflective, consider the combination with fabrics	
05	Plastic bags	Laminating, burning	Black smoke is generated during the firing process, and the firing pattern is not easy to manage	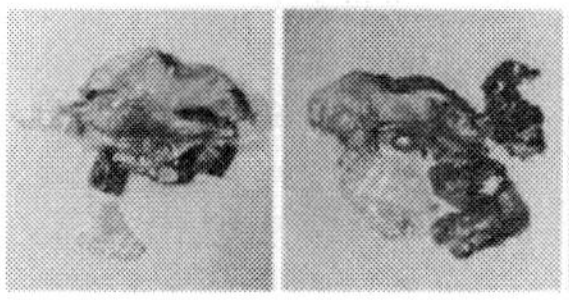
06	Strong reflective plastic, tinfoil	Laminating, stacking	Plastics are highly reflective, consider the combination with fabrics	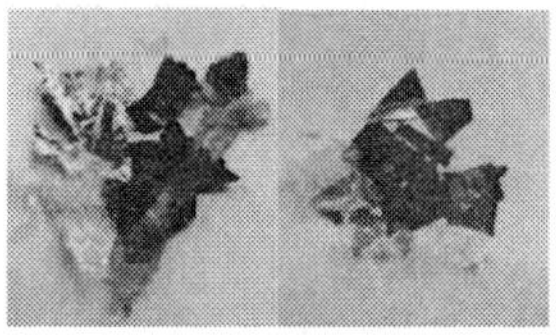
07	White Latex + Plastic bags	Laminating, stacking	The white latex covers the effect of the plastic itself	
08	Cotton thread + plastic TPU	Machine sewing	There are many varieties in the combination of lines and plastics	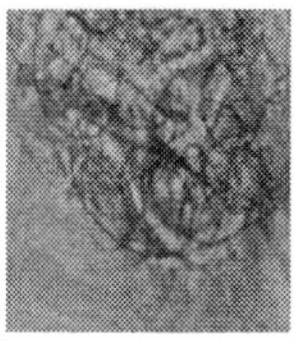

Based on the shortcomings and deficiencies of the above two schemes, in the third experiment, scattered and broken plastics were not used for lamination, and the characteristics of plastic waste materials were considered as the basic foothold of the design, and the waste packaging bags were used as the main experimental object, such as discarded plastic woven bags. In daily life, plastic woven bags are mainly used to carry heavy objects. The overall material is lighter, and its main component is a mixture of PP (polypropylene). The weaving structure of the woven bag in the warp and weft directions is similar to the structure of the fabric fiber, as shown in Fig.3. Therefore, in terms of technology, it was cut into strips, and the plastic tissue of the warp part was extracted in a destructive manner, and the weight and hardness of the warp was enhanced through the weaving method. At the same time, combined with the common characteristics of plastic materials, burning methods were avoided. The practical operation method is shown in Fig.4

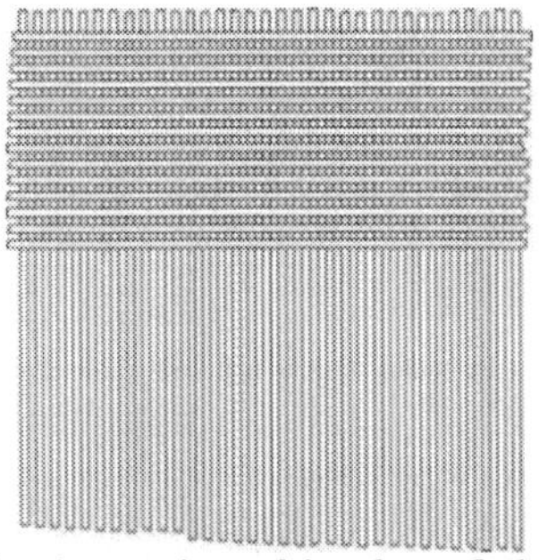

Fig.3 The organization and combination of plastic woven bags

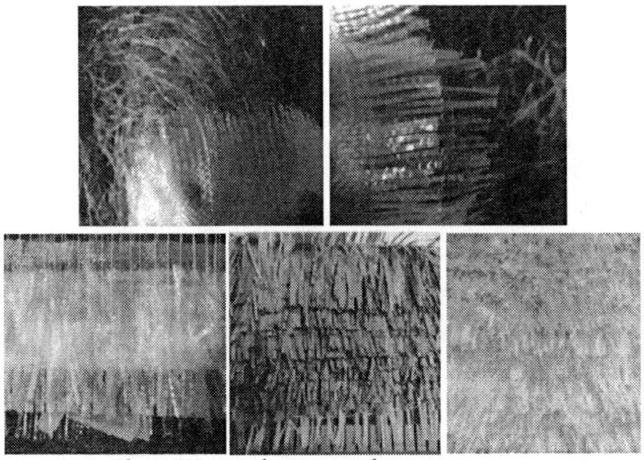

Fig.4 Woven bag weaving process

During the experiment, the woven bag has a mesh woven structure in the warp and weft directions and has unique toughness. It is suitable for strip-shaped tassel decoration in the shape of the innovative design of clothing, and the display is generally translucent and shiny. The fabric structure is organized, the cost of materials required for production is low, and at the same time it can be combined with clothing fabrics, and there is ample room for development in design. However, this solution is prone to detachment of the organizational structure during the cutting process, so the overlaid process method can be used to strengthen the integrity of the organizational structure and enhance its firmness during assembly.

4. The Application of Innovative Design Schemes in Clothing

In the entire experiment of sustainable practice of clothing, fabrics as one of the three elements of clothing usually need to go through clothing styling to show whether creative practical design can be implemented in the stage of garment production. Based on the visual display of innovative practical schemes, by providing three different process method as references, three design schemes can be creatively applied to clothing in the later design stage of the overall shape of the clothing. At the same time, it is not only limited to a single design practice, but also attempts to combine innovative schemes to show more possibilities and design methods of plastic waste in the sustainable design of clothing.

To better demonstrate the visual process of discarded plastics on clothing, this research offered a more detailed design of three design practice schemes combined with clothing modeling. In practice, the first two design schemes are suitable for displaying a small area on the clothing, so they are suitable for operation after the fabrics and garments are made. The last design scheme needs to be considered in the early stage of clothing plate making and tailoring. The area and method are suitable for combining large areas with clothing fabrics. After analyzing and comparation of the advantages and disadvantages of each experimental program, based on the sustainability of clothing fabrics and the application of discarded materials, this research combined the first two design innovation practices and used the complete fabric effect to demonstrate sustainable fashion design. The plastic waste blocks and printed fabrics were recombined by laminating and remanufacturing, thereby forming a superposition of two-dimensional plane and three-dimensional space, so that the surface of the fabric can form a raised lines and texture when viewed from a distance. It also created the effect of a whole piece of plastic waste piling up, enriching the visual and sensory experience, as shown in Fig.5.

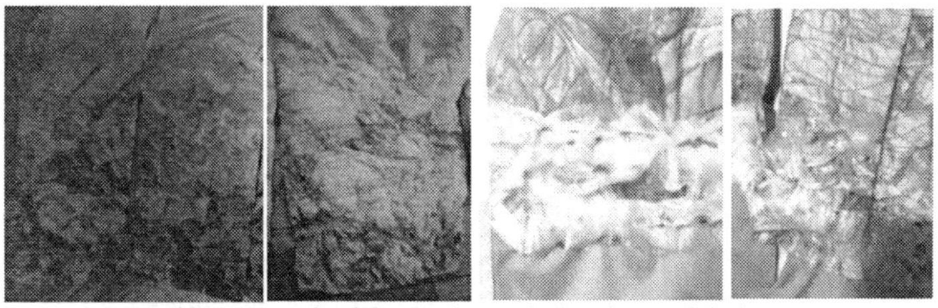

Fig.5 Sustainable design of plastic on fabric

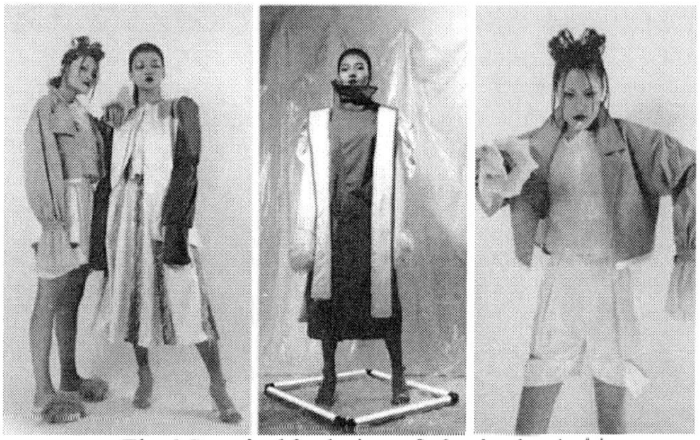

Fig.6 Sustainable design of plastics in clothing

The third innovative design practice can form an overall fabric texture effect. In order to facilitate the tailoring and production of clothing patterns, the design practice was directly displayed as a sustainable fabric, and the overall design was mainly applied to clothing modeling, as shown in Fig.6.

In view of the effects of visual display on the three design practices, it can be seen that plastic waste and clothing fabrics are organically integrated and applied to sustainable clothing design. Finally, the three innovative schemes can effectively solve the problems and difficulties in the recycling and classification of waste materials to a certain extent, but there are also certain limitations and shortcomings, which can be summarized as follows:

1) From the perspective of design, not only it must ensure that the theme and innovative thinking are closely related to the principle of selecting specific properties of plastic materials, but also it needs to be familiar with the requirements of various plastic products in clothing process and technology. In the selection of materials, with the consideration of the characteristics of plastic materials, the recycled plastic waste can be upgraded and re-used to enhance the added value of materials.

2) Compared with ordinary clothing fabrics, plastic waste is a special creative material. When it is used in clothing, detailed material screening and comparison are required to find the most suitable materials. The process methods appropriate for the overall design of clothing also need to be explored to meet the expected effects meanwhile.

3) During the overall design practice, this research provided three different process methods to show the innovative design of waste materials in clothing and fabrics, and carried out creative combinations by trying more waste material types in the possibility of sustainable design of clothing. Due to the numerous varieties and materials of plastic waste, in sustainable fashion design, we can consider the development of more new process and design methods to reorganize and derive and carry out sustainable exploratory research in combination of the overall shape of clothing.

4) With the gradual attention paid to the concept of sustainability, there is a large market and room for research and development in the sustainable design of various waste materials. How to creatively combine waste materials with clothing and improve the use value of clothing has become a subject and direction that can be studied in the future.

5. Conclusion

Beautiful and healthy wearing is always the goal pursued by mankind, but it should not be premised on breaking the ecological balance and destroying the environment. The fashion industry is one of the producers of environmental pollution. The serious ecological impact and the large amount of waste in the clothing industry have increased the call for "sustainable design". Clothing should have a certain awareness of sustainable design concepts in the early stages of design and strive to make breakthroughs. With the limitations of traditional materials, it is possible to find more sustainable materials in the application of clothing.

Through the conversion of recycled plastic waste into decorations and fabrics on clothing, the diversification of clothing styles and designs is enriched, awakening the public's sense of social responsibility, cultivating public awareness of environmental protection, and being able to actively protect the earth on which we live, and making eco-friendly wearing become a fashionable lifestyle. At the same time, it is hoped that this design practice can provide new ideas and methods for reference for the visual effects of waste materials in sustainable fashion design and promote the development of sustainable design in the fashion field.

References

[1] World Commission on Environment and Development. Our common future: Report 1987.
[2] Human perceptions of recycled textiles and circular fashion: A systematic literature review: Sustainability 2020; 12 (24): DOI: 10.3390/su122410599.
[3] Jambeck JR, Geyer R, Wilcox C, Siegler TR, Perryman M, Andrady A, Narayan R, Law KL. Plastic waste inputs from land into the ocean. Science: 2015; 347 (6223): 768-771.
[4] Foschi E, D'Addato F, Bonoli A. Plastic waste management: A comprehensive analysis of the current status to set up an after-use plastic strategy in Emilia-Romagna Region. Environmental Science and Pollution Research: 2020; 21 (1): DOI: 10.1007/s11356-020-08155-y.
[5] Gradus R. Postcollection separation of plastic recycling and design-for-recycling as solutions to low cost-effectiveness and plastic debris. Sustainability: 2020; 12 (20): 1-12.
[6] Hu SR. Analysis on the status of waste plastic recycling and import in China [J]. Renewable resources and circular economy: 2012; 5 (08): 24-27.
[7] Feil A, Pretz T, Jansen M. Velzen EUTV. Separate collection of plastic waste better than technical sorting from municipal solid waste? Waste Management & Research: 2017; 35 (2): 172-180.
[8] Diddi S, Yan RN. Consumer perceptions related to clothing repair and community mending events: A circular economy perspective. Sustainability: 2019; 11 (19): DOI: 10.3390/su11195306.
[9] Shim S, Kim J, Na Y. An exploratory study on up-cycling as the sustainable clothing life at home. Fashion and Textiles: 2018; 5 (1): 1-15.
[10] Information on https://www.ontimeshow.com/zh-hans/node/3482.
[11] Han S. 11 discarded plastic bottles to create eco-friendly clothing. Green Packaging: 2019; (10): 40-41.
[12] Foschi E, Zanni S, Bonoli A. Combining Eco-Design and LCA as Decision-Making Process to Prevent Plastics in Packaging Application. Sustainability: 2020; 12 (22): 9738.
[13] Zhang N. Innovative application of plastic material in popular clothing from space style in 1960s. Art Science and Technology: 2016; 29 (10): 143-144.
[14] Fu ZY. Research on national standard compilation of classification, crade and code for plastic waste. Tianjin University of Technology: Master's Degree Thesis: 2018.
[15] Classification and code of waste plastics, GB/T 37547-2019.
[16] Chen D, Huang XY, Wang P, Liang X, Zhou XH. Recycling effective ways of waste plastics. Engineering Plastics Application: 2012; 40 (9): 92-94.

Luggage Industry Sustainable Development Strategies Based on Triple Bottom Line Theory

Yu-Han Wei[1], You-Fang Chen[2*]

[1]*Chinese Academy of Lifestyle Design, Beijing Institute of Fashion Technology, No.2 East Yinghua Road, Chaoyang District, Beijing, China, 100029*

[*]Corresponding author's email:sxychenyf@163.com

Abstract

With the increasing severity of environmental issues, how to create benefits and reduce environmental pollution has become the common goal of every enterprise and brand. Therefore, how to achieve the sustainable development of enterprises and brands is a crucial problem. Based on the triple bottom line theory, this paper analyzes the relationship between the three dimensions of environment, society and the economy in the luggage industry chain and constructs a framework model for the sustainable development of the luggage industry by combining the literature method and the case analysis method. A sustainable development strategy of the luggage industry is put forward to provide ideas for the sustainable development of the luggage industry.

Keywords: Triple Bottom Line; Luggage Brand; Sustainable Development

1. Introduction

Consuming a large amount of resources to develop the social economy has brought about severe environmental problems in the 21st century. Global resource shortages, energy depletion, serious environmental pollution, global climate warming and other phenomena have forced companies to pay more attention to corrective actions while conducting business activities. Fashion is like a tire spinning at high speed on its own. According to consumers' consumption and purchase various fashion products. At the same time, the fashion industry is producing rapidly, and fashion products meet the needs of consumers, and the spirit is inflated with crazy profits [1]. As a result, a lot of old clothes and textiles are buried, causing our environment to deteriorate. With the protection of environment and resources, many companies have begun to implement sustainable development routes, which is also the future trend of the fashion industry.

Based on the triple bottom line theory, this article builds an analysis framework model around the luggage industry chain and proposes sustainable development strategies from three different dimensions: environment, society, and economy.

Sustainable luggage products not only need to satisfy public demands for material enjoyment, it also needs to protect the ecological environment. To achieve this goal, the fashion bags industry has to innovate to reduce resource consumption, encourage recycling, and meet social responsibility requirements, while still maintaining a profit.

Take luxury brands as examples of foreign luggage brands. The core of luxury brands is to create attractive but not essential goods and services, it is therefore a challenge to position luxury goods with sustainable development. The social responsibility of an enterprise lies in the obligation to protect and promote the overall interests of society. It integrates the protection of the earth, the survival and continuity of species, non-renewable resources, and respect for humans into the luxury goods industry.

Compared with the overseas luggage industry, China's luggage industry still has a long way to develop for it to be sustainable. There are Chinese brands that takes on the challenge of sustainability; their main focus is on how to solve the problem systematically, improve the overall sustainable development awareness of production and consumption, positively influence consumers' sustainable consumption concepts, and realizing sustainable lifestyles. This article will provide new ideas and feasible sustainable development strategies for domestic luggage companies and brands.

2. Domestic and Foreign Research Overview

2.1 Concept and Connotation of Sustainability

2.1.1 Sustainable Development

The World Commission on Environment and Development, which was established in 1984, published the Common Future in 1987 and formally defined sustainable development as "development that meets the needs of present generations without jeopardizing the ability of future generations to meet their needs." The sustainable development of enterprises is embodied in the protection of the environment, including the protection of biological resources, the responsibility to society and the responsibility to realize profit [2]. This section mainly combs the concerned domestic and foreign research results.

2.1.2 Chinese Scholars' Explanation of Sustainable Development

The first systematic article on the thought of sustainable development among Chinese is the article published by Professor Cheng Fukun, economist of the Chinese Academy of Social Sciences, in the sixth issue of Economic Research in 1983. In 1992, headed by Niu Wenyuan, China first established the Environment and Sustainable Development Research Office in the Institute of Science and Technology Policy of the Chinese Academy of Sciences.

On the basis of economics, sociology and ecology, the theoretical and empirical research of sustainable development in China has independently created the fourth theoretical direction of sustainable development: the direction of systematics. From the research results of domestic scholars, Zhang Li (2010) believes that the theory of sustainable development is based on the following basic concepts: the carrying capacity of the earth's ecosystem, decoupling and closed economy, life cycle assessment, and contemporary equity and intergenerational equity. Xu Ke and Wang Rui (2015) believe that the concept of sustainable development began at the end of the 20th century, and scholars at home and abroad began to pay attention to the relationship between economic development and environmental protection. Now the concept of sustainable development has developed into a systematic analysis of the whole society, involving five aspects: human society, natural resources, population development, economic growth and ecological environment. [3]

2.1.3 Foreign Scholars Explanation of Sustainable Development

In 1972, the concept of sustainable development was first formally discussed at the United Nations seminar on the human environment in Stockholm. By definition, sustainable development is that when we use the natural environment to develop economy and society, we should pay attention to the limitation of natural resources and ecological environment, the continuity of human development, and combine utilization with protection. It is the natural environment that makes a sustained contribution to human development. Further, it refers to the coordinated development of economic, social and natural environment protection.

After a long qualitative analysis, the research on sustainable development abroad has been carried out quantitatively, that is, the main body of sustainable development is all human beings, its representative is the governments of various countries, and mainly from the three aspects of urban, rural and enterprise.

2.2 Summary of Triple Bottom Line Theory

In 1997, British scholar John Elkington first proposed the concept of triple bottom line. He believed that corporate social responsibility can be divided into economic responsibility, environmental responsibility and social responsibility in terms of responsibility fields. Economic responsibility is also the traditional corporate responsibility, mainly reflected in the increase of profits, tax responsibility and dividends to shareholders and investors; Environmental responsibility is environmental protection; Social responsibility is the responsibility to other stakeholders in society. In the practice of CSR, enterprises must fulfill their responsibilities in the above three areas, which is the "triple bottom line theory" related to CSR. The following is a conceptual diagram of the triple bottom line theory (refer with: Fig.1). Based on the triple bottom line theory, this paper constructs the framework model of sustainable development of luggage industry.[4]

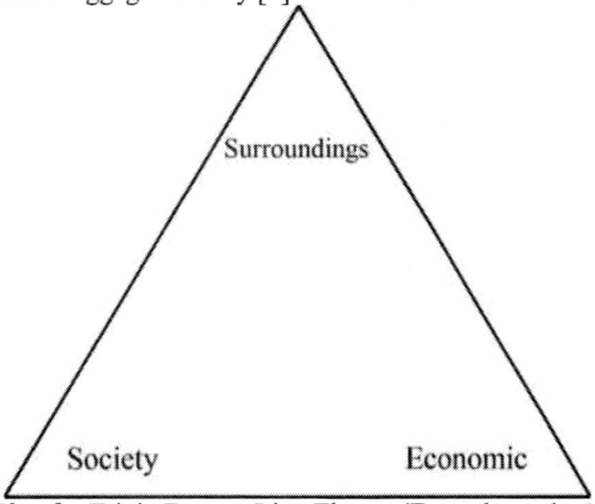

Fig.1 Concept Plan for Triple Bottom Line Theory (From the author's self-painted)

3. Build an Analysis Framework Model Based on the Luggage Industry Chain

3.1 The Luggage Industry Chain and its Connotation

The industrial chain of luggage category is mainly divided into three parts, including raw and auxiliary material suppliers, luggage manufacturers, and brands. The supplier of raw and auxiliary materials is mainly responsible for providing raw materials, including leather, fiber materials, metal materials and so on; the manufacturer of luggage is mainly responsible for contract production and manufacturing; the brand is mainly responsible for sales and brand management. This paper starts with the three main links of the luggage industry chain and combines the triple bottom line theory to construct the analysis framework model, which is convenient to clearly analyze the sustainable development strategy of the luggage industry. The following is a concept map of the luggage industry chain (refer with: Fig.2).

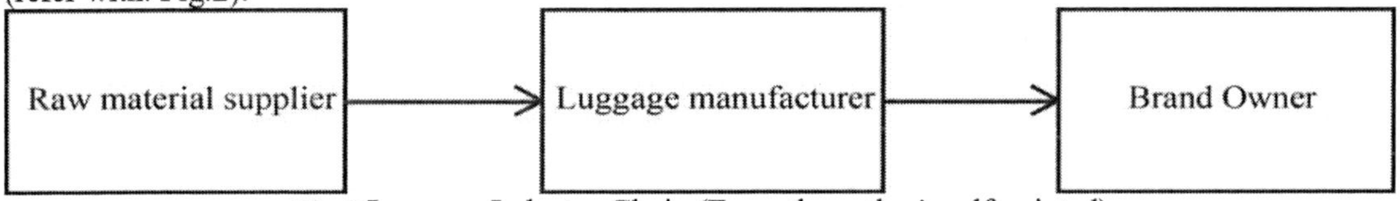

Fig.2 Luggage Industry Chain (From the author's self-painted)

3.2 Building an Analytical Framework Model for the Sustainable Development of the Luggage Industry

Fig.3 Sustainable Development Strategy Model of Luggage Industry (From the author's self-painted)

Based on the above two theoretical bases, taking the three dimensions of the triple bottom line theory as the ordinate, and taking the luggage industry chain as the horizontal coordinate, the analytical framework model of the following figure (refer with: Fig.3) is constructed and based on the model. Combined with case analysis, the sustainable development strategy of luggage industry is put forward.

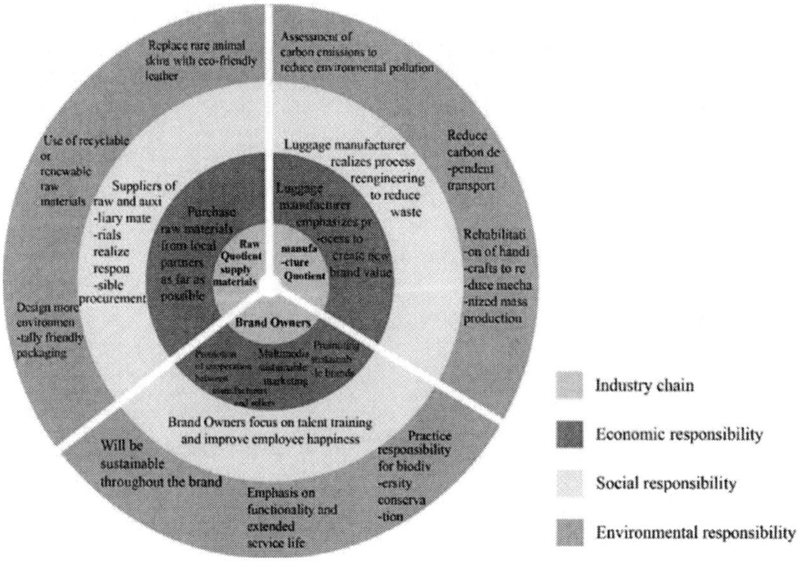

4. The Environmental Responsibility Dimension

4.1 Sustainable Strategies for Suppliers of Raw and Auxiliary Materials

In the whole industrial chain of luggage industry, the main role of raw and auxiliary materials manufacturers is to provide luggage raw materials and auxiliary materials, packaging supply, and so on.

4.1.1 Replace Rare Animal Skins with Eco-friendly Leather

Luggage brands such as Hermès, Louis Vuitton have designed many classic bags with precious animal skins as raw materials, which is common among luxury brands. However, as sustainability becomes a hot topic in society, luxury brands will also protect the earth, species survival, and continuation as the ultimate goal of enterprises. In 2018, following with Gucci, which gived up using animal fur, Michael Kors, Versace, Furla, Coach and Chanel also announced to abandoned using animal fur [5]. French luxury brand Hermès has launched a new version of sustainable leather for its "Victoria" travel bag. Hermès cooperated with Myco Works to create a leather derived from "mushroom mycelium" called “Sylvania” to replay traditional leather [6].

According to research, International Fur Association and China Leather Association Fur Economic Animal Breeding Professional Committee have reached a cooperation to carry out relevant certification and selection of fur animal farms in China and have launched the first batch of demonstration farms to promote the sustainable development of China's breeding industry. Not only that but the food processing industry fertilizer is used as the main feed in the breeding process. As a natural material, fur can decompose naturally in the natural environment without causing any burden to nature. It is a natural-friendly raw material for luggage. It is reported that the total output value of the fur industry chain has exceeded 40 billion U.S. dollars, and the retail link has reached 30 billion U.S. dollars. China is the world's largest producer and consumer of fur raw materials, as well as the world's largest fur trading and deep processing country. [7]

4.1.2 Use of Recyclable or Renewable Raw Materials

The American luggage brand Samsonite has launched more than 50 green eco-environmental protection series in the world, using recyclable materials, such as recycled PET, recycled nylon, recycled polypropylene, wood waste and cork. The British brand Elvis & Kresse collects the used fire hose from British fire brigade to make high-end luggage which has a strong resistance characteristics. Elvis & Kresse also recycles and redesigns coffee bags, recycled military aviation silk cloth, tea bags and shoe boxes, and all these efforts make the brand as a high-end sustainable brand.

Natural fiber is also a good choice. Sustainable concerns for natural fibers often include animal rights(how the animals are treated in the fiber generation and extraction processes),environmental effects with regards to grazing or growth processes, social lives of farmers, chemicals and water used in treatment and finishing processes,as well as recommended garment care methods.[8]

4.1.3 Design more Environmentally Friendly Packaging

Louis Vuitton shopping bags and gift boxes are 100% recyclable and 100% FSC® fiber, of which 40% are recycled fiber. At the same time, the gift box is foldable to reduce volume and carbon dioxide emissions during transportation to specialty stores. More environmentally friendly packaging is not only environmentally friendly but also can convey a sustainable brand concept for consumers in the process of imperceptibly, arousing consumers' awareness of environmental protection and sustainable consumption concept [9].

According to the Ellen MacArthur Foundation's 2019 "Global Commitment for a New Plastic Economy", British luxury brand Burberry plans to replace all its plastic packaging with reusable, recyclable or composting materials by 2025. Completely eliminate the use of ordinary plastic materials. The compost material eliminates the use of ordinary plastic materials. As of 2019, the company has eliminated plastic lamination from its retail bags and polyethylene bags for bags and garment jackets and has reduced the use of plastic by approximately 29 tons. In addition, Burberry also announced plans to replace all hangers, shields and plastic bags with environmentally friendly materials, and launched a hanger recycling program to recycle abandoned retail hangers. [10]

4.2 Luggage Manufacturer Sustainable Strategies

In the luggage industry chain, the main role of luggage manufacturers is OEM production, transportation, and so on. In view of these aspects, three strategies are proposed.

4.2.1 Assessment of Carbon Emissions to Reduce Environmental Pollution

Hermès uses specific digital tools to ® a global assessment of greenhouse gas emissions based on the bilan carbon (carbon footprint) method, in order to control carbon dioxide produced during production. MSCI, the world's largest index compilation company, will release a net-zero emission tracking report every quarter, aiming to promote global companies and industries that have made the most progress on the road to net-zero emissions, while spurring slow-moving backward companies and industries. The tracking report shows that listed companies need to keep their total emissions within the budget of 61.4 billion tons of carbon dioxide equivalent (CO2e) to avoid exceeding the 1.5°C temperature control threshold. As of May 31, 2021, Hermes' carbon emissions are shown in the table below [11]. Kering Group established the "environmental profit and loss "(EP&L) mechanism to achieve the goal of carbon neutralization. These enterprises and brands use various evaluation criteria to strictly control exhaust emissions in manufacturing and production. The use of evaluation tools is a reasonable and scientific method that can help enterprises and brands to reduce environmental pollution in the production process.

Table 1 Hermes' MSCI Net Zero Emission Tracking Report(As of May 31, 2021)

Issuer	Total reported emissions[tons CO2e]	Total emissions(reported and estimated)[tons CO2e]	Sum of reported emissions vs. MSCI estimated total emissions[%][2]
Hermès	520,996	2,824,119	18%

Data source: MSCI Net Zero Emission Tracking Report

4.2.2 Reduce Carbon-dependent Transport

Products often produce a large amount of greenhouse gases during transportation, which causes pollution to the environment. In order to reduce carbon emissions during transportation, companies and brands can use on-demand production and transportation, and alternative energy conversion (using electric vehicles, natural gas vehicles for transportation). Reduce environmental pollution caused by transportation.

4.2.3 Rehabilitation of Handicrafts to Reduce Mechanized Mass Production

The mechanized mass production process will inevitably produce a large amount of industrial waste and pollute the environment, and pure manual production can greatly reduce industrial waste and protect the environment. Hermès attaches great importance to the protection and inheritance of traditional handicrafts. In addition to respecting the in-service craftsmen, it also opens the Manufacto experience camp project. This project is aimed at primary and middle school students and aims to deepen the younger generation's understanding and interest in handicrafts.

4.3 Brand Owner Sustainability Strategy

In the whole industrial chain of luggage industry, retail brands mainly play the role of sales service, marketing promotion, brand management, design and development, and put forward three strategies for these aspects.

4.3.1 Will be Sustainable Throughout the Brand

Luxury brands Gucci, Hermès and other luggage brands have added sustainability to the brand story on the brand's official website, through these stories to show the brand's past life and sustainable development to consumers. It means that the brand regards sustainable development as the focus of brand future development and runs through the future development of brand.

4.3.2 Emphasis on Functionality and Extended Service life

In order to reduce waste generation, Samsonite provides durable luggage with long service life without compromising quality. Products are strictly tested to ensure that they do not affect consumers' travel, work use. Samsonite has established maintenance centers in many regions of the world, dedicated to provide the most timely services to consumers around the world. Prolonging service life is a manifestation of reducing waste. The longer the service life, the consumers will reduce the purchase of such items and achieve the purpose of energy-saving and environmental protection to the greatest extent.

4.3.3 Practice Responsibility for Biodiversity Conservation

According to statistics, more than half of the vertebrate species have become extinct in the past 40 years, and scientists believe this is the sixth major extinction. One of important reasons for this phenomenon is the great pressure on the earth caused by human activities, and the responsibility of environmental protection is on the shoulders of enterprises and brands. Hermès joined the French act4nature Plan to contribute to the achievement of the international community's biodiversity goals by integrating biodiversity conservation into one of its global development strategies in order to achieve Hermes' environmental commitment.

5. Society Feasible at the Responsibility Level

5.1 Suppliers of Raw and Auxiliary Materials Realize Responsible Procurement

The French luxury brand Louis Vuitton mostly uses natural materials to make and maintain its products, but not all of these materials are available. In 2017, LV conducted research on sustainability issues among consumers. The company formulated a sustainable development plan, which mainly focused on establishing a transparent and traceable system, and the top priority was to protect resources (including biodiversity, Water and energy), respect for animal welfare, avoid deforestation, prohibit the use of harmful chemicals, and reduce air and water pollution. In addition, the brand promises to achieve 100% responsible sourcing of raw materials by 2025. [12]

At present, many brands have successively realized responsible sourcing. As far as a single company is concerned, the purpose of responsible procurement is to ensure that the products and services provided by the company are "responsible" and avoid liability risks in the supply chain. From the perspective of society as a whole, the purpose of responsible procurement is to try to improve the sense of responsibility of all enterprises, promote all enterprises to improve the level of fulfilling social responsibilities, and jointly improve the welfare of the whole society, so that enterprises and brands can achieve sustainable development goals.

5.2 Luggage Manufacturer Realizes Process Reengineering to Reduce Waste

French luxury brand Hermch in order to reduce the waste of rare materials, in the manufacturing process, through improved technology and redesign, the remaining materials are used in other designs to achieve "zero waste" of raw materials. Moreover, Louis Vuitton established the "Cuirs Patrimoine" project, focusing on useless leathers and applying the remaining leathers scrap to window design. The upgrading and reengineering of waste materials is the art of "recovering from death", which can enhance its aesthetic value and function and accord with sustainable development.

5.3 Brand Owners Focus on Talent Training and Improve Employee Happiness

Hermès employs a total of 15417 employees, including 5200 artisans, the latter has achieved the brand's most outstanding craftsmanship. Hermès signed the first "Agreement on Assistance for People with Disabilities" in 2017, promising to increase the employment rate of people with disabilities and continue to increase the number of orders from supported industry organizations. In addition, the Hermès leather department recruits more than 200 craftsmen each year. The cobblers at the Hermès training center also chose a special course of study: learning sign language that takes four years to fully master. The purpose of this is to communicate with the hearing impaired or deaf-mute better, just like accepting other talents here for training, accepting them into the workshop training. This matter shows that Hermès is committed to setting up a pilot program for training people with disabilities in its workshops. The goal is to gradually establish customized training programs to help people with specific disabilities, such as dyslexia and spelling difficulties learn. The workshop further expands the diversity of talent recruitment. The goal is to train 80 people each year, including initial training and on-the-job training. As a result, the number of 750 craftsmen by the end of 2018 is expected to increase to 850 in the next three to four years [13]. The development of all companies and brands is inseparable from the joint efforts of employees, but at the level of corporate social responsibility, the well-being and happiness of workers are often overlooked. Therefore, companies should pay more attention to this aspect in the future and achieve sustainability development from multiple perspectives.

6. Feasible Strategies at the Level of Economic Responsibility

6.1 Purchase Raw Materials from Local Partners as far as Possible

German luggage brand RIMOWA selects high-quality raw materials supply as far as possible from local partners, and provides customer service to support charitable projects. Hermès contributes to local partners to promote the relationship between brands and suppliers of raw and auxiliary materials, and promotes the common development of various partners, interest symbiosis.

6.2 Luggage Manufacturer Emphasizes Process to Create New Brand Value

Hermès as the most well-known traditional luxury brand in the luggage industry, its products pursue luxury and pay attention to the exquisite craftsmanship. Only the brand series products always maintain high quality, can continue the classic, create new brand value, and bring benefits to the group.

6.3 Retail brand Sustainability strategies

6.3.1 Promotion of Cooperation between Manufacturers and Sellers

In the luggage industry chain, manufacturers and brand sellers are two independent parts. A craftsman in the Hermès leather workshop in the Sai Dôme and a local clerk completed a one-week cooperative sales, and the Tandem two-way exchange project was officially launched. The participation of manufacturing craftsmen in store sales can enable consumers to have a deeper understanding of the craftsmanship and design concepts of luggage and bags, thereby deepening the sense of identity and loyalty to the brand, which can increase brand sales to a certain extent.

6.3.2 Use Multimedia to Achieve Sustainable Marketing

Sustainable marketing is a relatively new concept that was put forward in the 1990s. It refers to the marketing activities in the development of sustainable economy to support the development of sustainable economy, but there is no complete system at present. Enterprises and brands are still trying. With the advent of multimedia era, some brands began to use multimedia to achieve sustainable marketing. Prada co-produced National Geographic in a series of short films, from Africa to the Americas, Asia, Oceania and Europe, bringing behind-the-scenes stories to the screen, presenting the goals, vision and reality of the Prada recycled nylon project, and demonstrating a transparent supply chain. [14]

6.3.3 Promoting Sustainable Brands

In order to promote sustainable development, luggage brands cooperate with other sustainable brands. The foundation Burberry Foundation announced in 2017 a five-year partnership with UK luggage brand Elvis & Kresse to recycle leather waste. The cooperation between these two brands is to convey the brand's product concept to consumers. When cooperating with sustainable brands, it naturally conveys the determination of sustainable consumption to consumers, which can make consumers who pursue sustainable consumption buy products at ease.

7. Conclusions and Prospects

This article first sorts out the concept and the explanation of sustainability by scholars at home and abroad, and then introduces the content of the triple bottom line theory. Based on the triple bottom line theory, combined with the luggage industry chain, including raw material suppliers, luggage manufacturers and brand owners, this article proposes sustainable development strategies. From an environmental perspective, the luggage industry should try to use environmentally friendly raw materials for example, avoiding rare animal skins, or using renewable and recyclable materials. The use of more environmentally friendly packaging means reducing plastics package; from a social perspective, raw materials Suppliers should achieve responsible sourcing. Manufacturers can achieve sustainable goals by reducing waste. What brand owners have to do is to improve the happiness of employees as much as possible; from an economic perspective, raw material suppliers should cooperate as much as possible from the local area. When purchasing raw materials from partners, manufacturers should emphasize that exquisite craftsmanship can bring new value to brands. Brand owners can promote cooperation between manufacturers and sellers as much as possible. They can use multimedia to achieve sustainable marketing and promote sustainable brands more locally. These strategies can further promote sustainable development for luggage companies and brands.

In the 21st century, the world is no longer just concerned about economic development, climate change and environmental pollution are also listed as the main focus, and fashion industry is the world's second largest environmental pollution industry after oil [15]. Fashion not only represents the pursuit of people's life, but also should guide people's thinking. The future fashion enterprises must also be able to take into account the environmental protection and sustainable development of enterprises. As a tributary of fashion industry, luggage industry also needs to keep up with the sustainable development of world fashion.

References

[1] Wei BY. Sustainable Fashion Development: Applying Transformational Design [D]. ProQuest Dissertations and Theses Full-text Search Platform: 2016.
[2] Dai YF. A Review of the Theoretical Literature on Sustainable Development [J]. 1 Shang: 2016 (13); 111.
[3] Li M. A Literature Review on Green Supply Chain and Sustainable Development in China's Textile and Garment Industry [J]. Contemporary economy: 2017 (03); 24-25.
[4] Sun BX. A Study on the Influencing Factors of Green Supply Chain Management Based on Triple Bottom Line Theory [J].]1 Inner Mongolia Science and Technology and economy: 2020 (21); 55-56+58.
[5] Global brands "green" are popular, and Chinese brands continue to release their potential [J].
[6] Hermès launches sustainable leather bags [J]. Textile Science Research: 2021 (04); 12.
[7] Su Y. Sustainable fashion [J]. Chinese Apparel: 2018 (07); 63.
[8] Begey, Melissa J. Green style: Discourses of sustainability among fashion industry professionals [D].ProQuest Dissertations and Theses Full-text Search Platform: 2016.
[9] Imformation on https://www.louisvuitton.cn/zhs-cn/magazine/sustainability
[10] Urberry announced the abandonment of plastics, and sustainable fashion has made new progress [J]. Rubber and Plastic Technology and Equipment: 2019; 45 (08): 61-62.
[11] Imformation on https://www.sohu.com/a/477588399_639898
[12] Su CY Wang LP. The first International Leather Sustainable Development Forum was held in Paris to outline the future sustainable development path [J]. Beijing Leather: 2020 (Z1); 30-31.
[13] Imformation on https://www.hermes.cn/cn/zh/story/158031-Sustainable-Development/
[14] Imformation on https://www.prada.com/cn/zh.html
[15] Xu H. "Sustainability" or the future of becoming a fashion industry [J]. China Fiber Inspection: 2020 (05); 100-102.

Cite as: N. Wang et al.,
TBIS 10.3993/tbis (2021).

On the Application of Traditional Tie Dyeing in Modern Women's Clothing Design

Ni Wang[1], Yu-Fei Dong[2], Xin Wei[3]

[1]*Wang ni of Wuhan Textile University of the first author, National Avenue, Wuhan, Asia, 430072, China*
[2]*Dong yufei of Wuhan Textile University of the second author, National Aveaue, Wuhan, Asia, 430072, China*

*Corresponding auther's email: 1137576590@qq.com

Abstract

Based on the background of the revitalization of traditional Chinese crafts, combined with relevant research literature and data, this paper aims to comprehensively sort out the characteristics of traditional handmade tie-dyeing skills and its own artistic characteristics, summarized through research methods such as literature investigation and field investigation. Research methods include innovative design methods such as integrated application, pattern extraction application, cross-border application, and comprehensive application. This research paper theorizes that traditional hand-made tie-dyeing techniques have a relatively large creative space, with extremely high originality and non-replicability in handwork, and there are contingencies caused by certain environmental factors in the production process, and resulting in unique products in each production cycle. Therefore, the design method proposed by the author is mainly based on element extraction and design structure.

Keywords: Traditional Skills; Hand Tie Dyeing; Textile Intangible Cultural Heritage; Fashion Design

1. Traditional Handmade Tie-dye

1.1 Fabric Color

The types of fabrics of the traditional handmade tie-dye technique can be traced back to ancient times and have a hierarchical nature. To show the royal family's hierarchical status and highlight their regality, there are many kinds of fabrics being used in their design. In February 2016, the author investigated "Shaanxi Provincial History Museum", "Tomb of Princess Tang Yongtai", "Tomb of Prince Tang Zhanghuai", "Qianling Museum" and "Tang Dynasty Furong Garden" respectively, as shown in Fig.1, to fully understand the gorgeous women's clothing of the Tang Dynasty. The textiles of the Tang Dynasty were mainly hemp and silk, with silk being often used by the royal family. Take the Jingchu culture as an example. In 1982, a large number of well-preserved silk weaving and embroidery articles were unearthed from the No. 1 Tomb of Mashan in Jiangling. Articles found include embroidery, brocades, Luo, gauze, silk, silk ribbons and other varieties, fine texture, all of which were well preserved. The clothes displayed a wide variety of color and patterns: with vermilion, crimson, eggplant purple, deep the, light green, tea brown, golden, brown and other colors of the silk line embroidery. There were also woven symmetrical patterns depicting dragons, phoenix birds, animal gods, dancers and geometric lines that alternate with various patterns, the color is soft. From showing Chu unearthed silk, it can be deduced that textile industry had reached quite high technological levels, undoubtedly a miracle of China dyeing technology [1]. Plant dyeing was an important technical achievement in this period. There are fewer kinds of fabrics used by ordinary people, and most of them wear clothes made of cotton and linen. The effect is different from that of dyeing.

Fig.1 Dyeing and Valerian Clothing in Tang Dynasty (Field Investigation: Shaanxi Provincial History Museum ; http://art.iqilu.com/bgsz/201510/20/679941.html)

2. Traditional Manual Tie-dye Technology

Table 1. Five categories of tiehua

The name	Binding method	Effect after strapping	Trend of final effect
Banding method	Will fabric according to the assumption in advance, pick up the fabric a bit, or along with the growth, or after doing all kinds of folding processing, with cotton thread or hemp rope for binding.		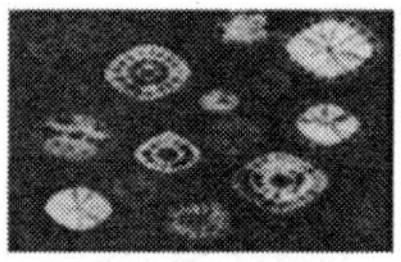
Suture method	The fabric is threaded or hung with a needle or thread to form a stained proof. Different effects are produced with different stitching methods. This is a convenient and freeflow way to tie flowers, which can fully express the designer's creative intention.		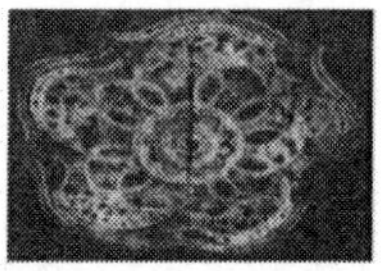
Splint fixing method	The folded fabric is clamped with round, triangular or hexagonal boards or bamboo pieces, bamboo clips or bamboo sticks, and then tied tightly with ropes to form anti-dyeing. The fabric between the splints produces a hard and straight "ice grain" effect.	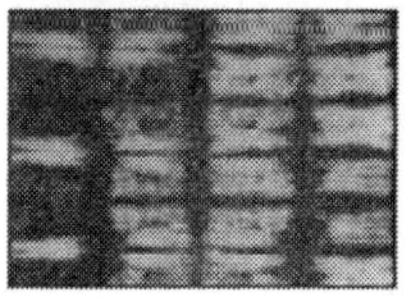	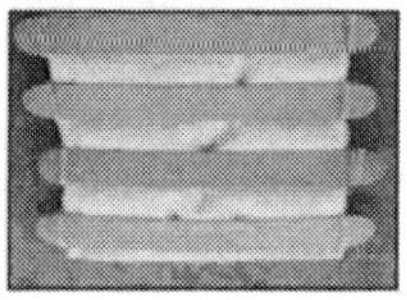
Arbitrary fold method	Arbitrarily wrinkled method, also known as the production of marbling pattern, is to create arbitrary wrinkles in the fabric after binding and dyeing; The fabric is then bundled and dyed again (or do a light to deep multiple bundling dyeing) to produce a marble-like texture effect.		
Comprehensive tie-dye method	Using a variety of skills such as binding, seam wringing and splinting, etc. to achieve different combinations for a rich and colorful effect.	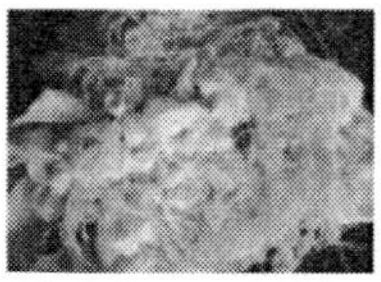	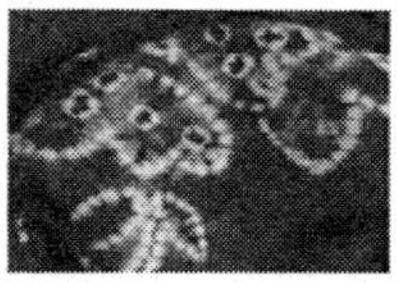

In the main steps of tie-dyeing, the key techniques of tie-dyeing and dip-dyeing are the binding process and dyeing process. Binding process is through yarn, thread, rope and other tools, the fabric is tied, sewed, clipped, and combined in other ways, and then dyed. (Table 1) This binding process can be divided into five categories: banding method, the suture method, splint fixing method, arbitrary fold method and comprehensive tie-dye method [2-3]. Its purpose is to prevent the part of the fabric bundled from dyeing, so that the bundled part retains its original color, while the part that is not tied is dyed uniformly. Thus, it creates a formation of uneven depth withrich layers of color through the formation of halo and wrinkles. The tighter and firmer the fabric is tied, the better the dye-proof effect is. It can be dyed into regular patterns of ordinary tie-dyed fabrics, but also can be dyed to show a concrete pattern of complex composition and a variety of gorgeous colors of exquisite handicrafts

There are many kinds of tying techniques in tie-dyeing, which can be roughly divided into three categories of banding, sewing, and splinting, each of which has different changes. In addition, there are the comprehensive application of the three tying methods and some free tying methods. Tie-dye is a special dyeing process, in which the academic term is referred to as the "anti-dyeing method" [4]. Before dipping, the artisan draws a pattern on the white cloth, and then uses a needle and thread to overlap or fold the part of the "flower" to form a "knot". Then, they adopt cold dyeing or cooking dyeing and other methods to dye the cloth, after repeated dipping and dyeing, they open the "knot" which has not been impregnated will form various flower shapes, the finished product is would have a blue or green background adorned with white flowers, creating a fresh, simplistic elegance with lasting appeal.. The quality of dyeing is not only related to the number of dips, but also to the distribution of dyes, dips and dyeing technology, as well as the use of dye medium, air drying, climate factors, etc. The rinsing process may seem simple, but it's a rule of thumb. The stitching process is not complicated, but extra care must be taken asif the fabric is broken, all the previous work will be wasted. Only when the process is iteratively improved upon can the quality of the product be guaranteed, otherwise there will be many problems such as uneven dyeing, lacking patterns and omission.

3. Application of Traditional Handmade Tie-dye Technique in ModernWomen's Dress

3.1 Pattern Element Extraction and Application Method

In the application and extraction of patterns, the traditional handmade tie-dye can be innovated by combining the cultural history of different regions. There are many ways to combine innovative methods from different regions and cultures. Taking the culture of Jingzhou as an example, part of the historical relics in the Jingzhou Museum with symbols related to Jingzhou and Chu were selected to conduct pattern extraction experiments. Based on modern design aesthetics, digital technology is used to extract the most obvious pattern elements (Table 2). For example, in Jingzhou Museum "A silk faces in the design of dragon and phoenix", utensils "A stone painted chime of phoenix pattern", "A jade pendant of man riding dragon" and other items in the Chu phoenix pattern.

Table 2. Extracting patterns of some historical relics in Jingzhou Museum

Name	Articles	Pattern element extraction
A silk faces in the design of dragon and phoenix	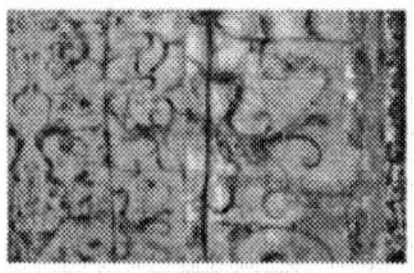	
A stone painted chime of phoenix pattern		
A jade pendant of man riding dragon		
A chopping board made of lacquerware technique.		

Pattern extraction is a step in understanding thetraditional manual tie-dye elements in their design, Before we conduct the pattern extraction, one must undergo field investigation: to establish an understanding of the pattern area that is extracted, as well asthe historical and cultural background of the development of the status quo and passing on state. This is to ensure that in the design, the final product's profound cultural connotation is effectively expressed, to ensure the effective design that inherits its cultural significance [5-7]. After the pattern extraction, design conception, style positioning, color orientation, detail positioning and other processes are carried out, and finally the production and adjustment of the garment. The author took the element extraction of the picture, as an example, and carried out the garment design experiment. The results are shown in the figure below. The design points are mainly applied to the back and shoulder, matched with relatively retro silhouette (Fig.2).

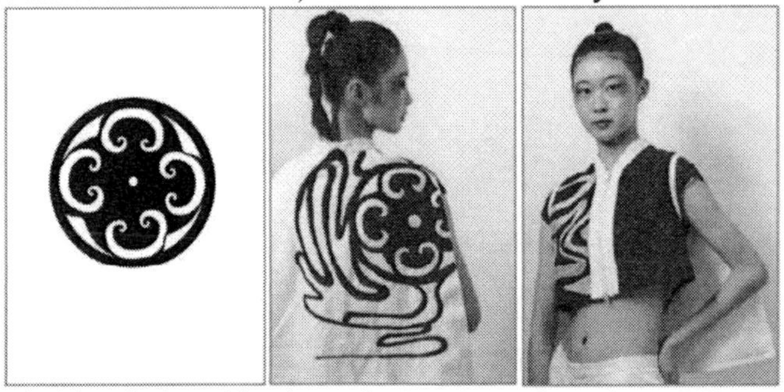

Fig.2 Clothing designed by the author's pattern element extraction method

3.2 Anatomy and Innovative Application Method of Traditional Manual tie-dye Technique

As mentioned above, traditional handmade tie-dye techniques can be divided into five categories: banding method, the suture method, splint fixing method, arbitrary fold method and comprehensive tie-dye method. The types of tools used in tie-dye are also constantly upgrading and changing with the times. The tools used in each stage are different. In modern times, there are some artists who innovate tie-dye tools and production methods. Today, with the continuous development of intangible cultural heritages, the tools of the tie-dye production process are also full of infinite possibilities [8-9]. With the development of science and technology, some simple tools, such as roller, wooden rod, splint, etc., can become more complex and interesting, and meet modern aesthetic standards based on inheritance. Anatomy refers to the design innovation of fixed tie-dyeing techniques by analyzing and summary, so as to explore the possibility of tie-dyeing to the maximum extent. For example, the following table (Table 3) takes the bundling method as an example to dissect and analyze the tie-dye effect through the microscopic experimental method. This analysis method can maximize the control of the general trend of the tie-dye effect, so as to carry out the product design (Fig.3).

Table 3. Dissecting the effect of tie-dye by using the bundling method as an example

Local refinement	Bundling effect	The final effect after dyeing
Take a piece of cloth and tie it with the rope clockwise, rotating as tightly as possible without any gap.	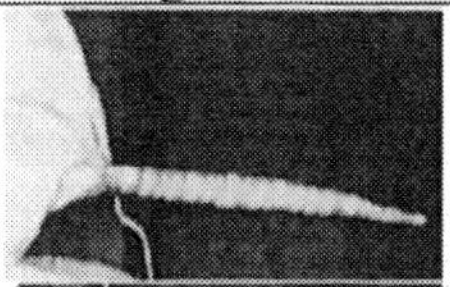	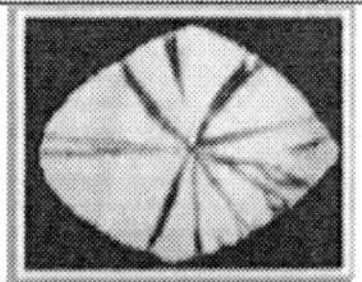
Take a piece of cloth and tie it with string only once at the bottom.	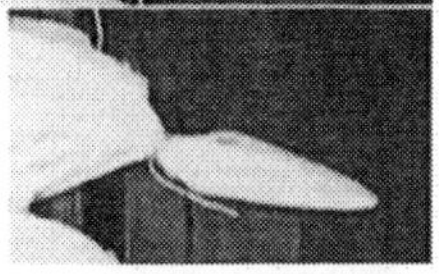	
Choose a piece of fabric that is tightly tied clockwise with no gaps at the bottom and top, with no strapping in the middle.	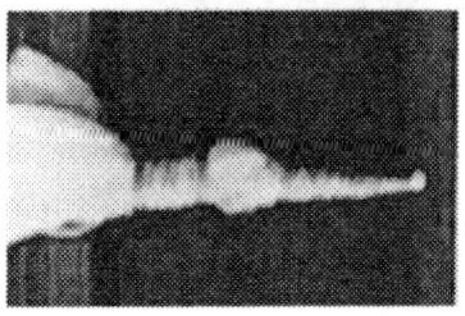	

Fig.3 Clothing experiment conducted by the author with the traditional manual tie-dye technique anatomy and innovative application method technique

3.3 Cross-border comprehensive innovative design method

Cross-industry first appeared in the marketing industry. Today, in the era of Internet economic development, the degree of importance and development speed of cross-industry are beyond imagination, especially in the process of multi-industry integration. It also plays an important role in the process of cultural industry integration. The cross-border comprehensive, innovative design method takes the cultural industry integration as the background and cross-border thinking as the basis for innovative design [10-11]. Taking "plane" as the basic point, explore common cross-border attributes. For example, the "flatable" attribute of textiles and the "flat" attribute of decorative paintings can achieve common attributes and cross-border innovations, such as the tie-dye decorative painting group creation series below [12-13]. (Fig.4) Due to the non-replicable nature of traditional handmade tie-dyeing, this series of cross-border combination products can be created forever.

Fig.4 Tie-dyed textile screen productstechnique

3.4 Integrated Application

The overall application of tie-dye refers to the application of patterns with more obvious characteristics and larger footprints to the more obvious parts of the clothing. However, in order to ensure that the patterns in the clothing are unbalanced due to the excessive concentration of the theme patterns, other patterns are often interspersed in the gaps for embellishment [14]. The characteristic of the overall application of the pattern is to help the clothing emphasize the style and reflect the personality of the wearer. The author first selects blue patterns and unknown edge decorative patterns for element extraction, and secondly breaks up and rearranges the selected patterns, and finally combines them to obtain two design schemes, style 1 and style 2. The innovative patterns are placed on the more obvious parts of the chest and waist of the garment, combined with a large area of tie-dye patterns (Fig.5).

Fig.5 The full-floor pattern is used in the rendering of the display

3.5 Fabric Matching Application

The integrity of fabric use is one of the factors that can affect the appearance of clothing [15]. From the fashion trend, elements such as hazy tulle, plant printing, and fabric stitching are quite popular. The author presents the plant pattern after replicating its pattern on the clothing. The form of digital printing can not only perfectly display the pattern design and reduce potential instability in the final product, but also provide a case for its use in subsequent development of traditional plant dyeing industrialization. The fabric carrier is homespun cloth with good quality cotton. With the development of technology, there are many kinds of cotton fabrics hand dyed in indigo, as shown in Figure (Fig.6). To complete this design, different materials were used such as splicing gauze fabrics, cotton jacquard fabrics, and plant printing and dyeing fabrics. Design fabrics are derived from a variety of plant blue-dyed cotton and linen fabrics withdigital printing fabrics as the main fabrics. The auxiliary stitching fabrics choose the Xiangxi Miao pattern as the edge stitching and the tulle fabric as the lining base. The collision of various ethnic elements makes the clothing more attractive and hierarchical.

Fig.6 The clothing designed by the author team using the fabric mixing method

4. Conclusion

The traditional hand-dyeing of intangible cultural heritage has a long history, and it can be used as a cultural resource in modern society to be widely used in design. their can also be useful to tracing back important aspects of self-confident Zeng Qiang our country's culture. The development of traditional handmade tie-dye in recent years, especially the promotion of Internet online sales during the epidemic, has expanded the development space of such crafts. This increase in demand also puts forward higher requirements for the innovation of traditional craftsmanship products. The research and promotion of innovative methods have enabled the development of traditional culture to adapt to the preferences in modern times. Some of the design methods summarized by the author are only for products currently on the market. There remain common problems, the efforts made in design can improve aspects of traditional culture and its stable inheritance. However, more industrial efforts need to be made to accommodate this improvement. Of course, innovative design is the fundamental priority in cultural inheritance, thus innovation must keep up with the pace of the times to preserve the intangible cultural history of our past

5. Acknowledgement

Fund Project: The research of this subject is the general project of the 2020 Open Fund of the Key Laboratory of Silk Culture Inheritance and Product Design,Digital Technology of the Ministry of Culture and Tourism (Project No.: 2020WLBO8) "Chu State Silk Pattern Classification and Digital Protection Path Research as an Example" Phased achievements; 2019 Hubei Provincial Colleges and Universities Philosophy and Social Science Research Major Project (Provincial Social Science Fund Preliminary Funding Project) (Project Number: 19ZD034) "Protection and Research of Traditional Cultural Heritage in Hubei-Using Textile Intangible Heritage Traditional Printing and Dyeing Techniques "Tianmen Blue Calico" as an example" phased results.

References

[1] Bronze Age · Tombs of Chu in Jiangling -- Archaeological China [EB/OL]
[2] Cao HY, Guan LS. Inheritance and Innovation of Silk Road Valerian Dyeing Art [J]. Young Writer: 2013 (20); 210-211.
[3] YU Q. Analysis of Chinese traditional folk dyeing and valerian technology [J]. Journal of Chongqing Three Gorges University: 2018; 34 (01): 50-56.
[4] Ma L. Inheritance, Protection and Development of Lu Embroidery [D]. Shandong University: 2010.
[5] Guo X. Hand-made tie-dyeing and raw material supply to the development of local agricultural economy [J]. Hubei Agricultural Mechanization: 2020 (07); 59.
[6] Guo LL. Brief Analysis of Traditional Hand-made Tie-dye Art and Modern Trend [J]. New West China: 2010 (09); 140-139.
[7] Li JR. A Brief Analysis of the Innovative Design of Cultural and Creative Products of Traditional Hand-made Tie-dye Technique [J]. Light Textile Industry & Technology: 2018; 47 (12): 36-38.
[8] Wu Q. Research on Cross-boundary Integration Development Mode of Exhibition Economy in the New Business Era [J]. Business Fair Economics: 2021 (08); 7-9.
[9] Zhang JP. Blue calico from Hubei Tianmen [J]. Decoration: 2006 (01); 91-92.
[10] Gui S, Sun ZY, Li XZ. Cultural space reconstruction and the path analysis of ethnic minority "intangible cultural heritage poverty alleviation" -- based on the field study of Enshi city in western Hubei [J]. Journal of Southwest University for Nationalities (Humanities and Social Sciences Edition): 2019; 40 (01): 29-35.
[11] Huang T. Positioning Error and Correction of Government Role in the Protection of Intangible Cultural Heritage in Recent Years [J]. Cultural Heritage: 2013 (03); 8-14.
[12] Shi WC. The research of textiles excavated from Han and Tang dynasties in eastern Xingjiang [D]. Northwest Normal University: 2017.
[13] Liu SQ, Liang HE, Gu M, Gao WD. Comparative study on Tang Dynasty pottery and tie-dye patterns [J]. Journal of Silk: 2012; 49 (02): 56-58.
[14] ZQ. Research on the application of blue calico auspicious patterns in modern women's clothing[J]. International Textile Herald: 2015; 43(04): 64-70.
[15] Bai YSS, Liang HE. The comparative study of Tang dynasty and unified silla's Male clothing [J]. Progress in Textile Science & Technology: 2009 (04); 93-95.

Cite as: Z.Y. Han et al.,
TBIS 10.3993/tbis (2021).

Factors Affecting Liquidity of Listed Companies in China's Textile Industry

Zhi-Yu Han, Lin Ma*

Beijing Institute of Fashion Technology, Beijing, 100029, China

*Corresponding author's email: malin_bift@126.com

Abstract

As a major producer and exporter of textile and garment, China's textile industry plays a leading and exemplary role. For a long time, the operation of listed companies in China's textile industry has not reached expectations, and many enterprises are facing the pressure of bankruptcy or transition. Liquidity is related to the survival and development of enterprises, textile industry should not ignore the liquidity problem. This paper discusses the main factors affecting the liquidity of textile industry through factor analysis, and puts forward feasible countermeasures on how to improve the liquidity of textile industry and realize the sustainable development of enterprises.

Keywords: Textile Industry; Liquidity; Factor Analysis

1. Introduction

Textile industry is a traditional pillar industry of China's national economy and is indispensable to people's livelihood. It is also an industry with clear international competitive advantages, and plays an important role in the booming market, expanding exports, promoting employment, increasing farmers' income, and promoting urbanization. [1]

However in recent years, due to the overcapacity and the intense competition of listed companies in China's textile industry, many enterprises are facing survival problems. This shows that competition of the textile industry is fierce, the development prospect is not optimistic, and there are complex issues with foreign trade.[2] Liquidity, as an important factor for the survival of enterprises. The liquidity of enterprise is affected by many factors, which requires enterprises to reasonably determine the overall asset liquidity on the basis of considering risks and profits, so as to realize the sustainable growth of enterprises. Only the reasonable allocation of asset liquidity can ensure the stable operation and long-term healthy growth of an enterprise. For example, the purchase department and the sales department are the two endpoints of the inflow and outflow of working capital. Therefore, enterprises can refine their responsibilities and strengthen the systematic management of inventory and accounts receivable. Managers also need to timely adjust sales and financial policies according to different operating conditions to ensure a reasonable asset structure.

2. Literature Review

Scholars from China and other countries have made some achievements in the study of corporate liquidity. Chen Xia (2009) argues the factors that affect the liquidity of listed companies in China are diverse and need to be comprehensively analyzed. [3] On the measurement of cash inflow, foreign scholars such as Opler (1999) pointed out in his research that the net cash flow from operating activities directly enhances the payment ability of enterprise's, which is a direct factor affecting the company's liquidity and the foundation of sustainable development of enterprises in the long run.[4] Chen Xia (2010) also researched the influencing factors of corporate liquidity, and found that the ratio of current assets, receivables turnover ratio, price-to-book ratio and other factors have a significant positive relationship with corporate liquidity, while the with-interest debt ratio and cash dividend per share have a significant negative relationship .[5] Hu Linqing, Zhu Xinlong and Wang Hong (2013) proposed that asset liquidity of different nature has a strong linear relationship with sustainable growth rate. [6] Liu Zhangsheng and Liu Hexiang (2015) believe that the key to the analysis of short-term liquidity is to analyze the liquidity of assets. [7] Baños-Caballero, García-Teruel, and Martínez-Solano (2010), Petersen and Rajan (1997) argued that firms with higher return receive significantly more credit from their suppliers and they, therefore, expected that firm returns inversely affects its liquidity. [8] Wasiuzzaman (2018) studied solvency of SMEs in Malaysia and concluded that "SMEs with high levels of liquidity do not rely on their profitability to improve their liquidity, instead increased profitability reduces liquidity". [9] Hristova, Stevcevska-Srbinoska, Mileva, and Zafirova (2019) believe that liquidity management plays a crucial role as the consequences of the insolvent risk can be very "expensive" and even lead to bankruptcy. [10]

According to the above research literature, it can be seen that there are many factors affecting corporate liquidity, and it is necessary to determine which factors directly affect corporate liquidity level, so as to avoid the multicollinearity problem in the empirical study.

3. Empirical Analysis of the Factors Affecting Liquidity of Textile Industry

3.1 Research Hypothesis

Liquidity, as a financial concept, is particularly critical in modern financial theory. Liquidity not only reflects the ability of enterprises to obtain cash, but also reflects the autonomy and vitality of enterprises. Liquidity manifests itself in the ability to trade quickly at reasonable prices and to sell assets quickly at predictable prices. The study of corporate liquidity can be started from the aspects of short-term solvency, long-term solvency, operation ability and profitability.

The measurement of relevant capabilities requires the cooperation of multiple indicators. Therefore, the indicators are related to corporate liquidity to a certain extent. [4] [5] In order to study the main factors affecting the liquidity of the textile industry, this paper puts forward the following four research hypothesises.

Hypothesis 1: Solvency is positively related to liquidity.

Hypothesis 2: Profitability is positively related to liquidity.

Hypothesis 3: Operating capacity is positively related to liquidity.

Hypothesis 4: Enterprises with good current asset management have stronger liquidity.

3.2 Research Design

3.2.1 Sample Selection

This paper selects 80 listed companies of the textile industry from Shanghai Stock Exchange and Shenzhen Stock Exchange as samples. Finally, 12 financial indexes from 2017 to 2019 are collected as the research variables which come from 60 textile industry listed companies including Hongdou Industrial, Ruyi Group and Peacebird after excluding the incomplete data. Data used in this study are both from China Stock Market & Accounting Research Database.

3.2.2 Define Variables

The research variables include a dependent variable and 11 independent variables.

The dependent variable (Y) is net working capital. Net working capital refers to the difference between the total current assets of an enterprise and various current liabilities. The net working capital of an enterprise is a key influencing factor for its internal management. It is also a widely used indicator in measuring the financial risk of an enterprise. It can affect the debt financing ability of an enterprise and measure the liquidity of its capital.

The current assets turnover ratio (X_1) is obtained by dividing the operating income by the average current assets. This index can fully reflect the use efficiency of enterprise assets, so as to observe the flow of enterprise current assets. Inventory turnover ratio (X_2), generally based on cost, is the ratio obtained by dividing the company's cost of sales by the average inventory balance in a specific period to analyze liquidity. It can explain the liquidity of enterprise inventory and the utilization rate of inventory, help enterprises to improve the utilization rate of funds in the short term, and improve the situation of debt financing. Current ratio (X_3) refers to the ratio between current assets and current liabilities, which can clearly show the ability of an enterprise to repay current liabilities by selling current assets in the current business cycle. Receivable turnover ratio (X_4) refers to the turnover frequency of accounts receivable in a specific time range. It generally plays an important role in measuring the turnover speed and management efficiency of accounts receivable. Net profit margin on sales (X_5) refers to the percentage of net profit divided by sales revenue. This ratio is necessary in measuring the sales ability of an enterprise. The financial situation of the enterprises whose net profit margin on sales keeps increasing should be good. Working capital turnover ratio (X_6), which can fully reflect the economic benefits of an enterprise, is usually obtained by dividing annual net sales by working capital. Through the working capital turnover ratio, we can know the working capital turnover in a year. Quick ratio (X_7), the ratio between quick assets and current liabilities, is used in measuring the ability of enterprises to repay current liabilities. The rate of return on net asset (X_8) is the ratio between profit and average shareholders' equity, which is crucial in measuring the profitability of listed companies. It is generally believed that the higher the rate of return on net asset, the higher the return on investment, and the stronger the ability of the company's own funds to obtain income. Asset-liability ratio (X_9), which is obtained by dividing the total liabilities by the total assets, is a key factor to measure the level of liabilities and the degree of risk. The cash ratio (X_{10}) is an indicator of solvency, which is obtained by dividing the sum of monetary capital and short-term investment by current liabilities. Cash ratio is a ratio that can accurately express the company's actual short-term solvency. When the cash ratio is more than 20%, the company has the ability to pay current liabilities, but the cash ratio is not the higher the better. The sales cash ratio (X_{11}) can reflect the net cash flow of an enterprise through sales. Generally speaking, the sales cash ratio which is an indicator of operating capacity is directly proportional to the capital utilization rate and income quality of the enterprise.[11]

3.3 Model

3.3.1 Mathematical Model

This paper is to study the influencing factors of textile industry liquidity and the standard index is obtained through Excel's basic operation of the financial data. The factor analysis method is used to simplify the original and complicated factors into several concise information factors. After all, we can finally obtain the factors that affect the liquidity of listed companies in China's textile industry.

Through data transformation, the original data index is expressed as a linear combination of common factors. The common factors are not related to each other. Therefore, suppose that the common factors selected in this paper are F_1, F_2... F_m ($m \leqslant 11$), and the 11 original indexes are X_1, X_2... X_{11} respectively, then the correlation between the original variables and the common factors is expressed as the following model (refer with Eq.1, Eq.2, Eq.3).

$$\begin{cases} X_1 = {}_{11}F_1 + 1_{12}F_2 + 2 \quad {}_{1m}F_m + \varepsilon_1 & 1) \\ X_2 = {}_{21}F_1 + 1_{22}F_2 + 2 \quad {}_{2m}F_m + \varepsilon_2 & 2) \\ X_{11} = 1_{111}F_1 + 1_{112}F_2 + 12 \quad {}_{11m}F_m + \varepsilon_{11} & 3) \end{cases}$$

In the above formula, the coefficient before the common factor is the factor load, ε_1, ε_2...... ε_{11} are special factors which are not included in the common factor and are not related to the common factor.

3.3.2 KMO and Bartlett's Test

In order to test whether the relationship between simple correlation coefficient and partial correlation coefficient meets the condition of factor analysis, KMO (Kaiser-Meyer-Olkin) and Bartlett test are needed. When the KMO value is closer to 1, the sample size is sufficient and the correlation between variables is stronger, and it indicates that the original variables selected are suitable for factor analysis. When KMO value is closer to 0, the correlation between variables is weaker, and the original variables are not suitable for factor analysis. It is generally believed that only when the KMO measures coefficient is greater than 0.5, it is meaningful to study. Bartlett test is a method to detect the correlation between variables in correlation matrix. When the p value of Bartlett's test is greater than 0.05, there is no significance of factor analysis, so the test should meet the condition of P value less than 0.05. The results of KMO and Bartlett tests are shown in Table 1. The data shows that KMO is equal to 0.610, the index is greater than 0.5, and the significance of test is less than 0.05. Each variable is independent to a certain extent, which indicates that the original variable selected in this paper is suitable for factor analysis.

Table 1. KMO and Bartlett's Test

		Calculated value
Kaiser-Meyer-Olkin Measure of Sampling Adequacy		.558
Bartlett's test	Approx. Chi-Square	1274.818
	df	55
	Sig.	.000

3.4 Data Analysis

The importance of the factors can be grasped from the size of variance contribution value. The common factor variance can be obtained by SPSS software (as shown in Table 2). It can be found from Table 2 that the common variances of the selected variables are all above 0.7, and the contribution values of current ratio and quick ratio are both above 0.9. Only the values of the two variables are lower than 0.5.

Table 2. Communalities

	Initial	Extraction
Current assets turnover ratio	1.000	.730
Receivable turnover ratio	1.000	.548
Current ratio	1.000	.903
Inventory turnover ratio	1.000	.741
Net profit margin on sales	1.000	.840
Working capital turnover ratio	1.000	.189
Quick ratio	1.000	.909
The rate of return on net asset	1.000	.860
Asset-liability ratio	1.000	.638
Cash ratio	1.000	.722
Sales cash ratio	1.000	.347

Table 3. Total Variance Explained

Component	Initial Eigenvalues			Extraction Sums of Squared Loadings			Rotation Sums of Squared Loadings		
	Total	% of Variance	Cumulative %	Total	% of Variance	Cumulative %	Total	% of Variance	Cumulative %
1	3.425	31.137	31.137	3.425	31.137	31.137	3.378	30.712	30.712
2	2.253	20.480	51.617	2.253	20.480	51.617	2.171	19.740	50.452
3	1.749	15.903	67.521	1.749	15.903	67.521	1.877	17.068	67.521
4	.909	8.260	75.780						
5	.843	7.661	83.441						
6	.612	5.561	89.003						
7	.493	4.482	93.485						
8	.381	3.461	96.946						
9	.224	2.038	98.985						
10	.080	.729	99.714						
11	.031	.286	100.000						

The degree of explanation of the total variance by the factor after "rotation" is shown in Table 3. According to the principle that the original feature value is greater than 1, three common factors are extracted. According to Table 3, the characteristic value of the first factor is 3.425, the characteristic value of the second factor is 2.171, and the characteristic value of the third factor is 1.877, which is greater than 1. The third column represents the factor contribution rate, while the fourth column is the cumulative factor contribution rate. The cumulative contribution rate of the three main components is 67.521%, and the cumulative contribution rate of the factors is consistent with that before rotation, which can reflect about 68% of the original 11 variables.

The component matrix after factor rotation can better explain the potential variables. The maximum variance method is used to factor rotate the initial factor load matrix. The results are shown in Table 4. It can be seen from Table 4 that after dimension reduction, there is an obvious gap in the load of variables after factor rotation. Common factor 1 includes current ratio, quick ratio and cash ratio, which reflects the solvency of textile enterprises, so it is named Debt Factor. Common factor 2 has the highest positive correlation with net profit rate and the rate of return on net asset, which reflects the overall profitability of textile enterprises, so it is named Profit Factor. Common factor 3 has higher load on inventory turnover rate and current asset turnover rate, which reflects the operation capacity of textile enterprises, so it is named Operation Factor. The component score coefficient matrix is shown in Table 5.

According to the score coefficient matrix, the regression method is used to construct the expression of each factor. F1, F2 and F3 are the expressions of Debt Factor, Profit Factor and Operation Factor respectively. The original variables in this paper are the standardized data. The three factor regression models are as follows (refer with Eq.4, Eq.5, Eq.6).

$$F_1=-0.049X_1+0.040X_2+0.281X_3+0.027X_4+0.009X_5-0.121X_6+0.284X_7-0.039X_8-0.229X_9+0.256X_{10}-0.022X_{11} \quad 4)$$

$$F_2=-0.007X_1+0.157X_2-0.005X_3+0.106X_4+0.421X_5-0.031X_6-0.010X_7+0.415X_8-0.046X_9-0.056X_{10}-0.278X_{11} \quad 5)$$

$$F_3=-0.435X_1+0.459X_2+0.005X_3+0.349X_4-0.049X_5+0.062X_6+0.038X_7+0.062X_8-0.001X_9+0.057X_{10}-0.059X_{11} \quad 6)$$

Finally, the contribution rate (common degree) of the above three factors is taken as the weight, and get the function of the total score of the influencing factors (refer with: Eq.7).

$$F=\frac{31.137}{67.521}F_1+\frac{20.480}{67.521}F_2+\frac{15.903}{67.521}F_3 \quad 7)$$

Table 4. Rotated Component Matrix

	Component		
	1	2	3
Quick ratio	.952		
Current ratio	.948		
Cash ratio	.846		
Asset-liability ratio	-.784		
Working capital turnover ratio			
Net profit margin on sales		.907	
The rate of return on net asset		.904	
Sales cash ratio		.587	
Inventory turnover ratio			.825
Current assets turnover ratio			.822
Receivable turnover ratio			.672

Table 5. Component Score Coefficient Matrix

	Component		
	1	2	3
Current assets turnover ratio	-.049	-.007	.435
Receivable turnover ratio	.027	.106	.349
Current ratio	.281	-.005	.005
Inventory turnover ratio	.040	-.157	.459
Net profit margin on sales	.009	.421	-.049
Working capital turnover ratio	-.121	.031	.062
Quick ratio	.284	-.010	.038
The rate of return on net asset	-.039	.415	.062
Asset-liability ratio	-.229	-.046	-.001
Cash ratio	.256	-.056	.057
Sales cash ratio	-.022	.278	-.059

3.5 Empirical Results

This paper uses the net working capital to represent the level of enterprise liquidity and measure the financial risk of enterprises. The results of the above model analysis show that: Debt Factor reflects the solvency of enterprises, and among the three factors Debt Factor has the greatest impact on net working capital. The Profitability Factor reflects the profitability of enterprises, and its impact on net working capital is slightly less than that of Debt Factor. The Operation Factor reflects the operation ability of the enterprise, and also has a certain impact on the net working capital. First of all, if we want to improve the liquidity of the textile industry, we should first consider improving the solvency of the textile industry. The empirical results show that the information extraction degree of quick ratio is high, so it can be achieved by improving the quick ratio of textile enterprises. The conclusion covers Hypothesis 1 and Hypothesis 4. Secondly, the Profitability Factor reflects the profitability of enterprises above scale in the industry. Enterprises can increase efforts to expand the scale of operation, strong alliance, and then reduce costs, improve sales. This is consistent with Hypothesis 2. Finally, the Operational Factor reflects the operational capability of an enterprise. Enterprises can reduce inventory, improve inventory turnover and enterprise liquidity by improving internal control system of inventory. This is consistent with Hypothesis 3.

4. Results

4.1 Improve Operation Ability and Strengthen Internal Control of Inventory

Textile and garment industry is faced with the problem of overcapacity due to the rapid upgrading of textile and garment industry. A large number of unsalable inventory increase the liabilities of the textile industry and cost a lot of money at the same time. In order to effectively deal with overcapacity, enterprises must maintain a solid cooperative relationship with suppliers, improve sales, grasp the market trend at the first time, and timely feed back the market demand. Enterprises should also timely reschedule their marketing plans, accelerate inventory turnover, reduce capital pressure, reduce debt repayment risk and operational risk.

4.2 Improve Profitability

With the acceleration of internationalization, the market competition is becoming increasingly fierce. Enterprises need to pay attention not only to solvency, but also to profitability. China's textile and garment industry can increase the production of enterprises, expand the business scale of enterprises, achieve economies of scale, and become more prominent in the aspects of cost, price, technology and service. China's textile and garment enterprises can expand their business scope when necessary, and can also adopt the way of strong alliance. Strong alliance can accelerate the speed of new product research and development, realize resource sharing, reduce the cost of market research and development, and improve the efficiency of market development.[12]

4.3 Speed up Scientific and Technological Innovation and go out Better

Only by focusing on quality and efficiency can new textile enterprises realize modernization. However, they still lack the independent innovation ability of industrial textile products and high-end precision clothing products.[13] From the perspective of product development, we should learn from foreign leading fashion design concepts and integrate technological innovation with brand building. For the development of textile products in daily use and high-end clothing products innovation is indispensable.[14] We should create brand, increase the investment of brand capital, form brand effect, and enhance the domestic and international reputation of enterprise brand.[15]

5. Acknowledgement

The author wishes to acknowledge the following programs:

1) Key project of education and teaching reform of Beijing Institute of Fashion Technology in 2019 ZDJG-1906: Research on the training program reform of fashion management specialty based on the idea of "curriculum thought and politics";

2) Key project of Beijing Institute of Fashion Technology 2014A-05: Research on the development path of Beijing's cultural and creative industry embedded in the global innovation network;

3) Youth top talent program of Beijing Institute of Fashion Technology 2014AL-54: Research on the influence of remanufacturing strategy of developed countries on innovation efficiency of innovation ecosystem of strategic emerging industries in China;

4) Talent introduction program of Beijing Institute of Fashion Technology 2013A-22: Research on the innovation mode of Beijing garment industry based on global innovation network.

References

[1] Liao Q. Analysis on the Status and Prospects of Textile Industry Development. Textile Reports: 2018; 5: 34-38.

[2] Xu WX, Xie Y. Empirical Analysis of the Factors Influencing the Liquidity of Listed Companies in China's Textile Industry. China Collective Economy: 2017; 4: 9-10.

[3] Chen X. Liquidity Holding Motivation of listed companies and sustainable development of Companies. Finance and Accounting: 2009; 06: 58-60.

[4] Opler T, Pinkowitz L, Stulz R, Williamson R. The determinants and implications of corporate cash holdings. Journal of Financial Economics: 1999; 1: 3-46.

[5] Chen X. Empirical Analysis on the Factors of Corporate Liquidity. Finance and Accounting Monthly: 2010; 23: 09-12.

[6] Hu LQ, Zhu XL, Wang H, Empirical Analysis on Relationship between Enterprise Asset Liquidity and Sustainable Growth Rate. Friends of Accounting: 2013; 1: 111-112.

[7] Liu ZS, Liu HX, Research on short-term solvency analysis index of enterprises based on Liquidity. Green Finance and Accounting: 2015; 8: 11-13.

[8] Baños-Caballero S, García-Teruel P, Martínez-Solano P. Working capital management in SMEs. Accounting and Finance: 2010; 50: 511-527.

[9] Wasiuzzaman S. Determinants of Liquidity in Malaysian SMEs: A quantile regression approach. International Journal of Productivity and Performance Management: 2018; 67: 1566-1584.

[10] Hristova S, Stevcevska-Srbinoska D, Mileva I, Zafirova A. Trade-Off Between Liquidity and Profitability: An Empirical Study of Pharmaceutical Sector in the Republic of North Macedonia. Timisoara Journal of Economics and Business: 2019; 12: 127-148.

[11] Hu SF. Financial Early Warning Analysis of Textile Listed Companies Based on Principal Component Logistic Model Market Modernization: 2015; 9: 194-195.

[12] Zhang JG, Zhang JR. Design of enterprise asset liquidity evaluation index. Journal of Shanghai Lixin Accounting College: 2010: 70-76.

[13] Liu ZY, Li WG. Analysis on the Development Path of China's Textile Export Trade under the Background of "the Belt and Road". Rural Economy and Science-Technology: 2018; 8: 75-76.

[14] Chen L. The Present and Development of Chinese Textile Industry. China Fiber Inspection: 2016; 5: 126-127.

[15] Zhang J. Existing problems and future development trend of textile industry. Modern Economic Information: 2017; 1: 385.

Cite as: M.X. Chen et al.,
TBIS 10.3993/tbis (2021).

Sustainable Development Strategy of Luxury Brands

Ming-Xi Chen[1], Guo-Xiang Yuan[1,2*]

[1]*College of Fashion and Design, Donghua University, West Yan'an Road, Changning District, Shanghai, 200051, China*

[2]*Key Laboratory of Clothing Design and Technology - Ministry of Education, Donghua University, West Yan'an Road, Changning District, Shanghai, 200051, China*

*Corresponding author's email: yuanguoxiang@gmail.com

Abstract

Sustainability plays an important role in today's fashion industry. The concept has permeated the design, production, and operation of clothing. As a pioneer in the fashion industry, luxury brands try hard to convey sustainability in design, achieve low energy consumption in production, and influence consumers in marketing. Based on the existing circumstances and sustainable fashion trend, this paper along with the sustainable cases of luxury brands, explains the sustainable development strategies from four aspects: material, production, design, and emotion. Through questionnaire surveys, the data are used to analyse consumers awareness, purchase intentions and decision-making of sustainable fashion. The paper combined literature reading, market research, questionnaire survey, data analysis and other research methods to propose meaningful sustainable development strategies for high-end fashion brands on the four perspectives of material, production, design and emotion.

Keywords: Sustainable Fashion; Luxury Brands; Sustainable Development Strategy

1. Introduction

In the 21st century, when climate issues are taking precedence over economic issues, sustainability has become the focus of the entire international community. Some reports relevant to the luxury fashion industry indicate that practitioners are paying more attention to sustainable practices to solve the challenges in balancing environmental, social and business needs [1, 2]. Due to the great benefit of fashion halo, luxury brands are always the primary target of criticism on environmental sustainability issues. Due to social responsibility and public pressure, luxury groups and high-end brands are developing strategies to improve environmental sustainability mainly reflected in the four aspects: materials, production, design, and emotion. How luxury brands truly make positive contributions to mankind and the planet, and give correct guidance to other brands, are the embodiment of top standing in the fashion industry.

2. Sustainability of Luxury Brands

In 1962, biologist Rachel Carson first proposed the term "sustainable development" in Silent Spring. Ten years later, American scholars Barbara Ward and Rene Dubos introduced the concept into the cognitive system of the human living environment. In the current definition of sustainable development, a generally agreed view is put forward by Gro Harlen Brundtland: to meet current needs without compromising the development of future generations to meet their needs. In April 2007, the World Intellectual Property Organization pointed out in the report Green Design: From Cradle to Cradle: The focus of today's design is sustainability. For the fashion industry, current study mainly focuses on the selection of raw materials, the repeatability of apparel structure, the carbon emissions and water footprint in production, the recycling and utilization of old clothing, and consumption awareness of sustainability. Due to the great influence of luxury brands, it is of leading significance for the promotion of sustainable fashion.

2.1 The Sustainable Status of Fashion Industry

Nowadays, the consumption frequency of clothing has increased a lot, and the pollution and emissions have also. According to United Nations forecasts, by 2050, the annual carbon emissions generated by fashion industry will account for 1/4 of the global total. At the current rate of growth, the natural resources needed in 2050 will be three times the natural intake in 2000. Therefore, international organizations, fashion groups and brands are changing the previous design techniques and production models, improving the eco-friendly industrial chain, and taking the concept of sustainability as the driving force for business development. Sustainability is not a marketing tool but a tenet to avoid pollution, improve material utilization and ensure energy cleanness. Because fashion industry is very closed to people's lives, the sustainability of fashion will benefit not only consumers, but also environment and society. There are many studies discussing sustainable fashion based on brand and clothing classification at the micro level. Chen Yuehan believes that fast fashion requires the organic combination of brand culture and sustainability to survive. Huang Ruqian thinks that consumers can only participate in sustainable fashion meaningfully if they maintain the same consciousness, cognition, and attitude. Yang Lu says that "slow fashion" is more in line with sustainable development. Based on the above studies and existing conclusions, the focus of this paper is the sustainable development strategy of high-end brands.

2.2 The Significance of Luxury Brands to Fashion Industry

Some literature proposes that sustainability is incompatible with the luxury industry [3]. But they also have rich economic capital to invest sustainable fashion. Due to the high quality and rarity, its sustainable development strategy is also different from common garments. Different from the speed of fast fashion brands and the functionality of sports brands, luxury brands have its aesthetic value, unique and stable brand tonality, and small but steady target market. Compared with fast fashion brands, the release frequency is relatively slow, so there is more time to consider the sustainability of the industrial chain, traceability of materials, and to ensure using clean energy and eco-friendly fabric. The target customers of luxury brands are insensitive to prices, so they can withstand the high costs caused by sustainable design. Besides, they also have a higher acceptance of the integration with sustainable design. In terms of great influence, luxury brands also receive a better influence in consumer communication. In general, compared with the contribution that common fashion brands can make to sustainability, high-end brands have greater advantages in terms of materials, production, design, and emotion.

3. Sustainability Cases of Luxury Brands

Luxury brand is the leader of fashion industry, and the correct leadership in sustainable design plays an essential role in the environment, economy and society. From the analysis of the current luxury industry, sustainable design is mainly embodied in four aspects: fabric, production, design and emotion.

3.1 Fabric

Sustainable fabric of high-end brands is mainly including degradable fabric and recycled materials. Degradable fabric includes bio-based fabric made of animal or plant fiber and synthetic fabric. The impact of sustainability can be clearly seen in latest fashion weeks.

Stella McCartney has practiced "environmentalism" since the beginning of the brand in 2001. The brand focuses on developing pure fur substitutes, that is, using pure plant resources, no animal raw materials and animal testing. Luxury goods with fine workmanship and high price do not have the problem of too short service cycle, but the use of animal fur has been controversial. In the "sustainable" boom, some luxury brands bans stop to use animal fur. Chanel has not used any animal fur since the spring and summer 2018 series. Prada also announced that fur will be discontinued from spring and summer 2020. In June 2020, Gucci released Gucci off the grid series, using many recycled and renewable materials, mainly renewable nylon. In addition, there are metal free tanned leather, recycled polyester yarn and lining, recycled brass, recycled gold and palladium plating accessories, solvent-free adhesives. Chloe's creative director is a practitioner of sustainable fashion. In 2021 autumn/winter collection (Fig.1), the striped poncho is made from recycled cashmere and redesigned with second-hand Chloe handbags purchased from eBay. Compared with last year, Chloe's sustainable composition this year is four times that of the past, and it has made a great breakthrough in the sustainability of material selection. In Vivienne Westwood's 2021 fall and winter series (Fig.2), she continued to cultivate the concept of sustainability and 90% of the materials is environmentally friendly, such as recycled denim and organic dyed cotton attractant silk, to create a longevity fashion that is never out dated and can match with any clothes in the closet. In addition, Marine Serre and Mytheresa cooperated and created a sustainable capsule (Fig.3). Some of the garments are made of recycled bed sheets and useless leather. Lacoste has also brought a sustainable twist to its classic Polo shirt. 30% of recycled Polo fabrics use recycled cotton, and the crocodile logo embroidery on the chest is made of renewable polyester fiber. Tommy Hilfiger promises to achieve 24 specific goals by 2030, including complete product recycling, 100% recyclable, reusable or compostable packaging, and eliminating oil-based disposable plastic packaging; Offices, warehouses and stores will use 100% renewable electricity.

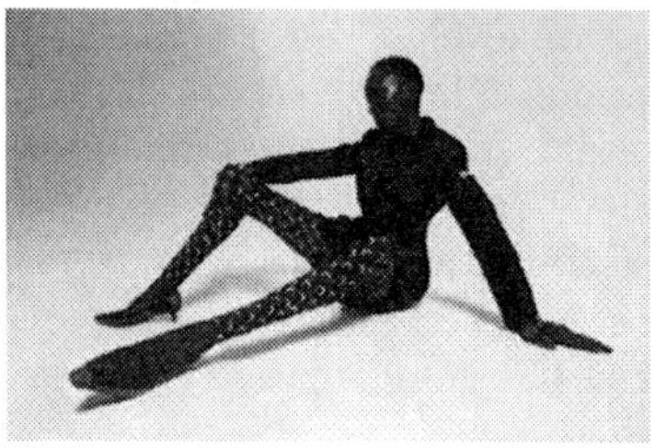

Fig.1 Chloe 2021 fall/winter Fig.2 Vivienne Westwood 2021 fall/winter Fig.3 Marine Serre × Mytheresa

3.2 Production

In 2018, Fashion and Sustainable Development Goals: The Role of the United Nations held in Geneva pointed out that 20% of the world's wastewater and 10% of carbon emissions are brought by the fashion industry. The source of marine plastic pollution is also considered to be caused by textile industry. As sustainability has been put on fashion industry agenda, many high-end brands have taken the lead in choosing "carbon neutrality" as the main sustainable path. The United Nations Fashion Industry Climate Action Charter also puts forward the vision of achieving net zero emissions by 2050. The economic capital owned by luxury group is a unique advantage to establish sustainable fashion. Kering group and LVMH Group invest a lot of money in the development of environmental protection clothing technology every year.

Burberry announced in 2019 that they planned to achieve carbon neutrality goal by 2022 completely and reduce greenhouse gas emissions in direct operations and supply chain operations for all stores in the UK, the US and the EMEA region. Burberry plans to reduce direct and indirect emissions by 95% between 2016 and 2022 and reduce 30% of other indirect emissions by 2030. In addition, the brand expected to achieve 100% use of renewable energy, which has already reached 58%. Gucci said that its brand operations and supply chain have already been carbon neutral. Stella McCartney promised to reduce 30% of total carbon emissions by 2030, and luxury retailer Farfetch launched a carbon neutral program to offset the carbon emissions caused by distribution and returns. Lacoste, a century-old French sports brand, uses a closed-loop production process, which consumes less water for production.

Fig.4 Burberry carbon neutral fashion show

3.3 Design

Fashion and sustainability are essentially opposites. Cause fashion industry perused the high-frequency and high-profit while sustainability emphasis on energy conservation and environmental protection [4]. Different from fast fashion, the update frequency of luxury brands is relatively stable and slow due to its exquisite design, time-consuming craftsmanship, and high price. However, sustainable fashion is not only about extending the product's life cycle and reducing the obsolescence rate of clothing, but also about improving the utilization of fabrics, which also called zero-waste design. Designers also use remodeling to redesign the old clothing and to make multi-functional clothes.

Zero-waste design refers to no waste in the process of pattern making and production. The use of fabrics is planned in detail from the initial design stage to maximize the utilization of fabrics [5]. Issey Miyake is one of the most typical brands using this design method. It advocates the concept of "a piece of cloth" and uses as little stitching as possible to complete a two-dimensional to three-dimensional transformation of a piece of cloth through manual modeling. The whole 132.5 ISSEY MIYAKE series (Fig.5) is sustainable. The four numbers represent one-dimensional fabrics, three-dimensional structure, two dimensions for storage and transcends time and space to five dimensions when dressing on. The series uses 100% light-weight recyclable polyester fiber knitted fabric, which is light, thin and easy to store. The character of this kind of fabric is greatly reduces the size of the clothing and also reduces the cost of transportation and waste of space resources.

Fig.5 132.5 ISSEY MIYAKE

3.4 Emotion

The sustainability of clothing depends not only on fabrics, production, and design levels, but also on consuming willingness, behavior and habits [6]. On the surface, the cause of serious environmental pollution in fashion industry is excessive or one-off clothing consumption, but the underlying reason lies in the unsustainable consumption awareness and behavior of consumers [7-9]. For luxury brands, promoting sustainability at the emotional level is mainly reflected in slogan print and marketing methods. In the process of transforming clothing from concepts to products, consumers' sense of participation and the empathy for brand stories is enhanced. Thereby deepening consumers' emotion for clothing can extend garments' life [10].

Vivienne Westwood is one of the representatives. She regards activism and design as equal importance, and her reputation as an environmental activist is better than that of a fashion designer. Her main point is to buy less, choose the better, and use for a longer time. Vivienne even implements her concept to consumers' mind through slogan-printed T-shirts (Fig.6). Vivienne's activism has undoubtedly deepened consumers' impressions of the brand. Lacoste has also worked with the International Union for Conservation of Nature to produce a limited-edition Polo shirt (Fig.7), each of which represents an endangered animal and aims to help people raise awareness of endangered animal protection.

Fig.6 Vivienne Westwood slogan T-shirt Fig.7 Lacoste Polo

4. Consumers' Perception of the Sustainability Concept of Luxury Brands

4.1 Questionnaire Survey

To understand consumers' awareness of sustainable fashion, questionnaires are used to investigate consumers' understanding, experience and acceptance of sustainability. The questionnaire used in this survey consists of three parts. The first part collects basic characteristics of consumers, including gender, age, and occupation. The second part investigates the respondent's understanding of sustainable fashion. The third part is about willingness and acceptance of sustainable fashion. After 8 pre-test questionnaires in this survey, 103 questionnaires were formally distributed, 103 valid questionnaires were returned, and the effective rate was 100%. A total of 103 valid sample data were obtained in this questionnaire survey. Through established facts and subjective wills, the questionnaire explores whether age, gender, and occupation have an impact on the understanding of sustainable fashion, as well as people's awareness of sustainable fashion in the current consuming environment, which will also reflect the effectiveness of corporate promotion of sustainable fashion.

4.2 Data Analysis

4.2.1 Sustainable Consumption Experience

Among the 103 valid questionnaires, one third of people have bought and know they have bought sustainable clothes, and more than half of them get in touch with this concept during shopping. Therefore, purchase willingness will be greatly enhanced based on existing knowledge. Among the remaining 66 people, some of them know a little about sustainable fashion. Most people engage with sustainable fashion through social media. Online channels are still an effective way to promote new concepts. From nothing-known to knowns is a fast process which only requires the push of promotion through internet, but from awareness to consumption is a slow development that requires the efforts from luxury industry and the change of consumers' mind. Table 1 is a cross-analysis of sustainability perception and purchase experience.

Table 1. Cross-analysis of sustainability perception and purchase experience

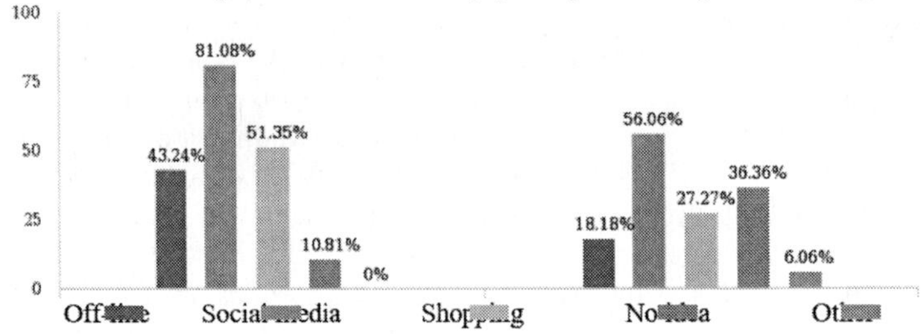

4.2.2 Sustainable Fashion Consumption Willingness

There is literature on sustainability in the luxury industry indicates that consumers' preference and demand for sustainability are highly varied and uncertain [11]. The result also shows 74.76% choose classic and durable styles, and the people choose fashionable and fast-changing styles are almost female consumers. Most male consumers update their clothing less frequently which means men's clothing consumption habit is more in line with sustainability concept. However, more than 30% of male consumers say that sustainability is not the dominant factor in purchase decision-making. Therefore, sustainability is still doubtful for male consumer and more positive guidance is needed for them. In contrast, female consumers are more likely to be driven in the shopping process, so, it is more important to strengthen the concept of sustainable consumption to increase the quality of orders and reduce the frequency of purchase.

4.2.3 The Relation between Sustainable Methods and Prices

From the perspective of sustainable methods, consumers have most knowledge about environmentally friendly materials (70.87%), followed by production emissions (53.4%) and clothing structure (47.57%). Recycling accounts for 18 %, promotion (14.56%) and environmental-friendly slogan (10.68%) are relatively few. The result shows that the implement of sustainable fashion has shifted from superficial promotion to fundamental issues such as materials, production and structure. Which also means luxury brands trust the society more and is willing to tell customer the essential aspects of sustainability. Meanwhile, whether the sustainability of fabrics, design, or craftsmanship will increase capital investment, which will lead to a higher price. From the result of the questionnaire, 85% of consumers can accept 5-10% price increase, which is completely affordable for high-end brands.

4.2.4 Perception of Sustainable Brand

The questionnaire lists 13 high-end brands that use sustainable fashion concepts. More than half consumers have no idea. Even if the brand tries best to use sustainable fabric or technology, it cannot be seen by consumers. The brand that conveys the concept of sustainability to consumers to the greatest extent is ISSEY MIYAKE. The reason is that the brand broadens their market through its concept. Lots of consumers know the brand concept before they get into contact with their product. What's more, the sustainable concept can be seen in shopping experience, from the interior of the store to the design details such as fabrics and cutting, as well as the clothing storage and shopping bags. Issey Miyake applies sustainability throughout the entire consumption process and passes this concept to consumers. Fig.8 shows the sustainable strategy cognition of high-end brands.

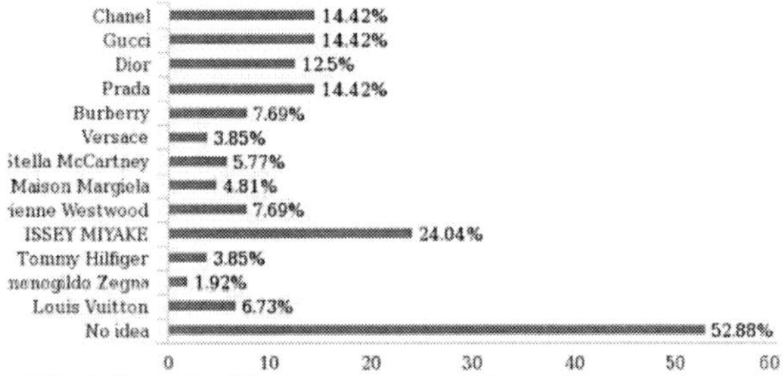

Fig.8 Sustainable strategy cognition of luxury brands

4.3 Summary

According to the results of the questionnaire, the knowledge of sustainable fashion is still deficient in China which mainly focused on fabrics and craftsmanship. The recognition is unclear that consumers cannot tell sustainable clothing during shopping. The main reason lies in the lack of breadth and depth in promotion of sustainability. In actual shopping experience, it is difficult for consumers to get in touch with this concept. It can be seen from market research that many brands are implementing sustainable plans, but it is not enough to promote internally only. What is fortunate is that most of customer have a positive attitude toward sustainable fashion, they are willing to know more about it but the shift from willingness to act still need time. Therefore, it is the duty of luxury brands to provide consumers better environment and opportunities to further understand sustainability and complete transformation from cognition to willingness to behavior.

5. Sustainability Strategies for Luxury Brands

The reason why luxury brands are different from other brands is that they have a smaller number but higher quality of products, wider social influence and an inescapable responsibility for environment protection. However, the research found that though many high-end brands have promoted sustainability, these efforts have not been seen by consumers. Whether fashion brands recognize the necessity of conveying sustainability to consumers or not, the sustainable development of mankind is undoubtedly not the only duty and responsibility of a certain industry or individual, but the joint efforts of everyone. The result of questionnaire also shows consumers' expectations for more participation in sustainable fashion. Now that the fashion industry has been implemented sustainability in practice, and consumers have the willingness to understand, there is no harmless to promote it more. Based on the social influence of luxury brands, combined with consumption psychology, behavior and willingness for sustainable fashion, it would be better for high-end brands implement sustainability throughout the entire industry chain, from raw material to consumers.

5.1 Fabric: Use Environmentally Friendly Fabrics

The effective way to fabric sustainability in material is realizing traceability of fabric production. Many fashion companies take recycled and natural textile as priority. At present, the most used method is to extract polyester materials from plastic waste. In addition, there are also recycled natural fibers for recycling, and some old clothes are physically cut into small pieces to make a new one. Fabric suppliers should have a comprehensive understanding of the source of fiber and fabric, monitor the production process to ensure the traceability of raw material and every process is environmentally friendly. The sustainability of fabric is mainly depending on the manufacture rather than designers. It is crucial to have a consistent sustainable concept among manufacturers and suppliers and to combined technology and production.

5.2 Production: Expand the Coverage of Carbon Neutral

Judging from the current situation, some companies have implemented the carbon neutral plan to reduce carbon emissions directly and indirectly. Clothing companies, especially high-end brands, have dozens of clothing factories and more than 100 stores around the world. To achieve carbon neutrality completely is not only covers the industrial chain, but also the supply chain.

5.3 Design: Improve Aesthetics of Sustainable Design

There are diversity ways to express sustainability in fashion design. One of the simple ways is slogan print, it is straightforward to consumer but more likely be used in fast fashion. For high-end brands should be advanced in not only fashion design but also application of sustainability. It is more valuable for ready-to-wear to make more efforts in clothing structure and functions, increase the utilization rate and extend the clothes longevity. Meanwhile, deepen consumers' understanding and aesthetics of sustainable design through advertising and promotion.

5.4 Emotion: Penetrate Sustainability Throughout the Entire Shopping Experience

Nowadays, the aim of shopping is not only to buy something, but to enjoy the service and experience. The combination of pre-promotion and shopping experience is an efficient way to strengthen consumers' awareness of sustainable consumption. It can be seen from the results of the research that social media is still the main method of online promotion. The high-end brands gain lots of attention from public, including direct and indirect consumers who read their tweets. Unfortunately, few consumers learn sustainability when shopping. The sales of high-end brands often have good selling skills and service spirit, but sustainability is always be neglected because it makes few influences to purchase decision. After visiting some high-end fashion shops, I found that many sales associates even did not know which certain apparel used sustainable craftsmanship or materials. It is gratifying for brands to promote sustainability a lot, but the involvement of consumers and implement of sustainability are equally significant.

6. Summary

This article mainly proposes four levels of sustainable development strategies: to use environmentally friendly materials, to achieve carbon neutrality, to improve sustainable aesthetics and consumer experience. Make sustainable fashion deeply rooted in people's mind, not only producers but also consumers. While strictly monitoring industry and supply chain, luxury brands should strengthen their connection with consumers to raise awareness of sustainable consumption and cultivate better consuming habits to promote the transition from awareness to behaviors. Finally, only when sustainability work throughout the entire industrial chain and root deeply in consumer awareness, can our only earth get better future.

References

[1] Caniato F, Caridi M, Crippa L, Moretto A. Environmental sustainability in fashion supply chains: An exploratory case based research. Int. J. Prod. Econ: 2012; 135: 659-670.

[2] Joy A, Sherry JF, Jr Venkatesh A, Wang J, Chan R. Fast fashion, sustainability, and the ethical appeal of luxury brands. Fash.: 2012; 16: 273-295. [CrossRef]

[3] Lochard C, Murat A. Luxe et Développement Durable: La Nouvelle Alliance; Eyrolles: Paris, France: 2011.

[4] An GY, Chu XL. Analysis on the social responsibility of the fashion industry under the sustainable background [J]. Western Leather: 2020; 42 (12): 64-65.

[5] Chen YH. The survival strategy of the fast fashion industry under the trend of sustainable development [J]. Textile Report: 2020; 39 (12): 29-30.

[6] Huang RQ, Li R. Consumer analysis of fast fashion brands under sustainable consumption mode [J]. Textile Science and Technology Progress: 2021 (02): 46-49.

[7] Liang JF, Cheng WY. Research status and dilemma analysis of sustainable clothing consumption behavior [J]. Silk: 2020; 57 (06): 18-25.

[8] Iran S, Geiger SM, Schrader U. Collaborative fashion consumption: a cross-cultural study between Tehran and Berlin [J]. Journal of Cleaner Production: 2019; 212: 313-323.

[9] Armstrong CM, Niinimaki K, Kujala S, et al. Sustainable product-service systems for clothing: exploring consumer perceptions of consumption alternatives in Finland [J]. Journal of Cleaner Production: 2015; 97: 30-39.

[10] Sun ZJ, Mao D. Overview and trend analysis of sustainable fashion design methods [J]. Textile Science and Technology Progress: 2020 (08); 40-44.

[11] Christopher M, Juttner U. Baker S. Demand chain management: Integrating marketing and supply chain management. Ind. Market. Manag: 2007; 36: 377-392.

A Comparative Study of Corporate Social Responsibility Between Overseas Investment and Non-Overseas Investment in the Manufacturing Industry

Peng-Fei Shao[1], Lei Yao[2*]

[1] *Beijing Institute of Fashion Technology, Beijing, 100029, China*
[2] *Beijing Institute of Fashion Technology, Beijing, 100029, China*

[*]Corresponding author's email: sxyyl@bift.edu.cn

Abstract

With the implementation of "the Belt and Road Initiatives", more and more enterprises are going abroad. Not only do the stakeholders such as the domestic government, society and the public, but also the stakeholders of host countries understand the importance of sustainable development, making corporate social responsibility a hot topic for theoretical research. Though the manufacturing industry is an important part of China's national economy, it causes a lot of environmental pollution problems and governance challenges. This paper takes listed manufacturing enterprises as the research object, in which they are evaluated based on the scores of corporate social responsibility (CSR) disclosed by Hexun rating agency in 2018, and the application of descriptive statistics. It is concluded that the performance level of corporate social responsibility of manufacturing enterprises with overseas investment is higher than that of non-overseas investment enterprises. Furthermore, through the analysis of the social responsibility of enterprises by industry, relevant suggestions are put forward.

Keywords: Overseas Investment; Corporate Social Responsibility; Manufacturing Industry

In September 2013, "the Belt and Road Initiatives" was proposed in China to improve the uneven economic development, unstable energy channels, passive trade and difficulties in expanding the enterprises. As Chinese enterprises continue going abroad, the international competitiveness of Chinese enterprises has been improved, and the economy continues to develop rapidly. However, China's manufacturing enterprises failed to fully balance the relationship among economy, society and environment due to the lack of relevant experience. As a result, the risk of corporate social responsibility increased, which brought enterprises huge losses.

There is a close relationship between overseas investment and the enterprise's performance of corporate social responsibility (CSR). However, at present, there are few studies on the impact of overseas investment on CSR in China, most scholars focus on the impact of overseas investment on corporate performance and the impact of CSR on its overseas investment. When exploring the impact of overseas investment on enterprises' CSR performance, scholars generally believe that it has a promoting effect, but mainly through case analysis and literature review. Few people conduct specific research through empirical analysis. Marano et al (2017) believes that under the background of globalization, under the dual pressure of the host and home countries, enterprises need to bear higher standards of CSR; Gao Wei et al (2018) believes that overseas investment has a significant role in promoting CSR, and the heterogeneity of host countries affects the effect of overseas investment on CSR. However, Zona et al (2013) thinks that when enterprises are eager to expand their business scope and field, their ability may not necessarily be sufficient to cope with the reality, which will cause them to act irresponsibly [1-3].

In this context, it is of great practical significance to explore the differences of CSR between overseas and non-overseas investment enterprises in China's manufacturing industry. The research object of this paper is listed manufacturing enterprises in 2018. Using statistical analysis, their CSR performance is explored and evaluated, which lead to some relevant suggestions. The limitation of this paper is that there is no perfect disclosure system of CSR in China, which leads to incomplete data and errors in the research process [4].

1. Total Score Analysis of Corporate Social Responsibility

1.1 Sample Selection

The research object of this paper is domestic A-share listed manufacturing enterprises. The selection of manufacturing enterprises is based on the c3-c43 categories of enterprises in the industry classification guidelines for listed companies (revised in 2012) issued by the China Securities Regulatory Commission. The year chosen was 2018, as the data up to that year is the most complete and up-to-date. According to the list of overseas investment enterprises (Institutions) of the Ministry of Commerce and the annual report of the company, one could determine whether the enterprise makes overseas investment. The CSR data of the enterprises are extracted from hexun.com, one of the largest third-party CSR rating agencies in China, and the above data are matched according to the enterprise code and year. After excluding the overseas investment events in tax havens such as the Caymans and Bermuda, 1215 valid samples were obtained, including 452 samples of foreign investment [5-7].

1.2 Result Analysis

Fig.1 depicts the CSR scores of 1215 manufacturing enterprises in 2018, in which the CSR scores of Chinese manufacturing enterprises are generally lower; the sample enterprises all have a score below 40 out of 100. This may be related to the industry characteristics of enterprises, and it is inevitable to cause pollution during the manufacturing processes. It is also possible that the disclosure of most CSR is insufficient to obtain accurate CSR data. As a result, in the CSR rating data of HeXun, there are a large number of missing social responsibility data from supplier, customer and consumer responsibility and environmental responsibility, denoted by 0 points. Among the sample enterprises, most of the manufacturing enterprises have a score of 16-30, with a total of 729. Only 20 enterprises scored between 34 to 39, and 80 enterprises scored below zero.

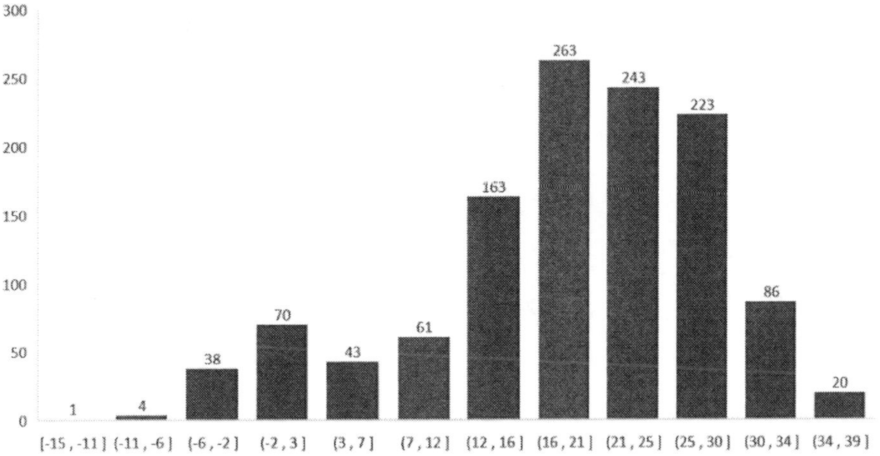

Fig.1 Total score of CSR of manufacturing enterprises in 2018
Data Source: From http://stockdata.stock.hexun.com/zrbg/Plate.aspx?date=2018-12-31

Fig.2 depicts the CSR scores of 452 manufacturing enterprises with overseas investment in 2018. The overall score is still low; most manufacturing enterprises have CSR scores of 17-32, a total of 729, 15 enterprises scored 32-37 and 30 enterprises have a negative score.

Compared with Fig.1 and Fig.2, although the overall score of CSR is low, the score of CSR of manufacturing enterprises with overseas investment is slightly higher. It can be concluded that in 2018, overseas investment improves the ability of manufacturing enterprises to fulfill corporate social responsibility, which provides a basis for data analysis of the research of overseas investment on corporate social responsibility. In the future, more years of data can be selected for empirical analysis.

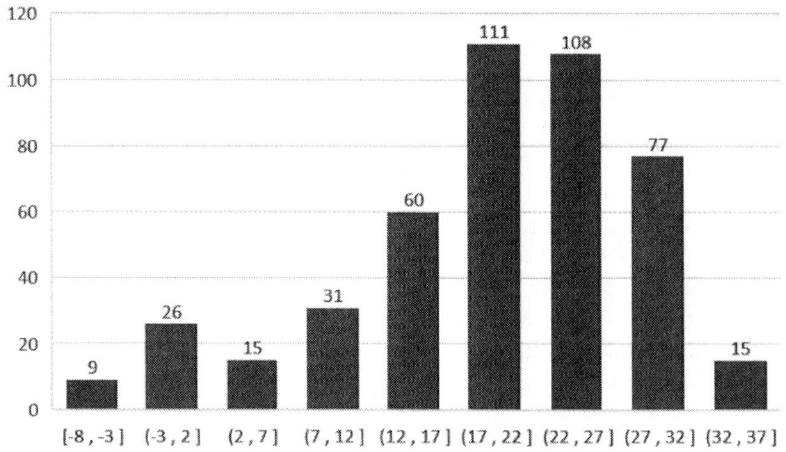

Fig.2 Total score of CSR of overseas manufacturing enterprises in 2018
Data Source: From http://stockdata.stock.hexun.com/zrbg/Plate.aspx?date=2018-12-31

2. Analysis on the Dimensions of Corporate Social Responsibility

2.1 Variable Selection

According to the CSR evaluation system released by Hexun, this paper measures the performance of CSR of manufacturing enterprises. Not only does the evaluation system reveal the total CSR, but also subdivides it into shareholder responsibility (SH), employee responsibility (EM), supplier, customer and consumer responsibility (SCC), environmental responsibility (ENV) and social public responsibility (SOC) to examine the performance level of CSR in detail. Specifically, SH includes the profitability, solvency and innovation ability of the enterprise; EM covers the level of care, training and income of employees; SCC includes product quality, customer satisfaction and supplier equity; ENV includes environmental protection investment, environmental protection awareness, environmental protection system, etc.; SOC includes public donation and income tax payment [8-10].

2.2 Result Analysis

Table 1. Descriptive statistics of CSR in manufacturing enterprises

Variable	Full sample					Non-overseas investment enterprises	Overseas-investment enterprise	T test of (2) -(3)
	Mean	Standard deviation	1/4 Quantile	Median	3/4 Quantile	Mean	Mean	
CSR	24.59	17.51	15.41	20.85	26.97	22.95	27.36	-12.08
SH	13.69	6.52	9.49	14.13	18.48	12.81	15.16	-17.41
EM	2.77	3.33	0.78	1.51	3.00	2.56	3.12	-7.97
SCC	1.95	4.99	0.00	0.00	0.00	1.69	2.37	-6.45
ENV	2.19	5.77	0.00	0.00	0.00	1.92	2.64	-6.00
SOC	4.01	3.79	2.27	3.56	5.52	3.97	4.07	-1.30

Data source: From http://stockdata.stock.hexun.com/zrbg/Plate.aspx?date=2018-12-31

Table 1 reports the descriptive statistical results of the dimensions of CSR. The overall performance level of CSR in manufacturing industry is not optimistic, with an average of 24.59 and a standard deviation of 17.51. The results show that there are great differences in the performance level of CSR among enterprises. The average score of social responsibility of overseas invested manufacturing enterprises is higher than that of non-overseas invested enterprises.

In the dimension of CSR, the fulfillment level of SH is the highest, which is 13.69, followed by SOC, with an average of 4.01. The fulfillment level of SCC, EM and ENV are very low, with an average of 1.95, 2.77 and 2.19, respectively. It shows that Chinese manufacturing enterprises are more inclined to fulfill shareholders' responsibility and social public responsibility, but not the other CSR.

To sum up, in terms of CSR segmentation, the performance level of overseas investment manufacturing enterprises is slightly higher than that of non-overseas investment enterprises.

3. Analysis of Corporate Social Responsibility by Industry

3.1 Industry Classification

According to the different industries, this paper divides enterprises with samples of C13, C14 and C15 into food and beverage industry; enterprises with samples of C17, C18 and C19 into textile and garment industry; enterprises with samples of C20 and C21 into wood furniture manufacturing industry; enterprises with samples of C22, C23 and C24 into paper printing entertainment industry, enterprises with samples of C26 and C28 are collectively referred to as the chemical manufacturing industry; enterprises with C36 and C37 samples are collectively referred to as the transportation industry; and some enterprises with small sample size are grouped under other industry, as shown in Table 2.

Table 2. Industry classification and distribution of manufacturing enterprises

Industry code classification	Industry reclassification	Amount	Industry code classification	Industry reclassification	Amount
C13	Food and beverage industry	25	C36	Transportation industry	39
C14			C37		
C15			C38	Electrical machinery and equipment manufacturing	50
C17	Textile and garment industry	15	C39	Manufacturing of computer communications and other electronic equipment	80
C18			C25	Other industry	101
C19			C29		
C20	Wood furniture manufacturing industry	2	C30		
C21			C31		
C22	Paper printing entertainment industry	10	C32		
C23			C33		
C24			C34		
C26	Chemical manufacturing industry	42	C40		
C28			C41		
C27	Pharmaceutical manufacturing	45	C42		
C35	Special equipment manufacturing	43	C43		

Data source: Guidance Note on Sector Classification of Listed Companies (Revised 2012)

It can be seen from table 2 that the computer, communication and other electronic equipment manufacturing industry, chemical manufacturing industry, pharmaceutical manufacturing industry and electrical machinery and equipment manufacturing industry comprise of the largest number of enterprises, with 80, 42, 45 and 50, respectively. The listed manufacturing enterprises are concentrated in high-tech fields such as machinery, equipment and medicine. In the sample, there are only 15 listed enterprises in the textile and garment industry. The reason is that although China is a big textile and garment country, most manufacturing enterprises chose not to be listed.

3.2 Result Analysis

The CSR score of manufacturing enterprises by industry is shown in Table 3. It can be seen from table 3 that for the whole sample of manufacturing enterprises, the performance of CSR of pharmaceutical manufacturing industry, textile and garment industry and food and beverage industry are relatively good, and their average and median are in the forefront, while that of special equipment manufacturing industry and electrical machinery and equipment manufacturing industry are the worst.

In terms of the sample of overseas investment manufacturing enterprises, the textile and garment industry has the best CSR, while the pharmaceutical manufacturing industry and food and beverage industry also perform well. The special equipment manufacturing industry, computer, communication and other electronic equipment manufacturing industry and electrical machinery and equipment manufacturing industry perform the worst.

Table 3. Descriptive statistics of CSR scores of manufacturing enterprises by industry in 2018

	Manufacturing enterprise				Overseas investment in manufacturing enterprises			
Industry	Mean	Ranking	Median	Ranking	Mean	Ranking	Median	Ranking
Food and beverage industry	21.357	3	23.365	2	22.375	4	26.610	2
Textile and garment industry	22.399	2	23.340	3	26.074	1	28.280	1
Paper printing entertainment industry	18.033	5	18.150	9	21.007	5	24.115	4
Chemical manufacturing industry	20.319	4	20.590	4	22.498	3	21.920	5
Pharmaceutical manufacturing	22.975	1	25.160	1	23.790	2	26.000	3
Special equipment manufacturing	15.937	10	17.130	10	16.366	10	17.230	10
Transportation industry	17.733	7	19.835	5	19.604	7	21.670	6
Electrical machinery and equipment manufacturing	16.676	9	18.550	8	18.670	8	19.985	8
Manufacturing of computer communications and other electronic equipment	17.250	8	19.270	6	17.409	9	19.465	9
Other industry	17.822	6	18.610	7	19.646	6	20.980	7

Comparing the two groups of data, we can see that the pharmaceutical manufacturing industry, textile and garment industry and food and beverage industry enterprises are doing well in fulfilling CSR, while the special equipment manufacturing industry and electrical machinery and equipment manufacturing industry enterprises need to be strengthened. The CSR score of overseas investment enterprises is higher than that of non-overseas investment enterprises.

4. Conclusions and Suggestions

4.1 Research Conclusion

Due to the particularity of cross regional operation, the fulfillment of CSR of overseas investment enterprises is not only affected by their own conditions, but is also affected by overseas investment policies. Therefore, this paper selects the listed manufacturing enterprises in 2018 as the research object, and through descriptive statistics and data analysis, tests whether there is a difference between the social responsibility of manufacturing overseas investment enterprises and non-overseas investment enterprises. The results show that overseas investment improved the total CSR of manufacturing enterprises; China's manufacturing enterprises are more inclined to fulfill the shareholder responsibility and social public responsibility, and lack other social responsibilities. In terms of performing CSR, pharmaceutical manufacturing industry, textile and garment industry and food and beverage industry have the best scores, while special equipment manufacturing industry and electrical machinery and equipment manufacturing industry have the worst scores. This paper lays a data foundation for the research of overseas investment on CSR, and proves that the research has academic value and practical significance, which can be followed by more in-depth empirical research.

4.2 Policy Implications

4.2.1 Enterprise Level

Manufacturing enterprises should actively take advantage of the policy advantages of "go global" and "the Belt and Road Initiatives", pay attention to investment cooperation with other countries, such as Lancang-Mekong River cooperation, and accelerate the pace of overseas investment. In addition to shareholder responsibility and social public responsibility, manufacturing enterprises should pay more attention to other responsibilities. Pharmaceutical manufacturing industry, textile and garment industry and food and beverage industry should play a leading role, actively carry out overseas investment activities, improve CSR standards according to the requirements of the host country and relay the feedback to domestic enterprises. Special equipment manufacturing and electrical machinery and equipment manufacturing enterprises need to learn from the experience of leading enterprises in fulfilling CSR to improve their own shortcomings. Enterprises should actively participate in the formulation of international standards of CSR in manufacturing industry, and strive to improve the international discourse power for the formulation of CSR standards.

4.2.2 Government Level

The government should encourage and support manufacturing enterprises to "go global", through a series of preferential means, including but not limited to certain tax incentives, government subsidies, and preferential financing incentives. The government should formulate more accurate CSR standards for employee responsibility, supplier, customer and consumer responsibility and environmental responsibility, to improve the performance of enterprises in these three subdivided dimensions. They should also establish a certain reward and punishment mechanism to encourage enterprises with excellent performance in CSR and enterprises with poor performance in CSR to pay attention to fulfill social responsibility; encourage capable enterprises to make overseas investment and strengthen their leading role; and lead the construction of overseas manufacturing park, forming the industrial scale effect under the leadership of leading enterprises. According to the different segments of manufacturing enterprises, the relevant departments should make corresponding adjustments in the CSR standards. Particularly for the pharmaceutical manufacturing, textile and garment industry and food and beverage industry, the government should actively encourage their overseas investment.

5. Acknowledgement

Beijing social science fund key project, construction and application of garment green consumption guidance system (19JDGLA010).

References

[1] Valentina M, Peter T, Tatiana K. Escaping the iron cage: Liabilities of origin and CSR reporting of emerging market multinational enterprises. Journal of International Business Studies: 2017; 48 (3).

[2] Gao W, Wang QJ. Overseas investment and corporate social responsibility: An Empirical Test Based on host country heterogeneity and corporate heterogeneity. Business research: 2018; (08): 88-98.

[3] Fabio Z, Mario M, Vittorio C. Antecedents of Corporate Scandals: CEOs' Personal Traits, Stakeholders' Cohesion, Managerial Fraud, and Imbalanced Corporate Strategy. Journal of Business Ethics: 2013; 113 (2).

[4] Gao W, Wang P. Overseas direct investment and Stock Liquidity: An Empirical Analysis Based on dynamic and heterogeneity. Journal of Southwest University of Nationalities (HUMANITIES AND SOCIAL SCIENCES): 2021; 42 (01): 98-108.

[5] Dang QM. Development trend and enlightenment of foreign corporate social responsibility. Gansu Social Sciences: 2019; (02): 229-236.

[6] Bian YM, Huang YY. Research on social responsibility system of foreign investment enterprises. Journal of Guizhou Party school: 2019; (03): 110-119.

[7] Fu YH. Research on social responsibility capacity building of Listed Companies in Guangdong. Guangdong Academy of Social Sciences: 2019.

[8] Song L, Bin B, Qiao XL. Research on the impact of institutional distance on corporate social responsibility of overseas investment in China: the moderating effect based on international experience. Journal of Beijing University of technology and industry (SOCIAL SCIENCE EDITION): 2019; 34 (02): 90-103.

[9] Wu JF, Qiao L. The relationship between corporate social responsibility and cross border M & A: Research Review and prospect. Economic management: 2018; 40 (11): 191-208.

[10] Huang LY, Zheng SF, Wang J, The cooperation and win win mechanism of foreign investment enterprises: Based on the perspective of social responsibility. Management review: 2018; 30 (02): 172-182.

Study on the Breast Shape of Young Women based on 3-D Body Scanning

Run-Yu Tian[1], Xue-Wei Jiang[1*]

[1] *Wuhan Textile and Apparel Digital Engineering Technology Research Centre, School of Fashion, Wuhan Textile University, No.1 Fangzhi Road, Wuhan, Hubei Province,430073, China*

*Corresponding author's email: xwjiang@wtu.edu.cn

Abstract

Through three-dimensional scanning of 39 women aged between 18 to25, 9 measurement items and 5 derived items related to breast shape were extracted. The basic breast shape of the sample women was divided according to the degree of breast sagging, breast fullness and breast expansion. The results showed that the ideal breast-to-body ratio of young women is about 0.713, and the ideal breast point spacing to breast diameter width ratio is about 0.69. When choosing bras, the ordinary type is most common, followed by the breast-lifting type. Fewer people need to wear bras with side-tightening functions. The results can provide reference for underwear enterprises to make bra production plans, as well as help women choose their bra sizes.

Keywords: Breast Shape; 3-D Body Scanning; Bra Size; Classification of Breast Shape

1. Introduction

With the development of science, technology and the improvement of people's living standards, as a functional garment for supporting, fixing, covering and protecting women's breasts, people have higher requirements for the fit, health and comfort of bras [1-3]. Breast is the only organ in a woman's body without bone support. As women age, the pressure of its own weight will result in changing the shape of the breast thus affecting the beauty of women's body. According to relevant research [4], the upper part of female breast is close to 1/2 cone, and the lower part is close to 1/4 ellipsoid. Different breast shapes need bras with different three-dimensional shapes for reasonable protection. If the bra does not match the breast shape, the bra cannot provide effective support and shaping of the breast causing pain in the breast, spine and other related parts [5]. Relevant research shows that about 70% of women wear bras that don not match their breast shape [6-7], due to the fact that some bra manufacturers do not have a good understanding of the breast characteristics of the target consumers, and the bra size is determined only by underbust girth and the size of the bust girth minus the underbust girth in bra design, which is very unfavorable to women's health. Therefore, the research and classification of breast shape will help to improve the relationship between bra structure design and breast shape, provide theoretical basis for bra size classification, and improve bra fit and wearing comfort.

2. Analysis of Breast Shape

Breast is one of the main characteristics of women's body shape and the basis of bra design. The appearance of female breast can be described by width, height, position and orientation [8]. Duan Xiaojuan [9] divided the breasts of 75B people into four categories according to breast point spacing, breast arc length and breast root circumference, described the shape of each type of breasts in the form of words, and gave suggestions on what bra structure to choose for different types of breasts. Wang Jianping et al. [10] obtained the size data of women's breasts through three-dimensional anthropometry and obtained the change rule of women's breasts with different cups through data analysis. They explained the morphological characteristics and related values of young women's breasts with different cups in detail. LIANG Suzhen et al. [11] divided women's breast shapes into nine categories based on the breast point spacing and breast height and combined the characteristics of each type of breast with bra structure design. Foreign scholars Pechter [12] obtained the breast circumference by direct measurement, and established the relationship between it and bra cup size; Rong Zheng et al. [13] used two multivariate statistical methods, principal component factor analysis and cluster analysis, to describe the shape of breast through eight factors; Reena Pandarum et al. [5] measured the relevant dimensions of women's breasts by three-dimensional body scanning method, and analyzed and compared the data to determine the relationship between various breast sizes and the size of bra cups; Yu Liu [14] et al. divided the female breast shape by the ratio of breast circumference to breast depth and breast diameter width, and developed the classification system of bra rims. These studies divided the breast shape from different angles, which provided the main basis for bra structure and design elements.

In order to divide the breast shape more carefully, this paper reflects the fullness of breast according to the ratio of breast diameter width to breast diameter thickness. For people with the same bust girth, the larger the ratio, the flatter the breast, and the smaller the ratio, the fuller the breast and the smaller the breast point spacing. The breast orientation is reflected by the ratio of breast point spacing to breast diameter width. The smaller the ratio, the more the breast is gathered. The relative height of breast is described by the ratio of neck den height minus breast high point to breast high point minus waistline height. The smaller the ratio, the higher the relative position of breast point.

3. Experiment

3.1 Measurement Requirements

Thirty-nine female college students aged between 18 and 25 were randomly selected and measured by German Anthroscan non-contact 3D body scanner. At this age, the breasts of women have just matured, and the selected subjects have no breastfeeding experience. Girls of this age don't fully understand what kind of bra they should wear, so it is of value to study their breast shape. The corresponding underwear market also has great potential. While scanning, testers are naked and wearing ordinary brief; Hair is tied up with rubber bands; Jewelry and watches are not allowed to be worn; The tester should stand naturally, her feet are placed on the foot mark on the scanning table, her arms are drooping, her elbows are slightly lifted and opened, her elbows are facing outwards; When measuring the subjects are requested to breathe naturally, look straight ahead, and not swing their bodies.

3.2 Parts to Be Measured

Table 1. Measurement items and codes

Sequence number	Codes	Items
1	A1	Bust girth
2	A2	Underbust girth
3	A3	Neck den height
4	A4	Breast height
5	A5	Waist height
6	A6	Height
7	I1	Width of breast diameter
8	I2	Thickness of breast diameter
9	I3	Breast point spacing
10	T1	The size of the bust girth minus the underbust girth
11	T2	Bra size
12	T3	Ratio of width of breast diameter to thickness of breast diameter
13	T4	Ratio of neck den height minus breast height to breast height minus waist height
14	T5	Ratio of breast point spacing to breast diameter width

According to the main factors affecting breast shape and bra size, 9 human parts including height, girth and width related to breast research were measured, including 6 automatic measurement items and 3 interactive measurement items. According to bra related research, 5 derived items and 14 statistical items were summarized. The measurement methods of width of breast diameter and thickness of breast diameter are shown in Fig.1. The measurement items related to breast shape in this study are shown in Table 1. Among them, A1-A6 are items obtained by automatic measurement, I1-I3 are obtained by interactive measurement of 3D human figure profile with Anthroscan software, and T1-T5 are derived items from calculation.

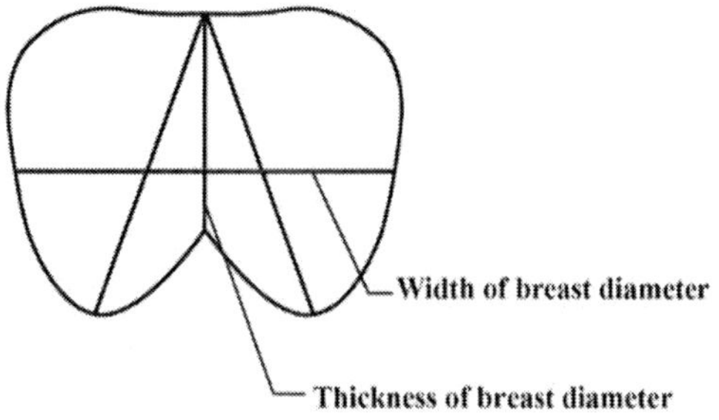

Fig.1 Schematic diagram of width of breast diameter and thickness of breast diameter

4. Data Processing and Analysis

4.1 Data Screening

The measurement data should be preliminary investigated first because of the occasional errors and systematic errors in measurement [15]. All the measurement items are analyzed and its graph can present the singular value, abnormal value, lost data and the characteristics of the data itself. The box diagram of relevant data is shown in Fig.2.

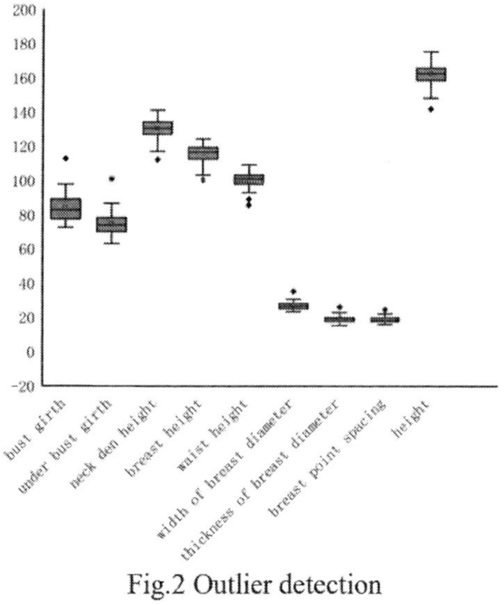

Fig.2 Outlier detection

It can be seen from Fig.2 that there are 10 abnormal data in this measurement data, and the errors of these data are more than 1.5 times standard deviation and less than 3 times standard deviation, which are mild singular values. By observing the samples, it was found that the 10 data came from the 15th tester, the 34th tester and the 35th tester. Among them, the fifteenth tester is 1.61 meters tall and weighs 68 kilograms, so the tester is obese, and its abnormal data is mainly reflected in the large bust girth and underbust girth, but the size of the bust girth minus the underbust girth belongs in the normal range; The width and thickness of breast diameter are large, but the ratio of width and thickness of breast diameter are within the normal range. The breast point spacing is large, but the ratio of breast point spacing to breast diameter width is within the normal range. Therefore, the abnormal data of the 15th tester does not affect the analysis of his breast shape, so the abnormal values of this tester are reserved. The 34th tester and the 35th tester are short in stature for their heights are 1.48 meters and 1.42 meters respectively, and their abnormal data are mainly reflected in the data of neck den height, breast height, height and waist line height, and the t4 index related to breast shape (the ratio of neck den height minus breast height to breast height minus waist height) is in normal range, which does not affect the analysis of their breast shape, so the abnormal values of these two testers are also retained.

4.2 Frequency Analysis

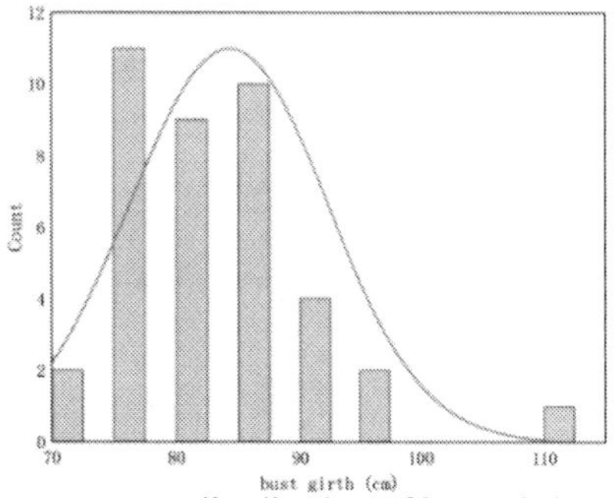

(a) Frequency distribution of bust girth

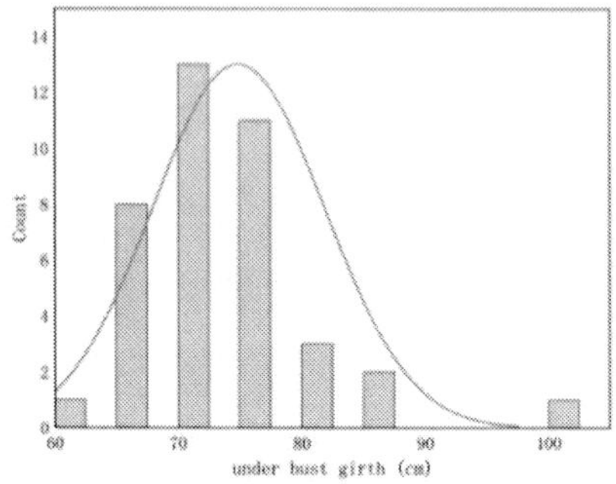

(b) Frequency distribution of underbust girth

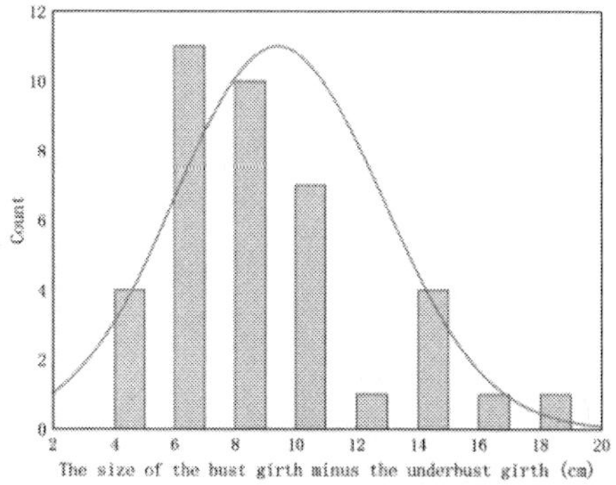

(c) Frequency distribution of the size of the bust girth minus the underbust girth

Fig. 3 Histogram of frequency distribution

The frequency distribution table can conveniently classify the data into groups, and form frequency distribution tables and graphs with different levels of each variable, so as to have a general understanding of the characteristics and observed distribution of each variable data. By analyzing the frequency of bust girth, underbust girth and the size of the bust girth minus the underbust girth, we can understand the overall characteristics of breast and the distribution of measured values. It can be seen from Fig.3(a) that the distribution curve of bust girth is biased to the left, and the peak value is smaller than the standard normal peak value, but the overall data obeys normal distribution. The bust girth of the population is concentrated between 75～90cm, with a total of 30 people, accounting for 76.9% of the total number of people measured; The average value is 84.3cm, which is close to the national standard intermediate 160/84A, indicating that the objects selected in this experiment conform to the standard body shape. The value of underbust girth in Fig.3(b) is concentrated between 65～80cm, with a total of 32 people, accounting for 82.1% of the total number, with an average value of 74.9cm. In Fig.3(c), the distribution curve of the size of the bust girth minus the underbust girth obeys normal distribution, and the value is concentrated between 6 cm and 12 cm, with an average value of 9.42cm. It can be seen from Figs.3(b) and 3 (c) that the bra sizes matching the breast shape of the sample are mainly 70A and 75A.

5. Classification of Breast Shape and Suggestion of Bra Selection

Breasts play a vital role in the aesthetic qualities of a women and its beauty is restricted by body proportion. [16], The breast height, breast expansion and breast fullness are the main factors affecting the beauty of breast shape.

The height of breast determines the up and down position of breast in human body, and the position of breast has an important influence on the structural design of bra [17]. Modern people think that the position of the bust line is the most suitable in the center of shoulder line and waistline. If the position is too high or too low, it will affect the breasts aesthetics. The degree of breast contraction determines whether the breast is gathered or divergent. Too wide or too narrow breast point spacing is not an ideal standard breast. The ideal breast spacing point is where the position of the two nipple points and neck dens form an equilateral triangle [18, 19]. The fullness of breast is the closest to the existing size standard, which is the most intuitive factor to reflect the size of breast. When the ratio of breast diameter width to breast diameter thickness is within a certain range, the breast adopts its most aesthetically pleasing shape [20].

5.1 Classification Basis and Method

Because of the special functions of bras, the basic shape of women's breasts directly determines the choice of bra types with different functions. According to the ratio of breast diameter width to breast diameter thickness, the basic breast shape can be divided into three types: flat, beautiful and plump. According to the ratio of neck den height minus breast height to breast height minus waist height, the breast shape can be divided into drooping type, middle type and higher type. According to the ratio of breast point spacing to breast diameter width, the breast shape can be divided into gathered type, middle type and outward type.

Let T3= ratio of width of breast diameter to thickness of breast diameter, then the value of T3 reflects the fullness of breast. For the same bust girth, the smaller the value of T3 indicates the fuller breast. T4= ratio of neck den height minus breast height to breast height minus waist height, which reflects the relative position of breast point between neck fossa and waist line. The larger T4 value, the greater the sagging degree of breast. T5= ratio of breast point spacing to breast diameter width, the value of T5 reflects the degree of breast expansion. The larger the value of T5, the more obvious the breast expansion.

General studies divide breast morphology into 3-6 categories. In this study, K-means clustering method was used to divide breast shape. Let K = 2, 3, 4, 5 and 6 respectively. When K=2, some parameters are the same, indicating that the clustered categories do not have good discrimination. When K>3, some categories are too small, which makes the categories unrepresentative, while when K=3, the divided categories have good discrimination. Therefore, the breast shape is divided into three categories , and the clustering results are shown in Table 2.

Table 2. Clustering results

category	T3	T4	T5	Number of people	Breast shape
1	1.38	0.64	0.71	17.9%	plump-outward higher
2	1.46	1.16	0.70	35.9%	flat-middle drooping
3	1.39	0.90	0.69	46.2%	beautiful- gathered middle

5.2 Overall Distribution Characteristics of Basic Breast Shape

It can be seen from Table 2 that the number of beautiful type in the sample population is the largest, accounting for 46.2% of the total sample proportion, followed by flat type, accounting for 35.9% of the total sample proportion, indicating that the breasts of women in this age have just matured and are in a beautiful and relatively flat state. In the sample group, the number of people with gathered type is the largest, accounting for 46.2% of the total, followed by those with breast in the middle position, accounting for 35.9% of the total, indicating that women's breast development is good at this stage, and outward expansion is not obvious; The number of people with breasts in the middle position =are the most, followed by the proportion of sagging breasts, and the number of people with higher breasts is the least, which indicates that the height of women's breasts at this stage has begun to change and each form is evenly distributed.

5.3 Characteristics of Height Ratio and Width Ratio of Various Breast Types

5.3.1 Breast-body Ratio Characteristics

Breast-to-body ratio is the ratio of breast height to height, and its size reflects the upper and lower position relationship of breast points in human body. Only considering the height direction, the average breast-to-body ratio of a higher type is 0.724, that of standard type is 0.713, and that of drooping type is 0.705. Among them, the breast-to-body ratio of standard breast is different from the human body ratio suitable for Chinese women proposed by Wacoal in 1979 (as shown in Table 3) [11]. This is related to the locality of the selected samples in this survey, and also to the changes of human body shape in China over the past 40 years.

Table 3. The proportion of human body suitable for Chinese women put forward by Wacoal in 1979

Age	Bust girth (cm)
20-29	Height*0.719
30-39	Height*0.715
40-49	Height*0.711

5.3.2 Width Ratio Characteristics

Width ratio is the ratio of breast point spacing to breast diameter width, which reflects the inward or outward expansion of breast peak to a certain extent. Only the circumference direction is considered in the analysis. The average width ratio of the gathered type is 0.68, the average width ratio of the standard type is 0.69 and the average width ratio of the outside type is 0.73. The ratio of breast spacing to breast width is commonly used in most studies regarding the definition of expansion degree, but the breast diameter width is used instead of breast width in this paper. This is because when measuring the human body, the breast width belongs to the body-fitted length, which easily causes errors to occur in the measurement process, while the breast diameter width is a straight-line distance, which can reduce the measurement errors. Compared with LIANG Suzhen 's research results [11], the value of width ratio is different due to the use of different indicators, however, the data distribution of gathered, standard and expanded breast types have the same characteristics, that is, the width ratio of gathered and standard breast is similar, but the ratio of expanded breast is quite different.

5.4 Selection of Bra Function Types

Bras are one form of plastic underwear, and its basic function is to protect women's breasts, maintain the ideal shape, position and height of breasts, prevent breasts from widening and sagging, and define the beautiful curves of women's breasts. Women strives to reach the standard ideal state by wearing bras. Based on the above reasons, bras are divided into ordinary type and adjustable type according to their functions. Among them, the adjustment type includes side-tightening type, breast-lifting type and side-tightening and breast-lifting type. Bras mainly correct the defects of sagging breasts by selecting the type of rims, determining the inclination angle of shoulder straps, adding padding at the lower part or side lower part of bra cups, etc., so as to improve the breast position or gathering the breasts. Hence, different types of breasts have different choices in bra function types, as shown in Table 4.

It can be seen from Table 4 that 46.2% of people in this group only need to wear ordinary bras, which shows that most of young women's breasts are in normal health and do not need to use adjustable bras to correct the breast shape; There are 17.9% of people who need to wear side-tightening bras; There are 35.9% of people who need to wear breast-lifting bras; There are 0 people who need to wear side-tightening and breast-lifting bras. These data show that only small number of women need to wear adjustable bras. Among them, most people need the lifting effect of bras to improve sagging; A small number of people need to reduce the breast spacing point through the side-tightening effect of the bra.

Table 4. Selection of bra types for breasts with different shapes

Breast type		Number of people	Bra type
Gathered type	Higher	0	Ordinary bra
	Medium	46.2%	Ordinary bra
	Mastoptosis	0	Breast-lifting bra
Medium type	Higher	0	Ordinary bra
	Medium	0	Ordinary bra
	Mastoptosis	35.9%	Breast-lifting bra
Expanded type	Higher	17.9%	Side tightening bra
	Medium	0	Side tightening bra
	Mastoptosis	0	Side-tightening and breast -lifting bra

6. Conclusion

In this paper, the breast size of 39 female college students aged 18-25 was measured by non-contact 3D body scanner. According to the ratio of breast diameter width to breast diameter thickness, the basic breast shape can be divided into three types: flat, beautiful and plump. According to the ratio of neck den height minus breast height to breast height minus waist height, the breast shape can be divided into drooping type, middle type and higher type. According to the ratio of breast point spacing to breast diameter width, the breast shape can be divided into gathered type, middle type and outward type. The overall distribution characteristics of breast morphology of young women in this survey are as follows: the number of beautiful women in the sample group is the largest, followed by the number of flat women, and the number of plump women are less. The phenomenon of breast expansion in sample population is not obvious; The number of people with breasts in the middle position accounts for majority of the population in relation to the height distribution of breasts, followed by the proportion of sagging breasts. From the aspect of breast-to-height ratio, the results of this study show that the average breast-to- height ratio of the standard type is 0.713. From the aspect of width ratio, the average width ratio of the standard type of this sample is 0.69 through calculation. The results of classification study on breast type show that: when choosing bras, the ordinary type is the most common, followed by the breast-lifting type. Fewer people need to wear bras with side-tightening functions. The results can provide reference for underwear enterprises to consider during the planning phase of bra production， as well as help women to choose correct bra sizes.

References

[1] Chen XN. Investigation and Study of Bra Purchasing Habits and Usage Status of Domestic Women. J Knitting Industries: 2018 (6); 63-67.

[2] Yu XK, Cai LX, Xiao P, et al. Requirements of Wearing Comfort for Bra Straps and Its Influence Factors. J Journal of Donghua University(Natural Science): 2014; 40 (01): 48-53,63.

[3] Peng H, Fu JF. Influent Factors Study of Shaping Effect of Steel Ring Free Molded Bra. J Knitting Industries: 2018 (11); 84-87.

[4] Zhang DY, Zheng LZ. Bra's Basic Pattern of Young Women in Shanghai Area. J Journal of Donghua University (Natural Science): 2009; 35 (01): 53-56.

[5] Reena P, Winnie Y, Lawrance H. 3-D breast anthropometry of plus-sized women in south Africa. J Ergonomics: 2011; 54 (9): 866-875.

[6] Zhang R, Liu C, Rao XY. Research and Application of Ergonomics in Nursing Bra Design Field. J Tianjin Textile Science & Technology: 2019 (03); 18-21.

[7] Li XT, Wang JP. Pressure Comfort and Gathering Effect of Bra Based on Breast Shape. J Knitting Industries: 2019 (03); 56-59.

[8] Wang MY, Zhou J. Research on the Relationship Between Different Breast Shapes and Bra Structure Design . J Tianjin Textile Science & Technology: 2019 (05); 18-21.

[9] Duan XJ. A Study on the Relationships of Breast Shapes and Cup Constructions of Bras. D Beijing Institute of Fashion Technology: 2007.

[10] Wang JP, Zhang WY, Gao SP. A primary research on the breast configuration of the young females. J Knitting Industries: 2007 (07); 23-26.
[11] Liang SZ, Zhang X, Zhou J. Basic breast shapes of female undergraduate in the west of China based on 3-D body scanning. J Journal of Textile Research: 2007; 28 (8) ; 75-78.
[12] Edward A, Pechter. A New Method for Determining Bra Size and Predicting Post augmentation Breast Size. J Plastic & Reconstructive Surgery: 1998; 102 (4): 1259-1265.
[13] Rong Z, Winnie Y, Jintu F. Development of a new Chinese bra sizing system based on breast anthropometric measurements. J International Journal of Industrial Ergonomics: 2007; 37 (8).
[14] Liu Y, Wang JP, Cynthia L. Istook. Study of optimum parameters for Chinese female underwire bra size system by 3D virtual anthropometric measurement. J Journal of The Textile Institute: 2017; 108 (6): 877-882.
[15] Zhao SS. Analysis and test of data's accidental errors and systematic errors in marketing research. J Journal of Liaoning Technical University: 2004; 23 (006): 835-838.
[16] Sun C, Zheng R, Li J. Current status of studies on female breast shape beauty. J Journal of Textile Research: 2009 (05); 142-146.
[17] Luo LL, Wang HF. Influence of breast location on structure of female upper garment. J Journal of Textile Research: 2015; 36 (11): 103-108.
[18] Li MJ. A Study on the Structure of Close-fitting Garment and its Virtual Design Based on the Analysis of Women's Body Shape. D DONGHUA UNIVERSITY: 2001.
[19] Chen ZF, Wang H, Wu HL, et al. Measurement on breasts of 100 female college students. J Medical Journal of Chinese Armed Police Forces: 1999; 8 (2): 95-98.
[20] Zhou JJ. Analysis of young women's breast shape and study on pressure comfort of seamless tube top. D Zhejiang Sci-Tech University: 2011.

Cite as: P.F. Shao et al.,
TBIS 10.3993/tbis (2021).

Study on the Potential of China's Textile and Garment Trade with the Five Lancang-Mekong Countries

Peng-Fei Shao[1], Lei Yao[2*]

[1] *Beijing Institute of Fashion Technology, Beijing, 100029, China*
[2] *Beijing Institute of Fashion Technology, Beijing, 100029, China*

*Corresponding author's email: sxyyl@bift.edu.cn

Abstract

With the development of economic globalization, the textile and garment industry, as a pillar industry of China's trade, has attracted more and more attention in the development of its foreign trade. Lancang-Mekong Cooperation is an important measure for Mekong countries to actively respond to the " the Belt and Road Initiatives " in the textile and garment industry. This paper selects the trade data of China and the five Lancang-Mekong countries from 2008 to 2018 and uses the one-step stochastic frontier gravity model to study the efficiency and potential of China's textile and garment trade with the five Lancang-Mekong countries. GDP, population of both sides, distance between both sides, economic freedom, tariff level and the signing of free trade agreements have different degrees of influence on trade efficiency. Finally, relevant suggestions are put forward to further improve trade efficiency. In order to promote the deep cooperation and development between China's textile and garment industry and the five countries, and build an industrial cluster.

Keywords: Lancang-Mekong Cooperation; Trade Efficiency; Trade Potential; Stochastic Frontier Gravity Model

The Lancang-Mekong River Cooperation (LMC) is a new sub-regional cooperation mechanism jointly initiated and built by China, Cambodia, Laos, Myanmar, Thailand and Vietnam. Since the official launch of the LMC mechanism in March 2016, the five Lancang-Mekong countries have gradually become important trading partners of China's textile and garment industry, with their share in the import and export trade constantly increasing, and their total scale surpassing that of Japan, Hong Kong China and other traditional main markets. In 2019, under the theme of "Building a Global Textile Community with a Shared Future", the six countries jointly issued the "Joint Statement on Capacity Cooperation on Textile and Garment in the Lancang-Mekong Region", which will further consolidate the textile and garment industry cooperation bond that has already been a close linkamong the Lancang-Mekong countries.

When analyzing the trade potential, most scholars will use the gravity model to demonstrate potential, and the focus is mostly on China and the countries along "the Belt and Road Initiatives". For example, Li Xiumin et al. (2006) used the gravity model to analyze the bilateral trade development of major countries in Northeast Asia, and proved that GDP, population and spatial distance of bilateral economies were the main factors affecting the bilateral trade volume. Fang Ying et al. (2018) identified that factors such as economic scale and spatial distance of bilateral trade bodies have different influences on bilateral trade volume. However, the assumption of frictionless trade in the gravity model is too idealistic, and the introduction of the stochastic frontier method can solve the defect in which trade resistance factors are ignored. At present, the stochastic frontier gravity model has been widely applied. Bano (2012) used the gravity model to measure the positive influence of ASEAN ANZ on ASEAN New Zealand trade. Ravishankar & Stack (2014) used the stochastic frontier gravity model to measure the trade efficiency of European countries. Wan Hongxian et al. (2019) have used this method to calculate the trade potential of China's traditional trade in goods [1-4].

To sum up, the traditional gravity model is mostly used in the study of trade potential by scholars, though the trade potential of textile and garment is rarely studied. The LMC mechanism under textile and garment has just been proposed, which makes for popular discussion. Based on this, this paper uses the stochastic frontier gravity model to analyze the potential of China's textile and garment export trade with the five Lancang-Mekong countries, and and the sample interval is 2008-2018. Through empirical analysis, this paper hopes to reveal the reasons for the Lancang-Mekong countries' trade growth, promote its further development and provide some reference for the improvement of the LMC mechanism.

1. Current Situation of Textile and Garment Trade Between China and Five Lancang-Mekong Countries

China's textile and garment trade with the five Lancang-Mekong countries from 2008 to 2018 is shown in Fig.1. It illustrates an overall upward trend and a continuous trade surplus, with the surplus value also rising year by year. In 2018, exports and imports were 16.44 billion US dollars and 4.69 billion US dollars respectively. Among the five countries, China has the closest textile and garment trade with Vietnam. China's exports to Vietnam have always accounted for more than 53% of the total exports, and even reached 61% in 2015. It can be seen that this trade-link between China and Vietnam is the main part of the textile and garment trade among the five Lancang-Mekong countries.

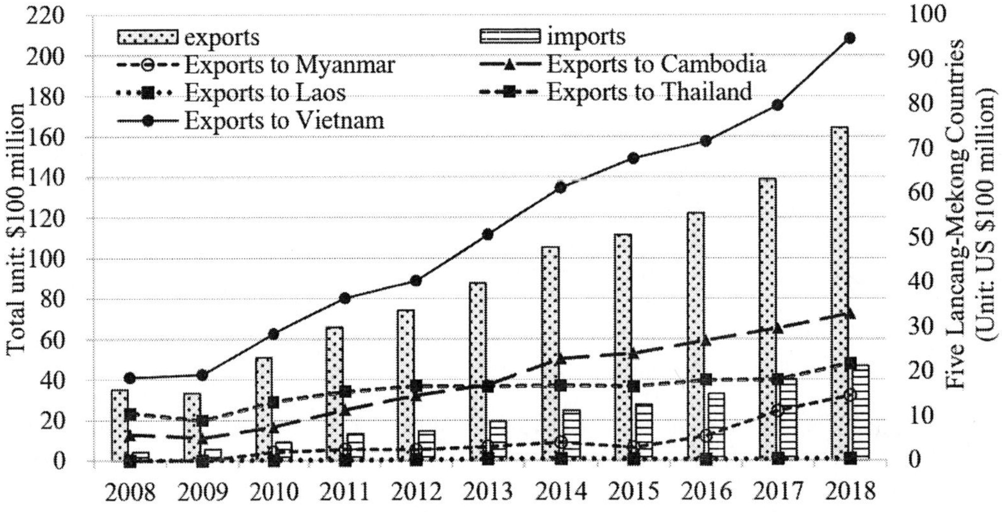

Fig.1 China's textile and garment trade with the five Lancang-Mekong countries (2008-2018)
Data Source: UN COMTRADE

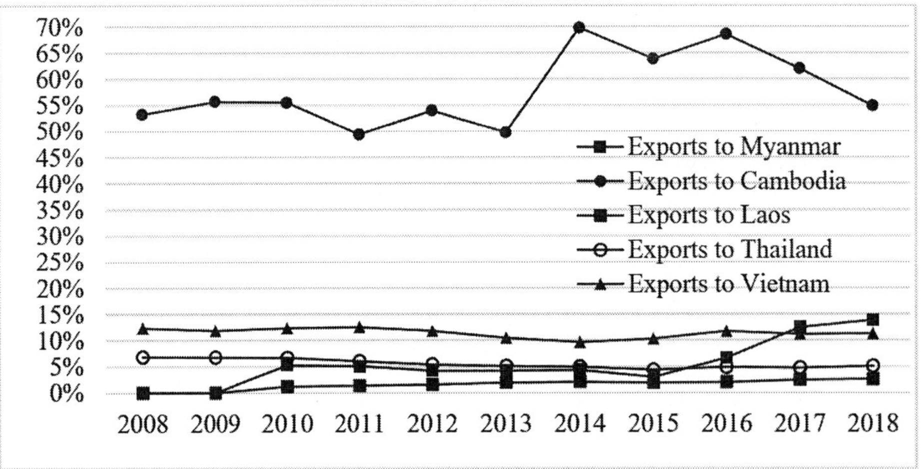

Fig.2 Ratio of China's textile and garment exports to the five Lancang-Mekong countries
Data Source: UN COMTRADE

In China's export trade to the five Lancang-Mekong countries, the proportion of textile and clothing is shown in Fig.2. Textile and garment play an important role in China's overall export trade to Cambodia, accounting for more than 50%. In the overall export trade between China and Laos, textile and garment accounted for a very small proportion, accounting for no more than 2.6%.

2. Models and Data

2.1 Model Setting

The traditional gravity model is a commonly used method to measure the trade potential. However, the traditional gravity model assumes that there is no trade friction between countries, and at the same time, the factors affecting trade efficiency are simply attributed to the residual term. The stochastic frontier analysis is used in the traditional gravity model to overcome the bias of trade potential measurement. In this paper, the Battese & Coelli (1995) model was adopted and relevant literature of Zhou Shudong (2018) stochastic frontier gravity model was referred to to build a model as follows [5]:

$$\mathrm{T_{ijt}} = f(x_{ijt}, \beta) exp(v_{ijt} - u_{ijt}) \qquad (u_{ijt} \geq 0) \tag{1}$$

$$\mathrm{T_{ijt}}^* = f(x_{ijt}, \beta) exp(v_{ijt}) \tag{2}$$

$$\mathrm{TE_{ijt}} = \frac{\mathrm{T_{ijt}}}{\mathrm{T_{ijt}}^*} = exp(-u_{ijt}) \tag{3}$$

Where, $\mathrm{T_{ijt}}$ represents the actual trade volume between country i and country j in the period of t; x_{ijt} represents the factors affecting the volume of trade; β is the parameter to be estimated; $\mathrm{T_{ijt}}^*$ denotes frontier trade level (The maximum level of trade that can be achieved in the existing environment); $\mathrm{TE_{ijt}}$ represents trade efficiency, which is equal to the ratio of the actual trade level to the frontier trade level. The closer the ratio is to 1, the closer the actual trade volume of the two countries is to the maximum trade level andthe higher the trade efficiency will be; or, vice versa and therefore, the lower the trade efficiency will be; v_{ijt} is the random error term, representing the unobserved factors affecting the trade volume; u_{ijt} is a trade inefficiency item, representing the degree of trade inefficiency, mainly including human factors to promote or restrain trade. $u_{ijt} = 0$ and $u_{ijt} > 0$ indicate that the trade volume between the two countries is equal to the maximum trade level and there is a gap.

In order to measure China's trade potential and trade efficiency with the five Lancang-Mekong countries, according to Armstrong, S. (2007), variables such as economic scale, total population, per capita income difference, distance and boundary of importing and exporting countries are introduced into the relevant influence variable x_{ijt} in Eq.1. After logarithms of both sides, the following equation can be obtained:

$$\ln \mathrm{T_{ijt}} = \beta_0 + \beta_1 \log POP_{jt} + \beta_2 \log POP_{it} + \beta_3 \log GDP_{jt} + \beta_4 \log GDP_{it} + \beta_5 \log AG_{ijt} + \beta_6 \log Dist_{ij} + \beta_7 Border_{ij} + v_{ijt} - u_{ijt} \tag{4}$$

After introducing artificial factors such as tariff level, institutional environment and trade agreement into the trade inefficiency item u_{ijt}, the one-step trade inefficiency comprehensive measurement model equation is:

$$\ln T_{ijt} = \beta_0 + \beta_1 \log POP_{jt} + \beta_2 \log POP_{it} + \beta_3 \log GDP_{jt} + \beta_4 \log GDP_{it} + \beta_5 \log AG_{ijt} + \beta_6 \log Dist_{ij} + \beta_7 Border_{ij} + v_{ijt} - (\alpha_0 + \alpha_1 MON_{jt} + \alpha_2 GES_{jt} + \alpha_3 SHP_{jt} + \alpha_4 TAF_{jt} + \alpha_5 TTI_{jt} + \alpha_6 FTA_{ijt} + \varepsilon_{ijt}) \quad 5)$$

2.2 Variable Description

Eq.4, Eq.5 are the stochastic frontier gravity model and the one-step inefficiency model respectively, where i is China, j is the five Lancang-Mekong countries, and t is the period. T_{ijt} represents the textile and garment export volume of China to country j in period t, and POP_{it} and POP_{jt} represent the total population of China and country j in period t respectively. GDP_{it} and GDP_{jt} represent the economic size of China and country j in period t respectively; AG_{ijt} represents the absolute difference of per capita income between the two parties in the period of t; $Dist_{ij}$ represents the distance between China and country j, reflecting the transportation cost between the two countries; $Border_{ij}$ denotes whether China and country j share a common Border. MON_{jt} is the economic freedom of country j; GES_{jt} reflects j 's investment in trade and transportation-related infrastructure; SHP_{jt} refers to the Liner Transport Connectivity Index of Country j; TAF_{jt}reflects the level of tariff facilitation in country j, and the higher the value is, the more serious the tariff barrier is and the lower the level of trade facilitation is. TTI_{jt} is the customs clearance time of country j, which is a trade inhibiting factor; FTA_{ijt} indicates whether country j implements the China-ASEAN Free Trade Agreement and belongs to the factor of promoting trade and is set as a dummy variable. Full implementation is defined as 1; otherwise, it is 0.

2.3 Data Sources

Data of China's textile and garment exports to the five Lancang-Mekong countries from 2008 to 2018 come from UN Comtrade database; GDP, per capita income and population are derived from the WDI database of the World Bank. GDP is adjusted in 2010 dollars to account for the impact of inflation. Distance from www.24timemap.com/distance/search; The boundary comes from the world map. TAF is derived from IMF reports; TTI comes from the World Bank; SHP data from UNCTAD STAT; GES comes from the World Bank's WGI database; Mon comes from the Wall Street Journal, Economic Freedom Index report; FTA for free trade agreement data from import and export trade documents [6-8].

3. Model Estimation Results and Analysis

This paper uses Frontier 4.1 software to measure the development level and potential of China's textile and garment export trade to five Lancang-Mekong countries by comparing the trade efficiency values obtained from the stochastic frontier gravity model and the trade inefficiency model.

3.1 Analysis of Results of Stochastic Frontier Gravity Model

This paper estimates China's textile and garment export trade to five Lancang-Mekong countries from 2008 to 2018 by stochastic frontier gravity model according to Eq.4. In order to compare the robustness of the results, the estimation results of the time-invariant model and the time-varying model are presented in this paper, as shown in Table 1. The γ values of the time-invariant model and the time-varying model are 0.986 and 0.966 respectively, and both are significant at the 1% level, indicating that trade inefficiency factors mainly affect China's export trade to the five Lancang-Mekong countries. The η of the time-varying model is less than 0 and significant at the 1% level, indicating that the trade inefficiency term increases with time, which proves that the time-varying model is more applicable [9].

Table 1. Estimation results of stochastic frontier gravity model

Variable	Time invariant model		Time-varying model	
	Coefficient	T value	Coefficient	T value
Constant term	96.310^{***}	93.141	109.231^{***}	96.660
$\log POP_{jt}$	3.026^{***}	15.728	3.026^{***}	13.439
$\log POP_{it}$	-10.707^{***}	-23.859	-10.521^{***}	-15.974
$\log GDP_{jt}$	-1.290^{***}	-6.635	-1.291^{***}	-3.273
$\log GDP_{it}$	2.482^{***}	9.098	2.484^{***}	4.220
$\log AG_{ijt}$	0.092	0.602	0.204	0.413
$\log Dist_{ij}$	-8.615^{***}	-9.607	-10.001^{***}	-9.361
$Border_{ij}$	-1.624	-1.632	-1.543^{***}	-16.452
σ^2	2.379^{***}	5.621	38.477^{***}	6.210
γ	0.986^{***}	124.385	0.966^{***}	135.965
η			-1.831^{**}	-2.044

Note: *, ** and *** indicate passing the test at the significance level of 10%, 5% and 1% respectively

It can be seen from Table 1 that the population size of the five Lancang-Mekong countries is significantly positive at the 1% level, indicating that the increase of population in the five Lancang-Mekong countries will promote China's textile and garment exports to them. However, the effect of China's population size is significantly negative. Possible reasons are as follows: with the increase of China's population, domestic demand will increase significantly and exports to other countries will, therefore, decrease; At the same time, the increase of population will increase the number of labor force, promote domestic textile and garment production, and reduce the import demand accordingly. The economic size of China has a significant positive impact on export trade, while the economic size of the five Lancang-Mekong countries has a significant negative impact. Both the geographical distance and the common border are significantly negative at the 1% level, indicating that the transportation cost will hinder the trade, and the proximity between the two countries will not necessarily expand the trade scale of textile and garment. This also illustrates the importance of other modes of China's export transportation to the five Lancang-Mekong countries. The difference of per capita income between the two countries is not significant at the level of 10%, which may be because the textile and garment are necessities of life and the price elasticity is low, so it has no obvious effect on export.

3.2 Result Analysis of Trade Inefficiency Model

This paper adopts a one-step method to estimate the trade inefficiency model, and the results are shown in Table 2. The overall estimation effect of the model is good, with LR value of 99.974 and γ value of 0.775, which are significant at the 1% level, indicating that trade inefficiency significantly exists and is the main factor hindering trade [10].

It can be seen from Table 2 that: firstly, economic freedom (MON_{jt}) is significantly negative at the 10% level, indicating that an open economic system with high freedom can effectively improve trade efficiency. Secondly, government expenditure (GES_{jt})has no significant impact on trade inefficiency, indicating that government investment in transportation and other infrastructure of partner countries is not prominent in textile and garment trade. It could also be that China's investment in transport in the One Belt and One Road countries has solved the transport problem, so partner countries do not need to make signficant investments. Thirdly, the Liner Shipping Connectivity Index (SHP_{jt}) has no significant influence on trade inefficiency, which is related to the geographical location of the five Lancang-Mekong countries and China, so maritime infrastructure is not necessarily used forfor trade transportation. Fourth, the tariff level (TAF_{jt}) is significantly positive at the 10% level, indicating that with the continuous improvement of regional integration level, the tariff can play an increasingly smaller role in trade and is no longer the main factor hindering the trade between China and the five Lancang-Mekong countries. Fifth, the impact of import customs clearance time (TTI_{jt}) is not significant, which is related to the level of public governance of the trading partners. The customs clearance process is artificially disturbed by more informal systems, which increases the uncertainty. Sixth, the impact of FTA (FTA_{ijt}) on trade inefficiency is significantly negative. FTA can promote trade development and offset the impact of trade inefficiency on trade. As China and the five Lancang-Mekong countries have signed the free trade agreement, their economic and trade relations will enter a new stage. Therefore, China should improve the corresponding trade policies and trade supporting facilities, so as to better promote the development of textile and garment trade between herself and the five Lancang-Mekong countries [11].

Table 2. Trade inefficiency model estimation results

Variable		Coefficient	T value
Stochastic frontier function	Constant term	96.297***	96.543
	$\log POP_{jt}$	3.260***	12.963
	$\log POP_{it}$	−9.826***	-15.778
	$\log GDP_{jt}$	−1.431***	-5.765
	$\log GDP_{it}$	1.733***	3.818
	$\log AG_{ijt}$	0.067	0.362
	$\log Dist_{ij}$	−8.168***	-9.060
	$Border_{ij}$	−1.485***	-11.479
Trade inefficiency function	Constant	0.481	0.818
	MON_{jt}	−0.019*	-1.674
	GES_{jt}	0.450	0.444
	SHP_{jt}	0.003	0.606
	TAF_{jt}	0.446*	1.807
	TTI_{jt}	0.127	0.420
	FTA_{ijt}	−0.295***	-2.815
σ^2		0.443***	6.285
γ		0.775***	3.709
log likelihood		-146.306	
LR inspection		99.974	

Note: *, ** and *** indicate passing the test at the significance level of 10%, 5% and 1% respectively

3.3 Trade Efficiency Analysis

In this paper, the time-varying stochastic frontier gravity model and Eq.2 are used to obtain the trade efficiency of China's textile and garment exports to Lancang-Mekong countries. The time span is from 2008 to 2018. Combined with Eq.3, the potential value of export trade can be calculated (Trade efficiency and trade potential value of inverse correlation, analysis can be one). The export potential of China and the five Lancang-Mekong countries from 2008 to 2018 is shown in Fig.3. There is a great variation in export trade potential between China and the five Lancang-Mekong countries, which indicates that the export distribution of textile and garment is unbalanced. Specifically, China's export trade potential to Vietnam has always been in the first place, and has been greatly increased since 2012, indicating that the cooperation in textile and garment trade between China and Vietnam has been continously deepening. However, the potential value of export fluctuated greatly from 2014 to 2016, and then bounced back rapidly after 2016. The potential value of export trade also increased steadily, making it an important export market for China's textile and garment trade. Secondly, the export trade potential between China and Myanmar, Cambodia and Thailand are all at a relatively weak level, but it also shows a trend of growth This growth rate of trade potential is significantly accelerated after 2016, which may be related to the conclusion of LMC mechanism. Laos has the lowest trade potential, and it has basically remained unchanged in the past ten years, which indicates that China faces great trade resistance with Laos. This is because Laos is still a traditional agricultural country, so China's trade advantages with other countries are not obvious. Non-tariff barriers, trade protectionism and other factors will hinder the development of trade.

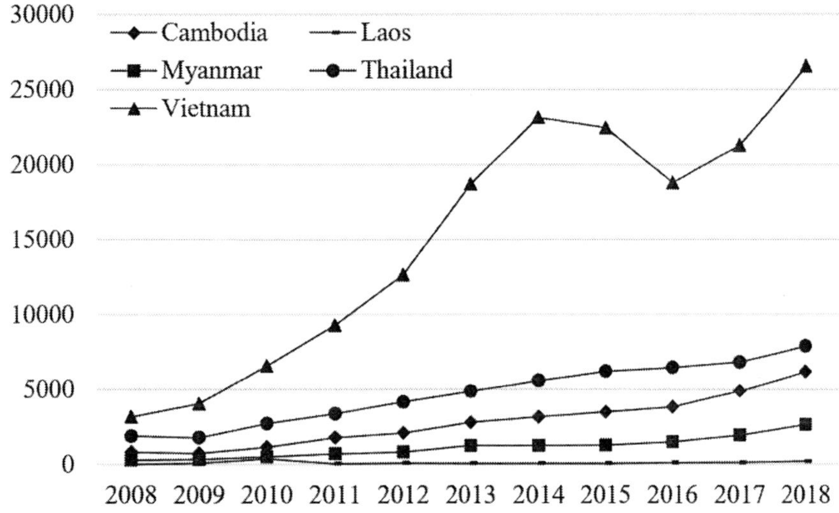

Fig.3 Export potential of China and Lancang-Mekong Countries from 2008 to 2018 (Unit: millions of US dollars)

4. Conclusions and Suggestions

4.1 Conclusions

Through empirical analysis, in terms of trade potential of textile and garment export between China and five Lancang-Mekong countries, Vietnam has the largest trade potential among them, while Laos has the lowest trade potential and is subject to large trade resistance due to tariff and non-tariff barriers together with other factors. Even though, until 2016,China's textile and garment exports for five Lancang-Mekong countries trade efficiency and potential are also relatively low, starting from 2016, in addition to Laos, other Lancang-Mekong countries' trade potential increase year by year, indicating that LMC plays a role in promoting textile and garment trade between China and the five Lancang-Mekong countries.

4.2 Suggestions

In addition, according to the estimation of export potential value, Vietnam is likely to continue to be the main trade partner of China's textile and garment products. Therefore, China should continue to dig deeper into the demand of Vietnam's textile trade and extend its industrial value chain. Next, it is necessary to further improve the efficiency of textile and garment trade between China and the five Lancang-Mekong countries and continue to tap the trade potential to promote the development of bilateral and multilateral trade.

LMC should give full play to its geographical advantages, focus on the construction of cross-border garment e-commerce cooperation zone and the overseas garment economic and trade cooperation zone, continuously optimize bilateral trade and in investment facilitation mechanism and, finally, defuse the fierce competition from other developed countries. To improve outside Pan-Asia railway sections, the Lancang - Mekong river port construction and upgrades, such as formation water, land, air, iron three-dimensional transportation network, the acceleration of construction infrastructure and other trade related facilities to consolidate the hardware foundation of LMC, to deal with the future development of the new opportunities and challenges.

China should enhance its scientific and technological content, itstechnological level, strengthen the competitive advantage of textile and garment export, develop the potential market brought by demographic dividend to fully stimulate its trade potential. At the same time, it should continue to promote the upgrading of cooperation in the textile and garment industry, and make full use of the Lancang-Mekong textile and garment cooperation mechanism. It should promote mutual recognition of standards and product certification, strengthen the construction of quality system, enhance the competitiveness of textile and garment products, and welcome the entrance of more textile and garment products from Mekong countries into the Chinese market.

5. Acknowledgement

Beijing social science fund key project, construction and application of garment green consumption guidance system (19JDGLA010).

References

[1] Li XM, Li SY. Empirical test and potential analysis of trade gravity model of Northeast Asian countries. Northeast Asia Forum: 2006; (02): 28-32.

[2] Fang Y, Ma R. The potential of cultural trade between China and countries along the "the Belt and Road Initiatives": An empirical study based on stochastic frontier gravity model. World Economic Research Journal: 2018; (01): 112-121+136.

[3] Wan HX, Feng TT. The efficiency and potential of trade in services between China and "the Belt and Road Initiatives" countries: Based on time-varying stochastic frontier gravity model. Productivity Research: 2019; (10): 1-7+161.

[4] Pierluigi M, Silvia N. Assessing the Trade Impact of the European Neighborhood Policy on EU-MED Free Trade Area[R]: 2012.

[5] Battese GE, Coelli TJ. A model for technical inefficiency effects in a stochastic frontier production function for panel data . Empirical Economics: 1995 (2).

[6] Zhou SD, Zheng J. Trade Efficiency and Influencing Factors between China and RCEP Partners: An Empirical Analysis Based on Stochastic Frontier Gravity Model. Research on Economic Problems: 2018; (07): 89-97.

[7] Hou M, Deng LL. Trade Efficiency and Potential between China and Central and Eastern European Countries: An Analysis Based on Stochastic Frontier Gravity Model. Shanghai Economic Research Journal: 2017; (07): 105-116.

[8] Zhang LJ, Zhang YF. Study on the potential of China's textile and garment trade with countries along the China Pakistan Economic Corridor. Price monthly: 2019; (11): 25-31.

[9] Wen SH, Zhang X. Trade potential of Indochina Peninsula and its influencing factors: An Empirical Analysis Based on Stochastic Frontier gravity model. International trade issues: 2017; (10): 97-108.

[10] Yao N. An Empirical Study on the trade scale and trade potential between Ningbo and central and Eastern Europe-Estimation Based on Stochastic Frontier gravity model. Enterprise economy: 2018; 37 (06): 154-162.

[11] Hou M, Deng LL. Trade efficiency and potential between China and central and Eastern European countries: an analysis based on Stochastic Frontier gravity model. Shanghai Economic Research: 2017; (07): 105-116.

Research on Innovation Performance of Textile and Garment Enterprises Based on Three-stage DEA Model

Ya-Nan Zhao, Zheng-Yan Shao*

Business School of Beijing Institute of Fashion Technology, Beijing, 100029, China

*Corresponding author's email: sxyshaozhengyan @bift.edu.cn

Abstract

Innovation driven industrial upgrades is the only way to develop the textile and garment industry. This paper constructs a Three-stage DEA model with innovative elements as the core, evaluates the innovative performance of textile and garment enterprises by using the sample data of listed Chinese textile and garment companies from 2016 to 2019, and studies the external environmental factors of innovation performance. The results show that: the level of innovative performance of Chinese textile and garment enterprises is low, and that the scale of innovative investment still needs to be increased; external environmental factors significantly affect the innovative investment of textile and garment enterprises, and that after eliminating environmental factors, the difference of pure technical efficiency of textile and garment enterprises is small, and the industry lacks unique enterprises with heterogeneous innovation.

Keywords: Innovation Performance; Textile and Garment Enterprises; Three-Stage DEA Model

1. Introduction

As an important traditional industry in China, the development of textile and garment enterprises has entered a mature period, facing many challenges, such as low-cost advantage fading, an insufficient domestic market demand, trade barriers in developed countries and so on [1]. In recent years, more and more textile and garment enterprises hope to break the bottleneck of industry development through innovation. According to the data of the China National Bureau of Statistics, the R&D experimental funds of China's textile and garment industry increased from 16.5 billion Yuan in 2011 to 37.15 billion Yuan in 2019, with an annual average compound growth rate of 9.43%. Whether the rapid growth of R&D investment in textile and garment industry can improve the innovation performance of enterprises and realize the high-quality development of the industry is a hot issue in academia and industry.

2. Literature Review

Enterprise innovation performance refers to the efficiency of the enterprise innovation process, the effect of output and its contribution to business performance [2]. As enterprise innovative activities involve many factors of the enterprise, the evaluation of innovation performance is complex. Many scholars have put forward a variety of opinions on how to evaluate innovation performance, but the standard index system and evaluation method have not yet been formed.

Based on the complexity of innovative performance evaluation, most scholars establish multi-index systems to i.: For example, Ma Ning et al. (2000) measured enterprise innovation performance with the number of enterprise patents, the number of innovative products and the number of new products separately [3] whilst Zhang Fanghua (2006) studies enterprise social capital and technological innovation performance. Zhang used the annual number of new products, annual number of patents applied, new product development speed, innovation success rate and new product output value rate to measure enterprise innovation performance [4]. Chen Jin (2009) chose to measure annual number of new products, new product sales rate, new product output value rate, development speed and patent application as indicators to evaluate innovation performance [5]. Xie Xuemei (2010) elected to use the new product sales revenue ratio, the product innovation ratio and the process innovation ratio to evaluate innovation performance on the basis of former scholars' research on innovation performance measurement to determine innovative performance [6]. Finally, Yi Jingtao et al. (2015) studied the relationship between enterprise heterogeneity, executive characteristics and enterprise innovation performance. He used the ratio of R&D expenditure to sales to measure innovative activity input, and the number of patent grants is used to measure innovation activity output [7].

As for innovation performance evaluation methods, stochastic frontier analysis (SFA) and data envelopment analysis (DEA) are the mainstream methods, especially with the former, which can not only build a multi-dimensional evaluation index system for the complexity of innovation activities, but also break through the limitations of previous evaluation by focusing on financial indicators [8-9]. Su Yi (2013) used stochastic frontier analysis to study and analyze regional innovation performance [10]; Chen Wei et al. (2013) used the DEA Malmquist index method to study professional innovation performance of knowledge intensive industries [11]; Mu Zhirui (2012) used a super efficiency DEA model to evaluate R&D investment performance of Beijing [12] and Han Bing et al. (2018) used a two-stage DEA model to study high-tech enterprises. In this paper, they analyzed the innovation performance of energy-saving and environmental protection listed companies in 2015 by using a Three-stage DEA model [13]; Zhou Lijuan et al. (2013) used the Three-stage DEA model to measure China's regional innovation efficiency [14] and Wang Yumei et al. (2019) used Three-stage DEA model to measure the innovation performance of 56 energy-saving and environmental protection listed companies in 2015, focusing on the impact of environmental factors on the innovation performance of energy-saving and environmental protection listed companies [15].

The current research focuses mainly on the innovation performance of high-tech enterprises and technology enterprises [16-19], and many scholars also study the innovation performance of manufacturing industry [20-22], but there are few articles that focus on the innovative performance of textile and garment enterprises. The uniqueness of this paper is to select textile and garment enterprises as the research object, and to use the Three-stage DEA model to evaluate the innovative performance of textile and garment enterprises. By comparing model results before and after excluding environmental factors and random interference factors, this paper speculates the impact of external environmental factors on the innovation activities of China's textile and garment enterprises. This will enrich the current research on the innovation and development of textile and garment enterprises, and provide a reference for the formulation of reasonable innovation and development policies.

3. Three-stage DEA Implementation Path

3.1 The First Stage: DEA-BCC Model

According to the difference of input-output relationship, the DEA model can be divided into input-oriented and output-oriented methods. Input variables are more operable than output, and the BCC model can measure the pure technical efficiency and scale efficiency of each decision-making unit which the CCR model cannot obtain. Therefore, this paper uses input-oriented DEA to evaluate the innovation performance of textile and garment enterprises. The equation is as follows:

$$\min[\theta-\varepsilon(\hat{e}^T s^- + e^T s^+)]$$

$$\begin{cases} \sum_{j=1}^{n} x_j\lambda_j + s^- = \theta x_0 \\ \sum_{j=1}^{n} y_j\lambda_j - s^+ = y_0 \\ \sum_{j=1}^{n} \lambda_j = 1 \\ \lambda_j \geq 0, \ j = 1,2,\dots,n,\theta \end{cases} \quad 1)$$

In the above formula, θ is the efficiency value of the j textile and clothing enterprise $DMU_{j0,}$ x_j is input variable, y_j is output variable, s^+ is input relaxation variable, s^- is output relaxation variable, and λ_j is weight coefficient of input and output index value. Suppose the optimal solution of BCC model is θ^*, λ^*, s^{*+}, s^{*-}. If $\theta^*=1$, $s^{*+}=0$, $s^{*-}=0$, then textile and clothing enterprises are DEA valid; if any of the conditions in $\theta^*=1=1$, $s^{*+}=0$ and $s^{*-}=0$, are not established, the textile and clothing enterprises are DEA invalid.

3.2 The Second Stage: SFA Model

In this stage, the external environmental factors and random errors are used to regress the input redundancy s^- obtained in the first stage. In this case, the interference of the two factors on the operation efficiency can be effectively stripped. The SFA regression model was as follows:

$$S_{ni} = f(Z_i;\beta_n) + v_{ni} + \mu_{ni}(\mathrm{i} = 1,2, \ \ldots, \ \mathrm{I};\ \mathrm{n} = 1,2, \ \ldots, \ \mathrm{N}) \quad 2)$$

In Formula (2), S_{ni} means the redundancy of the input of the textile and clothing enterprise; Z_i represents the environmental variable affecting the input; β_n represents the coefficient to be estimated for the corresponding environment, and $f(Z_i;\beta_n)$ indicates the influence of the environmental variable on the input excess, $f(Z_i;\beta_n) = Z_i\beta_n$. v_{ni} is the random interference term, $v_{ni}\sim N(0,\sigma_{ni}^2)$; μ_{ni} is the efficiency residual term, which reflect the inefficiency degree of the corresponding DMU, $\mu_{ni}\sim N(\mu_n,\sigma_{ni}^2)$. They are independent of each other. $v_{ni} + \mu_{ni}$ represents consistent error terms. $\gamma = \sigma_{uj}^2/(\sigma_{uj}^2 + \sigma_{vj}^2)$ stands for the ratio of management inefficiency variance to compound variance.If $\gamma \to 1$, then management inefficiency is dominant; If $\gamma \to 0$, then management inefficiency can be ignored.

Afterwards, according to the regression results, the excess part of investment is adjusted, and the corresponding measurement equation is as follows:

$$x_{ij}^* = x_{ij} + \left[\max\left(x_i \hat{\beta^n}\right) - \left(x_i \hat{\beta^n}\right)\right] + \left[\max\left(\hat{v_{ij}}\right) - \hat{v_{ij}}\right] \ i = 1,\ 2,\ \ldots,\ m;\ j = 1,2,\ \ldots,\ n \quad 3)$$

In formula (3), x_{ij}^* is the input variable adjusted by SFA regression, $\hat{v_{ij}}$ random error of input variable, which is needed to be seperated from $v_{ij} + u_{ij}$. According to the separation method of Jondrow et al. (1982) [23], the estimated value can be obtained as follows:

$$E(v_{ij}|v_{ij} + u_{ij}) = S_{ij}^- - z_i\beta^n - E(v_{ij}|v_{ij} + u_{ij}),\ i = 1,\ 2,\ \ldots,\ m;\ j = 1,2,\ \ldots,\ n \quad 4)$$

After the above calculation, the influence of different external environment and random noise on input is excluded, so that each textile and garment enterprise is in the same external environment, and then the input redundancy caused by inefficient management is obtained.

3.3 The Third Stage: Adjusted DEA Model

The input data adjusted by SFA is brough into the picture in the second stage and original input data into DEA-BCC model is again used to get the result of innovation performance of textile and garment enterprises excluding the influence of environmental factors and random noise.

4. Data Sources and Variable Selection

4.1 Data Sources

The Industry Classification Guidelines for Listed Companies (revised in 2012) issued by the China Securities Regulatory Commission selects the listed companies under the two categories of "textile industry" and "textile and clothing industry", and after excluding the enterprises that do not disclose relevant data, it obtains 25 enterprises in the textile industry and 24 enterprises in the clothing industry. The specific enterprise data comes from the RoyalFlush iFinD database and the relevant environmental variables data are from the websites of the China National Bureau of statistics and the Provincial Bureau of statistics.

4.2 Establishment of Innovation Performance Index System

According to the input-output characteristics of innovation activities, this paper constructs the innovation performance evaluation index system of textile and garment enterprises, as shown in Table 1.

Table 1. Index System of Innovation Performance Evaluation of Textile and Garment Enterprises

First Level Index	Second Level Index	Specific Index
Input	R&D investment scale	R&D expenditure
	R&D investment intensity	Proportion of R&D expenditure in operating revenue
	R&D personnel	Number of R&D personnel
	Intensity of R&D personnel	Proportion of R&D personnel in total employees
Output	Innovative achievements	Number of new patent applications
	Economic performance	Growth rate of operating revenue
		Return on total assets

4.3 Selection of External Environment Factors

Environmental variables have an impact on the innovation performance of enterprises, but they are not the factors that enterprises can subjectively control in innovation activities. The environmental variables selected in this paper are as follows:

1) Government's direct subsidy to enterprises

In this paper, the amount of government subsidy disclosed in the annual report of enterprises is defined as the direct subsidy from the government to enterprises. Government financial subsidy is a kind of affirmation and incentive for enterprise innovative activities. Enterprises with a strong innovative ability and have a remarkable innovative effect in the industry are more likely to get government subsidies.

2) Regional economic development level

This paper uses the per capita GDP of each region to measure the economic development level of the region where the enterprise is located. The level of regional economic development provides strong economic support for enterprise innovation, which will inevitably have an impact on enterprise innovation investment.

3) Regional openness

This paper uses the ratio of import and export volume to GDP to measure the degree of regional openness. The higher the degree of regional openness, the easier it is for all kinds of resources to gather, the greater the competitive pressure enterprises face, and the stronger the power to seek innovation and development.

4) Regional research environment

In this paper, the regional research environment is measured by regional research financial expenditure. The larger the research financial expenditure is, the more active the research activities are and the better the research environment is.

5) Regional education environment

In this paper, the regional education financial expenditure is used to measure the regional education environment. The larger the regional financial expenditure on education, the more attention the region attaches to education. At the same time, the human environment and population quality are also better, which provides a strong human support for innovation.

5. Innovation Performance Evaluation of Textile and Garment Enterprises Based on Three-stage DEA

5.1 The First Stage of Innovation Performance Analysis Based on DEA-BCC Model

Using the DEA-BCC model, this paper analyzes the sample data of 49 textile and garment listed companies from 2016 to 2019, and obtains the comprehensive innovation performance evaluation results of textile and garment listed companies, as shown in Table 2.

Table 2. Comprehensive Innovation Performance Measurement Results in The First Stage

Textile Listed Companies (25)					Clothing listed companies (24)				
Firm	2016	2017	2018	2019	Firm	2016	2017	2018	2019
SHSE:600220	1	1	1	1	SHSE:600177	0.858	1	1	1
SHSE:600232	0.164	0.318	0.331	1	SHSE:600398	1	1	1	1
SHSE:600370	0.187	0.237	0.409	0.553	SHSE:600400	1	0.465	0.398	1
SHSE:600448	0.034	0.129	0.109	0.132	SHSE:601566	0.607	0.811	1	0.714
SHSE:600493	1	1	0.79	0.695	SHSE:603518	0.348	0.357	0.23	0.158
SHSE:600630	0.293	0.356	0.468	0.431	SHSE:603555	0.655	0.924	0.501	0.244
SHSE:600987	0.451	0.464	0.25	0.206	SHSE:603808	0.341	0.41	0.35	0.212
SHSE:601339	0.111	0.132	0.201	0.274	SZSE:002003	0.336	0.854	0.58	1
SHSE:601599	1	1	1	0.349	SZSE:002029	0.16	0.219	0.495	1
SHSE:603558	0.803	1	0.612	0.264	SZSE:002154	0.13	0.317	0.614	0.38
SHSE:603889	0.484	0.573	0.318	0.178	SZSE:002269	0.099	0.08	0.237	0.273
SZSE:000726	0.18	0.19	0.235	0.278	SZSE:002404	0.475	0.577	0.606	0.451
SZSE:000850	0.329	0.868	0.606	0.26	SZSE:002486	0.293	0.391	0.653	0.545
SZSE:000955	1	0.533	0.651	0.597	SZSE:002503	0.321	1	0.879	1
SZSE:002083	1	1	0.65	0.717	SZSE:002563	0.271	0.207	0.245	0.276
SZSE:002087	1	0.643	1	0.726	SZSE:002569	0.952	0.946	1	1
SZSE:002144	0.255	0.357	0.315	0.203	SZSE:002612	0.159	0.203	0.272	0.194
SZSE:002193	0.119	0.181	0.28	0.135	SZSE:002634	0.34	0.703	0.607	0.53
SZSE:002293	0.316	0.503	0.329	0.435	SZSE:002656	0.251	0.548	0.916	0.121
SZSE:002327	0.96	1	0.586	0.61	SZSE:002687	0.47	0.618	0.894	0.493
SZSE:002394	0.303	0.307	0.255	0.245	SZSE:002740	1	1	1	1
SZSE:002397	0.255	0.38	0.294	0.215	SZSE:002762	1	0.998	0.692	0.325
SZSE:002516	0.378	0.472	0.414	0.287	SZSE:002763	0.453	0.498	0.56	0.952
SZSE:200726	0.18	0.19	0.235	0.278	SZSE:002832	0.5	0.498	0.506	0.251
SZSE:300577	1	1	1	0.402					
Mean	0.507	0.581	0.563	0.502	Mean	0.507	0.581	0.563	0.502
Effective=1	7	7	4	2	Effective=1	4	4	5	8
	28.00%	28.00%	16.00%	8.00%		16.67%	16.67%	20.83%	33.33%
Low	2	5	6	6	Low	5	9	12	4
Efficiency<1	8.00%	20.00%	24.00%	24.00%	Efficiency<1	20.83%	37.50%	50.00%	16.67%
Extremely	16	13	15	17	Extremely	15	11	7	12
Inefficient<0.5	64.00%	52.00%	60.00%	68.00%	Inefficient<0.5	62.50%	45.83%	29.17%	50.00%

It can be seen from Table 2 that, first of all, the number of enterprises that achieve effective innovation performance (efficiency value = 1) is very small in both the textile industry and the garment industry. In terms of time change, the number of listed companies that achieve effective innovation in the textile industry is decreasing year by year, and the number of listed companies that achieve effective innovation in the garment industry is increasing year by year. Among the listed companies with a low innovation performance (efficiency value < 1), the number and proportion of textile enterprises in 2016-2018 are lower than those in the clothing industry, and the situation of textile industry has improved in 2019. Among the listed companies with very low innovation performance (efficiency < 0.5), the number of textile enterprises is more, and the proportion is more than half for four consecutive years. The overall innovation performance of clothing enterprises is better than that of textile enterprises, and the innovation performance of the downstream industry is more prominent.

According to the results of the DEA-BCC model, the comprehensive performance, pure technical efficiency and scale efficiency of 49 decision-making units in each year are calculated, and the average values are calculated respectively, and a line chart is drawn, as shown in Fig.1.

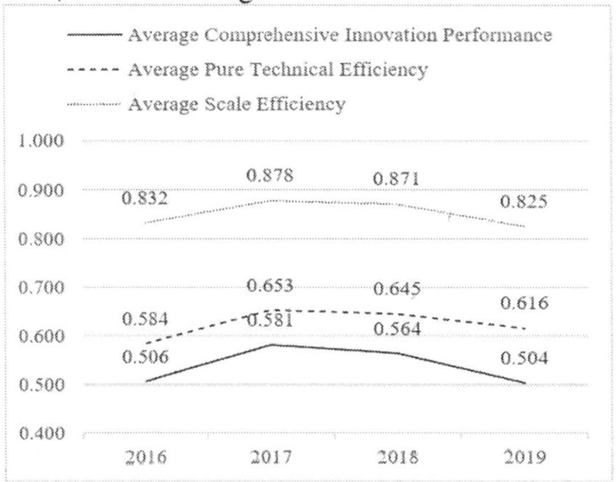

Fig.1 Line Chart of Average First Stage DEA Results of Textile and Garment Industry from 2016 to 2019

It can be seen from Fig.1 that the average comprehensive innovation performance of textile and garment enterprises remains between 0.5 and 0.6. During 2016-2019, the average comprehensive performance of textile and garment enterprises shows a trend of first, rising and then, declining. In 2017, the value reaches the peak and then begins to decline. This shows that the comprehensive innovation performance of China's textile and garment enterprises is generally low, and the promotion is weak; on the one hand, the lack of effective and reasonable allocation of innovation resources leads to the low conversion rate of innovation input to output. Furthermore, the poor ability of innovation output also affects the effective use of innovation input resources. The average pure technical efficiency of textile and garment enterprises fluctuates between 0.58 and 0.7, and it began to decline slowly after reaching the peak in 2017. The innovation efficiency driven by management and technology factors of textile and garment enterprises is not high and cannot be improved continuously. The average scale efficiency of textile and garment enterprises fluctuates between 0.8 and 0.9, and the average scale efficiency of innovation activities is better than pure technical efficiency.

The product of pure technical efficiency and scale efficiency is the comprehensive performance. The change in trend of the three variables is consistent from 2016 to 2019. Textile and garment enterprises generally have higher scale efficiency, which makes a great contribution to the improvement of comprehensive innovation performance. However, the lower level of pure technical efficiency restricts the overall improvement of comprehensive innovation performance of textile and garment industry.

Fig.2 shows the proportion of textile and garment enterprises in different returns to scale from 2016 to 2019. The proportion of enterprises with increasing returns to scale decreased year by year from 2016 to 2018 and increased slightly in 2019. The proportion of enterprises with decreasing returns to scale increased year by year from 2016 to 2018, and decreased slightly in 2019, but still maintained 57%. Meanwhile, the proportion of enterprises with increasing returns to scale increased year by year. The proportion of enterprises in a constant state has remained stable at about 20% for four consecutive years. The large scale of input relative to output of most textile and garment enterprises is the main reason for affecting the returns to scale.

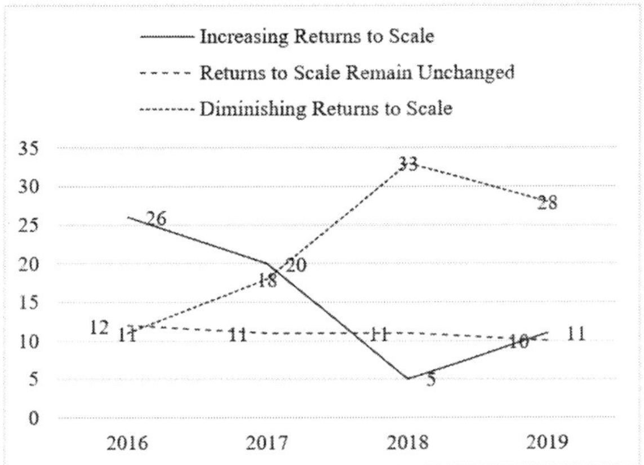

Fig.2 Line Chart of Scale Income of Textile and Garment Enterprises in The First Stage from 2016 to 2019

5.2 The Second Stage of Regression Analysis Based on SFA Model

A similar SFA model is established by taking the slack variables of four enterprise innovation investment variables, namely, R&D investment scale, R&D investment intensity, R&D personnel and R&D personnel proportion, as the explained variables. There are five environmental variables, namely, government direct subsidies to enterprises, regional economic development level, regional openness, regional scientific research environment and regional education environment, as the explained variables. The FRONTIER4.1 software is used to calculate and study the influence of five environmental variables on four input relaxation variables. The regression results are shown in Table 3. It can be seen from Table 3 that both σ^2 value and γ value pass the significance test, and the γ value is close to 1, which indicates that when measuring the comprehensive innovation performance of Listed Companies in textile and garment industry, the estimated values of regression parameters pass the significance test, and it is necessary to eliminate environmental variables and statistical noise.

Table 3. SFA Regression Results

	R&D Investment Scale	R&D Investment Intensity	Number of R&D personnel	R&D Personnel Ratio
Constant	9.66E+07*** (9.65E+07)	4.05E+06*** (1.17E+06)	-1.69E+08*** (-1.68E+08)	2.37E+07*** (2.37E+07)
Government's direct subsidy to enterprises	1.20E+00*** (8.01E+00)	3.58E+04*** (1.85E+02)	2.48E+04*** (8.91E+00)	-1.37E+05*** (-1.37E+05)
Regional economic development level	3.41E+03*** (6.42E+00)	-1.84E+05*** (-5.00E+03)	8.34E+03*** (6.70E+00)	1.19E+06*** (1.19E+06)
Regional openness	-5.20E+06*** (-5.19E+06)	-1.24E+06*** (-4.61E+04)	-3.21E+07*** (-2.69E+07)	-4.14E+06*** (-4.14E+06)
Regional research environment	5.47E+04*** (4.78E+04)	1.97E+04*** (1.20E+02)	9.24E+04*** (5.22E+03)	6.97E+04*** (6.97E+04)
Regional education environment	-1.97E+05*** (-7.04E+04)	-2.53E+04*** (-2.95E+01)	-3.78E+05*** (-1.42E+04)	-2.15E+05*** (-2.15E+05)
σ^2	7.90E+15*** (7.90E+15)	3.47E+14*** (3.47E+14)	2.28E+17*** (2.28E+17)	4.81E+15*** (4.81E+15)
γ	0.82*** (3.81E+01)	0.87*** (5.90E+01)	0.93*** (2.13E+02)	0.76*** (2.69E+01)
Log	-3.76E+03	-3.41E+03	-4.01E+03	-3.72E+03
LR	7.87E+01***	1.35E+02***	2.07E+02***	7.37E+01***

Note: "* *", "*", and "*" are significant at the levels of 1% (T > 4.541), 5% (T > 2.353), and 10% (T > 1.638) respectively. The values in brackets are the corresponding T values. The above models all passed the LR unilateral likelihood ratio test of 1% (T > 16.074).

According to the regression results, the effects of five environmental variables on four input relaxation variables were analyzed:

1) The results show that the larger the amount of direct subsidy, the more active the enterprises invest in these three aspects, which leads to the redundant investment of these factors to a certain extent. The direct subsidy from the government to enterprises has a significant negative impact on the proportion of R&D personnel, that is, the proportion of R&D personnel tends to be reasonable with the increase of the direct subsidy which indicates that the direct subsidy from the government promotes the optimization of the talent structure of enterprises. Moreover, direct subsidy from the government has obvious stimulating and promoting effect on the investment of enterprises in innovation activities. Although in the short term, enterprises will have redundancy in the use of innovation funds and innovative talents, the large-scale accumulation of innovation resources will improve the talent structure of enterprises in the long term and promote the adjustment and upgrading of product structure, raise the barriers to competition, and realize the high-quality development of enterprises.

2) The regression coefficients of relaxation variables of regional economic development level to R&D scale, R&D personnel and R&D personnel proportion are all positive, indicating that the higher the regional economic development level is, the more sufficient the capital investment, personnel investment, and personnel proportion of innovation activities of textile and garment enterprises will be, showing a certain degree of redundancy. The regression coefficient between the relaxation variables of regional economic development level and R&D investment intensity is significantly negative. This is because the higher the level of economic development, the larger the output scale brought by enterprise innovation, and the higher the efficiency of R&D investment.

3) The SFA regression coefficient of regional openness and the relaxation variables of four innovation inputs is significantly negative, which indicates that the higher the regional openness, the more it can reduce the redundancy of innovation resources and improve the innovation performance of enterprises.

4) There is a significant positive correlation between the regional research environment and the four slack variables of innovation investment, which indicates that the better the research environment of the region where the enterprise is located, the more abundant the research public resources provided by the local government for enterprise innovation. These results in a certain degree of redundancy in innovation resource investment. In the short run, a good regional scientific research environment cannot immediately improve the innovative output of enterprises and the innovative performance of textile and garment industry, but in the long run, it plays a role as a reservoir for enterprises to increase innovation investment.

5) There is a significant negative correlation between the regional education environment and the four slack variables of innovation investment, which shows that the better the regional education environment is, the better the allocation of innovation resources. After all, the higher the level of regional education, the higher the quality of regional talents, and the stronger the ability to attract foreign high-quality talents. A larger number of outstanding talents gathered in a certain region has played a positive role in promoting the innovation performance of enterprises in the region.

5.3 The Third Stage of Innovation Performance Analysis

Using the data of input variables and output variables adjusted in the second stage, the DEA-BCC model is used to recalculate the comprehensive innovation performance, pure technical efficiency and scale efficiency of textile and garment enterprises after eliminating environmental factors. The average value of the results is shown in Fig.3.

From Fig.3, it is obvious that the average comprehensive innovation performance, average pure technical efficiency and average scale efficiency of textile and clothing enterprises after adjustment have significantly improved compared with the first stage. The average comprehensive innovation performance shows an overall upward trend, and reaches the maximum value in 2019. The average pure technology efficiency is at the highest value of 0.983 in 2016, and then slightly decreases to 0.97. Compared with the fluctuation range of pure technical efficiency in the first stage, the fluctuation interval of pure technology efficiency has been significantly improved; the change trend of average scale efficiency of textile and garment enterprises is consistent with the average comprehensive performance. This means that scale efficiency becomes the main factor affecting the innovative performance of textile and garment enterprises after eliminating environmental factors and random noise.

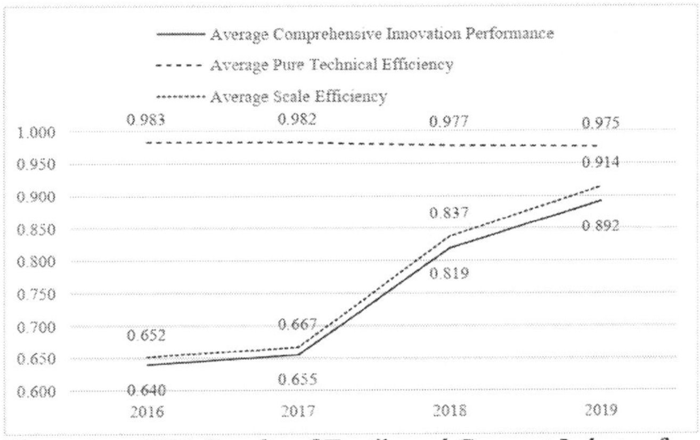

Fig.3 Average Three-stage DEA Results of Textile and Garment Industry from 2016 to 2019

On the other hand, this shows that, the comprehensive innovation performance evaluation of China's textile and garment enterprises based on the traditional DEA model is greatly affected by the environment, and the pure technical efficiency is generally high after removing the environmental factors, which indicates that the pure technical efficiency difference between enterprises is small, and the enterprises lacking of heterogeneous innovation lead the basic innovation of the industry. Nonetheless, after eliminating the influence of environmental factors, the comprehensive innovation performance of textile and garment enterprises shows a continuous and rapid rising trend from 2016 to 2019, and the main driving force for the rapid improvement of comprehensive innovation performance at the enterprise level comes from the improvement of scale efficiency. Therefore, the continuous improvement and optimization of the external environment, the individual investment in innovation activities of the enterprise is increasingly positive, and the increase of scale efficiency speeds up the improvement of enterprise level innovation performance.

The proportion of enterprises with different returns to scale after excluding environmental factors in the third-stage is shown in Fig.4.

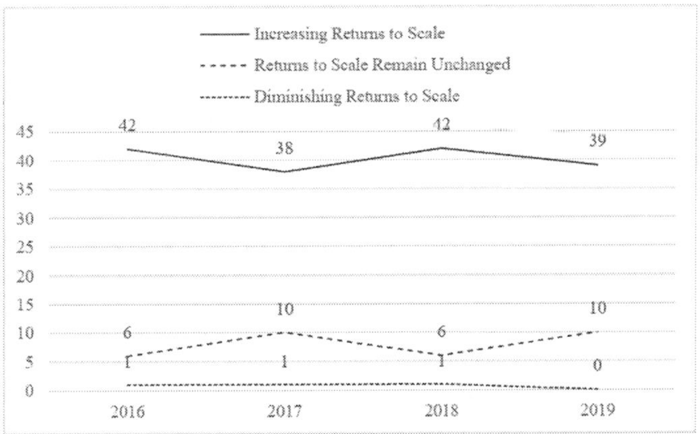

Fig.4 Line Chart of Three-stage Returns to Scale of Textile and Garment Enterprises in 2016-2019

Comparing Fig.4 with Fig.2, the proportion of enterprises in the increasing state has risen significantly, while the proportion of enterprises in the decreasing state has fallen significantly. It shows that environmental variables become the main reason of restricting the scale expansion of Listed Companies in textile and clothing industry. After eliminating the environmental variables, the innovation investment space of textile and garment enterprises is still very large. In conclusion, continuing to expand the scale of innovation investment will bring higher innovation income.

6. Conclusion and Suggestion

This paper uses a Three-stage DEA model to measure the innovation performance of 49 listed companies in China's textile and garment industry from 2016 to 2019, the conclusions are as follows:

1) From the comprehensive innovation performance results of textile and garment enterprises in the first stage, the innovation performance level of Chinese textile and garment enterprises is generally low. In contrast, garment enterprises in the downstream of the industry and and those that face the consumer market directly have stronger innovation ability and higher innovation performance than textile enterprises in the middle of the industry.

2) From the third stage, the DEA result which exclude environmental factors and random interference and the external environmental factors and random interference lead to the underestimation of innovation performance of textile and garment enterprises, and the degree of pure technical efficiency underestimation is higher than that of scale efficiency underestimation. Moreover, the difference of pure technical efficiency of textile and garment enterprises after eliminating environmental factors is small, and the main reason affecting the comprehensive innovation performance is the difference of enterprise scale efficiency. This indicates that the difference of technology and management level of textile and garment enterprises is low, and the unicorn enterprises lack of heterogeneous innovation. From the analysis results of the two stages of returns to scale, the proportion of textile and garment enterprises with increasing returns to scale after adjustment increased significantly. Therefore, textile and garment enterprises continue to increase innovation investment, which can effectively improve the scale efficiency of innovation investment.

3) From the regression results of five environmental variables to four innovation inputs obtained from SFA regression in the second stage, it can be seen that the government's direct subsidies to enterprises can significantly stimulate textile and garment enterprises to increase innovation investment and promotes the optimization of the structure of innovative talents in the industry. The improvement of regional economic development level is helpful for textile and garment enterprises to invest more actively in innovation funds and scientific research talents, so as to optimize the utilization rate of enterprise funds. The higher the degree of regional opening, the more competitive environment it brings, which is conducive to promote the innovative drive of textile and garment enterprises and improve the effective utilization of R &; D investment. Good research environment provides abundant public research resources for enterprise innovation, and promotes large-scale innovation investment of textile and garment enterprises. The higher the quality of the regional education environment, the more favorable it is to attract the inflow of all kinds of high-quality talents, so that the innovation input resources of enterprises can be reasonably allocated, and the comprehensive innovation performance of textile and garment industry can be improved.

Based on the above conclusions, combined with the current development of textile and clothing enterprises, the following suggestions are put forward:

1) At the government level, innovation and development is not achieved overnight, it must experience long-term investment. Scientific research innovation needs long-term development to bring economic benefits. Innovation activities have the characteristics of high investment and high risk. Enterprises must face the huge economic pressure brought by innovation activities. For textile and garment enterprises, a traditional industry, it is more difficult and challenging for enterprises to carry out technological innovation. The government should continue to increase support for innovation and development and further stabilize the determination of innovation and development of traditional industries.

2) At the enterprise level, the incentive mechanism for innovative activities should be formulated to activate the internal innovation development power and vitality of the enterprise, and enhance the enterprise's technological innovation ability and the ability to coordinate innovation resources efficiently. Considering the differences of external environment faced by textile and garment enterprises in different regions, these should make appropriate strategic arrangements according to different environmental conditions.

3) Industry level: relevant departments and industry associations need to take measures to provide impetus for the upgrading of the textile and garment industry. There should also be an increase to the technical differences which will activate the competition in the textile and garment market. The textile and garment enterprises should be stimulated to carry out technological innovation, strengthen the internal cooperation among regional enterprises, promote the flow and sharing of technology, and jointly promote the collaborative development of technological progress and innovation of textile and garment enterprises.

7. Acknowledgement

This work was supported by the education and teaching reform project of Beijing Institute of Fashion Technology (JG-1918) and Postgraduate Innovation Program at Beijing Institute of Fashion Technology (No.X2021-078).

References

[1] Xiao F, Yuan M. driving force and Inspiration of transformation and upgrading of China Textile and clothing industry. Modern economic information: 2020; 153-154.

[2] Gao J, Wang JF, Wei P. Performance indicators of technological innovation of enterprises: current situation, problems and new concept models [J]. Scientific research management: 2004 (S1); 14-22.

[3] Manning, official building. Key factors affecting the performance of technological innovation of industrial enterprises in China [J]. Science and technology management: 2000 (03); 16-20.

[4] Zhang FH. Corporate social capital and technological innovation performance: conceptual model and empirical analysis [J]. Research and development management: 2006 (03); 47-53.

[5] Chen YF, Chen J. Mechanism research on open innovation to promote innovation performance [J]. Scientific research management: 2009; 30 (04): 1-9+28.

[6] Jie XM. Empirical Study on collaborative innovation network and innovation performance of small and medium-sized enterprises [J]. Journal of management science: 2010; 13 (08): 51-64.

[7] Yi JT, Zhang XP, Wang HC. Heterogeneity of enterprises, overconfidence of senior executives and innovation performance of enterprises [J]. Nankai management review: 2015; 18 (06): 101-112.

[8] Farrellll MJ. The Measurement of productive efficiency [J]. Journal of the Royal Statistical Society: 1957; 253-290.

[9] Xiang J, Liu HW. A review of the performance evaluation of technological innovation [J]. Scientific and technological progress and Countermeasures: 2011; 28 (06): 161-166.

[10] Su Y, Li BZ. Performance analysis of regional innovation system innovation based on Stochastic Frontier [J]. Journal of system engineering: 2013; 28 (01): 125-133.

[11] Chen W, Sa R, Zhang YC, Tian SH. Research on the performance evaluation of patent innovation of knowledge intensive industries in China -- Based on DEA Malmquist index method [J]. Management review: 2013; 25 (08): 39-45+53.

[12] Mu ZR. Beijing R & D input performance evaluation based on super efficiency DEA model [J]. Scientific and technological progress and countermeasures: 2012; 29 (05): 117-120.

[13] Han B, Su Y, Li T, Wan M. Research on the performance of high-tech enterprise technology innovation based on DEA in two stages [J]. Scientific research management: 2018; 39 (03): 11-19.

[14] Zhou LJ, Xu M, Liu HD. The study of regional innovation efficiency measurement: Empirical Analysis Based on Three-stage DEA model [J]. Enterprise economy: 2013 (12); 160-164.

[15] Wang YM, Ji X, Wu HX. The evaluation of innovation efficiency based on Three-stage DEA model: Taking Listed Companies of energy conservation and environmental protection as an example [J]. Technology economy and management research: 2019 (03); 25-30.

[16] Zheng Y. Recognition of high-tech enterprises, external financing incentives and corporate performance: Empirical Study Based on the tendency score matching method (PSM) [J]. Research and development management: 2020; 32 (06): 91-102.

[17] Guo XQ, Sun YM. R & D investment, technology accumulation and market performance of high-tech enterprises [J]. Scientific research: 2020; 38 (09): 1630-1637.

[18] Lu X, Huang JY, Zhao Y. The research on the internal influence of innovation atmosphere of high-tech enterprises on Innovation Performance: Double intermediary role of tacit knowledge acquisition and organizational learning [J]. Technology economy and management research: 2020 (08); 49-53.

[19] Zheng SF, Tang FC, Ge AJ, Liu RJ. The influence of knowledge governance between high and new technology enterprises on collaborative innovation performance [J]. Scientific and technological progress and countermeasures: 2020; 37 (15): 107-115.

[20] Yu YM, East C. The influence of state-owned enterprises' introduction of non-state-owned capital on Innovation Performance -- Based on empirical evidence of state-owned listed companies in manufacturing industry [J]. Research and development management: 2020; 32 (05): 152-165.
[21] Zhao XY, Gao YF, Sun MD. Organizational flexibility, dynamic service innovation capability and service innovation performance of manufacturing enterprises [J]. Scientific and technological progress and countermeasures: 2020; 37 (15): 62-69.
[22] Li DH, Wu RH, Chen D. How "competition" affects innovation performance: Tracking Research on Chinese manufacturing enterprises choosing local and overseas competition [J]. Management world: 2020; 36 (02): 161-181+225
[23] Jondrow J, Lovell CAK, Materov IS, et al. On the Estimation of Technical Inefficiency in the Stochastic Frontier Production Function Model [J]. Journal of Econometrics: 1982; (19).

Cite as: Q.Q. Sun et al.,
TBIS 10.3993/tbis (2021).

Future of the Garment Industry: AI Promoted Mass Customization

Qian-Qian Sun[1], Xiao-Dong Sun[2*]

[1,2]*Beijing Institute of Fashion Technology, No. A2, East Yinghua Street, Chaoyang District, Beijing 100029, China*

*Corresponding author's email: sun_xiaodon@yahoo.com

Abstract

AI promoted mass customization is the promising solution for the traditional garment industry that is facing complex challenges caused by quarantine of global COVID-19 pandemic. However, there are still many open problems on how AI can be applied in mass customization. This paper describes the key technologies in accordance with: the knowledge management of customers, AI recommendation, AI measurement, flexible digital research, development technics and technics on the Internet of Things in the garment production process. These namely includes quality control as well as RFID and AI logistics products trace back, which will benefit the development of apparel industry 4.0.

Keywords: AI Promoted Mass Customization; AI Measurement; Apparel Industry 4.0; Flexible Manufacturing

1. Introduction

It is known that the traditional garment industry is labor intensive and time-consuming. The quarantine of COVID-19 pandemic have brought about many challenges to the traditional clothing industry. The AI promoted mass customization of garment industry is a promising solution for some of these problems, as it provides alternative online and digital production processes. The traditional garment industry requires many repeated processes of measurement, pattern making and garment fit. The traditional garment industry relies manually on measuring, pattern making and sewing sample cloth. In addition, there is a communication barrier between manufacturers, designers, and customers. Traditional garment production is not able to update the production progress in real time and is also not able to trace back the process for consumers. At the same time, product logistics mostly depend on third-party logistics companies, which has relatively higher logistics costs and has increased clothing production costs. The emergence of Industry 4.0 provides intelligent technology solutions for the apparel industry. Industry 4.0 adopts technologies such as the Internet of Things, big data, and artificial intelligence to achieve mass customization of clothing and provide personalized services to consumers [1]. Mass customization (MC) can realize the application of advanced computer technology with the apparel industry. The MC could provide online digital customization services without the presence of the customer. It could offer personalized garment for customers at a lower cost at the same time shorten production cycles. Flexible manufacturing in Industry 4.0 will accelerate the transformation trend of mass customization of clothing. However, there are many open problems on how AI could find its application in mass customization. This paper presents the key technologies of AI promoted mass customization.

This paper is organized as follows. Section 2 discusses the technologies of AI promoted mass customization of clothing in terms of the knowledge management of customers, AI measurement, AI recommendation, and flexible manufacturing for example RFID and AI logistics. It then proposes an optimized AI production method for mass customization. Section 3 provides conclusions and future research.

2. AI Promoted Mass Customization of Clothing

AI Promoted Mass customization can use AI knowledge management of customers and AI recommendation to obtain customer information. Apparel pre-sales, production, and logistics use the AI Cloud for data management and sharing. Consumers, manufacturers, and AI logistics interact with data to enable flexible manufacturing of clothing. Fig.1 shows the process of AI promoted mass customization. The MC could offer customization production on customer's demand. In addition, the MC could theoretically realize zero inventories of products.

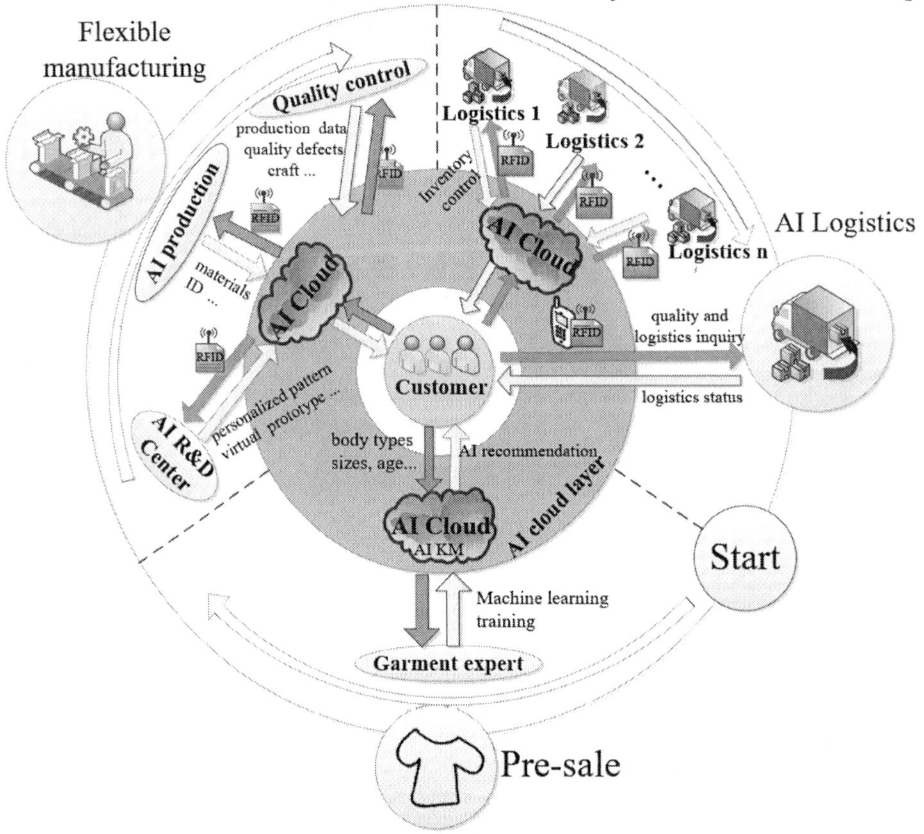

Fig.1 The process of AI Promoted Mass customization

2.1 The AI Knowledge Management of Customers

Traditional knowledge management obtains customer's size information at stores for customization. However, it does not accumulate user information or recommend the right clothing for the user. Therefore, the AI knowledge management of customers (AI KM) is very important in mass customization. The MC can obtain precise customer requirements for personalized garment production. For enterprises, the knowledge management of customers is customer-centered, with access to all information about apparel personalization. Fig.2 shows the AI knowledge management of customers. The AI KM with the priori knowledge of AI Expert library (AI EL) could apply deep learning technology to learn more about consumers' habits, tastes, purchase behavior, body types, sizes, fabrics, and positions, which will offer more intelligent recommendation to customer. The AI KM combines Information Technology (IT), Electronic data processing (EDP) and CAD technology could offer intelligent digital sample cloth R&D process. The AI KM can use the AI EL to learn designer's design style, which will design personalized clothing at lower cost. In addition, with the help of AI KM apparel companies could have quick response to consumers' demands and update of data from customer changes in real time. The target consumers of apparel companies are the center of all the activities of the companies. The knowledge management of customers is a prerequisite for mass customization of clothing. The MC will use AI KM to deeply explore customer factors: age, preferences, occupation, body type, and size. The AI KM will provide personalized customization for customers and real-time update of data according to customer changes, resulting in more timely design of new products and services, as well as early warning and competitive intelligence. And finally, The AI KM will improve customer commitment and loyalty, both of which will also benefit from AI promoted mass customization.

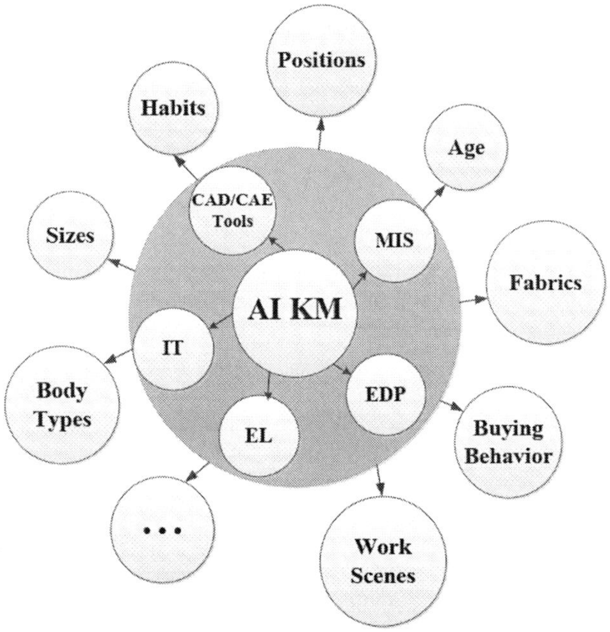

IT: Information Technology MIS: management information system
EDP: Electronic data processing EL: Expert library

Fig.2 The AI Knowledge Management of Customer

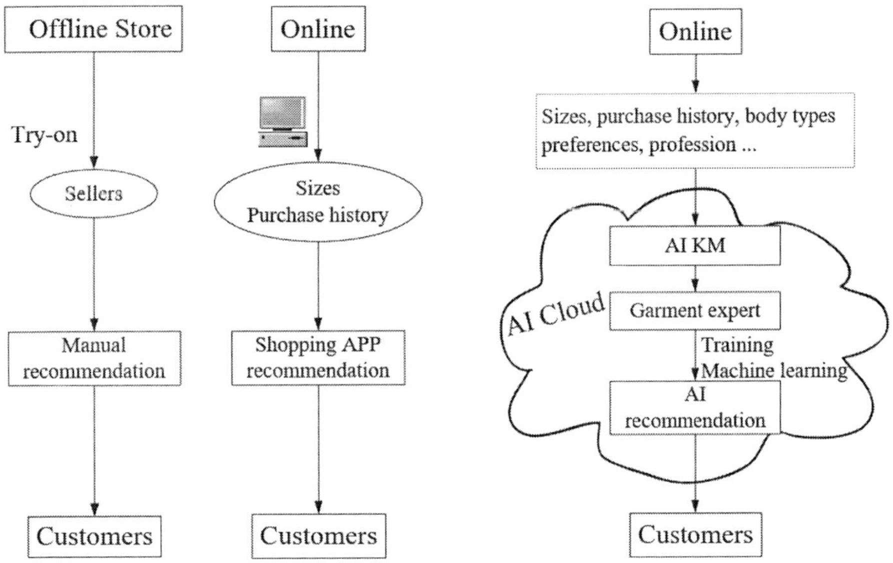

(a) Traditional clothing recommendation (b) AI recommendation

Fig.3 The comparison of traditional and AI recommendation

2.2 The AI Recommendation

Traditional recommendation uses warehouse or online app recommendations through purchase history, which is limited personalization. Garment recommendation information is provided by the customer. However, it cannot accurately recommend clothing suitable for customers. In addition, Traditional recommendation is time-consuming.

Fig.3 shows the comparison of traditional and AI recommendation. The AI recommendation can use the AI Cloud to highly match the customer's personalized data to recommend fit clothing. The AI cloud can learn from the experience of clothing experts to make new recommendation. In addition, the AI Cloud can use garment experts to train personalized styles, patterns, and make AI recommendations. In addition, AI measurement is a key technology to realize AI recommendation.

2.2.1 The AI Measurement

Traditional clothing measurement is done manually by measurers or consumers who fill in size information. However, the cost of manual measurement is high. Consumers generally do not know the body sizes, which will affect the entire process of clothing customization. Therefore, anthropometric methods needs to be intelligent. There are two smart measurement methods: 3D scanning and 2D image methods. The 3D scanning method can obtain accurate human body data. However, the human body scanners are usually expensive. It is inconvenient to measure at a fixed point with or without a strait jacket. The 2D image method can measure the human body quickly and inexpensively. Table 1 shows the comparison of human body measurement methods. The comparison shows that the existing measurement methods cannot fully extract the sizes required for mass customization. Therefore, The AI measurement can use big data to analyze consumer behavior and get customer' data based on sales records and the number of online orders. The AI measurement can use the AI KM and AI recommendation to make an accurate prediction of customers' data. At the same time, the MC can keep track of changes in consumers' body types, sizes and styles and respond quickly to meet the individual needs of consumers.

Table 1. The comparison of human body measurement methods

Method	Classify	Sizes	Category	Disadvantage
Based on 3D scan	Laser [2, 3]	1-9, 11-19, 21-23, 25	Suit shirt dress pants	High noise sensitivity, large space occupation, complicated size collection
	Structured light [4, 5]	1-9, 11-19, 21-23, 25		Great light influence, unmeasurable blind spot size, complicated data processing
	Depth map [6-8]	1-9, 14-17, 21-23, 25	Suit shirt pants	Low accuracy, high noise sensitivity, repeated depth map, large size error, point cloud holes, duplication
Based on 2D image	Single camera [9, 10]	1, 3-6, 9, 14-16, 22, 25	Suit shirt dress pants	Duplicate and missing point clouds, incomplete contour extraction; blind spots in measurement
	Multiple cameras [11, 12]	Camera coordinate distance		Complex equipment adjustment, disorder and missing point cloud

1-25: Bust, dress length, sleeve length, shoulder width, back length, neck circumference, upper arm circumference, cuff circumference, waist circumference, shoulder angle, shoulder slant length, chest width, back width, pants length, hips, upper length, feet mouth circumference, abdomen circumference, hip height, hip circumference, thigh circumference, inner and outer leg length, front and back girth length, back girth inclination, height.

2.3 Flexible Manufacturing

Flexible manufacturing is the key of mass customization. Traditional clothing production used manual cutting, sewing, and packaging. And the transactions conducted by stores. The traditional industrial chain has high labor costs and low efficiency and cannot achieve personalized clothing production. Flexible manufacturing can realize the transformation from factory to consumer model. QINGLIN QI [13] proposed that big data connects the information in the real and virtual world and realizes the information interaction and real-time sharing, which is conducive to the digital transformation of the manufacturing industry. The AI and big data are the key to flexible manufacturing. In the clothing industry, the AI can be used as an intermediary between consumers, production and logistics to realize the flexible manufacturing of clothing.

2.3.1 Flexible Digital Research and Development Technics

The AI pattern making is the key to flexible manufacturing of clothing. There are two traditional methods: manual pattern making and parametric pattern making. The manual pattern making requires the experience of pattern maker and requires repeated revisions. The patterns are placed in a specialized pattern warehouse. The second method is parametric pattern making. Parametric pattern making matched the appropriate pattern to CAD pattern making from the database. Traditional pattern making is time-consuming. In addition, traditional pattern making limited types of patterns in the existing database, which cannot provide personalized patterns of all customers. It is not really intelligent pattern making. Fig.4 shows the comparison of traditional and AI pattern making. The AI pattern making can use the AI Cloud pattern-making experts to automatically generate patterns. The pattern-making experts accumulate model design styles through machine learning and obtain customized patterns from customers. The AI Cloud will update the consumer's model data at any time and ensure the individualization of the pattern.

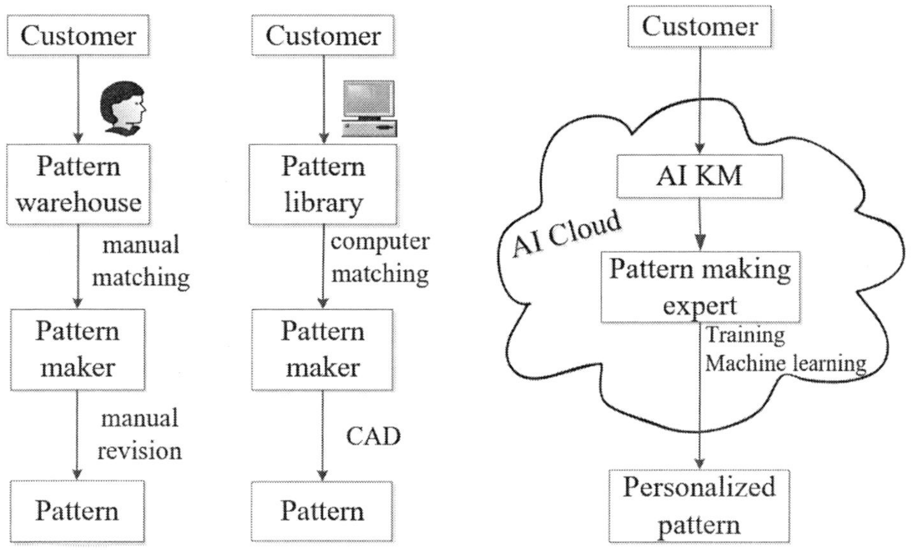

(a) Traditional pattern making (b) AI pattern making

Fig.4 The comparison of traditional and AI pattern making

In addition, flexible manufacturing of clothing is inseparable from virtual garment prototype. Traditional garment try-on is in offline shops and time-consuming. It is a repeatedly process in garment fit. However, virtual garment prototype can realize remote contactless clothing try-on, which will reduce cost and cycle of production. It can be tried on online and quickly modified. It can also evaluate the fit of virtual garment prototype to determine the fit of real garments. Virtual garment prototype will establish a virtual clothing library that includes all body types, fabrics and all clothing styles. And finally, virtual garment prototype will add virtual sewing process standards and establish rules for the evaluation the fit of the virtual garment prototype, which will enable the apparel industry to rapidly shift to mass customization.

2.3.2 Multiple Information Shared Via the RFID Technology

Internet of Things (IoT) is the key technology of flexible manufacturing. The RFID technology in the IoT can complete data collection and tracking when the IoT is applied to the apparel manufacturing industry. Traditional clothing information is recorded manually and easily lost. The garment production process is complicated and the product progress cannot be updated in real time. The RFID can quickly and accurately obtain all information about clothing customization. George Q. Huang [14] used RFID to synchronize real-time data in the manufacturing workshop to improve work efficiency. Ray Y. Zhong [15] used RFID to track and trace manufacturing objects and collect real-time production data. The garment with the smart RFID-label could offer not only tracing information of clothing design, production, and logistics for customers but also provide the easy methods for quality control and ensure the originality of clothing and realize smart clothing production in the clothing industry [16]. Mahir Oner [17] used the RFID application to track the production process of denim while managing the quality of the fabric and reducing the consumption of raw materials. Fig.5 shows the RFID technology in mass customization. The AI cloud can query and track production data from online digital R&D to workshop production. In flexible manufacturing, the AI Cloud can put data about design, pattern making, virtual garment fit, production and quality control into the RFID tags for AI sorting. In AI logistics, the AI cloud can import inventory data and logistics status into RFID tags and automatically match logistics vehicles by scanning RFID tags. Customers can scan the RFID tags with smart devices and get all the information about design, production and logistics. The AI Cloud will realize information interaction among customers, manufacturing, and logistics, which can save clothing customization cycle and realize flexible manufacturing. Meanwhile, the RFID tags will trace the brand and supplier information to ensure the brand originality. In addition, the MC will use RFID to connect clothing to AI homes, such as AI washing machines, which can automatically identify clothing fabric information for safe washing. The RFID can autonomously exchange information, trigger actions, and realize fully automated manufacturing.

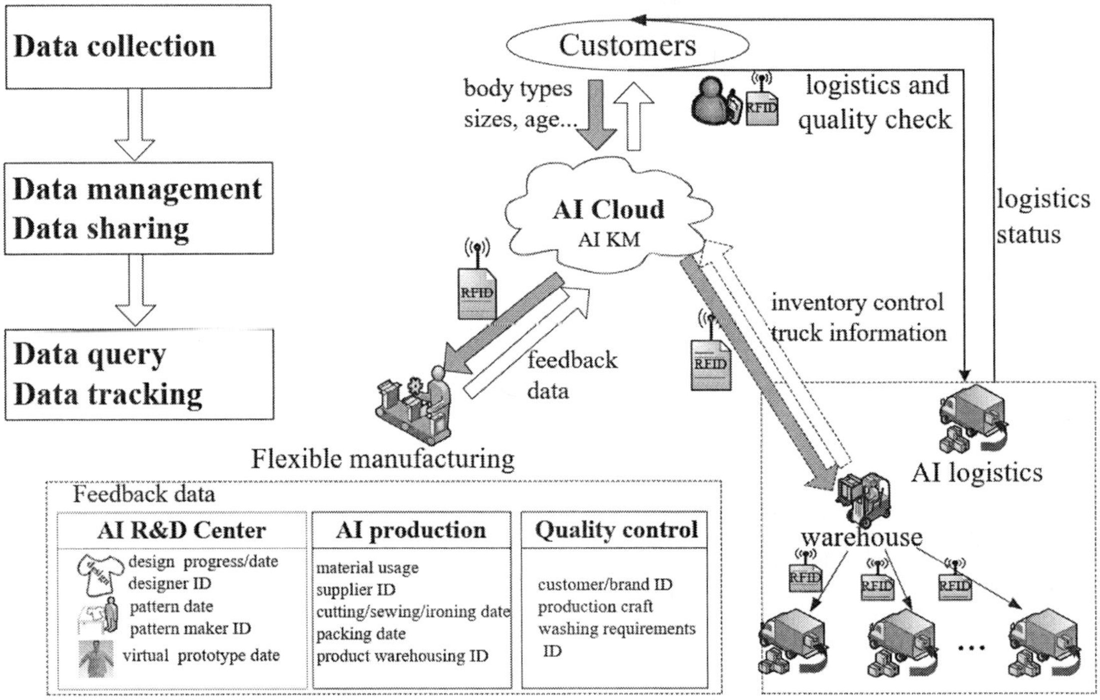

Fig.5 The RFID technology in mass customization

2.4 The AI Logistics

Traditional garment logistics includes the transportation and storage of materials throughout the production process, as well as the distribution process. Traditional logistics need manual sorting and rely on warehouse to deliver clothing to customers. Traditional logistics can be transmitted to retailers or clothing stores for sales through third-party logistics companies, which increases the cycle and cost of clothing customization. Therefore, the AI logistics can deliver garments directly to consumers, reducing cycle time and cost, and improving customer satisfaction. Lee, Zhang [18] used the IoT to realize the warehousing inspection and tracking of raw materials, semi-finished products and finished products. C.K.M. Lee [19] proposed an intelligent warehouse management system, which can provide support for the smart logistics of Industry 4.0. The AI logistics generally has functions such as perception, decision-making and feedback. The perception function can use RFID tags to automatically identify clothing packaging, storage, and vehicle information. Decision-making function can achieve logistics management and distribution. The feedback function can understand the status of logistics and transportation in real time [20]. Fig.6 shows the comparison of traditional logistics and AI logistics. The AI Logistics will use the AI Cloud to sort and match shipment. It can automatically match logistics vehicles for garment delivery by RFID tags data. The AI logistics can use the AI Cloud to obtain warehouse management information, customer addresses, and logistics status. Customers can inquire and track clothing quality and logistics by scanning RFID tags. With the application of AI logistics, The AI promoted mass customization will realize customer's personalized customization.

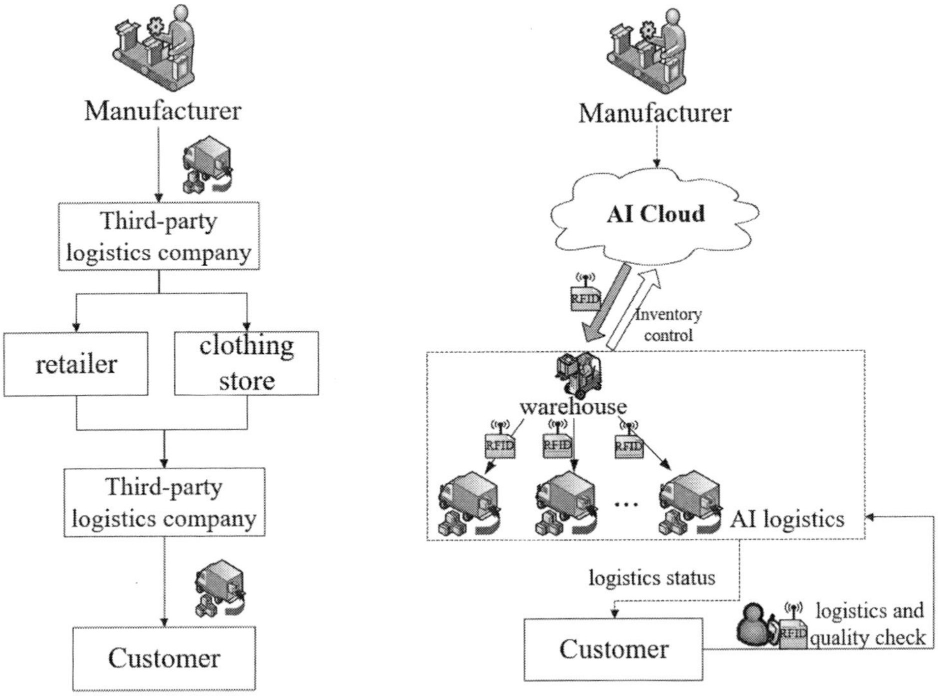

(a) Traditional logistics (b) AI logistics

Fig.6 The comparison of traditional logistics and AI logistics

3. Conclusion

The paper discusses the key technologies of AI promoted mass customization of clothing in terms of the knowledge management of customers, AI recommendation, AI measurement and flexible digital research and develop technics, which includes AI pattern making, virtual garment prototype. The AI Cloud, AI KM and AI recommendation by deep learning of customer data including body type, occupation, and preferences, which can recommend highly personalized garment for customers. The AI pattern making can use the AI cloud to learn from the experience of pattern experts to generate automatically personalized patterns. Virtual garment prototype establish virtual fit evaluation standards, which can reduce production cycles and costs. And the RFID is applied to the flexible manufacturing and AI logistics, including design, production date, quality control, and customer feedback, which will ensure the originality of apparel brands. The AI Promoted mass customization could offer online and digital production processes which are a promising option for garment industry to solving the challenges that COVID-19 had brought. It will be the crucial technology for promoting the competition ability of the garment industry in future industry 4.0.

4. Acknowledgement

The paper was supported by project A-2020-40 Fund for great project - Mass customization of garment development for Olympic game.

References

[1] Zheng T, Ardolino M, Bacchetti A, et al. The applications of Industry 4.0 technologies in manufacturing context: a systematic literature review. International Journal of Production Research: 2020; 59 (6): 1922-1954.

[2] Wu G, Zhong Y, Hu P, et al. Automatic body measurement based on slicing loops. International Journal of Clothing Science and Technology: 2018; 30 (3): 380-397.

[3] Nurunnabi A, Sadahiro Y, Laefer DF. Robust statistical approaches for circle fitting in laser scanning three-dimensional point cloud data. Pattern Recognition: 2018; 81: 417-431.

[4] Yue H, Yu Y, Chen W, et al. Accurate three dimensional body scanning system based on structured light. Optics Express: 2018; 26 (22): 28544-28559.

[5] Stančić I, Musić J, Zanchi V. Improved structured light 3D scanner with application to anthropometric parameter estimation. Measurement: 2013; 46 (1): 716-726.

[6] Xiao HT, Xiao YP, Li WL, et al. Automatic human body feature extraction and personal size measurement. Journal of Visual Languages & Computing: 2018; 47: 9-18.

[7] Rogge S, Hentschel C. A multi-depth camera capture system for point cloud library. 2014 IEEE Fourth International Conference on Consumer Electronics Berlin (ICCE-Berlin): 2014.

[8] Ren L, Lu J, Feng J, et al. Multi-modal uniform deep learning for RGB-D person re-identification. Pattern Recognition: 2017; 72: 446-457.

[9] Shah J, Shah C, Sandhu H, et al. A Methodology For Extracting Anthropometric Measurements from 2D Images. 2019 International Conference on Advances in Computing, Communication and Control (ICAC3): 2019.

[10] Gan XY, Ibrahim H, Ramli DA. A simple vision based anthropometric estimation system using webcam. Journal of Physics: Conference Series: 2020; 1529.

[11] Pesce M, Galantucci LM, Lavecchia F. A 12-camera body scanning system based on close-range photogrammetry for precise applications. Virtual and Physical Prototyping: 2015; 11 (1): 49-56.

[12] Peyer KE, Morris M, Sellers WI. Subject-specific body segment parameter estimation using 3D photogrammetry with multiple cameras. PeerJ: 2015; 3: e831.

[13] Qi Q, Tao F. Digital Twin and Big Data Towards Smart Manufacturing and Industry 4.0: 360 Degree Comparison. IEEE Access: 2018; 6: 3585-3593.

[14] Huang GQ, Zhang YF, Jiang PY. RFID-based wireless manufacturing for real-time management of job shop WIP inventories. The International Journal of Advanced Manufacturing Technology: 2008; 36 (7): 752-764.

[15] Zhong RY, Dai QY, Qu T, et al. RFID-enabled real-time manufacturing execution system for mass-customization production. Robotics and Computer-Integrated Manufacturing: 2013; 29 (2): 283-292.

[16] Lee CKH, Choy KL, Ho GTS, et al. A RFID-based Resource Allocation System for garment manufacturing. Expert Systems with Applications: 2013; 40 (2): 784-799.

[17] Oner M, Ustundag A, Budak A. An RFID-based tracking system for denim production processes. The International Journal of Advanced Manufacturing Technology: 2017; 90 (1): 591-604.

[18] Lee CKM, Zhang SZ, Ng KKH. Development of an industrial Internet of things suite for smart factory towards re-industrialization. Advances in Manufacturing: 2017; 5 (4): 335-343.

[19] Lee CKM, Lv Y, Ng KKH, et al. Design and application of Internet of things-based warehouse management system for smart logistics. International Journal of Production Research: 2018; 56 (8): 2753-2768.

[20] Tang X. Research on Smart Logistics Model Based on Internet of Things Technology. IEEE Access: 2020; 8: 151150-151159.

E-commerce Live Broadcast Influencing Factors on Impulsive Clothing Purchases

Hui-Ze Zhang[1], Qiu-Yue Wang[2*]

Business School, Beijing Institute of Fashion Technology, Beijing, 100029, China

*Corresponding author's email: sxywqy@bift.edu.cn

Abstract

The COVID-19 pandemic in 2020 has expedited the rapid development of the online economy. E-commerce live streaming has become a new way of online shopping. This paper studies consumer impulse buy factors of e-commerce live broadcasts. Based on SOR theory, the theoretical model and research hypothesis are put forward, and the conclusions are drawn through the empirical analysis of consumer impulse characteristics, preferential price, time pressure and live broadcast anchors. These features all have a positive impact on consumers' impulse buying intentions and heightens the perceived value of products. This paper proposes that we should attach importance to the training and operation of live anchors to enrich product information and promotional activities, improve the perceived value of consumers for effective marketing to enhance consumer's purchasing impulse.

Key words: Impulse Buying; E-commerce Live Broadcast; Purchase Intention

1. Introduction

The China Internet Network Information Center (CNNIC) released the 47th statistical report on the development of China's Internet network in February 2021. The report shows that as of December 2020, the number of online shopping users in China has reached 782 million, accounting for 79.1% of the total number of Internet users. The number of online live users reached 617 million, of which 388 million were e-commerce live users, accounting for 39.2% of the total number of internet users. [1] As a new way of online shopping, " live broadcast and e-commerce " has grown rapidly in the pandemic.

According to the survey data of China Consumer Association from January to March 2020, among the preferences of live shopping categories, 63.6% of consumers purchase clothes. [2] Clothing is sold on Tiktok, Taobao, Kwai and other mainstream sales platform and has a leading position in the live industry market share.

Scholars' research on impulse buying began in the 1950s, DuPont researchers defined impulse buying as "unplanned buying". Early scholars' research on impulse buying was based on this. With the deepening of research, scholars' understanding of impulse buying has deepened, and its definition has been revised. Many scholars such as Clover (1950), DuPont researchers and Applebesum (1951) all came to the conclusion that impulse buy is consumers' purchase behavior when they are stimulated in shopping stores. Scholars such as Wood (1998) further elaborated that impulse buying is unplanned, irrational, quick response, and strong emotional response shopping behavior.[3]

Wang Qingsen (2008) is one of the earliest scholars in China who began to pay attention consumers' impulse buying behavior in the online shopping environment, mainly based on the impulsive buying theory in the traditional environment, combined with the characteristics of online shopping environment. This paper studied the influence of website characteristics and consumer characteristics on consumers' online impulsive buying intention and behavior. [4] He Jianhua (2013) analyzed the influence of situational factors, marketing stimulation, personality traits and other factors on consumers' online impulsive buying behavior.[5] Chen Weidong (2018) and other scholars conducted research on the post purchase behavior of consumers' online impulsive buying.[6]

In 2020, under the influence of the COVID-19 pandemic, China's domestic enterprises normalized e-commerce live broadcasting. Domestic scholars have began to study consumer behavior and its influencing factors in the context of webcast. For example, Zheng cancan (2020) found that promotion discount, perceived interactivity and impulsivity had a significant role in promoting consumers' impulsive buying behavior in the context of e-commerce live broadcast. [7] Tian Xiuying (2020) believes that in the context of e-commerce live broadcast, price promotion, time pressure, interpersonal interaction and visual attraction will have a direct positive impact on consumers' impulse buying desire. [8] Wang Tong (2020) confirmed that the three variables of content quality, entertainment and opinion leader of e-commerce live broadcasting have a significant positive impact on consumer trust, perceived functional value and perceived emotional value.[9]

2. Hypothesis

The SOR (stimulus, organism and response) model was proposed in 1974 by psychologists Mehrabian and Russell. This model shows that consumers' purchase behavior is caused by stimulation, which includes both internal psychological and physiological factors as well as stimulation from external environment. These kinds of stimulation (S) will make the body (O) produce corresponding psychological changes or cognitive reactions, leading to corresponding reactions (R).[10]

Based on the SOR model and the characteristics of e-commerce live broadcast, this paper constructs a theoretical model of this study, taking into consideration of the impulsive characteristics of consumers, preferential price, time pressure and live broadcast anchor as incentive factors which are the independent variables; perceived value as physical response, namely intermediate variables; impulse purchase intention as behavioral response, namely dependent variables. It puts forward the following hypotheses:

1) The impact of each influence on perceived value

H1: Impulsive characteristics of consumers on the perceived value of clothing

H2: Preferential price has a positive impact on the perceived value of clothing

H3: Time pressure has a positive impact on the perceived value of clothing

H4: Live broadcast anchor has a positive impact on the perceived value of clothing

2) The impact of perceived value on impulse buying intention

H5: Perceived value has a positive impact on the impulse buying intention of clothing consumers

3) The influence of various factors on impulse buying intention

H6: Impulsive characteristics of consumers have a positive impact on impulsive buying intention of clothing consumers

H7: Preferential price has positive effect on impulse buying intention of clothing consumers

H8: Time pressure has positive effect on impulse buying intention of clothing consumers

H9: Live broadcast anchor has positive effect on impulse purchase intention of clothing consumers

4) The intermediary role of perceived value

H10: Perceived value plays an intermediary role in the influence of consumer impulse traits on impulse buying intentions

H11: Perceived value plays an intermediary role in the influence of price preference on impulse buying intention

H12: Perceived value plays an intermediary role in the influence of time pressure on impulse buying intentions

H13: Perceived value plays an intermediary role in the influence of live broadcast anchors on impulse buying intentions

3. Methodology

This paper adopts the method of questionnaire survey, referring to the relevant literature and the mature scale, according to the actual needs of the research. The variable measurement scale was formed, the specific scale items are shown in Table 1.

Table 1. Item of variable measurement scale

Variable	Scale Item	Source
Impulsive characteristics of consumers	IC1: Sometimes I buy things just to experience the fun of shopping, but I do not really need it. IC2: I rarely think too much about making a purchase decision. IC3: If I do not buy something I like, I always have a feeling of insatiability.	Jing Fengjie (2005)[11]
Preferential price	PP1: The discount promotion of the merchants in the live broadcasting room can attract my attention. PP2: The products sold in the live room are at exclusive special prices, which will have a certain impact on my impulse buying intention. PP3: In the process of e-commerce live broadcast, if I often issue coupons or draw prizes, I will be more attracted to watch. PP4: In the e-commerce live broadcasting room, the stronger the promotional activities, the stronger my impulse buying intention.	Wang Tong (2020)[9]
Time pressure	TP1: When watching the live broadcast, I feel that the promotional time stipulated by the host is generally short. TP2: When I watch the live broadcast, I feel like I have less time to decide whether to buy a promotional item. TP3: When watching the live broadcast, I feel that the second kill products in the live broadcast room will soon reach the deadline of promotion.	Tian Xiuying (2020)[8]
Live broadcast anchor	LA1: The more knowledge the anchor has on the product, the more I trust the quality of the product he recommends. LA2: The more influence an anchor has in this field, the more trust I have in the product. LA3: The anchor has a reputation in the society. I think the products he recommends will be more secure.	Wang Tong (2020)[9]
Perceived value	PV1: I think the product quality in the e-commerce live broadcasting room is trustworthy. PV2: I think the products recommended in the e-commerce live broadcasting room have better quality. PV3: Buying the products introduced in the e-commerce live broadcasting room can bring me joy. PV4: I am excited to buy the products introduced by the e-commerce live broadcasting room.	Wang Tong (2020)[9]
Impulse buying intention	II1: I have advance plan. I make my purchasing decision when I enter the studio. II2: I buy products without thinking it over. II3: During the shopping process, I will have strong purchasing intention for some goods that I did not intend to buy.	Tian Xiuying (2020)[8]

4. Results Analysis

In order to ensure that the results of the questionnaire are true and effective, a small-scale pre survey was conducted before the questionnaire was widely collected. The reliability and validity of the scale items were tested. The final questionnaire was formed and distributed. A total of 219 questionnaires were collected and the subjects were selected through the first two screening, In all samples, 48 invalid data were excluded, as they did not watch live broadcast of e-commerce or purchased clothing products through live broadcast of e-commerce. A total of 171 valid questionnaires were collected.

78.9% of the participants were female and 21.1% were male. In terms of age, the proportion of the people under 20 years old is 20.5%, 21-29 years old accounts for 45.8%, 30-39 years old accounts for 26.9%, and over 40 years old accounts for 5.8%; In terms of education background, high school and below account for 5.3%, junior college accounts for 9.4%, undergraduate course accounting for 77.8%, graduate students and above 7.6%. In terms of monthly average expenditure, 4.1% is less than 1000 yuan, 57.9% is 1000-3000 yuan, 3001-5000 yuan accounts for

23.4%, and over 5000 yuan accounts for 14.6%. In terms of the length of live broadcast per week, 56.7% of consumers watched live broadcast for 0-5 hours a week, 31.6% in 6-10 hours, 9.4% in 11-15 hours and 2.3% in 16 hours.

4.1 Reliability and Validity Analysis

It can be seen from the table below that only the reliability coefficient of time pressure is less than 0.7 (the specific value is 0.644), and the reliability coefficients of other variables are higher than 0.7. Therefore, the measurement of variables in the questionnaire has good reliability, and the KMO values of all variables are greater than 0.6. The factor load coefficient and common degree values of all research items are higher than 0.4, which indicates that the information of research items can be effectively extracted. Therefore, the questionnaire data also has good validity, which can be further analyzed.

Table 2. The results of reliability and validity analysis of research variables

variable	Question item	Factor load factor	Commonality	KMO	Cronbach α
Impulsive characteristics of consumers	IC1	0.871	0.484	0.614	0.714
	IC2	0.839	0.565		
	IC3	0.672	0.774		
Preferential price	PP1	0.784	0.615	0.780	0.778
	PP2	0.734	0.539		
	PP3	0.778	0.606		
	PP4	0.805	0.648		
Time pressure	TP1	0.802	0.643	0.645	0.644
	TP2	0.758	0.574		
	TP3	0.735	0.541		
Live anchor	LA1	0.804	0.646	0.682	0.757
	LA2	0.854	0.73		
	LA3	0.802	0.644		
Perceived value	PV1	0.842	0.709	0.810	0.835
	PV2	0.809	0.655		
	PV3	0.823	0.677		
	PV4	0.799	0.638		
Impulse buying intention	II1	0.799	0.638	0.688	0.739
	II2	0.814	0.663		
	II3	0.825	0.681		

4.2 Correlation Analysis

The results of correlation analysis among the influencing factors, perceived value and impulse buying intention are shown in the table below:

Table 3. The correlation analysis results between the influencing factors and impulse buying intention

Pearson relevance	Impulse buying intention	Perceived value	Live anchor	Time pressure	Preferential price	Product information	Impulsive characteristics of consumers
Impulse buying intention	1						
Perceived value	0.590**	1					
Live anchor	0.368**	0.606**	1				
Time pressure	0.428**	0.516**	0.588**	1			
Preferential price	0.343**	0.452**	0.612**	0.703**	1		
Impulsive characteristics of consumers	0.662**	0.515**	0.324**	0.446**	0.398**	0.409**	1

*p<0.05, ** p<0.01

The absolute value of the correlation coefficient between each research variable is less than 0.8, so it can be preliminarily derived that there is no collinear problem between each research variable. There is a significant correlation between impulse buying intention and perceived value, impulsive characteristics of consumers, preferential price, time pressure, live anchor, and the correlation coefficient values are 0.590,0.662, 0.343, 0.428, 0.368 respectively. The correlation coefficient values are greater than 0, which means that there is a positive correlation between variables. Therefore, hypothesis H1, H2, H3, H4, H5, H6, H7, H8 and H9 are preliminarily verified.

4.3 Regression Analysis

Each hypothesis was tested to explore the relationship between each independent variable, perceived value and impulse buying intention. The results obtained were sorted out as shown in Table 4. Among them, the regression results of influencing factors and perceived value are shown in the regression analysis (1); The regression results of influencing factors and clothing impulse buying intention are shown in the regression analysis (2).

Table 4. Regression analysis results

variable	Regression coefficient	R^2	F
Perceived value (2).	0.668**	0.348	90.311
Impulsive characteristics of consumers (1)	0.453**	0.265	60.867
Impulsive characteristics of consumers (2)	0.660**	0.439	132.017
Preferential price (1)	0.551**	0.204	43.358
Preferential price (2)	0.474**	0.118	22.563
Time pressure (1).	0.583**	0.266	61.321
Time pressure (2)	0.548**	0.183	37.897
Live anchor (1).	0.627**	0.367	97.878
Live anchor (2).	0.431**	0.135	26.422

*$p<0.05$, ** $p<0.01$

According to the analysis results, it can be concluded that the p value of each influencing factor variable is less than 0.01, and the correlation coefficient is positive, indicating that there is a significant positive relationship between them. In conclusion, H1, H2, H3, H4, H5, H6, H7, H8, and H9 are all valid in this study.

4.4 Intermediary test Analysis Results

In order to further test the mediating effect of perceived value, this paper uses bootstrap calculation method to test the significance of the mediating effect. If the 95% confidence interval of these path coefficients does not include 0, then the mediating effect exists. It can be seen from table 5 that the 95% confidence interval of the assumed path does not include 0. Therefore, the mediating effect of the four paths is verified.

Table 5. Intermediary test results

item	c	a	b	a*b	95% BootCI	c'	The conclusion of the test
Impulsive characteristics of consumers => perceived value => impulse buying intentions	0.660**	0.453**	0.384**	0.174	0.102 ~ 0.262	0.486**	Some intermediaries
Preferential price =>perceived value=> impulse buying intentions	0.474**	0.551**	0.619**	0.342	0.132 ~ 0.368	0.133	Full mediation
Time pressure => perceived value=> impulse buying intentions	0.548**	0.583**	0.570**	0.333	0.156 ~ 0.371	0.215*	Some intermediaries
Live anchor => perceived value=> impulse buying intentions	0.431**	0.627**	0.657**	0.412	0.227 ~ 0.497	0.019	Full mediation

5. Conclusion

The results of this study show that consumers' impulse characteristics, price concessions, time pressure and live anchor can positively affect clothing consumers' impulse purchase intention through consumers' perceived value.

Based on all the above conclusions of this study, the following four points are drawn about the marketing enlightenment of improving the sales of clothing products in e-commerce live broadcasting, to provide reference for clothing enterprises or brands carrying out e-commerce live broadcasting.

1) From the perspective of consumers it is important to introduce products thoroughly whether from style to fabric, or from the details to a general picture of the whole product. also It is important to answer consumers' questions in time, eliminate consumers' doubts, enhance consumers' reliability and trust in product information, and help consumers make better shopping decisions.

2) In the process of live broadcasting, it is important to make full use of time limited preferential activities such as second kill, rush purchase and lucky draw. The anchor repeatedly emphasizes the strength of price preference and time limit to stimulate consumers' desire to buy.

3) Focus on the training and operation of anchors. Improve the anchor's ability to explain products and control the field and adaptability in the live broadcasting room, so as to create a good staffing for the anchor and create a high popularity and good reputation.

References

[1] China Internet Network Information Center (CNNIC). The 47th China Statistical Report on Internet Development: 2021-02-03.

[2] First financial business data center (CBN Data). Research Report on live e-commerce in apparel industry: 2020-10-30.

[3] Zou LS. The Research on Emotion of Post-impulsive Buying online. Nanjing University: 2017.

[4] Wang QS. Research on online impulse buying based on website characteristics and individual characteristics of consumers. Zhejiang University: 2018.

[5] He JH. Analysis on Influencing Factors of consumers' online impulsive buying behavior. Consumer Economic: 2013; 029 (006): 46-50.

[6] Chen WD. Research on the behavior of consumers after online impulsive purchase. Tianjin University of Finance and economics: 2018.

[7] Zheng CC. Research on Influencing Factors of Consumer Impulsive Purchase Behavior Based on Online Shopping Live Situation. Anhui University: 2019.

[8] Tian XY. A Study on Impulse Buying Desire of Live E-commerce Consumers Based on SOR Model. Dalian Maritime University: 2020.

[9] Wang T. Research on Consumers' Purchase Intention in the Context of E-commerce Live Broadcast. Central University for nationalities: 2020.

[10] Bai YL. Consumer Psychology. Beijing: People's Posts and Telecommunications Press: 2018.

[11] Jing FJ, Yue HL. Research on Chinese Consumers' Impulsive Purchase Intention Scale. Financial Research: 2005 (05); 37-40

The Development of Virtual Fashion in the Global Clothing Industry

Shi-Yu Liu[1], Guo-Xiang Yuan[1,2*]

[1]*College of Fashion and Design, Donghua University, Shanghai, 200051, China*
[2]*Key Laboratory of Clothing Design and Technology - Ministry of Education, Donghua University, Shanghai, 200051, China*

*Corresponding author's email: yuanguoxiang@gmail.com

Abstract

Affected by the COVID-19 epidemic, the global economy has declined. However, the overall clothing industry has been hit hard. Virtual fashion has taken this opportunity to develop rapidly. Due to the pandemic, consumers behavior has changed to become more susceptible to the influence of social media and paying more attention to the social responsibility of brands. Virtual fashion represented by game collaboration, mobile apps, virtual avatars, and virtual clothing has gained traction in the post-epidemic era. In this paper, ten international brands related to virtual fashion were compared and analyzed by data collection. It indicates that the traditional clothing industry has ushered in a new transformation point with the help of modern cutting-edge technology, and the new virtual fashion will also enter our daily life from the experimental field.

Keywords: Virtual Fashion; Fashion Retailing; Digital Clothing; Virtual Economy; Fashion Industry

1. Introduction

Affected by the COVID-19 pandemic, the world economy has been gravely wounded and experienced a decline. According to the United Nations' World Economic Situation and Prospects 2020 and 2021, the global economy shrank by 4.3% in 2020, more than 2.5 times that of the 2009 global financial crisis. The latest report predicts that the global economy will rebound by 4.7% in 2021, offsetting the 2020 losses. Global GDP will shrink by nearly 1% in 2021. The United Nations indicates that due to the COVID-19 pandemic c. The global apparel market fell by 15.2% in 2020 compared to 2019, with a total loss of 297 billion US dollars. [1]

Based on the finding, it indicates that the research on virtual fashion either focuses on software such as CLO, VR, AR technology and studies virtual technology from the professional technical level. [2-4] focuses on the theoretical part, not combined with the latest reality background-COVID, and lacks actual case analysis. It is slightly inadequate in practical operations and applications. Others have mainly studied the key points such as [5-7], Or lack of support for the sudden economic downturn in the offline fashion industry [8-10].

Therefore, this paper includes virtual mobile phone applications, virtual fashion avatars, virtual clothing, E-commerce, virtual fashion branding. Each aspect will analyze ten international virtual fashion-related cases with examples, a total of 30 cases. Analysis of the clothing industry's development trend in the post-epidemic era is made through many cases and data analysis. The conclusions drawn are as follows: In the post-pandemic era, the fashion industry should pay attention to potential opportunities for cooperation with mobile game marketing and design; in-depth research and development of AR and VR technologies, increase investment in scientific research, and seize market opportunities; focus on broadening horizons. Companies are appropriately tilting their marketing focus to online channels, developing e-commerce, and developing offline simultaneously, positioning offline physical stores more as display and concept expression. Virtual fashion will be developed in the post-pandemic era. Online virtual methods can provide excellent display and promotion channels for more newcomers to design brands, and it can also make old international brands rejuvenated with the help of cutting-edge technology. Virtual fashion can also enable consumers and clothing industry players to rethink the entire clothing industry after experiencing the pandemic. In the future, everyone will consume more rationally and have a greater sense of social responsibility in the field of clothing. At the same time, they will be more susceptible to visual and online social media to stimulate consumption.

2. Method

Before the case analysis, it is necessary to define the virtual. Virtual fashion refers to the use of high-tech technologies, such as VR (Virtual Reality includes computers, electronic information, and simulation technology. The primary realization method is that the computer simulates the virtual environment to give people a sense of environmental immersion), CLO (3D clothing plate-making virtual test Clothing system), and other fashion activities that imitate real models, clothing, scenes, etc. It should be noted that "virtual fashion" is different from the concept of "online." "Virtual things" do not exist in reality but can be transformed into things that exist in reality, while "online" is through the Internet. Moreover, "online" uses the Internet as a channel to promote things that already exist in reality.

This part collects ten virtual fashion cases covering many countries worldwide, such as China, the United States, Japan, and other countries. It is divided into four parts: Cooperation with game industries mobile applications, virtual idols, avatars, and virtual clothing: the development and practical application of virtual fashion in the post-epidemic era.

2.1 Cooperation with the Games

2.1.1 Moschino × The Sims 4

Moschino collaborated with the lifestyle simulation game The Sims to launch a capsule series in April 2019. The Moschino x The Sims collaboration is now available to shop online. Creative director Jeremy Scott designed and produced ready-to-wear garments with Moschino's iconic patterns, such as teddy bears and vintage chains. Moschino-branded Freezer Bunny hoodie is the iconic flagship product in this collaboration

2.1.2 NIKE × Fortnite

Nike has turned the idea of exclusively selling sneakers in the field of online games into reality. Falling into the real world is known for causing long queues, causing a brand buzz, and driving high resale prices of the products offered, creating an exclusive and consumer hype engine. Nike hopes to reach with approximately 250 million registered players." "Fortnite" audience copy. [11] In May 2019, NIKE launched two Jordan virtual sneakers specially designed for the game. Players will change the game characters into Jordan brand clothing and start a racing parkour showdown on the street. Epic and Fortnite collaborated with Nike Air Jordans for an in-game collaborative campaign. The partnership included cosmetic items purchasable in in-game stores and loot drops that could contain branded apparel and Downtown Drop for a limited time. The exclusive Nike Air Jordan outfits are available in the Hang Time Bundle, purchased via the in-game shop for 1,800 V-Bucks. The bundle includes two outfits, Grind and Clutch, and includes the iconic Nike Air Jordan 1s trainers. Additionally, the skin also comes with a challenge pack, which can unlock different color options.[12]

2.1.3 Animal Crossing × Multiple Fashion Brands

On March 20, 2020, the simulation business game "Animal Crossing" broke the sales record for the first week in Switch game history as soon as it was launched. By the end of April, its global sales had reached 14.31 million copies. [13] At present, more than 20 fashion and non-fashion brands and platforms worldwide have built "Brand Island" in this game to promote themselves. For example, the Chinese brand JNBY, the internationally renowned cross-border e-commerce NET-A-PORTER, Valentino's luxury brand, and the famous GQ magazine. Among them, the e-commerce NET-A-PORTER has re-engraved the spring and summer new products of several Chinese designer brands in the game. These products not only appear in the game but can also be purchased in real life. The web designer Shel Orrock and Richmond Young, who has played this game since 2002, created the @crossingtherunway account. Their crossover works include Chanel, Balenciaga, JW Anderson, etc.

2.2 Mobile Apps

This part selects global clothing brands, including social media software, VR technology, virtual fashion bloggers, and online fashion week as case studies. Retailers need to understand how to integrate social media into CRM (customer relationship management) strategies and satisfy the needs of social networkers who engage brands in social media to maximize their profits and create keen brand awareness. [14] The clothing retail industry is facing the COVID-19 epidemic. Under such circumstances, it is necessary to change your marketing channels and focus on capital investment flexibly.

2.2.1 Gucci and other Brands × Drest

Drest and Gucci have cooperated to launch a modeling game application based on modeling models such as Unia Pakhomova. The user becomes a professional stylist in a fashion magazine and shoulders the mission of making a designer look suitable for his avatar. Besides, if people like what they see, users can purchase clothes through the program. The fashion games involve styles and allow gamers to participate in the entire creative process of fashion editorials. Provide users with dress codes, seasonal themes, and other information (such as colors), and then they must create a look at them. As players level up in the game, they can dress up their avatars and choose from various locations and backgrounds to purchase in the App or take photos as a reward. Players can also combine their appearances in a mood board form. Users can choose from five avatars of different sizes. The clothing selection covers about 130 high-end brands, including Burberry, Gucci, Prada, Stella McCartney, and Valentino.[16]

Fig.1 Unia Pakhomova, the face of Gucci, will appear as a game avatar in Drest

According to Drest, pieces from the company's fall 2019 collection will appear as cosmetic items within the game. Over the first four weeks, the center of the games on themes of time and place, including"Sixties London, " "Seventies Paris Rive Gauche, " "Eighties New York, " and "Nineties L.A. " Gameplay involves dressing up avatars for on-location photo shoots, with more backgrounds available for purchase or as rewards for progressing through the game. The so-called "still" games allow people to look together in a mood board using the in-game features.

2.2.2 Nike x Zepeto

Users can create their virtual cartoon images through the software, dress them up, and take photos with friends or even strangers to socialize. Nike has also cooperated with Zepeto; the store includes some co-branded shoes, Air Max 720, and even On Air's winning shoes from Tokyo, Japan. At the same time, Nike clothing and even accessories such as hats and socks were also readily available. Virtual clothes can be purchased by recharging virtual currency in the game with real money.

2.3 Virtual Avatars

Virtual avatars Lil Miquela, Noonoouri, Balmain, etc., participated in advertisement campaigns with Givenchy, Prada activities, among which blogger Lil Miquela shot Calvin Klein's advertisement. The well-known fashion magazine NYLON China also chose the mobile game "Glory of the King" virtual idol men's group Unlimited King as the cover character. In March 2020, the cover of "Vogue Italia" used an unusual model: Ida, a virtual model created by Computer Graphic. This is the first time in the magazine's history that the cover is dedicated to a "Nonexistent girl."

Fig.2 Instagram's popular virtual avatars

2.4 Virtual Clothing Brand

2.4.1 Catty Taylor

3D designer Catty Taylor focuses on creating digital clothing styles, and the clothing display effects produced will be promoted on social media. Many brands are rushing to cooperate with them, such as Balenciaga, Burberry, NIKE, a cold wall, Vetements, Off-White.

2.4.2 Fabricant, A Virtual Fashion Company

In May 2019, the world's first digital haute couture designed by The Fabricant was sold through the blockchain for $9,500. The Amsterdam virtual fashion company Fabricant has completed the design of the skirt. The way the dress is worn is a bit special. The company will take a photo of the customer and then use augmented reality technology to put a virtual dress on the customer to share the photo on social media.

Fig.3 Customers wearing virtual costumes made by Carling

2.4.3 Carlings

The Norwegian-based brand Carlings has made 3D design and employee training efforts to produce 3D products, with 120 million euros in 2018. They are well-known for their denim collections and are among the first brands to set foot to sell virtual 3D collections with prices ranging from 10 Euros to 30 Euros. [15] Consumers purchasing services need to submit personal photos, and then technicians will create personal 3D images in virtual costumes based on this.

Norway-based brand Carlings made 120 million euros in revenue in 2018, as it invests its efforts in 3D design and training employees to produce 3D collections. They were the first brands to stampede into selling virtual 3D collections and items from 10 euros to 30 euros apiece for their denim lines. Carlings CEO Ronny Mikalsen said he sees 3D collections as the future and a sustainable solution for the world, currently inundated with excess clothing waste. The digital clothing industry, currently driven by a legion of millennials interested in social networking, is expected to drive those sales and represent one percent of the fashion market share at $25 billion. Fortnite, the top money-making, free-to-play game, reportedly made an estimated $2.4 billion in 2018.

3. Results & Discussion

Based on the fashion brands listed above, the cooperation between the clothing industry and virtual reality technology has brought positive feedback from the market to the brands during the post-epidemic period.

First, because virtual fashion is operated virtually, it can be adjusted in real-time, which can reduce the waste of raw materials and practice sustainable clothing principles; second, virtual fashion can often predict a clothing's production outcome in advance, which is beneficial to a harmonious customer relationship through more efficient communication; thirdly, wearing different personal customized virtual clothes can meet the demands of modern people's social platforms.

3.1 Cooperation with the Games

First of all, according to the data provided by China Mobile Internet Database during the February 2020 epidemic, the industry with the highest increase in daily active users during the Spring Festival holiday in 2020 compared to the price, before accounting for costs, is the gaming industry, reaching a 20% growth. The 2020 Spring Festival The WeChat Mini Program, which has the most significant increase in daily active users during the holidays, is still in the game industry, reaching 38%. According to the data compiled by the Prospective Industry Research Institute, in 2019, the proportion of China Virtual Reality Industry's application in the game field reached 36.8%, and it is expected to account for 37.2% and 37.9% in 2020 and 2021, respectively, which is slowly increasing year by year. Compared with the second "video" field, which accounts for 20.5%-21.4%, the game field is about 16.3%-16.5% higher, and the Internet field is the field most frequently contacted by users.

According to the "2020 China E-sports Industry Research Report", the number of e-sports users in 2019 is 470 million, and in 2020 it will reach 520 million. In 2021, the number of this group will reach 550 million. The rise of mobile e-sports has made more and more female users come into contact with e-sports. In 2019, the proportion of female users in the Chinese e-sports group was 24%, increasing to 36% in 2020. The increase in the proportion of female users in the game field means that among e-sports users, the number of users willing to pay for clothing is increasing, and the potential customer group is gradually increasing. Cooperation with the games is a win-win business opportunity for the apparel industry and the game industry.

Table 1. Statistics of the proportion of applications in China's virtual reality industry

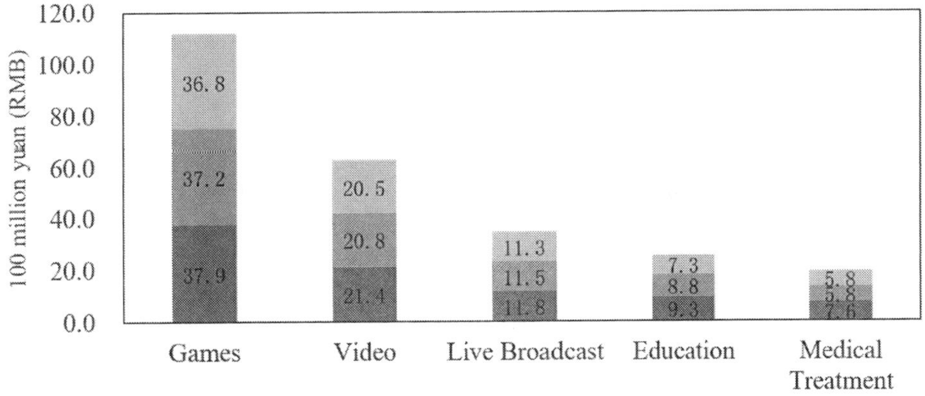

In China, cut to 2020, 63 percent of American mobile gamers are women, and women have always been the main force of consumption in the fashion industry. This once again proves the possibility of the fashion industry being combined with the game industry. Moreover, worldwide consumers' spending on mobile games reached 60billion in 2018. (Table.3) This figure will continue to grow in the future, bringing in more revenue.

Survey data from Statista shows that of 1,013 individuals aged 19-49, 44% were "very interested "in VR, while 26% of people were "quite interested." Right now, gaming seems to be the big winner with VR and AR. During the post-epidemic period, the real economy has been severely hit, and online digital technologies such as VR and AR are bound to develop. Young and middle-aged people have shown a strong interest in virtual technologies conducive to developing and promoting virtual fashion. Further data from Statista shows that revenues generated from VR gaming sales will reach an estimated $22.9 billion by 2020, up from the $660 million generated in 2015.

As for the brands mentioned in the previous method, the cooperation between the fashion industry and the game has increased sales and expanded its exposure during the post-pandemic period of the market downturn, especially for the younger generation of Z consumers. They did good publicity. The following table shows the economic benefits of brands through the design of virtual clothing for game characters.

3.2 Mobile Application Activities and Virtual Avatars

With the powerful functions of mobile phones and the development of the Internet, people spend more time on mobile phones than in previous years, and this phenomenon has been exacerbated during the pandemic. According to a survey conducted by Nielsen, an information and market measurement company, 54% of respondents said that their spending on unplanned shopping increased last year, and 80% of them said that social recommendations such as friend recommendations and social media groups had stimulated impulse buying. According to a McKinsey research report, among the WeChat users surveyed, most of the consumption of clothing and beauty, which accounts for 25% to 30% of the total consumption, is impulsive. Coincidentally, on January 9, 2020, the industry authoritative Tencent Advertising TMI Tencent Marketing Insight and Boston Consulting Group jointly released the "2020 China "Social Retail" White Paper," which shows that 56% of women are more likely to be interested in social media than men 45% of consumers; only 19% of consumers are loyal to the brand; 61% of consumers are easily planted by novel content.

Table 2. Worldwide consumers' spending on mobile games

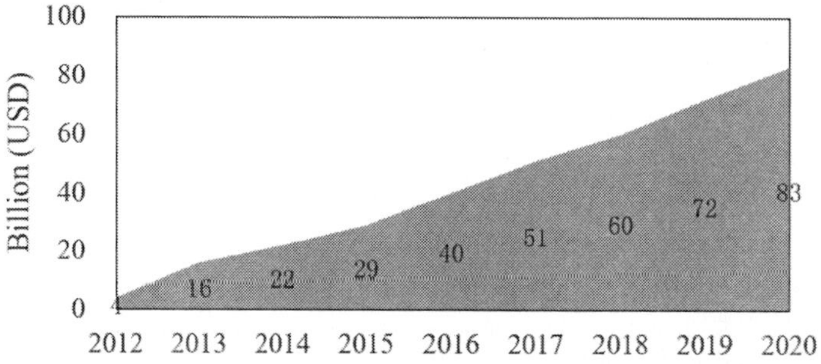

The above data shows that during the pandemic, governments worldwide have called on citizens to isolate themselves at home to reduce contact with others when they go out. Therefore, people's time to use mobile phones to socialize at home has been lengthened, and many offline activities have been changed to online activities. The original consumers have lost contact with the brand's physical stores, and browsing information on the Internet has become the primary way to communicate. Compared with offline content, online content is mainly media-based, making users more likely to be attracted by novel content and generate impulsive consumption, especially in clothing and beauty products with apparent visual effects, in the eyes. This also means that if some new brands or niche brands can grasp online marketing characteristics, they can also usher in a company's turning point.

Table 3. Cooperation with the game

Fashion Brand	Game	Results
Valentino	Animal Crossing	Tweet reposting rate has increased by 135% than usual
Moschino	The Sims	Wholesale sales posted a growth of 13.4 percent at constant exchange, contributing to 71.5 percent of consolidated sales, while sales at the company's directly-operated stores (DOS), representing 25.1 percent of consolidated sales, grew 4.8 percent at constant exchange rates. [16] 2019 The Sims sold over 200 million copies worldwide.
Nike	Fortnite	The game was downloaded more than 100 million times on iOS within five months of its launch. The Epic Games grossed $3 billion in profit in 2018, largely thanks to the popularity of "Fortnite."[17]

Moreover, with return rates for online retail at least 20% higher than brick-and-mortar store returns, virtual try-on is the perfect solution for foods and retailers struggling to get consumers into their stores until there is a coronavirus cure. Compared with physical store returns, the return rate of online retail is at least 20% higher. Virtual changing rooms are the perfect solution for brands and retailers to bring consumers into the store before in-person contact resumes.

3.3 Virtual Economy

According to China Electronic Information Industry Development Research Institute, the European virtual reality market will grow from 2.5 billion US dollars from 2017 to 2018 To US$4.73 billion. The Prospective Industry Research Institute predicts that by 2020, the European virtual reality market will reach US$11.35 billion. [18]

Whether it is China or Europe, where the virtual economy is currently at the top of the world, its market size increases year by year. In the future, under the influence of the new crown pandemic, the market size of the virtual economy will continue to expand. Virtual economy is manifested in clothing. There are probably development directions such as mobile games, mobile social applications, virtual catwalks, virtual avatars, virtual showrooms, and virtual clothing.

Table 4. Mobile application, virtual catwalks, virtual avatars, and virtual clothing

	Fashion Brands	Mobile Application Results
Virtual avatars	Virtual idol	NYLONChina used the virtual hero boy group of the Chinese mobile game "Glory of the King" to launch its first topic. The reading volume on Weibo soared to more than 60 million.
Virtual clothing	Carlings	Last November, Scandinavian retailer Carlings created a limited-edition, digital-only collection that sold out within one week. In May, a one-of-a-kind digital-only garment sold for $9,500 at a blockchain conference. The digital clothing industry is expected to drive sales and has the potential to represent one percent of the fashion market share at $25 billion.[19]

4. Conclusion

This global crisis has caused specific changes in consumers, brands, and apparel industries to varying degrees. Up to a point consumers are concerned, consumers will be more contradictory when they consume: on the one hand, they will reduce their consumption expenditure and remain rational, reducing the purchase of non-essential items; on the other hand, they will spend more time online, and relatively simple promotional channels make it easier for them to be attracted by new things on social media and make impulse consumption.

As far as the brand is concerned, in addition to focusing on the above points and resonating with consumers, the following measures are worth trying: smaller collections featuring more SKUs (Standard Product Unit) of trans-seasonal pieces; try to design for e-sports characters apparel may be co-branded with consumers to design apparel; shift the sales focus slightly from offline to online; if funds allow, try to develop beauty and apparel in the VR and AR fields; pay attention to the frontiers of trends and adjust marketing strategies according to younger consumers, particularly Gen Zs', market demographic in online spaces.

The entire clothing industry should redefine business models and build a more sustainable, progressive future to integrate all digitization forms from the supply chain and radically redefine their culture and operations to a more digital mindset.

5. Acknowledgment

This work was supported by the China Education Association for International Exchange (The China-CEEC Joint Education Project of Higher Education Institutions - Sustainable Surface Treatment for Garment and Design Innovation - 201911, ZX202005070044).

References

[1] Kaiser C, Fischer T, Schmeltzpfenning T, Stöhr M, Artschwager A. Case Study: Mass Customisation of Individualized Orthotics - The FASHION-ABLE Virtual Development and Production Framework. Procedia CIRP: 2014; 21: 105-110.

[2] Kawaf F, Stanbulluoglu DI. Online fashion shopping paradox: The role of customer reviews and facebook marketing. Journal of Retailing and Consumer Services: 2019; 48: p. 144-153.

[3] Lee SHN, Chow PS. Investigating consumer attitudes and intentions toward online fashion renting retailing. Journal of Retailing and Consumer Services: 2020; 52.

[4] Khandual A, Pradhan S. Fashion Brands and Consumers Approach Towards Sustainable Fashion, in Fast Fashion, Fashion Brands and Sustainable Consumption: 2019; p. 37-54.

[5] Armando E, Craparotta G. A Meta-Model for Fashion Retail Category Sales Forecasting, in Business Models and ICT Technologies for the Fashion Supply Chain: 2019; p. 79-93.

[6] Rovai S. Digitalisation, luxury fashion and "Chineseness": The influence of the Chinese context for luxury brands and the online luxury consumers experience. Journal of Global Fashion Marketing: 2018; 9 (2): p. 116-128.

[7] De Silva RKJ, Rupasinghe TD, Apeagyei P. A collaborative apparel new product development process model using virtual reality and augmented reality technologies as enablers. International Journal of Fashion Design, Technology and Education: 2018; 12 (1): p. 1-11.

[8] Herz M, Rauschnabel PA. Understanding the diffusion of virtual reality glasses: The role of media, fashion and technology. Technological Forecasting and Social Change: 2019; 138: p. 228-242.

[9] Porterfield A, Lamar TAM. Examining the effectiveness of virtual fitting with 3D garment simulation. International Journal of Fashion Design, Technology and Education: 2016; 10 (3): p. 320-330.

[10] Sayem ASM, Kennon R, Clarke N. Resizable trouser template for virtual design and pattern flattening. International Journal of Fashion Design, Technology and Education: 2012; 5 (1): p. 55-65.

[11] Young KE, Mattila P. Introduction: Fashion and smart virtual reality. Journal of Global Fashion Marketing: 2014; 5 (4): p. 267-268.

[12] Park HH, Jeon JO, Sullivan P. How does visual merchandising in fashion retail stores affect consumers' brand attitude and purchase intention? The International Review of Retail, Distribution and Consumer Research: 2014; 25 (1): p. 87-104.

[13] Nantel J. My virtual model: Virtual reality comes into fashion. Journal of Interactive Marketing: 2004; 18 (3): p. 73-86.

[14] Rich F, Maura G, Carson G, Aaron R. Redefining value and affordability in retail's next normal [J/OL]. (2020-06-15)[2020-06-18]. https://www.mckinsey.com/industries/retail/our-insights/redefining-value-and-affordability-in-retails-next-normal.

[15] The McKinsey Consumer Goods and Organization Practices, Ayush Agnihotri, Bryan Logan, Kristi Weaver, Oren Eizenman, Kate Lloyd George, and Lauren Ratner. Consumer organization and operating models for the next normal[J/OL]. (2020-05-19)[2020-06-18]. https://www.mckinsey.com/industries/consumer-packaged-goods/our-insights/consumer-organization-and-operating-models-for-the-next-normal.

[16] Maximilian B, Stefan E, Boris E, Stephanie P. Albert Serrano.A perfect storm for fashion marketplaces[J/OL]. (2020-06-8) [2020-06-18]. https://www.mckinsey.com/industries/retail/our-insights/a-perfect-storm-for-fashion-marketplaces.

[17] National data. Cumulative growth of total retail sales of consumer goods (%). [EB/OL].

[18] Redaccion EC. Negocios: E-commerce: Cu´anto gastan los peruanos que compran por Internet. Lima, Peru: El Comercio[J/OL]. (2018-11-19) [2020-07-14]. https://elcomercio.pe/economia/negocios/e-commercegastan-peruanos-compran-internet-noticia-nndc-559358.

[19] Cavero J. Nos vamos de Cybercompras? Peru: IPSOS[J/OL]. (2020-18-02)[2020-07-14]. https://www.ipsos.com/es-pe/nos-vamos-decybercompras.

Functional Values: How Usability, Quality, and Uniqueness Influence Consumer's Luxury Fashion Handbag Purchasing Intention

Tong Liu[1]; Dr.Carolina Quintero Rodriguez[1]

[1]*School of Fashion and Textiles, The Royal Melbourne Institute of Technology (RMIT) University, Melbourne, Victoria, Australia*

[*]Corresponding author's email: carolina.quinterorodriguez@rmit.edu.au

Abstract

Consumers use functional values to evaluate fashion handbags. Although previous research have determined that consumers purchase luxury goods for psychological satisfaction, other research have highlighted that a consumer's functional needs are also essential in making a purchasing decision. In this study, a handbag's functional values were measured through three specific values: usability, quality, and uniqueness. Usability value refers to an individual's evaluation of a handbag's functions as satisfying their needs, such as size and ease of use. Quality value is a consumer's evaluation of a handbag's physical qualities, such as material quality, craftsmanship, and lifespan. Uniqueness value is a consumer's need to be unique or distinct from others, which is evaluated in terms of a handbag's exclusivity. Thirty Chinese millennial luxury-fashion-brand consumers were invited to a survey that required participants to do a pairwise comparison of the importance of these three values on their purchasing decision of a luxury fashion handbag. This pairwise comparison method and data analysis were based on the analytic hierarchy process, which is a mathematically based, multi-objective, decision-making tool. The results demonstrate that quality value has the greatest impact on luxury fashion handbag purchasing decisions for Chinese millennial consumers, while the usability value has the lowest impact. This study can help luxury brands and designers better understand the influence of consumers' functional values on their intention to purchase a handbag and helps establish user-centered design criteria for developing luxury fashion handbags.

Keywords: Accessories Design; Luxury Handbags; Consumer Functional Values; Consumer Needs

1. Introduction

Designing a product to satisfy consumers' needs is key for designers and brands to position their product to consumers. Consumers' needs refer to consumers' desire to obtain benefits from a product. Previous research has demonstrated that luxury goods enable consumers to satisfy their psychological needs, such as hedonism and self-satisfaction [1, 2], which influence a consumer's purchase intention. Other research have demonstrated that a consumer's functional needs impact their purchasing decision of a luxury good [3, 4], leading designers and brands to study their consumers' functional needs in the process of product design.

Understanding consumers' functional needs requires understanding consumer functional value perception. In general, a business that understands consumer value perception gains a competitive advantage because consumer value perception influences consumers at every stage of the buying process [5, 6]. For example, consumers evaluate a luxury product on whether it meets their functional needs in order to make a purchasing decision. However, consumers' functional value perceptions are influenced by age, culture, educational background, and other factors [7]. There is little information on the influence of functional value perception of luxury handbags in design and product development. To better perceive consumers' functional evaluation of a luxury product, gain design-relevant information, and establish design criteria for developing luxury fashion handbags, luxury brands and designers should measure consumers' functional values.

Drawing from the conceptual framework of luxury value established by Wiedmann, Hennigs, and Siebels [4], the authors of this paper aimed to measure consumers' functional value perceptions of a luxury handbag. The researchers measured three key functional values: usability, quality, and uniqueness. The researchers used the analytic hierarchy process to analyze the data gathered from study participants and determine which functional value had the highest and lowest impact on luxury fashion handbag purchasing for participants, Chinese millennial consumers.

2. Literature Review

2.1 Study Theoretical Model

The multidimensional luxury framework established by Wiedmann, Hennigs, and Siebels [4] specifies the dimensions of consumers' orientations toward luxury-brand consumption. This luxury framework was widely used in previous researches that measured consumer luxury value perception. For example, a recent study [8] adjusted the individual and social value of this luxury framework to study the factors that impact Chinese millennials' luxury consumption, and also called for further research on Chinese millennials' other luxury values in order to understand the key factors that influence their purchasing intentions.

Functional value is defined as the individual evaluation of the physical benefits of luxury goods. Functional value [4] (Fig.1) encompasses a product's usability, quality, and uniqueness.

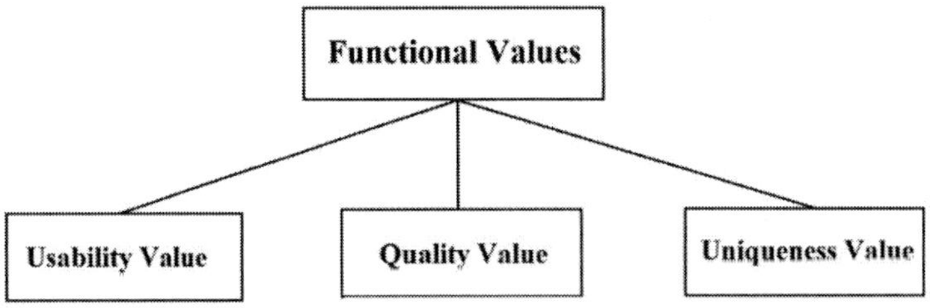

Fig.1 The luxury functional values framework

2.2 Usability Value

Every product, luxury goods included, is designed to perform specific functions [9], and usability is defined as a benefit of a product that satisfies consumer needs [4]. This benefit is often seen as the material value of luxury goods and is one of the motivations for consumers to purchase said goods [10]. In this study, usability value refers to a consumer's evaluation of whether a handbag's functions satisfy their needs. These functions include size and ease of use.

2.3 Quality Value

Quality is a factor used by consumers to evaluate the value of a product [7]. Quality value is a consumer's subjective perception of a luxury product as offering superior quality and performance [4]. Previous researches show that consumers believe luxury brands offer better quality than non-luxury brands due to their craftsmanship and material components [11, 12]. In this study, quality value is defined as the consumer's evaluation of the physical characteristics of a handbag's quality, such as material quality, craftsmanship, and product lifespan.

2.4 Uniqueness Value

Uniqueness is often associated with exclusivity, which stimulates consumers' desire to possess a rare or limited product [4]. Consumers use luxury goods to distinguish themselves from others because a luxury product is by definition not affordable to and owned by everybody [3]. Luxury-good consumers often avoid similar consumption as others to highlight their uniqueness and enhance their self-image and social image [13]. In this study, uniqueness value is an individual's evaluation of whether a handbag satisfies their need to appear distinct from others.

3. Methodology

3.1 Research Methods

A RMIT Qualtrics survey questionnaire was offered to participants. The participants completed the survey on an iPad device the researchers provided. The researchers offered face-to-face assistance if a participant had any concerns or confusion about a survey question. The researchers obtained ethics approval from the Design and Social Context College Human Ethics Advisory Network, a subcommittee of the RMIT University Human Research Ethics Committee.

3.2 Participants

The participants were Chinese millennials born between 1981 and 1995, which were identified as the valuable consumer for luxury consumption [14]. This group occupies over a quarter of the total population of luxury consumers, and they contribute to almost half of the luxury consumption in China [8].

3.3 Procedure

A questionnaire was prepared and saved in the software Qualtrics (Qualtrics, Provo, UT, version 2020). The questions were designed based on the functional values determined in the luxury functional values framework. This survey asked participants to conduct a pairwise comparison of usability, quality, and uniqueness values for luxury handbag purchasing to determine which value was most important in determining their purchasing intention. The survey was anonymous, and no personal information, such as name, phone number, and address, was retained. The questionnaire was offered in two languages, English and Mandarin.

Researchers went to Collins Street, Melbourne, Victoria, Australia, to recruit participants since this location was the site of many luxury fashion stores, and many of these stores' customers matched the participant criteria of this study. The researchers approached potential participants and invited them to participate in the survey. Potential participants were identified as individuals holding luxury shopping bags, such as Gucci, Chanel, Dior, LV, and others. Only the data of Chinese millennials who had luxury fashion shopping experiences at least once in the last six months were selected for analysis.

The researchers used the analytic hierarchy process (AHP) to do the data analysis in this study. The main calculations were carried out in Microsoft Excel.

3.4 Analytic Hierarchy Process

The AHP is a nonlinear framework for carrying out both deductive and inductive research considering several factors to arrive at a conclusion [15]. The main method used in the AHP is pairwise comparison, which can help researchers measure intangibles in relative terms, such as people perception [16]. Other researchers have successfully used the AHP to measure consumer values—for instance, measuring consumer values in luxury car selection [17], and measuring consumer values in traveling products shopping [18].

This study each part of the AHP process [19]:

3.4.1 Determine a Goal

In this study, the goal was to determine which functional values had the greatest and least impact on the luxury fashion handbag purchasing intention of participants.

3.4.2 Identify the Criteria or Subcriteria to Structure the Hierarchy

In this study, the criteria consisted of three main consumer functional values: usability, quality, and uniqueness values (Fig.2).

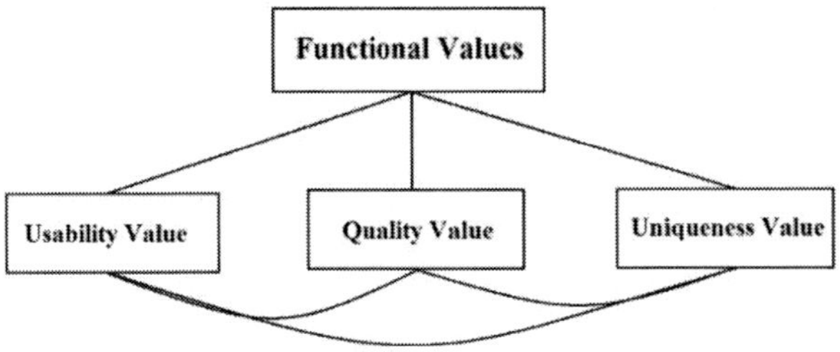

Fig.2 Pairwise comparison of usability, quality, and uniqueness values.

3.4.3 Survey Design

The questionnaire required participants to do a pairwise comparison of the three functional values (Table 1). Participants were asked the question, "Which value is more important for you when you consider purchasing a luxury fashion handbag?" Participants ranked each value on a scale from 1 to 9, where 1 = equally important, 3 = slightly more important, 5 = more important, 7 = much more important and 9 = absolutely important [19].

Table 1. The pairwise comparison of the values

Values	9	7	5	3	1	3	5	7	9	Values
Usability										Quality
Usability										Uniqueness
Quality										Uniqueness

Take one participant's response (Table 2) as an example. The participant believes the usability is more important than the quality, so the participant ticks the left box of the form (Usability, 5 points). Similarly, participant can tick the box with number 1 in the middle between the usability and uniqueness when they believe the usability is equally important as the uniqueness. And, the tick on the right box on the form (Uniqueness, 3 points) means the participant thinks the uniqueness is slightly more important than the quality.

Table 2. The example of participant's response

Values	9	7	5	3	1	3	5	7	9	Values
Usability			✔							Quality
Usability					✔					Uniqueness
Quality						✔				Uniqueness

3.4.4 Multiple Participant Data

Since the data came from a group rather than an individual, the researchers found the geometric mean of participant results to aggregate individual judgments (Fig.3).

$$w = \sqrt[n]{w_1 \cdot w_2 \cdot \cdots \cdots w_n}$$

Fig.3 The formula of geometric mean

3.4.5 Construct a Set of Pairwise Comparison Matrices

The researchers placed the pairwise comparisons in the matrix below (Fig.4).

$$A=\begin{bmatrix} 1 & a_{12} & \cdots & a_{1n} \\ \frac{1}{a_{12}} & 1 & \cdots & a_{2n} \\ \vdots & \vdots & \cdots & \vdots \\ \frac{1}{a_{1n}} & \frac{1}{a_{2n}} & \cdots & 1 \end{bmatrix}.$$

Fig.4 The pairwise comparisons matrix

Then, the researchers calculated the eigenvalue and the eigenvector (Fig.5).

$$W=\begin{bmatrix} w_1 \\ w_2 \\ \vdots \\ w_n \end{bmatrix},$$

$$W'=A\times W=\begin{bmatrix} 1 & a_{12} & \cdots & a_{1n} \\ \frac{1}{a_{12}} & 1 & \cdots & a_{2n} \\ \vdots & \vdots & \cdots & \vdots \\ \frac{1}{a_{1n}} & \frac{1}{a_{2n}} & \cdots & 1 \end{bmatrix}$$

$$\times\begin{bmatrix} W_1 \\ W_2 \\ \vdots \\ W_n \end{bmatrix}=\begin{bmatrix} W_1 \\ W_2 \\ \vdots \\ W_n \end{bmatrix},$$

$$\lambda \max=\frac{1}{n}\left(\frac{W_1}{W_1}+\frac{W_2}{W_2}+\frac{W_3}{W_3}+\cdots\frac{W_n}{W_n}\right),$$

Fig.5 The formula for eigenvalue and eigenvector

In this formula, n is the number of judgments; w1 ,w2 , ..., wn reflects the recorded value weights on the objectives. w is the eigenvector; and λmax is the largest eigenvalue of the pairwise comparison matrix.

The reason why the principal eigenvalue λmax of matrix W and its normalized eigenvector has been calculated is that the AHP derives priorities by using the eigenvalue method [20].

3.4.6 Calculate the Consistency Ratio (CR)

It is required to calculate the consistency ratio (CR) per Saaty [19] (Figure 7). This ratio demonstrates the consistency of subjective perception and the accuracy of the comparative weights if the value is 0.1 or less.

$CI=(\lambda_{max}-n)/(n-1)$ and $CR=CI/RI$

Fig.7 The CR and CI formula

CI is the consistency index.

The recommended random index (RI) values appear in Table 3 [19].

Table 3. The RI values

N	2	3	4	5	6	7	8	9	10
RI	0	0.52	0.89	1.11	1.25	1.35	1.40	1.45	1.49

3.5 Results and Discussion

A total of 30 participant responses were used in this study. Of these participants, 18 were female and 12 were male. Participants' age ranged between 25 and 39 years. All participants had luxury fashion brand shopping experiences at least once within three months. After the data was analyzed, the researchers proceeded with the AHP, first by calculating the geometric mean for each functional value per Table 4.

Table 4. The number calculated by the geometric mean method

Usability : Quality	0.596861146
Usability : Uniqueness	0.769495592
Quality : Uniqueness	1.493026979

Second, the researchers created a matrix of pairwise comparisons (Table 5).

Table 5. The matrix of pairwise comparisons

	Usability	Quality	Uniqueness
Usability	1	0.596861146	0.769495592
Quality	1.67543158	1	1.493026979
Uniqueness	1.29955251	0.66978027	1

Then, the researchers calculated the eigenvalue and the eigenvector based on the formulas in Figure 5. The results appear in Table 6.

Table 6.The result of the eigenvalue and eigenvector

	Usability	Quality	Uniqueness	Weight
Usability	0.251573334	0.263324027	0.235859086	0.2503
Quality	0.421493908	0.441181385	0.457629624	0.4401
Uniqueness	0.326932758	0.295494587	0.306511289	0.3096

Finally, the researchers calculated the CR to evaluate whether the results could be considered acceptable. The Table 7 shows the result of the CI and CR calculations.

Table 7. The result of CI and CR

CI	0.001196783	
CR	0.002301506	< 0.10

The normalized weights of usability, quality, and uniqueness values were 0.2503, 0.4401, and 0.3096, respectively. This CR was considered acceptable because the CR was 0.0023, which is less than 0.10.
These results show that quality value, which was 0.4401, was the most important value for participants, and the usability value had the least impact on their purchasing intentions of a luxury handbag (0.2503).

This study confirms the importance of quality value recognized by previous researchers [21-23]. The results further indicate that Chinese millennial consumers put greater emphasis on quality value than other functional values when determining purchasing intentions for luxury fashion goods.

Furthermore, researchers [13] [24] have highlighted that the uniqueness value of a luxury good is an important factor that distinguishes luxury goods from nonluxury goods, offering psychological satisfaction. For instance, Kim, Hsu, and Yuen [8] indicated that the need for uniqueness values influenced Chinese millennial's desire for status consumption, which eventually impacted their purchase intention of luxury fashion goods. However, this study demonstrated the uniqueness of a luxury handbag is attractive but not the most important functional value for Chinese millennial consumers. The need for quality is higher than a good's uniqueness.

On the other hand, while previous researcher [25] states that usability of a product can be considered one of the most important purchasing factors, the usability of a luxury handbag experienced less interest for Chinese millennial consumers.

4. Conclusion and Future research

Functional values such as usability, quality and uniqueness influence the consumer's luxury product purchase intention. The current study determined, through the use of the analytic hierarchy process, the influence of functional values on Chinese millennial consumers' purchasing intention of a luxury handbag. The results of this study show that quality value has the greatest impact on Chinese millennials' luxury fashion handbag purchasing decisions, while the usability value has the least impact. This information not only offers a better understanding of Chinese millennial consumers' functional values but also can be used by luxury brands to establish design criteria for developing luxury fashion handbags.

Admittedly, the total of participants was less and their data were limited to Chinese nationality only. But future research can involve more participants or expand the study into different participants, such as different ages (i.e Gen Z), and cultural backgrounds (i.e Australian and others). In addition, future research also can change the comparing values (i.e colour, pattern, style of design) to gain further understanding of consumer's value in a luxury handbag. This study model also can be considered applied in investigated different categories, such as shoes and clothing.

References

[1] Smith JB, Colgate M. Customer value creation: a practical framework. Journal of marketing Theory and Practice: 2007; 15 (1): 7-23.

[2] Jain S, Khan MN, Mishra S. Factors affecting luxury purchase intention: A conceptual framework based on an extension of the theory of planned behavior. South Asian Journal of Management: 2015; 22 (4): 136.

[3] Vigneron F, Johnson LW. Measuring perceptions of brand luxury. Journal of brand management: 2004; 11 (6): 484-506.

[4] Wiedmann KP, Hennigs N, Siebels A. Measuring consumers' luxury value perception: a cross-cultural framework. Academy of Marketing Science Review: 2007; 2007: 1.

[5] Spiteri JM, Dion PA. Customer value, overall satisfaction, end-user loyalty, and market performance in detail intensive industries. Industrial marketing management: 2004; 33 (8): 675-687.

[6] Lichtenthal JD, Wilson DT, Long MM. Scientific contributions to the field from the Journal of Business-to-Business Marketing. Journal of Business Research: 1997; 38 (3): 211-233.

[7] Agyekum CK, Haifeng H, Agyeiwaa A, et al. Consumer perception of product quality. Microeconomics and Macroeconomics: 2015; 3 (2): 25-29.

[8] Kim JH, Hsu MM, Yuen CA. Individual and social factors impacting Chinese millennialsics and Macroeconomicsconomicsmicsonomicsicssiness Researchial mar: 2020; 20 (1): 27-43.

[9] Chattalas M, Shukla P. Impact of value perceptions on luxury purchase intentions: a developed market comparison. Luxury Research Journal: 2015; 1 (1): 40-57.

[10] Alan A, Dursun I, Kabadayi E, et al. What influences the repurchase intention for luxury brands? The relative impacts of luxury value dimensions. International Business Research: 2016; 9 (5): 11-24.

[11] Shukla P, Purani K. Comparing the importance of luxury value perceptions in cross-national contexts. Journal of Business Research: 2012; 65 (10): 1417-1424.

[12] Faschan M, Chailan C, Huaman-Ramirez R. Emerging adults' luxury fashion brand value perceptions: A cross-cultural comparison between Germany and China. Journal of Global Fashion Marketing: 2020; 11 (3): 207-231.

[13] Tian KT, Bearden WO, Hunter GL. Consumers' need for uniqueness: Scale development and validation. Journal of consumer research: 2001; 28 (1): 50-66.

[14] Butcher L, Phau I, Shimul AS. Uniqueness and status consumption in Generation Y consumers: Does moderation exist? Marketing Intelligence & Planning: 2017.

[15] Saaty RW. The analytic hierarchy process—what it is and how it is used. Mathematical modelling: 1987; 9 (3-5): 161-176.
[16] Saaty TL. Decision making with the analytic hierarchy process. International journal of services sciences: 2008; 1 (1): 83-98.
[17] Apak S, Gö, G GG, Karakadkad. aS. An analytic hierarchy process approach with a novel framework for luxury car selection. Procedia-Social and Behavioral Sciences: 2012; 58: 1301-1308.
[18] Ku ECS, Fan YW. The decision making in selecting online travel agencies: An application of analytic hierarchy process. Journal of Travel & Tourism Marketing: 2009; 26 (5-6): 482-493.
[19] Saaty TL. Making and validating complex decisions with the AHP/ANP. Journal of Systems Science and Systems Engineering: 2005; 14 (1): 1-36.
[20] Costa CAB, Vansnick JC. A critical analysis of the eigenvalue method used to derive priorities in AHP[J]. European Journal of Operational Research: 2008; 187 (3): 1422-1428.
[21] Jiang L, Shan J. Heterogeneity of luxury value perception: a generational comparison in China. International Marketing Review: 2018.
[22] Zhang L, Zhao H. Personal value vs. luxury value: What are Chinese luxury consumers shopping for when buying luxury fashion goods? Journal of Retailing and Consumer Services: 2019; 51: 62-71.
[23] Perry P, Barnes L, Ye T. The Evolution of the Chinese Luxury Fashion Consumer: An Interpretive Study of Luxury Value Perception in Understanding Luxury Fashion. Palgrave Macmillan, Cham: 2020; 175-202.
[24] Chan WY, To CKM, Chu WC. Materialistic consumers who seek unique products: how does their need for status and their affective response facilitate the repurchase intention of luxury goods?. Journal of Retailing and Consumer Services: 2015; 27: 1-10.
[25] Han SH, Yun MH, Kim KJ, et al. Evaluation of product usability: development and validation of usability dimensions and design elements based on empirical models. International Journal of Industrial Ergonomics: 2000; 26 (4): 477-488.

Cite as: C.Z. Niu et al.,
TBIS 10.3993/tbis (2021).

Future-oriented Performance Measurement Method of the Textile Industry Based on Balanced Scorecard

Cheng-Zhi Niu[1], You-Gan Zhu[1]

[1] *Beijing Institute of Fashion Technology, Chaoyang District, Beijing, 100020, PRC*

*Corresponding author's email: macrov@yeah.net

Abstract

With the development of modern technology and economic globalization, the Textile Industry enterprises in China face fierce competition and operating pressure. Textile Industry enterprises are finding it more difficult to develop new competitive technologies. However, few enterprises pay attention to management innovation, which is an easier path to achieve competitive advantage. Therefore, this paper attempts to establish a future-oriented performance evaluation to innovate the management mode of textile enterprises, help textile enterprises to establish competitive advantages, and realize long-term development.

Keywords: Textile Industry; Analytic Hierarchy Process; Green Manufacturing; Manufacture

1. Introduction

Through the research summary of international researchers, we discovered many research results that in the effective integration of enterprise performance evaluation and balanced scorecard, but there are still some problems to be further addressed. [1]

Research on the former balanced scorecard system are only theoretical and is rarely applied in practice [2]. In fact, the success of different companies in this industry depends on their personality. Therefore, the evaluation system should be optimized for the characteristics of the textile industry. [3]

The Ministry of Environmental Protection of the People's Republic of China proposed cultivating new-type textile enterprises featuring circular energy saving and low carbon emission reduction. At the same time, the Chinese government will promote green technologies for the adjustment and transformation of the traditional textile industry, promote the green development of the traditional textile industry, and strengthen ecological and environmental protection. [4]

Most textile enterprises still use traditional methods to manage enterprises, which are difficult to motivate employees to create value for enterprises. In the traditional management mode, enterprise managers pay more attention to the short-term profitability of enterprises and do not address long-term healthy development. [5] The balanced scorecard is a new scientific performance appraisal system, wherein the indicators focus on the innovation driving factors.

Under the global trend of environmental protection and sustainable development, The current performance evaluation of textile enterprises has defects. [6] It is challenging to guide the green development of Chinese textile enterprises.

2. The Performance Measurement Method of the Textile Industry based on Balanced Scorecard

2.1 Characteristics of the Textile Industry Enterprises

China's textile enterprises have a long history of development, but with the impact of the new economy, Chinese textile enterprises are facing unprecedented shocks, and the development of Chinese textile enterprises is facing huge opportunities and challenges. [7]

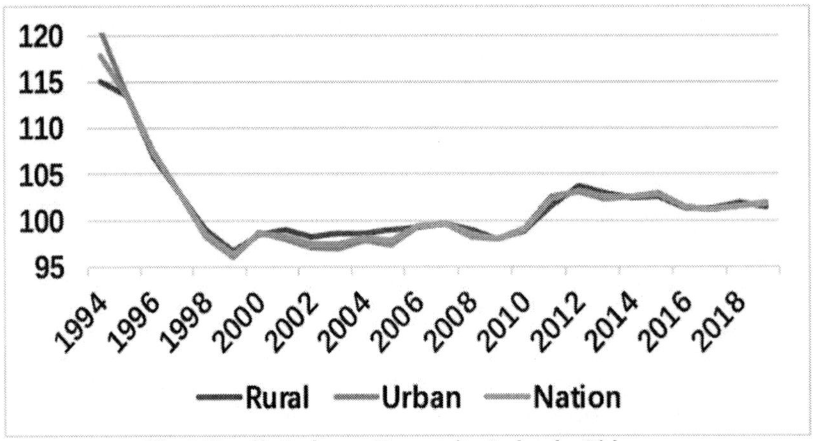

Fig.1 Textile Industry Prosperity Index in China

The picture above shows the prosperity index of China's textile industry calculated by the Chinese government. The prosperity index of China's textile industry is at a low point in the three dimensions of the nation, rural and urban. Chinese textile companies need to change their development thinking and make breakthroughs in industry development. [8]

In order to help textile companies to achieve their development and formulate a comprehensive index system that meets the characteristics of textile companies, the characteristics of textile companies have been summarized. [9]

2.1.1 Strong Industry Foundation

After years of development, the Chinese textile industry has a complete industrial chain globally, the highest level of processing facilities, and the ability to adjust to market risks. The Chinese textile industry has many developed industrial clusters. From the perspective of the international market environment, there is much room for expansion and opportunities in the international market. This has brought huge opportunities for China's textile trade. [10]

In the next few years, the world economy will still be on an upward trend, bringing beneficial international market environmental protection to the Chinese textile industry's exports. [11]

2.1.2 Traditional Industry, High-cost Pressure

The textile industry is traditional. The main materials of the textile industry are cotton, hemp fiber, wool, cashmere, cocoon silk, down feather, chemical fiber. The textile industry has a narrow source of raw materials, and the raw materials have large price fluctuations. Cost pressures are difficult to transmit to terminal consumption through the industrial chain, and the cost pressures faced by downstream textile companies are becoming increasingly prominent. [12]

2.1.3 Long Supply Chain, Uncertainty between Supply and Demand

The foreign textile industry is mainly divided into three sub-sectors: textile, clothing, and industrial textiles. But in China, the textile industry comprises dozens of sub-sectors, including chemical fiber, cotton textile, linen textile, wool textile, silk, printing, and dyeing, knitting, home textile, clothing and industrial textiles. The raw material suppliers, manufacturers, logistics companies, wholesalers, retailers, and final consumers involved in the production and circulation process constitute a vast supply and demand network and a long supply chain.

Companies usually take measures such as stocking excessive inventory or enlarging production capacity to buffer the uncertainty between supply and demand. These measures occupy a large amount of working capital, reduce equipment utilization, and easily cause inventory loss and high management costs. [13]

2.1.4 Labor-intensive Industries

Labor-intensive industries refer to industries that mainly rely on a large amount of labor for production. The textile industry is a typical labor-intensive industry, and employee wages account for a large proportion of the cost. [14][15][17]

2.1.5 Serious Pollution and High Energy Consumption

The textile industry is listed as a severe polluting industry. According to the State Environmental Protection Administration statistics in China, the total discharge of wastewater from the printing and dyeing industry ranks among the forefront of the national manufacturing, and this sewage is heavily polluted, difficult to treat, and low in reuse rate. In addition, the pre-treatment process of the wool, hemp, and silk industries will also discharge a large amount of sewage. [16]

In the production process of the chemical fiber industry, some products use a lot of acids and alkalis. These companies will eventually produce hazardous substances such as sulfur, sulfuric acid, and sulfate, causing severe pollution to the environment. Chemical fiber is not degradable, especially synthetic fiber, which has high waste recycling cost, causing soil environment's deterioration and pollutes the air after combustion.

In terms of energy consumption, textile and chemical fiber machinery's energy consumption is particularly huge. The total energy consumption of the chemical fiber industry in China is higher than the advanced level abroad. The textile industry has become a key industry in the energy-saving reforms of China.

2.2 Construction of a Performance Measurement Method for the Textile Industry based on the Balanced Scorecard

In the face of brutal market competition, textile companies should strengthen performance measurement method research to adapt to industrial upgrading and transformation. Constructing an effective evaluation index system is a prerequisite for the scientific evaluation of corporate performance. Balanced scorecard theory is based on financial indicators, combined with the enterprise's internal management to stimulate managers' creative thinking. The balanced scorecard also considers non-financial indicators such as learning and innovation capabilities, corporate development momentum, customer performance, and other non-financial indicators to optimize corporate performance. [1] It provides ideas for the establishment of a textile enterprise's performance evaluation index system.

Many companies neglect the sustainable development of the environment to maximize the interests of shareholders and cannot coexist harmoniously with the environment. This directly harms the interests of society, the basic lives of the public, and the corporate image and long-term benefits. Balanced development has become a new driving force for the progress of enterprises. Enterprises must fulfill their environmental responsibilities, and consider the public welfare social responsibilities of the government and the community. [3] For textile companies, the performance evaluation index system also needs to reflect the performance of corporate social responsibilities such as community responsibilities and environmental responsibilities, to conform to the development of the new environmentally friendly era.

Combined with the characteristics of textile enterprises, a performance indicator system for the textile industry has been established as follows:

Table 1. Textile Industry Enterprise Performance Measurement Index System

BSC Dimensions	Indicator
Financial data	Operating income growth rate
	Tangible asset turnover
	Net profit rate
	Estimated market value
Customer perspectives	customer satisfaction
	Brand social responsibility
	Customer retention rate
	New customer acquisition rate

Internal business process	Carbon emission reduction indicators Pollution reduction indicators Product return / rework rate After-sales service reaction cycle
Learning and growth	Staff training expense rate Key employee retention rate Employee recommendation adoption Employee productivity

3. Calculation of Index Weights based on the Analytic Hierarchy Process

The analytic hierarchy process is a complex multi-objective decision-making system. It decomposes the goal into multiple goals or criteria and further decomposes the goal into multiple indicators. It uses the quantification method to calculate the hierarchy order to optimize decision-making.

3.1 Establishing a Hierarchical Structure Model

The analytic hierarchy constructs a process model as follows. There are 3 layers in this model: the target layer, the criterion layer, and the indicator layer. The hierarchical model of the construction is shown in Table 2.

Table 2. The hierarchical model of Textile Industry Enterprise Performance measurement Index System

Target layer	Criterion layer	Indicator layer
Performance measurement of the Textile industry based on balanced scorecard A	Financial data B1	Operating income growth rate C1 Tangible asset turnover C2 Net profit rate C3 Estimated market value C4
	Customer perspectives B2	customer satisfaction C5 Brand social responsibility C6 Customer retention rate C7 New customer acquisition rate C8
	Internal business process B3	Carbon emission reduction indicators C9 Pollution reduction indicatorsC10 Product return / rework rate C11 After-sales service reaction cycle C12
	Learning and growth B4	Staff training expense rate C13 Key employee retention rate C14 Employee recommendation adoption C15 Employee productivity C16

3.2 Constructing a Judgment Matrix

Experts have been invited to revise the indicator system and give scores. These experts include professors and scholars from the Beijing Institute of Fashion Technology, and senior managers of Chinese textile companies.

The judgment matrix was constructed by using the Delphi Method. The Delphi method is based on systematic procedures, and experts use an anonymous way of expressing opinions. Experts can only communicate with investigators. The Delphi method will investigate experts' opinions on the questions raised in the questionnaire in multiple rounds. After repeated consultation, summarization, and modification, the results were finally summarized into the consensus views of the experts. This method is reliable in the research field.

Experts will score the indicator system. The score is between 1 and 9. The meaning of the score is as follows.

Table 3. AHP Scale Table

Scale	Meaning
1	Two elements have the same importance
3	The former element is slightly more important than the latter
5	The former element is obviously more important than the latter
7	The former element is significantly more important than the latter
9	The former element is exceedingly more important than the latter
2、4、6、8	Indicates the intermediate value of the above adjacent judgment
Reciprocal	Indicates the exchange of order of two elements

The A-B judgment matrix results of experts scoring the indicator system are as follows (Table 4).

Table 4.A-B Judgment Matrix

A	B1	B2	B3	B4
B1	1	2	3	3
B2	1/2	1	2	2
B3	1/3	1/2	1	1
B4	1/3	1/2	1	1

The A-B judgment matrix results of experts scoring the indicator system are as follows (Table 5-8).

Table 5. B1-C Judgment Matrix (Financial Dimensions)

B1	C1	C2	C3	C4
C1	1	2	3	3
C2	1/2	1	2	2
C3	1/3	1/2	1	1
C4	1/3	1/2	1	1

Table 6. B2-C Judgment Matrix (Customer Dimension)

B2	C5	C6	C7	C8
C5	1	3	7	5
C6	1/3	1	5	3
C7	1/7	1/5	1	1/3
C8	1/5	1/3	3	1

Table 7. B3-C Judgment Matrix (Internal Process Dimension)

B3	C9	C10	C11	C12
C9	1	7	5	7
C10	1/7	1	1/3	1
C11	1/5	3	1	3
C12	1/7	1	1/3	1

Table 8. B4-C Judgment Matrix (Learning and Growth Dimension)

B4	C13	C14	C15	C16
C13	1	1	3	5
C14	1	1	3	5
C15	1/3	1/3	1	3
C16	1/5	1/5	1/3	1

3.3 Calculating Weights and Conducting Consistency Tests

Before analyzing the indicators of the Analytic Hierarchy Process. The consistency index CI and the ratio CR are used to check the consistency. When CR<0.1, it indicates that the judgment matrix has passed the consistency test; otherwise, the judgment matrix should be corrected.

The calculation process of the consistency index is as follows:

$$X=(X1,X2,...,Xn)T. \tag{1}$$

$$W=(W1,W2,...,Wn)T. \tag{2}$$

$$AW=(AW1,AW2,...,AWn)T. \tag{3}$$

$$\lambda max = \frac{1}{n}\sum_{i=1}^{n}\frac{(AW)i}{Wi}. \tag{4}$$

$$CI = \frac{\lambda max - n}{n-1}. \tag{5}$$

$$CR = \frac{CI}{RI}. \tag{6}$$

Table 9. Standard values of the consistency indicator RI

Matrix order	1	2	3	4	5	6	7	8	9	10
RI	0	0	0.58	0.90	1.12	1.24	1.32	1.41	1.45	1.49

The maximum eigenvalue λmax=4.0109, CI=0.0036, RI=0.9, test coefficient is CR=CI/RI=0.004, CR<0.1 we can see that the result has passed the test. The weight of the four-dimensional matrix is (0.4547, 0.2631, 0.1411, 0.1411)T.

The maximum eigenvalues λmax obtained in the judgment matrix of finance, customer, internal process, learning, and growth are 4.0109, 4.1186, 4.0753, 4.0441, and the test coefficients CR are 0.004, 0.0439, 0.0279, 0.0163. The four CR are less than 0.1; we can see that all the judgment matrix passes the consistency test.

The weight of the indicator layer is shown in Table 10.

Table10. Textile Industry Enterprise Performance measurement Index Weight Table

Criterion Layer and Weight	Indicator Layer and Weight	Relative Weight
Financial Data 45.47%	Operating income growth rate 45.47%	20.67%
	Tangible assets turnover rate 26.31%	11.96%
	Net profit margin 14.11%	6.42%
	Estimated market value 14.11%	6.42%
Customer Perspectives 26.31%	Customer satisfaction 55.78%	14.67%
	Brand social responsibility 26.34%	6.93%
	Customer retention rate 12.19%	3.21%
	New customer acquisition rate 5.69%	1.50%
Internal Business Processes 14.11%	Carbon emission reduction rate64.74%	9.13%
	Pollution reduction rate7.82%	1.10%
	Product return / rework rate 19.62%	2.77%
	After-sales service reaction cycle 7.82%	1.10%
Learning and Growth 14.11%	Staff training expense rate 38.89%	5.49%
	Key employee retention rate 38.89%	5.49%
	Employee recommendation adoption level 15.35%	2.17%
	Employee productivity 6.87%	0.97%

The above table shows the financial dimension is still the focus of future textile enterprise performance, but other dimensions are also important. We can see from the scoring results of experts that textile companies should pay enough attention to their social responsibilities. With the increasing concern about environmental protection, those enterprises that rely on destroying the ecology, polluting the environment, and blindly consuming resources are now getting smaller living spaces. China is stepping up industrial upgrading, and outdated production capacity will be eliminated. Under China's policies, polluting companies will face high environmental fines and environmental taxes. Some seriously polluting enterprises were even forced to shut down. Pollution issues must be paid attention to in the transformation of Chinese textile enterprises, but this has been ignored in traditional performance evaluation. Chinese textile companies must transform to sustainable development and assume their social responsibilities.

4. Conclusion

4.1 From the Perspective of Strengthening Corporate Culture

We should strengthen the construction of corporate culture. As the core of an enterprise, corporate culture affects the operation and development of the enterprise. The corporate culture is embodied in all aspects of the company and affects the implementation of various company measures, including the performance management system。

4.2 Perspective of Employee Ability Improvement

At the same time, employees are a non-negligible part of it. Educating employees can improve the quality of the employees themselves, and effectively convey the corporate culture to the employees. We should conduct specialized training for enterprise employees, including managers and employees at all levels. We should also introduce advanced management methods to promote the development of the industry.

4.3 Green Development Perspective

The green development indicators and pollution reduction indicators in the internal process dimension are far more concerned than the traditional indicator system in this brand-new textile industry performance appraisal system. This should arouse the attention of textile companies. For the sake of protecting human health, governments in most regions of the world have issued laws and regulations to limit or prohibit toxic and hazardous substances in textiles. The development of green textiles is imperative. The impact of green development on textile companies is manifested in many aspects: textile companies' business concepts, raw material sources, specific operations and production, textile equipment and technology upgrades, brand planning, marketing, and international cooperation.
In the future, global textile companies will strengthen environmental protection in the production process of their products, and the products themselves will become more "green." The trend of the future development of the textile industry is a new technology, new processing, and green manufacturing. Green textiles and a sustainable economy will become the only way for future development.

4.4 Social Responsibility Perspective

As a future-oriented textile enterprise evaluation method, we have also paid due attention to the textile enterprises' green development and social responsibility. From the analytic hierarchy process results, Chinese experts and scholars have paid great attention to the green development of textile enterprises. Although from the analytic hierarchy process results, the current stage of green development indicators is not as important as financial indicators, but as a future-oriented development driving factor, green development indicators are also worthy of our attention. Therefore, our textile companies should actively respond to the green development trend of the textile industry.

4.5 Summary

In summary, we use the analytic hierarchy process to establish a comprehensive and scientific textile enterprise performance index evaluation system, which helps textile enterprise managers to see the future and gain competitive advantages. We hope that our research results can promote textile enterprises to achieve long-term green and sustainable development.

5. Acknowledgement

This research was financially supported by the Supporting Funds for Postgraduate research and innovation projects of the Graduate School of Beijing Institute of Fashion Technology (Grand NO. 120301990131/006).

References

[1] Hu HW. Research on the international competitiveness of China's textile industry and its influencing factors [D]. Heilongjiang University: 2020.
[2] Kaplan RS, Norton DP. Strategy Maps: Converting Intangible Assets into Tangible Outcomes, New York: Harvard Business Press: 2003.
[3] Liu JY, Meng Y, The Serviceability of the Balanced Scorecard: a experimental research, Accounting Research: vol.95; pp. 36-42: 2011.
[4] Deng LN. Study on the impact of Chinese textile enterprises' performance of environmental responsibility on exports [D]. Jiangxi University of Finance and economics: 2014.
[5] Hao J. Scientific and technological innovation has made China's textile industry a world leader [J]. Textile and garment weekly: 2021 (03); 12-13.
[6] Zheng JX. Research Report on China's textile industry's response to trade barriers [J]. Guangxi quality supervision guide: 2020 (01); 191-192.
[7] Chen N. Current situation and future of China's textile industry [J]. Textile science research: 2019 (12); 48-49.
[8] Zhou X, Wang D, Wang PF. Study on the evolution and development model of China's textile industry agglomeration pattern [J]. Research on technology, economy and management: 2019 (11); 89-95.
[9] LV ang. Transformation and upgrading of China's textile industry leads global industry change [J]. Financial world: 2018 (11); 126-127.
[10] Wang FB. Evaluation of input-output efficiency of textile industry in 14 provinces of China [J]. Western leather: 2018; 40 (18): 37.
[11] Liang JY. Empirical analysis of China's textile industry on export trade economy [J]. Mall modernization: 2018 (13); 65-67.
[12] Li J. Temporal and spatial evolution characteristics and influencing factors of interregional transfer of China's textile industry [D]. Henan University: 2018.
[13] Zhang ZY. Research on the relationship between Government R & D funding, R & D model and innovation performance of textile industry [D]. Zhejiang University of technology: 2021.
[14] Wang CN. Study on the impact of rising labor costs on the international competitiveness of China's textile industry [J]. Fortune era: 2019 (11); 154.
[15] Li J. Research on the status and influencing factors of international division of labor of China's textile industry under the global production network [D]. Hunan University: 2017.
[16] Wang DP. Study on the impact of labor cost on textile export trade [D]. Nanjing Agricultural University: 2012.
[17] Huang HJ. Study on the impact of environmental regulation on the international competitiveness of China's textile industry [D]. Hubei University: 2015.

Noise Reduction Algorithm for Point Cloud Reconstruction of Small Object based on SFM

Sui-Feng Yuan[1], Yue-Qi Zhong[1,2*]

[1] *College of Textile, Donghua University, Shanghai, China*

[2] *Key Laboratory of Textile Science & Technology, Ministry of Education, Donghua University, Shanghai, China*

*Corresponding author's e-mail: zhyq@dhu.edu.cn

Abstract

Three-dimensional point cloud acquisition with SFM (structure from motion) can generate point clouds that are 3D coordinated with color information however it induces additional noise to data. To deal with this issue, this paper introduces an outlier detection algorithm based on color clustering technique as the outliers are black when a black mask is used as the background. To reduce the outliers, our method uses K-means clustering to take black outliers with grayscale value lower than a threshold as the first initial clustering centre and separate the outliers from the real object. Experimental results proved that the proposed method can reduce the outliers to a very low level.

Keywords: Outliers Removal; Structure From Motion; Point Cloud; Clustering; RGB.

1. Introduction

SFM (structure from motion)-based 3D point cloud reconstruction uses image sequences to create a 3D model of the object, which has the advantages of fast reconstruction, easy operation and low cost. However, due to the influence of image background and the shape of the target object, many outliers are inevitable, which needs to be detected and removed. There are several point cloud processing methods that can be used to detect and remove outliers. For ordered or partially ordered point clouds, methods such as least square filter [1], Wiener filter [2] and smoothing filter [3] were used. For scattered point clouds, algorithm such as average curvature flow [4], bilateral filters [5, 6], moving least squares method [7-9] were used. For complex point clouds, Zhong [10] proposed a spatial outlier mining algorithm based on K- Nearest Neighbor (KNN). The KNN graph was built in terms of the K-nearest neighborhood to all objects, then the difference of non-spatial attribute values of adjacent objects was taken as the edge weight between two object points. Cutting edge strategy was employed to finally remove the edge with higher weight, thus identifying spatial outliers and outlier areas. Liu [11] proposed a statistical feature analysis method based on the distance distribution of the outliers' local domain to remove the outliers with sparse sampling. Then it proceeds to remove the isolated outliers according to the density within the optimal domain radius. Zhang [12] proposed a density-based outlier removal algorithm LSNOR that can estimate parameters autonomously and denoised on a large scale without determining parameters interactively. The aforementioned methods typically used the 3D coordinate values of data points to detect outliers. However, the point cloud obtained by the SFM method possesses not only 3D coordinates, but also color information stored in RGB format. Once the color of the outlier is different from the color of the object target, this difference can be used as the basis of color-based clustering. Therefore, an outlier detection method based on color clustering was proposed in this paper, which can effectively detect the outliers connected with the point cloud of the target object. This method provides a solution for noise reduction using specific color information rather than 3D coordinates.

2. Experimental Design

2.1 Preprocessing

The reconstructed model contains not only the point cloud of the target, but also the point cloud of the background. In order to reduce the interference of the background, a black flocking fabric with lower reflection is selected as the mask, which is attached to the other side of the turntable facing the camera to ensure the background taken in the image is always black, as shown in Fig.1. With this approach, the number of outliers can be greatly reduced. The remaining outliers can be further divided into two types in terms of the spatial distance: (1) the outliers floating around the target object; and (2) the outliers connected with the target object.

(a) (b)
Fig.1 Comparison of point cloud models before and after using masks.

Due to the fact that the points cloud generated by SFM approach requires two adjacent images possess similar features that can be used by feature comparison, it is not uncommon that highly overlapped area occurred in image sequences. In practice, this is one of the major reasons that outliers were produced [13]. Assuming that the point cloud model in Fig.1(b) is generated from n images, at every interval for these n images, one image is deleted to get n/2 images, two images are deleted to get n/3 images, and three images are deleted to get n/4 images. The point cloud models are generated from these different number of images as input are shown in Fig.2. According to our observations, the optimal number of images is n/3, with which the number of outliers is minimized while still keeping a watertight point cloud of target without gaps.

(a) n (b) n/2 (c) n/3 (d) n/4
Fig.2 Comparison of the reconstructed point cloud model after deleting different numbers of images.

2.2 Outlier Detection based on Color Clustering

After preprocessing, a new model with minimum outliers is generated. To further remove the remaining outliers, a method based on color clustering is employed.

2.2.1 K-means Clustering

The principle of the traditional k-means clustering is to divide a given data set into K clusters according to the distance between the data.

The mathematical expressions are as follows: the input data set is D=$\{x_1, x_2, \cdots, x_m\}$, the number of clusters is k, the maximum number of iterations is N, and the output cluster set is C=$\{C_1, C_2, \cdots, C_k\}$. Firstly, randomly select k data points from the data set D as the initial k centroid vectors: $\{u_1, u_2, \cdots, u_k\}$. Secondly, initialize the cluster set C as $C_t = \emptyset, t = 1,2 \ldots k$ and calculate the distance between each data point $x_i (i = 1,2 \ldots m)$and each centroid vector $u_j (j = 1,2 \ldots k)$,

$$d_{ij} = \|x_i - u_j\|_2^2 \quad 1)$$

Thirdly, mark the category δ_i corresponding to the smallest d_{ij} of x_i, and update $C_{\delta i} = C_{\delta i} \cup \{x_i\}$. Finally, recalculate new centroids for all data points in $C_j (j = 1,2 \ldots k)$,

$$u_j = \frac{1}{|C_j|}\sum_{x \in C_j} x \quad 2)$$

When the k centroid vectors no longer change, the cluster set $C = \{C_1, C_2, \cdots, C_k\}$ is regarded as the final output.

Traditional k-means clustering which randomly selects the initial cluster center converges slowly. To improve it, an optimization scheme is proposed:

1) Use eq.3 to compute the grayscale value of each data point. The grayscale refers to dividing the white and black into several levels according to the logarithmic relationship. The range is generally from 0 to 255, where white is 255 and black is 0. When the grayscale value is equal to 30, the color is very close to black. Therefore, the one with a grayscale value lower than 30 is randomly considered as the cluster center u_1.

$$\text{Gray} = 0.3R + 0.6G + 0.1B \quad 3)$$

2) Calculate the distance between each data point x_i in the data set and the nearest cluster center among the selected cluster centers:

$$D(x_i) = \text{argmin}\|x_i - u_r\|_2^2, r = 1,2, \ldots, k \quad 4)$$

3) Select the data point with larger $D(x)$ as the new cluster center;

4) Repeat steps (2), (3) until k cluster centers are selected;

5) Run the traditional k-means clustering with the selected k cluster centers as the initial centroids.

2.2.2 Determining the Initial Clustering Center

The first initial clustering center is randomly selected among all the points with grayscale value lower than 30. Based on the RGB value, the point cloud farthest from the first initial clustering center is a certain point in the point clouds of target. Therefore, the second initial clustering center is selected from the point clouds of target.

2.2.3 Determining the Number of Clusters

When k is equal to 2, parts of the point clouds of target and outliers are divided into a cluster, as shown in Fig.3. To separate these point clouds of target, the number of clusters has to be increased. When k is equal to 3, it is observed from Fig.4 that the point clouds of target is divided into three clusters. Some point clouds of target are still divided into the same cluster as outliers, but the number is reduced. When k is equal to 4 or 5, some point clouds of target and outliers are still classified into the same cluster, but as the value of k increases, the number of these point clouds of target gradually decreases, as shown in Fig. 4. When k is equal to 6, it can be found from Fig.5 that the point clouds contained in one of the clusters is almost all black outliers. The point clouds of each cluster are saved in one file, and the five files containing only the point clouds of target are merged into one file. A model containing only point clouds of target and a model containing outliers are shown in Fig.6.

Fig.3 Classified point clouds when k=2.

k=3

k=4

k=5

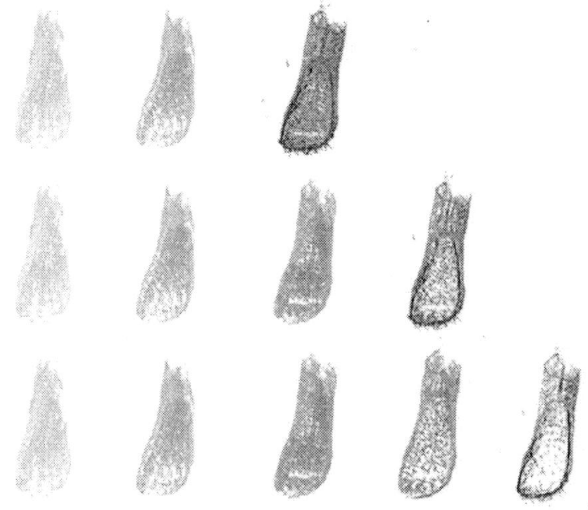

Fig.4 Classified point clouds when k=3,4,5.

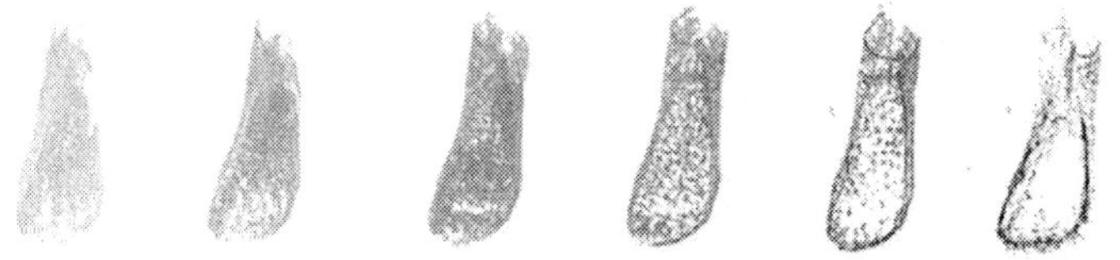

Fig.5 Classified point clouds when k=6.

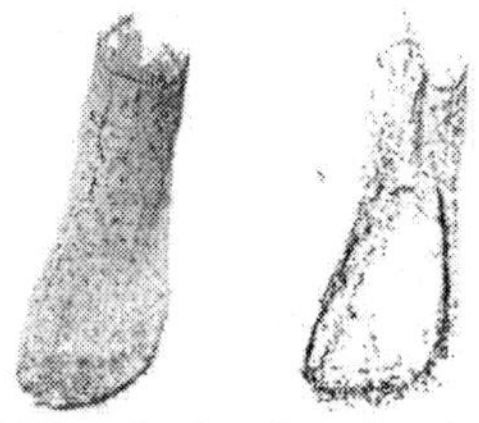

Fig.6 The final point cloud models

3. Experimental Results and Discussion

In order to verify the effectiveness of the algorithm proposed in this paper, two different subjects are scanned. The corresponding raw points cloud are shown in Fig.7. The distribution of the point clouds of target in Fig.7(a) is uniform, and the distribution of the local point clouds of target in Fig.7(b) are sparse.

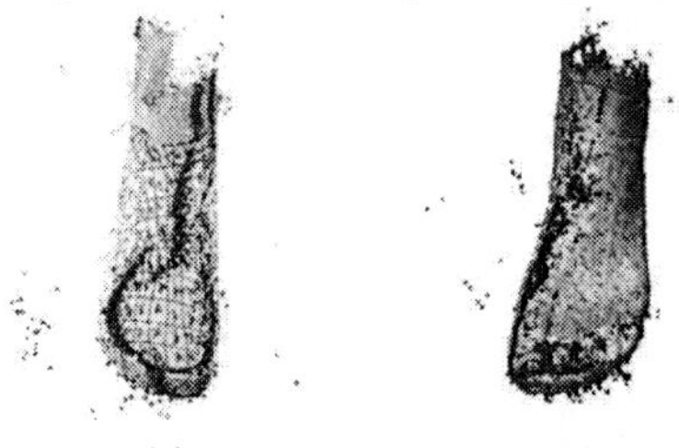

(a) (b)

Fig.7 Raw points cloud of two different scans.

3.1 Methods Comparison

Three methods are employed for outlier removal: the statistical analysis, the proximity analysis and the k-means method proposed in this paper. In the following, pct represents the point clouds of target, sa represents statistical analysis, and pa represents proximity analysis, oc represents the outliers connected with target.

For the point cloud model shown in the Fig.8(b), it can be found that compared with the model shown in Fig.8(a), although the black outliers floating around the target are removed, the black oc have hardly been removed, but the pct in the upper part of the model have been removed. In Fig.8(c), it can be clearly found that outliers are partially removed. Compared with other point clouds, the model shown in the 8(d) is the best since the oc are almost fully removed. For the point clouds shown in Fig.9(b) and Fig.9(c), the outliers floating around the target were completely removed, and the oc are almost fully removed as well. However, there are obvious gaps on the surfaces of the two models, as shown in Fig.9(b) and (c), which can introduce unwanted problems in the downstream treatments including normal estimation, surface reconstruction, etc. In Fig.9(d), the sparse pct in the point cloud model has not been removed to avoid the generation of huge gaps.

The execution time and point cloud removal rate can be found in Table 1 and 2. The point cloud removal rate r is obtained by calculating the percentage of the number of the removed point clouds N_r to the number of the raw point clouds N.

$$r = \frac{N_r}{N} \times 100\% \qquad 5)$$

As shown in Table 1, the r from the first row to the last row is represented by r_1, r_2, r_3 respectively. The r_1 is 1% and the r_2 is 2.1%, neither of which exceeds 3% and is much lower than r_3 which is 15.3%. The distribution of outliers and the different detection mechanisms cause this huge difference in the r. In Fig.8(a), although pct are different from all outliers in color, most of pct have similar spatial features to some outliers such as the density and the number of neighboring point cloud. In the three methods, the methods based on sa and pa can only remove unqualified point clouds through spatial features. Therefore, some outliers that have similar spatial features to the pct not be removed. Our method based on color clustering only considers the color difference between point clouds. The point clouds with relatively close colors are divided into a cluster. Therefore, when the number of clusters is reasonable, most of black outliers are divided into same cluster and removed. As shown in Table 2, the r from the first row to the last row is represented by r_4, r_5, r_6 respectively. Contrary to the results in Table 1, the values of r_4, r_5, r_6 are very close. In Fig.9(a), the most of pct are tightly distributed, and a small part of pct are sparsely distributed. These sparse pct are lower than the set threshold in density and the number of neighboring point clouds. Therefore, the sparse pct are identified as outliers and removed by two methods based on sa and pa.

According to the above analysis of the figures and tables, it can be proved that the methods based on the sa and the pa can detect and remove sparse outliers, but have a poor performance in detecting the dense outliers. In Fig.9, it can be proved that if the pct have sparse parts, these sparse pct are regarded as the outliers and wrongly removed. Obviously, the K-means cluster approach has better performance.

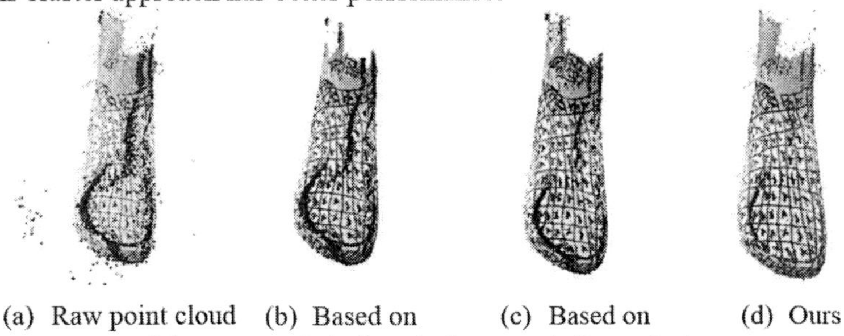

(a) Raw point cloud model (b) Based on statistical analysis (c) Based on proximity analysis (d) Ours

Fig.8 Comparison of outlier detection effects of various methods on No.1 point cloud model.

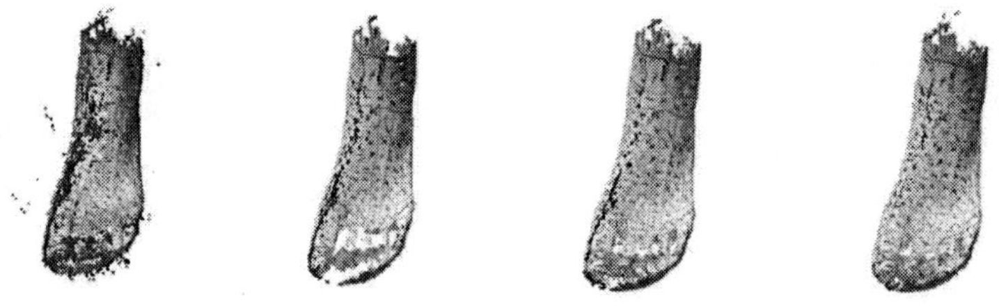

(a) Raw point cloud model (b) Based on statistical analysis (c) Based on proximity analysis (d) Ours

Fig.9 Comparison of outlier detection effects of various methods on No.2 point cloud model.

Table 1. Comparison of each method on No.1 point cloud model.

Methods	Number of original point clouds	Number of point clouds after processing	Point cloud removal rate/%	Time /s
Based on statistics	391776	387878	1.0	112
Based on proximity	391776	383586	2.1	153
Ours	391776	331823	15.3	118

Table 2. Comparison of each methods on No.2 point cloud model.

Methods	Number of original point clouds	Number of point clouds after processing	Point cloud removal rate/%	Time /s
Based on statistics	235529	229625	2.5	40
Based on proximity	235529	227229	2.6	112
Ours	235529	229308	2.6	84

3.2 Differences in Point Cloud Removal Rates

As shown in Table 3, after our method is employed, r_3 is more than five times higher than r_6. But it is observed from Table 4 and 5 that r_1 and r_2 are close to r_4 and r_5. The reason is that the methods based on sa and pa are essentially to identify a few point clouds that do not conform to the mainstream as the outliers. Assuming that a point cloud model is divided into two parts by setting a threshold, and the number of point clouds in the two parts is very different, the part with a small number of point clouds will be regarded as outliers through these two methods. Therefore, the r obtained by these two methods is always at a low level. Our method classifies point cloud by color information. The N_r is only related to the color of point clouds themselves. According to this standard, the point clouds with similar RGB are classified into one cluster. Among these clusters, one cluster contains all black outliers, and the number of point clouds in this one cluster is the N_r. Therefore, the robtained by our method have significantly different results for different models.

Table 3. Comparison of the point cloud removal rate of the method proposed in this paper on different models.

Point cloud models	Number of original point clouds	Number of point clouds after processing	Point cloud removal rate/%	Times/s
Model No.1	391776	331823	15.3	118
Model No.2	235529	229308	2.6	84

Table 4. Comparison of the point cloud removal rate of the method based on statistical analysis on different models.

Point cloud models	Number of original point clouds	Number of point clouds after processing	Point cloud removal rate/%	Times/s
Model No.1	391776	387878	1.0	112
Model No.2	235529	229625	2.5	40

Table 5. Comparison of the point cloud removal rate of the method based on proximity analysis on different models.

Point cloud models	Number of original point clouds	Number of point clouds after processing	Point cloud removal rate/%	Times/s
Model No.1	391776	383586	2.1	153
Model No.2	235529	227229	2.6	112

4. Conclusion

After using the mask, the color information of outlier changes from multiple to a single. Then using the difference in colors between the outliers and the target object, an outlier detection approach based on k-means clustering is proposed in this paper. From our observation, when the number of clusters (k) equals to 6, the outliers can be removed automatically. It can be found that on the No.1 point cloud model, the proposed approach in this paper achieves an outlier detection rate of up to 15.3%, which is seven times higher than other approaches, and on the point cloud model in Fig.9(a) which have sparse parts in point clouds of target, it can accurately detect outliers without wrongly removing the point clouds of target as outliers.

5. Acknowledgments

This work was supported by National Natural Science Foundation of China (Grant No.61572124).

References

[1] Jung J, Hong S, Yoon S, et al. Automated 3D Wireframe Modeling of Indoor Structures from Point Clouds Using Constrained Least-Squares Adjustment for As-Built BIM [J]. Journal of Computing in Civil Engineering: 2015; 04015074.

[2] Diwakar M, Kumar M.CT. Image noise reduction based on adaptive wiener filtering with Wavelet packet thresholding [C]. International Conference on Parallel Distributed and Grid Computing: 2014.

[3] Mazhari M, Hasanzadeh RPR. Suppression of noise in SEM images using weighted local hysteresis smoothing filter [J]. Scanning: 2016; 634-643.

[4] Colding TH, Minicozzi WP, Pedersen EK. Mean curvature flow [J]. Bulletin of the American Mathematical Socicty: 2015; 52 (2): 1.

[5] Fleishman S, Drori I, Cohen-Or D. Bilateral Mesh Denoising [J]. ACM Transactions on Graphics: 2003; 22 (3).

[6] Jones TR, Frédo D, Desbrun M. Non-Iterative, Feature-Preserving Mesh Smoothing [J]. ACM Transactions on Graphics: 2003; 22 (3): 943-949.

[7] Alexa M, Behr J, Cohenor D, et al. Point set surfaces [J]. IEEE Visualization: 2001.

[8] Fleishman S, Cohen-Or D, Silva, Cláudio T. Robust moving least-squares fitting with sharp features[J]. Acm Trans Graph: 2005; 24 (3): 544-552.

[9] Oztireli AC, Guennebaud G, Gross M. Feature Preserving Point Set Surfaces based on Non-Linear Kernel Regression[J]. Computer Graphics Forum: 2010; 28 (2): 493-501.

[10] Zhang ZP, Xu XY, Wang P. Spatial outlier mining algorithm based on KNN graph [J]. Computer Engineering: 2011; 37 (04): 37-39.

[11] Liu JJ. Research on denoising algorithm in 3D point cloud reconstruction [D]. Beijing Jiaotong University: 2019.

[12] Zhang BB. Research and Implementation of 3D Model Reconstruction and Optimization Technology [D]. Beijing University of Posts and Telecommunications: 2018.

[13] Li XG. Research on Outlier Detection Algorithm of Reconstruction Point Cloud Based on Image [D]. Yunnan Normal University: 2017.

Cite as: X.Y. Han et al.,
TBIS 10.3993/tbis (2021).

Interactive Design Platform for Pant Denim Development

Xiao-Yu Han[1], Qi Lin[2]

Quanzhou Normal University, Donghai Street No398, Fengze Distict, Quanzhou 362000,China

*Corresponding author's email: hxyjy123@126.com, 740662924@qq.com

Abstract

Nowadays, people are increasingly seeking out personalized clothing, breaking the past pattern where designer offers a design and customers passively accept it. Customers are demanding to be more involved in the design process and to express themselves in their everyday wear. This paper summarizes and analyzes the common styles and shapes of pant denim, explores the changes of various components and the applicable occasions of each element based on modularization, and eventually simulates an interactive design platform for pant denim using interactive genetic algorithm.

Keywords: Interactive Design; Pant Denim; Genetic Algorithm.

1. Data Analysis of Pant Denim

Interaction refers to the direct or indirect communication among people, equipment, and systems [1]. Compared with the traditional design discipline, interactive design pays more attention to the essence and connotation of a product. Some scholars define interactive design as "interactive product design to assist people's daily life and work" [2]. Interactive design roughly involves the following process: understand customer requirements, design options, create interaction pattern, and establish evaluation system.

With the development of society, customers pay more attention to personalized apparel. As such, many personalized customization systems are emerging, such as YUNYI [3], which provides customization of men' suit, and Cotte [4], choosing international famous fabric and developing an app to provide high-class customization service [5]. They all provide online figure measurement, design, and fitting service. However, most of them are complicated and at high cost, so the price of the apparel is too expensive for consumers. Meanwhile, as a symbol of fashion, pant denim is one of the most popular apparel for all ages, while there is no customization platform. It is therefore necessary to develop a design platform with high efficiency.

This paper focuses on how to build an interactive design system or platform for pant denim. Since users are often not clothing experts, they may usually type some descriptive words into the system. These words input by users are classified and summarized as below: leisure, casual, faddish, personalized, sexy, intellectual, etc. Obtaining the keywords on user requirements helps to identify the needs of users for clothing products. If any user wants to customize a pair of jeans for work on the interactive platform, it is difficult know exactly what kind of jeans the consumer needs with keywords alone. Therefore, the interactive design platform should collect various features of office clothing, such as simple, elegant and intellectual style; then the system should convert these keywords into the elements of clothing such as style, color, fabric and technique. For example, silhouette options include straight, tapered, and flared; the zip closure options include front, side and back. Users can select jeans according to their personal preferences.

The implementation of the interactive design system for pant denim creates the need to incorporate the data on style, color, fabric, and technique of corresponding pants into the system. Among them, style of clothing is the dominant element of clothing design and determines the overall apparel styling, and therefore forms the focus of this research. Specifically, the effect drawings of a wide range of pant denim are collected to analyze the element characteristics and classification of pant denim more thoroughly.

1.1 Style Analysis

The style of women's jeans studied herein is composed of length, looseness, silhouette, rise, pocket, closure, etc. Jeans can be roughly divided by length into short, middle and long ones and by silhouette into skinny, slim, medium-loose and loose ones. Silhouette options mainly include straight, tapered, and flared. Jeans rise includes low rise, mid-rise and high rise. Jeans pocket mainly includes side pocket and hip pocket, the former of which includes cross pocket, curved pocket and no pocket, and the latter includes patch pocket and patch pocket with a flap. Jeans closure includes zip, button and strap. Jeans button options include 1-button, 2-button and multi-button, and strap options include front, side and back straps. Wearing occasions of women's jeans include: work, business trip, traveling, shopping, dating, parties, performances, etc.

1.2 Color Analysis

Selection of clothing color depends much on the occasion of dressing. Usually, dark jeans are the best for formal occasions like work and business trip, including black, dark gray and navy blue ones; brighter, colored and white jeans are a good option for dating and parties; and jeans of super bright and dazzling colors such as gold and red are suitable for performance activities.

1.3 Fabric Analysis

The common fabrics of jeans include elastic weft denim, warp, or weft slub denim, ring spun yarn denim, overdyed denim, colored denim, super indigo dyed denim, etc.

1.4 Technique Analysis

There are two main techniques of jeans making, namely washing and trimming. Washing techniques with special effects include pickling, sanding, monkey washing, scratch washing, etc. Trimming techniques include burr, embroidery, pleat, rip, spray dye, rivet, leather patch, graffiti, zip, etc.

2. Process of Interactive Design

2.1 Overall Framework of Interactive Design

This paper is intended to build a web-based customization and interactive design platform for women's jeans by which users can customize their jeans using the available system data extracted from the database of style, color, fabric and technique. After the selecting available data, users need to evaluate the design renderings out of the combinations output by the system. Fig.1 shows the flow chart of human-computer interactive design for customizing women's jeans. A user should register with the system and fill in the leg size information before typing in her desired keywords based on the occasion and style of dressing. The system will then analyze the keywords input by the user according to the reasoning principles, and screen out the clothing style elements matching the keywords from the database for the user to select. If the user is satisfied with the options, she may click OK and go to the next step where she can select the system recommendations of combinations among fabric, color and trimming technique generated via the keywords. Afterwards, the user is required to evaluate the final design of clothing. If satisfied, she can click Confirm to submit the design, or otherwise the system will replace the jeans components through crossover or mutation in the interactive genetic algorithm, and execute a new round of combination among style, fabric, color and technique until the user is satisfied with the final design.

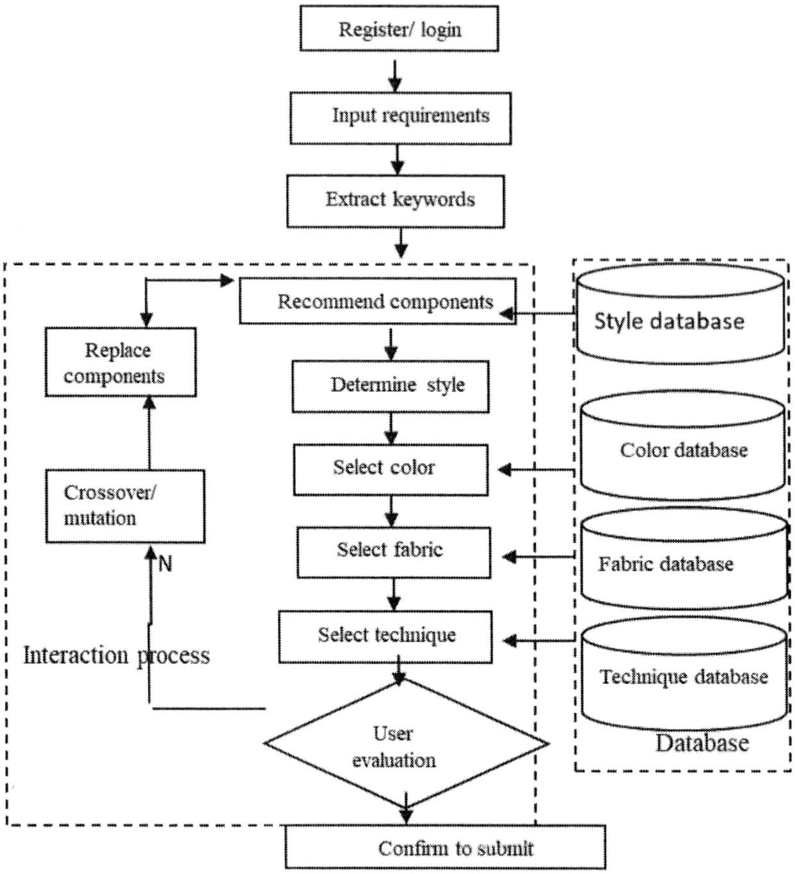

Fig.1 Flow chart of human-computer interactive design for customizing women's jeans

2.2 Database Establishment and Data Coding

Database is a warehouse that organizes, stores, and manages data based on a given structure, that is, the management mode of all kinds of data required by users [6]. The database of women's jeans studied in this paper is divided into style database, color database, fabric database and technique database, and then the database is constructed by texts, pictures and tables. The concept of "modular design" is also integrated into the database, which is a design theory and practice that subdivides a piece of clothing into a series of components called modules. Various final products are created through the selection and combination of these components so as to meet the different needs of customers for clothing [7].

The normal running of the interactive design platform necessitates the coding of each component for customers to choose. Genetic algorithm has the advantages of simple binary encoding and decoding, in addition to easy implementation of crossover and mutation steps, so it is selected for this research. Genetic algorithm uses binary coding to represent each component of clothing, that is, to use binary symbols 0 and 1 to form a coded set. The output of interaction results is achieved by decoding, i.e. the inverse process of encoding, for the purpose of generating the final product [8].

2.2.1 Style Database and Coding

There are many styles of jeans, with the main variations found in details. Jeans can be divided by details into six modules, including: Length module: short, middle and long; 2. Looseness module: skinny, slim, medium-loose, loose; 3. Silhouette module: straight, tapered and bell-bottomed; 4. Rise module: low rise, mid rise and high rise; 5. Pocket module: cross side pocket, curved side pocket, no side pocket, patch back pocket, patch back pocket with a flap, no back pocket; 6. Closure module: front zip, side zip, back zip, 1-button, 2-buttons, multi-button, front strap, side strap and back strap. However, in light of the design principle of jeans, some of the specific elements in each module cannot be configured, as shown in Table11. Further, based on the various occasions of dressing mentioned above, it is necessary to match the feature elements of each component of women's jeans with keywords for each occasion, including simple, personalized, ladylike, intellectual, etc. The specific component elements of each module have corresponding 1-4 keywords. The components of jeans style are coded by modules. For example, in the Length module, short jeans is coded 00, middle jeans 01, and long jeans 10; in the Looseness module, skinny jeans is coded 00, slim jeans 01, medium-loose jeans 10, and loose jeans 11. In the Silhouette module, straight jeans is coded 00, tapered jeans 01, and bell-bottomed jeans 10. The remaining modules are coded in the same way, using coding numbers 0 and 1.

Table 1. Configuration of components beyond the design rules

Short	Tapered, bell-bottomed
Middle	Bell-bottomed, 5-button
Skinny	Straight, bell-bottomed
Loose	Bell-bottomed, no side pocket, no back pocket, side zip, back zip
Low rise	2-button, multi-button

2.3 Other Database Establishment and Coding

Color database and coding. Colors include scarlet, pink, light yellow, dark yellow, medium green, dark green, greenish black, navy blue, sky blue, dark blue, and deep purple; neutral colors include black, white, and dark gray. All the colors must be coded. The corresponding codes for 19 colors are: 00000 light blue, 00001 medium blue, 00010 dark blue, 00011 watermelon red, 00100 purplish red, 00101 pink, 00110 orange, 00111 light yellow, 01000 dark yellow, 01001 light green, 01010 medium green, 01011 dark green, 01100 lavender, 01101 dark purple, 01110 black, 01111 white, 10000 light gray, 10001 medium gray, 10010 dark gray.

Fabric database and coding. The common jeans fabrics are coded and stored in the database, including 000 elastic weft denim, 001 warp or weft slob denim, 010 ring spun yarn denim, 011 overdyed denim, 100 colored denim and 101 super indigo dyed denim.

Technique database and coding. There are two main techniques of jeans making, namely washing and trimming. When coding jeans techniques, since the two techniques can co-exist on a pair of jeans, the washing technique and trimming technique are coded separately as follows: 00 pickling, 01 sanding, 10 monkey washing, 11 scratch washing; 0000 burr, 0001 embroidery, 0010 pleat, 0011 rip, 0100 spray dye, 0101 rivet, 0110 leather patch, 0111 ironing applique, 1000 graffiti, 1001 zip.

3. Application of Interactive Genetic Algorithm

The essence of genetic algorithm is to search the content of requirements directly. When solving a problem about genetics, the system should first code the problem as an entirety. At the same time, the set of individuals is called population, and the individuals in a population set represent the size of the population [9] [10]. After the individuals are coded, the system must feedback the fitness of each individual, and the fitness must be determined according to the problem seeking the optimal solution. The system generates many different populations in iteration, from which the optimal solution can be obtained. Individuals in each generation will cross and mutate with the help of genetic operators and fitness values, thus generating a new population set [10] [11]. This algorithm is an exact match for the ideas of jeans interactive design.

3.1 Fitness Evaluation

Interactive genetic algorithm (IGA) is a technique that combines human and genetic algorithm for optimization of the target, which encompasses subjective factors such as user's personal preference, emotion and psychology. The individual fitness is not obtained from the established evaluation function but derived out of the user's subjective intentions input to the system [13-14]. When operating interactive design, the use knows little about the relevant expertise, so she needs a handy method for individual fitness input that is easy to operate. In this paper, slider is used to allow users to drag the slider with the mouse to the evaluation range that she deems appropriate, so that the whole evaluation process can be completed quickly and provide feedback for interactive design.

The slider helps users with fitness evaluation on the clothing keywords (like simple, personalized, ladylike, intellectual) and the final rendering of clothing. In this paper, English letters are used to express the correlation, as shown in Fig.2. After a user completes evaluation with the slider, the interactive system will iterate and generate new individuals through selection, crossover and mutation, so as to output the user evaluation with the highest fitness.

A No correlation B	**Weak correlation C**	**Medium correlation D**	**Strong correlation**

Fig.2 Schematic diagram of slider

3.2 Control of Probability

There are two probability parameters in interactive genetic algorithm: crossover probability and mutation probability. Crossover probability is a measure of the probability that crossover occurs between individuals, so the selection of crossover probability is also very important. An excessively high probability will bring about a random crossover to disturb the fitness of each individual and affect the evolution and iteration of the population. An excessively low probability will lead to the slow generation of new individuals and affect the efficiency of the interactive system. Therefore, the crossover probability Pc typically ranges from 0.41 to 0.98, which is set to 0.65 in this paper [12]. The resulting outcome of mutation probability is same with that of crossover probability, so mutation probability also cannot be too high or too low. Generally, mutation probability Pm is set to 0.055[13][14].

3.3 Results Output and Feedback

On the interactive platform, the system will absorb the keywords input by the user, work out the corresponding sketch of style using the optimal solution of the algorithm, and then display it on the interface of the interactive system for users to choose. The user can choose her favorite one and then make feedback evaluation on the results. The system will calculate the corresponding fitness value according to the user's evaluation, eliminate the clothing components and elements with low fitness, and allow the clothing components and elements with high fitness to appear in the search results with same keywords.

Taking the requirement "pant denim for dating" input by the user as an example, we illustrate the interface of a simple interactive platform for design of pant denim, as shown in Fig.3. After entering this interface, the user can select desired style, color, fabric, and technique from the four modules on the left side of the interface, and then select her preferred elements from the specific classification of elements on the right side according to the style diagram, and finally view the selected elements in the list of used elements displayed on the bottom right of the interface. At last, the user can preview the final rendering in the effect display area. If satisfied, she can submit it to complete the customization of jeans, or otherwise she can clear it and select new elements to combine.

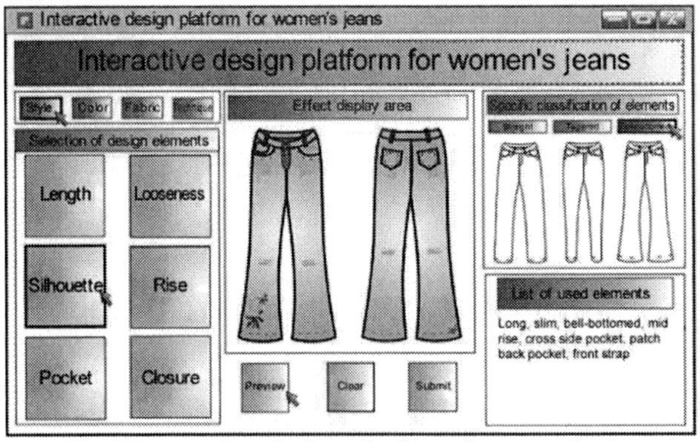

Fig.3 Interactive design platform for pant denim

Basing on analyzing the style of pant denim and using the genetic algorithm, the article has developed an Interactive design platform for pant denim. Customers can choose the style, color, fabric, and technique and evaluate the result to get the optimal pant for themselves.

4. Funding

The authors disclosed receipt of the following financial support for research, authorship, and/or publication of this article: This research project is funded by Young and Middle-aged Teacher Educational Research Project of Fujian Province. The project number is JAT190529.

References

[1] Huang Q, Bi ZW. Interactive Design [M]. Zhejiang University Press: 2012; 25-26.
[2] Li J, Li SG. Exploring the Emotional Communication Between Human and Products from the Perspective of Interaction Design [J]. Packaging Engineering: 2008; 29 (9): 151-153
[3] Song Y. Interactive design of cheongsam personalized customization and display system [J]. Textile Journal: 2021; 42 (04): 144-148
[4] Song Y, Guo Y. Interactive design of men's shirt customization system based on virtual display [J]. Wool Textile Journal: 2021; 49 (06): 70-74
[5] Shang HL, Chen YY. Influencing factors of online apparel customization for customers [J]. Wool Textile Journal: 2018; 46 (5) : 62-66.
[6] Wang S, Sa SX. Introduction to Database System (5th Edition) [M]. Beijing: Higher Education Press: 2014.
[7] Zhu XJ, Li XF, Chen XY. Interactive genetic algorithm based on typical style for clothing customization [J]. Journal of Engineered Fibers and Fabrics: 2020 (15); 1-9.
[8] Yang XH, Shen ZY. Intelligent Algorithms and Their Applications in Resource and Environment System Modeling [M]. Beijing: Beijing Normal University Press: 2005.
[9] Cheng G, Li JY. Research on the Clothing Style Preference Model Based on Interactive Genetic Algorithm [J]. Computer Applications and Software: 2011; 28 (2): 229-231: 238.
[10] Zhu XJ, et al. Interactive genetic algorithm based on typical style for clothing customization [J]. Journal of engineered fibers and fabrics: 2020 (15); 1-9
[11] Run LD, Zong C, Guo FN. Multi-stage interactive genetic algorithm for collaborative product customization [J]. Knowl Based Syst: 2016; 92: 43-54.
[12] Guan CY, Qin SF, Long Y. Apparel-based deep learning system design for apparel style recommendation. International Journal Of Clothing Science And Technology [J]: 2019; 3 (31): 376-389.
[13] Qu LD, He DX. Artificial Fish Swarm Algorithm Based on Hybrid Mutation Operator [J]. Computer Engineering and Applications: 2008; 44 (35): 50-52.
[14] Yang LM. Research and Application of Interactive Genetic Algorithm Based on Gene Classification [D]. Qingdao University: 2011.

Evaluation of Radiant Heat Transmission of Biopolymer Based Flame Retardant Treated Cotton Fabrics

Sajid Faheem[1*], Nazia Nahid[2], Jakub Wiener[1], Vijay Baheti[3], Adnan Mazari[4], Jiri Militky[1]

[1]*Department of Materials Engineering, Faculty of Textile Engineering, Technical University of Liberec, Studentska 1402/2, Liberec-46117, Czech Republic*
[2]*Department of Bioinformatics and Biotechnology, Government College University Faisalabad, Faisalabad-38000, Pakistan*
[3]*Department of Textile Technology, Indian Institute of Technology, New Delhi, Dehli-110016, India*
[4]*Department of Clothing, Faculty of Textile Engineering, Technical University of Liberec, Studentska 1402/2, Liberec-46117, Czech Republic*

*Corresponding author's email: msajidfaheem@gmail.com

Abstract

The main objective of this research work is to investigate the improvement in thermal resistance performance of the flame retardant cotton fabric treated with milk casein, a biopolymer-based flame retardant. Milk casein (a phosphoprotein) was applied, as a green flame retardant finishing material, on the cotton fabric through aqueous solutions by using an industrially applicable method. For estimating the thermal protective performance, casein treated cotton fabrics were exposed to a radiant heat source at a specific radiant heat flux density Q_o (kW/m^2). The lower values of transmitted heat flux density Q_c (kW/m^2) and heat transmission factor TFQ_o (%) were observed for the casein treated cotton fabric samples as compared to untreated cotton fabric.

Keywords: Cotton Fabric; Milk Casein; Green Flame Retardant; Thermal Resistance Performance

1. Introduction

Modern technologies and developments have contributed to a vast increase in the many types of hazards to which human beings are exposed. The protection from flame, fire, heat radiation sources, etc., is a principal requisite for both civil and defense applications. Most of these hazards are confronted at workplaces. The workplace hazards can be categorized into chemical, mechanical, radiation, biological, and thermal hazards [1]. The thermal resistance behavior of textiles provides protection from the thermal hazards that comprise of exposure to high-temperature radiant sources, flame impingement, hot gases and liquids, molten substances, hot solids and surfaces [2]. The heat transfer from a thermal hazard can be ensued by different phenomena such as radiation, convection or conduction, and/or a combination. Three general classifications are used to categorize the common thermal environmental conditions, namely routine, hazardous, and emergency [3, 4]. Many researchers have performed work on the category of emergency conditions [5-10], which are characterized by high-intensity exposures for short durations. The exposure levels among these emergency conditions normally range from 20-160 kW/m^2, which can be experienced during large fires and explosions. Severe thermal problems and life-threatening injuries are connected with these conditions. The total heat energy from a flame/fire can cause textile clothing to ignite and/or melt, subsequently fissuring and causing severe burns to the skin. Skin burns occur when the human tissue temperature is raised from a normal blood temperature of 36.5°C to 44.0°C at a rate that depends on the level of raised temperature [11-13]. Thermal resistance textiles should provide insulation against heat, have ignition resistance, remain intact, shrink and melt resistance as well as resistance from forming brittle char residues that may rupture and expose the wearer [14]. Flame retardant textile apparel (clothing/garments) and fabrics are a crucial requirement and are currently used as personal protective equipment (thermal resistance substrates), in order, to ensure people's survivability and to protect the structure against a thermal exposure. In many industrial settings, workers face potential exposure to an intense radiation heat flux. The thermal exposure may cause skin burns even on wearing a flame retardant clothing or garment. Therefore, the significance of thermal performance of flame retardant textile fabrics is indisputable [15].

Cotton; a natural cellulosic fiber, is one of the most important biopolymers in the world. It is commonly used in wide application areas, indeed, due to its availability in large quantities and plentiful diverse advantages such as good mechanical properties, hydrophilicity, physiological comfort, biodegradability, absence of static charge problems, etc. [16]. However, the thermal properties such as; the high flammability, ease of ignition, low thermal stability, and quick burning/combustion of cotton textiles [17], restricted many of their specific applications in transportation, automotive, protective garments, military, home/furniture upholstery, apparels, bed/night-wears, etc. Thus the demands for flame retardant cotton textiles possessing good radiant heat resistance have become an urgent exigency and have had steady growth for the past decades. The flame retardant treatments of organic and/or inorganic halogen, nitrogen, and phosphorus compounds are generally applied on cellulosic materials for the prevention of fire and protection of human life [18, 19]. These compounds act as flame retardants because they influence the pyrolysis reaction, prevent the formation of levoglucosan and flammable volatiles, and increase the formation of char [18, 20-22]. Nowadays, the search for environment-friendly flame retardants that ensure fire performances comparable to the conventional flame retardant compounds has gained significant importance [23-25]. This is because of the persistent, bio-accumulative, and environmentally toxic nature of some molecules employed in the formulation of conventional flame retardants [26]. In this context, the effectiveness of different bio-macromolecules (whey proteins, caseins, hydrophobins, and deoxyribonucleic acid) has been assessed for textiles recently [27]. Among them, the application of casein as flame retardant provided an attractive solution for the valorization of the dairy industry by utilization of its by-product/waste. Casein is the major milk protein (80%) and is obtained as a co-product during the production of skim milk. The casein macromolecules can be considered as polyamino acids bearing several phosphate groups in their micellae structure [28], which decompose similarly to the ammonium polyphosphate salt. In addition, their moisture adsorption features could partially dissipate the heat evolved during the combustion of the fabric, and dilute the produced combustible volatile species.

It is evidenced, from the data results of our previous work that the casein treatments on cotton fabrics promoted an increase in total burning time, thus exhibiting their ability for the delayed burning and partial flame protection [29]. Furthermore, the casein treated cotton fabric samples burnt at a slower rate and resulted in stronger char residue while preserving the original texture of the fabrics. Hence, in the present work, casein solutions of different concentrations and casein treated cotton fabric samples were prepared according to described and used methodology in our previous work [29]. Moreover, the thermal resistance performance of the developed flame retardant casein treated cotton fabrics has been investigated by exposing them to a radiant heat source at the specific radiant heat flux density.

2. Experimental Part

2.1 Materials

A bleached 100% cotton woven fabric having 145 g/m2 area density, purchased from Licolor, Czech Republic, was used in this research work. Bovine milk casein (technical grade) was received from Sigma-Aldrich, Czech Republic with approximate composition (g/l) of: α-s1 12-15, α-s2 3-4, β 9-11 and κ 2-4, as specified in the product information sheet. Sodium hydroxide (analytical grade $\geq$98%) was obtained from Lach-Ner, Czech Republic.

2.2 Preparation of Casein Solution

The aqueous solutions of casein were prepared by dissolving the casein powder in distilled water at 5, 10, 15, 20, 25, 30 w/v % concentration under magnetic stirring (300-600 rpm). The solutions were subsequently heated at 80°C in a thermostatic bath and the pH was adjusted to 9 with dropwise addition of 1M NaOH, under continuous magnetic stirring. The process was later stopped when the casein was completely dissolved. Finally the solutions were cooled down to almost 30°C before applying to cotton fabrics.

2.3 Application of Casein Solution onto Cotton Fabric

The lab-scale two roller padder (Mathis AG, Switzerland), was used to apply the casein solutions on the cotton fabrics according to the padding method, as shown in Fig.1. The prepared casein solutions of different concentrations were applied on cotton fabric samples by adjusting the pressure and speed of the padding rollers to achieve 80% wet pick-up of the casein solutions. The treatment of cotton fabrics was carried out in a climatic chamber at 30°C and 30% RH atmosphere. The casein treated samples were then dried at 100°C in a lab-scale hot air stenter-frame dryer (Mathis AG, Switzerland). Subsequently, the uptake (i.e. total dry solids or add-on percentage) of casein by cotton fabrics was calculated using Eq.1, according to the gravimetric principle from the oven-dry weight of the fabric samples before and after treatments.

$$\text{Add on (wt.\%)} = \frac{w_f - w_i}{w_i} \times 100 \qquad 1)$$

where 'w_i' and 'w_f' are the oven-dry weights of the untreated and casein treated cotton fabric samples respectively.

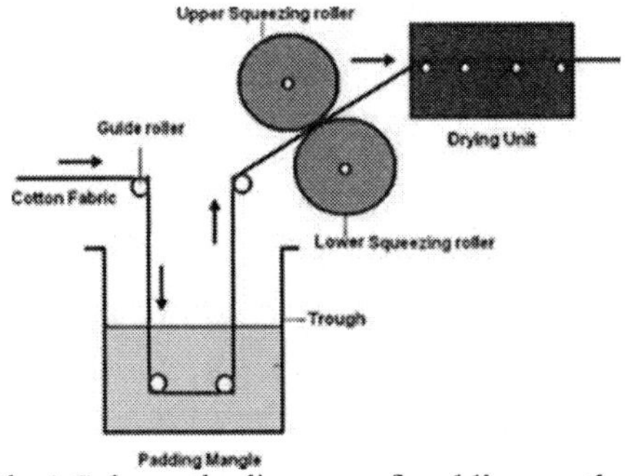

Fig.1 Schematic diagram of padding method

3. Characterization and Testing

The untreated and casein treated cotton fabric samples were conditioned, in a standard atmosphere with 65±2% relative humidity and 20±2°C temperature for at least 24 hrs, before performing any characterization and testing.

3.1 Surface Morphology Analysis

The surface morphology, of untreated and casein treated cotton fabric samples was examined by scanning electron microscope (TS5130-Tescan SEM) at an accelerating voltage of 20-25 kV. The samples were mounted rigidly on a specimen holder/stub using a conductive adhesive. Prior to SEM analysis, the surface of all the samples was sputter-coated with gold (a conductive layer). A sputter-coating was done to improve the electrical conductivity and stability of the samples. So that, they can withstand the high vacuum conditions and the high energy (voltage) beam of electrons, generated by a heated tungsten filament electron gun/source (cathode). The images were taken at different suitable accelerating voltages and magnifications through slow-scanning speeds to obtain higher quality images.

3.2 Radiant Heat Transmission Analysis

The transmission of radiant heat (heat flux/heat flux density) through untreated and casein treated cotton fabrics was evaluated by employing the radiant heat transmission analyzer (combustion behavior test equipment X637 B) [30], according to the standard test method ISO 6942-B. The equipment consists of a heating arrangement, curved copper plate calorimeter (with an area of 50 mm x 50.3 mm and mass of 35.9-36.0 grams), and a moveable frame assembly as shown in Fig.2. The heating arrangement comprises six carbide rods serving as a radiation heat source, which can generate heat flux density up to 80kW/m^2. The calorimeter is mounted on a non-combustible block and determines the heat flux. Its face is coated with a thin film of an optically black paint having a coefficient of absorption greater than 0.9. The moveable frame assembly is constantly cooled with water flowing in cooling pipes. It contains a holder for the calorimeter and a moveable screen to expose the calorimeter (without or with fabric sample, mounted over it) to radiant heat.

Fig.2 Radiant heat transmission analyzer

At first, the calibration was performed by exposing the calorimeter without mounting the fabric sample on its face side. The moveable screen was withdrawn and reverted to the point when the rise of temperature was reached 30°C and incident heat flux density Q_o (kW/m^2) was evaluated from Eq.2.

$$Q_{\circ} = \frac{C_p RM}{a.A} \qquad 2)$$

where,

C_p is the specific heat of the copper metal (0.385 kJ/kg°C),
R is the rising rate of the calorimeter temperature in the linear region in (°C/s)
M is the mass of copper plate (kg)
a is the absorption coefficient of the painted surface of calorimeter
A is the area of the copper plate (m^2)

Then the fabric sample of specified size, having length 230 mm and width 80 mm, was clamped in such a way that it was in contact on the face side of the calorimeter applying a mass of 200 grams. The calorimeter, with a mounted fabric sample over it, was subjected to heat radiations (medium heat flux density of 40kW/m^2) by lifting the moveable screen in its open position. It was reverted to its closed position as the temperature reached 30°C. Then the time in seconds for temperature acceleration/rise of 12°C and 24°C in the calorimeter was recorded by a copper constantan thermocouple, mounted on its back and expressed as $RHTI_{12}$ and $RHTI_{24}$. The transmitted heat flux density Q_c (kW/m^2) was evaluated by Eq.3.

$$Q_c = \frac{C_p M 12}{A.(t_{24} - t_{12})} \qquad 3)$$

where,

$12/t_{24} - t_{12}$ is the mean rising rate of the calorimeter temperature (°C/s) in the region between 12 and 24°C
t_{12} is the time to attain (12±0.1) °C rise in calorimeter temperature
t_{24} is the time to attain of (24±0.2) °C rise in calorimeter temperature

Heat transmission factor TFQ_o (%) for incident heat flux density level was elucidated by Eq.4.

$$TFQ_{\circ} = \frac{Q_c}{Q_{\circ}} \times 100 \qquad 4)$$

The test was performed in a room that was free of any air currents and any system capable of producing stray heat radiations. The temperature of the test room was maintained between 15°C and 35°C. All the samples were conditioned for at least 24 hours at a temperature of (20 ± 2)°C and relative humidity of (65 ± 2) %. At least five replicas were tested for each fabric sample. The output results of this test were concluded and expressed in the form of two threshold times i.e., the radiant heat transfer indexes $RHTI_{12}$ (threshold time in seconds (t_{12}), when the temperature of calorimeter increase in 12°C) and $RHTI_{24}$ (threshold time in seconds (t_{24}), when the temperature of calorimeter increase in 24°C) respectively, incident heat flux density (Q_o), transmitted heat flux density (Q_c) and the percentage heat transmission factor (% TFQ_o as the mean % heat transmission over the range between 12°C and 24°C) [30].

4. Results and Discussion

In textile chemical finishing processes and/or after-treatments, the different desired end-use aesthetic, as well as, the functional properties presented by the textiles, depend on the chemical/finish amount deposited on the textile substrates. Therefore, the casein add-on/weight-gain (wt. %) of cotton fabrics was assessed, after treatment with casein solutions of different concentrations (w/v %) through roller padding method at fixed/constant wet pick up percentage. The effect of casein solutions of different concentrations, on the casein add-on of cotton fabrics is shown in Fig.3. The add-on was found to increase with an increase in the casein solution concentration. It was also observed that the higher casein add-on was obtained beyond the 20 w/v % solution. It was due to the increased viscosity of the casein solution usually at higher concentrations [31], beyond 20 w/v % that results in maximum wetting and binding of individual fibers and yarns, and ultimately retaining in the fabric structure after squeezing and drying as evidenced from the SEM images of casein treated cotton fabric samples (see Fig.3) .

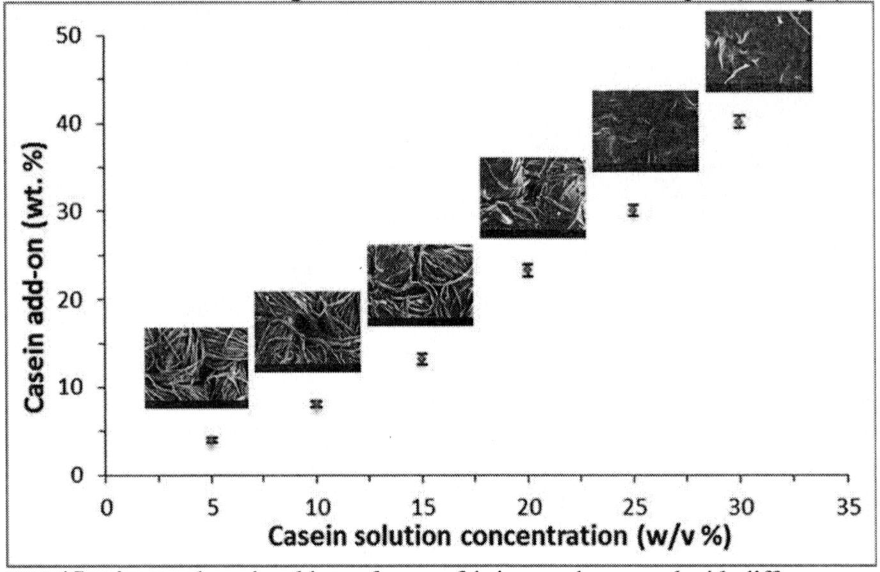

Fig.3 SEM magnifications and casein add-on of cotton fabric samples treated with different casein solutions

4.1 Radiant Heat Transmission through Untreated and Casein Treated Cotton Fabrics

The thermal resistance behavior, of untreated and casein treated cotton fabrics, as protection against radiant heat, was evaluated by exposing the fabric samples to a radiant heat source at the medium heat flux density (Q_o) of 40 kW/m^2. The rise of temperature was measured at the back of cotton fabric samples by calorimeter, which resulted in; two threshold times / radiant heat transfer indexes, i.e., time for the rise of 12 °C ($RHTI_{12}$) and time for the rise of 24°C ($RHTI_{24}$), transmitted heat flux density (Q_c) and percentage heat transmission factor (% TFQ_o), as listed in Table 1. The lower values of Q_c and ultimately an analogous pattern in the values of percentage TFQ_o was noticed for the casein treated cotton fabric samples as compared to untreated cotton fabric. The drop in Q_c and percentage TFQ_o values was attributed to the greater difference between $RHTI_{12}$ and $RHTI_{24}$ values, owing to the thicker casein layer/films as well as the blocking of spaces/interstices between the cotton fibers and yarns [32-34], due to the higher casein add-on as corroborated from the SEM micrographs (see Fig.2). The higher values of radiant heat transmission indexes and their difference ($RHTI_{24}$-$RHTI_{12}$), for the casein treated fabrics, as compared to untreated cotton fabric, demonstrated that the rate of temperature rise occurred at a lower rate with respect to time, indicating the dilatory exchange of radiant heat towards calorimeter (subsequently the slower transmission of heat exchange). The time differences ($RHTI_{24}$ - $RHTI_{12}$) provided a good indicator of the skin pain alarm time. In different studies, it has been indicated that the pain threshold of human skin is around 44°C. When the temperature of skin exceeds this threshold, the absorbed energy by the skin determines if and how much severe burns will be instigated and received. Subsequently, the skin receives second-degree burns, when the temperature of the skin approaches 55°C. The time gap between these two points/thresholds, i.e., the time when the skin starts to feel pain and when it receives irreversible burns, is called the skin pain alarm time [12, 35, 36].

Table 1. Radiant heat resistance parameters of untreated and casein treated cotton fabrics

Sample	Q_0 (kW/m^2)	$RHTI_{12}$ (sec)	$RHTI_{24}$ (sec)	$RHTI_{24}$-$RHTI_{12}$ (sec)	Q_c (kW/m^2)	TFQ_0 (%)
Untreated	40	3.8±0.04	6.7±0.05	2.9±0.04	23.10±0.37	57.76±0.92
5% casein conc.	40	3.9±0.08	6.9±0.08	3.0±0.07	22.19±0.52	55.46±1.31
10% casein conc.	40	4.1±0.07	7.2±0.08	3.1±0.04	21.33±0.30	53.32±0.75
15% casein conc.	40	4.2±0.07	7.4±0.09	3.2±0.05	20.54±0.35	51.35±0.86
20% casein conc.	40	4.5±0.09	7.9±0.11	3.4±0.08	19.46±0.48	48.66±1.20
25% casein conc.	40	4.7±0.08	8.3±0.08	3.6±0.07	18.49±0.36	46.21±0.91
30% casein conc.	40	5.1±0.10	9.0±0.11	3.9±0.09	16.89±0.39	42.23±0.98

These higher values of ($RHTI_{24}$ - $RHTI_{12}$) for the casein treated fabrics endorsed the reason that infrared radiation that performed a pertinent part in heat transference was also absorbed by casein layers/films due to which casein layers/films provided the improved thermal resistance and insulation. That might be helpful to prevent the wearer from acquiring any severe burn injuries. Simultaneously, the reduction in the value of transmitted heat flux density, i.e., the lower quantity of radiant heat transmitted, was also beneficial for the textile substrate, i.e., cotton fabric, resulting in its later on slower burning. More accurately, it can be inferred that the resistance to a radiant heat flux of casein layers/films can also dramatically reduce the time to ignition of casein treated fabrics as compared to untreated cotton fabric. This can be accredited to the degradation/decomposition of casein to release phosphoric acid exerting a deleterious effect that sensitizes the early ignition of cotton fabric and favors the cellulose dehydration to produce more thermally stable char residues instead of volatile release [37].

As, the investigation of thermal resistance performance of single-layer flame retardant woven fabrics is highly relevant, in order, to procure and further develop the thermal protective multi-layer clothing and their overall thermal protective performance [2, 6, 14, 33, 38-40]. The general obtained fallouts from this research study/work were supported and in close agreement with the results that were already reported in the literature. Such as, Mandal et al. evaluated the thermal resistance behavior and/or thermal protective performance of single-layer flame retardant woven fabrics by calculating the skin pain alarm time ($RHTI_{24}$ - $RHTI_{12}$) [33]. These single-layer woven fabrics were constructed in different weave designs from the combination of natural and synthetic flame retardant fibers. However, the value of skin pain alarm time for the single layer flame retardant woven fabrics was found in the range from 2.1-3.1 sec, at an incident heat flux density of 40kW/m^2. It was also concluded that in the case of single-layer woven fabrics systems, the weight and/or thickness fundamentally control the thermal resistance of the fabric.

5. Conclusion

The casein protein was applied on cotton fabrics to provide an environment-friendly flame retardant and the thermal resistance performance of the casein treated cotton fabrics was evaluated in this research work at a medium level of radiant heat flux density. The shift of transmitted heat flux density (Q_c) and percentage (TFQ_o) towards the lower values was observed, due to the greater difference between $RHTI_{12}$ and $RHTI_{24}$ indexes. The drop of heat transmission factor (TFQ_o) was found (-26%) for the cotton fabric treated with 30 w/v % casein solution at the medium heat flux density of 40kW/m^2. Consequently, the single-layer casein treated flame retardant cotton fabrics can be considered as an attractive approach for thermal insulation and subsequent prevention against different types of radiant heat flux accidents.

6. Acknowledgement

This work was supported by the Ministry of Education, Youth and Sports of the Czech Republic, and the European Union - European Structural and Investment Funds in the frames of Operational Programme Research, Development, and Education - project Hybrid Materials for Hierarchical Structures (HyHi, Reg. No. CZ.02.1.01/0.0/0.0/16_019/0000843).

References

[1] Bajaj P, Sengupta AK. Protective clothing. Text Prog: 1992; 22: 1-110.
[2] Song G, Paskaluk S, Sati R, et al. Thermal protective performance of protective clothing used for low radiant heat protection. Text Res J: 2011; 81: 311-323.
[3] Abbott NJ, Schulman S. Protection from fire: Nonflammable fabrics and coatings. J Coat Fabr: 1976; 6: 48-64.
[4] Hoschke BN. Standard and specifications for firefighters' clothing. Fire Saf J: 1981; 4: 125-137.
[5] Lee YM, Barker RL. Thermal protective performance of heat-resistant fabrics in various high intensity heat exposures. Text Res J: 1987; 57: 123-132.
[6] Sun G, Yoo HS, Zhang XS, et al. Radiant protective and transport properties of fabrics used by wildland firefighters. Text Res J: 2000; 70: 567-573.
[7] Shalev I, Barker RL. Analysis of heat transfer Characteristits of fabrics in an open flame exposure. Text Res J: 1983; 53: 475-482.
[8] Shalev I, Barker RL. Protective fabrics: A comparison of laboratory methods for evaluating thermal protective performance in convective/radiant exposures. Text Res J: 1984; 54: 648-654.
[9] Crown EM, Ackerman MY, Dale JD, et al. Design and evaluation of thermal protective flightsuits. Part II: instrumented mannequin evaluation. Cloth Text Res J: 1998; 16: 79-87.
[10] Song G, Barker RL, Hamouda H, et al. Modeling the thermal protective performance of heat resistant garments in flash fire exposures. Text Res J: 2004; 74: 1033-1040.
[11] Buettner K. Effects of extreme heat and cold on human skin. II. Surface temperature, pain and heat conductivity in experiments with radiant heat. J Appl Physiol: 1951; 3: 703-713.
[12] Stoll AM, Chianta MA. A method and rating system for evaluation of thermal protection. Naval Air Development Center Warminster Pa Aerospace Medical Research Dept: 1968.
[13] Stoll AM, Chianta MA. Heat transfer through fabrics as related to thermal injury. Trans N Y Acad Sci: 1971; 33: 649-670.
[14] Perkins RM. Insulative values of single-layer fabrics for thermal protective clothing. Text Res J: 1979; 49: 202-212.
[15] Kilinc FS. Handbook of fire resistant textiles.
[16] Bashar M, Khan M. An Overview on Surface Modification of Cotton Fiber for Apparel Use. J Polym Environ; 21. Epub ahead of print 1 March: 2012. DOI: 10.1007/s10924-012-0476-8.
[17] Wakelyn PJ, Rearick W. Cotton and flammability: overview of new developments. Am Dyest Report: 1998; 87: 13-21.
[18] Horrocks AR. Flame-retardant Finishing of Textiles. Rev Prog Color Relat Top: 1986; 16: 62-101.
[19] Lu SY, Hamerton I. Recent developments in the chemistry of halogen-free flame retardant polymers. Prog

Polym Sci: 2002; 27: 1661-1712.
[20] Beninate JV, Trask BJ, Drake GL. Durable Flame-Retardant Treatments for Blends of Cotton, Wool, and Polyester. Text Res J: 1981; 51: 217-224.
[21] Wu W, Yang CQ. Comparison of different reactive organophosphorus flame retardant agents for cotton: Part I. The bonding of the flame retardant agents to cotton. Polym Degrad Stab: 2006; 91: 2541-2548.
[22] Wu W, Yang CQ. Comparison of different reactive organophosphorus flame retardant agents for cotton. Part II: Fabric flame resistant performance and physical properties. Polym Degrad Stab: 2007; 92: 363-369.
[23] Alongi J, Carosio F, Malucelli G. Current emerging techniques to impart flame retardancy to fabrics: An overview. Polym Degrad Stab: 2014; 106: 138-149.
[24] Basak S, Ali SW. Sustainable fire retardancy of textiles using bio-macromolecules. Polym Degrad Stab: 2016; 133: 47-64.
[25] Pan H, Song L, Hu Y, et al. An Eco-friendly Way to Improve Flame Retardancy of Cotton Fabrics: Layer-by-Layer Assembly of Semi-biobased Substance. Energy Procedia: 2015; 75: 174-179.
[26] Nguyen TMD, Chang S, Condon B, et al. Development of an environmentally friendly halogen-free phosphorus-nitrogen bond flame retardant for cotton fabrics. Polym Adv Technol: 2012; 23: 1555-1563.
[27] Malucelli G, Bosco F, Alongi J, et al. RSC Advances Biomacromolecules as novel green fl ame retardant systems for textiles : an overview †. RSC Adv: 2014; 4: 46024-46039.
[28] Phadungath C. Casein micelle structure : a concise review. J Sci Technol: 2005; 27: 201-212.
[29] Faheem S, Baheti V, Nahid N, et al. Flame retardancy, physiological comfort and durability of casein treated cotton fabrics. Fibers Polym: 2019; 20: 1011-1020.
[30] Mess GW. Combustion Behavior Test Equipment according to ISO 6942 -Manual. Dr-Ing Georg Wazau Meß- und Prüfsysteme GmbH: 2002; 1-9.
[31] Zoller HF. Casein viscosity studies. J Gen Physiol: 1921; 3: 635-651.
[32] Zheng Z, Wang H, Zhang N, et al. Heat transfer through glassfiber fabrics coated with silicone resin. J Text Inst: 2017; 108: 743-749.
[33] Mandal S, Annaheim S, Camenzind M, et al. Characterization and modelling of thermal protective performance of fabrics under different levels of radiant-heat exposures. J Ind Text: 2019; 48: 1184-1205.
[34] Zhang M, Denning R, Zhong Z, et al. Investigation of Heat Transfer Properties of Plasma-Treated and Silicone-Elastomer Coated Basalt Fabric. Coatings: 2019; 9: 292.
[35] Hardy JD, Wolff HG, Goodell H. Pricking pain threshold in different body areas. Proc Soc Exp Biol Med: 1952; 80: 425-427.
[36] Stoll AM, Greene LC. Relationship between pain and tissue damage due to thermal radiation. J Appl Physiol: 1959; 14: 373-382.
[37] Alongi J, Carletto RA, Bosco F, et al. Caseins and hydrophobins as novel green flame retardants for cotton fabrics. Polym Degrad Stab: 2014; 99: 111-117.
[38] Rossi RM, Zimmerli T. Influence of humidity on the radiant, convective and contact heat transmission through protective clothing materials. In: Performance of Protective Clothing: Fifth Volume. ASTM International: 1996.
[39] Rossi RM, Bolli W, Stämpfli R. Performance of firefighters' protective clothing after heat exposure. Int J Occup Saf Ergon: 2008; 14: 55-60.
[40] Mandal S, Song G, Ackerman M, et al. Characterization of textile fabrics under various thermal exposures. Text Res J: 2013; 83: 1005-1019.

Impact of Morphology and Content of Fly Ash Filler on the Mechanical Properties of Epoxy Resin Composites

Shi Hu[1], Dan Wang[1], Şebnem Sözcü[1], Aamir Mahmood[1], Dana Křemenáková[1], Jiří Militký[1], Yuan-Feng Wang[1]

[1]*Department of Material Engineering, Faculty of Textile Engineering, Technical University of Liberec, studentska 1402/2, Liberec, Czech Republic*

*Corresponding author's email: shi.hu@tul.cz

Abstract

Fly ash (FA) is a particulate waste generated from the combustion of coal in thermal power plants. The reuse of fly ash as filler in the epoxy matrix has been studied by many researchers. In this review article, the impact of fly ash morphology and content of fly ash filler on the epoxy matrix's mechanical property has been discussed from several angles. The difference in fly ash particle size will directly affect fly ash content added to improve epoxy resin composites' mechanical properties. Fly ash with uniformed particle size performed better on epoxy composites than mixed-sized fly ash filler. To improve the adhesion characteristics at the interface, surface modification of fly ash such as silane treatment, could also improve the mechanical property.

Keywords: Fly Ash; Composites; Morphology; Mechanical Property

1. Introduction

Coal fly ash (FA) is a by-product of coal from the combustion process, mainly generated in thermal power plants [1]. FA has a complex heterogeneous mixture of amorphous and crystalline phases. Depending on the difference of the coal, the FA gematric characterization and components can vary. Generally, FA is a fine powdered ferro aluminosilicate material with Al, Ca, Mg, Fe, Na, and Si as the predominant elements. Coal fly ash contains significant amounts of toxic metals such as As, Ba, Hg, Cr, Ni, V, Pb, Zn, and Se, characteristically enriched in coal fly ash particles [2] [3]. Exposure to FA without protections is a great risk to human health, as it may induce cancer and other respiratory diseases [4]. The pollution from the disposal of FA in landfills and groundwater poses an environmental risk [5].

Research on the recycling and utilization of fly ash is of great significance. According to the data published on the American Coal Ash Association (ACAA) 2019 [6], the production and utilization of coal combustion products (CCP) are illustrated in Figure 1. More than 85% of FA was reused for concrete products and blended cement. Due to the several sustainability benefits, fly ash-based geopolymer concrete (FA-GPC) is gaining popularity in the construction industry. It is estimated that approximately 470 kg/m^3 of FA is usually required to produce a metric ton of GPC [7]. However, some researchers presenting critical reviews on FA-GPCs found that the mechanical properties of the FA-GPC could be equivalent or even much higher compared to the ordinary Portland cement (OPC) concrete. Also, the employment of FA-GPC technology could provide economic and environmental benefits [8] [9]. Depending on the research of Chand [10] and Kishore et al. [11] about the effect of volume fraction of fly ash on the mechanical properties of unsaturated polyester composites, both of the research found that increasing the volume fraction of fly ash reduced the tensile and impact strengths of the composites for epoxy composite system. Another study from Kishore et al. [12] found that in fiber epoxy composites, the addition of fly ash led to a reduction of the density and an increase in the composites' modulus; the latter effect is caused by an improved dispersion of the fiber in the matrix.

According to literature, there have been several published papers on using fly ash as filler for the epoxy composites [9-12]. In this review paper, comprehensive information on fly ash morphology's impact on the epoxy matrix's mechanical properties is discussed.

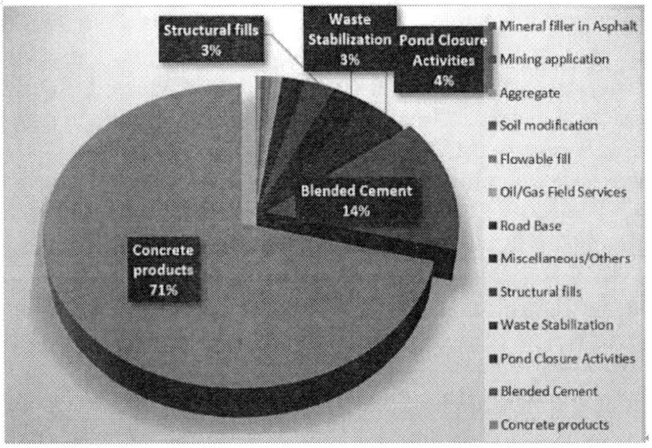

Fig.1 Utilization of fly ash in 2019 by category. (data source ACAA. www.acaa-usa.org)

2. Fly Ash Characterization

2.1 Fly Ash Morphology

Depending on the temperature and cooling rate of coal combustion, a fly ash particle's morphology is variant. The study from Kutchko and Kim [13] shows the sizes of the particles observed ranged from less than 1μm to greater than 200μm. The majority of the particles ranged in size from approximately 1μm to 100μm and consisted of solid spheres (Fig.2A). Hollow cenospheres (Fig.2B) and irregularly shaped unburned carbon particles tended to be in the upper end of the size distribution. Minerals and mineral aggregates often showed surface melting. Agglomerated particles and irregularly shaped amorphous particles may have been due to inter-particle contact or rapid cooling. To select demanded particle size as the composite's filler, normally sieve can be used for the selection process.

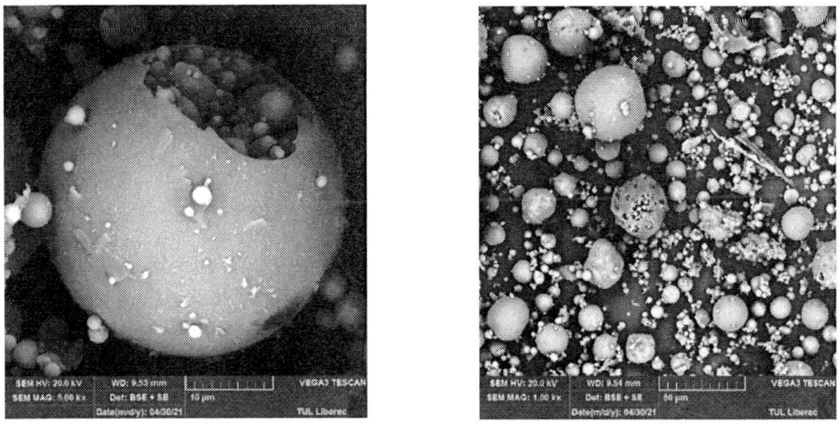

(A) (B)

Fig.2 Typical fly ash morphology SEM images of (A) typical fly ash spheres; (B) hollow cenospheres in cross-section

2.2 Fly Ash Chemical Characterization

The chemical composition of FA mainly depends on the coal source. The main component is silicon dioxide (SiO_2), aluminum oxide (Al_2O_3), iron oxide (Fe_2O_3), sulfur trioxide (SO_3), and calcium oxide (CaO). Table 1 showed one typical FA particle composition supplied from the power station of Opole, Poland [14]. According to the ASTM C618, three fly ash classes are Class N, Class F, and Class C [15]. The difference for the three classifications of the

FA depends on the containment percentage of silicon dioxide (SiO_2) plus aluminum oxide (Al_2O_3) plus iron oxide (Fe_2O_3); maximum containment of sulfur trioxide (SO_3), and calcium oxide (CaO). A. The primary difference between Class C and Class F fly ash is the chemical composition of the ash itself. While Class F fly ash is highly pozzolanic, meaning that it reacts with excess lime generated in the hydration of Portland cement, Class C fly ash is pozzolanic and also can be self-cementing. ASTM C618 requires that Class F fly ash contain at least 70% pozzolanic compounds (silica oxide, alumina oxide, and iron oxide), while Class C fly ashes have between 50% and 70% of these compounds. Typically, Class C fly ash also contains significant amounts of calcium oxide - over 20%. Most Class F fly ash contains little calcium oxide; however, some Class F fly ash sources may contain intermediate levels (8% to 16%) of calcium oxide. While both classes of fly ash greatly reduce concrete permeability as compared to the cement only mixes, Class F tends to give proportionately greater permeability reduction.

Table 1. Chemical composition of FA from the power station of Opole, Poland [14]

Components	Content (%)
SiO_2	40 - 55
Al_2O_3	22 - 30
Fe_2O_3	8 - 15
SO_3	Max 2.8
CaO	Max 1

FA is typically used for its binding properties, resulting from the chemical reaction with other reactive constituents. In general, silica in FA reacts with alkaline compounds like lime to cause cementing actions. Silica, reacting in the form of silicic acid, is used for characterizing pozzolanic materials. Often, high silica content is referred to as acidic and is preferred. However, over time, this characterization has shifted from chemical analysis to mineralogical identification due to developing an understanding of reactive phases and their role in pozzolanic activity [16] [17].

3. Mechanical Property of Fly Ash Filled Epoxy Matrix

These fly ash characteristics qualify it as a good alternative for use as a filler in polymers, especially in epoxy. Epoxy resins are the most commonly used thermoset plastic in polymer matrix composites. The FA-filled epoxy composite's good mechanical properties and excellent corrosion resistance have carved for itself a niche in various engineering applications.

In this context, the adhesion between the particles and the matrix gains importance as it influences the system's strength. The size of the interface is generally dependent on the specific surface area of the filler. Srivastava et al. [18] investigated the ultimate tensile strength, modulus of elasticity, and fracture properties of epoxy resin filled with FA particles by the tensile test. The fly ash mesh size is 105 μm, and epoxy resin FA was mixed by hand and then cured. The tensile strength of an epoxy resin composite filled with FA decreases as the proportion of FA increases. At 6.5% FA by volume, the maximum tensile strength of epoxy resin filled with FA is achieved. With an increase in FA's proportion, the modulus of elasticity of epoxy resin filled with FA particles increases. At 6.5% FA by volume, the fracture surface energy of epoxy resin is filled with FA composite peaks.

Compare to the work from researcher T. Chaowasakoo and Sombatsompop [19], more than 90% of the FA particle size they use is between 50-100μm. It was found that the increases in the tensile and flexural moduli of the composites due to the addition of fly ash can be considered in two zones, one being 0-40% FA and the other being 40-80 % FA. The addition of fly ash had a more pronounced effect on the tensile and flexural moduli below 40 %. Beyond these concentrations, the effect became minimized. The increased tensile and flexural moduli were associated with the higher modulus of fly ash. The tensile and flexural strengths decreased slightly with increasing fly ash content, but these properties did not change much after 40% FA. The decreases in these ultimate mechanical strengths of the composites were probably caused by the ash particles' incompatibility and the epoxy matrix, leading to poor interfacial bonding. Specifically, it was found that the fly ash particles formed a cluster or agglomerate among themselves, resulting in a filler–filler interaction and a poor interfacial bonding due to the strong polarity of hydroxyl groups on the fly ash surfaces [20].

Regarding to the impact strength of composites with FA as a filler, the improvement is significant compared to pure epoxy composites. Researcher Jana Novotna made the research of the impact strength of fly ash filled epoxy composite. The results show that, the best result of all percentages was achieved by a sample of 4wt% filler. The impact strength is about double that of the pure resin. This amount of filler best absorbs the energy developed by the test hammer. A similar sample was obtained with 3wt% filler. While composites with 2wt% concentration FA have a lower impact strength compared to the strength of unrefined resins, maybe the presence of the voids is the main reason for the reduction of the impact strength of the composites containing fly ashes, these voids work as stress concentration centers [21] (Fig.3.).

As the above works have approved, FA's morphology significantly influences the epoxy composite's mechanical properties. To improve the adhesion characteristics at the interface and reduce the formation of debonds, a silane treatment on the fly ash filler was attempted by S. M. Kulkarni and Kishore [22]. The strengths for composites with silane-treated fly ash show an improvement over the untreated filler composites. An increase of the strength of about 26% (maximum) was observed from such an exercise.

The microscopy revealed improved interface compatibility between treated particles and the epoxy matrix.Regarding the FA particle size, the composites with uniform particle size, no matter smaller or larger, registered improvement of modulus over the fly ash used in the bimodal particle distributed condition. The smaller average of the FA particle size values showing improvement in strength. The microscopic observation of these samples revealed lesser debonds at the epoxy interface and the smaller particles, compared to that witnessed for corresponding positions in composites containing the larger ones.

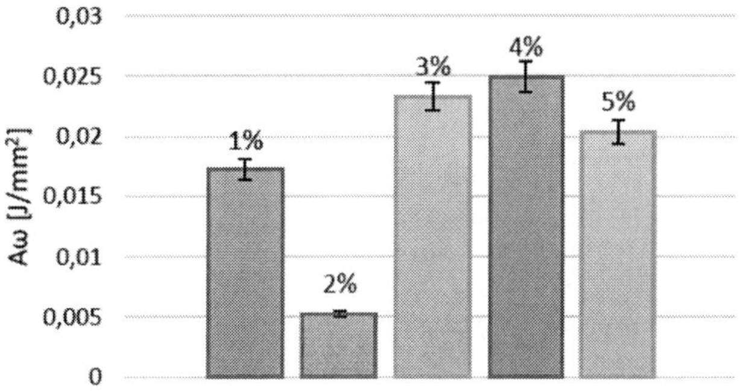

Fig.3 Impact strength of specimens with fly ash. [20]

4. Conclusion

Fly ash is a very valuable industrial waste recycling material. For epoxy matrix composites, filler as FA particles has significant interaction between FA and polymers. The morphology of FA has an impact on the mechanical properties. Depending on the average size of the FA particle, FA's optimized content for improving the epoxy matrix's mechanical property can be different. Accurate surface modifier agents can improve the mechanical properties. It is also evident that the FA content in the polymer matrix has to be optimized to achieve the desired properties.

5. Acknowledgement

The authors are grateful for the financial support from the Student research project 2021 No. SGS-2021-6025 supported by the Czech Ministry of Education and Faculty of Textile, Technical University of Liberec.

References

[1] Sahu SK, Bhangare RC, Ajmal PY, Sharma S, Pandit GG, Puranik VD. "Characterization and quantification of persistent organic pollutants in fly ash from coal fueled thermal power stations in India," Microchem: J., vol. 92, no. 1; pp. 92-96: May 2009; doi: 10.1016/j.microc.2009.02.003.

[2] Adriano DC, Page AL, Elseewi AA, Chang AC, Straughan I. "Utilization and Disposal of Fly Ash and Other Coal Residues in Terrestrial Ecosystems: A Review," J. Environ. Qual: vol. 9; no. 3: pp. 333-344; Jul. 1980: doi: 10.2134/jeq1980.00472425000900030001x.
[3] Jain M, Dwivedi A. "Fly ash - waste management and overview : A Review Fly ash -nwaste management and overview: A Review," Recent Res. Sci. Technol: 2014; vol. 6 (1), no. january: pp. 30-35; 2014, [Online]. Available: http://recent-science.com/.
[4] Munawer ME. "Human health and environmental impacts of coal combustion and post-combustion wastes," Journal of Sustainable Mining: vol. 17; no. 2. Central Mining Institute in Katowice: pp. 87-96; Jan. 01: 2018; doi: 10.1016/j.jsm.2017.12.007.
[5] Yao ZT. "A comprehensive review on the applications of coal fly ash," Earth-Science Reviews: vol. 141. Elsevier; pp. 105-121: Feb. 01; 2015: doi: 10.1016/j.earscirev.2014.11.016.
[6] The American Coal Ash Association, "Production & Use Reports - ACAA." https://acaa-usa.org/publications/production-use-reports/ (accessed Mar. 27, 2021).
[7] Assi LN, Carter K, Deaver E, Ziehl P. "Review of availability of source materials for geopolymer/sustainable concrete," J. Clean. Prod: vol. 263; p. 121477: Aug. 2020; doi: 10.1016/j.jclepro.2020.121477.
[8] Amran M, Debbarma S, Ozbakkaloglu T. "Fly ash-based eco-friendly geopolymer concrete: A critical review of the long-term durability properties," Constr. Build. Mater: vol. 270; p. 121857: 2021; doi: 10.1016/j.conbuildmat.2020.121857.
[9] Baheti V, Wang Y. "Ohmic heating and mechanical stability of carbon fabric/green epoxy composites after incorporation of fly ash particles," Mater. Today Commun: vol. 26; pp. 2352-4928: Mar. 2021; doi: 10.1016/j.mtcomm.2020.101710.
[10] Manimaran R, Jayakumar I, Mohammad Giyahudeen R, Narayanan L. "Mechanical properties of fly ash composites-A review," Energy Sources, Part A Recover. Util. Environ. Eff: vol. 40; no. 8: pp. 887-893; 2018: doi: 10.1080/15567036.2018.1463319.
[11] Kishore, Kulkarni SM, Sharathchandra S, Sunil D. "On the use of an instrumented set-up to characterize the impact behaviour of an epoxy system containing varying fly ash content," Polym. Test.: vol. 21; no. 7: pp. 763-771; Jan. 2002: doi: 10.1016/S0142-9418(02)00008-9.
[12] Kulkarni SM, Kishore. "Effect of filler-fiber interactions on compressive strength of fly ash and short-fiber epoxy composites," J. Appl. Polym. Sci: vol. 87; no. 5: pp. 836-841; Jan. 2003: doi: 10.1002/app.11501.
[13] Kutchko BG, Kim AG. "Fly ash characterization by SEM-EDS," Fuel: vol. 85; no. 17-18: pp. 2537-2544; Dec. 2006: doi: 10.1016/j.fuel.2006.05.016.
[14] Powerment. "Downloads - Powerment." https://www.powerment.de/downloads (accessed Mar. 29: 2021).
[15] ASTM Internationa. "ASTM C618 - 19 Standard Specification for Coal Fly Ash and Raw or Calcined Natural Pozzolan for Use in Concrete," West Conshohocken, PA: 2019; doi: 10.1520/C0618-19.
[16] Medina A, et al. "Fly ash from a Mexican mineral coal I: Mineralogical and chemical characterization," J. Hazard. Mater: vol. 181; no. 1-3; pp. 82-90: Sep. 2010; doi: 10.1016/j.jhazmat.2010.04.096.
[17] Ustabaş İ, Kaya A. "Comparing the pozzolanic activity properties of obsidian to those of fly ash and blast furnace slag," Constr. Build. Mater: vol. 164; pp. 297-307: Mar. 2018; doi: 10.1016/j.conbuildmat. 2017. 12. 185.
[18] Srivastava VK, Shembekar PS. "Tensile and fracture properties of epoxy resin filled with flyash particles," J. Mater. Sci: vol. 25; no. 8: pp. 3513-3516; 1990: doi: 10.1007/BF00575379.
[19] Chaowasakoo T, Sombatsompop N. "Mechanical and morphological properties of fly ash/epoxy composites using conventional thermal and microwave curing methods," Compos. Sci. Technol: vol. 67; no. 11-12: pp. 2282-2291: 2007; doi: 10.1016/j.compscitech.2007.01.016.
[20] Choi SS, Nah C, Lee SG, Joo CW. "Effect of filler-filler interaction on rheological behaviour of natural rubber compounds filled with both carbon black and silica," Polym. Int: vol. 52; no. 1: pp. 23-28: Jan. 2003; doi: 10.1002/pi.975.
[21] Novotna J, Baheti V. "Impact strength of Epoxy composites with fly ashe as the fillers," no. November: 2018.
[22] Kulkarni SM, Kishore, "Effects of surface treatments and size of fly ash particles on the compressive properties of epoxy based particulate composites," J. Mater. Sci: vol. 37; no. 20: pp. 4321-4326; 2002: doi: 10.1023/A:1020648418233.

Study on Silk Fibroin Non-knotted Suture Containing Barbs

Zu-Qiang Yin[1], Yu-Min Zhang[1], Kang Cheng[1], Fang Zhao[1], Zhen-Zhen Qi[1], Shen-Zhou Lu[1*]

National Engineering Laboratory for Modern Silk, College of Textile and Clothing Engineering, Soochow University, Suzhou, 215123, China

[*]Corresponding author's email address: lushenzhou@suda.edu.cn

Abstract

Surgical sutures can achieve wound suture and tissue closure. However, common sutures require complicated knotting and complicated operation. This paper developed a kind of silk suture with barbs. Infrared absorption spectroscopy and X-ray diffraction were used to test and analyze the structure of the barbed sutures. The results showed that ethanol immersion and high pressure treatment changed the crystal structure of silk fibroin barbs and improved the crystallinity of silk fibroin proteins. The dissolution and mechanical properties of silk fibroin barbs showed that the dissolution rate of silk fibroin barbs was lower but the mechanical properties were stronger. The microscopic morphology of the barbed suture lines was observed by microscope, and it was shown that the silk fibroin protein barbs were closely bound to the silk and formed a regular barb structure on the silk suture lines. In this paper, 30°, 45°, 60° different barb angle suture lines were prepared, the mechanical properties test results show that 45° barbed suture lines have a better suture anchoring effect. This kind of barbed silk sutures is expected to be used in tissue suture.

Keywords: Silk Fibroin; Barbed Suture; Silk Suture; Medical Textiles

1. Introduction

Surgical sutures are a kind of widely used medical textile, which can close the wound and tissue. The surface of ordinary surgical sutures is mostly smooth, and the suture and fixation of biological tissue wounds can only be completed by tying knots after surgical suture. The complicated knot tying work needs a lot of time, and ligation clips and other auxiliary fixation are needed when suturing wounds with surgical sutures [1].

With the development of medical technology and material science, surgical sutures with barbed structures have been developed. Warner, JP et al. [2] used the barb suture technique to perform abdominal surgery, and the results proved that the use of barbed sutures were a safe and effective method, which could significantly reduce the operation time and retain the suture effect of ordinary suture. Barbed sutures make suturing faster, easier and knot-free. The stress distribution of the barbed sutures in the tissue is uniform, which can effectively relieve wound pain. Compared with traditional surgical sutures, barbed sutures have excellent postoperative wound healing effect and reduced postoperative complications [3]. The barbed sutures are safe and effective, and the wound strength and tissue response score are comparable to knotted sutures. [4]

Silk sutures are a kind of natural protein material with good biocompatibility and no adverse effects on the human body [5]. Many clinical experiments have shown that after complete degumming, silk sutures will not cause any inflammatory reaction after implantation into human tissues and will not cause biological rejection [6]. Viju et al. [7] combined tetracycline with silk sutures to prepare silk sutures with antibacterial properties while retaining the original mechanical properties. Silk protein, as a natural protein with excellent properties, can promote and speed up wound healing and repair. Silk sutures have good mechanical properties and do not need to be treated chemically to avoid the residue of harmful chemicals.

Silk fibroin protein is a kind of natural protein material with excellent biocompatibility. Silk fibroin protein is composed of heavy chain and light chain. It is a kind of structural protein that can give silk strong mechanical strength, softness and elasticity. Silk fibroin protein has two crystalline structures: Silk Ⅰ crystalline structure and Silk Ⅱ crystalline structure [8, 9]. The conformation of β-folded molecules in silk fibroin is relatively stable [10].

Many studies have chosen various ways to transform silk fibroin protein from random coil structure to β-folded structure to improve the crystallinity of silk fibroin protein materials, so as to improve the properties of materials. However, regenerated silk fibroin protein materials have the problems of low mechanical properties and high water solubility in wet state [11].

In this paper, silk barbed sutures were developed by pouring silk fibroin protein solution on the surface of silk sutures. It is hoped that they will have excellent mechanical properties while retaining the excellent biocompatibility of silk. In order to achieve this goal, the following difficulties need to be solved: 1. Reduce the water solubility of silk fibroin protein, and avoid the silk barbed sutures from dissolving due to water absorption in wet environment; 2. Improve the mechanical properties of barbs in wet state, and solve the problem of mechanical properties decline caused by hygroscopic silk fibroin barbs; 3. Optimize the combination of sutures and barbs to avoid barbs falling off due to the weak combination of silk fibroin barbs and silk sutures. In order to solve the problem of low mechanical properties and water solubility of barbs in wet state, ethanol immersion and high temperature and pressure treatment were used. In order to increase the binding force between barbs and sutures, silk sutures degumming and vacuum defoaming were used. This kind of barbed silk sutures has good stability and excellent mechanical properties through the related performance tests, and it is expected to be used in the field of knot free tissue suture.

2. Materials and Methods

2.1 Experimental Materials

Table 1. Experimental materials and reagents

Type	Specification	Manufacturer
Silk sutures	0.265 mm	Wuxi ideal silk thread co. LTD
Silicon substrate	120×120mm	Meidi Family co. LTD
Bombyx cocoon	Fresh cocoon	Suzhou Xian Can Silk Biotechnology Co. LTD
Lithium bromide	Analytically pure	Tiancheng Chemical Co. LTD
Sodium bicarbonate	Analytically pure	Suzhou Keqing Biological Reagent Co., Ltd
Sodium carbonate	Analytically pure	Suzhou Keqing Biological Reagent Co., Ltd
Anhydrous ethanol	Analytically pure	Suzhou Keqing Biological Reagent Co., Ltd
Deionized water	Conductivity < 10μS	Self-made
Dialysis bag	Molecular cut-off 14kDa	Jiangsu Bomeida Life Science Co., Ltd
New Zealand Rabbits	Conventional-CV animal	Laboratory Animal Center, Soochow University

2.2 The Experimental Method

2.2.1 Preparation of Silicon Template

Suture lines with different barb angles of 30°, 45° and 60° were drawn respectively. The barb negative molds were carved on the silicon substrate by laser engraving machine, and then cleaned and dried for standby.

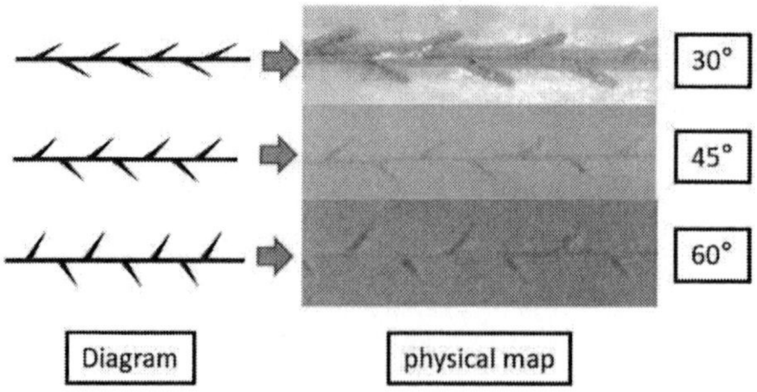

Fig.1 Schematic diagram of preparation of barb molds

2.2.2 Preparation of Silk Fibroin Solution

Weigh 80g Bombyx cocoons for use. Measure 4000ml of deionized water and heat it. After the water was heated to boiling, 1g of sodium carbonate and 3g of sodium bicarbonate were added. After stirring evenly, the Bombyx cocoons were added and treated in a slightly boiling state for 30min. Then, the Bombyx cocoons were kneaded and washed with deionized water to remove the sericin on the surface. Repeatedly degummed three times. The degummed wet silk was dried in the oven at 60°C to get the dry and degummed silk for use.

A lithium bromide solution with a molar concentration of 9.3mol/L was prepared for later use. The degummed silk was mixed with lithium bromide solution at the mass ratio of 15:100, and the silk was fully dissolved for 1h under the condition of heating and stirring in a water bath at 65°C to obtain the mixed solution of silk fibroin lithium bromide. After cooling, the mixture was placed in dialysis bags (with an intercepted molecular weight of 14kDa) and dialysed with deionized water for 3 days. Change dialysis water every 2 hours during dialysis. After dialysis, pure silk fibroin protein solution was filtered and stored in a refrigerator at 4°C for later use.

2.2.3 Degumming Treatment of Silk Sutures

Weighed 80g silk sutures, measured 4000mL of deionized water, heated to boiling, added 1g of sodium carbonate and 3g of sodium bicarbonate, then added silk sutures and kept it slightly boiling for 30min. Remove the sutures and rinse them with deionized water to remove the coating, sericin and impurities. After washing, the suture lines were put in the oven at 60°C to dry to get the degummed silk suture lines, and put in a dry and clean environment for standby.

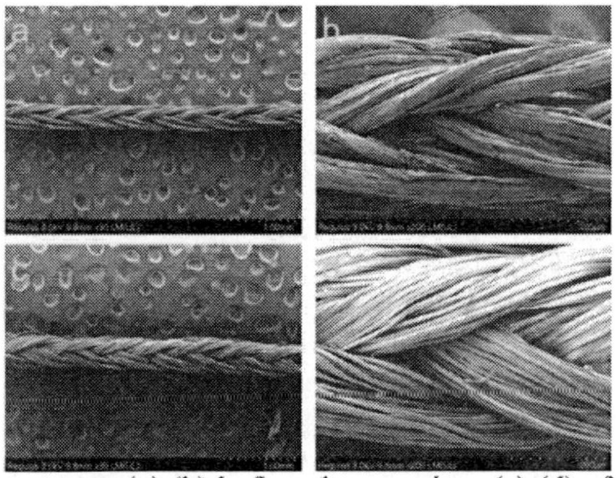

Fig.2 SEM images of silk sutures: (a) (b) before degumming; (c) (d) after degumming treatment. The magnification is 30 times and 200 times respectively.

2.2.4 Roller Treatment of Degummed Silk Sutures

In order to remove the air inside the silk sutures and optimize the binding effect of barbs and sutures. The degummed sutures were immersed in silk fibroin solution at 60°C for 5min, then the sutures were taken out and treated with pressure rollers for three times.

For the first rolling, the rolling pressure is 0.4kg/cm^2~2kg/cm^2, and the residual rolling rate is 10%-20%. For the second rolling, the rolling pressure is 0.2kg/cm^2~1kg/cm^2, and the residual rate is 20%-40%. For the third rolling, the rolling pressure is 0.1kg/cm^2~0.5kg/cm^2, and the residual rate is 40%-60%.

2.2.5 Preparation of Barbed Silk Sutures

The rolled sutures were placed in the barb molds, and the silk fibroin protein solution was poured until the barb molds and sutures were completely covered. Then the cast molds were placed in a vacuum drying oven to remove the bubbles inside the suture line by vacuum, and finally the solution was completely immersed in the barb molds and suture lines. After taking them out, they were naturally dried under constant temperature and humidity (temperature: 25±2°C, humidity: 65±5%), and then the samples were taken out of the mold to obtain the barbed silk suture lines.

2.2.6 Post-treatment of Barbed Silk Sutures

The barbed silk sutures were soaked in anhydrous ethanol solution for 6h and then taken out and dried to obtain the barbed sutures treated with ethanol. The barbed sutures were placed in a autoclave for 20min with high pressure steam at 120°C, and then dried to obtain the barbed sutures treated with high pressure. Untreated barbed silk sutures served as the control group.

2.2.7 Observation of Microscopic Morphology of Barbed Silk Sutures

The surface morphology of original silk sutures, degummed silk sutures, roller processed silk sutures and barbed silk sutures were observed under a type microscope.

2.2.8 Infrared Absorption Spectrum Analysis

The barbs of the three groups of barbed silk sutures mentioned in 2.2.6 were peeled off and ground to produce barb powder samples. The barb powder and potassium bromide powder were mixed and ground, then pressed into sheet samples. The absorption spectra of barb powder were measured by Fourier transform infrared spectrometer, and the infrared absorption spectra were obtained.

2.2.9 Wide-angle X-ray Diffraction Analysis

The samples of barb powder were put into the automatic X-ray diffractometer to scan the X-ray diffraction curve of 2θ=5°~45°. The scanning speed is 2°/min.

2.2.10 Dissolution Loss test of Barbed Silk Sutures

The barbs of the three groups of barbed silk sutures with different post-treatment were peeled off, with 5 parallel samples in each group. The barbs were baked in an oven at 100°C to constant weight, and the weights were recorded as M_1. PBS buffer solution was added to the peeled barbs according to the bath ratio of 1:100, and the water bath oscillated at 37°C for 24h at constant temperature. Take the barbs, place them on the filter gauze, rinse them with deionized water for 3 times, and then wipe the water on the surface of the barbs with absorbent paper. The barbs were baked in an oven at 100°C to a constant weight, and the weights were recorded as M_2. The dissolution rates of each barbed material were calculated according to the following formula.

$$\text{Dissolved rate}=\frac{M_1\text{-}M_2}{M_1}\times 100\%\#(1)$$

2.2.11 Tensile Mechanical Properties of Barbed Silk Sutures

Prepare untreated original silk sutures, degummed silk sutures, roller processed silk sutures, and take 5 of each suture. Each group of sutures was placed on a universal tensile tester for axial unilateral tensile test. The stretching distance was 50mm and the stretching speed was 50mm/min.

2.2.12 Simulated Tissue Tensile Test

Fig.3 shows the experiment of pulling out the barbed sutures from the simulated tissue to explore the anchoring ability of the barbed sutures from different barb angles to the tissue. The simulated tissue tensile material was fresh rabbit skin. The surface hair of experimental New Zealand white rabbits was removed, and then the abdominal skin of rabbits was removed with a scalpel. The surface villi of rabbit skin were removed by sodium sulfide treatment. Using a scalpel to take the simulated tissue tensile samples (length 30mm, width 10mm, thickness 1.5~2mm).

Uniaxial tensile test of simulated tissue/barbed sutures was performed by using a texture analyzer. The upper chuck holds the barbed suture, the lower chuck holds the simulated tissue, and the suture line punctures the tissue, and 10mm of suture line is reserved outside, and stretch along the axis (test parameters: tensile speed 50mm/min, pretension 0.05N, spacing 50mm). At the same time, in order to test the force of the barbed sutures in the process of penetrating and pulling out the tissue, two test methods of pulling out the suture from the tissue in the direction of the barbs and in the direction of the inverse barbs were used to determine which barb angle is more suitable for the tissue suture.

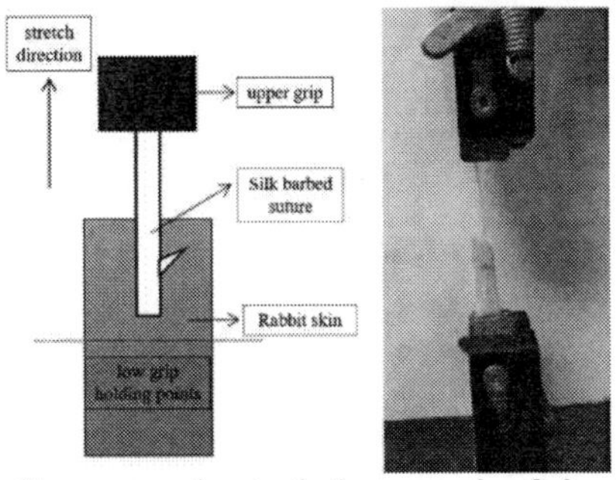

Fig.3 Schematic diagram and actual photograph of simulated tensile test

3. Results and Discussion

3.1 Morphological Characterization of Barbed Silk Sutures

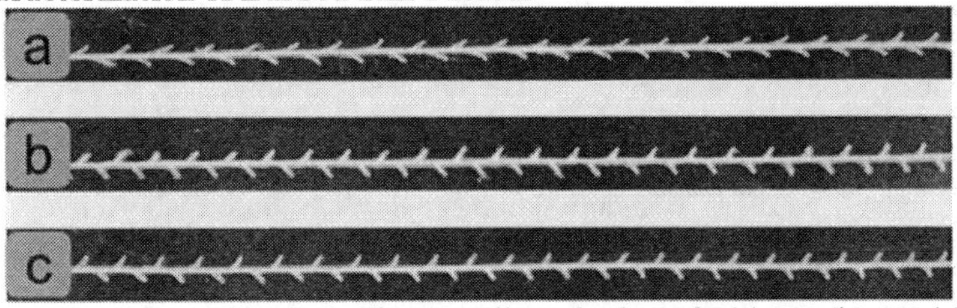

Fig.4 Actual image of barbed silk suture lines: barb angle (a) 30° (b) 45° (c) 60°

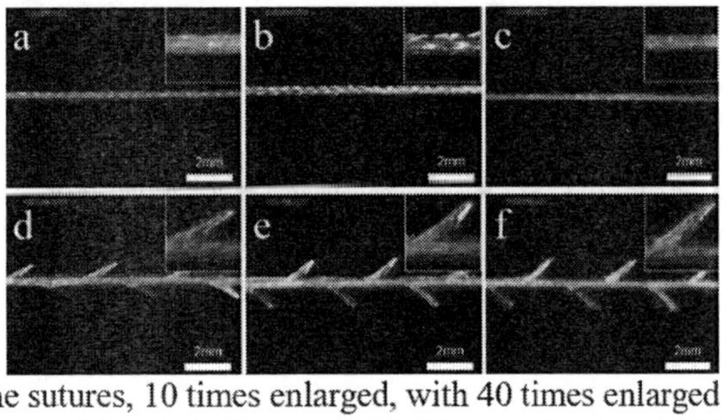

Fig.5 Microscopic image of the sutures, 10 times enlarged, with 40 times enlarged upper right corner (a) original silk suture (b) degummed silk suture (c) rolled silk suture (d)30° barbed suture (e)45° barbed suture (f)60° barbed suture

It can be seen from Fig.4, the total diameter of the barb fitting line body of the 30° barbed suture line did not increase too much. The barbs with angle of 45° and 60° barbed sutures increased the diameter of the whole suture greatly. For tissue suture, too large barb angle hinder the penetration of the suture into the muscle tissue, and at the same time, the greater penetration resistance also have a greater impact on the mechanical properties of the barbs.

By Fig.5 (a) (b) (c) comparison, the gluing of the surface of the suture line was more bright and clean, white as a whole. The overall appearance of the shape was basically unchanged, each woven monofilament could be clearly observed. The sutures treated by press roller were coated with silk fibroin evenly on the surface and inside of the sutures after being soaked by silk fibroin solution. It can be observed in Fig.5 (d)(e)(f) that the silk barbed suture lines were perfectly formed, the connection between the barbs and the line body was excessively natural, without obvious defects, and the overall suture lines had no defects such as bubbles and defects.

3.2 Infrared Absorption Spectrum Analysis

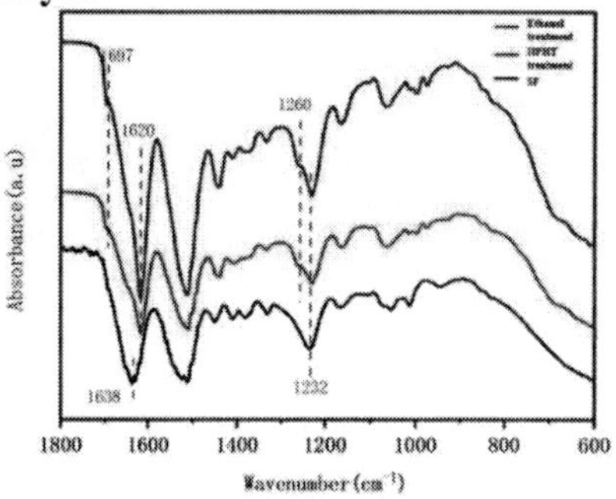

Fig.6 FTIR of suture barbs

It can be seen from Fig.6 that the barbs treated by high-pressure steam or ethanol immersion have a sharp absorption peak shape at 1620cm^{-1} in the Amide Ⅰ region. The barbs treated by high pressure steam or ethanol immersion at 1697cm^{-1} and 1260cm^{-1} are the absorption peaks of silk fibroin in Amide Ⅰ and Amide Ⅲ respectively, and the above three peaks all show β-folded molecular conformation [12]. This is because the random curled structure inside silk fibroin molecules was changed after high-pressure steam treatment, and a large number of β-folded structures were generated in silk fibroin [13]. While the ethanol immersion can induce the interentanglement of silk fibroin molecular chains, a large number of entanglements made the internal aggregation phenomenon and produce β-folded structure [14]. The original silk fibroin barbs had an absorption peak (Amide Ⅰ) at 1638cm^{-1} and a certain β-folded molecular conformation. The results indicate that some β-folding structures appeared in the preparation process of pure silk fibroin barbs. The three kinds of barb sutures all have an obvious absorption peak at 1232cm^{-1}, which is a random coiling molecular conformation, indicating that there are many random coiling structures in silk fibroin even after the treatment of high pressure steam and ethanol immersion.

3.3 Wide-angle X-ray Diffraction Analysis

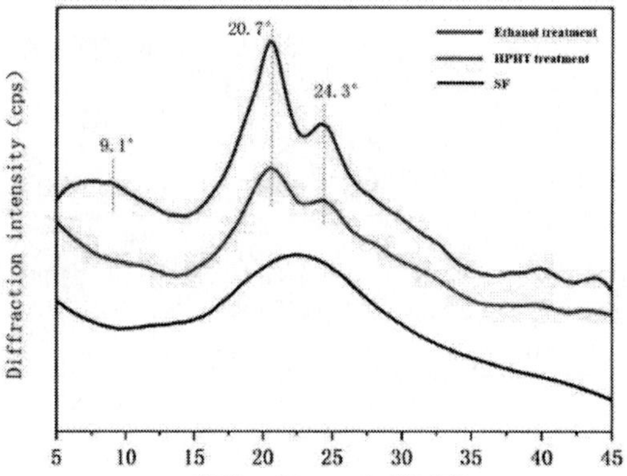

Fig.7 X-ray diffraction curve of suture barbs

It can be seen from Fig.7 that the X-ray diffraction curve of pure silk fibroin barbs is a large and gentle peak, showing a typical amorphous structure. While the barbs treated by ethanol immersion have strong absorption peaks at 9.1°, 20.7° and 24.3°, as well as the barbs treated by high-pressure steam at 20.7°and 24.3°, showing the crystal structure of Silk fibroin protein Silk Ⅱ [15]. This is because most pure silk fibroin materials are amorphous structures. Under the treatment of high temperature and high pressure or ethanol, the internal molecules of silk fibroin appeared a large number of crystal nuclei and neatly arranged and stacked into an orderly crystal structure. At the same time, it can be found that the crystallization peak after ethanol treatment is larger than that of high pressure steam, indicating that ethanol soaking treatment has a stronger effect on promoting the crystallization of silk fibroin protein.

3.4 Solubility Test and Analysis of Barbed Silk Sutures

As can be seen from Fig.8, the solubility of silk fibroin materials treated by high pressure steam and ethanol immersion is significantly different from that of original silk fibroin materials. Due to the existence of a large number of amorphous structures, the binding force between molecules is weak, and there is a large dissolution loss in water. After high pressure steam and ethanol immersion treatment of silk fibroin material, due to the production of a large number of β-folded structures inside the silk fibroin, between the molecules stacked into an orderly crystalline structure, water molecules are difficult to break the structure between molecules, making the dissolution rate of the material greatly reduced. Compared with the high pressure steam treatment (11.5%), the ethanol treatment (4.4%) showed a better modification of silk fibroin crystal structure than the high temperature and high pressure treatment.

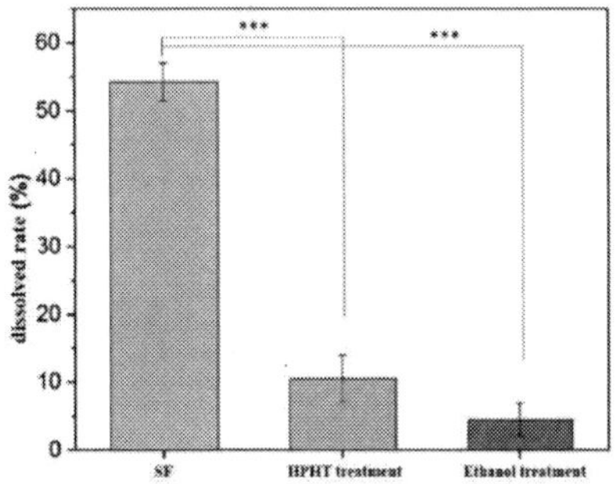

Fig.8 Dissolution loss rate of barbed suture
*p<=0.05, **p<=0.01, ***p<=0.001

3.5 Test Analysis of Mechanical Properties

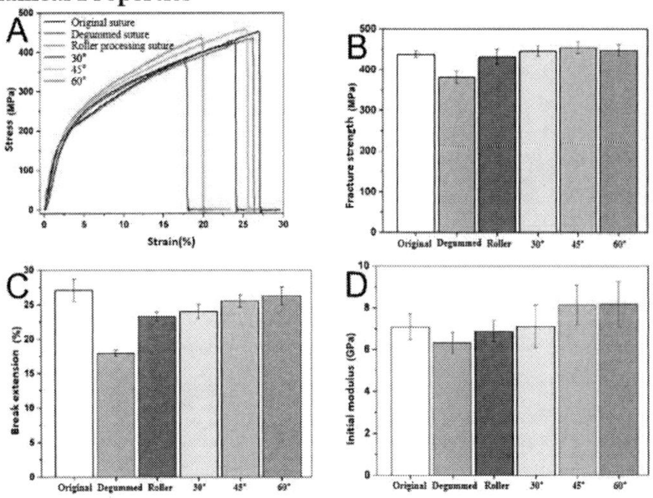

Fig.9 Tensile mechanical properties of silk sutures
(A) stress-strain curve (B) tensile breaking strength (C) elongation at break (D) initial modulus of tension

It can be intuitively observed from Fig.9(A) that the strain of the silk sutures decreased significantly after degumming, and the elongation at break of the degummed sutures also decreased in Fig.9(C). This is because the silk sutures were damaged by alkali solution, and the coating and sericin on the surface of the sutures fell off, which weakened the overall mechanical properties of the sutures. In the absence of sericin, silk fibroin sutures fracture at a relatively small tensile displacement, showing a small elongation at break. As shown in Fig.9(A) and (C), the strain and elongation at break of the rolled sutures and the barbed sutures were increased after the degumming sutures were treated by rolling and the barb solution was poured, almost reaching the strain and elongation at break of the original silk sutures. This is because after the pressure roller and vacuum defoaming treatment, the degummed sutures were completely infiltrated by silk fibroin protein solution, so that the sutures were finally wrapped by silk fibroin, which increased the overall mechanical properties of the sutures, so that the sutures could maintain good strain and elongation at break.

It can be seen from Fig.9(B) that the fracture strength of the original silk sutures is as high as 438MPa and the initial modulus is over 7GPa, indicating that the silk suture lines have excellent mechanical properties. Compared with the above barbed silk sutures, the mechanical properties of the degumming thread were reduced due to the removal of sericin. While mechanical properties of barbed sutures were improved due to the addition of silk fibroin. Their tensile fracture strength was almost close to that of the original silk sutures. As can be seen from Fig.9(C) (D), the elongation at tensile break and initial modulus of the barbed sutures obtained by pouring silk fibroin protein on the basis of the soaked and pressed thread body increased. The elongation at break and the initial modulus also increased slightly with the increase of the barb angle. It may be due to the existence of silk fibroin protein crystal structure, the outer layer of silk fibroin wrapped in the suture was more rigid, so the initial modulus of the barbs thread was stronger than the original silk suture to a certain extent. The tensile fracture strength of the sutures with barbs was slightly increased. Barbs from different angles had no significant effect on the strength of the sutures. This indicates that the barbed sutures have excellent mechanical properties similar to the original silk sutures.

3.6 Tensile test Analysis of Simulated Tissue

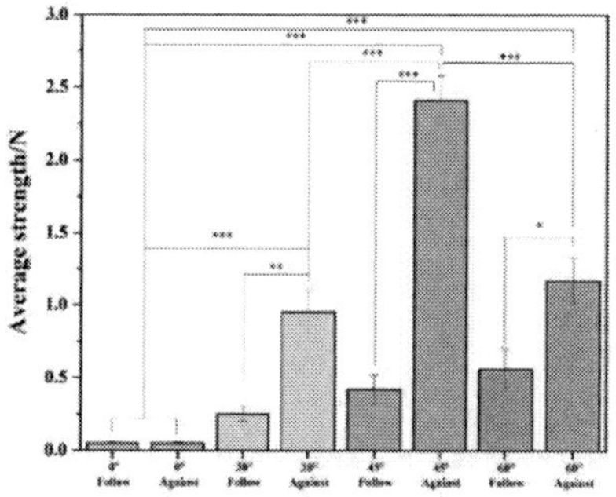

Fig.10 The mechanical properties of barbed silk sutures with angles of 0°, 30°, 45° and 60°
*p<=0.05, **p<=0.01, ***p<=0.001

The anchoring effect of barbed sutures is directly related to the effect of tissue suture. The stronger the ability of the barbs to anchor the tissue, the less likely the wound to crack after tissue suture. The key to avoiding knotting sutures is that their barbs can have a strong anchoring effect on the tissue, which means the sutures require a large force to pull out the tissue.

As can be seen from Fig.10, the force of the sutures with barbs in reverse stretching is significantly different from that of the original suture. There is a significant difference in the force required for the barb direction and the reverse barb direction of the barbed sutures. With the increase of the barb angle, the force stretching the simulated tissue along the direction of the barb gradually increased. It indicates that the larger the barb angle is, the greater the force requires for the barbed sutures to penetrate the tissue during the process of tissue suture. So choosing a smaller barb angle is essential. At the same time, it can be found that with the increase of the barb angle, the force of stretching the simulated tissue against the barb direction increases first and then decreases. It shows that the increase of the barb angle can increase the anchoring effect of the suture on the tissue to some extent. However, when the barb angle increases to a certain extent, the barbs with a large angle are more likely to break away from the line due to the increase of the extension length of the barbs under the application of force.

Through the comparison of the results, 45° barbed sutures have a more significant difference in inverse/alignment force compared with other angle of barbed sutures, and they have a better anchoring effect and superior suture mechanical properties, so it is the better choice.

4. Conclusion

In this paper, a kind of silk sutures with barbs was prepared by mold casting method in order to realize the operation suture without knotting. In order to reduce the water-solubility of silk fibroin protein and improve the mechanical properties of silk barbs, this paper studied the treatment of silk fibroin barb sutures by ethanol immersion and high temperature and high pressure treatment. Through the dissolution property test, infrared spectrum and XRD analysis, it was found that the two post-treatment methods could promote the formation of silk fibroin silk II crystal structure. The crystallinity of silk fibroin protein was significantly increased, and the solubility of silk fibroin barbs was decreased. The higher crystallization effect and lower dissolution rate of silk fibroin protein were obtained by ethanol immersion. The anchoring effect of barbed sutures at 30°, 45° and 60° barb angles on the tissue was investigated, and the results showed that the anchoring effect of the 45° sutures was better.

In conclusion, silk fibroin protein barbed sutures with barb angle of 45° and treated with ethanol has excellent mechanical properties, low water solubility and good suture anchoring effect. This kind of barbed sutures is expected to be used in tissue suture field.

5. Acknowledgement

This work was supported by College Nature Science Research Project of Jiangsu Province, China (Grant No. 20KJA540002), PAPD and Six Talent Peaks Project in Jiangsu Province (Grant No. SWYY-038).

References

[1] Abhari RE, Martins JA, Morris HL, et al. Synthetic sutures: Clinical evaluation and future developments [J]. Journal of Biomaterials Applications: 2017; 32 (3): 410-421.
[2] Warner JP, Gutowski K. A. Abdominoplasty With Progressive Tension Closure Using A Barbed Suture Technique [J]. Aesthetic Surgery Journal: 2009; 29 (3): 221-225.
[3] Murtha AP, Kaplan AL, Paglia MJ, et al. Evaluation of a novel technique for wound closure using a barbed suture [J]. Plastic and Reconstructive Surgery: 2006; 117 (6): 1769-1780.
[4] Zaruby J, Gingras K, Taylor J, et al. An In Vivo Comparison of Barbed Suture Devices and Conventional Monofilament Sutures for Cosmetic Skin Closure: Biomechanical Wound Strength and Histology [J]. Aesthetic Surgery Journal: 2011; 31 (2): 232-40.
[5] Altman GH, Diaz F, Jakuba C, et al. Silk-based biomaterials [J]. Biomaterials: 2003; 24 (3): 401-16.
[6] Vepari C, Kaplan DL. Silk as a biomaterial [J]. Prog Polym Sci: 2007; 32 (8-9): 991-1007.
[7] Viju S, Thilagavathi G. Characterization of tetracycline hydrochloride drug incorporated silk sutures [J]. Journal of the Textile Institute: 2013; 104 (3): 289-294.
[8] Wang Q, Yang YH, Chen X, et al. Investigation of Rheological Properties and Conformation of Silk Fibroin in the Solution of AmimCl [J]. Biomacromolecules: 2012; 13 (6): 1875-1881.
[9] Liang CX, Hirabayashi K. IMPROVEMENTS OF THE PHYSICAL-PROPERTIES OF FIBROIN MEMBRANES WITH SODIUM ALGINATE [J]. Journal of Applied Polymer Science: 1992; 45 (11): 1937-43.
[10] Hu X, Kaplan D, Cebe P. Determining beta-sheet crystallinity in fibrous proteins by thermal analysis and infrared spectroscopy [J]. Macromolecules: 2006; 39 (18): 6161-6170.
[11] Sashina ES, Bochek AM, Novoselov NP, et al. Structure and solubility of natural silk fibroin [J]. Russian Journal of Applied Chemistry: 2006; 79 (6): 869-876.
[12] Pelton JT, McLean LR. Spectroscopic methods for analysis of protein secondary structure [J]. Analytical Biochemistry: 2000; 277 (2): 167-716.
[13] Rnjak-Kovacina J, DesRochers TM, Burke KA, et al. The Effect of Sterilization on Silk Fibroin Biomaterial Properties [J]. Macromolecular Bioscience: 2015; 15 (6): 861-874.
[14] Zheng ZH, Wei YQ, Yan SQ, et al. Preparation of Regenerated Antheraea yamamai Silk Fibroin Film and Controlled-Molecular Conformation Changes by Aqueous Ethanol Treatment [J]. Journal of Applied Polymer Science: 2010; 116 (1): 461-467.
[15] Tanaka C, Takahashi R, Asano A, et al. Structural analyses of Anaphe silk fibroin and several model peptides using C-13 NMR and X-ray diffraction methods [J]. Macromolecules: 2008; 41 (3): 796-803.

Cite as: U. Reischl et al.,
TBIS 10.3993/tbis (2021).

Assessment of Electrostatic Potential Resulting from Friction between Fabric Samples made of Natural and Synthetic Fibers

Uwe Reischl[1*], Budimir Mijovic[2]

[1]*Boise State University, Boise, Idaho, USA*
[2]*University of Zagreb, Zagreb, CROATIA*

Corresponding author's email: ureischl@boisestate.edu

Abstract

Synthetic fibers are known to be a major source of electrostatic potential occurring on clothing. The electrical properties of fibers are determined by their chemical composition and polymer structure. Environmental factors such as humidity, temperature, and friction intensity can influence these properties. Due to the insulative characteristics of most fabrics, fibers are able to keep their charge for relatively long periods. This can result in clothing to cling to each other, attract dust, and create discharge shocks. To reduce these issues, suggestions have been proposed to mix garment layers made of natural fibers with garment layers made of synthetic fibers. The purpose of this study was to identify combinations of commercially available natural and synthetic fabrics that may result in reduced electrostatic build-up. Sixteen combinations of four fabric types were evaluated under controlled laboratory conditions. The fabric samples included 100% Polyester, 100% Nylon, 100% wool and 100% cotton. All combinations were evaluated using the same friction test protocol. The results showed that wool and cotton both exhibited the lowest electrostatic charge build-up while the polyester sample created the highest electrostatic charge build-up, even when paired with fabric samples of cotton and wool. The results suggest that layering synthetic fabrics with fabrics made of natural fibers will not significantly reduce the electrostatic potential created by the friction with synthetic fibers.

Keywords: Triboelectric Effect; Electrostatic Potential; Synthetic Fibers; Natural Fibers

1. Background

1.1 Static Electricity

Static electricity is generated when different materials rub against each together. Under static conditions, the atoms within a material are neutral as negatively charged electrons are being balanced by the positively charged protons. However, when one of the atoms comes in contact with another atom through friction, electrons are attracted to the nucleus of the other atom. Some of the electrons may be removed altogether through this friction. The material which loses the electrons then becomes positively charged while the other material will become negatively charged. Fabrics used in clothing are known to differ significantly in their propensity to lose such electrons due to friction [1].

1.2 Textile Materials

In general, textile materials used for clothing are poor conductors of electricity. Clothing layers will develop an opposite charge as they rub against each other. When the layers are separated, one will retain its positive charge and the other a negative charge, resulting in static cling. Charged clothing can also induce a charge on the entire human body [2].

1.3 Triboelectric Charging

The electrostatic charge created by friction between two materials is called "triboelectric charging". The amount of charge created by the triboelectric effect depends on the area of contact between two materials, the moisture in the material, the water vapor pressure in the air, the molecular structure of the materials, the friction intensity imposed on the two materials, and other factors [3, 4]. Once a charge is created, it becomes an electrostatically charged material. This charge may be transferred to another material resulting in an electrostatic discharge event.

Virtually all materials, including water and air, can be triboelectrically charged. How much charge is being generated, where that charge goes, and how quickly, depends on a material's physical, chemical, and electrical conductive characteristics.

1.4 Laboratory Tests

Research in triboelectric charging has relied on three main measurements approaches:

1) Measurement of fabric surface resistivity
2) Measurement of electrical capacitance
3) Measurement of electrostatic potential

With the availability of new instrumentation, it is now possible to accurately measure the electrostatic potential resulting from friction between different materials including textiles. This measurement method was chosen for the current study.

1.5 Research Objective

The objective of this study was to determine whether combining textiles made of natural fibers with textiles made of synthetic fibers can reduce the electrostatic charge build-up on clothing since natural fibers are known to exhibit a lower propensity for electrostatic charge build-up than synthetic fibers.

2. Methods and Procedures

2.1 Triboelectric Generator

A friction generator was constructed consisting of a non-conductive support frame, a 9cm diameter variable-speed rotating cylinder (rotor), and a 77g fabric tension weight. A True-RMS Multimeter was used to record the electrostatic potential generated by the fabric materials exposed to the friction caused by the rotating cylinder. The negative terminal of the Multimeter was placed on the rotor while the positive terminal was attached to the fabric sample. The design of the triboelectric friction generator is illustrated in Fig.1.

2.2 Fabric Samples

Four types of fabric material were evaluated including 100% Cotton, 100% Wool, 100% Polyester and 100% Nylon. Samples were obtained from clothing worn previously. Fabric structural specifications such as yarn count, yarn density and weave density could not be examined. Two sets of each were made. The dimensions were 6 cm in width and 42 cm in length. While one sample was mounted onto the rotating cylinder, the other was suspended vertically toughing the rotating cylinder at a constant pressure. The geometry and dimensions of the fabric samples are illustrated in Fig.2.

2.3 Test Protocol

All fabric samples were exposed to the same friction level created by the rotating cylinder (rotor). 20 rotations during a 15-second period were used for each test. Fabric pressure on the rotor was created by a 77 gram weight attached to the bottom of the fabric sample. Each combination was tested five times and the average value calculated. Laboratory air temperature was maintained at 24°C with a relative humidity of 20%

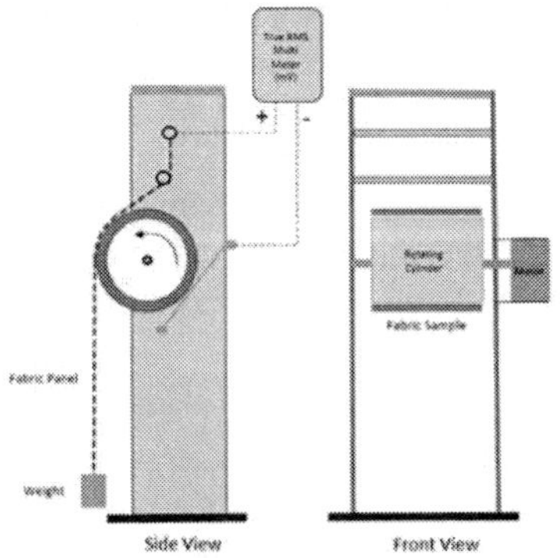

Fig.1 Triboelectric generator used in assessing friction induced electrostatic charge on fabric samples

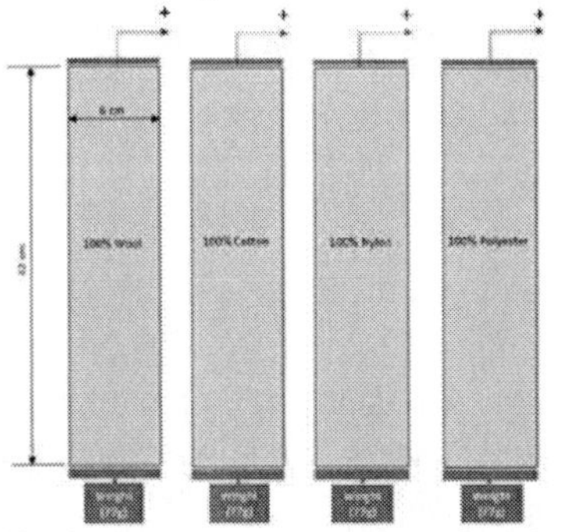

Fig.2 Geometry of fabric samples used in the tests. Two samples of each were made. One was mounted onto the rotor and the other was suspended externally as illustrated in Fig.1

2.4 Fabric Combinations

Sixteen combinations of fabric pairs were evaluated including Nylon, Polyester, Wool and Cotton. The combinations are illustrated in Table 1.

Table 1. Combination of fabric sample pairs tested

Rotor Mounted Fabric Sample	Suspended Fabric Sample
Nylon	Nylon
Nylon	Polyester
Nylon	Wool
Nylon	Cotton
Polyester	Polyester
Polyester	Nylon
Polyester	Wool
Polyester	Cotton
Wool	Wool
Wool	Nylon
Wool	Polyester
Wool	Cotton
Cotton	Cotton
Cotton	Nylon
Cotton	Polyester
Cotton	Wool

3. Results

Table 2 summarizes the electrostatic charge build-up observed for the sixteen fabric sample combinations exposed to the controlled triboelectric charging. The table lists separately the fabric samples mounted onto the rotating cylinder and the fabric samples contacting the rotor externally.

Table 3 summarizes the average electrostatic charge observed for the Wool and Cotton fabric samples interacting with other Wool and Cotton samples, with Nylon and other Polyester fabric samples.

Table 4 provides a statistical comparison between synthetic fabrics and natural fabrics in the reduction of the electrostatic charge build-up. The reductions are relative to the electrostatic build-up exhibited by the synthetic fabric combinations.

Table 2. Summary of electrostatic charge observed for sixteen fabric combinations

	Fabric Panel	Fabric Panel	Fabric Panel	Fabric Panel
Rotor Mounted	100% Wool	100% Cotton	100% Nylon	100% Polyester
100% Wool	200 mV (SD=22)	125 mV (SD=18)	230 mV (SD=24)	180 mV (SD=17)
100% Cotton	60 mV (SD=7)	20 mV (SD=8)	80 mV (SD=6)	55 mV (SD=5)
100% Nylon	100 mV (SD=20)	100 mV (SD=20)	80 mV (SD=11)	100 mV (SD=15)
100% Polyester	240 mV (SD=16)	200 mV (SD=20)	270 mV (SD=23)	200 mV (SD-22)

Table 3. Summary of average electrostatic charge build-up observed for fabric combinations tested.

Fabric Sample Combination	Electrostatic Build-up
Wool/Cotton + Wool/Cotton	101 mV (SD=78)
Wool/Cotton + Nylon/Polyester	145 mV (SD=72)
Nylon/Polyester + Nylon/Polyester	152 mV (SD=88)

Table 4. Statistical comparisons of electrostatic charge reductions observed for all combinations of fabric samples tested.

Fabric Comparison	Reduction in Electrostatic Static Buildup	Statistical Significance
Natural vs. Synthetic	$\Delta = 51$mV	$p<0.05$
Natural / Synthetic vs. Synthetic	$\Delta = 7$ mV	$p<0.05$

4. Discussion

Differences between the electrostatic charge build-up observed for the natural fiber textiles and synthetic fiber textiles were seen. Higher electrostatic levels were observed when a fabric sample was mounted onto the triboelectric rotor in comparison to the fabric sample when it was placed externally, touching the rotor. Although the length of both fabric samples was the same, i.e., 42 cm, the contact area between the external fabric sample and the rotor was different, i.e., 7 cm. While the fabric sample mounted on the rotor was exposed a total of 28 cm for each rotation, the surface contact area for the external fabric panel remained at 7 cm. This may explain the differences in electrostatic charge build-up observed between the two panel locations. However, averaging the two values together provided an overall value as summarized in Table 3.

In this study, the electrostatic potentials were documented for natural/natural fiber pair, the natural/synthetic fiber pair, and the synthetic/synthetic fiber pairs separately. While the natural/natural pair exhibited a substantial lower electrostatic charge than the synthetic/synthetic pair, the difference between the natural/synthetic pair and the natural/natural pair was only 14%. The results of this study, therefore, suggest that combining fabrics made of synthetic fibers with fabrics made of natural fibers may not be able to eliminate or substantially reduce the electrostatic potential created by this combination of textiles.

5. Limitations

The fabric samples evaluated in these experiments were obtained from previously worn clothing and may have been impacted by their previous use. Furthermore, the structural fiber characteristics such as yarn count, weave type and weave density could not be determined. However, such variables may have influenced their electrostatic characteristics. Controlling such variables in the future will allow a more precise differentiation of a fabric sample's propensity to create an electrostatic charge due to friction with other fabric samples.

6. Conclusions

Combining natural textiles with synthetic textiles has shown to reduce electrostatic charge build-up by approximately 14%. Although this reduction appears "small", it can contribute to an overall reduction of fabric cling when combined with other methods such as antistatic surface agents and methods of integrating metallic or highly conductive fibers into fabrics [4, 5]. To date, these alternative approaches have been successful in lowering the electrostatic charge on garments. However, future preferences for natural fibers in clothing may reduce the demand for anti-static chemicals and use of conductive Nano-materials.

References

[1] Verschave R, Firmin R. Hazards of Static Electricity in Work Clothing, J. Oil and Color Chem. Assoc: 1980; 63: 7: 287-289.

[2] Crow RM. Static Electricity: A Literature Review. Canadian Defence Research Establishment, Ottawa: 1991; PCN: 051LC.

[3] Gonzales JA.Development of a Qualitative Method for Fiber Idantification using Electrostatic Characterization. AATCC International Conference Poreceedings: 2001; 247-254.

[4] Hersh SP, Montgomery DJ. Static Electrification of Filaments:Theoretical Aspects. Textile Res. Journal: 1956; 26: 903-913.

[5] Lobel W. Antistatic Mechanism of Internally modified Synthetics and Quality Requirements for Clothing Textiles. Proceedings of the Institute of Physics Conference: 1986; 8: 2: 183-186.

Peel Property of 3D Printed Re-entrant/Textiles Composite Manufactured by Hot Pressing Conditions

Imjoo Jung[1], Sunhee Lee[1,2*]

[1]*Dept. Fashion and Textiles, Dong-A University, 37 Nakdong-daero 550 beon-gil, Saha-gu, Busan, 49315, South Korea*

[2]*Dept. Fashion Design, Dong-A University, 37 Nakdong-daero 550 beon-gil, Saha-gu, Busan, 49315, South Korea*

*Corresponding author's email: shlee014@dau.ac.kr

Abstract

Composite fabrics are fabrics made from two or more materials, which can improve the performance of fabrics. For manufacturing, the adhesion of said materials is important. Thus, in this study, 3DP-RE fabrics were printed using FDM-based TPU filament, and 3 types of textiles were used to laminate the composite fabric. The hot-pressing process was performed under three separate temperature and pressure conditions. To evaluate this, the morphology, thickness change rate, and peel strength were measured and analyzed. Therefore, to improve the adhesion between the two materials of composite fabrics, this study was attempted to determine the most suitable hot pressing process conditions.

Keywords: FDM 3D Printing Process, Re-entrant, 3D Printed Composite Fabric, Peel Strength, Morphology

1. Introduction

Composite fabrics are materials of composite structure consisted of two or more components that have different physical and chemical properties. Through combining various components, each material can improve the overall performance of its composites. Furthermore, this process can provide specific characteristics for particular use [1]. It can also be applied to a wide variety of fields. In recent years, many studies have been reported on composite fabrics' capability in enhancing performance in strength and durability in the textile field. A method mainly used to manufacture such a composite fabric is through lamination. Lamination is a process that creates composites by bonding different layers together with or without adhesive. As laminating layers are made with different processes, it can bring many advantages as seen invarious studies conducted in the textile and fashion related fields. Woven, knitted and non-woven fabrics can also be used as layers [2]. In particular, current research has been focusing on making composite fabrics with 3D printed textiles [3-10].

The manufacturing of composite fabrics using 3D printed materials is being carried out in various ways, either by using coating and adhesives or by printing them directly on the fabric. Several studies have compared the results according to the 3D printing parameters or the structure of each fabric as a manufacturing method for composite fabrics by 3D printing directly onto a fabric. Loe et al. (2021) have reported on a peel strength corresponding to different materials and printer settings by 3D printing PLA on nylon and polyester based fabrics [5], Mpofu et al. (2019) have compared the peel strength of 3D printed composite fabrics with 15 types of fabrics [6], Narula et al. (2018) have evaluated separation test and tensile test of various pretreatment methods and infill angles of cotton fabrics [7], and Kozior et al. (2018) provided a comparison of peel strength according to z-distance and 3D printing thickness of 3 filaments and 4 fabrics [8]. In addition to this, a study on the peel strength according to printing width ratio, the pretreatment for the fabrics and structure of fabric by 3D printing PLA on knit using coating or adhesive [9], and polymer coating to strengthen the adhesion of fabric/3D printing composite fabric were all relevant factors in its report on peel strength [10].

Our research team has compiled research on manufacturing 3D printed textiles and composite fabrics based on FDM 3D printing technology to develop fabrics suitable for application in the field of safety protection clothing products [11-15]. By applying FDM 3D printing technology and thermoplastic polyurethane (TPU) filament, a composite fabric was manufactured and evaluated by combining 3D printed patterns on various fabrics. Lee. (2018) evaluated the mechanical performance and washability of these composite fabrics by making a composite fabric printed on a polyester foil fabric. Kabir et al. (2020) produced composite fabrics by printing auxetic sinusoidal patterns with different density conditions on nylon fabrics and analyzed auxetic and mechanical properties. In addition, shape memory TPU and sinusoidal patterns were used to evaluate physical properties by printing them on nylon fabrics, and the possibility of their application as shoe uppers was confirmed by changing the pattern under different temperatures. Kim et al. (2021) confirmed auxetic characteristics of poisson's ratio and bending property, compression property, tensile property by incorporating a re-entrant pattern into neoprene for each of five tilting angles. Previous study confirmed the possibility of manufacturing composite fabrics based on 3D printed auxetic pattern, but it also showed that more research on adhesion between two fabrics for fabrication of composite fabrics was needed. Therefore, in this study, the possibility of using the hot-pressing process as a method to manufacture a 3D printed re-entrant pattern/textile composite fabric applied to the shoe upper and improve its adhesion is conducted. First, 3 kinds of cotton, polyester, and aramid knit were selected as the substrate textile, and a 3D printed composite fabric was laminated by combining it with a 3D printed auxetic re-entrant pattern based on a TPU filament. The prepared 3D printing auxetic re-entrant pattern/textile composite fabric was hot pressed under 3 temperature and pressure conditions to improve adhesion. This was to confirm the optimal hot-pressing process through analysis of morphology, thickness change rate and peel strength.

2. Experimental

2.1 Materials

In this study, the filament used is TPU filament (Ninjatek®, Fenner Inc., USA). The filament has a diameter of 1.75 mm and a hardness of 85 A. As for the manufacture of composite fabrics, Table 1 shows 3 types of textiles used. Used textiles were 2×1 basket weave cotton fabric (Sombe, Korea), plain weave polyester (Sewing fabric, Korea) and aramid knit (Mirae advanced material Co. Ltd., Korea). For the lamination process of the 3D printed auxetic re-entrant pattern/textile, spray adhesive (3M 77, 3M, USA) was used.

Table 1. Characteristics of textile used in this study

Type	Cotton	Polyester	Aramid knit
Sample code	CO	PET	ARNT
Surface image			
Weave	Plain	Plain	Knit
Thread counts (threads/inch)	76 × 42	56 × 54	-
Weight (g/cm^2)	0.03 ± 0.02	0.02 ± 0.01	0.02 ± 0.01
Thickness (mm)	0.58 ± 0.01	0.36 ± 0.01	1.0 ± 0.01

2.2 CFDM 3D Printing Conditions

The auxetic re-entrant (RE) pattern structure used in the experiment was used in accordance with previous studies [9]. The repeating unit of RE is 10 mm × 10 mm in width and length, and the inward angle is 70°. 3D modeling was performed with the 123D Design program (Autodest Inc., USA) so that the width × length × thickness was 100 mm × 1000 mm × 2 mm. This was converted into a .stl file, and converted into a .g-code file in the Blackbelt Cura 3.6.0 program. The conveyor-type FDM (CFDM, conveyor fused deposition modeling) process 3D printer (Blackbelt, Netherland) with 0.6 mm diameter nozzle was used. Following printing condition of nozzle temperature of 240°C, printing speed of 15 mm/s, gantry angle of 15°, 100% of infill density and zigzag of infill pattern were used as the setting.

2.3 Fabrication of Hot Pressed 3DP-RE/Textile

To prepare hot pressed 3D printed auxetic re-entrant pattern/textile(3DP-RE/textile) composite fabric, first cut 3D printed auxetic re-entrant pattern into 100 mm × 120 mm × 2 mm. And on the back side of the 3DP-RE, 3M strong adhesive spray was applied to 100 mm × 120 mm × 2 mm of 3DP-RE, except 100 mm × 45 mm × 2 mm and laminated onto each textile in the machine direction (MD). After then, it was pressed for 3 minutes using a press machine and dried for 24 hours at room temperature. The laminated sample was hot-pressed under three conditions of temperature and pressure. Based on the conditions set through the preliminary test, the hot-pressing process condition of 3DP-RE/textile shown under Table 2. Hydraulic press machine (12-10H, Carver Inc., USA) and stainless steel plate of 200 mm × 300 mm × 2 mm was used. The hot pressed 3DP-RE/textile was cut into 25 mm × 120 mm × 2 mm and used in the characterization.

Table 2. The hot pressing conditions of fabrication of hot pressed 3DP-RE/Textile

Temperature (°C)	Pressure (psi)	Times (s)
20	0	
175	50	60
200	100	

2.4 Characterization

2.4.1 Morphology

Morphology was collected through fabric image analysis microscopy (NTZ-6000, Nextecvision Co. Ltd., Korea) used magnifications were x6.5. Lateral sectional of sample was taken. And the 3D printed re-entrant pattern and changes in textiles were confirmed.

2.4.2 Thickness Variation

In this experiment, the thickness reduction rate was confirmed to determine the change in thickness of the fabric after the hot press process. The thickness of specimens was measured using a digimatic caliper (CD-20CPX, Mitutoyo corp., Japan). After then, the measurement value of the decrease in thickness (percentage %) was calculated in comparison with untreated samples.

2.4.3 Peel Strength

Peel strength was measured in 3DP-RE/textile composite fabric with different hot-pressing conditions, according to the KS M 6522 standard. It was measured using a universal mechanical testing machine (AGS-X, SHIMADZU Co. Ltd., Japan). After that, samples were prepared in width × length of 20 mm × 115 mm. Further tests were conducted at a tensile speed of 25 mm/min and gauge length 5mm using a tensile tester. And then, the sample's non-adhesive part of 20 mm × 45 mm of 3DP-RE was clamped to the upper clamp and textile was fixed the lower clamp. And peel off up to 50 mm from the start point. As a result, the peel strength was measured with the average of the 5 highest peak values of Strain-Stress (S-S) curve. Three samples were tested for each type Fig.1. shows the schematic of peel strength test.

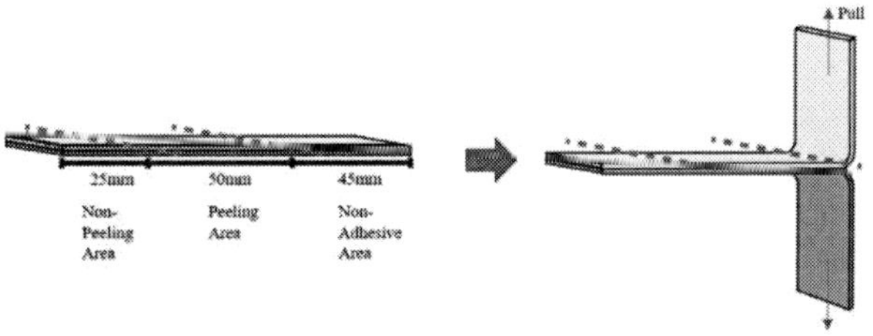

Fig.1 Schematic of hot pressed 3DP-RE/textile peel test

3. Result and Discussion

3.1 Morphology

Table 3~5 shows the morphology of hot-pressed 3D printed auxetic re-entrant 3D pattern/textile composite fabric with various hot pressing conditions. The lateral section can be seen to result in a change in thickness. All samples showed a decreased thickness when applying increased temperature and pressure

Compared to temperature, it is confirmed that: at 20°C, even when the pressure was increased, there was little change in the thickness of the composite fabric. At 175°C, even at 0psi, the thickness decreased slightly, and as the pressure increased, the thickness gradually decreased. At 200°C, it could be seen that the thickness decreased at a larger rate than when exposed to temperatures of 175°C. In case of 3DP-RE/CO and 3DP-RE/PET, thickness of 3D printed auxetic re-entrant pattern decreases when compared to other textiles. Comparatively, there was little change in the thickness of the textile. Also, 3DP-RE/PET at 200°C, 3DP penetrated the textile structure. On the other hand, thickness of 3DP-RE/ARNT reduced 3DP-RE and textile when applied temperature and pressure. In the case of 3DP-RE/ARNT, it was confirmed that the 3D printing sample penetrated the fabric structure when pressure was applied compared to the CO and PET fabric.

Table 3. Morphology of hot pressed 3DP-RE/CO

Temperature (°C) / Pressure (psi)	20	175	200
0		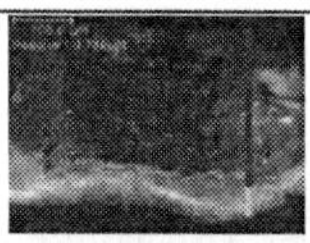	
Thickness (mm)	2.57 ± 0.12	2.45 ±0.04	2.43 ±0.06
50			
Thickness (mm)	2.58 ± 0.04	2.25 ±0.04	2.25 ±0.04
100	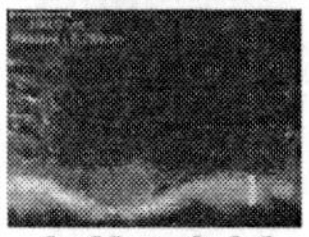		
Thickness (mm)	2.63 ± 0.06	2.17 ±0.02	2.16 ±0.02

Table 4. Morphology of hot pressed 3DP-RE/PET

Temperature (°C) / Pressure (psi)	20	175	200
0		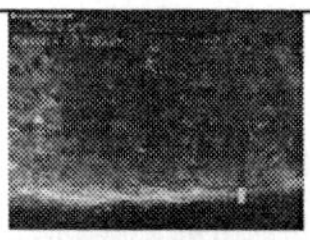	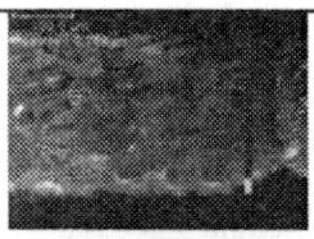
Thickness (mm)	2.12 ± 0.13	2.14±0.07	2.11±0.09
50	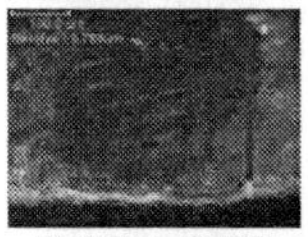		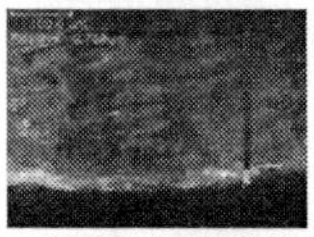
Thickness (mm)	2.18 ± 0.07	2.03 ± 0.05	1.93 ± 0.07
100	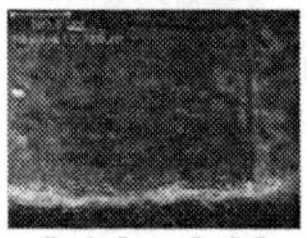	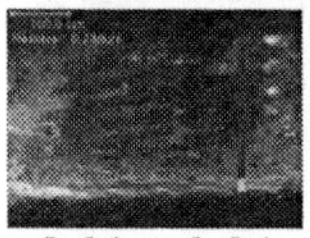	
Thickness (mm)	2.16 ± 0.08	2.01 ± 0.04	1.89 ± 0.03

Table 5. Morphology of hot pressed 3DP-RE/ARNT

Pressure (psi) \ Temperature (°C)	20	175	200
0		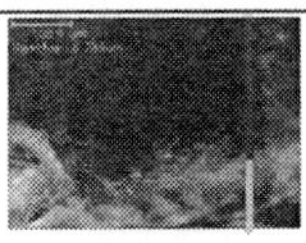	
Thickness (mm)	2.78 ±0.10	2.82 ±0.04	2.64 ±0.05
50			
Thickness (mm)	2.78 ±0.13	2.66 ±0.01	2.50 ±0.03
100			
Thickness (mm)	2.90 ±0.16	2.57 ±0.10	2.39 ±0.03

3.2 Thickness Variation

The thickness variation of samples with various hot-pressing condition is compared by thickness reduction rate. Fig.2 shows the results according to their rate of change.

It is confirmed that as temperature and pressure increased, the reduction rate of thickness change tends to increase. A increased range is organized in the order of 3DP-RE/PET < 3DP-RE/CO < 3DP-RE/ARNT. In the case of 3DP-RE/CO, when the fabric was applied pressure at 20°C, the thickness reduction rate was less than 5%. At 175°C and 200°C, thickness of 3DP-RE/CO decreased 3.4 % ~ 16.5 % and 7.3 % ~ 18.8 %. 3DP-RE/PET viewed by pressure at 20°C and 175°C, the thickness reduction rate was less than 5 % even when 100 psi was applied. When pressured at 200°C, no pressure was applied, only 5% of the thickness was changed, but when 50 psi and 100 psi pressures were applied, the thickness decreased by 11.7% and 12.8%. On the other hand, 3DP-RE/ARNT showed a 5.9% reduction in thickness even at 20 degrees and 0 psi pressure conditions. On the other hand, 3DP-RE/ARNT showed that the change in thickness decreased more than 3DP-RE/PET. It was found that the thickness decreased by 5.9% even at 20°C and 0 psi pressure conditions. At 175°C, the thickness decreased by 8.0 ~ 18.9 %. And at 200°C, the thickness decreased to 13.6 ~ 21.7 %, the thickness reduced more than 20%. It seems that aramid knit. This is because, as shown Table 3-5, cotton and aramid knit has a bulkier structure than plain woven PET, and it seems that the thickness is further reduced when pressure is applied.

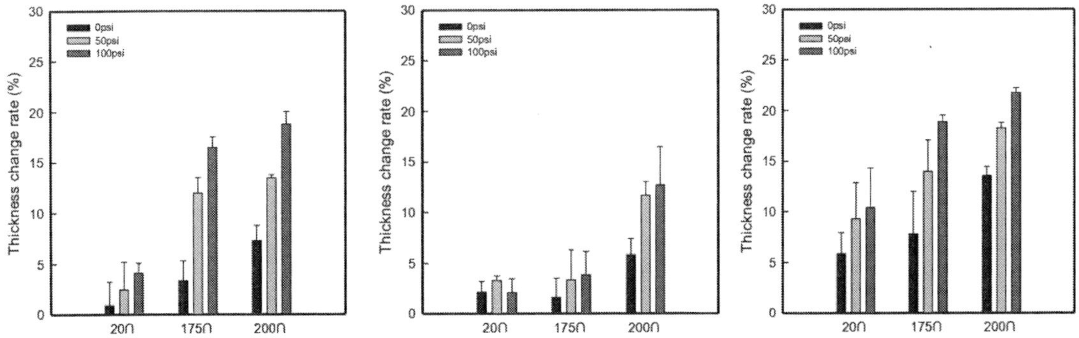

(a) 3DP-RE/CO (b) 3DP-RE/PET (c) 3DP-RE/ARNT

Fig.2 Thickness change rate of hot pressed 3DP-RE/textile composites

3.3 Peel Strength

The peel strength of samples with various hot-pressing condition is showed Fig.3 that shows peel strength graph by fabric.

As a result of the peel strength, it tends to increase 3DP-RE/CO < 3DP-RE/ARNT < 3DP-RE/PET. In the case of 3DP-RE/CO, it does not appear to be significantly affected by the hot-pressing process, as the value of 3DP-RE/CO does not differ significantly even when the pressure temperature increases. 3DP-RE/PET, the peel strength was large in the order of 100psi < 0psi < 50psi at 20°C and 200°C. However, at 100psi, the peel strength was found to be the greatest at 175°C. On the other hand, 3DP-RE/ARNT showed more increase in peel strength applied at all temperatures under pressures of 50 psi and 100 psi were applied compared to no pressure. In particular, the peel strength at 200°C was the best value when 100psi pressure was applied.

Like the previous results, these results seem to be heavily influenced by the fabric structure. the case of 3DP-RE/PET, which has the highest peel strength, PET is made of filament yarn and has a smooth surface, so it seems that when heat and pressure are applied, the adhesion between 3DP and textile is further increased. 3DP-RE/ARNT showed the lowest peel strength when hot-pressed at 20°C without heat. However, it was found that the peeling strength becomes stronger as the temperature was applied. It seems that when heat and pressure are applied, 3DP penetrates the pores between the fiber weaves and the strength increases. 3DP-RE/CO has a high density, so it is not significantly affected even when the hot press process is applied.

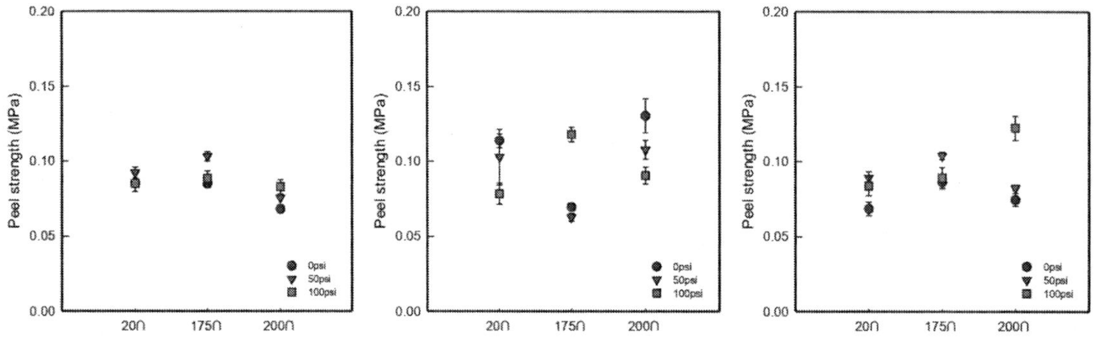

(a) 3DP-RE/CO (b) 3DP-RE/PET (c) 3DP-RE/ARNT

Fig.3 Peel strength of hot pressed 3DP-RE/textile composites

4. Conclusion

In this study, to fabricate composite fabrics, 3D printed auxetic re-entrant pattern/textile composite fabrics were custom manufactured for each textile. And the useability of the hot press process is studied to strengthen the adhesion between the two materials of the composite fabric. 3DP-RE pattern was printed using FDM printer based TPU filament, and three types of fabrics were used. The hot-pressing process was performed under three temperature and pressure conditions. To evaluate these samples, the morphology, thickness change rate, and peel strength were measured and analyzed.

As a result of the experiment in morphology, it was confirmed that the thickness of all samples decreased when the temperature and pressure were increased under the hot-pressing. For 3DP-RE/CO and 3DP-RE/PET, it was confirmed that the 3DP-RE's thickness decreased. In 3DP-RE/CO and 3DP-RE/ARNT, the thickness of the fabric was further reduced, which was confirmed to be due to the bulky structure of the aramid knit. With regards to its reduction rate, it was found that the thickness decreased in the order of 3DP-RE/PET < 3DP-RE/CO <3DP-RE/ARNT. This was also confirmed by the same trend as the result of the morphology. As for the peel strength, the peel strength increases in value in the order of 3DP-RE/CO <3DP-RE/ARNT < 3DP-RE/PET. It was confirmed that the peel strength was also greatly influenced by the surface structure of the textile. 3DP-RE/CO was confirmed that the 3DP-RE was little affected by increasing the temperature and pressure conditions of the hot-pressing process due to the high density of the fabric. On the other hand, 3DP-RE/PET was found to increase the adhesion between 3DP-RE and textile as temperature and pressure were applied. In 3DP-RE/ARNT, it was confirmed that

3DP-RE penetrated the fabric as hot press conditions were applied due to the pores created by the structure of the aramid knit.

Therefore, in this study, even under hot pressing progress conditions, 3DP-RE/PET, which has the smallest deformation, the lowest thickness reduction ratio, and the strongest peel strength, was found to be the most suitable for manufacturing 3DP-RE/textile composite fabrics. In addition, it was confirmed as the best hot pressing process condition at 200°C and 0psi. In addition, through this study, it was confirmed that the structure of the textile greatly influenced the manufacturing process of the composite fabric. Therefore, in a later study, we intend to establish a hot-pressing process to apply 3DP auxetic pattern/textile composite fabric as a shoe upper and conduct performance evaluation.

5. Acknowledgement

This research was funded by the National Research Foundation of Korea (NRF) grant funded by the Korea government (MSIT) (No. NRF- 2019R1A2C2084041).

References

[1] Yosomiya R, Morimoto K, Nakajima A, Ikada Y, Suzuki T. Adhesion and Bonding in Composites. In: US: CRC Press: 2020; p. 1-42.
[2] Novak N, Duborovski PD, Borovinsek M, Vesenjak M, Ren Z. Deformation behaviour of advanced textile composites with auxetic structure. Comp Struct: 2020; 252.
[3] Kozior T, Blachowicz T, Ehrmann. Adhesion of three-dimensional printing on textile fabrics: Inspiration from and for other research areas. J Eng Fiber Fabr: 2020; 15: 1-6.
[4] Spahiu T, Al-Arabiyat M, Martens Y, Ehrmann A, Piperi E, Shehi E. Adhesion of 3D printing polymers on textile fabrics for garment production. IOP Conf. Series: Materials Science and Engineering : 2019; 459: 012065.
[5] Loh GH, Sotayo A, Pei E. Development and testing of material extrusion additive anufactured polymer-textile composites. Fash text : 2021; 8: 2.
[6] Mpofu NS, Mwasiagi JI, Nkiwane LC, Njuguna D. Use of regression to study the effect of fabric parameters on the adhesion of 3D printed PLA polymer onto woven fabrics. Fash text: 2019; 6: 24.
[7] Narula A, Pastore CM, Schmelzeisen D, Basri SE, Schenk J, Shajoo S. Effect of knit and print parameters on peel strength of hybrid 3-D printed textiles. Journal of Textiles and Fibrous Materials: 2018; 1: 1-10.
[8] Kozior T, Dopke C, Grimmelsmann N, Junger IJ, Ehrmann A. Influence of fabric pretreatment on adhesion of three-dimensional printed material on textile substrates. Advances in mechanical engineering: 2018; 10 (8): 1-8.
[9] Meyer P, Dopke C, Ehrmann A. Improving adhesion of three-dimensional printed objects on textile fabrics by polymer coating. J Eng Fiber Fabr: 2019; 14: 1-7.
[10] Grimmelsmann N, Kreuziger M, Korger M, Meissner H, Ehrmann A. Adhesion of 3D printed material on textile substrates. Rapid Prototyp J: 2018; 24 (1): 166-170.
[11] Lee SH. Evaluation of mechanical properties and washability of 3D printed lace/voil composite fabrics manufactured by FDM 3D pirnting technology. Fash. Text: 2018; 20 (3): 353-359.
[12] Kabir S, Kim HR, Lee SH. Characterization of 3D printed auxetic sinusoidal patterns/nylon composite fabrics. Fiber. Polym: 2020; 21 (6): 1372-1381.
[13] Kabir S, Kim HR, Lee SH. Physical property of 3D-printed sinusoidal pattern using shape memory TPU filament. Text. Res. J: 2020; 90 (21): 2399-2410.
[14] Kabir S, Lee SH. Study of Shape Memory and Tensile Property of 3D Printed Sinusoidal Sample/Nylon Composite Focused on Various Thicknesses and Shape Memory Cycles. Text. Res. J: 2020; 90 (21): 2399-2410.
[15] Kim HR, Kabir S, Lee SH. Mechanical properties of 3D printed re-entrant pattern/neoprene composite textile by pattern tilting angle of pattern. J. Korean. Soc. Cloth. Text: 2021; 45 (1): 106-122.

Cite as: C.C. Li et al.,
TBIS 10.3993/tbis (2021).

Influence of Test Conditions on the Properties of Optical Heat Storage Fabrics

Cong-Cong Li, Li-Hua Chen*

School of Material Science and Engineering, Beijing Institute of Fashion Technology, Beijing, 100029, China

*Corresponding author's email: wclh@sina.com

Abstract

In this paper, the warp is made of polyester, the weft is made of Thermoheat-Solorx polyester filament, and the fabric is woven with twill structure design. At the same time, polyester fabrics with the same structure were selected for comparison. The heat storage and heating properties of the fabric were tested at different irradiation heights and ambient temperatures, the effects of irradiation height and ambient temperature on the heating capacity and heating rate of the fabric and their significance were analyzed. The test results show that, Thermoheat-Solorx fiber fabric has strong heat storage and temperature rise properties, the effects of irradiation height and ambient temperature on heat storage and heating of fabrics are significant. The research results provides a reference value for the design and development of optical heat storage fiber fabrics.

Keywords: Solar-thermal Conversion; Functional Textiles; Temperature; Thermoheat-SolorX

1. Introduction

Textiles products have primarily been made for warmth. Commonly used textiles can retain warmth by using fibers with high thermal resistance as raw materials to prevent body temperature from escaping and low temperature sources from coming into direct contact with the body. [1] [2] With the emergence of new technology and new materials, textiles have evolved from their traditional function of heat preservation, also it has also grown to play a more active role in heat storage and heating. The technology and materials of heat storage and heating can improve the utilization rate of resources and achieve sustainable development. [3] [4] Optical heat storage fiber is a new type of active insulation material that converts the energy of solar radiation to heat energy for heat storage and emission. [5] [6]

In this paper, the warp is made of polyester, the weft is made of Thermoheat-Solorx polyester filament, and the fabric is woven with twill structure design. At the same time, polyester fabrics with the same structure were selected for comparison. The heat storage and heating properties of the fabric were tested at different irradiation heights and ambient temperatures, the effects of irradiation height and ambient temperature on the heating capacity and heating rate of the fabric and their significance were analyzed. The research results have certain reference value for the design and development of optical heat storage fiber fabrics.

2. Test and Analysis of the Performance of Light Heat Storage Fabrics

2.1 Test of Performance of Light Heat Storage Fabric

2.1.1 Test Standard

GB/T 18319-2019. Test method for light heat storage of textiles.

2.1.2 Test Instrument

Equipment: infrared thermometer; Camera obscura.

Light source: incandescent lamp 300W. [7] [8]

2.1.3 Sample Preparation

Thermoheat-Solorx twill fabric, 10×10cm, 3 pieces, placed in constant temperature and humidity room for balancing for 24 h. [7] [8]

2.1.4 Test Parameters

Nitial ambient temperature: 15°C, 20°C, 25°C.

Light height: 20cm, 30cm, 40cm.

Lighting time: 10 minutes.

Turn off the power and test for 10 minutes.

2.1.5 Test Indicators

1) Temperature rise at time point (ΔTi) Calculate according to Equation (1). [9]

$$\Delta T_i = T_i - T_{0i} \quad 1)$$

ΔTi: Temperature rise at time point i, unit is Celsius (°C).

Ti: Temperature value of sample at time point i, unit is Celsius (°C).

T0i: Ambient temperature value at time point i (surface temperature value of reference sample), unit is Celsius (°C).

2) Heating rate: on the curve before the inflection point when the temperature change on the temperature-regulating curve slows down obviously, make the tangent line of the part closest to the straight line and find the slope of the tangent line.

2.2 Significant Analysis of the Performance of Light Heat Storage Fabrics

Unary variance was used to analyze whether the environmental conditions had significant influence on the heating performance of the optical heat storage fabrics. In the experiment, the irradiation height factor has three levels to test whether the irradiation height has significant influence on the heating performance of the optical heat storage fabric. There are three levels of ambient temperature to test whether the influence of ambient temperature on the heating performance of the optical heat storage fabric is significant. [1]

3. Influence and Analysis of Test Conditions on the Properties of Optical Heat Storage Fabrics

3.1 Influence of Irradiation Height on Absorption and Heating Properties of Photothermal Storage Fabrics

3.1.1 Influence of Irradiation Height on Heating Capacity of Optical Heat Storage Fabrics

The heat storage and heating capacity of the optical heat storage fabric was tested under different irradiation heights. The results are shown in Table 1, Table 2

Table 1. The influence of irradiation height on the heating capacity of optical heat storage fabric

irradiation height (cm)	Highest temperature (°C)	ΔT_{10} (°C)
20	49.4	7.2
30	46.0	6.4
40	44.4	5.2

It can be seen from Table 1, 10 minutes later, when the irradiation height was 20cm, 30cm, and 40cm, the surface temperature of the reference sample increased to 49.4°C, 46°C, and 44.4°C. The change of the surface temperature of the reference sample represents the change of the ambient temperature, so it can be seen that the irradiation height has a certain influence on the ambient temperature. During the same irradiation time, With the increase of irradiation height, the change of ambient temperature decreases. [10]

ΔT_{10} represents the performance of the fabric's light absorption and heating. It can be observed from the data in Table 1 that the temperature rise value ($\Delta T_{10)}$ is 7.2°C, 6.4°C and 5.2°C under the same light source for 10 minutes at the irradiation height of 20cm, 30cm and 40cm, respectively. The results show that the irradiation height has a great influence on the heat storage and heating capacity of the optical heat storage fabric. In the same irradiation time, with the decrease of the irradiation height, the temperature rise value (ΔT_{10}) becomes larger, and the heat storage and heating capacity of the optical heat storage fabric plays better. [11] [12]

Table 2. Influence of irradiation height on effective irradiation time

irradiation height (cm)	effective illumination time (min)	Highest temperature (°C)
20	12	7.6
30	14	7.0
40	17	6.6

It can be seen from Table 2 that the effective illumination time is different under different illumination heights. When the illumination height is 40cm, the effective illumination time is the longest, which is 17 minutes. However, its corresponding final temperature is still the lowest, which is 6.6°C. When the illumination height is 20cm, the effective illumination time is the shortest. Therefore, the higher the illumination height, the longer the effective illumination time ,but its corresponding final temperature is still the lowest.

3.1.2 Effect of Irradiation Height on Temperature Change Rate of Optical Heat Storage Fabrics

The effect of irradiation height on the heating rate of the light-heat storage fabric is shown in Fig.1.

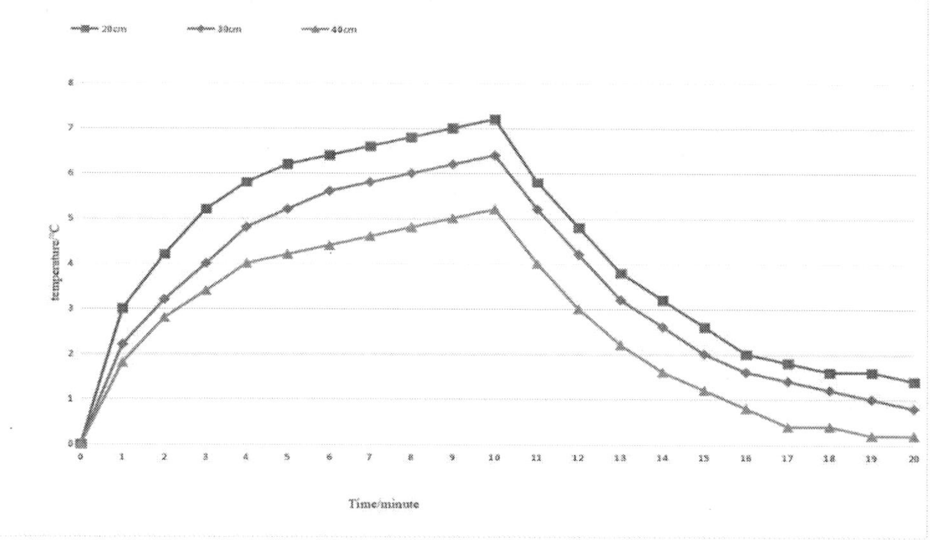

Fig.1 Temperature curve

As can be seen from the curve of the temperature rise value in Fig.1, the temperature rose sharply in the first minute of irradiation and then increased gently. The rate of temperature rise was relatively high in two minutes, and then the rate of temperature rise was stable. At the same time, the heating rate is different under the three irradiation heights. It can be seen from Figure 1 that the heating rate is the highest when the irradiation height is 20 cm, it shows that the irradiation height has an effect on the absorption and heating rate of the optical heat storage fabric, the smaller the irradiation height, the greater the heating rate. After the power is turned off, the cooling trend of the three irradiation heights is basically consistent with the rate, indicating that the irradiation height has little effect on the heat dissipation rate of the optical heat storage fabric.[13]

3.1.3 Significant Analysis of the Effect on the Temperature Rise of Optical Heat Storage Fabric

Analyse whether the irradiation height has significant influence on the temperature rise of the optical heat storage fabric, and calculate its F value. The results are shown in Table 3

Table 3.the effect of irradiation height on the temperature rise of the optical heat storage fabric

independent variables	dependent variable	F
irradiation height	ΔT_i	5.0

$F_{0.05}(3,27) = 2.96$. As can be seen from Table 3, F value of temperature rise under the influence of irradiation height is 5, greater than $F_{0.05}(3,27)$, indicating that exposure height has a highly significant impact on temperature difference.

3.2 Influence of Initial Ambient Temperature on Absorption and Heating Properties of Photothermal Storage Fabrics

3.2.1 Influence of Initial Ambient Temperature on the Heating Capacity of Optical Heat Storage Fabrics

The heating capacity of the optical heat storage fabric was tested at different ambient temperatures. The results are shown in Table 4, Table 5.

Table 4. Influence of the initial ambient temperature on the heating capacity of the optical heat storage fabric

initial ambient temperature (°C)	Highest temperature (°C)	ΔT_{10} (°C)
15	41.6	5.0
20	50.4	6.4
25	60.2	7.6

As shown in Table 4, when the initial ambient temperature was 15°C, 20°C and 25°C respectively, the temperature of the reference sample increased to 41.6°C, 50.4°C and 60.2°C. The temperature of the reference sample indicates the ambient temperature of the test under the same conditions. The ambient temperature in the darkroom will rise under the light. It can be seen that the initial ambient temperature has a certain influence on the test ambient temperature, and the greater the initial ambient temperature is in the same irradiation time, the greater the ambient temperature will be. [10]

After 10 minutes, ΔT_{10} represents the performance of the fabric to absorb light and heat. According to the data in Table 3, under the same light source, the initial ambient temperature is 15°C, 20°C, 25°C, ΔT_{10} is 5.0°C, 6.4°C, 7.2°C.The results show that the initial ambient temperature of the same kind of optical heat storage material has a great influence on the heat absorption and temperature rise performance of the optical heat storage fabric. The higher the initial ambient temperature is, the higher the rising temperature of the fabric will be, and the better the heating performance of the optical heat storage fabric will be. [11] [12]

Table 5. Influence of test temperature on effective irradiation time

initial ambient temperature (°C)	effective illumination time (min)	Highest temperature (°C)
15	9	5.0
20	11	6.8
25	13	8.2

It can be seen from Table 5 that the effective illumination time is different under different test temperatures. When the test temperature is 25, the effective illumination time is the longest, which is 15 minutes, and the final temperature is the largest, which is 8°C. When the test temperature is 15, the effective illumination time is the shortest. Therefore, the higher the test temperature, the longer the effective illumination time and the greater the final temperature.

3.2.2 Effect of Initial Ambient Temperature on the Temperature Change Rate of Optical Heat Storage Fabrics

The influence of the initial ambient temperature on the heating rate of the photo-heat storage fabric is shown in Fig.2

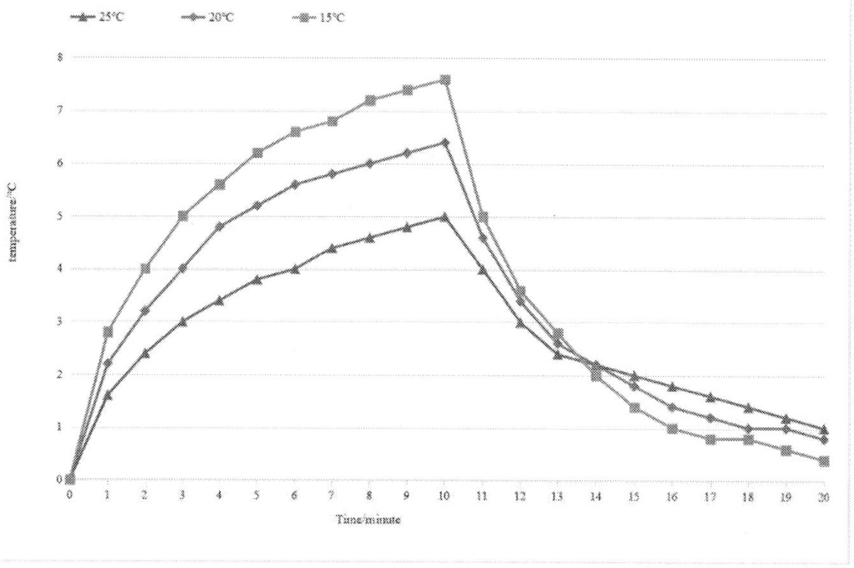

Fig.2 Temperature curve

As can be seen from the curve of temperature rise value in Fig.2, the temperature rose sharply in the first minute of irradiation and then increased gently. The rate of temperature rise was relatively high in the two minutes, and then was stable. At the same time, the heating rate is different at the three test temperatures, it can be seen from Fig.2 that the temperature rises fastest when the test temperature is 15°C, indicating that the test temperature has an influence on the temperature rise rate, The lower the test temperature is, the higher the temperature rise rate will be. When the power is turned off, the cooling rate is different under the three test temperatures. The temperature drops fastest when the test temperature is 15°C, It shows that the initial ambient temperature has a certain influence on the heat dissipation rate of the optical heat storage fabric. The smaller the initial ambient temperature is, the greater the heat dissipation rate will be. [13]

3.2.3 The Significance Analysis of the Effect of Initial Ambient Temperature on the Temperature Rise of Photoregenerative Fabrics

Analyse whether the influence of the initial ambient temperature on the temperature rise of the optical heat storage fabric is significant, and calculate its F value. The results are shown in Table 4.

Table 6. the effect of initial ambient temperature on the temperature rise of the optical heat storage fabric

independent variables	dependent variable	F
initial ambient temperature	ΔT_a	6.6

$F_{0.05}(3,27) = 2.96$. It can be seen from Table 6 that the temperature rise F value under the influence of ambient temperature is 5.9, which is greater than $F_{0.05}(3,27)$, indicating that ambient temperature has a significant impact on the heat dissipation temperature difference. Therefore, it can be inferred that ambient temperature has a significant impact on the heat dissipation performance of the optical heat storage fabric.

4. Conclusions

The textile can absorb the light energy in the environment and convert it into heat energy, so as to increase the surface temperature of the sample and achieve the effect of heat storage and warmth. [14] [15] This paper carried out tests under different conditions, and the test results showed that:

1) The lower the irradiation height is, the higher the rising temperature of the fabric is, and the better the heat absorption and rising performance of the fabric is, the higher the heating rate is. The irradiation height has a significant effect on the heating performance of the fabric.

2) The smaller the initial ambient temperature, the lower the fabric temperature, but the faster the heating rate. The initial ambient temperature has a significant effect on the temperature rise of the optical heat storage fabric.

References

[1] Huang SR, Chen LH, Zhou HQ. Effect of temperature on the temperature-regulating property of phase change temperature-regulating fiber fabric [J]. Journal of Beijing Institute of Fashion Technology (Natural Science Edition): 2018; 38 (02): 16-23.

[2] Zhang H. Research and application of cold-resistant textiles and clothing [J]. Journal of Textile Science: 2003; 24 (5): 499.

[3] Shi HF. Development of photothermal conversion heat storage and temperature regulating fiber [D]. Tianjin Polytechnic University: 2002.

[4] Hosokawa H, Wang N. Polyacrylonitrile fiber Thermocatch ~ (TM) -W [J]. Foreign Textile Technology: Chemical Fiber Dyeing and Finishing of Textile Knitting Garment: 2001 (6): 8-11.

[5] Li L, Meng JG. Development and application of thermal fiber and thermal fiber [A]. China Textile Information Center, National Textile Product Development Center, Information Committee of China Textile Engineering Society. Proceedings of the 16th National Symposium on Technological Progress of Fancy Yarns and Fabrics [C]. China Textile Information Center, National Textile Product Development Center, Information Committee of China Textile Engineering Society: 2010; pp: 6.

[6] Gutian CS, Development of Sunlight Thermal Storage and Insulation Fiber [J]. Functional Materials: 1992; 12 (1): 37-44

[7] Liao YL, Zhan Q, Zhang YQ, et al. Test Method for Light Absorbing and Heating Property of Textile [J]. Quality Technical Supervision Research: 2016; (5): 11-17.

[8] Xia Y. Research progress on testing methods of thermal storage performance and thermal insulation performance of textiles [J]. Jiangsu Textiles: 2014; (04): 47-48.

[9] Zhang HY, Xia Y, Wei XY. Preliminary study on a new test method for heat storage and thermal insulation performance of textile [J]. Wool & Textile Science and Technology: 2015; 43 (03): 56-58.

[10] Xiao CF. Properties and preparation of a new type of heat storage and thermal insulation fiber [J]. Journal of Tianjin Textile Institute: 1995; 14 (1): 44-47.

[11] Chen JY, Zhu G. Briefly describe the heating mechanism and development of heating and thermal insulation fiber Application [J]. Shandong Textile Science and Technology: 2012; (5): 38-40.

[12] Li LL. Hair development and application of hot fiber dimension [J]. Printing and dyeing: 2015; (21): 49-51.

[13] Weng L, Wu ZM. Heat storage and thermal insulation finishing of fabrics [J]. Dyeing and Finishing echnology: 2005; 27 (8): 6-9.

[14] Wang M, Li J. Development status and development of heating and thermal clothing materials Trend [J]. industrial textiles: 2009; (4): 6-9.

[15] Li JS, Ma XG, Zhang XL, et al. Photothermal conversion heat storage materials and their heat storage Textiles [J]. journal of textile science: 2007; 28 (2): 76-80.

Cite as: Z.Z. Qi et al.,
TBIS 10.3993/tbis (2021).

Stability of Silk Fibroin Blend Films with Silk Ⅰ Structure

Zhen-Zhen Qi, Mei-Hui Zhao, Xiao-Sheng Tao, Ge-Yu Jin, Shen-Zhou Lu*

National Engineering Laboratory for Modern Silk, College of Textile and Clothing Engineering, Soochow University, Suzhou, 215123, China

*Corresponding author's email: lushenzhou@suda.edu.cn

Abstract

The crystalline structure of Silk I fibroin is a metastable structure that may transition to silk II crystalline structure. It is unclear under what conditions is crystalline structure of silk Ⅰ stable and under what conditions when the transition will occur. In this paper, Silk Ⅰ crystalline structure blend film was prepared with a glycerol/silk fibroin mass ratio of 20: 100, and a pure silk fibroin solution was used to prepare a random coil structure film. Different concentrations of methanol and ethanol were used to soak the above two materials to investigate the influence of monohydric alcohol on the crystalline structure of silk fibroin materials. X-ray diffraction (XRD), Fourier transform infrared spectroscopy (FTIR), Raman scattering spectroscopy (Raman) and Thermogravimetric analysis (TGA) were used to characterize the structure of silk fibroin before and after the treatments. The results showed that after methanol and ethanol treatment, the random curled structure film was transformed into Silk Ⅱ crystal structure, while the glycerin/silk fibroin protein film remained with the crystal structure of Silk Ⅰ without obvious transformation. To investigate the time stability of Silk I crystal structure, glycerin/silk fibroin blend film and pure silk fibroin protein film were placed in an oven at 60°C for 32 days. The results showed that the glycerin/silk fibroin film had no obvious structural changes and had good humidity, heat stability and time stability. The random coil structure of pure Silk film gradually changed to the more regular crystal structure of Silk Ⅰ and Silk Ⅱ. Silk Ⅰ film has considerable stability and can be used as biomaterials.

Keywords: Silk Fibroin; Film; Silk Ⅰ Crystalline Structure; Crystal Structure; Stability; Glycerol

1. Introduction

Silk fibroin is a kind of fibrin, that has excellent biocompatibility and mechanical properties. Based on these properties, silk fibroin has moved from the traditional textile industry to the biomedical field. Silk protein can be processed into different forms with a very broad application prospect [1]. At present, the silk fibroin materials have been developed and applied to mainly 3D printing [2], tissue repair [3], drug controlled release [4], disease model [5], bone tissue scaffold and skin wound dressing [6].

Silk fibroin has a polypeptide chain, composing of hydrophobic heavy chain (H-chain, 390 kDa), hydrophilic light chain (L-chain, 25 kDa) and P25 glycoprotein chain (P25, 30 kDa) [7]. H-chain and L-chain are connected by disulfide bond. One P25 glycoprotein and six H-L chain aggregates are assembled by hydrophobic interaction [8]. Silk fibroin contains crystalline region and amorphous region. H-chain runs through the crystalline region and amorphous region, while L-chain only exists in the amorphous region. The amorphous region is composed of non-repetitive sequences. Amino acids such as tryptophan, phenylalanine and tyrosine with large side chain groups exist in an amorphous state. The GAGAGS repetitive sequence in the crystal region forms an antiparallel β-sheet structure by intermolecular and intramolecular forces [9].

The exploration of the crystal structure of silk fibroin can be traced back to Shimizu Masanori. He obtained two crystal states of silk fibroin through experiments, which were named silk fibroin α and silk fibroin β [10]. Later, Kratky [11] and others found that silk fibroin had two types of crystals, Silk I and Silk II. Subsequently, the researchers [12] found that Silk I crystal structure model was a repeat unit dipeptide, and the molecular chain was a crankshaft type, belonging to the orthorhombic system. The Silk II crystal structure model is a layered structure formed by β-antiparallel folding, belonging to the monoclinic crystal system [13]. Regina [14] et al found a new crystal on the gas-liquid interface, which was called Silk III, similaring to triple helix. The crystal structure of Silk III is similar to that of polyglycine II, belonging to hexagonal system.

At present, silk fibroin materials of Silk II crystal structure have been studied and applied in many fields, while studies on the crystalline structure of Silk I is slowly developing. At present, it is found that silk fibroin mainly transforms to Silk I when the film forming temperature is lower than 40° 0 and the concentration is low; Lu [15] and others successfully induced silk fibroin to form Silk I crystal structure by adding polyols with specific hydrophile-lipophile value. Compared with Silk II crystalline materials, Silk I crystalline materials have better hygroscopicity, flexibility and higher degradation rate [12]. Generally speaking, the crystal structure of Silk I is a kind of metastable structure with poor time stability, which is easy to be induced to Silk II crystal structure. In order to explore the change of silk fibroin Silk I crystal structure, the stability of silk fibroin blend film with Silk I crystal structure was discussed. The stability after soaking in methanol and ethanol and time stability of glycerol / silk fibroin blend film were studied, to expand the application of Silk I as biomaterials.

2. Materials and Methods

2.1 Experimental Materials

Fresh mulberry silkworm cocoons of Bombyx mori (Suzhou Soriano Biological Technology Co. , Ltd. , China), sodium carbonate, sodium bicarbonate, lithium bromide (Tiancheng Chemical Co. , Ltd. , China), Anhydrous methanol (Aladdin reagent Shanghai Co. , Ltd. , China), Anhydrous ethanol (Aladdin reagent Shanghai Co. , Ltd. , China), Dialysis bag (MWCO: 8-12kd, Glycerol (Aladdin reagent Shanghai Co. , Ltd. , China).

2.2 Preparation of Silk Fibroin Solution

80g of silkworm cocoon shells were weighed and steamed in a 0.3% (w/v) Na_2CO_3/0.1% (w/v) $NaHCO_3$ solution at 100°C three times, each for half an hour, stirring once every ten minutes. The cocoon shell was washed with deionized water to remove sericin. The silk was dried in an oven at 60°C. 15g degummed silk fibroin fiber was dissolved in 9.3M LiBr solution at 65°C for 1h. After cooling to room temperature, the above solution was put into dialysis bag and placed in 4°C deionized water for dialysis for 3 days. Particulate impurities in the solution were then filtered to remove the resulting silk fibroin solution.

2.3 Preparation of Silk Fibroin Film

2.3.1 Preparation of Random Coil Structure Film

The silk fibroin solution prepared in 2.2 (5.0%, w/v) was injected into a 60mm polyethylene petri dish and dried for 24h in a constant temperature and humidity room (25°C, relative humidity 65%) to obtain a pure silk fibroin film with random coil structure (later referred to as SF-P) (Fig.1A).

2.3.2 Preparation of Silk I Structure Film

Set the mass ratio of glycerin to silk fibroin as 20:100, mix the silk fibroin solution obtained from 2.2 with corresponding mass of glycerin, pour it into a 60mm polyethylene petri dish, and let it stand for 24h in constant temperature and humidity room (25°C, relative humidity 65%). The glycerin/silk fibroin blend film (later referred to as SF-G) (Fig.1B) can be obtained by this preparation method, which is a silk fibroin film with Silk I crystal structure [16].

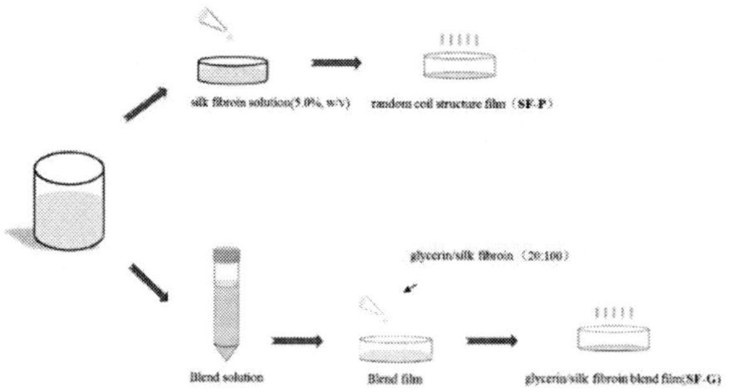

Fig.1 Flow chart of sample preparation: (A) pure silk fibroin film (SF-P), (B) glycerol/silk fibroin blend film (SF-G)

2.4 Determination of Stability of Monohydric Alcohol Treatment

2.4.1 Methanol Treatment

The samples of SF-G and SF-P were immersed in 75%, 90% and 99% (v/v) methanol solution respectively, and the gradient time was set to 1, 2, 4 and 8 h. After soaking, the samples were cleaned with deionized water and placed for 24 hours in a constant temperature and humidity room (25°C, relative humidity 65%).

2.4.2 Ethanol Treatment

The two samples were soaked in 75%, 90% and 99% (v/v) ethanol solution respectively. After 1, 2, 4 and 8h of time gradient treatment, the samples were dried for 2h in a constant temperature and humidity room (25°C and 65% relative humidity).

2.5 Thermal Stability Determination

TA Differential Thermometer SDT Q600 was used to measure the mass stability of the silk samples with increasing temperature. The thermogravimetric curve was obtained at a gas flow rate of 50 mL/min under a nitrogen atmosphere. 4 mg of each sample was heated from 25°C to 450°C at a heating rate of 10 °C/min. During heating the percentage of mass change was recorded.

2.6 Time Stability of Silk Fibroin Materials

Based on the principle of time temperature equivalence, the accelerated aging temperature TAA is set at 60°C, and the accelerated aging factor AAF is calculated by Arrhenius Equation (1).

$$AAF = Q_{10}^{[\frac{(T_{AA}-T_{RT})}{10}]} \quad 1)$$

Where TAA is the accelerated aging temperature (°C), TRT is the ambient temperature (°C), Q_{10} is the reaction rate coefficient, usually Q_{10}=2 is a conservative method to calculate the aging factor.

The accelerated aging time AAT needs to be established by the expected aging time and the accelerated aging factor AAF, which is calculated according to Equation (2).

$$AAT = \frac{RT}{AAF} \quad 2)$$

2.7 Structural Characterizations

A certain number of treated samples and untreated materials was cut with scissors and sieved with 80 mesh sieves.

2.7.1 X-ray Diffraction Analysis

The crystal structure of the sample was detected by the fully automatic X'PERT PRO MPD ray diffractometer (Panalytical Company, Netherlands). The diffraction intensity curve was recorded between 5°C and 45°C under the conditions of 10°/min scanning speed, 40kV, and 30mA.

2.7.2 Fourier Transform Infrared Spectroscopy

Each sample was thoroughly mixed with KBr, crushed, and then compressed into transparent small discs with a tablet press. A Nicolet 5700 infrared spectrum analyzer was used to analyze the infrared absorption spectrum of the sample with a scanning step range of 400~4000 cm^{-1}

.

2.7.3 Raman Scattering Spectroscopy

Raman spectrum was measured using a Japanese HORIBA Raman Microscopy. The excitation wavelength was 532 nm, slit width was 100 μm, 1200 gr/mm grating was selected. The scanning time was 20s, and the step range of Raman scattering spectrum was 200~2000 cm^{-1}.

3. Results

3.1 Chemical Stability

3.1.1 X-ray Diffraction Analysis

The X-ray diffraction peak of silk fibroin Silk I crystal mainly appeared at 12.2°(ms), 19.7°(s), 24.7°(m), 28.2°(m), 32.3°(w), 36.8°(mv), 40.1°(mw); and the diffraction peak of Silk II crystal mainly appeared at 9.1°(ms), 18.9°(ms) and 20°(VS) [17]. Fig.2 shows that the pure silk fibroin film mainly exhibited an amorphous structure, without no crystallization peak. After 8 h treatment with different concentrations of methanol and ethanol, the crystallization peaks appeared at 9.1°and 20.7°. The random coil structure was transformed into Silk II crystal structure. This is because small molecule monohydric alcohols are very easy to enter the interior of silk fibroin materials, making silk fibroin molecules easier to move and changing to a more stable crystalline structure, which will lead to the transformation of silk fibroin from an amorphous state to a crystalline structure [13]. However, the crystal structure of Silk I in SF-G material remained Silk I crystal and no significant difference appeared in the XRD analysis before and after treatments. This indicated that the Silk I crystal structure in silk fibroin material had good stability, and will not transform into Silk II crystal structure.

3.1.2 FTIR Analysis

The test results in Fig.3 shows the characteristic peaks SF-P at 1644 cm^{-1} (amide I), 1531 cm^{-1} (amide II) and 1235 cm^{-1} (amide III) [18], indicating that silk fibroin molecule is mainly random coil structure. Silk I material (SF-G) has infrared absorption peaks near 1651 cm^{-1} (amide I), 1531 cm^{-1} (amide II) and 1234 cm^{-1} (amide III). With the increase of ethanol and methanol concentration, the random coil structure gradually transformed to the β-sheet. In this paper, the samples treated for 8h were selected for infrared detection. The results showed that the absorption peak position of SF-G in amide I and amide II remained unchanged with the increase of methanol and ethanol concentration. The absorption peak of SF-P in amide II did not change significantly with the increase of treatment concentration. However, the absorption peaks in amide I and amide III shifted, especially the absorption peak in amide I region shifted to a lower wavenumber. The results showed that with the increase of methanol ethanol content, the rearrangement of hydrogen bonds in random coil materials caused more silk fibroin molecules to transform from random coil to β-sheet [18]. Generally, it is difficult to distinguish random coil and α-helix by FTIR [15]. The combination of FTIR spectra and XRD curves enables us to determine that the shift of the absorption peak in the amide I region should be caused by the coexistence and superposition of α-helix, random coil and β-sheet peaks. The results indicate that the random coiling materials treated with methanol and ethanol can promote random coil structure to β-sheet structure. However, the crystal structure of Silk I still has good stability after methanol and ethanol treatment, which is consistent with the results of XRD.

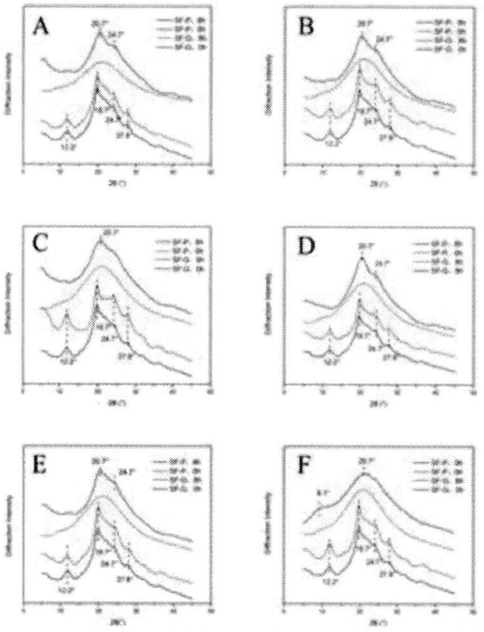

Fig.2 XRD analysis of SF films before and after alcohol treatment: (A) 75% methanol, (B) 90% methanol, (C) 99% methanol, (D) 75% ethanol, (E) 90% ethanol, (F) 99% ethanol

3.1.3 Raman Scattering Spectrum

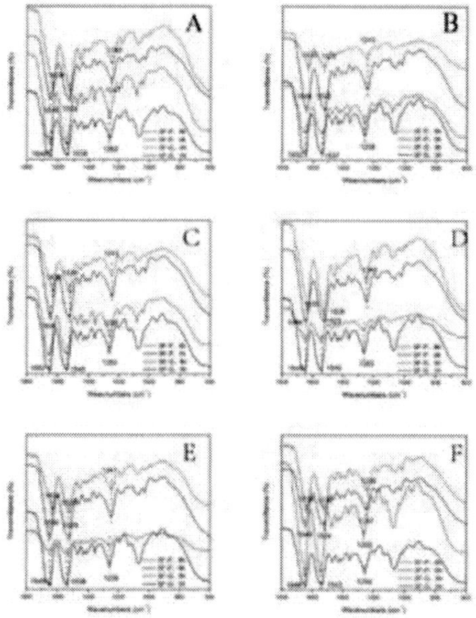

Fig.3 FTIR spectra of SF films before and after alcohol treatment: (A) 75% methanol, (B) 90% methanol, (C) 99% methanol, (D) 75% ethanol, (E) 90% ethanol, (F) 99% ethanol

In order to further explain the structural changes of silk fibroin materials, the Raman spectra of silk fibroin materials were detected. In Raman spectrum(Fig.4), SF-P has scattering peaks near 1665cm^{-1} (amide I), 1263cm^{-1}, 1234cm^{-1} (amide III) and 1104cm^{-1}, which are typical characteristic peaks of random coil conformation [19]. It can be seen from Fig.5(A) that after the treatment with 75% methanol for 8h, the amide I band moved to 1669 cm^{-1} and the width became narrower. In amide III region, the peak at 1264cm^{-1} weakened and the peak at 1231cm^{-1} became sharp; A sharp scattering peak at 1085cm^{-1} appeared and the scattering peak at 1104cm^{-1} disappeared. This indicated that the treatment of SF-P with 75% methanol will result in the production of β-sheet structure. In contrast, before and after treatment, the scattering peaks of SF-G around 1661cm^{-1} (amide I), 1273cm^{-1}, 1243cm^{-1} (amide III) and 1106cm^{-1} remained unchanged, which indicated that the Silk I crystal structure was stable after 75% methanol treatment. Fig.4 (B-F) shows the same test results as Fig.4 (A). These results are consistent with the results of XRD and FTIR, which proves that the crystal structure of Silk I can exist stably after being treated with methanol and ethanol.

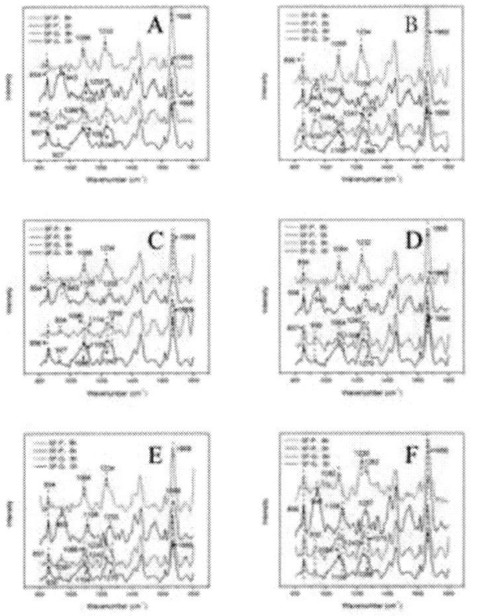

Fig.4 Raman spectra of silk fibroin materials before and after alcohol treatments: (A) 75% methanol, (B) 90% methanol, (C) 99 % methanol , (D) 75% ethanol, (E) 90% ethanol, (F) 99% ethanol.

3.3 Thermal Stability

In order to evaluate the thermal stability of the materials, the weight loss rates of silk fibroin fiber (mainly Silk II crystal), SF-G and SF-P were determined by thermogravimetric analysis. Fig.5 showed the mass percentage change of each sample during heating from room temperature to 450°C. The mass loss percentage curve in the figure reveals each sample experienced three main stages of weight loss. In the first stage, there was a slight quality drop at 100°C, the weight loss of random coiled material (SF-P) was 9.36%, Silk I material (SF-G) lost 6.61%, and silk fibroin fiber lost 7.02%. This part of weight loss was caused by moisture volatilization. The second stage was between 200°C and 300°C, which was a slow thermal decomposition stage. The initial decomposition temperature of three samples is 286.37°C, 243.26°C and 265.68°C respectively. The results showed that three kinds of silk fibroin material had good thermal stability, and there was no thermal degradation before 200°C. In the third stage, the weight loss was between 300°C and 400°C, and the three samples all had very fast mass loss, showing rapid weight loss. The temperature corresponding to the maximum degradation rate was 313.73°C (silk fiber), 276.87°C (SF-G) and 280.54°C (SF-P) respectively. The weight loss was mainly caused by the cleavage of peptide bond and side group [20]. Silk I crystal is a metastable structure, making its thermal decomposition slightly easier than that of Silk II crystal. When the temperature reaches 450°C, the weight remaining rate of random coiled material is similar to that of the Silk I material.

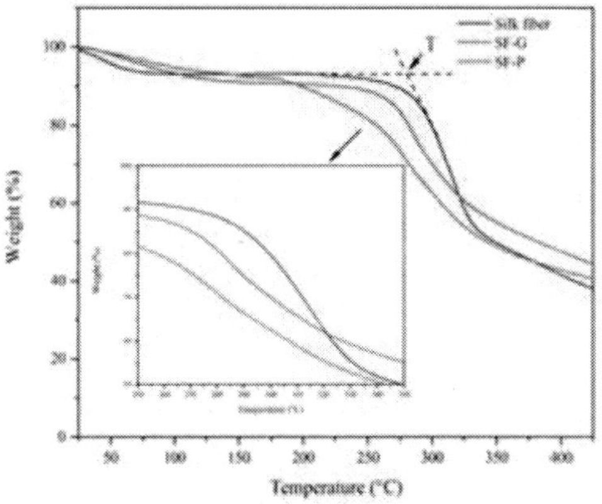

Fig.5 TGA curves of different silk fibroin materials

3.4 Time Stability

According to the Equation (1) and (2), the accelerated aging time is about 32 days. The glycerol/silk fibroin blend film (SF-G) and pure silk fibroin film (SF-G) were stored in the oven at 60°C for 32 days. Fig.6(A) showed that SF-G was almost no change in the XRD curve before and after placement, and the X-ray diffraction peak mainly appeared at 12.2° (ms), 19.7°(s). This indicated that the Silk I crystal structure material can maintain stable structure at room temperature for one year. In contrast, the random coil material (SF-P) showed a new Silk I crystal diffraction peak at 12.2° after being placed for 32 days. These results indicated that the random coil structure can change to a more regular crystal structure. Fig.6(B) showed that SF-G material had absorption peaks at 1650cm^{-1} (amide I), 1541cm^{-1} (amide II) and 1243cm^{-1} (amide III). After being placed in the oven at 60°C for 32d, the position of the absorption peak did not change significantly, just shifted from 1243cm^{-1} to 1250cm^{-1} in amide III. On the contrary, after the above treatment, the amide I region shifted from 1642cm^{-1} to 1650cm^{-1}, and the amide II region shifts from 1534cm^{-1} to 1528cm^{-1}, which indicates that the SF-P material changed from random coil structure to α-helix structure and β-sheet structure.

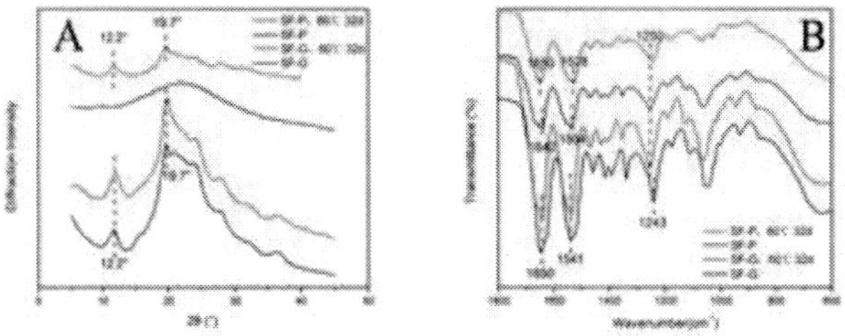

Fig.6 Structure of silk fibroin materials before and after placed at 60°C for 32d. (A): XRD; (B): FTIR

4. Discussion

Previous studies had shown that glycerol can induce the conformational change of silk fibroin from amorphous structure to crystal structure [21]. Joseph E. Brown [22] et al. mentioned that the interaction of protein fragments with different molecular weights in the mixture may be different (through peptide-peptide or peptide glycerin interaction). In this paper, the buffer solution (sodium carbonate/sodium bicarbonate) with less damage to silk fibroin was used for degumming, and lithium bromide solution was used for degumming at 65°C. When the solution is dissolved under different conditions, the obtained silk fibroin solution has high molecular weight, and the polypeptide chain of silk fibroin is less damaged, and the polypeptide chain remains more complete [23]. It can be seen from Fig.2 that the SF-G prepared in this paper had α-helix. The crystal structure of Silk I was observed by XRD. The formation mechanism of Silk I structure in SF-G has been considered as the following five types [22]: (1) peptide bond and typical of hydrogen bond β-sheet crosslinking; (2) The glycerol bridge interacts with two peptide chains on each glycerol molecule by hydrogen bonding; (3) When a large amount of glycerol is added to the protein matrix, it may lead to hydrogen bonding with silk fibroin polypeptide and other glycerol molecules. This results in phase separation of glycerol rich regions between peptide chains; (4) It occurs during the immersion of glycerin plasticized silk material in aqueous medium. Hydrogen bonding interaction between glycerol and water molecules may occur more frequently. These interactions may contribute to the extraction of glycerol from silk matrix; (5) Due to the destruction of physical crosslinking and hydrogen bond between peptides, protein folding occurs.

However, our previous studies showed that the addition of 20% wt glycerol can effectively induce the formation of characteristic silk fibroin with rich α-helix. At the same time, the solid content of silk fibroin in the glycerol/silk fibroin blend solution before drying is 50mg/ml. The higher solid content of silk fibroin and the retention of intact silk fibroin peptide chain make the entanglement of silk fibroin polypeptide chain more intensive and the movement more difficult. The corresponding Fig.3 and 4 showed that the aggregation mechanism of SF-G after methanol ethanol treatment was not as the same as SF-P. It can be considered that more dense polypeptide chain entanglement leads to less glycerol rich domains in the blend film, and the hydrogen bond between glycerol and polypeptide chain is not easy to be destroyed by methanol and ethanol, making glycerol difficult to be replaced and having good chemical stability. At the same time, SF-G can maintain the aggregation structure well, and is not easy to volatilize, because of large inter chain entanglement and low affection by water molecules. It can be seen from Fig.6 that compared with SF-P, SF-G had good time stability. Because there are few glycerol rich domains, the binding between glycerol and silk fibroin is more uniform and dense, and there is no obvious glycerol evaporation at 222°C [22], and the thermal degradation peak appears near 276°C, so it has good thermal stability.

5. Conclusions

In this paper, the stability of silk fibroin films with Silk I crystal structure was discussed. Compared with silk fibroin film with random coil structure, the glycerol/silk fibroin film with Silk I crystal structure had good thermal stability, hydrothermal stability and time stability, which can be stored at room temperature for one year keeping the structure unchanged. Methanol and ethanol soaking treatment can promote the transformation of random coil structure to Silk Ⅱ crystal structure, but it does not induce the change of crystal structure of silk Ⅰ. This paper proves that the crystal structure of Silk I remains stable after alcohol disinfection, which provides a good practice basis for its application in medical and health fields.

Acknowledgments

This work was supported by College Nature Science Research Project of Jiangsu Province, China (Grant No. 20KJA540002), PAPD and Six Talent Peaks Project in Jiangsu Province (Grant No. SWYY-038).

References

[1] Wei W, Liu J, Peng ZB, et al. Gellable silk fibroin-polyethylene sponge for hemostasis [J]. Artificial Cells: 2020; 48 (1): 28.
[2] Lee, Hyeongjin, Yang, et al. Fabrication of micro/nanoporous collagen/dECM/silk-fibroin biocomposite scaffolds using a low temperature 3D printing process for bone tissue regeneration [J]. Materials science & engineering, C. Materials for Biogical applications: 2018; 84: 140.
[3] Melke J, Midha S, Ghosh S, et al. Silk fibroin as biomaterial for bone tissue engineering [J]. Acta Biomaterialia: 2016; 31: 1.
[4] Wang S, Zhu M, Zhao L, et al. Insulin-loaded silk fibroin microneedles as sustained release system [J]. ACS Biomaterials Science and Engineering: 2019; 5 (4): 1887.
[5] Zhu M, Liu Y, Jiang F, et al. Combined Silk Fibroin Microneedles for Insulin Delivery [J]. ACS Biomaterials Science and Engineering: 2020; 6 (6): 3422.
[6] Angelica, Panico, Federica, et al. Development of regenerative and flexible fibroin-based wound dressings [J]. Journal of Biomedical Materials Research Part B Applied Biomaterials: 2019; 107 (1): 7.

[7] Rockwood DN, Preda RC, Yücel T, et al. Materials fabrication from Bombyx mori silk fibroin [J]. Nature Protocols: 2011; 6 (10): 1612.
[8] Inoue S, Tanaka K, Arisaka F, et al. Silk Fibroin of Bombyx mori Is Secreted, Assembling a High Molecular Mass Elementary Unit Consisting of H-chain, L-chain, and P25, with a 6:6:1 Molar Ratio* [J]. Journal of Biological Chemistry: 2000; 275 (51): 40517.
[9] Jin HJ, Kaplan DL. Mechanism of silk processing in insects and spiders [J]. Nature: 2003; 424 (6952): 1057.
[10] Liang CX, Hirabayashi K. Improvements of the physical properties of fibroin membranes with sodium alginate [J]. Journal of Applied Polymer Science: 1992; 45 (11): 1937.
[11] Kratky O, Schauenstein E, Sekora A. An Unstable Lattice in Silk Fibroin [J]. Nature: 1950; 165 (4191): 319.
[12] Hu X, Shmelev K, Sun L, et al. Regulation of Silk Material Structure by Temperature-Controlled Water Vapor Annealing [J]. Biomacromolecules: 2011; 12 (5): 1686.
[13] Cebe P, Partlow BP, Kaplan DL, et al. Silk I and Silk II studied by fast scanning calorimetry [J]. Acta Biomaterialia: 2017; 323.
[14] Valluzzi R, Gido SP, Muller W, et al. Orientation of silk III at the air-water interface [J]. International Journal of Biological Macromolecules: 1999; 24 (2-3): 237.
[15] Lu S, Li J, Zhang S, et al. The influence of the hydrophilic–lipophilic environment on the structure of silk fibroin protein [J]. Journal of Materials Chemistry B: 2015; 3 (13): 2599.
[16] Lu S, Wang X, Lu Q, et al. Insoluble and Flexible Silk Films Containing Glycerol [J]. Biomacromolecules: 2010; 11 (1): 143.
[17] Riekel C, Burghammer M, Rosenthal M. Nanoscale X-Ray Diffraction of Silk Fibers [J]. Frontiers in Materials: 2019; 6.
[18] Yang H, Yang S, Kong J, et al. Obtaining information about protein secondary structures in aqueous solution using Fourier transform IR spectroscopy [J]. Nature protocols: 2015; 10 (3): 382.
[19] Monti P, Taddei P, Freddi G, et al. Raman spectroscopic characterization of Bombyx mori silk fibroin: Raman spectrum of Silk I [J]. Journal of Raman Spectroscopy: 2001; 32 (2): 103.
[20] Müller V, Cestari M, Palácio S, et al. Silk fibroin nanofibers electrospun on glass fiber as a potential device for solid phase microextraction [J]. Journal of Applied Polymer Science: 2015; 132 (13): n/a.
[21] Yin Z, Liang X, Zhou K, et al. Biomimetic Mechanically Enhanced Carbon Nanotube Fibers by Silk Fibroin Infiltration [J]. Small: 2021; 17 (19): 2100066.
[22] Brown JE, Davidowski SK, Xu D, et al. Thermal and Structural Properties of Silk Biomaterials Plasticized by Glycerol [J]. Biomacromolecules: 2016; 17 (12): 3911.
[23] Bucciarelli A, Greco G, Corridori I, et al. A Design of Experiment Rational Optimization of the Degumming Process and Its Impact on the Silk Fibroin Properties [J]. Acs Biomaterials Science & Engineering: 2021; 7 (4): 1374.

Performance Evaluation and Application of Hot-Spot Silver Fabrics

Xue-Li Lin[1], Li-Min Shi[2*], Qi Wang[2], Jin-Hui Ren[2], Jing Guo[2]

1 School of Material Design and Engineering, Beijing Institute of Fashion Technology,
2 Universal Fashion Lab, Beijing Institute of Fashion Technology, No.2 Yinghua Road Chaoyang District, Beijing, 100029, China

[*]Corresponding author's email: Shilm@uflbift.com

Abstract

Hot-spot silver fabric has become a popular material in textiles during recent years, used as the lining of cold-proof and warm clothing. It warmth, moisture permeability, and air permeability properties. In order to objectively understand the comprehensive performance of the material and better apply it to thermal clothing, this article investigates and selects 11 commonly used hot-spot silver fabrics from the market for performance testing and analysis; then based on traditional mathematical analysis, mathematical fuzzy synthesis is used. The analysis method systematically and objectively evaluates the comfort performance of the samples. The results show that most of the hot silver fabrics have soft texture, anti-wrinkle, and abrasion resistance; 9# and 11# samples can meet the far-infrared emissivity and far-infrared radiation temperature rise value. From the fuzzy comprehensive analysis, the 5# and 8# samples have better comfort performance, and the 9# sample has better far-infrared radiation performance, which can be used in the local warmth of clothing. The thesis research can provide reference for the development and application of new fabrics.

Keywords: Hot-spot Silver Fabric; Far-infrared Performance; Performance Evaluation; Warm Clothing Lining; Fuzzy Comprehensive Analysis

1. Introduction

Warm clothing must ensure people's comfort in a cold environment. In order to maintain the body's heat balance, warm clothing must be able to balance the heat generated by the body's normal metabolism. Ambient air may vary from mild and moist to dry and extremely cold and so thermal protective clothing has to vary accordingly. [1] In a cold and humid environment, the human body will feel uncomfortable due to clothing that is not warm or relatively warm and not breathable. To better achieve warmth retention, it is necessary for the selection of surface accessories, sewing technology and finishing of warm clothing, wherein the process requirements are very high. Lining is the material of the innermost layer of clothing, which can enhance the appearance, smoothness, thermal insulation and three-dimensional effect of clothing, prevent the filler from being exposed, protect the fabric, and support the fabric. [2] In addition, it improves thermal and moisture sensations on the skin and, therefore, may significantly affect sensorial comfort. [3] There have been much research on the fabrics of cold and warm clothing at home and abroad, but little research on the lining. This paper selects 11 representative hot-spot silver fabrics, through assessing basic parameters such as heat retention, air permeability, far infrared and other performance tests. Through fuzzy comprehensive analysis the comprehensive performance of hot-spot silver fabrics were analyzed and evaluated. A reasonable application is then proposed according to different linings and the structure of the garment.

2. Experimental Materials and Methods

2.1 Experimental Materials

After investigation, 11 kinds of fabrics with different coating shapes were collected. The fabric specifications and structural parameters are shown in Table 1.

Table 1. Fabric specifications and structural parameters

Sample No.	Raw material	Density [root/10cm] warp	Density [root/10cm] weft	Yarn fineness [Tex]	Organization	Mass per unit area [g/m^2]	Thickness [mm]	Base cloth color	Hot-spot shape
1	Polyester	748	413	5.56*5.56	plain weave	69.08	0.082	black	gradient dots
2	Polyester	770	420	5.56*5.56	plain weave	65.38	0.075	black	herringbone
3	Polyester	724	406	5.56*5.56	plain weave	64.46	0.091	black	dots
4	Polyester	717	406	5.56*5.56	plain weave	68.72	0.076	black	rectangle
5	Polyester	709	445	5.56*5.56	plain weave	82.22	0.143	black	type T
6	Polyester	480	322	8.33*8.33	plain weave	60.34	0.067	purple	dots
7	Polyester	740	413	5.56*5.56	plain weave	77.16	0.105	black	dots
8	Nylon	480	300	7.78*7.78	plain weave	71.8	0.105	black	square
9	Polyester	/	/	/	knitted film	81.2	0.311	knitted mesh	whole silver film
10	Polyester	453	322	7.00*7.00	plain weave	58.8	0.091	blue	dots
11	Polyester	712	448	5.56*5.56	plain weave	89.88	0.156	black	type T

According to the data in the above table, the warp and weft density of the fabric is similar, the mass per unit area is 60-85 g/m^2, and the thickness is within 0.2mm. Most of the hot silver fabrics are made of filament yarn. Filament yarns are useful in windbreaker jackets and in the shell and lining of skiwear, where a combination of dense weave and low surface coefficient of friction is desirable. [4]

Way.

2.2 Fabric Performance Measurement

1) The appearance performance test was as follows: 1. Bending rigidity was determined according to GB/T 18318.1-2009. The Textile Bending Performance was determined with the Inclined Plane Method, using LLY-01 electronic stiffness meter with the unit as mN·cm; 2. Wrinkle recovery was determined according to GB/T 3819-1997 "Determination of recovery angle of textile fabric crease recovery", using YG541L digital fabric fold elasticity meter, unit: °, ambient temperature and humidity was set as 25°, 55%.

2) The durability test was conducted as follows: 1. Tensile performance was determined according to GB/T 3923.1-2013 Textile Fabric Tensile Properties Part 1: Determination of Breaking Strength and Breaking Elongation (Strip Method)", using YG065C electronic fabric Powerful tester. 2. Abrasion resistance was determined according to GB/T 21196.2-2007 "Determination of the abrasion resistance of textile Martindale fabrics Part 2: Determination of sample damage", GB/T 21196.3-2007 "Determination of the abrasion resistance of textile Martindale fabrics" Determination Part 3: Determination of Mass Loss", the YG401L fabric flat grinder and electronic balance are used to test the wear-resistant mass loss of the fabric. The ambient temperature and humidity was set as 25°C, 55%.

3) Comfort performance test was conducted as follows: 1. Air permeability was determined according to GB/T 5453-1997 Determination of Air Permeability of Textile Fabrics, using YG461Z automatic air permeability tester, unit: mm/s; 2. Moisture permeability was determined according to GB/ T 12704.2-2009 "Textile Fabric Moisture Permeability Test Method Part 2: Evaporation Method", using YG601H computerized fabric moisture permeability tester, moisture permeability unit: g/(m^2·h). 3. Thermal resistance and moisture resistance was determined according to GB/T 11048-2008 "Textiles-Physiological comfort-Determination of thermal and moisture resistance under steady-state conditions", the M259B thermal and moisture resistance tester is used. 4. Electrostatic performance was determined according to GB/T 12703.1-2008 "Evaluation of electrostatic performance of textiles Part 1: Static voltage half-life", using YG(B)342E fabric induction electrostatic tester, unit: s, ambient temperature and humidity was set as 25°C , 55%.

4) Far-infrared performance test was determined according to GB/T 30127-2013 "Testing and Evaluation of Far-infrared Performance of Textiles", the samples were sent to a third-party testing agency for testing. Ambient temperature and humidity was set as 25°C, 55%.

3. Experimental Results and Analysis

Performance tests on the samples were conducted according to relevant standards and the experimental test results are shown in Tables 2, 3, 4, and 5. As the fabric has the same performance and different degrees of performance in various aspects, it is difficult to use a single index as a comprehensive evaluation of the fabric. In addition, because the regular distribution of data is not obvious and the amount of data is small, it is difficult to draw effective conclusions using traditional deterministic mathematics analysis. [5] Therefore, on the basis of traditional description and analysis, this paper adopts the mathematical fuzzy comprehensive evaluation method to objectively evaluate the comfort of 11 kinds of fabrics.

3.1 Experimental Test Rresults

3.1.1 Appearance Performance

The appearance performance test results of 11 kinds of hot silver fabrics are shown in Table 2.

Table 2. Test value of fabric appearance performance

Sample No.	Bending stiffness [mN·cm]		Acute elastic wrinkle recovery angle [°]		Slow elastic wrinkle recovery angle [°]	
	warp	weft	warp	weft	warp	weft
1	0.897	0.380	73	94	89	111
2	1.065	0.327	65	75	80	90
3	0.722	0.340	62	69	80	88
4	1.163	0.384	77	74	91	89
5	0.568	0.767	129	148	143	156
6	1.036	0.690	54	76	70	92
7	0.546	0.590	113	120	126	129
8	1.367	0.660	102	104	123	130
9	0.462	0.410	77	106	91	128
10	1.016	0.576	89	107	101	113
11	0.789	0.656	107	136	127	145

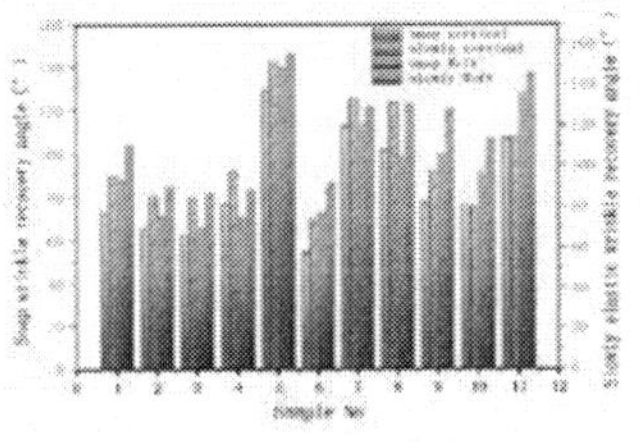

Fig.1 Fabric wrinkle recovery angle

The bending stiffness test results of samples 1-11 are shown in Table 2 and Fig.1. The test results of No.3, No.7 and No.9 linings are relatively small, and they are softer as linings. The test result of No.6, No.8 and No.10, linings are larger, so the fabric is relatively stiff. The initial modulus of the fiber is an important factor in determining its bending performance. Generally, the lower the initial modulus, the softer the fiber, and the more suitable the fabric for wear. [6]

3.1.2 Durability

The durability performance test results of 11 kinds of hot silver fabrics are shown in Table 3.

Table 3. Test values of fabric durability

Sample No	Breaking strength[N]		Breaking strength[N/cm²]		Break work[n·mm]		Elongation at break [%]		Quality difference of wear-resistant 20000 times[g]
	warp	weft	warp	weft	warp	weft	warp	weft	
1	637.00	394.33	12.74	7.89	8547.21	7134.98	25.30	31.53	0.0032
2	523.00	369.33	10.46	7.39	5591.03	3088.70	19.53	16.67	0.0012
3	508.67	285.33	10.17	5.71	4902.03	2191.22	17.98	14.75	0.0033
4	581.33	36.67	11.63	6.73	6503.17	2692.35	21.25	15.38	0.0011
5	631.00	351.67	12.36	7.03	10649.97	7890.82	37.43	33.80	0.0013
6	522.33	417.00	10.45	8.34	8127.87	6860.58	26.73	27.42	0.0011
7	544.33	369.00	10.83	7.38	7538.77	5135.40	27.93	26.10	0.0014
8	676.00	415.67	13.52	8.31	16073.13	10275.32	38.97	42.32	0.0036
9	190.33	153.67	3.81	3.07	4500.88	5227.48	55.25	89.18	0.0181
10	593.00	374.33	11.86	7.49	10379.83	4748.48	29.90	21.97	0.0012
11	730.00	439.00	14.60	8.78	9554.76	7917.01	42.48	30.60	0.0014

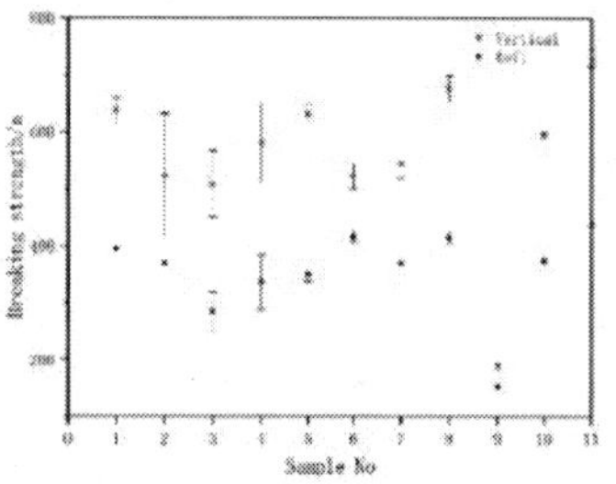

Fig.2 Fabric breaking strength

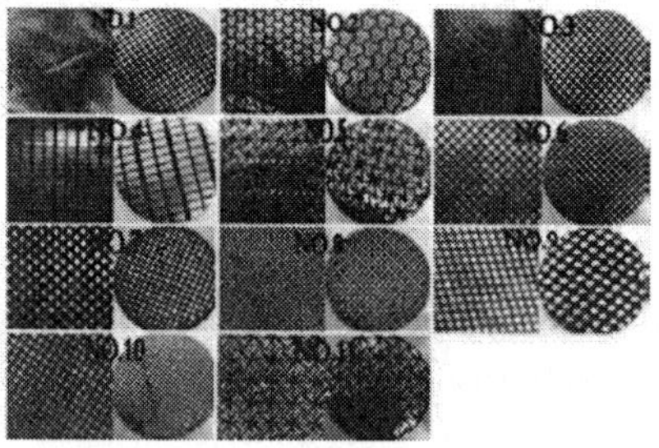

Fig.3 Comparison of samples of 20000 experiments

It can be seen from Table 3 that the longitudinal and latitude breaking strength test results of No.11 sample is the largest, and the latitude and longitude breaking strength test results of No.9 sample is the smallest. As can be seen from Fig.2, the breaking strengths of No.1, 8, and 11 are relatively large. The breaking strength of No.9 sample is the smallest; the breaking elongation of No.3 is the smallest in the warp and weft directions, and the breaking elongation of No.9 is the largest in the warp and weft directions. This is because the No.9 sample is covered with a silver film as the base fabric. The sake of layer knitting loops. The mass loss rate of No.1-11 samples after 20000 times is 3.9%. It can be seen from Fig.3 that after 20000 times, the appearance of No.3 and No.11 samples is not obvious, the appearance of No.1, 2, 4, 6 sample is slightly worn, the appearance of No.8, 10 is slightly worn, and the appearance of No.5, 7 and 9 is more obvious. Therefore, No.3 and No.11 linings can be used on elbow joints, hips, knee joints, cuffs, trouser openings, collars and other areas that are easily worn.

3.1.3 Comfort Performance

The comfort and antistatic test results of 11 kinds of hot silver fabrics are shown in Table 4.

The air permeability of No.6 and No.8 samples is relatively good, the air permeability of No.5, 10 and 11 samples are relatively general. The air permeability of No.1,2,3,4, and 7 is not good. No.5 and No.8 samples have good moisture permeability, No.1, 2, 3, 4, 6, 7, and11 samples have general moisture permeability, and No.9 and No.10 samples have poor moisture permeability. No.1, 4, 5, 8 and 11 samples have relatively large thermal resistance and crow values, while No.2, 3, 9, and 10 samples have relatively high thermal resistance and crow values. The thermal resistance of No.6 and No.7 is too low to be ignored. The larger the crow value is, the greater the obstacles the sample encounters when heat transfer occurs: the less heat is lost, the better the warming performance. The converse corresponds to a worse warming performance. [7] The moisture resistance of No.9 and No.7 is relatively large, while that of No.10 and No.11 is relatively small. The static decay period of No.1, 3 and 4 samples are small, and its antistatic performance is better; No.7 and No.10 samples have poor antistatic performance.

Table 4. Test values of fabric comfort and antistatic properties

Sample No.	Air permeability [mm/s]	Moisture permeability [g/(m²·24h)]	Thermal resistance [m²·K/W]	Wet resistance [m² Pa / W]	CLO value [clo]	Static decay period [s]
1	2.85	3024.170	0.0119	3.200	0.077	1.06
2	6.54	2701.484	0.0039	2.761	0.025	50.05
3	16.31	2888.481	0.0048	3.289	0.031	2.97
4	5.13	3323.110	0.0079	4.501	0.051	2.08
5	30.08	5378.799	0.0125	2.275	0.081	48.50
6	67.02	3065.300	0.0007	2.176	0.045	13.65
7	3.41	3136.537	0.0030	6.831	0.019	66.21
8	69.59	4770.742	0.0078	2.890	0.050	54.64
9	0.00	1382.756	0.0053	12.402	0.034	31.00
10	44.36	2256.678	0.0047	0.359	0.030	67.09
11	33.13	2768.481	0.0084	1.541	0.054	22.53

According to the test results, No.9 has poor air permeability and moisture permeability. The cause of the lower crow value of the No.9 sample can be attributed to the film composition containing far-infrared properties and knitted loops. The film is not breathable and has poor moisture permeability. It can be used on the body, knees and other parts of the garment, which is beneficial to improve the warmth of the clothing; No.5 and No.8 have a large air permeability, moisture permeability, and crow value. It can be used on the underarms and lower legs, as well as other parts that are prone to sweating; No.1, 3 samples have good antistatic performance and can be used in sleeves and other parts that are prone to static electricity.

3.1.4 Far Infrared Performance

Since far-infrared fiber is added with far-infrared radiation material with high emissivity, its thermal insulation performance can make full use of the heat radiation of the organism through absorbing and storing the energy radiated from the outside to the organism. This helps the organism by creating a "greenhouse effect" through preventing heat loss with good thermal insulation. [8] The hot silver fabric is a fabric with a finishing agent containing far-infrared functional powder particles, which is a new type of active thermal insulation material. The new type of active thermal insulation material can insulate or reduce the heat loss of the human body on the basis of the negative thermal insulation material. It also has the ability to absorb and store external heat and transfer heat to the human body through an additional thermal effect. [9] Therefore, the use of far-infrared textiles to make clothing (especially when the underwear is made into direct contact with the human body) can prevent the body's heat from radiating to the outside and provide a highly effective thermal insulation effect. [10]

According to the standard GB/T 30127-2013 "Testing and Evaluation of Far-Infrared Performance of Textiles", testing the far-infrared performance of experimental materials (third-party testing institutions), No.9 and No.11 samples can meet far-infrared emissivity and far-infrared radiation for the temperature rise value, the test results are shown in Table 5.

The heat storage and heat preservation function of far-infrared textiles is mainly because it can absorb the energy of external electromagnetic wave radiation and emit far-infrared rays, as well as reflect the far-infrared rays emitted by the human body to obtain a warmth preservation effect. [11] Therefore, No.9 and No.11 linings are suitable for use on the body and knee structures of warm and cold-proof clothing to obtain warmth.

Table 5. Test values of fabric comfort performance

Sample No.	Far infrared emissivity	Far infrared radiation temperature rise value [°C]
9	0.88	1.4
11	0.93	1.5

3.2 Comprehensive Evaluation of Mathematical Fuzzy

The fuzzy comprehensive evaluation method was used to evaluate the comfort of hot silver fabrics. Including air permeability, moisture permeability, crow value, moisture resistance and static decay cycle five indicators.

3.2.1 Establish a Set of Factors for Judging Objects

The set of factors U, U= {air permeability, moisture permeability, Crowe value, moisture resistance, electrostatic decay period} were determined, corresponding to U={u_1,u_2,u_3,u_4,u_5}, the related experiments in Table 6 above are filled in with the test data from the factor set U:

$$U=\{u_{ij}\}\begin{Bmatrix} 2.85 & 6.54 & 16.31 & 5.13 & 30.08 & 67.02 & 3.41 & 69.59 & 0.00 & 44.36 & 33.13 \\ 126.007 & 112.562 & 120.353 & 138.463 & 224.117 & 127.721 & 130.689 & 198.781 & 57.615 & 94.028 & 115.353 \\ 0.077 & 0.025 & 0.031 & 0.051 & 0.081 & 0.045 & 0.019 & 0.050 & 0.034 & 0.030 & 0.054 \\ 3.200 & 2.761 & 3.289 & 4.501 & 2.275 & 2.176 & 6.831 & 2.890 & 12.402 & 0.359 & 1.541 \\ 1.06 & 50.05 & 2.97 & 2.08 & 48.50 & 13.65 & 66.21 & 54.64 & 31.00 & 67.09 & 22.53 \end{Bmatrix}$$

Table 6. Test values of fabric comfort and antistatic properties

Sample No.	Air permeability u_1	Moisture permeability u_2	CLO value u_3	Wet resistance u_4	Electrostatic decay period u_5
1	2.85	126.007	0.077	3.200	1.06
2	6.54	112.562	0.025	2.761	50.05
3	16.31	120.353	0.031	3.289	2.97
4	5.13	138.463	0.051	4.501	2.08
5	30.08	224.117	0.081	2.275	48.50
6	67.02	127.721	0.045	2.176	13.65
7	3.41	130.689	0.019	6.831	66.21
8	69.59	198.781	0.050	2.890	54.64
9	0.00	57.615	0.034	12.402	31.00
10	44.36	94.028	0.030	0.359	67.09
11	33.13	115.353	0.054	1.541	22.53

3.2.2 Establish a Comprehensive Evaluation Transformation Matrix

Due to the different physical meanings of fabric properties, the influence of various index test results on fabric properties is inconsistent [12], such as the factors in matrix U, the test of fabric air permeability, moisture permeability, and crow value. The larger the result, the better its performance; while the wet resistance and static decay period of the fabric is the smaller the test result, the better its performance. It is therefore necessary to standardize the test results of all factors in the matrix, so that all test results are within the same order of magnitude; in this way, establish a comprehensive evaluation matrix R based on the standardized test results. [12] The test data of air permeability, moisture permeability and crow value are processed according to formula (1) for standardized data, and the test data of moisture resistance and static electricity decay period are processed according to formula (2).

$$B=A\cdot R=\{0.227\ 0.178\ 0.283\ 0.187\ 0.125\}\times \begin{Bmatrix} 0.04 & 0.09 & 0.23 & 0.07 & 0.43 & 0.96 & 0.05 & 1.00 & 0.00 & 0.64 & 0.48 \\ 0.41 & 0.33 & 0.38 & 0.49 & 1.00 & 0.42 & 0.44 & 0.85 & 0.00 & 0.22 & 0.35 \\ 0.94 & 0.10 & 0.19 & 0.52 & 1.00 & 0.42 & 0.00 & 0.50 & 0.24 & 0.18 & 0.56 \\ 0.76 & 0.80 & 0.76 & 0.66 & 0.84 & 0.85 & 0.46 & 0.76 & 0.00 & 1.00 & 0.90 \\ 1.00 & 0.26 & 0.97 & 0.98 & 0.28 & 0.81 & 0.01 & 0.19 & 0.55 & 0.00 & 0.67 \end{Bmatrix}$$

$$=\{0.6152\ 0.2896\ 0.4370\ 0.4962\ 0.7507\ 0.6717\ 0.1769\ 0.6857\ 0.1367\ 0.4224\ 0.5818\}$$

$$r_{ij} = \frac{u_{ij} - u_{min}}{u_{max} - u_{min}} \quad 1)$$

$$r_{ij} = \frac{u_{max} - u_{ij}}{u_{max} - u_{min}} \quad 2)$$

According to formulas (1) and (2), the data in the matrix U is standardized and then a comprehensive evaluation matrix R is established.

$$\mathbf{R} = \{r_{ij}\} = \begin{Bmatrix} 0.04 & 0.09 & 0.23 & 0.07 & 0.43 & 0.96 & 0.05 & 1.00 & 0.00 & 0.64 & 0.48 \\ 0.41 & 0.33 & 0.38 & 0.49 & 1.00 & 0.42 & 0.44 & 0.85 & 0.00 & 0.22 & 0.35 \\ 0.94 & 0.10 & 0.19 & 0.52 & 1.00 & 0.42 & 0.00 & 0.50 & 0.24 & 0.18 & 0.56 \\ 0.76 & 0.80 & 0.76 & 0.66 & 0.84 & 0.85 & 0.46 & 0.79 & 0.00 & 1.00 & 0.90 \\ 1.00 & 0.26 & 0.97 & 0.98 & 0.28 & 0.81 & 0.01 & 0.19 & 0.55 & 0.00 & 0.67 \end{Bmatrix}$$

3.2.3 Determine the Weight Coefficient Matrix

Different performance indexes of fabrics have different effects on the wearability of fabrics, so the concept of weight coefficient is introduced. [13] In the weight coefficient matrix A, the larger the index value, the greater the influence of the index on the overall wearability of the fabric. According to professional knowledge, expert opinions and references, the weight coefficient of each performance index is summarized: A= {0.227 0.178 0.283 0.187 0.125}.

3.2.4 Comprehensive Evaluation

The weight coefficient matrix A and the comprehensive evaluation matrix were used in the formula B=A·R to acquire matrix B. The data in the B matrix can be used for the comprehensive evaluation of the lining wear performance. The larger the value, the better the overall comfort performance of the fabric, making it more suitable for the lining of the warm clothing.

The larger the value in matrix B, the better the comprehensive comfort performance of the fabric. From the data of matrix B, the comprehensive comfort performance of 11 kinds of linings is ranked as 5#> 8#> 6#> 1#> 11#> 4#> 3#> 10#> 2#> 7#> 9#.

According to the experimental research and analysis, hot-spot silver fabrics do not have excellent comfort performance due to different factors. According to the structural characteristics of clothing, hot-spot silver fabrics can be precisely developed and applied, as shown in Fig.4 and Fig.5.

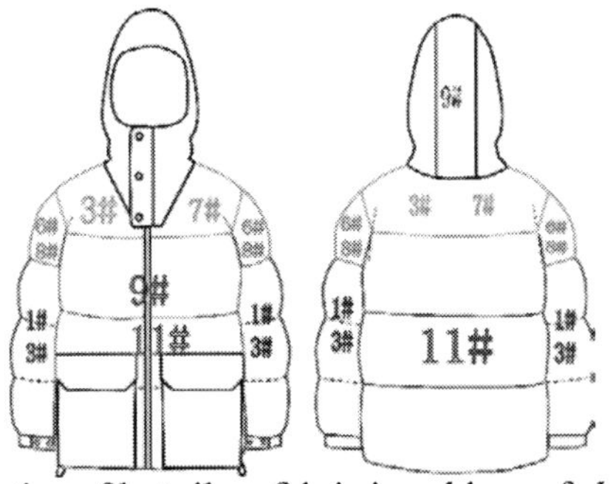

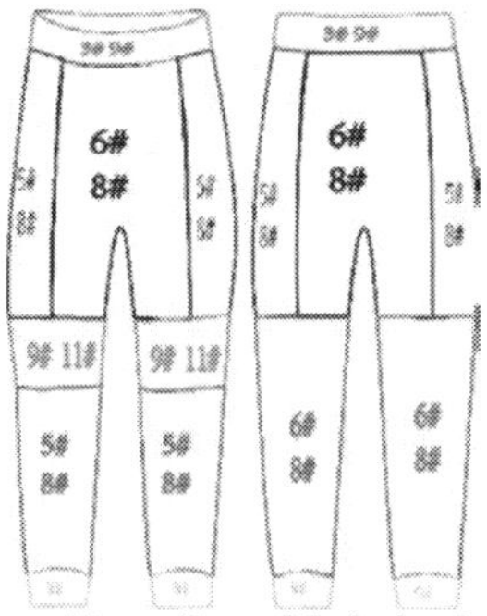

Fig.4 Application of hot silver fabric in cold-proof clothing Fig.5 Application of hot silver fabric in thermal pants

4. Conclusion

When considering the performance of different hot silver fabrics, through experimental research on appearance, durability, comfort and far-infrared performance, the following conclusions are drawn:

1) After the investigation, it became apparent that the hot silver fabric is made of ordinary fabric with the base fabric material processed to achieve far-infrared performance through the hot stamping process. It is often used in the innermost layer of cold and warm clothing.

2) In terms of appearance performance and durability, the flexural rigidity of the hot silver fabric is 0.4-1.2 mN·cm, and the fabric is soft; only a few samples have a larger wrinkle recovery angle, and the fabric has better wrinkle resistance. The tensile breaking strength of the hot silver fabric is not much different in 500-800N. The lower breaking strength of No.9 sample is due to the film layer of the base fabric; the surface of the fabric can achieve an abrasion-resistant of up to 20,000 times.

3) In terms of far-infrared performance, 9# and 11# samples can meet far-infrared emissivity and far-infrared radiation temperature rise.

4) From the fuzzy comprehensive analysis, it can be seen that the comfort performance of 5#, 8#, 6#, 11# samples is better, the comfort performance of 4#, 3#, 10#, 2# samples is general, 7#, 9 #Sample comfort performance is not ideal, while 9# sample has far-infrared radiation performance, which can be used for local warmth of clothing.

5. Acknowledgement

The authors would like to express their sincere appreciation to the financial support received from Funding project for National Key R&D Program of China: Construction and demonstration on accessible, convenient and intelligent life service system (No.2019YFF0303300) Subject IV: Universal apparel and accessories research and practice based on body and sporting features of disabled people (No.2019YFF0303304).

References

[1] Paul R, Harlin A, Jussila K, Ilen E. (2019). Sports Textiles and Comfort Aspects. In High Performance Technical Textiles, R. Paul (Ed.).

[2] Pu W. Clothing fabrics and accessories [M]. Beijing: China Textile Publishing House: 2004; 184.

[3] Stanton JH, Speijers J, Naylor GRS, et al. (2014). Skin comfort of base layer knitted garments: part I: Description and evaluation of wearer test protocol. Textile Research Journal: 84 (13); 13.

[4] Anand SC. (2013). Recent Advances in Textile Materials and Products for Activewear and Sportswear. Faisalabad, Pakistan, Covitex13: Plenary Lecture, National Textile University.

[5] Wang LS, Ma XH, Meng H, Feng AF. Analysis and prediction of thermal and moisture comfort of knitted fabrics based on grey correlation model [J]. Wool Spinning Technology: 2020; 48 (05): 82-86.

[6] Chen LH. Testing and Evaluation of Garment Accessories[M]. Beijing: China Textile Publishing House: 2015; 52.

[7] Wu B, Wang ZX, Li S, Li WH, Wu F, Zhu WJ. Research status and development prospects of thermal insulation materials [J]. Journal of Hunan Institute of Engineering (Natural Science Edition): 2021; 31(01): 74-79.

[8] Shi JS. The application of far infrared rays in textile fabrics [J]. Jiangsu Silk: 2013 (04); 40-42.

[9] Su JM. Research on the application of new thermal insulation materials in clothing [D]. Beijing Institute of Fashion Technology: 2012.

[10] Zhang XD, Sheng JY. Research on testing methods of far-infrared textile thermal insulation performance [A]. China Conference: 2007; 2.

[11] He ZP, Ping Y. Far-infrared textiles and their testing and evaluation [J]. Dyeing and Finishing Technology: 2014; 36 (06): 50-52.

[12] Liu YT. Research and development of moisture-absorbing and quick-drying textiles based on environmentally friendly recycled polyester fibers [D]. Zhejiang Sci-Tech University: 2015.

[13] Wang YJ. Performance research and product development of comfortable flame-retardant apparel fabrics[D]. Zhejiang Sci-Tech University: 2016.

Effect of Proline and its Derivatives on the Properties of Silk Fibroin Microneedles

Bin Tian, Yu-Min Zhang, Zhen-Zhen Qi, Le-Hao Zhang, Zu-Qiang Yin, Shen-Zhou Lu*

National Engineering Laboratory for Modern Silk, College of Textile and Clothing Engineering, Soochow University, 215127, *Suzhou, China*

*Corresponding author's email address: lushenzhou@suda.edu.cn

Abstract

In this paper, the effects of proline and its derivatives prolinamide and hydroxyproline on the properties of silk fibroin microneedles were studied. The composite silk fibroin microneedles were obtained by pouring the amino acid/silk fibroin mass ratio of 0/10, 1/10, 2/10, 3/10 and 4/10 into a polydimethylsiloxane mold, after vacuum defoaming and drying. The length of the microneedles was about 600μm. The aggregation structure of amino acid/silk fibroin microneedles was measured by X-ray diffraction (XRD) and Fourier transform infrared spectroscopy (FTIR). The mechanical properties of the microneedles were measured by texture analyzer. The results showed that: (1) The silk fibroin microneedles prepared by adding proline and its derivatives had predominately Silk Ⅰ crystal structure; (2) When the mass ratio of proline and its derivatives to silk fibroin reached 2/10, it had a higher swelling degree and a lower dissolution rate; (3) The silk fibroin microneedles prepared by proline and its derivatives have good mechanical properties. The following conclusion was drawn: with the addition of proline and its derivatives, silk fibroin microneedles with higher swelling degree and lower dissolution rate can be obtained. The crystal structure of Silk Ⅰ is formed inside the microneedles, which has good penetration and fracture properties. It is expected that the microneedles can be used as swelling microneedles for drug transdermal delivery.

Keywords: Silk Fibroin; Microneedle; Proline; Silk Ⅰ Crystalline Structure

1. Introduction

In the past few decades, transdermal delivery has been an attractive route for drug delivery [1-3]. Skin is the largest organ in the human body. It receives one-third of the blood supply of the entire body and was not used as a drug delivery route until the end of the 20th century. The mechanism of microneedle transdermal drug delivery is to use microneedles to penetrate the tightly arranged stratum corneum of the skin [4], and directly deliver the drug to the dermis. Microneedle penetration strengthens the drug delivery channel into the skin, and the microneedle will not penetrate too deep, which can protect the human skin. The administration process is almost painless, which can improve the compliance of patients. In addition, the dosage of drugs delivered by skin is usually lower than oral drugs, which can avoid the side effects caused by unstable absorption and metabolism of drugs in gastrointestinal tract [5]. In the case where oral drug delivery is difficult, transdermal drug delivery technology can be easily applied on the skin, and provide effective blood concentration level [6].

Different forms of microneedles have different ways of releasing drugs. There are five types of microneedles: solid microneedles, hollow microneedles, swelling microneedles, soluble microneedles, and coated microneedles [7]. The main advantage of solid microneedles is their firmness, which makes it easier to penetrate the skin and can be used to pretreat the skin. Henry et al [8]. first demonstrated the increase of transdermal flux of calcein after the silicon microneedles were prepared by ion etching. Dissolving microneedles are mainly made of polymers or polysaccharides, which release drugs through dissolution after piercing the skin [9]. Hollow microneedles have the same empty cavity as a traditional hypodermic needle to deliver drugs to the skin or through the skin into the blood, but the microneedles are shorter and the flow rate can be controlled by a micropump or syringe [10]. Coated microneedles are made from a substrate coated with drugs that can be delivered quickly to the skin and increase the long-term stability of the active drug, although the amount of drug can be uncontrolled. Swelling microneedles can absorb the interstitial fluid of the skin and provide channels for drug delivery in the microneedles. When used, the integrity can be retained without residual accumulation in the skin, and it is convenient for patients to use and improve patient dependence. Swelling microneedles have the function of rate control and prolong the time of drug administration by adjusting the swelling performance [11]. Now, the most widely used is the use of microneedles to carry drugs, after the microneedle enters the epidermis, the needle body swells and the drugs enter the human skin.

Silk fibroin protein is a kind of natural polymer material with good biocompatibility and natural degradation, and rarely has sensitization reaction [12]. Silk fibroin protein can also maintain the biological activity of the drugs it carries for a long time, so it is extremely suitable for the preparation of biological materials [13]. Based on previous studies by scholars, it can be known that pure silk fibroin protein microneedles have good mechanical properties and can successfully achieve the purpose of drug release by microneedles. However, the untreated pure silk fibroin is random coil structure, its molecular chain arrangement is disordered, and water molecules can enter smoothly, which makes the microneedles dissolve quickly when contacting body fluid. In this paper, we explored the effects of proline and its derivatives on the structure and properties of silk fibroin microneedles, and explored a more balanced addition strategy to prepare the swelling silk fibroin microneedles with low solubility, high swelling and good mechanical properties. It provides new methods and ideas for the improvement of microneedle transdermal drug delivery materials.

2. Materials and Methods

2.1 Experimental Materials and Instruments

2.1.1 Experimental Materials

Silkworm cocoon shell (Suzhou siruibao Biotechnology Co., Ltd., China), sodium carbonate, sodium bicarbonate, lithium bromide (Tiancheng Chemical Co., Ltd., China), dialysis bag (MWCO: 8-12kd, Pierce), L-proline (Shanghai Yuanye Biotechnology Co., Ltd.), L-proline amide (Shanghai Aladdin Biochemical Technology Co., Ltd.), L-hydroxyproline (Shanghai Aladdin Biochemical Technology Co., Ltd.) etc.

2.1.2 Experimental Instruments

FA2004 electronic balance, SHY-2 digital display water bath constant temperature oscillator, DT5-2 low speed table centrifuge, DHG-9246A electrothermal constant temperature blast drying oven, 84-1A magnetic stirrer, SmartSpec Plus spectrophotometer, X'PERT-PRO MPD X-ray diffractometer, Nicolet is 5 intelligent Fourier transform infrared spectrometer, TMS-PRO texture analyzer.

2.2 Experimental Part

2.2.1 Preparation of Silk Fibroin Solution

Weigh 80g silkworm cocoon shell, put it into 100°C sodium carbonate/sodium bicarbonate buffer solution and cook for half an hour. Wash the cocoon shell with deionized water to remove the surface degraded sericin. Repeat three times. Dry the silk in an oven at 60°C. Weigh 15g degummed silk fibroin and dissolve it in LiBr solution (9.3M) at 65°C for 1h. After cooling, the solution was dialyzed in deionized water at 4°C for 3-4 days, and the particulate impurities in the solution were removed by filtration. The obtained silk fibroin solution was stored at 4°C for standby.

2.2.2 Preparation of Microneedles by Die Casting

The concentrations of proline, hydroxyproline and prolinamide were 150 mg/ml. The mass ratio of amino acid to silk fibroin was set to be 0/10, 1/10, 2/10, 3/10 and 4/10. The amino acid and silk fibroin solution were mixed in proportion, and then the mixed solution was cast into the PDMS microneedle mould. After vacuumizing, the formed silk fibroin microneedles were dried in a constant temperature and humidity room for 24 hours.

2.2.3 X-ray Diffraction

X-ray diffraction sample preparation: the prepared microneedles were cut into pieces with scissors.

X-ray diffraction analysis: Using automatic X'PERT PRO MPD X-ray diffractometer. The diffraction intensity curve was recorded between 5° and 45° under the conditions of 10°/min scanning speed, 40kV, and 30mA.

2.2.4 Infrared Detection

The prepared silk fibroin film was tested on the Nicolet is 5 intelligent Fourier transform infrared spectrometer. The scanning range was 400 ~ 4000 cm^{-1}, and the infrared absorption spectrum was obtained.

2.2.5 Detection of Dissolution and Swelling of Microneedles

The prepared microneedles were weighed and recorded as M_1. The microneedles were put into a centrifuge tube with PBS buffer (pH = 7.4) according to the bath ratio of 1:100 (W/V). The microneedles were shaken in 37°C water bath constant temperature oscillator for 24h. Take out the soaked microneedles, wash them with deionized water, then absorb the surface water with absorbent paper, weigh them and record them as M_2. After centrifuging the solution in the centrifuge tube, the supernatant was taken out, and the absorbance A was measured at 278nm with ultraviolet spectrophotometer. The swelling rate Q (%) and silk fibroin dissolution rate C (%) of microneedles were calculated by formula (1) and (2) respectively.

$$Q = \frac{M_2 - M_1 \times (1-F)}{M_1 \times (1-F)} \times 100\% \qquad 1)$$

$$C = \frac{K \cdot A \cdot V}{M_1 \times (1-F)} \times 100\% \qquad 2)$$

Q - Silk fibroin swelling degree (%)
M_1 - The initial weight of the sample (g)
M_2 - Weight of sample after soaking (g)
F - Water content (%)
C - The dissolution rate of c-silk fibroin (%)
K - UV absorption constant of silk fibroin solution
A - Absorbance
V - Volume of PBS solution (ml)

2.2.6 Measurement of Compression Strength of Microneedles

Different amino acid/silk fibroin microneedles were evenly cut into 3 × 3 array, 5 parallel samples in each group, the needle tip was placed upward under the TMS-PRO texture analyzer, and the breaking strength of the needle tip was detected. Test conditions: initiation force 0.02N, maximum detection range 25N, deformation 80%, compression speed 10mm/min.

3. Experimental Results and Analysis

3.1 Morphology of Microneedles

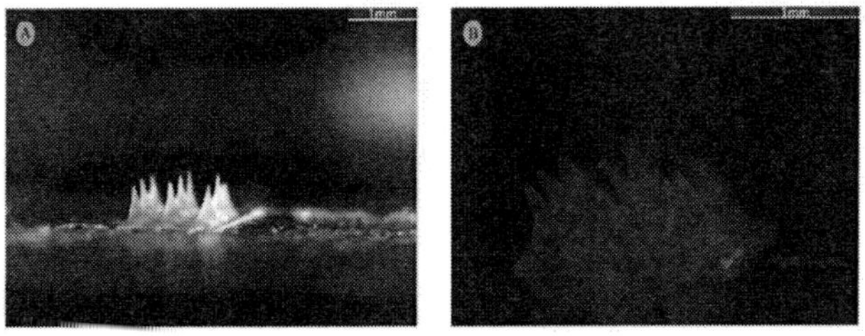

Fig.1 Morphology of microneedles

As can be seen from Fig.1, the obtained microneedles are highly homogeneous, with a height of about 600μm. The microneedle arrays were prepared for further research.

3.2 X-ray Diffraction Analysis

In the X-ray diffraction pattern, the crystal peaks of Silk I appeared at 12.2°, 19.7°, 24.7°, 28.2°, 32.3°, 36.8° and 40.1°, while the crystal peaks of Silk II appeared at 9.1°, 18.9°, 20.7° and 24.3° [14].

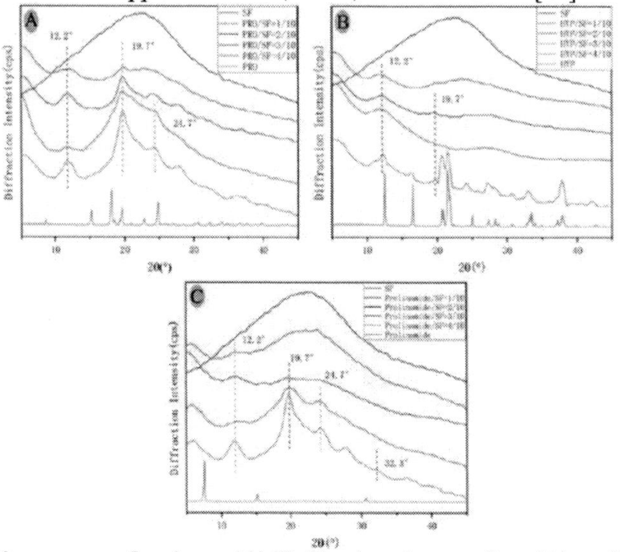

Fig.2 X-ray diffraction curves of amino acid/silk fibroin microneedles: (A) proline/silk fibroin, (B) hydroxyproline/silk fibroin, (C) prolinamide/silk fibroin

As shown in Fig.2(A), the X-ray diffraction curve of pure silk fibroin microneedles shows a steamed bread shape without obvious absorption peak. The results showed that the molecular chain of silk fibroin was in random coil state, and no crystal was formed in the microneedles. Therefore, pure silk fibroin microneedles are easily soluble in water. When the ratio of proline to silk fibroin is 1/10, 2/10, 3/10 and 4/10, the crystal peaks appear at 12.2° and 19.7°. With the increase of the proportion of proline, the peak pattern becomes more acute, indicating the formation of the crystal structure of Silk I. Proline mainly induces silk fibroin protein to form the crystal structure of Silk I. The addition of proline greatly reduced the dissolution rate of silk fibroin microneedles.

In Fig.2(B), when the ratio of hydroxyproline to silk fibroin was 1/10, the curve showed no significant difference from that of pure silk fibroin, presenting a state of random coiling. At 2/10, 3/10 and 4/10, the crystal peaks of 12.2° and 19.7°. As the proportion of hydroxyproline increases, the peak shape becomes sharper. When the ratio of hydroxyproline to silk fibroin was 4/10, a large number of crystallization peaks of hydroxyproline appeared. The above results indicate that when hydroxyproline/silk fibroin is at 1/10, hydroxyproline does not induce silk fibroin to form Silk I crystallization. When the ratio reached 2/10, 3/10 and 4/10, the silk fibroin protein mainly formed the crystal structure of Silk I under the induction of hydroxyproline. When hydroxyproline/silk fibroin is 4/10, hydroxyproline begins to crystallize, which indicates that the compatibility between hydroxyproline and silk fibroin protein is poor at this time.

In Fig.2(C), 1/10 prolinamide/silk fibroin barely changed compared to pure silk fibroin. At 2/10, 3/10 and 4/10, the crystal peaks of Silk I appeared at 12.2° and 19.7°, and gradually became sharp with the increase of prolinamide dosage. When reaching 3/10, the crystal peaks of 24.7° and 32.3° appeared, indicating that a large number of crystal structures of Silk I were formed at this time. It can be seen from the figure that prolinamide has good compatibility with silk fibroin, and induces silk fibroin to mainly form the crystal structure of Silk I.

3.3 Infrared Detection

In order to further explore the aggregation structure of modified silk fibroin microneedles, Fourier transform infrared spectroscopy was used to test the secondary structure.

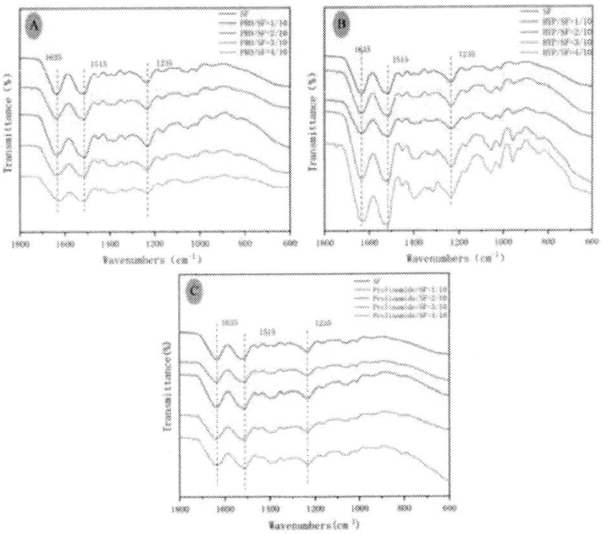

Fig.3 FTIR of proline and its derivatives/silk fibroin microneedles: (A) proline/silk fibroin, (B) hydroxyproline/silk fibroin,(C) prolinamide/silk fibroin

In the infrared spectrum, it can be seen that the silk fibroin microneedles have no other obvious absorption peaks compared with pure silk fibroin microneedles after adding proline and its derivatives. Absorption peaks were observed at 1635cm^{-1} (amide I), 1515cm^{-1} (amide II) and 1235cm^{-1} (amide III) [15]. With the increase of amino acids, the absorption peaks of silk fibroin in the amide I, the amide prism and the amide prism remain basically unchanged. In general, it is difficult to distinguish random coil from α-helix directly by FTIR. By combining FTIR spectra with XRD data, we can determine that the absorption peak should be the coexistence and superposition of α-helix, random coil and β-folding peak.

3.4 Swelling and Dissolution Properties of Microneedles

In order to obtain silk fibroin microneedles with high swelling rate and low dissolution rate, three kinds of amino acids, proline, hydroxyproline and prolinamide, were added. The swelling and dissolution properties of silk fibroin microneedles were compared by controlling the dosage of drugs. To simulate the internal environment, we used PBS solution to test. Tukey test was used for significant difference analysis in Fig.4 and Fig.5.

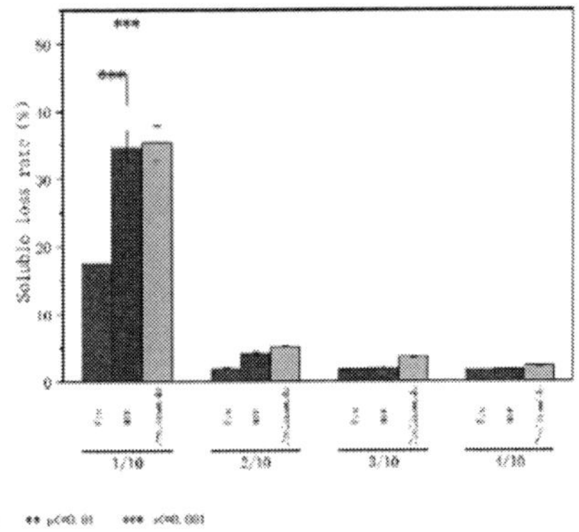

Fig.4 Dissolution property of silk fibroin microneedles

It can be seen from Fig.4 that when the ratio of proline and its derivatives to silk fibroin is 2/10, 3/10 and 4/10, the dissolution loss of the three microneedles is very small, less than 6%, and there is no significant difference. But at 1/10, hydroxyproline and prolinamide were significantly different from proline. At this time, the dissolution rate of hydroxyproline and prolinamide is as high as 35%, while that of proline is only 18%. This is because when hydroxyproline/silk fibroin and prolinamide/silk fibroin are 1/10, their structures are basically random coil, they are easily soluble in water, and the loss of protein is large, so the dissolution rate is relatively large.

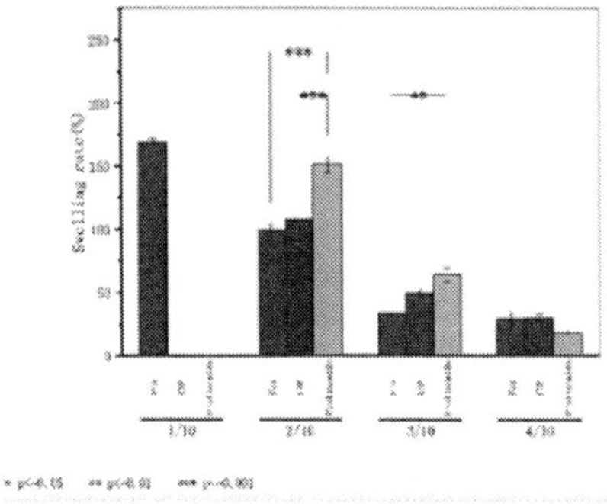

Fig.5 Swelling properties of silk fibroin microneedles

Pure silk fibroin microneedles dissolve rapidly in PBS solution, so it is impossible to measure its swelling degree. When the ratio of hydroxyproline/silk fibroin and prolinamide/silk fibroin was 1/10, the silk fibroin microneedles dissolved rapidly in solution, so the swelling degree could not be measured. It can be seen from Fig. 5 that when proline and its derivatives/silk fibroin is 4/10, the swelling of the three microneedles is about 25%, and there is no significant difference. In the case of no significant difference in dissolution rate, there was a significant difference in swelling rate when amino acid/silk fibroin was 2/10. When the amino acid/silk fibroin ratio was 2/10, the swelling degree of silk fibroin microneedle was up to 150% when prolinamide was added, while the swelling degree of other microneedles were only about 100%. It can be considered to use the additive amount of 2/10 to get the higher swelling rate of silk fibroin microneedles. Prolinamide microneedles at 2/10 were significantly different from other microneedles. In addition, when the ratio of amino acid to silk fibroin was 3/10, the swelling degree of microneedles with prolinamide was 60%, while that of proline was only 34%.

With the increase of amino acid content, the swelling and dissolution of micro needle decreased gradually. This is because the small molecules of proline, hydroxyproline and prolinamide promote the formation of a small amount of silk fibroin crystallization, thus reducing the loss rate of silk fibroin, so it has certain swelling and insoluble. When PBS solution is added, the non-crystalline region expands, and a small amount of crystal region suppresses its infinite expansion. With the increase of the small amino acid molecules, silk fibroin was induced to produce more crystallized regions. The results showed that the dissolution rate of microneedle decreased with the increase of amino acid small molecule. With the increase of addition dose, hydroxyl, peptide bond and amino group in hydroxyproline, proline and prolinamide molecules can interact with polar groups such as amide bond, Unbound hydroxyl group or carboxyl group in silk fibroin molecular chain to form hydrogen bond. The formation of these hydrogen bonds further increases the interaction of silk fibroin proteins, so the swelling properties of microneedles also tend to decrease.

3.5 Mechanical Properties of Microneedles

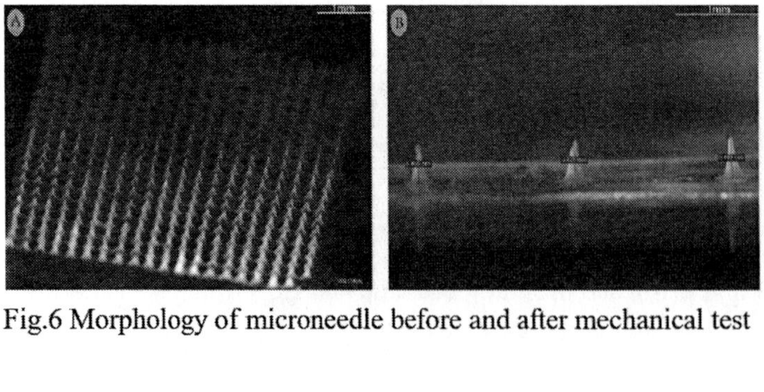

Fig.6 Morphology of microneedle before and after mechanical test

As shown in Fig.6(A), the 3*3 microneedle array is not compressed. The microneedles in Fig.6(A) have full tips and similar lengths. Fig.6(B) shows the compressed 3*3 microneedle array, in which the microneedle tips are broken.

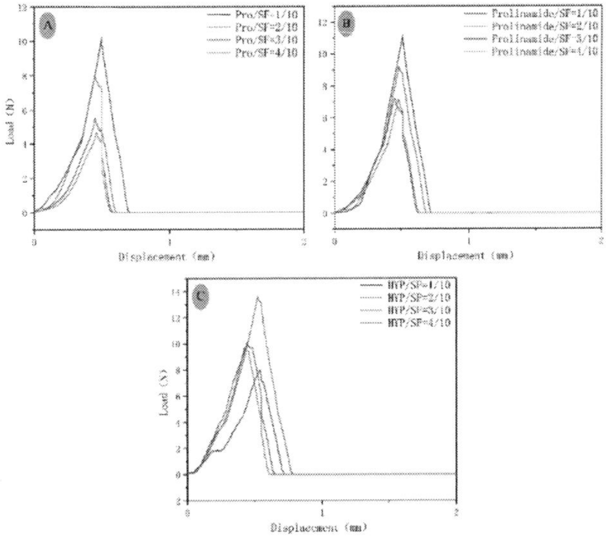

Fig.7 Displacement load curve of proline and its derivative 3*3 microneedles

In Fig.7(A), it can be seen that when the proline/silk fibroin ratio is 1/10, the displacement load of the microneedle is the largest, which is 10.1N; when the proline/silk fibroin ratio is 4/10, the displacement load of the microneedle is the smallest, which is only 4.7N.

In Fig.7(B), when prolinamide/silk fibroin is 1/10, the maximum displacement load of the microneedle is 11N, and when 4/10, the minimum displacement load of the microneedle is 7N.

In Fig.7(C), when hydroxyproline/silk fibroin is 2/10, the maximum displacement load of the microneedle is 13.3N, and when it is 1/10, the minimum displacement load of the microneedle is 7.8N.

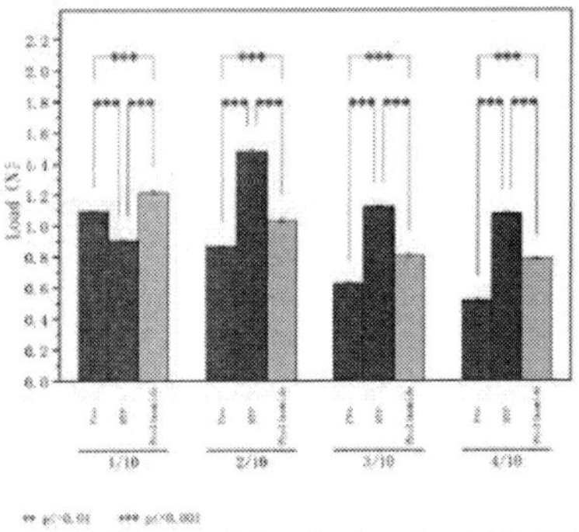

Fig.8 Fracture strength of proline and its derivative/silk fibroin single microneedle

Fig.8 Significant difference analysis was performed using Tukey test. As can be seen from Fig.8, there are significant differences among the three microneedles in any proportion. At 2/10, the maximum compression strength of hydroxyproline/silk fibroin single needle reached 1.4N. According to the observation, except for 1/10, the compressive strength of single needle was hydroxyproline > prolinamide > proline. Therefore, the mechanical properties of hydroxyproline were the best, followed by prolinamide, and proline was the worst. In general, the breaking strength of microneedles decreased with the increase of amino acids, which was due to the fact that more small molecules were inserted between the molecular chains of silk fibroin to play the role of lubricant. When the external force acts on the microneedles, the relative slip of silk fibroin molecular chain is more likely to occur due to the action of small molecules, resulting in the decrease of its breaking strength.

4. Conclusion

In this paper, proline, hydroxyproline and prolinamide, which are nontoxic amino acids and their derivatives, were selected to modify silk fibroin to prepare swelling microneedles. The results showed that: (1) The silk fibroin microneedles with proline and its derivatives were mainly Silk I crystal structure, (2) When the mass ratio of proline and its derivatives to silk fibroin was 2/10, both had higher swelling degree and lower dissolution rate, (3) Both hydroxyproline and proline have good mechanical properties. When the ratio of prolinamide/silk fibroin is 2/10, the swelling degree is as high as 150%, and the dissolution rate is only 5%. The breaking strength of a single needle can reach 1 N, which has good mechanical properties. It is expected that it can be used as a swelling microneedle for drug transdermal delivery.

References

[1] Abiandu I, Ita K. Transdermal delivery of potassium chloride with solid microneedles [J]. Journal of Drug Delivery Science and Technology: 2019; 53: 101216-101223.

[2] Waghule T, Singhvi G, Dubey SK, et al. Microneedles: A smart approach and increasing potential for transdermal drug delivery system [J]. Biomedicine & Pharmacotherapy: 2019; 109: 1249-1258.

[3] Wiedersberg S, Guy RH. Transdermal drug delivery: 30+ years of war and still fighting! [J]. Journal of Controlled Release: 2014; 190: 150-156.

[4] Prausnitz MR, Mitragotri S, Langer R. Current status and future potential of transdermal drug delivery [J]. Nature Reviews Drug Discovery: 2004; 3 (2): 115-124.

[5] Lee H, Song C, Baik S, et al. Device-assisted transdermal drug delivery [J]. Advanced drug delivery reviews: 2018; 127: 35-45.

[6] Delgado-Charro MB, Guy RH. Effective use of transdermal drug delivery in children [J]. Advanced Drug Delivery Reviews: 2014; 73: 63-82.

[7] Rzhevskiy AS, Singh TRR, Donnelly RF, et al. Microneedles as the technique of drug delivery enhancement in diverse organs and tissues [J]. Journal of Controlled Release: 2018; 270: 184-202.

[8] Henry S, McAllister DV, Allen MG, et al. Microfabricated Microneedles: A Novel Approach to Transdermal Drug Delivery [J]. Journal of Pharmaceutical Sciences: 1998; 87 (8): 922-925.

[9] Ita K. Dissolving microneedles for transdermal drug delivery: Advances and challenges [J]. Biomedicine & Pharmacotherapy: 2017; 93: 1116-1127.

[10] Ita K. Ceramic microneedles and hollow microneedles for transdermal drug delivery: Two decades of research [J]. Journal of Drug Delivery Science and Technology: 2018; 44: 314-322.

[11] Donnelly RF, Singh TRR, Alkilani AZ, et al. Hydrogel-forming microneedle arrays exhibit antimicrobial properties: Potential for enhanced patient safety [J]. International Journal of Pharmaceutics: 2013; 451 (1): 76-91.

[12] Wu XL, Mao L, Qin DK, et al. Impact of Sterilization Methods on the Stability of Silk Fibroin Solution [J]. Advanced Materials Research: 2011; 311-313: 1755-1759.

[13] Altman GH, Horan RL, Lu HH, et al. Silk matrix for tissue engineered anterior cruciate ligaments [J]. Biomaterials: 2002; 23 (20): 4131-4141.

[14] Li Y, Yuen CWM, Hu JY, et al. Analysis of the structural characteristics of nanoscale silk particles [J]. Journal of Applied Polymer Science: 2006; 100 (1): 268-274.

[15] Yang H, Yang S, Kong J, et al. Obtaining information about protein secondary structures in aqueous solution using Fourier transform IR spectroscopy [J]. Nature protocols: 2015; 10 (3): 382-396.

Effect of Looseness on the Style of Lapel Collar in Men's Suit

Kun Zhang[1], Wei Cheng[2], Gui-Yu Jin[2 *]

[1]*Jiangxi Institute of Fashion Technology, No.103 Lihuzhong Blvd, Xiangtang Economic Development District, Nanchang, Jiangxi Province, 330201, China*
[2]*Wuhan Textile University, No.1 Fangzhi Road, Hongshan District ,Wuhan, Hubei Province, 430073, China*

[*]Corresponding author's email: 13006519979@163.com

Abstract

There are many factors affecting the style of lapel collar. Among them, looseness is the most important one. To find the best value of looseness, the main factors affecting looseness are discussed using CAD technology. Under certain conditions: the length of stand collar for 2.5cm, top collar for 3.5cm, the sizes of notch are 3.5cm and 4cm, the position of bar point is at the waistline and extends 2cm away. It can be concluded that Angle A decreases 3.46°, and the length of collar edge line increases 0.33cm for every 0.5cm increase of looseness. The correlation between them is analyzed and the regression equation is established. 2.5cm is the best value of looseness on the style of lapel collar in men's suit, confirmed by sample trial-production and sensory evaluation test. This paper can enrich the knowledge of looseness and provide reference for designers.

Keywords: Clothing Structure; Lapel Collar; Looseness; Style

1. Introduction

Style, color, and fabric are the three elements of clothing design. The style of collar have a direct impact on the beauty and comfort of garments [1]. The design methods of lapel collar mainly have two categories: the first is to directly add the length of collar edge line to complete the turn-over collar structure; the second is to increase the value of looseness by increasing the angle of collar inclination [2]. The latter is generally used at home and abroad. Looseness is considered to be the most influential factor in the style of lapel collar [3, 4]. Zhu [5] et al. discovered with the CLO3D digital design system that when looseness is insufficient, there will be a strong pressure on the back neck. It is necessary to change the value of looseness from "X+1" to "X+1+0.7". The design of looseness not only affects the comfort of garments according to the design philosophy of lapel collar, it also plays a vital role in the style and the beauty of suit. Lee [6] put forward that the best shape of collar and lapel appeared when the value of looseness was 3cm and 3.5cm. Tian [7] established a linear relationship between the thickness of lapel collar fabric and the value of looseness on a double-sided woolen suit, that is, $y=0.223x+2.945$. However, the best value of looseness depends on several factors. The main factors that affect looseness are learned through literature reading method. They are the position of bar point, the width of overlap, the slant of shoulder and the degree of muscular development, the difference between the length of top collar and stand collar, the plasticity of fabric etc. [8] Therefore, the best value of looseness can be found only when the factors affecting the looseness are clear.

2. Methods and Procedures

The main method of this paper is drawing lapel collar patterns and recording the value of different variables through CAD technology. The best value of looseness can be found by making cotton cloth samples and conducting evaluating test. The control group in this paper is based on the basic paper pattern of Liu Ruipu's third-generation upgraded men's standard (Fig.1), because there is less information on men's pattern making, but more on women's in Japanese cultural prototypes [9]. The basic principle of Liu Ruipu's prototypes is the same as Japanese cultural prototypes. In terms of Wuhan local garment enterprises, the design method compiled by Liu Ruipu is more common. A standard half men model is set as 170/88A with the following sizes: chest circumference for 88cm, waist circumference for 74cm, hip circumference for 90cm referencing to standard sizing systems for garments-Men (GB/T 1335.1-2008). In addition, the length of stand collar is set as 2.5cm, top collar is set as 3.5cm, the sizes of notch are set as 3.5cm and 4cm [10]. In order to record the data more conveniently, the angular distance is set between the neckline auxiliary line and vertical line generated by the neck side point X cm, and the value of looseness is equal to the X value plus the difference between the length of top collar and stand collar as (X+1) cm (Fig.2).

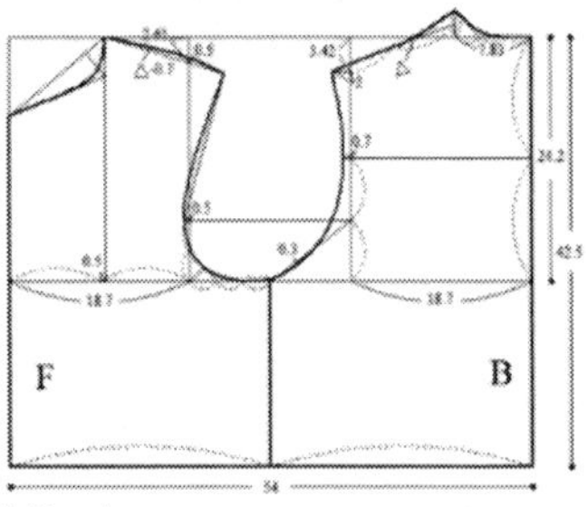

Fig.1 Basic paper pattern

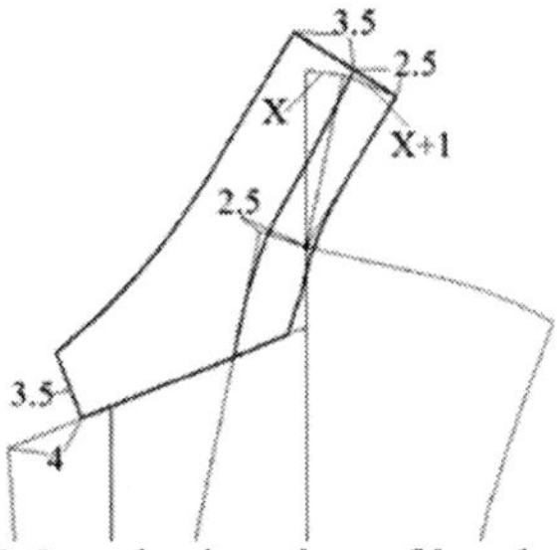

Fig.2 Standard setting of lapel collar

3. The Main Factors that Affect the Value of Looseness

3.1 The Position of Bar Point

The bar point is the basic point of fold line for lapel, the position of bar point will affect the value of looseness. In general, the highest position of bar point should not exceed the chest line, and the lowest position should not be more than 8cm below the waist [11]. Thus, the first experimental group A is aimed to study the changes of looseness when the position of bar point is at the waist line, 5cm above and below the waist line (Fig.3). The higher the bar point, the greater the value of X, and the greater the value of looseness (Table 1). Since the position of bar point is the basis to determine the fold line, the higher the bar point, the greater the inclination. This leads directly to an increase in the length of collar edge line.

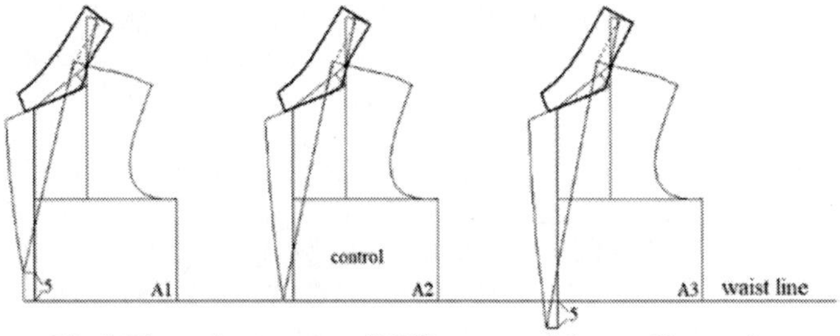

Fig.3 The suit patterns of different positions of bar point

Table 1. The effect of the position of bar point on the value of looseness

The position of bar point	X+1(cm)	Difference(cm)
A1	3.10	——
A2	2.87	-0.23
A3	2.68	-0.19

3.2 The Width of Overlap

The overlap is where closure overlaps with under lap. The width of overlap depends on the size of button-hole and button stay. The suit can be divided into single breasted and double breasted according to the width of the overlap [12]. The width of the overlaps are set to 2cm, 4cm, 6cm, respectively in the second experimental group B (Fig.4). The study found that the wider the overlap, the greater the X value and the greater the value of looseness and vice versa (Table 2). With the widening of the overlap, the bar point goes outward at the waist-line, the inclination of the fold line changes, and the length of collar edge line naturally increases.

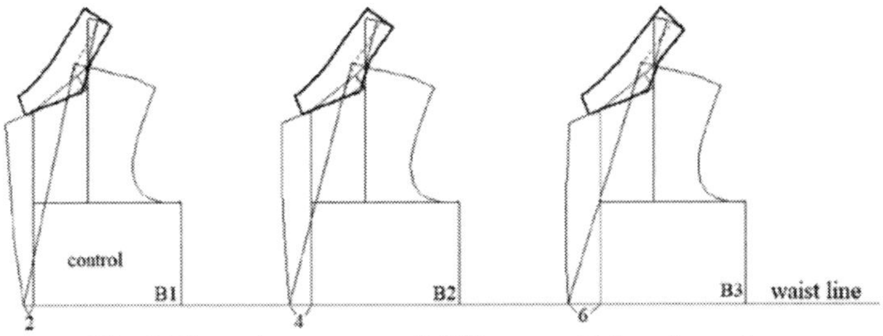

Fig.4 The suit patterns of different widths of overlap

Table 2. The effect of the width of overlap on the value of looseness

The width of overlap	X+1(cm)	Difference(cm)
B1	2.87	——
B2	3.24	0.37
B3	3.61	0.37

3.3 The Slant of Shoulder and the Degree of Muscular Development

The third experimental group C focuses on exploring the influence of the slant of shoulder on looseness. The slant of shoulder has a certain influence on the value of looseness when shoulder point extends 2cm up the armhole line, extends 2cm down the armhole line, and remains in the original position (Fig.5). Whereas the difference in looseness is almost similar (Table 3).

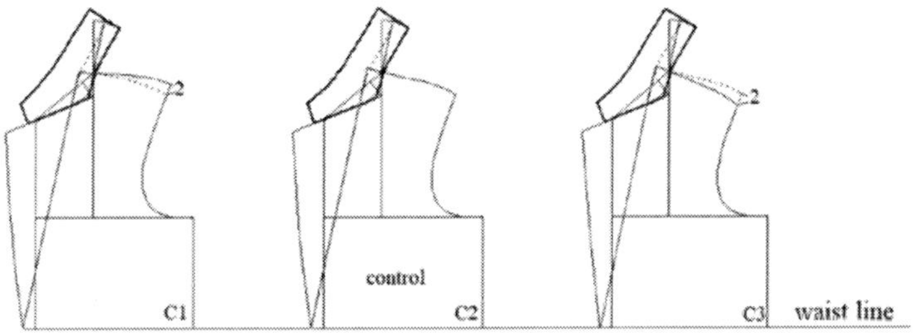

Fig.5 The suit patterns of different slants of shoulder

Table 3. The effect of the slant of shoulder on the value of looseness

The slant of shoulder	X+1(cm)	Difference(cm)
C1	2.87	——
C2	2.87	0
C3	2.88	0.01

3.4 The Difference between the Length of Top Collar and Stand Collar

The difference between top collar and stand collar is an important factor in determining the value of looseness. The top collar is the part of a lapel collar that turns out. The stand collar is upright and tightly wrapped around the neck. The lapel collar is smooth and beautiful when the difference between the length of top collar and stand collar is 1cm, and the stand collar is covered with top collar with 2.5cm looseness. Increasing the difference over 3cm, collar edge line is slanting and longer than before. The arc of fold line and back collar will increase. When the value of looseness is increased, the lapel collar will cling to the back and shoulders [13]. No matter how wide the total collar is, the length of stand collar will not change much, because it is limited by the height of the human neck.

Thus, the length of top collar is longer when the total collar is wider and the value of looseness is larger, and vice versa.

3.5 The Plasticity of Fabric

Looseness is related to the plasticity of fabric. Clothing fabrics include cotton, hemp, wool, silk, chemical fiber fabrics. Among them, the plasticity of wool is higher than other fabrics. This kind of fabric can be treated by the process of blocking during sewing. Hence, the value of looseness will be smaller. Wool fabric, also known as woolen fabric, can be divided into coarse woolen fabric, worsted woolen fabric, long woolen fabric, and camel fabric according to its production process and appearance characteristics. Using worsted fabric to make suit can increase the value of looseness, because the worsted fabric is made of worsted yarn which has smooth surface, clear texture, fine and dense yarn, smaller plasticity when compared to other woolen fabrics [14]. In general, for fabrics with greater plasticity such as woolen fabric, the value of looseness can be smaller. While for fabrics with poor plasticity such as man-made fiber and worsted fabric, the value of looseness should be larger.

4. Experimental Verification

4.1 The Patterns of Lapel Collar

Based on the study of the factors affecting looseness, the variables are controlled. The position of bar point is set at the waistline, the width of overlap is 2cm, the slant of shoulder is normal, the difference of top collar and stand collar is 1cm, cotton cloth is used. The value of looseness is set to be the independent variable from 1 to 5 with an interval of 0.5cm. Particularly, a horizontal line is made over the side neck point, so that the angle between the horizontal line and the acute angle of the looseness auxiliary line is Angle A (Fig.6). In this section, the patterns of lapel collar are drawn with CAD technology and the value of looseness is recorded (Table 4).

SPSS 22.0 is used for data analysis in this paper. Angle A decreases by 3.46°, and the length of the collar edge line increases by 0.33cm for every 0.5cm increase of looseness. Through One-Sample Kolmogorov-Smirnov Test, the progressive significance of the value of looseness and the length of collar edge line is both 0.2, greater than 0.05. It is considered that these two variables in Table 4 is subject to normal distribution. The correlation coefficient between the value of looseness and the length of collar edge line is 0.999, and the correlation coefficient between the value of looseness and Angle A is -0.999. They are extremely significant at the level of 0.01 through the Pearson Correlation Coefficient. It can be indicated that the value of looseness and the length of collar edge line is significantly positive, the value of looseness and Angle A is significantly negative. This demonstrates that the larger the value of looseness, the longer the length of collar edge line and the smaller the Angle A. Furthermore, the linear regression equations are established respectively (Eq.1, Eq2). The adjusted R^2 in these equations are above 0.9 and the image of line chart is shown in Fig.7.

$$\text{the length of collar edge line} = 0.653 \times \text{the value of looseness} + 19.354 \quad 1)$$

$$\text{Angle A} = -6.9 \times \text{the value of looseness} + 65.596 \quad 2)$$

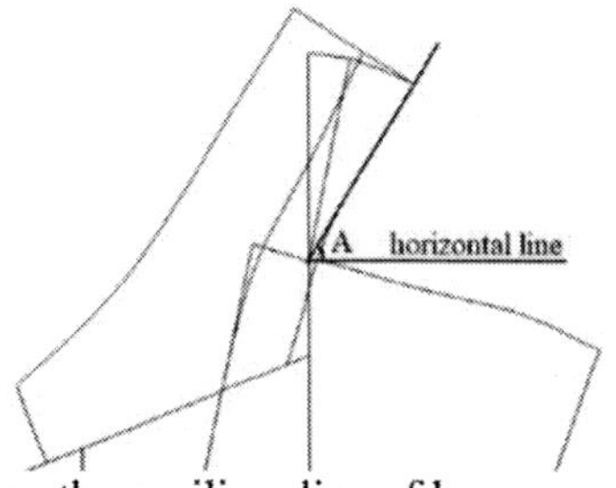

Fig.6 Angle A is between the auxiliary line of looseness and the horizontal line

Table 4. Measurements of the degrees of Angle A and the length of collar edge line

The value of looseness(cm)	Angle A (°)	Difference of degrees(°)	The length of collar edge line(cm)	Difference of lengths(cm)
1	58.59	——	19.94	——
1.5	55.17	-3.42	20.33	0.39
2	51.92	-3.25	20.66	0.33
2.5	48.23	-3.69	21.04	0.38
3	45.06	-3.17	21.38	0.34
3.5	41.57	-3.49	21.65	0.27
4	38.09	-3.48	21.98	0.33
4.5	34.56	-3.53	22.27	0.29
5	30.88	-3.68	22.57	0.30
Mean	——	-3.46	——	0.33

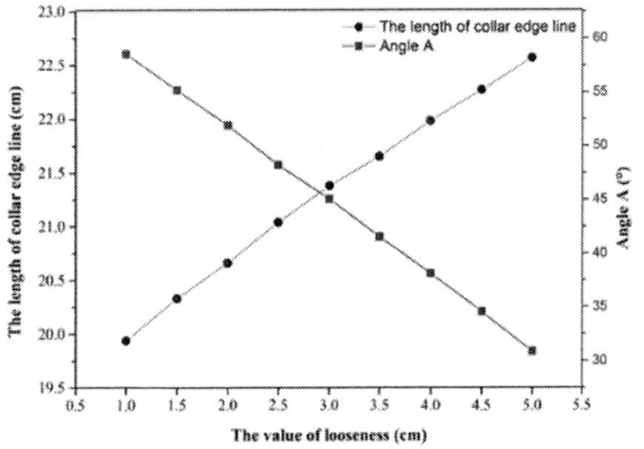

Fig.7 The image of line chart

4.2 Practice Research

In this section, we selected the white cotton cloth to make samples for experiment verification, and to study the best value of looseness. According to the manufacturing requirements of garment, to reduce the influence of external factors such as processing conditions, process flow and processing quality on the quality of collars, all samples of lapel collars are made by the same sample maker through the same process flow, so as to minimize the errors. The sewing criteria for judging lapel collars is the following: collar shape should be regular, neat, firm; neckline should be smooth; lapel collar should be symmetrical; bar point, overlap, side line and shoulder line should be straight at the same time. Then, the samples are put on the mannequin, the points of side neck, shoulder line, waistline and the position of bar point are fixed, then fold line is marked.

The images of different lapel collars are shown in Fig.8. The value of looseness is 1cm(a), 1.5cm(b), 2cm(c), 2.5cm(d), 3cm(e), 3.5cm(f), 4cm(g). When the value of looseness is 4cm, it can be found that the back collar is too large to make the structure reasonable. Hence, the samples with looseness of 4.5cm and 5cm are deemed unnecessary. The samples are evaluated via a sensory evaluation test by means of observing images including the front, side, back of different lapel collars [15]. Evaluation items and corresponding scores are shown below (Table 5). Tightness refers to the fit between the back and neck. Participants are 43 college students majoring in fashion design.

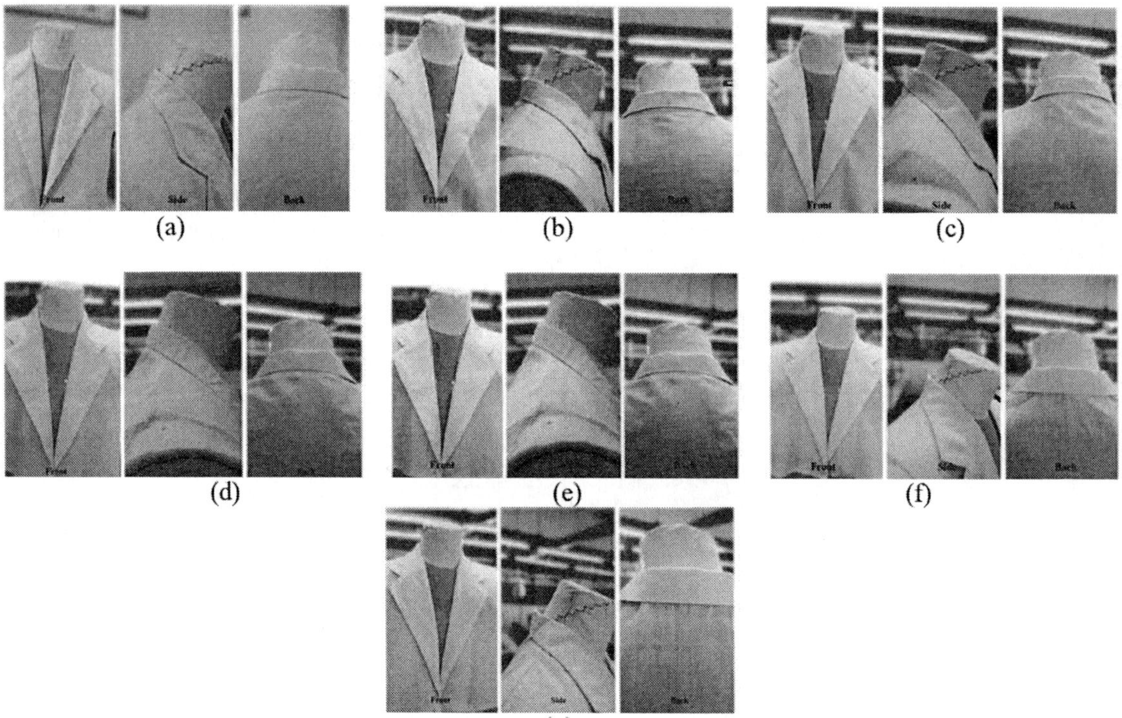

Fig.8 The images of different lapel collars

Table 5. Evaluation items and scores in the sensory evaluation test

Evaluation items	Scores(1→5)
Beauty	Not beautiful→Beautiful
Smoothness	Not smooth→Smooth
Fitness	Not fitting→Fitting
Tightness	Not tight→Tight

5. Results and Discussion

Seven samples are evaluated in terms of beauty, smoothness, fitness and tightness. The average score of sensory evaluation is shown in Fig.9. When the value of looseness is less than 2cm, the length of collar edge line is insufficient. As a result, the top collar is not able to cover the stand collar and stitch line. The lapel collar will stick close to the back of neck, and there will be wrinkles in the collar and shoulders. This leads to a loss of beauty in the overall style. When the value of looseness fluctuates within the range of 2cm to 3cm, the style of lapel collar is performs better. However, the average score of smoothness is only 3.58 with 3cm looseness. This problem can be attributed to sewing and ironing, but it does not affect the overall effect. Particularly, when the value of looseness is 2.5cm, four evaluation items have high scores. The sample is not only beautiful and smooth, it also fits the neck. The top collar can well cover the stand collar when the value of looseness is greater than 1.5cm. However, the larger the value of looseness, the longer the length of the collar edge line. When the value of looseness is greater than 3cm, the length of collar edge line becomes longer, and the back collar becomes loose. This kind of lapel collar is neither fitting nor beautiful.

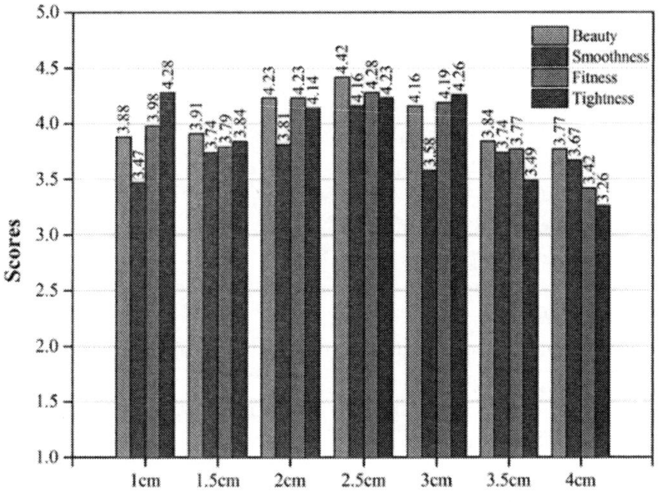

Fig.9 The results of sensory evaluation test

6. Conclusion

The main factors that affect the looseness were discussed in this paper. It is found that Angle A decreases by 3.46°, and the length of collar edge line increases by 0.33cm for every 0.5cm increase of looseness. Under the condition that the length of stand collar for 2.5cm, top collar for 3.5cm, the sizes of notch are 3.5cm and 4cm, the position of bar point is at the waistline and extends 2cm away. The value of looseness and the length of collar edge line is significantly positive, and the length of collar edge line=0.653×the value of looseness+19.354. The value of looseness and Angle A is significantly negative, and Angle A=-6.9×the value of looseness+65.596. Through the experiments of samples and data analysis, it can be concluded that when the value of looseness fluctuates within the range of 2cm to 3cm, the style of lapel collar in men's suit is reasonable and beautiful, and that 2.5cm is the best value of looseness.

In addition, the effective way to find the most suitable lapel collar is to change the value of looseness according to the above research findings. Therefore, the suggestion for beginners is to increase the value of looseness or the length of collar edge line if the collar is tight in the process of design. On the contrary, if the loose amount of the back collar is large, the value of looseness and the length of collar edge line should be appropriately reduced. These experimental results provide reference data for structural designers.

References

[1] Wang HG, Shao XW, Xu JH. The Research of Slopeness and the Looseness of the Structural Designing of Turndown Collars. Journal of Henan Institute of Engineering: 2009; 21 (01): 11-13.
[2] Zhang H. Design method of lapel collar structure based on structure model of lapel collar looseness. Journal of Textile Research: 2020; 41 (11): 128-135.
[3] Li Z. Calculation method of the gradient data of lapel collar. Journal of Textile Research: 2008; (07): 83-86.
[4] Wang J, Hou DY, Jin Y. Exploring the establishment of mathematical model based on the structure of garment lapel collar. 2010 International Conference on Computer Application and System Modeling (ICCASM): 2010.
[5] Zhu QY, Yang YR, Cao Y. Digital modification of lapel structure based on virtual simulation system. Textile Dyeing and Finishing Journal: 2021; 43 (01): 37-40+44.
[6] Lee SY. A Study on the Character of Collar and Lapels According to Collar Laying Amount -Focused on Tailored Jacket-. Journal of the Korean Society of Clothing and Textiles: 2011; 35 (04): 421-430.
[7] Tian HW. Factors about lapel collar structure influence on double-sided woollen women's dress. Wool Textile Journal: 2017; 45 (02): 50-53.

[8] Zhou LH. Comparison and Applicatoin of Lapel Collar Structure Design Methods. Journal of Zhejiang Fashion Institute of Technology: 2014; 13 (03): 34-41.
[9] Liu RP. The principles & practices of pattern design. Beijing: China Textile & Apparel Press: 2005.
[10] Wang XY, Xu D. Research and improvement of structure design methods of turndown collars. Journal of Tianjin Institute of Textile Science and Technology: 2000; 9 (05): 74-76.
[11] Xue M. The handling skills of collar looseness in lapel collar. Journal of the Staff and Worker: 1999; (03): 84-89.
[12] Min AM. Analysis of the influence factors of the change of looseness on the design of lapel collar structure. Tianjin Textile Science & Technology: 2015; (03): 38-39.
[13] Yang J. Analysis of the correlative factors affecting the looseness of lapel collar. Science & Technology Information: 2013; (10): 472-473.
[14] Song YH. Analysis of the influence of looseness on the shape of lapel collar. Neijiang science and technology: 2005; (05): 54.
[15] Ai M, KyoungOk K, Masayuki T. Effect of the difference between body dimensions and jacket measurements on the appearance of a ready-made tailored jacket. International Journal of Clothing Science and Technology: 2017.

Cite as: X.H. Sui et al.,
TBIS 10.3993/tbis (2021).

An Empirical study on Communication Content and Effect of Chinese Traditional Costume Culture

Xiao-Hui Sui[1], Ping Zhao[1*]

[1]*Beijing Institute of Fashion Techology, Beijing, 100029, China*

*Corresponding author's email: zhaobaiou@139.com

Abstract

Based on the elaboration likelihood model and information adoption behavior model in communication theory, a theoretical model of the communication content against communication effect was established for Chinese traditional costume culture moderated by engaging with the audience. With questionnaires and through regression analysis and verification of the moderating effect, the results suggested: (1) more effective communication result was generated by high quality traditional costume culture, wherein completeness, accuracy and practicability of communication content produces significant influence on communication result; (2) the relation between communication content and effect was regulated by the costume engagement with the audience. The research results provide references and suggestions for the preparation of differentiated communication strategies from the angle of communication content targeting at audience with different professionalism and fashion involvement.

Keywords: Communication Effect; Communication Content; Traditional Costume Culture; Fashion Involvement

1. Introduction

Chinese traditional costume culture reflects many aspects i.e. nationality, society, mentality and aesthetics. Its communication and inheritance are helpful not only for in-depth mining of traditional costume culture, it also brings vitality and vigor. Existing research on Chinese traditional costume culture mainly concentrated on characteristics [1], influencing factors [2] and inherited design of Chinese ancient costumes [3-5]. There is a lack of research conducted from the angle of communication. Furthermore, most of them are qualitative researches [5-7].

In the communication process, the communication content and its driving effect on transmitters and audience are the prime power of communication activities. Communication is a message-centered process built on information [8]. Therefore, a research on the communication effect of Chinese traditional costume culture from the angle of communication content has, in essence, seized the important factor that influences the communication effect. Currently, the orientation of research of traditional costume culture focuses on the effective promotion of audience's cognition of traditional costume culture, improvement of their cultural identification and promotion of constume consumption with traditional elements through communication. In this article, a model of the influence on communication content on communication effect of Chinese traditional costume culture was constructed based on communication theory, Chinese traditional costume culture and other relevant literature studies, accompanied by an in-depth research of the influence between them through empirical analysis.

2. Overview of Relevant Theories

2.1 Culture and Traditional Costume Culture

Culture is a complicated concept which is commonly defined in broad and narrow terms [9]. In narrow terms, it refers to literature and art; in broad terms, it refers to everything created by human in social life, including all material and intellectual production, covering religions, philosophy, science, technologies, literature, art, social psychology and folk custom, etc. FU Xiaocong, a researcher of costume culture, divided costume culture into hidden culture and explicit culture based on the ways how culture elements were presented on costumes [10]. The specific content of the two categories of culture differs to some extent. In broad terms, the explicit costume culture includes costume, decorations and wearing ways; in narrow terms, it includes the style, colors and fabric, etc. of costumes. The hidden costume culture includes aesthetic taste, values, social custom and morals, etc.

2.2 Communication and Communication Effect

American scholar Cooley, from the angle of sociology, defined communication as a mechanism on which relationship among people was established and developed; the scholar Peirce, from the angle of semiology, believed that communication was a process of passing concepts or significance [11]. The birth of information science expanded the concept of communication. Ayer, the scholar in communication, pointed out that communication was passing of information in broad terms [12], including sentiment or concepts, etc. Based on the research purpose herein, communication is defined as a process in which information is exchanged through symbols and medium for the purpose to influence the attitude or behaviors of others. The communication effect refers to any conscious or unconscious result of communication. As the communication effect on audience is researched herein, the communication effect of Chinese costume culture is defined as changes of attitude and behaviors of audience caused by communication behaviors relating to Chinese traditional costume culture.

2.3 Communication Effect of Chinese Traditional Costume Culture

According to the hierarchy of effects model created by Lavidge and Steiner [13] in 1961, the communication effect of Chinese traditional costume culture can be also summarized at three levels: cognition (the knowledge about traditional costume culture), sentiment (the sentiment toward traditional costume culture) and willingness (the willingness to consume costumes with traditional elements). Therefore, in this article, the identity and willingness to consume Chinese traditional costume culture are taken as the criteria for assessing communication effect. The identity [14] of Chinese traditional costume culture refers to cognition of and sentiment toward traditional costume culture, as a result of which people develop a certain behavior mode by acting under such guidance. Based on research of relevant literatures on Chinese traditional costume culture and cultural identity, etc., WANG Yu [15] prepared a identity scale of Chinese traditional costume culture. ZUO Yuping [16] made a modification to the scale and proposed one that consisted of three dimensions i.e. individual sentiment, cognitive pride and behavioral intention which were consistent with the dimensions of communication effect. Among the factors that influence communication effect, the influence of content on effect is the most direct. In 1986, psychologists proposed an elaboration likelihood model of information(ELM), pointing out the two routes of information processing by audience and their change of attitude [17]. Based on ELM theory, Sussman and Siegal proposed the information adoption model(IAM), believing that the quality of communication content was the core factor that influenced usefulness and adoption of information [18]. Therefore, in this article, the discussions on the quality of communication content on communication effect of Chinese traditional costume culture will be conducted based on ELM model and IAM model.

3. Research Design

3.1 Research Model and Assumptions

In this article, based on ELM and IAM models, a research model of communication effect of Chinese traditional costume culture is established with the quality of communication content of Chinese traditional costume culture as a dependent variable that influences communication effect and the costume professionalism and fashion involvement of audience as the moderator variable, as shown in Fig.1.

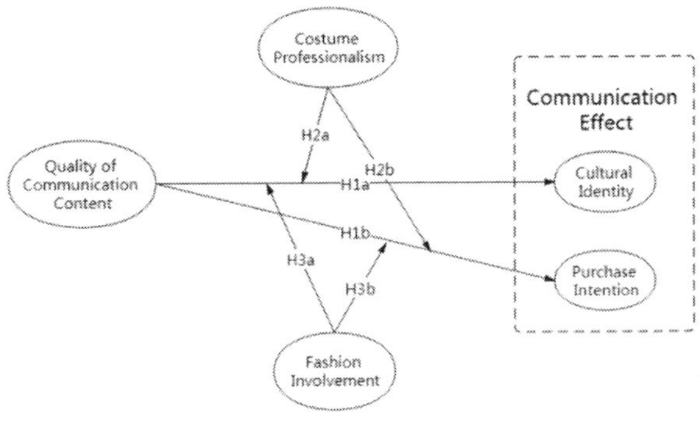

Fig.1 Research model

3.1.1 The Influence of Information Quality of Chinese Costume Culture on Communication Effect

Information quality is an overall concept and scholars have researched its evaluation dimensions. ZHANG Yi [19] believed that interestingness, practicability and vividness of communication content would affect information quality, which further influenced audience's attitude. Chen and Rodgers [20] proposed that there were three characteristics of network information i.e. intelligence, interestingness and organized form and that the more interesting the information, the better the communication effect. McMillan et al. [21] through empirical study proved that the interestingness, vividness, usefulness and richness of communication content would affect communication effect. Therefore, in this article, the communication content of Chinese traditional costume culture is measured from three dimensions: the dimension of completeness and accuracy; the dimension of vividness and interestingness; and the dimension of practicability. The dimension of completeness and accuracy means that the communication content should not only give accurate and complete elaboration on the style, colors, fabric and patterns etc. of ancient costumes but also expound the connotation and cause of formation. The expression of communication content is an important dimension for evaluating information quality. Since some obscure terminologies are involved concerning the style, colors and patterns etc. of traditional costumes, the dimension of vividness and interestingness means vivid description of traditional costumes to facilitate audience's better understanding of communication content. The communication content is practical if it offers help and knowledge to audience and promotes the inheritance of traditional culture. The higher the quality of communication content of Chinese traditional costume culture (i.e. information that is more complete, more accurate, more vivid and more practical), the better the communication effect. Therefore, the following research assumptions are proposed:

H1: The communication content of Chinese traditional costume culture has positive influence on communication effect.

3.1.2 The Regulating Effect of Audience's Costume Professionalism and Fashion Involvement

Audience are receivers of information in different communication activities. According to ELM [17] theory, audience's motive and capacity of information processing will influence the route of information processing, thus affecting communication effect. The research by Sussman and Siegal [18] proved that audience's involvement and professionalism had important regulating effect on the relation between information quality and information adoption intention. Therefore, it's believed in this research that, during communication of Chinese traditional costume culture, audience's fashion involvement and costume professionalism have regulating effect on communication effect. The fashion involvement refers to the extent to which the audience give attention to fashion. Costume professionalism refers to individual's understanding of the knowledge about costume and traditional costume. The following assumptions are proposed:

H2: Audience's costume professionalism produces a regulating effect on the influence of communication content on communication effect of traditional costume culture.

H3: Audience's fashion involvement produces a regulating effect on the influence of communication content on communication effect of traditional costume culture.

3.2 Design and Collection of Questionnaires

The research was conducted through questionnaires which consisted of 3 parts: the first part was the survey of communication content quality of traditional costume culture; the second part was the measurement of communication effect and the measurement scale was from the WANG Yu's cultural identity scale [15] and JIN Gaoxia's Purchase Intention scale of Chinese costume [4]; the third part was the survey of costume professionalism and fashion involvement. The measurement scale of costume professionalism was by reference to LIU Huihui's [22] and Bansal and Voyer's [23] scale of characteristics of costume professionalism; and that of fashion involvement was by reference to CHEN Wenpei's [24] and Ho Jung Choo's [25] research scale. As to the form of questionnaires, the five-level Likert scale was adopted for measurement. The questionnaires were distributed on the Wenjuanxing Platform. A total of 342 questionnaires were recovered, including 331 effective questionnaires which accounted for 96.8%. In terms of gender, males accounted for 49.8% and females accounted for 50.2%; in terms of age, people aged 18-30 accounted for 59.5%, people aged 31-45 accounted for 37.0% and people aged 46-60 accounted for 3.5%. The recovered samples were reasonable in terms of both gender and age.

4. The Result of Empirical Study

4.1 Verification of Reliability and Validity

For sample data, SPSS22.0 and Amos 22.0 were adopted for verification of its reliability and validity. The verification result concerning the reliability and validity of scales is shown in Table 1 below. The Alpha and Composite Reliability (CR) of latent variables in each scale exceeded the standard threshold value 0.7. Meanwhile, all AVE value exceeded 0.5, indicating that the variables in this research demonstrated satisfactory reliability and validity and conformed to the research requirements.

Table 1. Verification result concerning the reliability and validity of scales

Variables		Alpha	AVE	CR
Content Quality	Completeness and Accuracy	0.850	0.567	0.867
	Vividness and Interestingness	0.720	0.534	0.820
	Practical	0.709	0.522	0.813
Costume Professionalism		0.799	0.507	0.836
Fashion Involvement		0.793	0.492	0.828
Communication Effect	Cognition	0.786	0.507	0.837
	Pride	0.787	0.494	0.853
	Interest	0.730	0.514	0.808
	Purchase Intention	0.725	0.513	0.840

4.2 The Relation between Communication Content and Communication Effect of Chinese Traditional Costume Culture

To understand whether communication content quality of Chinese traditional costume culture directly predicts the communication effect, a regression analysis was performed with the three measurement dimensions of communication content as independent variables and with the recognition of traditional costume culture and willingness to buy Chinese costumes as dependent variables. The regression results are shown in Table 2 below. The completeness, accuracy and practicability of communication content of traditional costume culture contributed significantly to cultural identity and all standard regression coefficients were positive, indicating the two factors' influence on traditional costume culture were positively significant. Meanwhile, the three factors of communication content of traditional costume culture had positive contributions to willingness to purchase Chinese costumes and all standard regress coefficients were positive, indicating the influence of the three factors of communication content on the willingness to purchase Chinese costumes were positively significant.

Table 2. Regression result

dependent variable	independent variable	unstandardized coefficients	S.E	standardization coefficient	t	VIF	R^2	adjusted R^2	F
Cultural identity	constant	1.28	0.10		12.78**				
	Practical	0.42	0.03	0.47	15.51**	1.47	0.49	0.49	393.78**
	Completeness and Accuracy	0.28	0.03	0.32	10.54**	1.47			
Purchase intention	constant	0.98	0.13		7.12**				
	Practical	0.38	0.04	0.38	11.47**	1.14			
	Completeness and Accuracy	0.27	0.03	0.27	7.50**	1.07	0.41	0.41	170.18**
	Vividness and Interestingness	0.07	0.04	0.07	2.83**	1.15			

4.3 Verification of Regulating Effect

4.3.1 Verification of Regulating Effect of Audience's Costume Professionalism

With adoption of Model8 in SPSS macro by Hayes which was consistent with the theoretical model herein, the audience's professionalism was entered into the regression equation. The professionalism had regulating effect in direct prediction of cultural identity with communication content. The results are shown in Table 3 below. The result of further simple slope analysis is shown in Fig.2 below. With improved costume professionalism of individuals, the role of communication content quality in prediction of cultural recognition was gradually reduced.

Table 3. Results of regulating effect of professionalism

Model	coeff	S.E	t	significance	R^2 change	p
constant	4.080	0.013	324.113	0.000		
communication content	0.539	0.028	19.237	0.000	0.0084	0.0002
audience's professionalism	0.222	0.020	11.156	0.000		
communication content*professionalism	-0.139	0.037	-3.739	0.000		

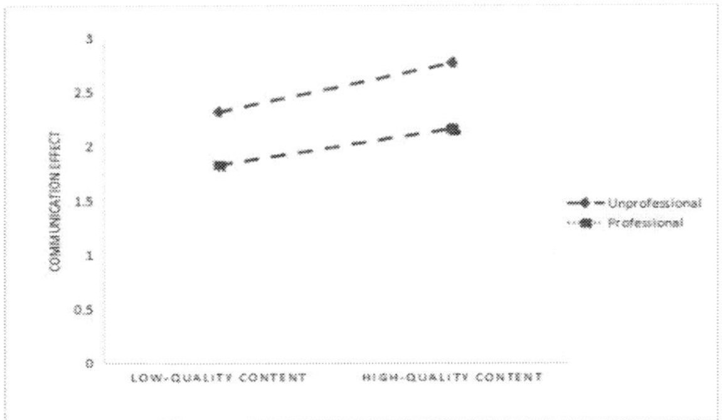

Fig.2 Verification of regulating effect of audience's costume professionalism

4.3.2 Verification of Regulating Effect of Audience's Fashion Involvement

The regulating effect of audience's fashion involvement was verified with the same method. The fashion involvement had regulating effect in direct prediction of cultural identity with communication content. The results are shown in Table 4 below. The result of further simple slope analysis is shown in Fig.3 below. With improved fashion involvement of individuals, the role of communication content quality in prediction of cultural recognition was gradually reduced.

Table 4. Results of regulating effect of fashion involvement

Model	coeff	se	t	significance	R^2 change	p
constant	0.037	0.028	1.305	0.192		
communication content	1.116	0.061	18.256	0.000	0.0171	0.000
fashion involvement	0.289	0.036	8.016	0.000		
communication content*fashion involvement	-0.340	0.071	-4.816	0.000		

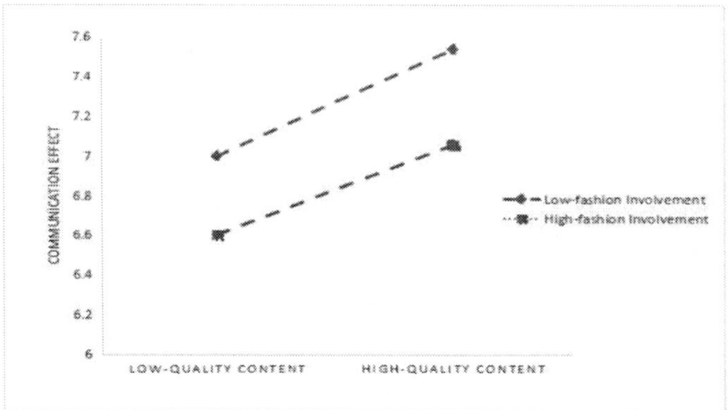

Fig.3 Verification of regulating effect of audience's fashion involvement

5. Conclusion

In this article, based on elaboration likelihood model and information adoption behavior model in communication theory, a theoretical model of the communication content of traditional costume culture against communication effect was established. Then, the model was verified through empirical study. The research result expands the application mechanism of communication theory on the communication effect of Chinese traditional costume culture and is of important theoretical significance to research of traditional costume culture. According to the analysis, the comprehensiveness and accuracy of communication content are key factors that affect communication effect, which is consistent with the conclusion of SHI Ke's research of influencing factors of traditional cultural communication [26]: whether the communication content is accurate and complete will directly affect audience's knowledge about traditional costume culture and further influence the communication effect at cognition stage where audience change their sentiment toward traditional costume culture and consumption willingness. Second, the practicability of communication content also had significant influence on communication effect, which was consistent with Davis's research conclusion [27]. The traditional costume cultural information of high practicability will increase audience's knowledge, prompt them to feel the value of communication content to themselves and social development and help them develop more positive attitude, thus achieving better communication effect. The vividness and interestingness of communication content only produce significant positive influence on willingness to purchase Chinese costumes, because audience will better combine it with present living style when they encountered vivid content of traditional costume culture, thus developing the willingness to consume Chinese costumes.

Audience's costume professionalism has a regulating effect in the relation between communication content and communication effect. Audience with lower professionalism are easily affected by high-quality communication content. On one hand, audience with lower costume professionalism will develop comprehensive understanding and cognition of traditional costumes, thus changing their recognition significantly; on the other hand, with richer professional knowledge about clothing and traditional costumes, information receivers will analyze the content based on their own professionalism and their attitude won't be easily changed. The audience's fashion involvement has a regulating effect in the relation between communication content and communication effect. Audience with higher fashion involvement will develop their own ideas about fashion and pay more attention to custom trend. Since traditional costume culture is focusing more on ancient costumes and not quite matching modern fashionable life style, the audience with high fashion involvement do not have higher focuses on its communication content and the communication effect is not significantly improved.

Since audience's understanding of and interests in costume culture vary from one to another, differentiated communication may be prepared. As to audience who pay less attention to fashion and lack professional costume knowledge, the quality of content of traditional costume culture should be improved to demonstrate the evolution of style, fabric, colors etc. of ancient costumes in easily understood manner, so as to improve audience's cognition and emotional recognition. Meanwhile, rich means of expression i.e. pictures and videos may be adopted to make the communication content more affectionate to the people and more interesting, thus making costume culture more infectious. For example, in the TV program of National Treasure, the silk costume of painted wooden figurine in Tang Dynasty was restored and displayed, thus prompting audience's interests in traditional costume culture and their willingness to get understanding of traditional costumes. In this way, the communication effect is enhanced. As to audience who have certain professionalism and like to follow fashion, the communication content should give more focuses on innovative design which integrated the modern life style. For example, some Chinese clothing brands blend the elements of traditional costumes with current fashion and aesthetics, so that the traditional costume culture is able to satisfy the fashion demand of such audience while maintaining its substantial characteristics.

References

[1] Zhang PF. The culture implication and aesthetic meaning of Chinese traditional costume. J Zhengzhou Univ Light Ind Sci: 2009; 10 (4): 8-11.
[2] Zhang XL, Yang CY. The sedimentary accretion of Chinese tradition clothing culture. J Zhengzhou Inst Light Ind Sci: 2005; (6): 18-19+23.
[3] Xu X. The influence of Chinese traditional dress culture on modern fashion. Silk Mon: 2001; (12): 36-8.
[4] Jin GX. A study on the influence of life style on Chinese clothing consumer behavior. Beijing: Beijing Institute of Fashion Technology: 2015.
[5] Qiu ZY, Song J. A brief discussion on the predicament of Chinese traditional clothing culture and the media communication path. Liaoning Tussah Silk: 2014; (4): 23-24.
[6] Wang NN. The intercultural transmission of Chinese traditional clothing. Press: 2016; (16): 129-130.
[7] Li YM. Dissemination of Chinese clothing culture under the new media environment. J News Res: 2015; 6 (1): 66.
[8] Littlejohn SW. Encyclopedia of communication theory. Beijing: China Social Sciences Press: 1999.
[9] Zhang DN. A brief analysis of cultural elements. Beijing: Educational Science Press: 1988.
[10] Fu XC. Appearance contains obvious culture and invisible culture. Art Res: 2010; (4): 36-37.
[11] Guo QG. The course of communication studies. Beijing: Renmin University of China Press: 2011.
[12] Alfred J. What is communication? Studies in communication. London: Martin and Warburg: 1955.
[13] Lavidge RJ, Steiner GA. A model for predictive measurements of advertising effectiveness. J Mark: 1961; 25 (6): 59-61.
[14] Zhang CX. Zhang's dictionary of psychology. Shanghai: Shanghai Dictionary Publishing House: 1992.
[15] Wang Y. The construction of chinrse traditional costume culture identity scale. Beijing: Beijing Institute of Fashion Technology: 2015.
[16] Zuo YP. Research on the influence mechanism of Chinese traditional clothing culture identity on Chinese style clothing brand loyalty. Beijing: Beijing Institute of Fashion Technology: 2018.
[17] Petty RE, Cacioppo JT, Petty RE, Cacioppo JT. The elaboration likelihood model of persuasion. Commun Persuas: 1986; 19 (4): 1-24.
[18] Watts Sussman S, Schneier Siegal W. Informational influence in organizations: An integrated approach to knowledge adoption. Inf Syst Res: 2003; 14 (1): 47-65.
[19] Zhang Y. The Research of microblog users' forward-based on the theory of persuasion communication. Guangzhou: South China University of Technology: 2014.

[20] Chen Q, Rodgers S. Development of an instrument to measure web site personality. J Interact Advert: 2006; 7 (1): 4-46.
[21] McMillan SJ, Hwang JS, Lee G. Effects of structural and perceptual factors on attitudes toward the website. J Advert Res: 2003; 43 (4): 400-409.
[22] Liu HH. Research on the influence of net red opinion leaders on apparel consumers' willingness to buy-Taking micro-biog as an example. Hangzhou: Zhejiang Sci-Tech University: 2019.
[23] Harvis SB, Peter AV. World-of-mouth processes within a services purchase decision context. J Serv Res: 2000; 3 (2): 166-177.
[24] Chen WP. Relationships of consumer lifestyle, consumer innovativeness and new products buying behavior-An empirical analysis based on chinese context. Chongqing: Chongqing University: 2014.
[25] Choo HJ, Sim SY, Lee HK, Kim H Bin. The effect of consumers' involvement and innovativeness on the utilization of fashion wardrobe. Int J Consum Stud: 2014; 38 (2): 175-182.
[26] Shi K. The Study on the problems and countermeasures existing in the current traditional Chinese cultural dissemination. Jinan: Shandong Normal University: 2018.
[27] Davis FD. Perceived usefulness, perceived ease of use, and user acceptance of information technology. MIS Q: 1989; 13 (3): 319-339.

Restoration of Niya's Brocade Robe with Yellow and Blue Checkered Pattern Based on Digital Technology

Hui Zhang[1], Man Zhou[1], Guang-Tu Zhang, [2]Xiao-Yu Xin[1*]

[1] *College of textile and clothing, Xinjiang University, Urumqi, 830046, China*
[2] *Northeastern University, Shenyang, 110819, China*

Corresponding author's email: 479790302@qq.com

Abstract

In order to better inherit and protect the cultural relics of traditional clothing and shorten the research and production cycle of clothing cultural relics restoration work, a method for virtual restoration of traditional clothing based on digital technology is proposed. Firstly, it is necessary to fully analyze detailed information regarding the shape, pattern and color of the restored clothing. On this basis, digital software is used to complete the processes of drawing, cutting, pattern restoration, texture effect production and other processes, and then realize traditional clothing virtualization based on CLO3D software Try on the display, and finally evaluate the restoration effect and pressure comfort of the restored clothing. The results show that this method can effectively complete the restoration of unearthed costumes.

Keywords: Niya; Digital Restoration; CLO3D; Virtual Display

Introduction

The rapid development of information technology has expanded the methods of clothing design and production. Many digital softwares on the market can simulate series of processes of clothing from design to production. In the field of unearthed clothing cultural relics protection and heritance, combination of digital technology and clothing restoration has been a new demand. Unearthed clothing are fragile relics therefore not easy to restore. Part of the physical objects can only be seen in museums. Some of the more precious clothing are collected by museum and are not on display. At the same time, researches on unearthed costumes are only carried out by professionals, universities and museums, which is not conducive to the inheritance and development of costumes. A method of restoration of unearthed costumes based on digital technology is proposed in this paper in order to show the charm of traditional Chinese costumes and provide new ideas for protection and restoration of unearthed costumes.

The brocade robe with yellow and blue checkered pattern is unearthed at the Niya site in Xinjiang. Based on the analysis of its shape, we restored a digital technology version and completed the virtual try-on, hoping that this method can provide a reference to the virtual restoration of clothing.

1. Clothing Shape Analysis

Brocade robe with yellow and blue checkered pattern has two layers. The outer layer is made of square plaid brocade and inner layer is made of silk. Sleeves are tight and narrow, Hem is wide, Side seams are open, Body length is 122cm and Sleeve length is 225 cm [1].The method of weaving brocade consist of two colors. A two-way brocade is knitted with the intertwining of two sets of warps and a set of weft. 2 warps will be blue/green and yellow/deep red and show figures with the distance of 1 cm. Brocade pattern will appear yellow, blue, green, red four different colors of squares, which makes brocade more colorful [2], as shown in Fig.1.

1.1 Styles

Brocade robe with yellow and blue checkered pattern has the costume characteristic of western regions which has left lapel, tight and narrow sleeves, open side seams. "Left gusset" has the typical clothing characteristics of nomads in Han Dynasty [3]. Nomads in Han Dynasty had lifestyles that involved riding and archery to fulfill their everyday needs .Smaller left lappet saves the Normad's some effort as they held their bow with their left hands and drew with right hands [4]. Narrow sleeves are convenient to ride as well as provide better warmth. Wide hem and side seam opening design increases the amount of activity of the lower limbs, which is beneficial whenriding horses, so that they can live a better nomadic life.

Fig.1 Yellow and blue checkered brocade robe [1]

1.2 Structure

Structurally, although there are some differences between Central Plains and the Western Regions, they mainly adopt a cross-shaped flat structure. Due to limited fabric width, the Central Plains clothing is the embodiment of moral concept, while the Western Region clothing is the embodiment of nomadic way of life and practicality [5]. Archaeological data shows that the out layer is 42 cm wide and the inner layer is 41cm wide.A concluson can be drawn after a closer look at the picture that there are obvious trace of seam between left cuff and right shoulder of sleeve part, lappet and body, hem and body, side seams and body. The robe should be cobbled together from a clothing piece. Side seam cutting is originally rectangular. Its upper portion is removed to form a trapezoid when stitching. During the Han and Jin Dynasties, due to the use of scissors [6], diversified clothing cutting designs were introduced. The sleeve roots, underarms and waist circumference appeared arc shapes, and the armpits of the brocade robe were cut using this design.

Based on the information above, a diagram is drawn and fixed. The robe adopts traditional "cross-shaped flat structure", using horizontal sleeve line and front and rear clothing center line as the axis as the cross coordinates. Limited by the width of cloth, front and rear pieces of cloth and sleeves are stitched together. There are triangular inserts under the armpits. There are triangular inserts under armpits. Collar and lappet are connected together. A trapezoidal insert is stitched to side seams. As shown in Fig.2.

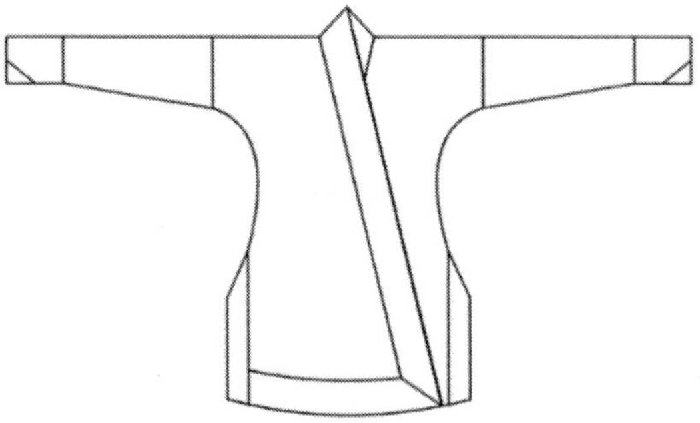

Fig.2 Structure chart

2. Digital Recovery

2.1 Man-model Establishment

CLO3D software model library provides male and female models of black, yellow, white skin. Users can create mannequins according to their own needs and size of each part of model can be adjusted. According to the excavation report, the male body were 174 cm tall, so we select an adult male model from the model library, adjust model's height to 174 cm, other size as shown in Tab 1.After selecting the gender and size of model, the posture of the person model is also needed to be adjusted to facilitate the virtual trial wear of garments. Then open the X-Ray binding point of the 3D window model, adjust the shoulder joint point to flatten the arm and create the positions that satisfy the needs.

Table 1. Mannequin size

Parts	Dimensions /cm
Height	174
Neck circumference	40
Bust	96
Waist	84
Hips	95

2.2 Clothing Pieces Drawing

2D window interface tool of CLO3D software can be used to draw clothes [7]. However, CLO software plate making takes a long time to generate the model compared to professional pattern making software. Therefore garments are always drawn by other professional pattern making software, which are saved as DXF format files, then imported it into the CLO3D software. According to the data and scale provided in the excavation report, the length of the garment and the length of the sleeve can be obtained, and the remaining dimensions can be determined according to the scale and the number of squares. The dimensions of each part are shown in the table 2. Using the data, the patterns are drawn according to the size of the garment and then saved as a DXF format file. By importing the saved file into the CLO3D software, the garment pieces will appear in the 2D window.

Table 2. Brocade size

Parts	Size /cm
The clothes are long	122
Long sleeves	225
Sleeve width	15
Waist width	60
The hem is wide	95

2.3 Clothing Arrangement

Clothing arrangement is to arrange cloth pieces around the mold reasonably based on location [8]. If the clothing piece is not arranged reasonably, the resulting model will be broken and clothing pieces will fall off, which is a highly probable phenomenon. Due to the large number of patterns of the brocade robe, it is necessary to use the arrangement point and free drag to complete the arrangement of the patterns at the same time, rotate the 3D window to check whether the position of the patterns is reasonable, and adjust the pieces with unreasonable positions.

2.4 Virtual Stitching

The stitching relationship between the pieces of the Brocade robe is relatively simple. All the patterns can be stitched using only thread sewing. This process requires the selection of sewing tool to sew the arranged patterns, while paying attention to the stitching sequence during the sewing process, and adjust the perspective of the three-dimensional window to check the patterns whether the suture relationship is correct and whether there is suture crossing, adjust the wrong suture in time. As shown in Fig .3.

Fig.3 Virtual try on

2.5 Fabric Simulation

2.5.1 Pattern Restoration

The motifs on the Brocade robe is mainly square grid with a side length of 1cm and adopt a relatively simple shape. The color of them are mainly yellow, blue, purple and green, which sometimes appear on the front of the brocade robe, making the brocade robe more colorful. Choosing Coreldraw to draw motifs [9]. First, draw a square with a length and width of 1cm.Then use the transform function to arrange it horizontally 70 times, next arrange it 111 times in the weft direction, finally we get a 70×110cm rectangular grid, select the grid and fill it with four colors of yellow, blue, magenta, and green. The obtained motifs is as shown in the figure4, save it as a jpg format file.

After making the motifs, use PS software to add texture effect on the picture to make the restoration effect more realistic [10]. As shown in Fig.4.

Fig.4 Part of texture making

2.5.2 Fabric Simulation

Select the silk fabric in the fabric library [11], import the prepared texture, adjust the surface roughness and reflection strength of the fabric to make the surface texture better, and then adjust the physical properties of yarn strength, diagonal tension, bending strength, deformation rate, etc, making the fabric more real.

2.6 Virtual Try-on Effect

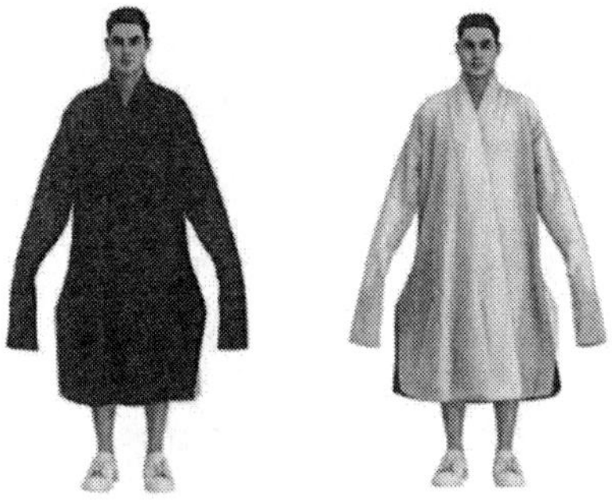

Fig.5 Figure of dress

The restored brocade robe with yellow and blue checkered pattern has simple motifs and rich colors which sleeves are longe and the collar is tightly wrapped around the neck, forming a unique style of clothing. In order to further study the effect of digital restoration, we evaluated the comfort of the brocade. Adjust the clothing to a translucent state. After observation, it is found that the clothing and the human body have a large amount of looseness, so the looseness meets the wearing requirements. As shown in Fig.5.

Pressure comfort is a important item in the evaluation of clothing comfort indicators [12]. The distribution of pressure points and the pressure values of different parts of the body can be observed in the three-dimensional window, which is convenient for us to evaluate the pressure comfort [13]. Observe the distribution of pressure points on the surface of the Brocade robe, the more concentrated the pressure points, the more pressure points are generated in this part, the more pressure points indicate the stronger the feeling of pressure [14]. For this reason, recording the densely concentrated parts of the pressure points, and then Open the pressure value tool, different colors indicate different pressure values. The pressure increases from blue to red. Click any part of the brocade to display the pressure value of that point, as shown in table 3, the pressure value of the concentrated area of the pressure point is small, and the pressure on the human body is small, which meets the requirements of pressure comfort [15], As shown in Fig.6 and 7.

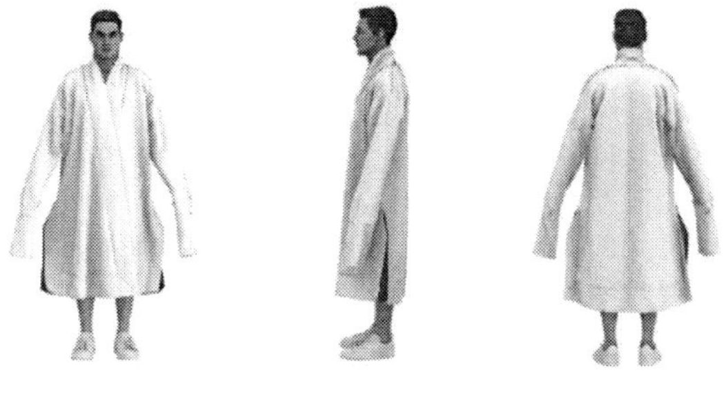

Fig.6 Press point

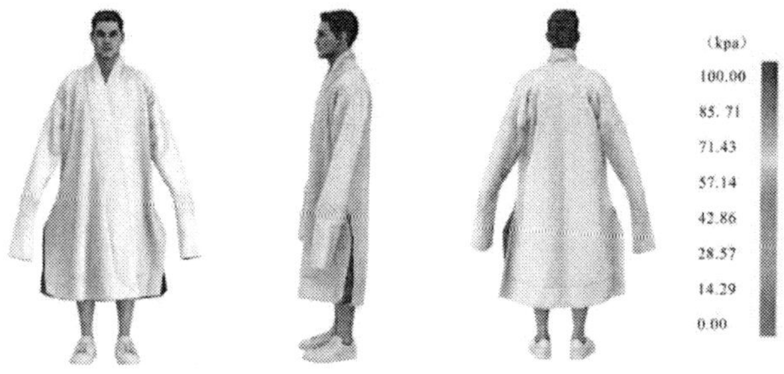

Fig.7 Press

Table.3 Pressure value

Shoulder point (left)	Shoulder point (right)	Nape point	Elbow point (left)	Elbow point (right)
10.11	9.84	10.92	4.11	4.32

3. Conclusion

In this paper, a method of restoration of unearthed clothing and cultural relics based on digital technology is proposed. Firstly, the paper analyzes the dress shape, and then proceeds to the drawing, cutting, pattern restoration and texture effect production based on digital softwares. Finally, the virtual stitching and trial dressing of clothing are finished by CLO3D. The results show that the pattern, color, style and other information of unearthed clothing can be recovered with this method, while also can appreciate the dress wearing status. Our method does not only protect the unearthed costume cultural relics, but also better display and promote the Chinese national costume culture, and provide a reference for the unearthed costume protection and restoration.

References

[1] Zhao F, Yu ZY. Legacy of the Desert King[M]. Hongkong: ISAT: 2000; 55.
[2] China-Japan Japan-China Joint Academic Investigation Team of Niya Remains. China-Japan Japan-China Joint Academic Investigation Report on Niya Remains [M]. China-Japan Japan-China Joint Academic Investigation Team of Niya Remains: 1996; 91-100.
[3] Wang TB, Liang HE. Exploration of Ancient Zuojun Clothing Based on the Protection and Inheritance of Cultural Heritage [J]. Silk: 2010; 10: 51-59.
[4] Dong LL. Re-understanding of the "Zuo Ren" dress custom in the Han Dynasty [J]. Folklore Research: 2019; 3; 50-55+158.
[5] Liu RP, Chen JJ. Construction Patterns of Chinese Costume [M]. Beijing: China Textile&Apparel Press: 2013; 45.
[6] Zhou JL, Li WY. Excavation of Yingpan graveyard at Yuli Xinjiang [M]. Cultural Relics: 2002; 6: 4-45.
[7] Li J. Virtual clothing design based on CLO3D [J]. Science&Technology Information: 2019; 4: 22-23.
[8] Liu DS, Shi H, et al. Research on the shape and structure of Mongolian gowns from the Ujumqin tribe and virtual fitting [J]. Silk: 2020; 57 (8): 50-56.
[9] Yao J, Zhang YJ. The Application of CorelDRAW in Graphic Editing [J]. Ocean Surveying and Mapping: 2004; 3: 47-48.
[10] Adobe Company. Adobe Photoshop CS5 Chinese Edition Classic Course [M]. The People's Posts and Telecommunications Press: 2014.
[11] Zhang H, Guo RL. Overview of clothing simulation methods in network virtual display [J]. Shandong textile science and technology: 2018; 59 (5): 52-56.
[12] Zhan H, Xu J. Research and application of clothing pressure comfort [M]. Beijing Textile: 2004; 5: 58-60.
[13] Zhang Wm, Ma F. Analysis of Cross Structure of Han Chinese Clothing Based on CLO3D Platform [J]. Silk: 2021; 58 (2): 131-136.
[14] Tian BQ, Xu ZB, Hu SZ. Evaluation of Dress Fit based on CLO3D Virtual Fitting Technology [J]. Journal of Donghua University (Natural Science): 2018; 44 (3): 397-402.
[15] Chen Y, Zhou L. Tibetan style clothing design and display based on CLO3D platform [J]. Progress in Textile Science and Technology: 2018; 11: 50-53.

Research on Color Digitization of Chinese Traditional Patchwork Costumes: Taking the NCS Analysis of Traditional Patchwork Baby Carriers of the Zhuang Ethnic Group as an Example

Ru Zhang[1,2], Qi Liu[3*], Shuang Zheng[4]

[1]*Graduate School, Beijing Institute of Fashion Technology, No.2, East Yinghua Road, Chaoyang District, Beijing, 100029, China*
[2]*Institute of Jiangsu Intangible Cultural Heritage, Nanjing Museum, No.321 East Zhongshan Road, Nanjing, 210016, China*
[3]**Institute of Art and Culture, Beijing Institute of Fashion Technology, No.2, East Yinghua Road, Chaoyang District, Beijing, 100029, China*
[4]*Beijing PuYuan Culture&Art Co., Ltd, Beijing Institute of Fashion Technology, No.2, East Yinghua Road, Chaoyang District, Beijing, 100029, China*

*Corresponding author's email: fzylq@bift.edu.cn

Abstract

The qualitative research method is usually used to analyze the color features of ethnic costumes. Differently, this paper adopts the most state-of-the-art quantitative research method in chromatics and the modern research theory and method for color digitalization to analyze the color styles and characteristics of patchwork baby carriers of the Zhuang. This paper aims to allow people to better understand the digital ways of traditional ethnic colors. This research contributes to the establishment of a color database of traditional ethnic costumes. Such color database can serve as the intuitive and referenceable guideline for other future studies on color digitalization of traditional ethnic costumes, thus having strong significance.

Keywords: Patchwork; Traditional Costumes of the Zhuang; Modern Chromatology; Color Digitization; NCS

1. Introduction

The Zhuang has a long history. It is an ethnic minority with more than 16.9 million people, the most population in China [1] in 2010. The Zhuang has distinctive features, including the rich and colorful clothing culture. The Zhuang has developed skills in weaving and embroidery. A distinctive apparel process of the Zhuang is patchwork, also known as Piecine or Quilting. Patchwork is a technique for connecting and stitching the regular or irregular pieces of cloth. It is also a unique art form and a common way for decoration. Patchwork is often used for clothing and household supplies [2]. Patchwork is often deemed as a kind of embroidery. However, we do not think patchwork should be classified as embroidery because it overlaps with embroidery only to a certain extent [3]. There are 56 ethnic groups in China, 31 of which are good at using patchwork to make costumes and daily necessities, including the ethnic Han [4]. Patchwork is also time-honored. It was recorded that "a native bride wore the clothing made with many collaged cloth of different colors and patterns in a Qinzhou village" in the Ling Wai Dai Da written by Qu-Fei Zhou in the Song Dynasty. [5] therefore, the typical patchwork process can be summarized as follows: collage many pieces of cloth of different colors and patterns according to the designer's intention, and then nail the corners with silk threads. Patchwork products look simple but very beautiful. The Zhuang women favor them for their diligent and thrifty.

Baby carriers were called "swaddles" in ancient times. Now it is still popular in the minority areas of Southern China. Generally, a baby carrier is T-shaped. Through packing and tying, a mother can fix her child on her back with a baby carrier. In this way, the mother can take care of her child while working. A baby carrier is regarded as an "amulet" for a child. Mothers like decorating their baby carriers with various patterns or ornaments to pray for their children's good luck and health. According to the folk custom of the Zhuang, a grandmother will present a beautiful baby carrier to her one-month-old grandson, hoping that he can grow up safely and happily. Therefore, patchwork baby carriers the highest craftsmanship level of the Zhuang people.

The application of colors reflects the culture and aesthetic psychology of an ethnic group in the most intuitive manner. The Zhuang people are fond of black, so they are mainly dressed in black. They look calm and dignified. The Zhuang has many collateral ethnic branches that vary significantly in the styles of costumes and decorations. In addition to black Zhuang clothing, there is a wide range of clothings in the folk of the Zhuang, including blue Zhuang clothing, cyan Zhuang clothing, red Zhuang clothing, colorful Zhuang clothing, White Sand clothing, and Black Sand clothing. According to the Chronicles of Guangnan County, written in the Qing Dynasty, "the White Sand Zhuang People liked white clothes" [6]. All costumes of the Zhuang people are wonderful. The Zhuang's traditional patchwork baby carriers are very distinctive in the application and matching of colors.

This paper analyzes the color styles and characteristics of Zhuang's patchwork baby carriers with the quantitative research method of modern chromatology. It aims to allow people to better understand the matching ways of traditional ethnic colors more digitally and rationally and benefit people in enjoying their color-related value. This research contributes to the establishment of a color database of traditional ethnic costumes. It can be used to reference the design of clothing colors, and also serve as the intuitive and referenceable guideline for other future studies on color digitalization of traditional ethnic costumes. This research has strong guiding significance, and comes up with some new ideas and ways for the research on digitization and standardization of traditional costumes in terms of inheritance and innovation.

2. Technical Roadmap

2.1 Innovation in Technical Roadmap

In the past, the qualitative research method was usually used to research on the color features of ethnic costumes. For example, such qualitative research method was adopted in the subsection "Analysis of Color Features of Minority Women's Costumes" of the monograph Design of Ethnic Minority Women's Wear [7], Color Languages of Ethnic Minorities-From Totem Symbols to Social Symbols written by Jing-Yu Zhu and Jia-Quan Li [8], the paper The Symbol of Ethnic Costume Colors written by Ding-Hong Pan [9], and the paper The Colors of Chinese Ethnic Costumes written by Ping Dai [10], etc. These qualitative researches are very beneficial for understanding the color culture of ethnic costumes, but some details are still unclear. For example, what kind of red do some branches of the Miao people like? What kind of black do the Yi people prefer? What is worse, chromatic aberrations exist between the image data and the costumes in kind. There are massive uncertainties due to the lack of unified standards for information expression. In addition, the unique charms of ethnic costumes in color matching, and the relations between colors cannot be interpreted through qualitative research.

This paper adopts the most state-of-the-art quantitative research method, NCS color theory and tool, and the modern research theory and method for color digitalization to comprehensively analyze the colors of the Zhuang's patchwork baby carriers in combination with the results of qualitative research and from the perspectives of both ethnic costumes and modern chromatology.

2.2 What is NCS?

The Natural Color System (NCS) is a color system based on the human color perception mechanism. It can define colors that people intuitively perceive in daily life. The NCS enables everyone to convey color concepts, define colors, and describe any imaginable surface colors in a simple and easy-to-understand way. The NCS is an international universal color system that applies cross-disciplinary and across-industry [11]. NCS is a symbolic language that provides various possibilities for describing colors and relationships, both by graphical illustrations and by letter-digit notations. Furthermore, colorimetry provides a physical representation of the psychometric attribute of NCS for color notation. [12]

NCS color solid: The NCS is based on six reference colors, including four chromatic ones: yellow (Y), red (R), blue (B), and green (G), and two colorless ones: white (W) and black (S). It is believed that all visible colors are related to these six reference colors.

A three-dimensional color solid is formed based on the color gradient between the six reference colors and the four chromatic colors. In this color solid, each color visible to naked human eyes has a unique number and position in the color space.

NCS color ring: The NCS color ring is a cross-section of the color solid above, reflecting the color hue. The colors on its four vertexes are yellow (Y), red (R), blue (B), and green (G) clockwise in turn, respectively. Each quarter of the color ring is equally divided into 100 parts.

NCS color triangle: The NCS color triangle is a vertical cross-section of the color solid above, reflecting the nuance, namely the relationship between white (W), black (S), pure chromatic color (C, gray), and also the attributes of the colors, namely, their whiteness (gradient), blackness (depth) and grayness (saturation). The three attributes of colors are divided into 100 equal parts in the color triangle. As in the color ring, the attributes of color can be represented by the percentage values, accordingly (Fig.1).

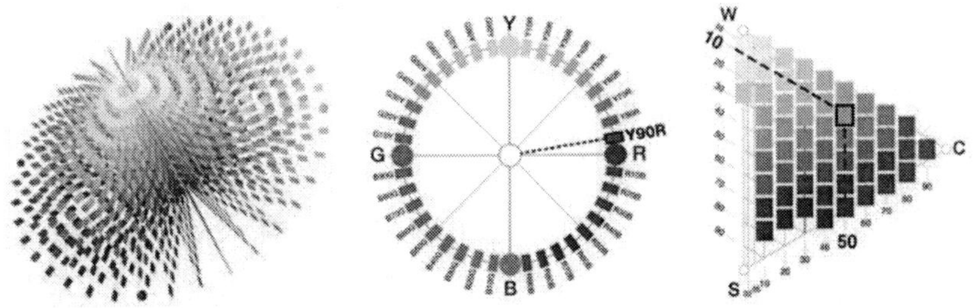

Fig.1 NCS color solid, NCS color ring and NCS color triangle [13] (Provided by Shuang Zheng)

2.3 Reasons for Choosing NCS

Currently, the NCS color system and Munsell color system are the two main color systems commonly used in the world. The Munsell color system was designed and launched by the American painter A.H.Munsell in 1905 and scientifically validated by the Committee on Colorimetry of the Optical Society of America in 1943. It has been widely used in the world after several rounds of modification and improvement. The Munsell color solid determines color classification and marks the surface colors of matters based on the Hue, Chroma, and Value of the colors, and also in combination with the visual characteristics of the colors [14]. The Munsell color system requires the complex background knowledge of physics for interpretation.

In contrast, the color naming of the NCS color system is in line with people's visual habits. The NCS color system makes it possible to clearly show the relationship between colors in a more intuitive and easy-to-understand manner. For example, for S 1050-Y90R, S means the revised standard color swatch, 10 indicates the blackness, 50 denotes the chroma, and -Y90R refers to the hue (90% red and 10% yellow) (Fig.2). Due to the convenience and easiness to understand, the NCS color system is favored by many color designers and researchers, becoming an indispensable tool for color professionals.

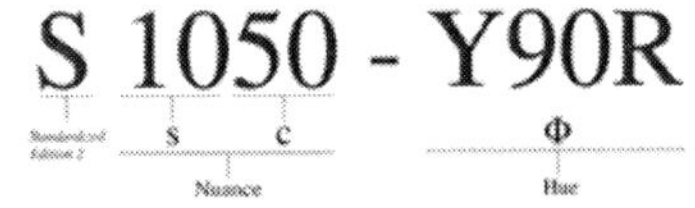

Fig.2 NCS color numbering [15] (Provided by Shuang Zheng)

3. Workflow

The color measurement workflow: Color measurement of samples→ Result recording of color measurement of samples→ Color analysis by sample→ Data summary and analysis.

3.1 Color Measurement Tools: Color Swatch and Colorimeter

Color swatch is an indispensable tool for measurement, recording, calibration, selection, and spectrum production of colors. As a universal language, it conveys color information based on precise values. The NCS Mini

Colorimeter 2.0 was used to measure the sample colors. It can assign an NCS number to each surface color selected and display it quickly. The surface colors of the embroidery, cotton, linen, silk, satin, and other items are identified.

3.2 Colorimetry Based on Visual Sense

The standard color swatches for the NCS color system are usually used for visual color measurement. According to the objective numerical values of color conversion, the common points in the hue, value, and chroma of colors, and the tendency characteristics of colors, can be analyzed. The fundamental data obtained through the research and analysis are conducive to the inheritance and improvement of the local patchwork skills. (Fig.3 and 4)

Fig.3 A Color professional performing color measurement (Photographed by Liu Qi)

Fig.4 Manuscripts of color measurement records (Photographed by Qi Liu)

3.3 Data Collation and Analysis

The color measurement data of the samples are collected and analyzed based on the coordinates of the NCS color ring and the NCS color triangle. In the NCS color ring, the color hues such as yellow, red, blue, green, black, white and others are indicated. In the NCS color triangle, the attributes of value and chroma are acquired. The overall color status of the samples is analyzed based on the color triangle and color ring.

4. The NCS Analysis of Measurement Samples of the Zhuang Ethnic Group

4.1 Selection of Color Measurement Samples

The Zhuang has many ethnic branches and different dialects. As we know, different ethnic groups differ in their languages and costumes. However, the language and costumes of an ethnic group interact with each other [16]. Different branches of the Zhuang vary greatly in their costumes' styles, patterns, colors, and craftsmanship. Some branches of the Zhuang have gradually lost their traditional costume features due to the impact of modernization, but most of the Zhuang people still maintain their traditional costume features in Western Guangxi, Northwest Guangxi, Southwest Guangxi, and Southeastern Yunnan. We selected five well-preserved baby carriers from the San-Jie Liu Culture Impression Museum as the color measurement samples, including two from Yunnan Province and three from Guangxi Province. They are different in style and are typical representatives of the high craftsmanship level of the Zhuang. (Table 1)

Table 1. List of Color Measurement Samples (Drawn by Qi Liu)

No.	Name	Time	Origin
Sample 1	Magatama baby carrier with the cloud dragon patterns	1940s	Xichou County, Yunnan Province
Sample 2	Patchwork embroidery baby carrier	1940s-50s	Northern Guangxi
Sample 3	Embroidery baby carrier	1930s	Nandan County, Guangxi Province
Sample 4	Applique and embroidery baby carrier	1920s	Nandan County, Guangxi Province
Sample 5	Applique and embroidery baby carrier (partial)	1950s	Qiubei County, Yunnan Province

Data source: Provided by the San-Jie Liu Culture Impression Museum

4.2 Analysis of Measurement Samples

4.2.1 Sample 1: The Magatama Baby Carrier with the Cloud Dragon Pattern

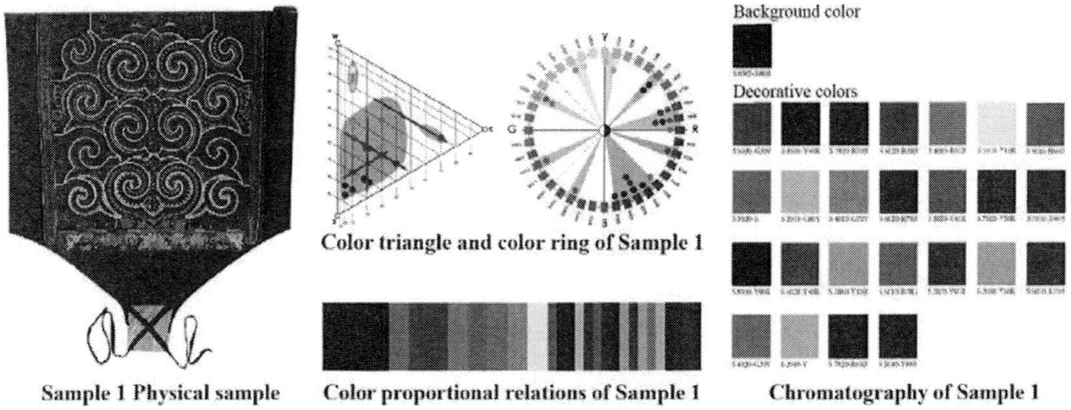

Fig.5 Color measurement results of Sample 1 (Photographed by Qi Liu and drawn by Shuang Zheng)

With black as the background color, the sample contains the factors of the colorful patchwork and embroidery. The dominant red and purple blue (Y80R-R90B) are supplemented by bluish green, yellowish green, and orange yellow in small areas. With various hues, similar colors, contrasting colors and complementary colors are combined to form a rich color contrast. A wide range of colors are used. Color harmony is achieved through area control. In addition, the convergence in blackness, chroma, and whiteness are used repeatedly to reflect the original aesthetic wisdom of the Zhuang. The color value of the sample lies mainly in the medium range, supplemented by colors with an appropriately high value and low value. The black with a low value is adopted to harmonize all the colors (Fig.5).

4.2.2 Sample 2: The Patchwork Embroidery Baby Carrier

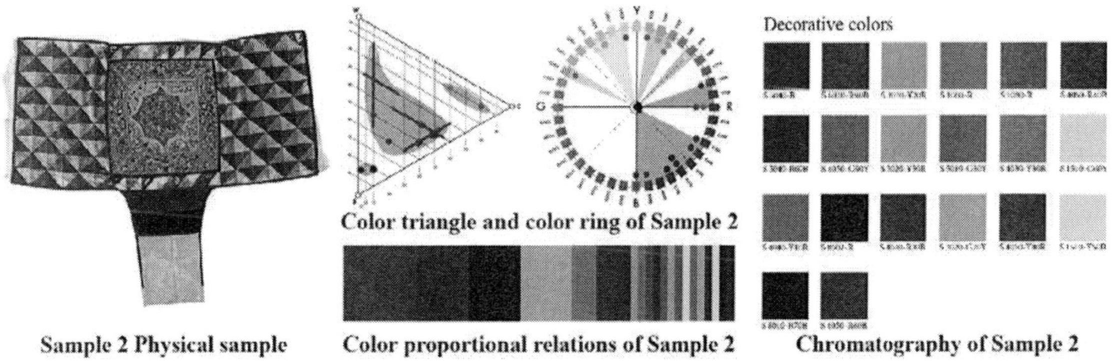

Fig.6 Color measurement results of Sample 2 (Photographed by Qi Liu and drawn by Shuang Zheng)

The sample is purple overall. The R30B-R70B purple occupies the most significant proportion and is supplemented by the light warm orange with a high value. The four basic colors of the sample present the obvious gradient changes in value and chroma. Due to the relatively strong color contrast of basic patterns, the sample looks distinct and vivid.

The embroidery part at the center of the sample is decorated with yellow, red, and blue with medium to high chroma. The delicate colors, and the convergence in blackness, chroma and whiteness makes the sample look so harmonious and beautiful (Fig.6).

4.2.3 Sample 3: The Patchwork Embroidery Baby Carrier

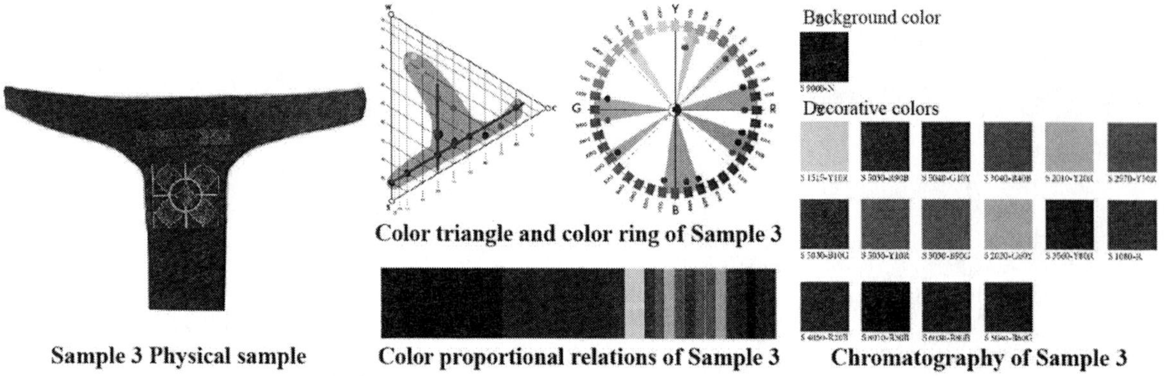

Fig.7 Color measurement results of Sample 3 (Photographed by Qi Liu and drawn by Shuang Zheng)

Black is the background color of the sample. The colors of the patchwork and embroidery parts are mainly red and blue (Y80R-R and R80B-B10G), decorated with cyan, green, yellowish green, and orange. Despite the complex colors, the sample with delicate patterns looks proportionable, harmonious and beautiful by virtue of the excellent convergence in whiteness and chroma and the harmonization of black as the background color. (Fig.7)

4.2.4 Sample 4: Applique and Embroidery Baby Carrier

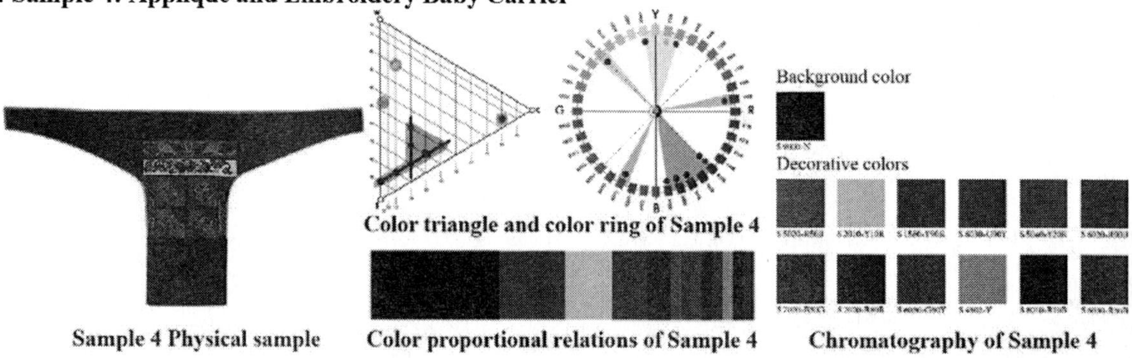

Fig.8 Color measurement results of Sample 4 (Photographed by Qi Liu and drawn by Shuang Zheng)

Black is the background color of the sample. The colors of the patchwork and embroidery parts are mainly red and blue (Y90R and R70B-R90B), decorated with yellow, purple, cyan and yellowish green. Despite the complex colors, the sample with delicate patterns looks proportionable, harmonious and beautiful by virtue of the excellent convergence in whiteness and chroma and the harmonization of black as the background color (Fig.8).

4.2.5 Sample 5: Applique and embroidery baby carrier (partial)

The sample looks very colorful due to many colors, mainly pink and black, two colors occupying the largest proportion in color. Some supplementary colors are adopted for decoration, including purple blue, cyan, and yellowish green, and yellow and orange for the embellishment. The colors are mostly medium and high chroma, and there are also many contrasting and complementary colors. However, the sample looks harmonious due to a small proportion of each contrasting and complementary color, and the convergence in blackness, chroma, whiteness, and Nuance (Fig.9).

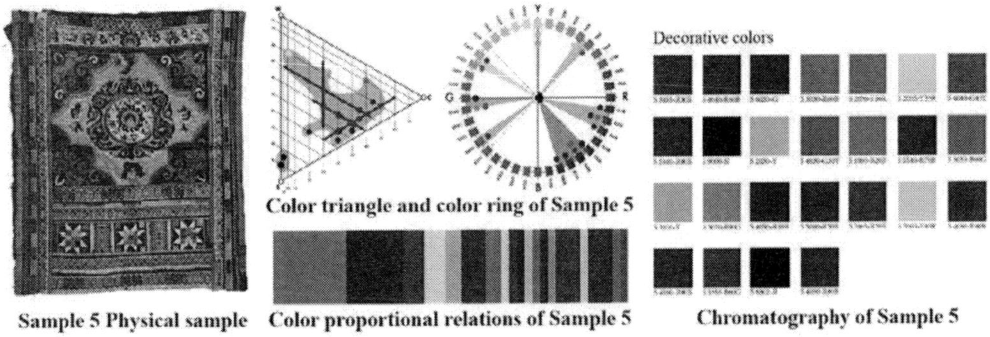

Fig.9 Color measurement results of Sample 5 (Photographed by Qi Liu and drawn by Shuang Zheng)

5. Conclusion

From the color measurement results above, we can see the following characteristics of the traditional patchwork baby carriers of the Zhuang.

1) Most patchwork baby carriers of the Zhuang have black as their background color. Black is the most popular color of the Zhuang people, which is generally reflected in the ethnic costumes of various ethnic branches of the Zhuang. As their background color, a large NCS S 9000-N black area harmonizes the complicated colors brought by the complex patchwork and embroidery in those patchwork baby carriers. Their colors are harmonious and beautiful overall.

2) Multiple colors are used for patchwork baby carriers. Patchwork technology means to use the remaining cloth again, so colors are not preplanned deliberately. The producers are good at selecting and arranging various colors, however, so the overall color effect is rich and gorgeous.

3) White lines are used to outline patchwork baby carriers so that their colors are not chaotic. In addition, many patchwork textiles have an obvious division of areas, and the edges of these areas are often embroidered with white. These white lines play the role in harmonizing colors.

4) As for hue, red, purple, and blue hues are often used within the hue range from red to blue. This is also related to the tradition of dyeing with vegetable dyes. Vegetable dyes, known as "dyeing grass" in ancient China, are used to color fibers by using pigments contained in wood, root, stem, leaf, bark, fruit, bud, and flowers of plants [17]. Colors in this range are easy to obtain, and no taboo is considered. Dark blue is mainly used in the range of R60B-R90B, 40-70 blackness, and 20-50 chroma and with a medium to high chroma and a medium to low value. Grape purple and aubergine are mainly used in the range of R40B-R60B, 30-50 blackness, and 30-50 chroma, and with a medium to high chroma and a medium to high value. For red series colors, pink is mainly used in the range of R10B-R20B, 10-50 blackness, 20-65 chroma, and with medium to high chroma and value; bright pink and rose-red in the range of R interval, 10-30 blackness, and 60-80 chroma, and with high chroma and value; and also, bright red in the range of Y80R-Y90R range, 15-50 blackness, 40-80 chroma, and with medium to high chroma and value.

5) Colors used for patchwork baby carriers generally have a higher chroma (brighter) and mainly a medium to high value. They form the bright and vivid visual effect of those samples. Their commonly used color matching method follows the principle of convergence in chroma, blackness, and whiteness. As a result, a lot of colors are adopted for the samples, but they are not chaotic, reflecting the primitive aesthetic tendency of the local people of the Zhuang.

6. Acknowledgement

We would like to express our gratefulness for the support received by the following programs:

1) 2018 Major Art Project under the National Social Science Fund of China: Research on Chinese Clothing Culture (Approval No.: 18ZD20)

2) Key Scientific Research Project of Beijing Institute of Fashion Technology: Research on Traditional Patchwork Skills and Color Application of Costumes of Ethnic Minorities in Southwest China" (Approval No.: 2017A-04).

3) Doctoral Training Project of Beijing Institute of Fashion Technology for National Special Needs: the Urgent Inheritance and Design Innovation of Traditional Chinese Clothing Culture.

References

[1] The Office of the State Council for Population Census, P.R.C., Department of Population and Employment Statistics of National Bureau of Statistics, P.R.C. (ed.). China Report on 2010 Census (First Half Part). Beijing: China Statistics Press: 2012.
[2] Xu W, Liu Q (ed.). The Art Exhibition of Patchwork. Beijing: China Textile & Apparel Press: 2011.
[3] Liu Q. Classification and Orientation of Patchwork——Best Handicrafts of China. Fashion Color: 2020; 08.
[4] Jin YS. Mother's Fragrance: Patchwork Art of Jin Yuanshan. Beijing: China Textile & Apparel Press: 2019.
[5] Zhou QF (the Song Dynasty), Yang WQ (proofread). Annotations to Ling Wai Dai Da. Beijing: Zhonghua Book Company: 1999.
[6] Li XL (the Qing Dynasty). Chronicles of Guangnan County.
[7] Wei L. Design of Ethnic Minority Women's Wear. Beijing: China Minzu University Press: 2014.
[8] Zhu JY, Li JQ. Color Languages of Ethnic Minorities-From Totem Symbols to Social Symbols. Kunming: Yunnan People's Publishing House: 1993.
[9] Zhang YF (EIC), Chinese National Museum of Ethnology (ed.). Research on Chinese Ethnic Costumes. Beijing: The Ethnic Publishing House: 2003.
[10] Dai P. Discussion about Colors of Chinese Ethnic Costumes. Theatre Arts (Journal of Shanghai Theatre Academy): 2000; 1.
[11] Information on http://ncsColor.com/
[12] Hård A, Sivik L (SWE). NCS—natural color system: a Swedish standard for color notation. Color Res Appl: 1981; 6.
[13] Information on https://ncscolour.com/ ncs/
[14] Wang XW, Fan H (EIC), Li D, Chen M (AE). Principles of Color Design. Beijing: Beijing Institute of Technology Press: 2014.
[15] Information on https://ncscolour.com/ncs/
[16] Li YJ (EIC), Hu DZ (written), Liang HC and Hu QS (PHO). Beautiful Costumes of the Zhuang. Nanning: Jieli Publishing House: 2012.
[17] Wang YP. Return to Nature. Beijing: China Textile & Apparel Press: 2019.

Cite as: L.H. Ren et al.,
TBIS 10.3993/tbis (2021).

Aesthetic Culture and Art Exploration of Chinese Kunqu Opera Costumes

Li-Hong Ren[1], Dong-Sheng Chen [2], Jia Lyu[2*]

China Changshu Institute of Technology, Changshu, Jiangsu, 215500;
Engineering Research Center of Textile and Clothing (Minjiang University), Fujian Province University, 350108

[*]Corresponding author's email: 61012023@qq.com

Abstract

Kunqu Opera is known as the ancestor of all operas. On May 18th 2001, it was declared as the representative work of human oral and intangible cultural heritage by UNESCO, emphasizing its importance in the world. Kunqu costumes is the most important external representation of Kunqu Opera, which highlights its historical and cultural connotation. Kunqu costumes are the most common costumes among costumes for other dramas in China, and are especially known for their profound and distinct artistic characteristics and strict program specifications. These artistic characteristics and styles are developed in the long-term development process. To protect the treasures passed down from our ancestors, this paper analyzes and explores the aesthetic culture of Kunqu costumes from various aspects, including the formation of artistic characteristics, stylization, artistry and aesthetics of Kunqu costumes.

Keyword: Kunqu Opera Costumes; Aesthetic Culture; Art Exploration

Kunqu opera costumes, also known as "costumes"[1], are an important means of shaping the modeling characteristics and external images of opera characters. Kunqu costumes are mainly divided into five series, namely python, robe, rely on, pleat, and clothing [2]. These costumes cover almost all kinds of opera characters across different occasions and regions. Kunqu costumes are elegant, exquisitely made and aesthetic, which contain the unique spiritual values and way of thinking of China. the dress rules of Kunqu opera have achieved the perfect role shaping of Kunqu Opera Art and are an important symbol of the inheritance of national spirit and the continuation of context. This paper explores the artistic and aesthetic characteristics and culture of Kunqu costumes from the following aspects.

1. The Formation of Artistic Characteristics of Kunqu Opera Costumes and the Origin of Jiangnan Culture in the Ming and Qing Dynasties

During the historical evolution of Kunqu Opera, it has been known as " Kunshan tune", "Kunqu Opera", "Kundiao", "Nanqu opera" and "Yabu". These names carry a long history and rich humanistic spirit [3]. The Kunqu Opera era began at the turn of Longqing and Wanli in the Ming Dynasty, and lasted till the early years of Jiaqing in the Qing Dynasty. Though short-lived, there were numerous works of high quality created in this period.Kunqu Opera made great achievements in this period. It can be said that Kunqu Opera is the most prosperous drama in the history of Chinese classical opera.

As the most important representation of Kunqu Opera, "Kunqu costumes" has a long history, which demonstrates the Chinese's admiration of Kunqu Opera. These costumes carry the humanistic spirit of the Ming and Qing Dynasties, and remain a bright pearl in the history of Chinese art and culture. The term "Kunqu costumes" covers more than just the garment, and also includes accessories, hats and headgear. In the dictionary of Chinese Kunqu Opera, "costume" is clearly defined, which includes costumes, beards, boots, helmets, headgear and facial makeup [4]. During the Ming and Qing Dynasties, Kunqu costumes gradually formed what is known today as the suitcase system. The suitcase system was founded in Yuan Dynasty, developed in Ming Dynasty and matured in Qing Dynasty. It divides the costumes into five categoriess: overcoat box, second suitcase, third suitcase, helmet box and flag box. Together, they form the "five suitcases" [5].

Kunqu Opera originated in the Kunshan area at the end of the Yuan Dynasty. Therefore, there are reports of influence from the Jiangnan culture and folk customs on the Kunqu Opera suitcases. Kunqu Opera costumes exhibit Jiangnan taste and aesthetics explicitly. There were a large number of talents and prosperous culture in Suzhou during the Ming and Qing dynasties. Due to economic development, urban prosperity and the growing power of the new class in Suzhou, literati usually regard Kunqu opera as the best way to express their feelings. They added their own cultural perception to Kunqu Opera and costumes [6]. For one, they embedded spiritual culture into Kunqu costumes; for another, they integrated into the elegant culture and aesthetic style of Jiangnan literati. Kunqu Opera costumes are deeply marked with the brand of Jiangnan literati culture, with unique artistic features and cultural connotations. Kunqu Opera is elegant, which requires its clothing to adapt to it, integrate into the scene, and reflect the elegant artistic style. They also need to accurately convey the emotions of the characters and realize the combination of Kunqu Opera and stage practice. For instance, in Kunqu Opera The Peony Pavilion (Fig.1), the character Du Liniang is dressed in a pink embroidered cape with a moonlight colored goffer. Her singing and dancing display the emotional process during and after the dream, reflecting the character's state of mind and charm. The pink Ruyi embroidered cape is decorated with plum blossom, orchid, bamboo, chrysanthemum and other ornaments, showcasing the character's inner purity. Clothing design embodies the combination of internal and external beauty in art philosophy, and eventually enables the costume to reflect the plot, and in the meantime, embodies the aesthetic taste of the literati in the south of Yangtze River [7].

Fig.1 Kunqu Opera the Peony Pavilion Du Liniang Dressed in Pink Robe
(the picture is taken from the stills of Peony Pavilion)

The pursuit of simplicity and elegance of the literati in the south of Yangtze River is reflected in the clothing of Kunqu Opera costumes. Kunqu opera costumes are constantly improved and innovated over time, but the basic and inherent cultural characteristics of traditional Kunqu opera suitcases remain unchanged. The Kunqu opera suitcase system is becoming increasingly substantial, which can be called a very precious national cultural heritage in the world.

2. The Aesthetic Principle of Freehand Brushwork of Kunqu Costumes and the Virtual and Real Expression of Procedural Characteristics

The design of Kunqu opera costume is closely related to the dress system in ancient China. Not only does it retain the basic shape and scope of coronal clothing, but it also inherits the tradition of Chinese clothing in pursuing beauty [8]. Freehand brushwork is a commonly used form of artistic expression in Kunqu Opera costumes design. For example, Kunqu Opera costumes originated from python clothes of the Ming Dynasty. "Python clothes", also known as "elephant dragon clothes", is originally a special gift from the emperor to aristocratic eunuchs and meritorious ministers, which do not belong to the series of official clothes.

The embroidered robe is the same as the emperor's dragon robe, except the dragon in the embroidered robe is missing one claw. The difference between the python and the dragon lies in the python's four claws (Fig.2) and the dragon's five claws (Fig.3) [9]. The "python robes" worn by military generals and civil officials in Kunqu Opera costumes are all consistent with the actual python clothes in the Ming Dynasty, with a few exceptions. For example, the collar is changed into a round collar, and the python embroidered on the python robe has four claws instead of five. First of all, this form of expression is restricted by the feudal hierarchical social system; second, under the influence of the form of "freehand brushwork", python clothing is used to represent the dragon robe, achieving the artistic effect that carries the spirit but not the shape of the dragon robe. The beauty lies between the resemblance and the unlikeness, which is a representative form of imaginative clothing.

Fig.2 Python Robe (Four claws of boa robe and partial enlargement)
(photos taken from Nengfu Huang, Yuanjuan Chen, China Dragon Robe)

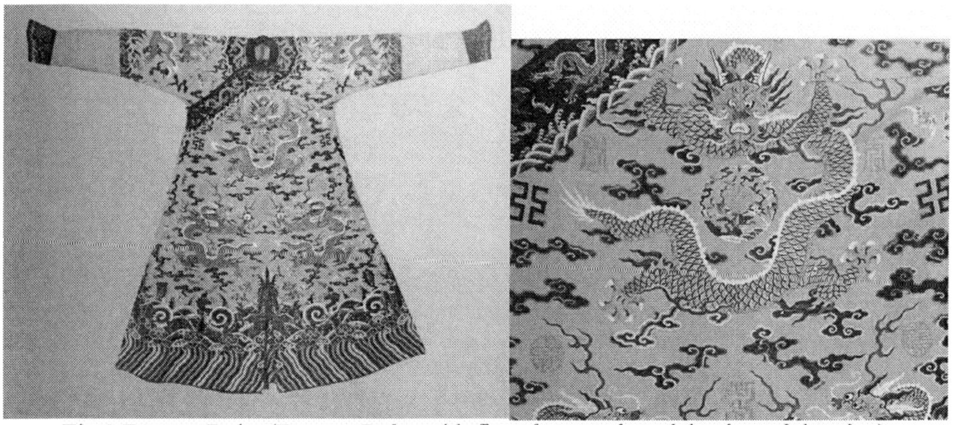

Fig.3 Dragon Robe (Dragon Robe with five claws and partial enlarged drawing)
(photos taken from Nengfu Huang, Yuanjuan Chen, China Dragon Robe)

Kunqu Opera costumes, like Kunqu Opera, have been passed down from generation to generation, and have experienced years of perfection. Traditional Kunqu Opera costumes advocate fine carving in the aspects of fabrics, patterns, colors, embroidery and craftsmanship. Kunqu Opera costumes are diverse in variety and exquisite craftsmanship, creating colorful and hierarchical clothing, forming a complete and meticulous system in the dress code [10].

There is a Kunqu opera jargon called "better worn than wrong" [11], which reflects not only the actors' serious attitude towards the costume regulation, but also embodies the formulaic form of Kunqu opera costumes. The stylization of Kunqu Opera costumes is shown in several aspects, such as style, color and pattern, which requires actors to obey the rules of Kunqu opera costumes to bring out the essence of the characters in the play.

For example, the stylized use of color in Kunqu costumes reflects the hierarchy of China opera characters. Yellow (true yellow or bright yellow) can only be worn by emperors, which is reflected in the bright yellow Python Robe (Fig.4) and robe (Fig.5) worn by Emperor Tang Ming in Kunqu Opera. Red is also the more precious color, followed by green and blue, and black is the humblest color. Color expresses folk customs as well as the occasion: red is generally used as an auspicious color, often used in festive scenes. Generally, poor students dress in green pleats; housekeepers and the servants in dark blue; poor women in plain colors or in black. Colors express spirit and temperament; for example, young characters on the stage usually wear brighter colors. Mei Lanfang once said, "Du Liniang in A Dream in the Garden is a young girl from a wealthy family, and her clothes should be beautiful. And she is also a talented woman. 'In the midst of beauty, the color is light and elegant' [12]. The color of the costumes and the collocation of each color are the formulaic artistic expression of Kunqu Opera costumes, which is also the rule that actors or characters follow when playing on stage [13].

Fig.4 The Python Robe Worn by Emperor Tang Minghuang in Kunqu Opera The Hall of Everlasting Life (Picture taken from the stills of Kunqu Opera " The Hall of Everlasting Life ")
Fig.5 The robe by Emperor Tang Minghuang in Kunqu Opera The Hall of Everlasting Life (Picture taken from the stills of Kunqu Opera " The Hall of Everlasting Life ")

Kunqu Opera costumes are perfectly integrated with its singing and tunes, reaching a higher state of beauty. Kunqu Opera costumes inherited the traditional aesthetic ideas and pursue the aesthetic realm of "coexistence of virtual and reality". Under the aesthetic principle of "freehand brushwork", they pursue likeness in spirit rather than in shape. However, in terms of institutional requirements, they are very strict. For example, putting Zhang's hat on Li's head in the performance of Kunqu Opera causes a ridicule and ruins the play. The serious attitude of actors towards costume regulation forms an important foundation for the development of Kunqu Opera art. Costumes of Kunqu Opera have absorbed many excellent elements in its long history of development, in which the regulations were also formed throughout the many plays and rehearsals. The beauty of these costumes remains on the stage of Kunqu Opera, like a dream for hundreds of years.

3. The Aesthetic Characteristics Contained in the "Exaggerated" Art Form of Kunqu Opera Costumes

Kunqu Opera costumes follow the principles of stage beauty and create a basic formula, while integrating Chinese traditional philosophy, Jiangnan literati thoughts and profound cultural connotations, creating an extreme and gorgeous sense of beauty. Kunqu Opera costumes are a re-creation based on costumes from the Ming Dynasty. Different from life costume, Kunqu Opera costumes pursue stage effect, which requires it to be more decorative, colorful and prominent than life costume. The artistic technique of "exaggeration" is often used in the design of Kunqu costumes, which brings out the personality of the actors, creating a stage that is richer, more gorgeous and charming [14].

Among all Kunqu Opera costumes, nothing embodies the brightest and boldest artistic features of exaggeration and deformation than the costume Kao. Kao, formerly known as "armor-binding", is commonly used by military generals in Kunqu Opera performances. From top to bottom, it is composed of a tiger's head and shoulder, sleeves, breastplate, belly, hanging fish and Shang Jia. The attachment is composed of Kao flag, Kao collar and Kao leg. Kao on the stage reflects the strongest exaggeration, deformation performance of Kunqu Opera costumes.

For example, the design of "Kao belly" uses measurements that are the opposite of the regular shape of the human body. Instead of a wide shoulder and a thin waist, the "Kao belly" is relatively fat and full. The exaggerated waist means "a fat man with a round waist" and "strong as a bear in the hips and with a back supple as a tiger", which changes the actor's figure and implies the generals' bravery and burly spirit [15].

Fig.6 Kao's Static Display

Fig.7 Kao's Dynamic Display

Fig.8 Dynamic Display of Kao Flag on the Costume
(Fig.6-8 pictures taken from the Internet)

Kao is the upper coat joined with the lower coat (the armors are connected), Similar to ancient "deep clothes". The front and the rear of Kao is separated; like clothes but not exactly the same as clothes, like armor but not the real armor. This does not exist in the ancient military clothing, but the design of the upper and lower parts shows the overall beauty of Kao properly (Fig.6). The armpit of Kao is decorated with a "bird's nest" (armpit), which has the function of connecting the front and the back of the body; like link but not the actual link. The separation of the front from the back of Kao is possible to allow free movement of the actors' legs, doing great dance movements such as kicking and pedaling. This kind of "separated" Kao, with the actors' fierce martial arts movements spinning and flying, create a fierce fighting atmosphere on the stage. It shows the heroism of generals and the dynamic beauty of costumes (Fig.7). Kao flag on the martial arts costume (Fig.8) is not only colorful and dazzling in the character's martial arts performance. The mood of the characters can also be reflected by the swing of the Kao flag, which has a strong artistic and aesthetic function. The design of "Kao" on the stage of Kunqu Opera completely gets rid of the original form of life and carries out extreme exaggeration and deformation. This separate style of clothing, the static component, gives the characters majestic spirit and embodies the adornment of Kunqu Opera costumes. On the other hand, the dynamic component facilitates exaggerated dance movements, which separates the characters, reflecting the danceability of Kunqu Opera costumes [15]. Through the exaggerated artistic means of Kunqu Opera costumes, it brings out the shape of the characters, which appear to be more distinct and stronger, leaving a deep impression on the audience.

4. Conclusion

Kunqu costume brings together the quintessence of our national traditional culture. It is the wisdom and crystallization of many generations of artists, as well as the sedimentation and accumulation of art, forming the unique artistic style and profound cultural connotations of Kunqu costume. It can be said that Kunqu costume is a bright pearl in the treasure house of traditional Chinese culture and art. Because of its symbolic and stylized aesthetic characteristics, it has become the most complete continuation and embodiment of ancient Chinese costume culture in today's society.

5. Acknowledgement

The authors wish to acknowledge their funding sources: Engineering Research Center of Textile and Clothing (Minjiang University), Fujian Province University (Project number: MJFZ17109) and Art Science Project of National Social Science Foundation of China (2020bh00911)

References

[1] Chen S, Li RS. Traditional Chinese Costume. Beijing: People's Art Press: 2006.
[2] Liu YM. China Kunqu Suitcase [M]. Shanghai: Shanghai Dictionary Press: 2010.
[3] Su Z. Municipal Bureau of Culture (Suzhou Opera Records [M]. Suzhou: Guwuxuan Press: 1998.
[4] Wu XL. The Dictionary of Kunju in China [M]. Nanjing: Nanjing University Press: 2002.
[5] Tan YJ. Costume Design of Opera [M]. Beijing: Beijing Cultural and Art Press: 2000.
[6] Zhu L. Kunqu and the Social Life in the Jiangnan in the near Future - An Investigation of the Audience Group of Kunqu Music [D]. Suzhou University: 2006.
[7] Bai XY. Colorful Peony Pavilion: A Dream of Four Hundred Years of Youth [M]. Guilin: Guangxi Normal University Press: 2004.
[8] Zhu HF. On Kunqu Xingtou and Wearing Principles [J]. Sichuan Drama: April 2019.
[9] Liu RP, Chen JJ. A Study of the Structure of Chinese National Dress [M]. Beijing: China Textile Press: 2013.
[10] Zhou XB. The History of Ancient Chinese Dress [M]. Beijing: China Drama Press: 2002.
[11] Zeng CS. Kunju Wear [M]. China: Compiled by Suzhou Municipal Bureau of Culture, Radio and Television: 2005
[12] Mei LF. Xu Ji's Biography of 40 Years of Stage Life, [M]: 1952, Pingming press.
[13] Su WH. Analysis of the Secularization of Kunqu Opera Clothes Based on the Youth Version of Peony Pavilion Costume [J]. Clothing Guide, August: 2020.
[14] Sun LH. On the Basic Characteristics of Traditional Chinese Opera Dress [J]. Opera Art: 1991; 7.
[15] Chen J. Study on the Function and Aesthetic Characteristics of "Relying on" of the Costume [J].Textile Industry and Technology: August 2013.

Innovative Design Study of New Chinese Style Sportswear

Jun-Shan Xu[1], Zhao-Qing Li[2*]

[1]*Xiamen Academy of Arts and Design, JiMei Street, Xiamen, FuJian, 361000, China*
[2] *Xiamen Academy of Arts and Design, JiMei Street, Xiamen, FuJian, 361000, China*

*Corresponding author's email: 961226603@qq.com

Abstract

This paper is a research and practice on the innovative design of New Chinese style in sportswear. Through research of New Chinese style clothing, we explore the application characteristics of New Chinese style in sportswear from various aspects such as style, color, and pattern. Using the new perspective of sportswear design, we extract and reorganize the New Chinese style and redesign it with the contemporary characteristics of sportswear, aiming to give a new spiritual connotation to sportswear, try to solve the integration of traditional culture and modern trends and provide development ideas for the domestic clothing design with the elements of New Chinese style.

Keywords: New Chinese Style; Sportswear; Traditional Culture; Innovative Design

The New Chinese style is not a replica of the traditional Chinese style in the complete sense. It is the extraction of the Chinese style characteristics, restating its cultural background, digging deeper into its cultural heritage as well as using modern design techniques to express the pursuit of a unique oriental civilization and spiritual connotation. The New Chinese style was born in the new period of the revitalization of Chinese traditional culture, with the rebound of China's international status, Chinese culture has been widely spread around the world, the cultural confidence of the people has also been enhanced, people's national consciousness gradually increased, and began to change from " replicate " to " create [7]. The emergence of the New Chinese style is not only an inheritance of the aesthetic elements of the five thousand years of Chinese civilization, it is also has a high cultural connotation and aesthetic value, which can reflect the current aesthetic trend of consumers. It is also a reflection of the temperament of a great nation [8]. Nowadays, the design research of New Chinese style domestically and internationally is mostly combined with dresses, casual and formal wear, and other categories. Sportswear is one of the main categories of modern clothing design. In the general environment exercise consciousness, people's demand for sportswear design are growing. In the current clothing design environment, only a few combines the New Chinese style with sportswear. This paper combines the New Chinese style with modern sportswear, taking traditional culture as the background and sportswear as the carrier; the two collide with each other, fitting the clothing trend and designing sportswear with both traditional cultural connotation and the spirit of the times [2].

1. Overview of the New Chinese Style Clothing

The emergence of the New Chinese style is closely related to the lengthy traditional Chinese culture, to dig deeper and explore the long history of Chinese clothing, to use a more fashionable expression, combined with the carrier of sportswear, to inherit and recreate.

1.1 From the Perspective of the Evolution History

The historical background of the use of new Chinese style in clothing can be traced back to the traditional "Chinese clothing." As an important part of the clothing category, "Chinese clothing" first appeared in opposition to "Western clothing," referring to the traditional Chinese style of clothing categories. During the Republic of China, the emergence of both the Zhongshan suit and the modified Cheongsam was a typical "combination of the East and the

West," a collision of traditional elements and current fashion. The former was designed based of military uniforms, incorporating the characteristics of Chinese clothing and suits, making it easier and more elegant, and giving its clothing style a unique political meaning [14]; the latter was a vertical, loose Cheongsam that had been worn for nearly three hundred years during the Qing Dynasty, adopting the waist-saving and chest-saving concepts of Western-style clothing, breaking the Cheongsam's plate-like style and tightening the waist to highlight the beauty of a woman's body curves [5]. Both of them can become the representative of "national costume" in the Republic of China period; both styles cannot be separated from the use of traditional Chinses costumes or traditional patterns at the same time, combined with the current fashion to recreate, rather than to replicate [6]. Looking back at the process of modern national dressing evolution, whether it is learning the dressing style of the Soviet Union in the 1950s and 1960s or the flared trousers and fitness pants that became popular in the 1980s and 1990s, these purely "imported products" lack the characteristics of the beauty of the five thousand years of the dress. With the increasing international influence of China, both domestic and international designers have made Chinese style inherited and evolved; a new style symbol has gradually formed - New Chinese Style, which inherits the excellent traditional culture of the Chinese nation and also caters to the current aesthetic trend, and appears as a new style symbol, in public's vision. [10]

1.2 From the Perspective of International Development

As for the international development of the New Chinese style, as early as the 1970s and 1980s, Chinese elements were already fancy by Western fashion designers' creativity. They mostly used Cheongsam, the coiled buckle, as a visual symbol to represent the oriental elements. In Christian Dior's 1997 fashion show (Fig.1), John Galliano used the typical Chinese red as the main color with oriental style - Cheongsam to express the whole idea of series design. In Prada's Spring/Summer 2017 collection, designer Misia Prada incorporated the classic slant collar and stand-up collar details into the design of Cheongsam, combined with the modern plaid prints, so that the entire collection has a modern feeling while retaining a unique Chinese flavor. Gucci, a luxury clothing brand famous for its pattern design, has used oriental elements in its designs in recent years, showing his obsession and love for traditional Chinese culture. The development of the Chinese style has been promoted. The collision between the two allows more consumers to understand Chinese culture, appreciate the beauty of Chinese style and give unique artistic value. [9]

Fig.1 Christian Dior's 1997 fashion show (From http:// www.vogue.com)

2. The Current Appreciation of Sportswear

With the improvement of awareness of public health and the change of lifestyle, the increasing improvement of people's living standards, China has set off a trend of national sports. This trend is a benefit for the development of sportswear. Nowadays, sportswear is the leader of the garment market, a large part of sports brands in the innovative design is not in place, facing problems such as single style, lack of spiritual connotation. Young consumers have a strong demand for innovative design of sportswear, especially the age group of young men between 18-30 years old, they have independent consuming power, also have a pursuit of fashion, with "seeking beauty, seeking name, seeking new" and other characteristics [1]. This consumer group hopes that sportswear is not only simple to meet the daily sports function but also has strong uniqueness and aesthetics, and a certain spiritual and cultural connotation. The New Chinese style men's sportswear design combines the traditional Chinese straight and loose fit with modern plate making based on people's modern lifestyle and aesthetic trends. The traditional

Chinese elements are summarized, integrated, and recreated, and applied to the sportswear design with a new element so that the clothing in terms of shape, color, and pattern are in line with the basic attributes of sportswear requirements, to create a series with Chinese elements design. [11]

3. The Embodiment of New Chinese Style in Clothing Design

Applying the New Chinese style in the design of clothing is not a pure restatement of traditional Chinese clothing but focuses on the refinement of traditional Chinese cultural elements, blending traditional culture with modern trends to collide with the products of the times.

3.1 Clothing Design Concept of New Chinese Style

The essence of apparel design is to serve consumers, and whether it is to do practical design or visual design, it should cater well to or guide the market [1]. The new Chinese design style is a design style that inherits and carries forward the traditional Chinese culture, grasps the connotation of Chinese style, refines Chinese design elements, and innovates with current materials and technology. The new Chinese style is an improved design based on traditional Chinese clothing, which usually has a clear and simple expression, adhering to a particular visual symbol of traditional Chinese clothing and drawing on the shape of modern clothing to give it a cultural connotation. The New Chinese style discards the impractical design points of traditional Chinese clothing, such as complicated handicraft embroidery and poor wearability, and satisfies modern people's usage needs and aesthetic habits. Therefore, the New Chinese style can be said to be a new improvement and optimization of traditional aesthetics by using modern clothing design concepts, which is good for promoting traditional Chinese culture and is of great significance for enhancing the international influence of Chinese culture.

3.2 The Current Situation of the Application of New Chinese Style in Garment Design

More and more Chinese designers have entered the international arena, and the New Chinese style with Chinese characteristics has become synonymous with many domestic and foreign brands in recent years. More and more brands and industry experts believe that "national attributes are a necessary highlight for the international development of Chinese clothing brands."

Fig.2 Li Ning autumn and winter 2018 series (From http:// www. lining.com)

Fig.3 Gaia legend spring and summer 2019 series (From https://www.ellechina.com)

Fig.4 MUKZIN autumn and winter 2017 series (From https://www.ellechina.com)

In 2008, Tmall International cooperated with New York Fashion Week to introduce "CHINA DAY." Many Chinese brands appeared in New York, among which Chinese sports brand Li Ning appeared in the New York Fashion Week Fall/Winter 2018 show. As one of the first Chinese sports brands to appear at the four major fashion weeks, China Li Ning took the theme of "Enlightenment" (Fig.2), combining the traditional Chinese virtues of "self-reflection, self-enlightenment, and self-creation" and inspired by the Chinese culture of "unity of heaven and man", integrating Chinese traditional culture and retro trends into its apparel design and giving it a unique spiritual connotation. The famous domestic New Chinese style brand Gaia Legend has appeared in major fashion weeks with traditional Chinese elements as its design inspiration and expression. The brand's Spring/Summer 2019 collection (Fig.3), with the theme of "painting the wall, a thousand years," combines traditional Chinese elements such as Bodhisattva and flying sky in Dunhuang frescoes with traditional handicrafts such as Su embroidery and woof as forms of expression, combined with a certain modern sense of dress silhouette, to elevate the beauty of the East from meek and gentle to majestic and fairy-like. In addition, MUKZIN, a "hip Chinese style" brand established less than six years ago (Fig.4), deconstructs the visual symbols of traditional Chinese culture with modern designs that are not clichéd and blends a new dimension of New Chinese style symbols by using clothing as a carrier. The brand designer has the experience of studying in the West and is deeply influenced by Western aesthetics and design concepts but is also well versed in the essence of traditional Chinese aesthetics. From the mythological stories in the Shanhaijing to the Guinan Tibetan embroidery, a non-heritage handicraft, are all applied as visual symbols in the pattern design by MUKZIN, showing a professional sense of collision between traditional culture and modern trends. When the national trend style is prevalent, New Chinese clothing with a unique sense of responsibility and mission has emerged repeatedly in the large market of clothing design. Through the analysis of the existing New Chinese clothing brands on the market, it is found that the clothing containing the New Chinese style is mostly dresses and casual clothing for women, formal clothing for men, and only a few sports and casual clothing designs with the New Chinese style.

4. Design Path of Sportswear in New Chinese Style

Except for a very few sports brands such as Li Ning, most of the sports brands are still decorated with simple and random patterns, and their designs lack artistic and cultural connotations. In the context of the prevalence of the "national trend," it has become necessary to integrate Chinese traditional culture in sportswear design and give it up-to-date characteristics. In the design practice, we should pursue the "new" while reflecting the "Chinese," follow the trend of the times, and better adapt to the shopping psychology of the contemporary consumer mainstream of seeking new and not the same. The series design will extract the elements of the New Chinese style and carry out an innovative redesign with sportswear as the carrier.

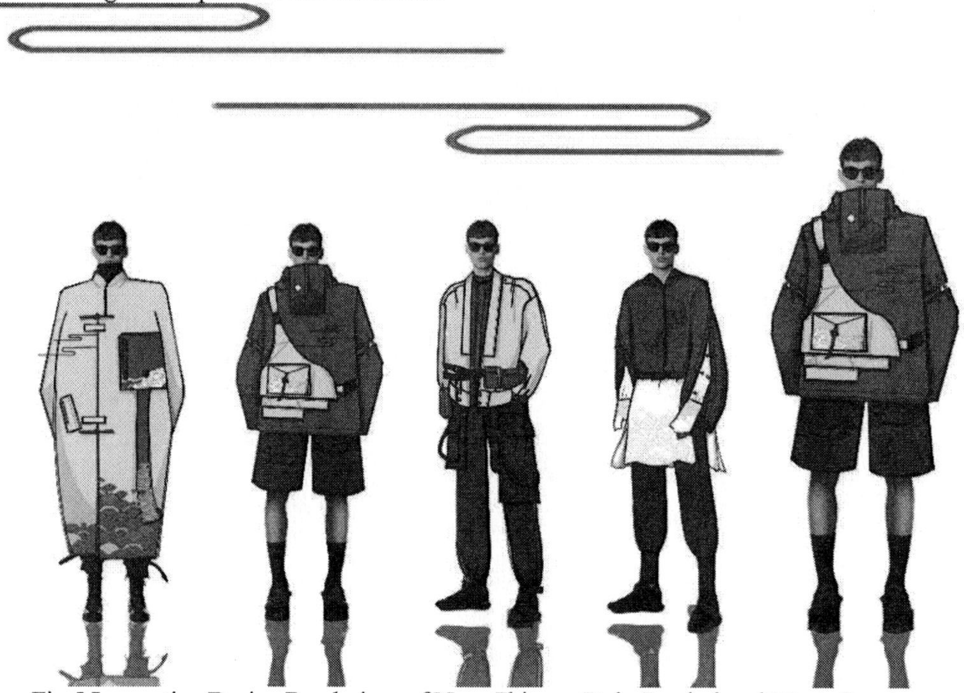

Fig.5 Innovative Design Renderings of New Chinese Style Symbol and Men's Sportswear

4.1 Inheritance and Improvements in Style Design

As one of the three elements of clothing design, style design can directly influence the style positioning of clothing. There are many traditional Chinese clothing shapes, from ancient robes, lapels, the Han Chinese dress to the modern flag dress, horse face dress, and even the Cheongsam formed by the influence of Western thinking, all in line with the characteristics of Chinese national dress. The way it is styled mainly adopts loose and linear cutting method, flat structure processing, which seems simple but contains the wisdom of Chinese philosophy.

The overall silhouette of the series is mainly loose and casual, in line with the basic needs of sports. The top shape is mainly an "H" shape, similar to the effect presented by the straight-line cutting method, which has strong wearability and can very well modify the body line. The bottom style structure is relatively simple, drawing on the traditional clothing design of the mouth, both to meet the current aesthetic and in line with the style of sportswear positioning. In the style's details, the common elements of traditional Chinese clothing, such as coiled buckle, lapel, and slant lapel, are modified and reused. The design is based on the features of the lapel style and the ancient waistband, and the functional backpack is used as a form of phenotype, which enriches the clothing category and highlights the currently popular concept of overlapping.

4.2 Refinement and Differentiation in Color Design

The color design of clothing is the most intuitive design, which can distinctly and strongly give a person's vision to the "first impression." [4] The traditional Chinese culture has given birth to unique Chinese colors, such as the mysterious Dunhuang colors, colors that give character to theatrical faces, and the majestic Forbidden City, which are widely used in all major categories of modern clothing design. Nowadays, the status level affects the color scheme no longer exists, and the traditional color symbolism is increasingly forgotten, leaving people with a wide variety of color schemes.

As the most classic and high-profile color in traditional Chinese clothing, red is the most typical Chinese traditional color, and the high saturation of Chinese red can also enhance the eye-catching degree of the garment and meet the psychological demand of the mainstream consumers. Black has a dual cultural meaning in traditional color culture, with mixed reviews. At the same time, in contemporary clothing design, black represents understated mystery and cold, and then uses different shades of gray to harmonize, increasing the overall harmony and fashion sense of clothing [3]. The color design of modern sportswear is not just a direct reinterpretation of the traditional Chinese color scheme in contemporary times but should be combined with the characteristics of the times, in line with modern trends, to make a very New Chinese style of modern sportswear design.

4.3 Inheritance and Continuity in Pattern Design

The pattern design of the garment is the embodiment of the details of the garment. A good pattern design can enrich the garment, just like the finishing touch. At the same time, as the most direct expression of pattern design, the pattern can be the fastest way to convey design elements as visual symbols. There are many kinds of traditional Chinese patterns, among which the most modern is the landscape pattern, which is similar to the geometric pattern, mostly abstract graphics, including auspicious clouds, and water patterns.

The series design uses Xiang Yun pattern and Shui Bo pattern, among which Xiang Yun pattern originates from Shang and Zhou Dynasty, the most representative design of Chinese style elements. It is reflected in many ancient bronze vessels, lacquerware, and clothing, deconstructing them and treating them asymmetrically. The Shui Bo pattern is reconstructed and redesigned, re-coloring some color blocks and changing the arrangement to be more fashionable. New Chinese style symbols of sportswear in the design of the pattern, to find both innovative points, rather than copy the traditional pattern, learn to use deconstruction, reconstruction, and other methods of pattern innovation and redesign, while maintaining the traditional beauty should be the greatest extent to meet the modern trend of aesthetics.

4.4 Innovation and Transformation in Design Process Presentation

In the form of traditional clothing craft expression, mostly embroidery, dyeing, weaving, and other handicrafts, the process is exquisite and complicated, difficult to learn, and even more difficult to refine. Nowadays, the garment market is dominated by industrial production, and most of the common processes are digital printing, which can quickly achieve the ideal design effect at a lower cost.

The design of the series combines traditional embroidery, and modern digital printing, the collision of tradition and modernity, the use of both in the design of men's sportswear with New Chinese style symbols can give a unique value to the series design, making the series design more traditional beauty and characteristics of the times. In the details of the presentation process, the designer also tried to use the plate buckle to do the splicing design, the plate buckle and modern work buckle with the use of each other, so that the clothing more design sense at the same time also enrich the clothing level. The target of modern sportswear is young people with certain aesthetic abilities, so the design process should not be limited by the predecessors, and they should try to integrate various ways to explore infinite possibilities.

5. Conclusions

Under the trend of "traditional culture,", domestic and foreign brands have been trying to design New Chinese style symbols, but the innovation in sportswear design is not prominent enough, and there are few representative brands. Even if a few brands have designed series, there are problems such as single expression, coincidental ideas and repetition. Therefore, in the design of sportswear in the New Chinese style, how to do popular design is a problem that every designer should think about, try to explore new angles, find innovative points of design elements, start from various aspects such as style, color, pattern, and technology, broaden the design ideas and enhance the spiritual connotation of the series design. [13]

Modern clothing design must be a design with the background and soul of the times and cannot pursue the superficiality of losing the inner spirit of formality. The New Chinese style sportswear design is not a simple duplicate of a symbolic element of traditional culture, nor is it simple splicing of Western elements, but should support the design itself with a rich cultural connotation and the spirit of the times. Thus, it can better support modern sportswear design.

Regarding the innovation and design of traditional culture, we cannot talk about inheritance in isolation, and we cannot lose the spiritual connotation of traditional culture to pursue superficiality. It should be combined with innovation to inherit and develop and give it a distinctive spiritual image and characteristics of the times; only in this way, the "national trend" will prevail, and traditional culture will be liked and accepted by more people. [12] The innovative design research of the New Chinese style sportswear contains the long-standing traditional Chinese culture, and reflects the current fashion trend, and gives a new definition to the New Chinese style by integrating and recreating the traditional and modern trends. The design examples further prove the feasibility of using the New Chinese style sportswear in innovative design and better open up the reader's mind. The designers should adhere to the innovative design of traditional culture and give the future Chinese sportswear design a robust cultural connotation and the spirit of the times to highlight the influence of Chinese design. [15]

6. Acknowledgment

Fund: The National Social Science Fund of China: History of Modern Fashion Design (1900-1949) in China, 18CG192.

Reference

[1] Liu S. Design psychology [M]. Beijing: Tsinghua University Press: 2006.
[2] Li CD. Design Aesthetics [M]. Hefei: Anhui Fine Arts Publishing House: 2004.
[3] Jia JS. Costume Color [M]. Beijing: Higher Education Press: 1999.
[4] Pang . Fashion Design Color [M]. Beijing: China Youth Publishing Group: 2008.
[5] Bao MX. Real records of modern Chinese women's clothing [M]. Shanghai: Donghua University Press: 2004.
[6] Shi ST, Zhao F. Fashions of the Civil War [M]. Taiyuan: Xi'an Education Press: 2015.
[7] Chen H. "Neutral and Beautiful": the Aesthetic Care of New Chinese Design Style [J]. Journal of South China Normal University (Social Science Edition): 2016 (4); 187.

[8] Wang Q. New Chinese Knitted Garment Design Features and Its Path [J]. Wool Textile Journal: 2019 (11) 45-50.
[9] Zhou X. Practice, Accommodation and Openness of "Chinese Clothing" (I) [J]. Journal of Clothing Research: 2018 (2): 59-66.
[10] Zhou X. Practice, Accommodation and Openness of "Chinese Clothing" (II) [J]. Journal of Clothing Research: 2018 (6): 247-255.
[11] Tang YL. On the innovation Application of the Traditional Cultural Elements and Artistic Values in Modern Clothing Design [J]. Art Criticism: 2017 (8); 171-174.
[12] Xu RR. Research on Inheritance and innovation in modern clothing design [J]. Shanghai Textile Science and technology: 2006; Vol. 34 (6): 67-69.
[13] Li ZQ. On Promoting the Role of Enhancing Design Culture on Popular Clothing [J]. Science and Innovation Herald: 2015; 12 (12): 241.
[14] Li ZQ. Research on old Shanghai Fashion (1910-1940s) [D]. Shanghai Academy of drama: 2015.
[15] Li CD. New Perspectives Needed for Non-Heritage Research and Innovative Design [J]. Hunan Packaging: 2020/1 Issue 24-27.

Study on Structures of Ancient Knitted Fabrics Based on Unearthed Relics

Da-Wei Liu [1]

[1] *Institute of Archaeology, Chinese Academy of Social Sciences, 1 National Stadium North Road, Chaoyang District, Beijing 100101, China*

*Corresponding author's email: 291949054@qq.com

Abstract

By taking unearthed objects as the research object and the coil string structure as the basic condition for selecting fabric samples, time and location as the axes of the spatial distribution for the fabric, fabric data in the excavation report was screened out in this paper through literature analysis. The data analysis indicated that the empirical data of ancient knitting technology were mainly distributed in the middle and lower reaches of the Yangtze River and along the ancient Silk Road. In terms of the fabric structure, from the perspective of fabric contour, structure and based on the fabric structure and the layout of loop and columnar, these samples were classified. Additionally, the development of ancient knitting technology was also illustrated by analyzing and sorting out the relationship between fabric structure and its origin from the perspective of real objects.

Keywords: Ancient Knitting; Hand Knitting; Looping Knitting Technology

1. Introduction

Knitting technology is both artistic and practical. At present, there have been many research on the artistry of knitting technology but few on its "practicability" from the historical perspective. Lots of problems on knitting technology in early history remain unclear. In addition, there is no unified name for ancient-knitted fabrics and craftsmanship, as they were passed down mostly by oral instruction from masters to apprentices, resulting in the unclear definition of knitting technology. In literatures on textiles, hand-knitting technology was not clearly defined. The definition of knitted fabric is usually based on the study of textile machinery, that is to use knitting needles to form a coil of yarn, and then the coils are interlaced to form a fabric where the coil is the basic unit of knitting fabric [1]. The coil is composed of column, sinker loop and coil loop. Hand knitting is more flexible than mechanical knitting; as for the definition of knitted fabric, its scope of application is not generalized and has certain limitations as it often ignores the changes caused by the combination of column, sinker loop and coil loop in neighbouring coils. Therefore, it is critical to summarize and classify the structure of unearthed knitted fabrics on the basis of empirical data and discuss the structural correlation and technological origin.

2. Collection and Analysis of Fabric Data

By taking unearthed ancient knitted fabrics as the research object and coil string structure as the basic condition for selecting fabric samples, data on ancient fabrics in the excavation report were screened out for this paper through literature analysis from the perspective of spatial distribution, and the knitted fabrics were purposefully investigated, as shown in Fig.1. The author participated in the onsite emergency protection of unearthed ancient knitted fabrics and laboratory cleaning and collected the data on ancient knitting technology through field measurement and observation. The structural characteristics of the fabrics are shown in Table 1.

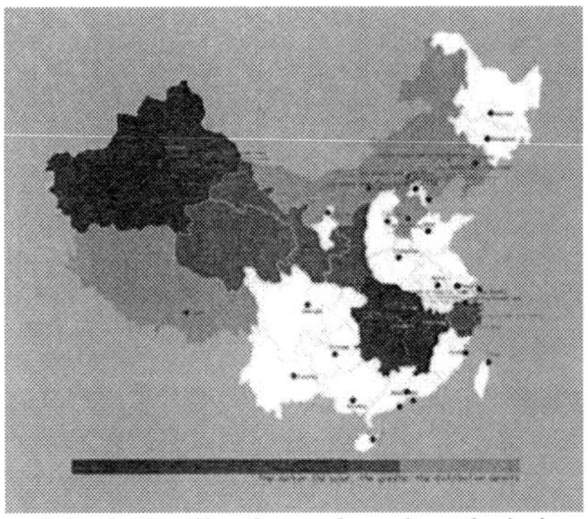

Fig.1 Geographical distribution of ancient knitting technology

Table 1. A summary of the structure and chronology of unearthed knitted fabrics

No.	Item	Chronology	Tomb	Category	Coil Characteristics	Coil Layout
1	M1:13	The late Shang Dynasty to the early Western Zhou Dynasty	Shanshan Yanghai cemetery [2]	Hairnet	Coil twisted with column	Coil sleeved in sinker loop
2	M8:7	The late Shang Dynasty to the early Western Zhou Dynasty	No.1 cemetery in SUBASHI, Shanshan [3]	Head hood	Coil intersected with column	Coil sleeved in sinker loop
3	M8:7	The late Shang Dynasty to the early Western Zhou Dynasty	No.1 cemetery in SUBASHI, Shanshan	Plait net	Coil twisted with column	Coil sleeved in sinker loop
4	1992 SASM	The Warring States Period to the Western Han Dynasty	No.3 cemetery in SUBASHI, Shanshan [4]	Hairnet	Coil intersected with column/Coil twisted with column	Coil sleeved in sinker loop
5	M13:19	The Jin Dynasty to the middle of the Southern and Northern Dynasties	Three-bridge tomb in Shanshan [5]	Head hood	Coil intersected with column	Coil sleeved in sinker loop
6	M13:19	The Jin Dynasty to the middle of the Southern and Northern Dynasties	Three-bridge tomb in Shanshan	Plait net	Coil twisted with column	Coil sleeved in sinker loop
7	M151:41	The late Shang Dynasty to the early Western Zhou Dynasty	Wubao cemetery in Hami [6]	Hat	Coil intersected with column	Coil sleeved in coil
8	T11:3	The early Western Zhou Dynasty to the middle and late Spring and Autumn Period	Yanbulake cemetery in Hami [7]	Fragment of cloth	Coil intersected with column	Coil sleeved in coil
9	M5:2	The late Western Zhou Dynasty to the late Warring States Period	Aikesi Xiaer cemetery in Hami [8]	Hat	Coil intersected with column	Coil sleeved in sinker loop
10	M5:3	The late Western Zhou Dynasty to the late Warring States Period	Aikesi Xiaer cemetery in Hami	Hat	Coil intersected with column / Coil separated from column	Coil sleeved in sinker loop/coil sleeved in coil

11	C:2	The late Western Zhou Dynasty to the late Warring States Period	Aikesi Xiaer cemetery in Hami	Hat	Coil intersected with column	Coil sleeved in sinker loop
12	M6:5	The Warring States Period (770 B.C.- 221 B.C.)	Tomb No.1 of Qiemozhagunluk [9]	Hat	Coil intersected with column	Coil sleeved in coil
13	M67: 2	The Warring States Period (770 B.C.- 221 B.C.)	Tomb No.1 of Qiemozhagunluk	Hat	Coil intersected with column	Coil sleeved in sinker loop
14	M4:8	The Warring States Period (770 B.C.- 221 B.C.)	Tomb No.1 of Qiemozhagunluk	Hat	Coil intersected with column	Coil sleeved in sinker loop/coil sleeved in coil
15	M55: 13	The Warring States Period (770 B.C.- 221 B.C.)	Tomb No.1 of Qiemozhagunluk	Hat	Coil twisted with column	Coil sleeved in sinker loop
16	N7	The Warring States Period (770 B.C.- 221 B.C.)	Chu tomb No.1 in Mashan, Jiangling, Hubei [10]	Silk Braid	Coil intersected with column	Coil sleeved in coil/ heavy looping/intarsia
17	N9	The Warring States Period (770 B.C.- 221 B.C.)	Chu tomb No.1 in Mashan, Jiangling, Hubei	Silk Braid	Coil intersected with column	Coil sleeved in coil/ heavy looping/intarsia
18	N14	The Warring States Period (770 B.C.- 221 B.C.)	Chu tomb No.1 in Mashan, Jiangling, Hubei	Silk Braid	Coil intersected with column	Coil sleeved in coil/heavy looping/dotted jacquard
19	N19	The Warring States Period (770 B.C.- 221 B.C.)	Chu tomb No.1 in Mashan, Jiangling, Hubei	Silk Braid	Coil intersected with column	Coil sleeved in coil/ heavy looping/dotted jacquard
20	N25	The Warring States Period (770 B.C.- 221 B.C.)	Chu tomb No.1 in Mashan, Jiangling, Hubei	Silk Braid	Coil intersected with column	Coil sleeved in coil/ heavy looping /dotted jacquard
21	43	The Warring States Period (770 B.C.- 221 B.C.)	Chu tomb No.1 in Mashan, Jiangling, Hubei	Silk Braid	Coil intersected with column	Coil sleeved in coil/ heavy looping/dotted jacquard
22	/	The Warring States Period (770 B.C.- 221 B.C.)	Chu tomb No.1 in Guojiagang, Jingmen, Hubei	Silk Braid	Coil intersected with column	Coil sleeved in coil/ heavy looping/dotted jacquard
23	M40 6:05 4	The Warring States Period (770 B.C.- 221 B.C.)	Tomb No. M406 in Wulipai, Changsha [11]	Fragm ent of cloth	Coil intersected with column	Coil sleeved in coil/ heavy looping /dotted jacquard
24	/	The Yuan Dynasty(1271A.D.-1368 A.D.)	Pigeon cave in Longhua, Hebei [12]	Face cloth	Coil intersected with column	Coil sleeved in coil/ heavy looping /dotted jacquard
25	/	The Yuan Dynasty(1271A.D.-1368 A.D.)	Pigeon cave in Longhua, Hebei	Face cloth	Coil intersected with column	Coil sleeved in coil/ heavy looping/dotted

						jacquard
26	008300	The Song Dynasty(960A.D.-1279 A.D.)	Zhao boshu's Tomb of the Southern Song Dynasty in Huangyan, Zhejiang Province [13]	Sock	Coil intersected with column	Coil sleeved in coil/with a weft line in the weft direction
27	/	The Liao Dynasty(907A.D.-1125 A.D.)	Joint burial tomb of the Chen's princess and emperor's son-in-law of the Liao Dynasty [14]	Net cloth	Coil intersected with column	Coil sleeved in coil
28	/	The Liao Dynasty(907A.D.-1125 A.D.)	Tombs of the Liao Dynasty in bugugou, Longhua, Hebei Province	Net cloth	Coil intersected with column	Coil sleeved in coil
29	M2A b:11	The Ming Dynasty(1368A.D.-1644 A.D.)	Tomb of the Wang Luo family in the Ming Dynasty in Changzhou [15]	Fragment of cloth	Coil intersected with column	Coil sleeved in sinker loop

In terms of the geographical distribution of knitted fabrics, Figure 1 showed that in the early stage, knitted fabrics were mainly distributed in Hunan Province and Hubei Province of the Central Plains, south of the Tianshan Mountains, north of the Qilian Mountains, and along the Silk Road along the Tarim Basin. And after the Song Dynasty (960A.D.-1279A.D.), the distribution scope shifted to eastern China, mainly gathered in the influential areas of the minorities such as Liao, Jin and Yuan in the north of the Yanshan Mountains in northern China and the Yangtze River Basin and southwestern coastal areas in Southern China. Ying, the capital of the State of Chu, was in the first place in terms of both quantity of knitted fabrics and technological difficulty.

From the aspect of time distribution of knitted fabrics, Table 1 indicated that the knitting technology could be traced back to the Bronze Age more than 3,000 years ago, mainly distributed in the northwest of China. The time of unearthing ancient knitted fabrics which were mostly made of wool in the northwest of China mainly concentrated from the middle and late Shang Dynasty (1600B.C.-1064B.C.) to the middle of the Northern and Southern Dynasties. The empirical data of ancient knitted fabrics unearthed in the Central Plains could be traced back to the Warring States Period (770 B.C.- 221 B.C.) and continued to the Ming Dynasty(1368A.D.-1644A.D.), and these fabrics were made of silk. After the Song Dynasty(960A.D.-1279A.D.), fabric materials and technologies became more diversified, and knitted fabrics in the Central Plains changed from functional to decorative. Embroidery technology, as the main carrier, was applied to garment edges and Buddhist ornaments. And in the northern minority areas, fabrics made of metals (silver and copper) and horsetails successively emerged, of which the former was mostly seen in the collection of clothes worn by the tomb owner, and it was often accompanied by the structure of twisted knitted fabrics [16], which had the same technological property in the knitting process.

3. Fabric Classification

Fabrics can be classified in many ways, for example, by material, technology, purpose and structure. This paper focused on the structural type of fabrics. Therefore fabric structure and production mode are the main factors for classification. As the tombs where the fabric was unearthed were rarely accompanied by the unearthed textile tools, and the appearance of such textile tools was slender and cylindrical, it was usually difficult to distinguish the production and living materials such as hairpins and sticks, so it was difficult to judge the production mode, which therefore will not be discussed here. The research scope of fabrics is limited by modern knitting science, therefore they are categorized into twisted looping, crossed looping, simple looping, knit looping and heavy looping based on coil structure and layout.

The simple looping fabric, which appeared early with wide distribution and numerous unearthed evidence, possess the following characteristics: the yarn moves and bends to form a coil, and the two columns of the coil intersect to form a closed coil structure like a Chinese character '乂'. According to the position relationship of the two columns, it can be divided into S and Z types, that is, when the left column is on the top while the right one is at the bottom, then it is S type; otherwise, it is Z type. The coil structure is the smallest unit that constitutes the simple looping fabric, and each newly formed coil passes through the sinker loop between the two existing coils, with only one coil passing through each sinker loop, as shown in Fig.2. In the Song Dynasty (960A.D.-1279A.D.), with richer fabric structure, weft-knitted fabric emerged, namely there is a weft string in the weft coil to improve the recovery capacity of the fabrics after weft stretching, as shown in Fig.3. After the Song Dynasty (960A.D.-1279A.D.), this fabric structure emerged mostly together with embroidery technology and took advantage of the position relationship between coil and sinker loop to form meshes and holes effect.

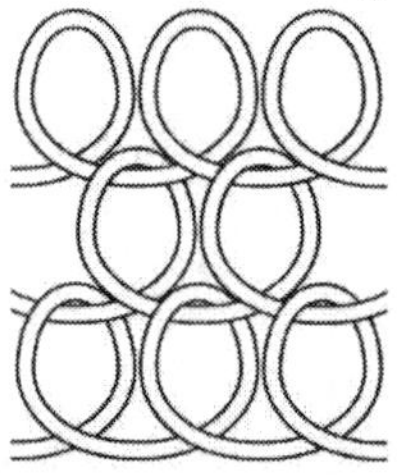

Fig.2 Organization chart of simple looping

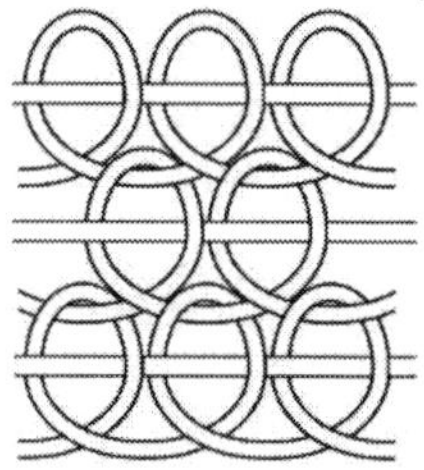

Fig.3 Simple looping on an identical element

Twisted looping and simple looping fabrics are basically identical in emergence time, distribution range and sample size, but they differ slightly in fabric structure, in which multiple twisted loops are formed after the two columns of the coil are intersected. And the column of the coil in the twisted looping fabric is generally longer than that of the coil in the simple looping fabric, and its height of the coil is higher or even several times higher than that of the simple looping fabric, thus forming more mesh and hole effects, as shown in Fig.4.

The fabrics with an crossed looping structure can be mostly traced back to the Spring and Autumn Period and the Warring States Period (770 B.C.- 221 B.C.). The coil structure of such fabrics is exactly in line with that of the simple looping fabric, but there is a significant difference in the coiling method between the two. The coils of crossed looping fabrics are circulated laterally along the fabric to form horizontal rows, with new coils passing through the existing coils, as shown in Fig.5. Among crossed looping fabrics unearthed from tombs such as Zharunluk No. 1 Cemetery in Qiemo, Xinjiang and Aiske Xiar Cemetery in Hami, Xinjiang, there have been a wide range of knitting methods, such as moving to stitch, forward and reverse stitching, etc. The aforementioned fabric structure can be completely simulated by a modern computerized flat knitting machine for weft knitting.

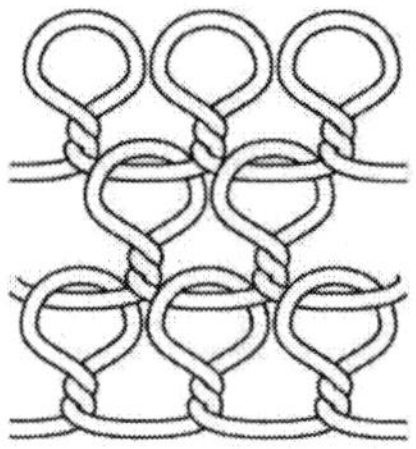

Fig.4 Organization chart of twisted looping

Fig.5 Organization chart of crossed looping

Knit looping fabrics could only be found in the Aiske Xiar Cemetery in Hami, Xinjiang, which dates back to the Bronze Age about 3,000 years ago. The enriched looping and simple looping fabrics were also unearthed from this cemetery. The structure of knit looping fabric is identical with that of the modern knitted fabric, that is, the coil is open, the two columns of the coil are separated, the sinker loop is outward without being intersected to form an omega-like structure, and the coils are mutually intersected. Each coil on the front side of the fabric is provided with two columns arranged at a certain angle along the vertical line of the coil, and each coil on the reverse side of the fabric is equipped with a circle arc with the same configuration as the horizontal line of the coil, as shown in Fig.6. It is named as weft flat knitted structure in modern knitting [17].

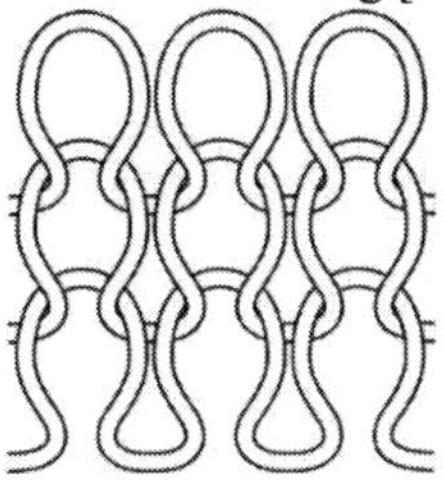

Fig.6 Organization chart of knit looping

Heavy looping fabrics were mainly distributed in Hunan Province and Hubei Province of Central China such as Mashan No.1 Tomb of Chu in Jiangling City, Hubei Province and Guojianggang No.1 Tomb in Jingmen City, Hubei Province. The technology is mostly used in silk braid and clusters. Each newly formed coil passes through two existing coils in the same longitudinal line to form a heavy loop. According to the modern knitting theory, the pattern principles can be divided into intarsia fabric and jacquard fabric. The technological difficulty of the latter one is extremely complex, each coil horizontal row is alternately woven with two different colors of silk threads, the configuration of coils with different colors forms a pattern, and there is a loose thread in the reverse side of the coil that is not involved in knitting. The knitting principle of pattern mode is in line with that of the dotted jacquard fabric in modern knitting.

4. Analysis and Comparison of Fabric Structure

The above five categories of coil string are not isolated, nor are they affiliated with each other, but there is a certain connection between the technology. The above coils usually appear in the structure of the same fabric sample. For example, in the fabric numbering M46: 2 unearthed from the No. 1 Tomb of Zhagunruke in Qiemo County, Xinjiang, in order to exhibit width seam, the fabric structure combined with twisted looping and simple looping methods was used in the joint between cap ear and cap body. In the knitted fabric numbering M5: 3 unearthed in the Aiske Xiar Cemetery in Hami, Xinjiang, crossed looping was used for round cap top and simpling loop for cap body. The simple looping, crossed looping and twisted looping were used in the knitted fabrics numbering 1992SASM unearthed in the No. M6 Male and Female Co-burial Tomb, Subashi No.3 Cemetery, Shanshan, Xinjiang. The fabrics have the same technical attribute and should be derived from ancient hand-knitted fabrics. According to the appearance sequence of the above four types of fabrics and the technological difficulties thereof, it can be deduced that the simple looping fabrics appeared first, followed by twisted looping fabrics, the crossed looping fabrics and knit looping fabrics emerged almost at the same time.

Coil string structure fabric was not only found in the northwest of China, but also widely distributed in Hunan Province and Hubei Province of the Central Plains. It can be traced back to the Warring States Period (475 B.C. to 221 B.C.), such as Mashan No.1 Tomb of Chu in Jingzhou City, Hubei Province, Guojiagang No.1 Tomb in Jingmen City, Hubei Province, and M406 Wuli Pai Tomb in Changsha City, Hunan Province. During this period, not only crossed looping fabrics emerged, but also heavy looping fabrics were unearthed. With a complex structure, the heavy looping fabric is composed of coils and floating threads. The structure of unit coil is the same as that of the crossed looping fabric. However, each coil passes through two adjacent coils in the same longitudinal line, and each complete coil is composed of coils of two colored yarns in the horizontal line. Therefore, with pattern changes, the fabric structure is similar to that of modern weft knitted dotted jacquard fabric, and it shall belong to the category of the coil crosscut fabric excavated in Northwest China. The coil structure fabric unearthed in this area is later than fabric with the same structure unearthed in Northwest China. In addition to above areas, ancient relics were also found along the Silk Road, and the time of their emergence was later than that of the fabrics with the same structure unearthed in Northwest China. For example, the ruins of the palace in Nineveh, northern Iraq, and the ancient city of Dula, Syria, and three pieces of textiles unearthed from the ancient city of Dula, Syria date back to the third century A.D. The fabric, which is now in the Yale University Gallery of Art was woven with the same technology as the simple looping fabrics unearthed from tombs of the Eastern Zhou Dynasty (770 B.C.-256 B.C.), such as Zharunluk No. 1 Cemetery in Qiemo, Xinjiang and Aiske Xiar Cemetery in Hami, Xinjiang. They all adopted lifting, and front and back needle knitting methods. The former utilized the above knitting methods to form a flower pattern with a concave and convex appearance, which made the weaving method more complex. However, there is no direct evidence to prove that the ancient knitting technology in Northwest China is necessarily related to the textile technology in other regions.

5. Conclusion

The three characteristics of the unearthed ancient fabrics are as follows: First of all, from the perspective of location and time distribution, ancient fabrics mainly distributed in the middle and lower reaches of the Yangtze River and along the ancient Silk Road, and concentrated in the two Lake Basin and along the eastern section of the ancient Silk Road. Its long period can be traced back to the Bronze Age 3000 years ago. After the Song Dynasty(960A.D.-1279A.D.), the application of knitting technology changed from functional to decorative; secondly, analyzing from the angle of fabric structure, it can be divided into five categories, which were developed from the same textile technology, and according to their appearance sequences, they are respectively simple looping, twisted looping, crossed looping, knit looping and heavy looping. The structure of knit looping is identical with that of modern weft knitted fabric, and there is an inevitable connection among textile technologies, therefore, the above five categories of fabrics are knitted fabrics; Finally, by comparing with fabrics with the same coil structure unearthed in different areas, it can be discovered that the ancient knitting technology unearthed in Northwest China is much earlier than that in other areas.It still needs more archaeological data to verify the origin and development of technology.

6. Acknowledgement

The authors wish to acknowledge the Youth research project of Chinese Academy of Social Sciences-"Research on the organizational structure of unearthed textile relics"

References

[1] Chen WJ. Encyclopedia of China, Textile Volume [M], Beijing: Encyclopedia of China Press: 1984; pp. 324-325.
[2] Cultural Relics Bureau of Turpan City, Yanghai Cemetery in Xinjiang [M], Beijing: Cultural Relics Publishing House: 2019; pp. 458-459.
[3] Institute of Cultural Administration in Turpan region, Ancient Tombs in SUBASHI, Shanshan, Xinjiang [J], Archaeology: 1984; pp.41-50.

[4] Xinjiang Institute of Cultural Relics and Archaeology, No. 3 cemetery of Su Beixi Tombs in Shanshan County [J], Xinjiang Cultural Relics: 1994; pp. 1-20.
[5] Xinjiang Institute of Cultural Relics and Archaeology, Excavation Bulletin of Sangeqiao Tomb in Shanshan, Xinjiang [J], Cultural Relics: 2002; pp. 46-56.
[6] Xinjiang Institute of Cultural Relics and Archaeology, Tombs 151 and 152 in Wubao Cemetery, Xinjiang [J], Xinjiang Cultural Relics: 1992; pp. 1-9.
[7] Xinjiang Institute of Cultural Relics and Archaeology, Excavation of Aiskexia'er Cemetery in Hami, Xinjiang [J], Archaeology: 2002; pp.30-41.
[8] Department of Cultural Relics, Department of Culture, Xinjiang Uygur Autonomous Region,Yanbulake Cemetery, Hami, Xinjiang [J], Journal of Archaeology: 1989; pp. 325-362.
[9] Museum of Xinjiang Uygur Autonomous Region, Excavation Report of No.1 Cemetery of Zhagonluke in Qiemo, Xinjiang [J], Journal of Archaeology: 2003; pp. 89-136.
[10] Jingzhou Regional Museum of Hubei Province, Chu Tomb No.1 in Mashan, Jiangling [M], Beijing: Cultural Relics Publishing House: 1985; pp. 50-56.
[11] Institute of Archaeology, Chinese Academy of Sciences, Excavation Report of Changsha [M], Beijing: Science Press: 1957; pp. 63-64.
[12] Longhua National Museum, 600 Years of Splendid Cave Collection - Yuan Dynasty Cultural Relics in Pigeon Cave Collection in Longhua, Hebei Province [M], Beijing: Cultural Relics Publishing House: 2015; pp. 95-104.
[13] The China Silk Museum, Silk Mansion with Song Rhyme - An Exhibition of Clothing Excavated from the Huangyan, Zhao Bo Tomb in the Southern Song Dynasty [M], Hangzhou: China Silk Museum: 2017; pp.44-45.
[14] Inner Mongolia Institute of Cultural Relics and Archaeology, et al., Tomb of Princess of Chen in Liao Dynasty [M], Beijing: Cultural Relics Publishing House: 1993; pp.75-77.
[15] Hua Q, Luo Q, et al. Tian Sun Zhu - Tomb of Excavated Textiles from Wang Luo Family Tomb in Changzhou Ming Dynasty [M]: 2017; pp. 92-95.
[16] Zhao F. The Chronological Development of Needle-looping Embroidery [J], Hong Kong: Orientations: Feb 2000; pp.44-53.
[17] Zhao F. Fabric Types and Texture Elements [J], Journal of China Textile University: 1996; pp.27-32.

Research on the Construction of the Protection and Inheritance System of Chinese Traditional Costume Skills

Zhuo Chang[1], Bei-Bei Zhang[2*]

[1]*School of Textile and Clothing Engineering, Soochow University, Soochow, Jiangsu, 215000, China*
[2]*School of Art, Soochow university, Soochow, Jiangsu, 215000, China*

[*]Corresponding author's email: lovingbaby210@163.com

Abstract

Based on fieldwork, this paper constructs the protection and inheritance system of Chinese traditional costume skills from three aspects: authenticity, productivity, and cultural space. Finally, based on the holistic protection theory, this paper puts forward the holistic protection measures of traditional costume skills from the three dimensions of point, line, and area, provides a theoratical protection framework according to local conditions for the protection and inheritance of Chinese traditional costume skills and explores more scientific and efficient protection methods.

Keywords: Chinese Traditional Costume Skills; Protection System; Cultural Heritage; Protection and Inheritance; Cultural Space

1. Introduction

The rich and colorful Chinese costume culture in history has left us many costume skill treasures, including weaving skills, manufacturing skills, printing and dyeing skills, and decoration skills. For example, embroidery categories (Suzhou embroidery, Gu embroidery, Xiang embroidery, and Shu embroidery) have been listed in the "human intangible cultural heritage list" by UNESCO. Their long history and brilliant artistic achievements show the pluralistic and integrated Chinese culture, and confirm the distinctive characteristics of Chinese costume culture in its artistic structure. These costume skills themselves reveal the essence of Chinese culture and have become the dress art symbols, representing the image of China in international communication and cross-cultural communication. Traditional costume skills are one of the core contents of traditional costume culture, and it is also an essential part of Chinese excellent traditional culture. The State Council issued the "Opinions on the Implementation of the Inheritance and Development Project of Chinese Excellent Traditional Culture" clearly proposed to "implement the revitalization plan of traditional crafts." In the appendix of the notice of the Ministry of Culture and Tourism, Ministry of Industry and Information Technology on releasing the first batch of the national traditional crafts revitalization catalog in May 2018, there are 283 items in the first batch of national traditional crafts revitalization catalog, among which 81 items are involved in the textile, printing, embroidery, and dyeing. Under the strategic layout of national cultural rejuvenation and traditional craft revitalization, more and more attention has been paid to the inheritance of traditional costume skills by the society and academia.

After China acceded to the Convention for the protection of intangible cultural heritage, the law of the People's Republic of China on intangible cultural heritage has been issued, which provides directional guidance for the protection, inheritance, and dissemination of intangible cultural heritage, including traditional costume techniques. At present, the protection of intangible cultural heritage at home and abroad has achieved some theoretical and practical achievement, forming a basic theoretical system framework.

Foreign research on intangible cultural heritage started earlier and achieved rich results. Since Japan put forward the concept of "intangible heritage" in the 1950s, scholars' research on intangible cultural heritage has shifted from the early concept and value discussion on how to protect and develop intangible cultural heritage resources. Among them, the research on how to protect intangible cultural heritage abroad focuses on using digital means and the combination of museum protection. For example, Cary Karp puts forward that "Building a network, virtual and invisible digital museum to collect digital resources including intangible cultural heritage can provide a sharing platform for the whole cultural heritage." In addition, the research on intangible cultural heritage policies and regulations has also attracted the attention of scholars all over the world. For example, Jeon chil soo analyzed the new intangible cultural heritage policies, investigated the problems, and sought ways to improve them in A Study on the Convergent Change Revitalization of Paradigm for Intangible Cultural Heritage Transmission Support Policy.

In addition to theoretical research, foreign practical projects related to the protection of intangible cultural heritage have also been gradually carried out, of which the most representative is the European cultural heritage online (ECHO) sponsored by the European Commission. The ECHO framework consists of six parts: current situation research, technology development infrastructure, content supply, AGORA network construction, AGORA management structure, communication ,and development. According to these theories and structures, five research examples have been built, involving art history, science history, language learning, social culture, and anthropology. ECHO is of great significance in publicizing and utilizing Europe's rich cultural heritage resources and provides a practical reference for Chinese cultural institutions to integrate resources.

The research on intangible cultural heritage in China started late, but it has developed rapidly and achieved remarkable results in recent years. The representative is The Introduction to Intangible Cultural Heritage, edited by Wang Wenzhang, which summarizes the basic protection methods of intangible cultural heritage into the following five items: first, establish a protection list system; Second, transform intangible cultural heritage into tangible form; Third, keep its vitality in the authentic atmosphere of its generation and growth; Fourth, the protection of productive mode; Fifth, protect the inheritors. Taking the subject of environmental protection and investigation of folk craft culture as an example, Tang Jialu proposed that the sustainable development of folk art should start with its life form and cultural ecology and discuss the related production and lifestyle, folk creation concept, folk belief, technical factors, and organization form. In addition, many scholars have investigated the individual traditional costume skills and discussed the protection mode of the case from different angles. For example, Hu Jianfang and Yan Hua used the theory of sharing economy to analyze the inheritance and development mode of embroidery intangible cultural heritage and explored the ways of resource integration, optimization, and promotion of embroidery intangible cultural heritage. In The Study on the Salvage Protection of Hubei Tianmen Blue Printed Cloth, Liu Yanqi put forward some protection ways, such as government support and digital protection, establishing publicity and education mode between colleges and communities, brand building, and experiential market management mode.

From the standpoint of present researches at home and abroad, there is no discussion on the protection and inheritance system of traditional costume skills as a whole. This paper not only combs and summarizes the theory of the protection and inheritance of Chinese traditional costume skills, but also makes specific analysis and demonstration in combination with a single category, to further supplement the lack of systematic research on costume culture.

2. The Construction of the Protection Path of Chinese Traditional Costume Skills

First, there is the authentic protection and inheritance system represented by museums. We believe that the authenticity in protecting cultural heritage is to attach importance to the development of the material composition of cultural heritage, protect the native, original, and accurate historical relics, and protect all the historical and cultural information it has left. The display and dissemination of costume skills take the production process of inheritors as the core content. Digital display can comprehensively show the expression form and internal charm of costume skills, creating more possibilities for the authenticity protection of costume skills [1]. We investigated more than ten textile and costume museums, including the National Costume Museum of Beijing Institute of Fashion Technology, the Han Folk Costumes Museum of Jiangnan University, Nanjing Yunjin Museum, China National Silk Museum, and Jingchu textile Intangible Cultural Heritage Museum. These museums adopt the form of material protection, and on this basis, they integrate static and dynamic display methods. The static aspect is mainly realized through the static display of intangible cultural heritage physical objects or physical models, words, labels, drawings, pictures, photos and other material elements, in which pictures and photos can be displayed digitally on the screen; Dynamic display includes stage performance, on-site manual guide, inheritor performance display, visitor participation activities, and other non-digital methods, as well as digital display through video play, voice introduction, slide show, website, 3D, VR, AR, human-computer interaction and artificial intelligence (as shown in Fig.1). Taking Jingchu textile intangible cultural Heritage Museum as an example, the museum focuses on the inheritance, protection, research, and innovation achievements of 20 textile and costume intangible cultural heritage projects in the Jingchu area. The interactive multimedia equipment displayed throughout the museum includes multi-touch screen, LED screen, AR(Augmented Reality)equipment, and interactive ground projection, especially the multi-touch screen, which not only allows the audience to have a more comprehensive understanding of these textile and costume skills, It can also form a database of various textile and costume skills through network management, to meet the needs of both exhibition viewing and academic research and development [2].

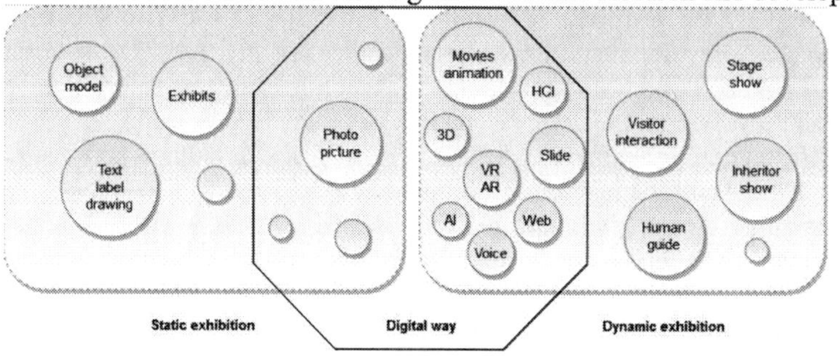

Fig.1 Exhibition mode of Intangible Cultural Heritage Museum in China(drawn by the author)

The second is to carry out effective protection and inheritance. Productive protection is to transform intangible cultural heritage into commodities through production, circulation, sales and other means, to achieve the protection of intangible cultural heritage projects in the production process [3]. Effective protection is based on transforming intangible cultural heritage into productive forces or products, and its production process itself is a kind of cultural production. On the premise of effective protection, we should give full play to the resource advantages of cultural heritage, revive its original functions, activate its vitality in the market, drive the development of relevant cultural industries, and promote economic growth and harmonious social development [4]. At the same time, we should make full use of traditional costume skills to create cultural products and develop intellectual property, highlight the multiple functions of traditional costume skills in the development of contemporary society to make it produce greater social and economic benefits [5]. According to our field research, at present, most of the Chinese traditional costume skills adopt the productive protection mode of "production workshop + transmission institute + cultural tourism line," as well as other production modes, such as the traditional underwriting model of "company + base + farmers," as well as the mode of "School-enterprise cooperation". Taking plant printing and dyeing skills as an example, there has been an upsurge of traditional crafts in some Chinese cities. Emerging dyeing workshops have opened handicraft experience courses such as "plant dyeing" and "blue dyeing", organized manual enthusiasts to experience plant dyeing and sold daily-use products with a sense of design. At present, the handicraft workshop with plant dyeing as the carrier is trendy in China, which enriches urban consumers to experience intangible cultural heritage. The development of traditional plant dyeing skills in Japan started early and mature, and the brand recognition is high. For example, VISVIM, KUON and Porter Classic are famous blue dyeing brands in Japan. The designer combined the blue dyeing skills with Boro, Kimono, Sashiko and other Japanese cultural processes to realize the redesign and branding of traditional costume skills. At present, the product R&D, market development, and expansion of Chinese traditional costume skills are not mature enough. In particular, the textile and garment industry in ethnic minority areas is underdeveloped. They can seek cooperation with eastern enterprises to improve their design level and product quality and promote the inheritance and development of intangible cultural heritage with the power of brand [6].

The third is the protection and inheritance of cultural space. In costume cultural heritage protection, its attached cultural space and ecological environment also have signigificant protection value. Cultural space provides the soil for the survival of intangible cultural heritage and allows for the inheritance, development, and protection of intangible cultural heritage. The significance of cultural space protection lies in the integrity and living inheritance of intangible cultural heritage [7]. So, we should protect the specific traditional costume skills, and pay attention to the environment they rely on. That is to say while protecting the material form, we must protect the cultural-ecological environment of cultural heritage, which is often ignored in the process of cultural heritage protection and inheritance. To ensure that the protected objects are not only preserved but also to maintain "vitality". The ethnic minorities in Guizhou, Yunnan, Guangxi, and other remote mountainous areas still retain relatively complete cultural space [8]. Effective measures must be taken to protect and promote the sustainable development of these cultural-ecological spaces. Certainly, there are a few traditional costume skills inheritance ecological protection areas in China, such as Zhenhu embroidery town is a typical case. Zhenhu is famous for its "hometown of Suzhou embroidery." In ancient times, the women here were very good at Suzhou embroidery [9]. In the trend of inheritance and innovation of intangible cultural heritage, Zhenhu embroidery town plans to integrate Suzhou embroidery design, production, exhibition and sales, experience interaction, learning and training of Suzhou embroidery, tourism and intangible cultural heritage protection base within a few years, and create an industrial integration development belt around the three cores of tradition, regeneration and future. The traditional costume skills are grafted with the industry to enter people's lives in the form of cultural and creative products, to adapt to the changing trend of costume culture.

Therefore, the protection and inheritance of folk costume cultural heritage can be divided into three ways, as shown in Fig.2.

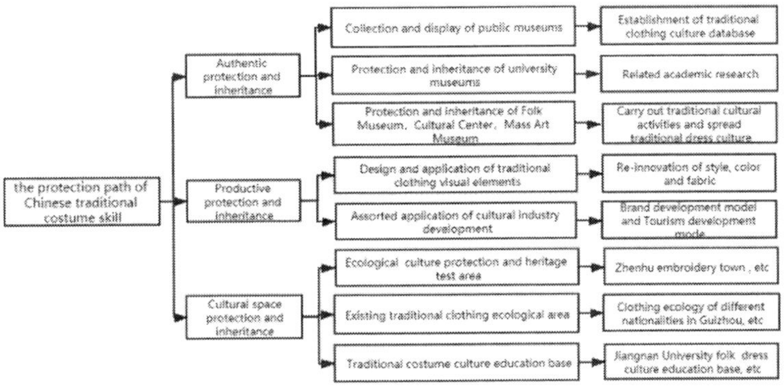

Fig.2 The construction of the protection path of Chinese traditional costume skills(drawn by the author)

3. Specific Protection Measures of Chinese Traditional Costume Skills

3.1 Punctiform Level: Individual Protection and Inheritance

A single traditional costume technique is composed of many materials and non-material cultural factors. At the individual level, a single project's protection and inheritance also need comprehensive and multi-level protection [10]. Taking "blue calico printing and dyeing skill" as an example, as a national intangible cultural heritage, it is a complex combination. It is a systematic and complete cultural expression composed of multiple processes and multistep tasks: engraving, scraping, uncovering, sizing, and dyeing. The blue calico also conveys rich folk cultural connotations: "endless generation," the world view of "false or true complement," and the aesthetic view of "simple and elegant." Blue calico has been handed down from generation to generation in the colorful folk culture from wedding ceremony to funeral ceremony. Every living cultural factor constitutes the soul of the blue calico. Cutting off any part of it will directly affect the vitality of the blue calico. In addition, the production of blue calico needs to rely on such essential material factors as cowhide engraving, indigo dye, and stepping stone. These materials and tools are necessary factors to highlight the characteristics of handicrafts. Through the research on the transformation of handcraft tools and innovation, we can realize the innovative inheritance of traditional costume skills [11].

The innovation of costume skills carries the changes of costume and promotes the development of costume. In the natural evolution, especially in the alternation of the old and new culture, the traditional costume skills have been integrating, improving, and innovating. For example, the improved modern slant opening female coat symbolizes typical Chinese symbols given on the framework of western-style cutting. It abandons traditional crafts such as starching and water line and introduces western technology such as dart and shoulder pad, but it still retains the Chinese characteristics of euphemism and connotation. It can be seen that "tradition" is not a "past tense." It is an eternal "present tense." The traditional costume skills are gradually formed by constantly absorbing and integrating the excellent parts of foreign forms based on inheritance according to the actual needs. As a kind of historical existence, the national characteristics of traditional skills are naturally generated. As long as the nation and its independence exist, it will not disappear. I am afraid that this characteristic becomes solidified and sanctified. Only breaking through the rigid dogma, giving the traditional form new ideas, making it active and flowing, embodies the vitality of national culture [12]. In the new era, the protection and inheritance of traditional costume skills should not stay in the study of theoretical literature, mechanically copied or completely accepted. It must be interpreted and verified from design and technology, combined with modern industry, new materials, and new technology experiments. Based on the perspective of design art and industrialization, using modern technology and design methods to transform and enhance the material gene of traditional costume skills is also an essential embodiment of the contemporary value of traditional costume skills. For example, the popular 3D printing technology, VR technology, and artificial intelligence technology provide reference and reflection for stimulating the modern transformation of traditional costume skill value [13].

3.2 Linear Level: Connection and Construction of Cultural Industry Chain

The essence of an industrial chain is to describe the structure of an enterprise group with some internal connection. The upstream link delivers products or services to the downstream link, and the downstream link feeds back information to the upstream link. The connection and construction of the industrial chain on the online level refers to connecting the closely related industries in the industrial chain to realize the overall linkage development. Taking "grasscloth weaving skill", which is listed as the second batch of national intangible cultural heritage list in China, as an example, the planting industry is the source of its industrial chain. Only when the source is guaranteed can the industrial chain develop healthily and continuously. However, due to the high intensity and low benefit of ramie planting, the quantity of ramie planting is less and less in the areas with less government support, and there is a shortage of raw materials, only through government policy support, increasing the income of growers and output, can we break through the bottleneck of the lack of raw materials. The complex production process determines that the production enterprises need more labor and longer process flow in the manufacturing process. In addition, the process technology and equipment update are slow, which further restricts the improvement of the production efficiency of grasscloth. Only by strengthening the research and development of grasscloth technology equipment and introducing advanced technology can we reduce production costs and improve output efficiency. In the downstream link, the emerging enterprises of grasscloth embroidery, grasscloth handicraft development, creative workshops, and grasscloth culture industry communication have widely used the grass cloth in clothing, household goods, handicrafts, tourism products, and health care products. By relying on the transformation and upgrading of the industry, adjusting the product structure and increasing the brand cultivation, the overall economic benefits of the fabric industry can be improved and the regional economic growth can be promoted.

In addition, for the same kind of traditional costume skills distributed in different regions, the connection and construction of the industrial chain at the linear level should break the regional restrictions [14]. Take the overall distribution of costume skills resources in China as an overview, integrate resources, and promote coordinated development. For example, Nantong is known as the "hometown of blue printed cloth". It took the lead in making the "printing and dyeing technology of Nantong blue printed cloth," which was selected into the first batch of national intangible cultural heritage batch [15]. However, there are many other areas in China where blue printed cloth exists, and the artistic style and production skills with local characteristics are declining because they are not valued. If the blue printed cloth skills in China are composed into an overall system to form a special category, all parts of the country can use the driving role of their own regional resource advantages to realize complementary advantages and develop the industrial chain of blue printed cloth. Such cultural heritage protection and the inheritance will have diversity and universality, and the heritage system of traditional costume skills will be more perfect and sound [16].

3.3 Areal Level: Regional Planning and Construction

At the areal level, areal planning and construction mainly include relevant productive bases, cultural and ecological protection bases, living cultural protection and inheritance experimental areas, ecological museums, Expo parks, and creative cultural villages (towns), which is a new model of intangible cultural heritage protection after human beings pay more attention to the relationship between human and environment [17]. UNESCO emphasizes in the Convention on the protection of intangible cultural heritage that intangible cultural heritage has the characteristics of spirituality, liveliness, initiative, and practicality [18]. Therefore, to ensure the vitality of intangible cultural heritage requires the inherent characteristics of intangible cultural heritage to be fully and truly reflected in the practice of human intergenerational inheritance. Therefore, only by maintaining the stability of traditional costume ecosystem can its vitality be ensured.

As the carrier of human's lifestyle and spiritual culture, traditional costume skills involve many fields, such as politics, economy, geography, religion, etiquette, custom and so on. The protection and inheritance of traditional costume skills can not be separated from its original cultural ecosystem, and cultural space is also included in this structure, which plays a crucial restrictive role [19]. However, in the context of today's globalization, cultural space, the cultural matrix for the survival of traditional costume skills, has undergone dramatic changes. For example, the Shengpu rural area in the Wudong watery area of Suzhou has become a modern industrial park, and the clothing attached to the rice-growing culture in this area, such as cardigan splicing shirt, kerchief, and apron, has disappeared. The traditional costume culture system under the construction of the original farming civilization is disappearing and dying out, meanwhile, the living space of the traditional costume skills heritage has been occupied, which has brought the fault of the inheritance of the traditional costume skills. Therefore, in implementing the protection of traditional costume skills, we should not be limited to the scope of traditional costume skills, but should deeply excavate and analyze the types of cultural heritage, various cultural factors, and cultural performances in the whole cultural ecosystem [20]. In Chinese remote minority areas, there are many successful cases of cultural space protection, such as Matang village in Guizhou Province, a traditional village inhabited by the Ge Nationality. It is in a remote mountainous area. Its economy has lagged for a long time. There are more than 100 households and more than 500 people in the village. The Ge Nationality belongs to a branch of the Miao Nationality and has its language, clothing, and living customs. Their batik, step-by-step dance, Unique costumes have been listed in the protection list by the local government. Since the 1980s, Matang has been identified as one of the ethnic, cultural tourism villages by the People's Government of Guizhou Province. Up to now, Matang has received more than 150000 tourists per year; there are an endless stream of tourists at home and abroad every year. Many experts, scholars, painters, and photographers who study national costumes come here for field investigation and art collection. This small village has been lifted out of poverty through the protection and development of the cultural area, with an average annual comprehensive tourism income of more than 1 million Chinese Yuan and a per capita annual tourism income of more than 5000 Chinese Yuan.

4. Concluding Remarks

Based on the field investigation and case analysis of Chinese traditional costume skills, we constructed the protection and inheritance system of Chinese Traditional costume skills. We put forward three ways: authenticity protection and inheritance, effective protection and inheritance, and cultural space protection and inheritance. First of all, we especially emphasize the application of modern digital technology to interpret the traditional costume skills based on authenticity protection to ensure the complete, true, reliable, and comprehensive protection and inheritance of the traditional costume skills process. Secondly, we should carry out productive protection of traditional costume skills, make them keep pace with the times and integrate into modern production and daily life, to give full play to the "productivity" of traditional costume skills in the real sense. Finally, the protection of the cultural space of traditional costume skills aims to protect the cultural diversity and the tradition, personality of different cultural groups to truly realize the overall protection and inheritance of the living state of traditional costume skills.

Based on the holistic protection theory, we further put forward the holistic measures for the protection and inheritance of Chinese traditional costume skills: personal protection and inheritance, cultural industry chain integration, regional planning and construction. From point to line to area, broadening the protection boundary and adopting an integrated protection perspective can better activate many cultural factors, and improve the protection efficiency, so that the overall cultural ecosystem, including people, environment, and cultural project individuals, can achieve a symbiotic balance.

5. Acknowledgments

Thanks for the support from "the Art project of the National Social Science Fund of China"

6. Funding

The work described in this paper was fully supported by a grant from " the Art project of the National Social Science Fund of China " named "A Study of Ancient Chinese Women's Costume with Multi Integration(From Qin and Han Dynasties to Qing Dynasty)" (No.20BG119)

References

[1] Cui RR, Liang HE, Niu L. The protection and inheritance of han folk costume cultural heritage from the perspective of cultural circle. Creativity and Design: 2012; 3: 19-23.
[2] Li T. Reflection on the current situation of Chinese traditional national costume culture. Textile Technology Progress: 2013; 1: 78-80.
[3] Du HY. "Intangible cultural heritage" value characteristics and its realization. Social Scientist: 2018; 5: 98-102.
[4] Qi Y, Li Q, Su DY. Research on the spiritual value of intangible cultural heritage resources from the perspective of contemporary culture. Hundred Schools in Arts: 2012; 4: 222-224.
[5] Huang LG. The protection of intangible cultural heritage in China and the practice path of public folklore. National Art: 2020; 5: 65-74.
[6] Yang JJ. Cultural heritage through legislation. Chinese Jurisprudence: 2020; 5: 127.
[7] Jiang XW. Visual design and development of Xifu intangible cultural heritage. Packaging Engineering: 2020; 12: 227-235.
[8] Mao QH. Evolution of folk art research in the context of intangible cultural heritage. National Art: 2020; 4: 47-54.
[9] Hu M. Research on the path of protecting the private rights of ethnic minorities' intangible cultural heritage from the perspective of public property theory. Journal of Central University for Nationalities (Philosophy and Social Science): 2020; 4: 130-140.
[10] Yu YS. The dilemma and outlet of intangible cultural heritage inheritance. Chinese Market: 2020; 18: 13.
[11] Li YB, Gao XY. Research on the commercial development and authenticity protection of intangible cultural heritage. Yunnan Archives: 2007; 5: 10-11.
[12] Zhao YF. Practice and thinking on the recording and protection of intangible cultural heritage in the new era. Folk Studies: 2019; 6: 47-55.
[13] Han HY. Application of artificial intelligence in intangible cultural heritage protection and innovative design: taking Inner Mongolia as an example. Art and design (Theory): 2020; 8: 73-75.
[14] Jin C, Liang HE. Research on the value identification path of Jiangsu traditional costrime handicraft:from "productive protection" to "lifestyle inheritance". Hundred Schools In Arts: 2020; 3: 77-81.
[15] Niu L, Yan L, Cui RR. Inheritance and development of traditional clothing techniques in Jianghuai Area of Jiangsu Province. Journal of Inner Mongolia Art College: 2019; 01: 133-137.
[16] Li KJ, Tian HW. Research on heritage of Zhongyuan traditional clothing fabric and handicraft.Journal of silk: 2013; 3: 44-45.
[17] Information on http://www.ihchina.cn/zhengce_details/11566.
[18] Wang WZ. Research on the protection of intangible cultural heritage.Beijing:Culture and Art Press: 2013; p. 15-30.
[19] Zeng MY. Research on the living inheritance strategy of intangible cultural heritage of traditional skills. Reform and Opening-up: 2020; 4: 75-79.
[20] Zhang M. Exploration on the innovative mode of intangible cultural heritage protection.Cultural Industry: 2020; 2: 65-67.

Cite as: D.L. Chen et al.,
TBIS 10.3993/tbis (2021).

Perceptual Image of Zhangpu Paper-cut Pattern

Dao-Ling Chen[1*], Peng-Peng Cheng[2]

[1]*Clothing and Design Faculty, Minjiang University, Fuzhou, China*
[2]*College of fashion and design, Donghua University, Shanghai, China*

*Corresponding author's email: 316910027@qq.com

Abstract

To understand consumers' perceptual cognition of the intangible cultural heritage Fujian Zhangpu paper-cut patterns, accurately grasp the innovative design and application direction of Zhangpu paper-cut patterns as well as to meet the perceptual needs of modern consumers, we recruited young consumers aged 20-30 to carry out the aesthetic evaluation of Zhangpu paper-cut patterns based on Kansei Engineering. We then analyzed the experimental data by SPSS to identify design elements that are significantly related to consumer perception. Through the mathematical model, the corresponding relationship between the design elements of paper-cut patterns and the perceptual image vocabulary is clearly expressed, providing a reference for the realization of personalized paper-cut pattern designs based on different perceptual needs and providing new ideas for the development and inheritance of traditional paper-cut culture.

Keywords: Zhangpu Paper-cut Patterns; Consumption Perception; Design Elements; Correlation Analysis

1. Introduction

Zhangpu paper-cut in Fujian has its unique local characteristics due to the influence of China Central Plain Culture, Minyue culture and Ocean culture. As a representative of southern Chinese paper-cut, it is loved by people for its exquisite, primitive and naive beauty [1]. However, in the process of social development, Zhangpu paper-cut demand has decreased since the reduction of traditional custom activities [2]. Although the current protection of Zhangpu paper-cut has allowed it to obtain better conditions for survival and inheritance, the design and application of Zhangpu paper-cut patterns mostly relies on the designer's personal experience. It lacks consideration of the psychological needs of modern consumers, which has resulted in paper-cut pattern design and product style to be homogenized, wherein the aesthetic needs of consumers cannot be met [3]. With the enrichment of material products and the improvement of design level, people's demand for commodities has gradually changed from the pursuit of quality to perceptual consumption with more attention on the aesthetic experience [4-5]. Modern consumers tend to take spiritual satisfaction as the focus of their choice and purchase of products, and they pursue products that can show their personality, hobbies, values and other personal characteristics [6]. Therefore, grasping the specific psychological trend of consumers and taking it as the focal point of design will become an important way to design and develop paper-cut patterns.

With the coming of the era of perceptual consumption, the research on consumer sensibility promotes the development and application of Kansei Engineering. Kansei Engineering technology is to quantify consumers' perception and discuss the relationship between perceptual cognition and design elements in a clear way, so as to provide scientific and rational reference for product design [7-8]. Since 1995, when Mitsuo N systematically introduced the concept, classification, application field and application method of Kansei Engineering, Japan and other countries have successfully applied it in product design, architectural design, design of a webpage, fashion design, service design, and so on [9-13]. In recent years, some scholars have also carried out research on the combination of pattern design and Kansei engineering, while it only focus on simple geometric elements such as stripes, lattice, wave dot and diamond pattern in patterns field [14-15]. Even if there are a few researches on complex patterns, they mainly evaluate the application effect of patterns on clothing, such as the application effect of blue and white patterns in the Ming Dynasty on the cheongsam, perceptual research on print patterns of men's shirts and so on

[16-17]. However, the research on consumers' perception of paper-cut patterns has not yet been carried out, which plays a vital role in the innovation application of traditional paper-cut patterns to meet the needs of modern consumers.

Therefore, this research put forward the theory and methods of Kansei Engineering to explore the relationship between the design elements of paper-cut patterns and consumers' perception. Realizing the perceptual research on complex patterns is helpful for designers to apply and combine design elements according to different consumer's perceptual needs, and provide support for the innovative design, development and application of Zhangpu paper-cut patterns.

2. Methodology

2.1 Screening of Zhangpu Paper-cut Patterns and Stimulus Diagram Production

Collect Zhangpu paper-cut patterns by the internet, literature and fieldwork, and invite local paper-cutting experts and professors in the art field to screen the representative samples from 80 samples. And it is necessary to ensure that each sub-element has a corresponding pattern sample in every representative sample, 12 representative samples were selected at last. At the same time, in order to avoid the interference in color on the experimental results, we use Photoshop software to decolorize the selected paper-cut patterns, and then perform noise and brightness processing to achieve the basic aesthetic effect. And finally we obtained 20 stimulus diagrams, as shown in Fig.1.

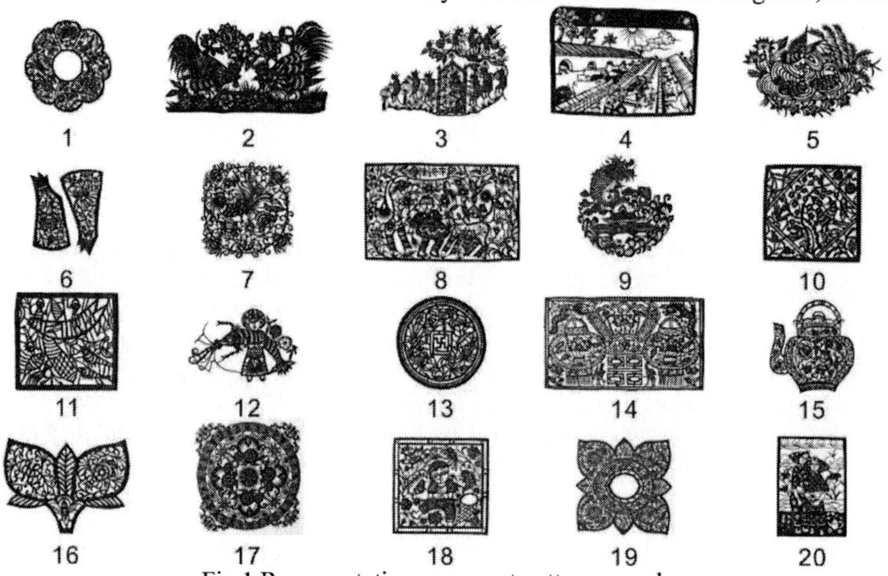

Fig.1 Representative paper-cut pattern samples

2.2 Morphological Analysis and Extraction of Design Elements of Paper-cut Patterns

The morphological analysis method is used to analyze things based on morphology, Whose basic idea is to divide the product into several independent items, find the variable elements of each independent item after decomposition, and establish a product design element system. This research proposes morphological analysis and invites 3 experts in the design field to decompose the decisive elements of the paper-cut sample's overall shape design style, and finally divide it into 5 design elements (namely composition method, outer contour shape, pattern expression technique, pattern content, line type), which were coded as X1, X2, X3, X4, and X5 respectively. In addition, every design element contains several sub-design elements, and every element is coded, as shown in Table 1.

Table 1. Paper-cut pattern design elements deconstruction and coding

Design elements	Sub-design elements 1	2	3	4	5	6
Composition method (X1)	central composition	frame composition	guide line composition	symmetrical composition	arc composition	scattered composition
Outer contour shape (X2)	square	circle	rectangle	shape according to the object	no contour	
Pattern expression technique (X3)	realism	exaggeration	symbolism			
Pattern content (X4)	auspicious pattern	daily life scene	folk legend			
Line type (X5)	thick line	thick and thin lines				

2.3 Selection of Perceptual Adjectives

Get perceptual adjectives describing patterns through multiple channels such as the market, the Internet, newspapers, books, magazines, and websites, and combine these adjectives with their opposite semantics to form 15 pairs of perceptual adjectives. And then 3 experts in the design field were invited to choose the adjectives which were most suitable for describing the paper-cut patterns. And finally, 5 pairs of perceptual adjectives were selected which are Vivid-Rigid, Modern-Tradition, Unique-Ordinary, Plain-Noble, Delicate-Rough respectively, and they are numbered A1, A2, ..., A5 in sequence.

2.4 Questionnaire Preparation and Distribution

This research adopts the Semantic-Differential method and the Likert scale with 5 levels to conduct evaluation experiments. Take the perceptual adjective Vivid-Rigid as an example. 1-level is very vivid, 2-level is a bit vivid, 3-level is neither vivid nor rigid, 4-level is a bit rigid, 5-level is very rigid. Then, consumers aged between 20 and 30 years old are selected as the objects of the survey, because this group of consumers are active in thinking, chasing fashion, and has its own unique opinions on aesthetics and consumption. In order to ensure the respondents' accurate understanding of the intention of this survey, face-to-face questionnaires were adopted in the experiment. A total of 100 questionnaires were issued, and the unqualified questionnaires were removed, 95 valid questionnaires were recovered at last. The recovery rate was 95%, which met the sample size requirements of this survey.

3. Research Results and Analysis

3.1 Perceptual Evaluation Scores of Paper-cut Patterns

The subjects graded perceptual adjectives on 20 paper-cut pattern samples in turn, and every sample will have 475 results (95 questionnaires ×5 pairs of perceptual adjectives). Excel software was used for the mean processing of the survey results, and the perceptual evaluation results of 20 paper-cut patterns were finally obtained. In order to further analyze the main feelings generated by every sample and the differences between them, the highest and lowest scores of every sample and every pair of adjectives were taken as the main emotional psychology of the sample. The specific results are shown in Table 2.

According to Table 2, we can clearly get the main feeling of each paper-cut sample. For example, Sample 2 and Sample 3 give people a vivid feeling, while Sample 16 and Sample 19 give people a rigid feeling. Sample 1, Sample 3, Sample 7, Sample 10, Sample 12, Sample 13 and Sample 14 give people a traditional feeling, Sample 4 and Sample 9 give people a modern feeling, but this feeling is not particularly strong, because the evaluation score is relatively close to 3. Sample 2, Sample 4, Sample 5, Sample 6, Sample 8, Sample 14, Sample 15, Sample 16,

Sample 17, Sample 18, Sample 19, Sample 20 give people an ordinary feeling, only Sample 3 and Sample 12 give people a unique feeling. Sample 1, Sample 4, Sample 6, Sample 8, Sample 9, Sample 10, Sample 11, Sample 12, Sample 13, Sample 15, Sample 16, Sample 19, Sample 20 give people a plain feeling, and no sample gives people a feeling of noble. Sample 5, Sample 7, Sample 9, Sample 14, Sample 17, Sample 18 give people a delicate feeling, and Sample 6, Sample 11 give people a rough feeling.

Table 2. The main perceptual evaluation and scores of Zhangpu paper-cut pattern samples

Sample number	Main perceptual evaluation and score	Sample number	Main perceptual evaluation and score
1	Tradition (4.532) Plain (1.674)	11	Plain (1.373) Rough (4.362)
2	Vivid (1.236) Ordinary (4.597)	12	Tradition (3.102) Unique (1.463) Plain (1.425)
3	Vivid (1.297) Tradition (4.323) Unique (1.463)	13	Tradition (4.525) Plain (2.674)
4	Ordinary (3.548) Plain (1.231) Modern (2.735)	14	Tradition (4.612) Ordinary (4.672) Delicate (1.795)
5	Ordinary (4.389) Delicate (1.832)	15	Ordinary (4.537) Plain (2.778)
6	Ordinary (4.464) Plain (2.343) Rough (4.362)	16	Rigid (4.312) Ordinary (4.675) Plain (2.875)
7	Tradition (3.766) Delicate (2.101)	17	Ordinary (4.489) Delicate (1.231)
8	Ordinary (4.142) Plain (1.874)	18	Ordinary (3.972) Delicate (1.231)
9	Modern (2.311) Plain (2.964) Delicate (1.357)	19	Rigid (4.312) Ordinary (4.573) Plain (2.669)
10	Tradition (4.532) Plain (2.561)	20	Ordinary (3.612) Plain (1.787)

From the above results, it can be seen that although the paper-cut patterns are different, some samples give people the same feeling. After comparing the samples, it is found that the samples with the same feeling have some same design elements, which shows the paper-cut pattern design elements affecting the consumers' perceptual evaluation. In order to more accurately find out the relationship between design elements and consumers' perception, their correlation is analyzed in the following.

3.2 Correlation Analysis of Consumption Perception and Design Elements

Establish the mapping relationship between paper-cut pattern design elements and consumers' perception by SPSS19.0, regarding design elements as independent variables and perceptual vocabulary evaluation values as dependent variables. And correlation analysis was carried out to find out the design elements significantly related to perceptual vocabulary, as shown in Table 3.

Table 3. Correlation between perceptual adjectives and design elements

Perceptual adjectives	Correlation	Design elements				
		X1	X2	X3	X4	X5
A1	Pearson correlation	-0.437	-0.523*	0.371	-0.861**	-0.411
	Significance (two-tailed)	0.054	0.018	0.136	0.000	0.072
A2	Pearson correlation	-0.730**	-0.151	0.722**	-0.174	-0.355
	Significance (two-tailed)	0.000	0.525	0.000	0.464	0.125
A3	Pearson correlation	-0.530*	-0.426	0.475*	-0.832**	-0.350
	Significance (two-tailed)	0.042	0.061	0.034	0.000	0.130
A4	Pearson correlation	-0.010	0.061	0.686**	-0.516*	-0.230
	Significance (two-tailed)	0.967	0.798	0.001	0.039	0.330
A5	Pearson correlation	-0.205	-0.312	0.105	-0.436	-0.929**
	Significance (two-tailed)	0.386	0.181	0.660	0.054	0.000

**. At the 0.01 level (two-tailed), the correlation is significant.
*. At the 0.05 level (two-tailed), the correlation is significant.

It can be seen from Table 3 that the design elements that are significantly related to A1 are X2 and X4, and the Pearson correlation coefficients are -0.523 and -0.861, respectively. The design element X4 is the most relevant, with a significant value 0.000<0.01. The design elements related to A2 are X1 and X3, and both of these design elements are significantly related to A2, with a significant value of 0.000. The design elements X1, X3, X4 are significantly related to the perceptual adjective A3, and the design element X4 is the most significant, followed by X3 and X1. The design elements related to A4 are X3 and X4, and X3 is the most relevant, with a significant value 0.001. Among the five design elements, only X5 is significantly related to A5, and the Pearson correlation coefficient is -0.929.

In order to more accurately express the mapping relationship between design elements and consumers' perception, the data is further analyzed by multiple linear regression and a correlation model between consumers' perception and design elements is established, as shown in Table 4.

Table 4. Regression model between consumer perception and design elements

Perceptual adjectives	Regression model		
A1	Y=59.592-0.129X2-0.883X4	R^2=0.872	P=0.000
A2	Y=12.817-0.471X1+0.438X3	R^2=0.830	P=0.003
A3	Y=39.122-0.045X1+0.168X3-0.996X4	R^2=0.858	P=0.001
A4	Y=-13.736+0.497X3+0.153X4	R^2=0.842	P=0.009
A5	Y=139.923-2.3X5	R^2=0.954	P=0.000

It can be seen from Table 4 that all the models have good fitting degree, with $R^2 \geq 0.830$. It can be seen from the regression model that both X2 and X4 have an impact on the perceptual dimension of A1, and both of these two design elements make the scoring result biased to the adjective on the right, namely "rigid". It can be seen from the absolute value of the regression coefficient that X4 has a greater impact on A1. The design element that has a greater impact on the perceptual dimension of A2 is X1 and the scoring results are biased to "traditional". There was a positive correlation between design element X3 and perceptual adjective A2, that is, the scoring result was biased to "modern". Among the design elements related to A3, X4 has the greatest influence on it, and X1 makes the scoring result biased to "ordinary", and design element X3 makes the scoring result "unique". Both X3 and X4 are positively correlated with A4, which makes the scoring results are biased toward "plain". The design element X5 significantly influences A5 and makes the scoring result towards "rough".

4. Conclusion

Based on Kansei Engineering, this research recruited young consumers to explore the perceptual cognition of Zhangpu paper-cut patterns, the result show that although the paper-cut pattern samples are different, some samples could give people the same feeling. And proposed multiple linear regression analysis method to analyze the experimental data, the corresponding relationship between consumers' perceptual cognition and the design elements of Fujian Zhangpu paper-cut pattern is found out. It not only provides the direction for the innovative design and application of Zhangpu paper-cut patterns, but also helps to improve the working efficiency of designers and consumers' satisfaction with the design.

5. Acknowledgements

This paper was supported by 2020 Fujian Social Science Planning Project (No. FJ2020C049).

References

[1] Chen YH. An analysis of Zhangpu's paper-cut art. Modern Decoration (Theory): 2014; 12: 135-136.
[2] Wu ZH. An analysis of the local cultural characteristics of Zhangpu's paper-cut auspicious patterns. Popular literature and art: 2014; 5: 49-50.
[3] Ma QL. Zhangpu paper-cut inheritance from the perspective of community participation. Journal of Xuzhou Institute of Technology (Social Science Edition): 2017; 2: 9-13.
[4] Wang TX, Zhou MY. A method for product form design of integrating interactive genetic algorithm with the interval hesitation time and user satisfaction. International Journal of Industrial Ergonomics: 2020; 76: 1-15.
[5] Wang ZQ, Zhang M, Sun HY, Zhu GL. Effects of standardization and innovation on mass customization: an empirical investigation. Technovation: 2016; 48: 79-86.
[6] Su JN, Fan YF, Zhang ST, Ou Y, Lin Z. Product modeling design based on perceptual engineering and neural network. Journal of Lanzhou University of Technology: 2011; 4: 47-50.
[7] Li Y, Wu J. The correlation between the consumer perception and design elements of clothing wave dot pattern. Journal of Textile Research: 2020; 10: 132-136.
[8] Steinbrg F, Tursch P, Woll R. The improvement of Kansei engineering by using the repertory grid technique. Management & Production Engineering Review: 2015; 3: 77-82.
[9] Mele M, Campana G. Prediction of Kansei engineering features for bottle design by a Knowledge Based System. International Journal on Interactive Design and Manufacturing: 2018; 12 (4): 1201-1210.
[10] Mahsal B, AbasAli S. Effect of opening architectural shapes on users' emotion with Kansei method. Intelligent Buildings International: 2018; 10 (2): 103-121.
[11] Guo F, Liu WL, Cao Y, Liu FT, Li ML. Optimization design of a webpage based on Kansei engineering. Human Factors and Ergonomics in Manufacturing & Service Industries: 2016; 26 (1): 110-126.
[12] Zhou XX, Liang HE, Dong ZY. A personalized recommendation model for online apparel shopping based on Kansei engineering. International Journal of Clothing Science and Technology: 2017; 29 (1): 2-13.
[13] Hartono M. The modified Kansei engineering based application for sustainable service design. International Journal of Industrial Ergonomics: 2020; 79: 1-14.
[14] Shi JP, Zhu JN. Research on striped knitted garments based on the wearer’ s perceptual knowledge. Knitting Industry: 2011; 1: 53-56.
[15] Wang HY, Liu GL. Lattice pattern recognition based on consumer's perceptual needs. Journal of Textile Research: 2014; 11: 152-156.
[16] Gu ZH, Wang YJ. The application effect of Qing pattern on cheongsam in Ming Dynasty. Journal of Textile Research: 2018; 1: 126-132.
[17] Zheng HZ, Wang JP. Perceptual research on design of garment printing pattern. Journal of Textile Research: 2020; 8: 101-107.

A Study on the Figure of "Lotus Flower" on the Tibetan Buddhist Vestment "Da Gang"

Wei-Min Chang[1*], Le Chang[2]

[1]BeiJing Institute of Fashion Technology, Ying Hua Road Jia 2nd, Chao Yang district, Bei Jing, China, 100029,
[2]BeiJing Institute of Fashion Technology, Ying Hua Road Jia 2nd, Chao Yang district, Bei Jing, China, 100029

*Corresponding author's email: 842414810@qq.com

Abstract

Monk's clothes are one of the most important cultural carriers of Tibetan Buddhism, it includes a lot of cultural information. The article posits that Da Gang is an important vestment in Tibetan Buddhism as research objects. Combined with the characteristics of shape and pattern, and the presentation state in the scene of Da Gang to make an in-depth analysis and interpretation of the "Lian Hua " pattern on it. Da Gang adopts the Buddhist meaning through lotus's shape. It shows the regulation of Tibetan Buddhism during study everyday. It emphasizes the determination of Tibetan Buddhism's religious reform. It reproduces the cultural characteristics of the pluralistic integration of Han and Tibetan Buddhist of monk's clothes.

Keywords: Tibetan Buddhism; Monk's Clothes; Da Gang; Lotus "Lian Hua"

1. Introduction

According to 《Mahabharata》, at the beginning of the universe, Brahma is self-born out of a lotus flower which grew from the navel of Vishnu. He sat and created all things in the world. Lotus flower was used to represent Vishnu and his spouse God. Today, the presentation of lotus flower in Tibetan monk clothes has to be said to be the essence of Buddhist culture.

《The Wonderful Dharma Lotus Flower Sutra》 is a classic that uses lotus flower as a metaphor for the right Dharma. In the pure land of bliss, there are cyan lotus flower, yellow lotus flower, red lotus flower and pure white lotus flower. According to Great Treatise on the Perfection of Wisdom (Volume 8) (Da Zheng 25.116a), "the lotus flower in human world is not big enough, but in Mantuqi Ni pool and Anaboda pool, it is as big as a car cover.

Great Treatise on the Perfection of Wisdom in Volume 8 (Da Zheng25.116a):"Lotus flowers that grow up in the man's world are less than one foot. Lotus flowers that grow up in Mandākinī and Anavatapta, are as big as a car cover". The size of the cherished lotus flower in heaven is larger than this." It is easy to make full- lotus posture in it. Buddha's lotus flower for sitting is ten million times larger" [1]. It can be seen that lotus flower is the representation of Buddhism and lotus flower culture of Buddhism has a long history.

The Tibetan monk's vestment known as the "Da Gang" is one of the important dharma clothes formed during the design of a monk's garment, which is characterized by the "conceptualization" of lotus shape.

The article uses the field research methods, specimen drawing, as well as exclusive interviews with eminent monks and textual analysis. Drawing from different stereoscopic states, plane expansions and elements analysis, this paper aims to summarize the Buddhist significance and function of the lotus pattern in Da Gang.

2. Shape and structure of Da Gang

In the 7th century AD, Indian Buddhism and Han Buddhism were introduced into Tibet at the same time. Tibetan Buddhism, the mainstream religion of Tibet, was formed by blending with the Bon religion of Tibet [2]. With the introduction of Buddhism, monk's clothes were also introduced into Tibet, forming the basis of Tibetan monk's clothing [3]. The Tibetan Buddhist monk's clothing system is a unique clothing system formed by the integration of Central Plains culture and local secular culture after Buddhism was introduced into the plateau. The figure of "lotus flower" on the clothing forms the iconic cultural connotation of Da Gang. Da Gang is divided into Da Gang in Sutra Hall and Da Gang in monk's dormitory, which is worn when chanting sutras or attending dharma meetings. Da Gang is worn by monks when they practice sutras in monks' houses. It can be seen that the two kinds of Da Gang are closely related to the study of Buddhist scriptures (Fig.1).

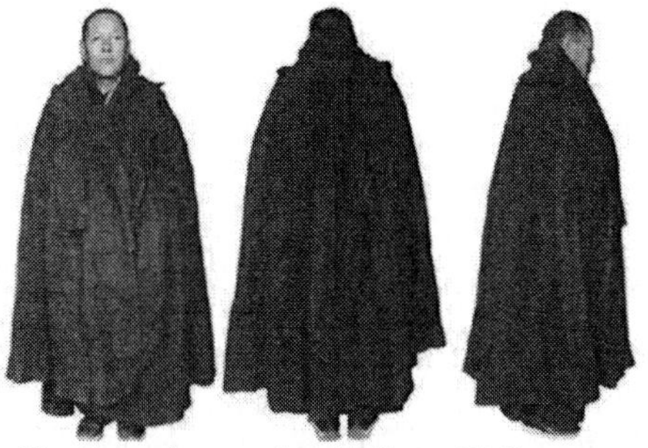

Da Gang is wearing in Sutra Hall
(From Rongbuk Monastery shot in 2017.10)
(From Wei-Min Chang drawing in 2018.1)

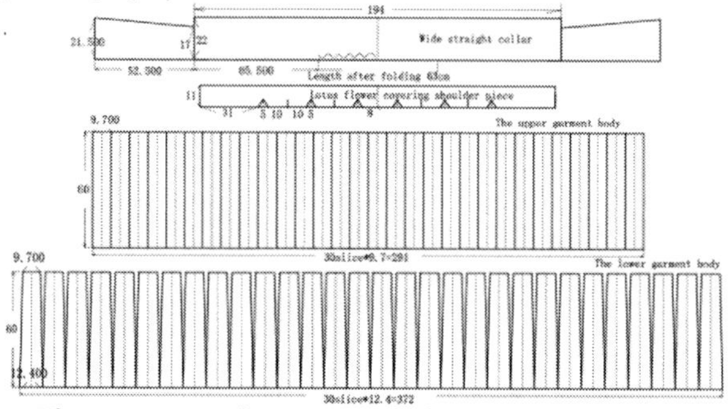

The structure of Da Gang clothes: upper is collar; middle is shoulder cover; lower is body with pleats

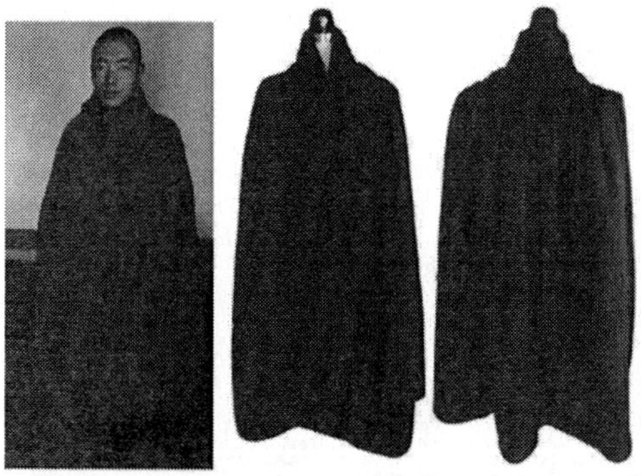

Da Gang is worn in monk's dormitory
(From Gaden Temple shooting in 2017.10)

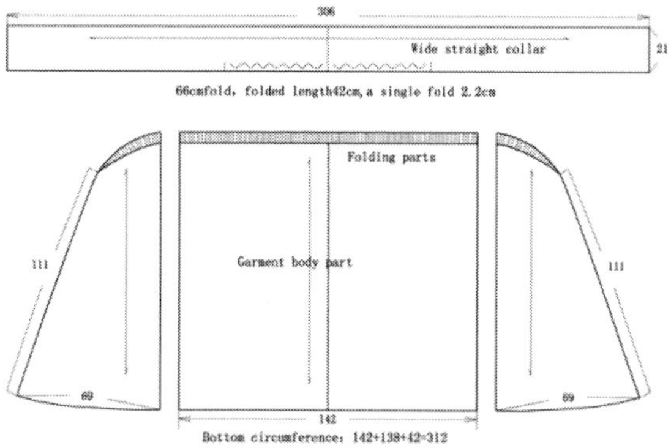

The structure of Da Gang that wearing in monk's dormitory
(From Wei-Min Chang drawing in 2018.1)

Fig.1 The shape and structure of Da Gang

3. The Presentation of Lotus Flower's Shape and Structure of Da Gang

Da Gang is the unique clothing of Tibetan Buddhism. It is also a symbolic garment worn by Tibetan Buddhist monks when they study scriptures and attend major Dharma meetings [4]. From the silhouette to inner structure, each aspect of the clothing's design embodies certain characteristics of the lotus flower. From the whole shape to the local form of Da Gang, it shows the figure of "lotus flower". Whether it is coincidental or inevitable, this synergistic design is obviously the embodiment of the internal relationship between the connotation of lotus flower and Buddhist doctrine.

With the help of lotus flower form, the dogma of Buddhist precepts can be perceived in the design. It is the image requirement of the wearer's spiritual practice.

3.1 The Lotus Bud Posture

Da Gang is displayed in fixed posture like a lotus bud when studying scriptures and resting (Fig.2). At the end of studying Sutra everyday, monks would take off Da Gang and fold it in the shape of a lotus bud. An esteemed language teacher named Luosong Ciding in Xiangda Temple explains this:" Da Gang is placed in this way to show a sense of wholeness, which makes people feel comfortable when they look at it." Judging from this interpretation, harmony is the first rule for monks to follow, which inevitably results in a reverence for the Buddha. Li Shangyin, a famous poet in the Tang Dynasty, wrote in his poem "Gift of Lotus" that "flowers and leaves in the world are not compared, flowers into the golden basin leaves as dust. Only the lotus red bud, roll open and open as naive."

Lotus Bud jumps over the flowers, and becomes a sublime perception of life. This poem borrows the lotus flower, actually builds the harmony between things. In addition, this form of lotus bud often appears on Buddhist temple architecture (Fig.3). When the Rinpoche died, he wore Da Gang and made a posture of a lotus bud (Fig.4), showing that Da Gang's appearance is cyclical and timeless.i.

The sample of lotus bud and abstract figure (From www.redocn.cn)

Daily state of Da Gang Khenpo Sutra Monks learn Sutra Making the shape of Da Gang looks like a bud after learn sutra

(From Bayon temple shooting in 2019.9) (From Tradruk Temple shooting in 2019.1)
(From Zhadong Templeshooting in 2019.1)

The monk was folding Da Gang shape liked bud of lotus
(From Guxiu Temple in Nyingchi shooting in 2019.9)
Fig.2 Posture of Da Gang

Indian Lotus Temple Pillar head of Huayan Temple guardrail
(From https://mydecorative.com/lotus-temple-of-delhi/) (From Huayan Temple shooting in 2021.2)
Fig.3 Lotus bud pattern in architecture

Gongkhar Ningbu bronze statue of Sakya (The founder of a Sakya) [5]

Statue of the first Dalai Lama dge-vdun-vgrub-pa Translator the late development period [6]

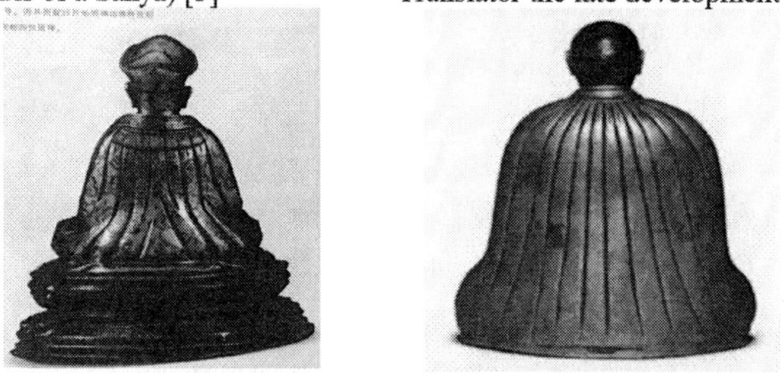

The statue of forth Panchen Erdeni LUOSANGQUEJIJIANZAN [7]

Vbrog mi shākya ye shes (The late development period) (From Zhiguan Art Museum)

Fig.4 The leaders statues wear Da Gang in Tibetan Buddha

3.2 The Round and Broad Shape of Lotus Leaves

The base structure of Da Gang takes the form of a wide round type of lotus leaf. The center of the lotus leaf is located in the neck. The edge of the lotus leaf is flat and the shape is similar to a "circle". The number of dorsal folds around the neck fundmentally determines the degree of perfection and satisfaction of the Da Gang. Generally, there are 36-80 dorsal folds in each Da Gang. The number of dorsal folds is different, as the degree of perfection required is different. Da Gang's level is also different. Even for Da Gang, which has the largest back fold, the circular edge is not fully expanded. The expansion range of the round edge is often not a full circle, and the radial dorsal folds are similar to the meridians of lotus leaves.

Da Gang for different sects and regions in Tibet follows lotus's shape (Fig.5). The shape of Da Gang presents a round wheel. This is inseparable from the aesthetic ideal of omnipresent perfection advocated by Buddhism.

Da Gang with 80 dorsal folds (From Qiangbarin Temple shooting in 2019.9)

The structure of Da Gang with 80 dorsal folds (From Wei-Min Chang drawing in 2019.10)

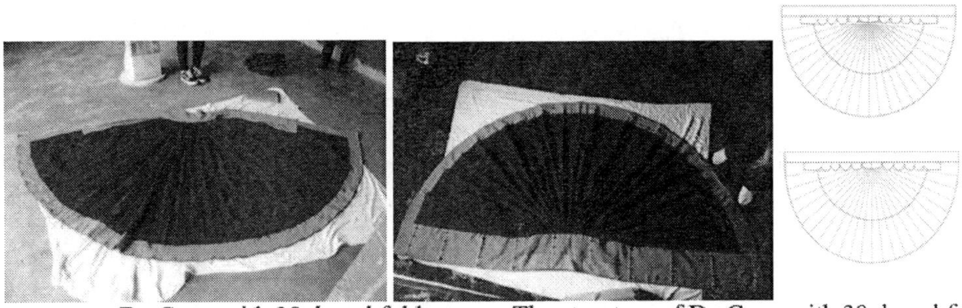

Da Gang with 38 dorsal folds The structure of Da Gang with 38 dorsal folds
(From Rongpu Temple shooting in 2018.9 and Longga Temple shooting in 2019.9) (From Wei-Min Chang drawing in 2019.10)

Lotus leaf (From https://gimg2.baidu.com/image)
Fig.5 Silhouette of Da Gang

3.3 Shoulder Piece Covered with Lotus Petal Pattern

Da Gang is composed of three parts: the collar piece, the body piece and the shoulder piece. The shoulder piece is a rectangular piece of cloth at the junction of the collar piece and the body, and the rectangular side is in the shape of lotus petal pattern. One side of which adopts the shape of the lotus petal pattern. The shoulder piece covers the pleats of the body completely. The center of the back collar of the lotus petal pattern is symmetrical. There are mainly three forms, among which the lotus petal pattern of the shoulder piece of Da Gang of Bon religion is more elaborate than that of Tibetan Buddhism (Fig.6). According to Gyung Drung Khenpo, introduced in Long Ka Temple, there is no Da Gang in the six pieces of clothes of Bon religion. During the integration with Tibetan Buddhism, Da Gang appeared in the Bob religion, so the local form of Da Gang is not as that of Tibetan Buddhism. The lotus petal pattern on Da Gang's shoulder piece emphasizes the integration of Tibetan maternal culture with Buddhism.

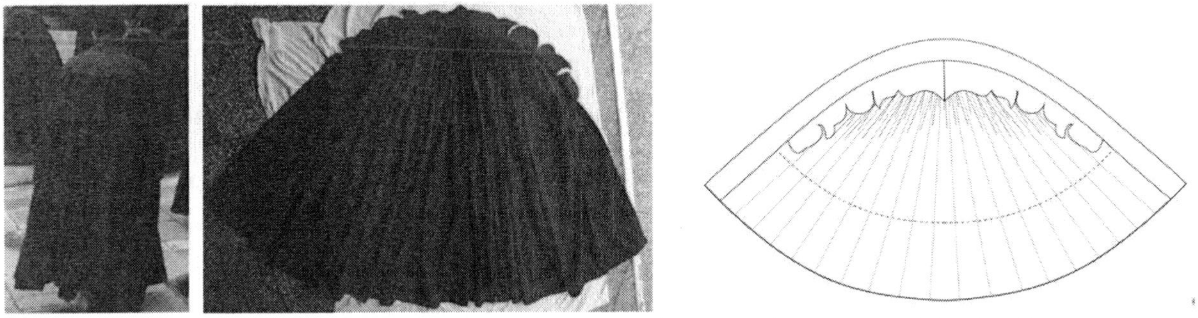

The monk wear Dang Gang attend the class in the morning the structure of the shoulder piece
(From Qiangbarin Temple shooting in 2019.9) (From Wei-Min Chang drawing in 2019.10)
Fig.6 shoulder piece covered with Lotus petal pattern

3.4 Lotus Stems Like Dorsal Folds

According to the body structure of Da Gang in Sutra Hall, there are 35-80 folds on Da Gang's shoulder and back. The number of folds would form different fold at different heights. The height of each fold resembles the stem of a lotus flower. The upper part of the stem is covered by the shoulder piece representing the lotus petal which forms a three-dimensional abstract illustration of the "lotus flower". The three-dimensional stem takes on different forms according to the garment's shape. These three-dimensional stemmed frills not only enhance the wrapping of Da Gang, but also resonates strongly with the Da Gang's religious significance (Fig.7).

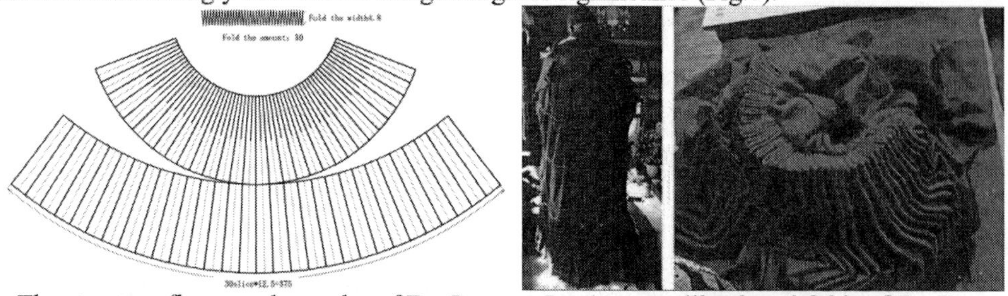

The structure figure and samples of Da Gang (From Wei-Min Chang drawing in 2019.10)
Lotus stems like dorsal folds of Da Gang (From Bayon temple shooting in 2019.9)

Red and red with yellow of the "stem" of Da Gang
(From Qiangbarin Temple shooting in 2019.9) (From Longga Temple shooting in 2019.9)

Lotus stem (From https://gimg2.baidu.com/image)
Fig.7 the dorsal fold and lotus leaf

4. The Religious Meaning of Da Gang's "Lotus" Figure

Lotus, also known as Zheng Jue flower [8], symbolizes purity, as it comes out of the mud yet remains clean and unstained. It represents the pure heart in all living beings. Both the whole shape and its portrayal in the Da Gang's designs conveys the internal relationship between the quality of lotus flower and Buddhist doctrines.

4.1 Symbolism of Objects

The symbolism of Da Gang lotus pattern represents a way of thinking inspired by a series of lotus patterns. The Wonderful Dharma Lotus Flower Sutra is short for Lotus Sutra in which the lotus is used as a metaphor, symbolizing purity and elegance of religious doctrine. Lotus represents the pure Dharma body of Buddha. Lotus is also regarded as the symbol of Buddhism and Dharma in Buddhist classics [9]. For example, the Buddhist kingdom is called "Lotus Kingdom", and the temple is called "Lotus House". As we can see, the lotus is the de facto symbol and holy flower of Buddhism. Buddhism regards life as a sea of suffering, hoping that people could get rid of the misery. The lotus grows in the muddy water and is otherworldly, not contaminated by the mud, and blossoms into incomparably beautiful flowers. According to the theory of great wisdom, the lotus comes from the mud; its beauty persevering against unclean sources. The "Sutra in Forty" says: "I am sramana, in the turbid world, and should be like a lotus flower without pollution". In short, Buddhism uses lotus as a metaphor, to point out to us that we should be as clean as the lotus in life and achieve purity and non-action in Buddhism and Taoism. The Avatamsaka Sutra, Brahmajala Sutta say that lotus hides the world, and the eight-leaves lotus flower is also the cradle of the Foetus that hides the Mandala Esoteric Buddhism" [10] (Fig.8). In terms of modeling, most of them are square shapes with eight petals, which symbolizes the concept of "four directions and eight positions" of the universe [11]. Arnheim once said that "all art is symbolic" [12]. Da Gang expresses the flower's celestial and dogmatic significance through its lotus flower inspired design.

Chakrasamvara——padma Mandala (From The Palace Museum)
Fig.8 Padma Garbha-dhatu-mandala

In conclusion, Buddhism uses lotus as a metaphor everywhere. In life, we emphasize to be like one lotus, without pollution and achieve the way to be Buddha that lets everything to be it is, not forced to confront worldly troubles.

4.2 Life Regeneration

Da Gang's bud-like posture emphasizes the feeling of rebirth when practitioners chant scriptures to keep out the cold weather. It implies that lotus can breed new life and practitioners can experience rebirth, emphasizing the unique maternal culture of Tibet [13]. The merits and wisdom of lotus are respected by Buddhist disciples and respected by the common people in the world.

As a result, the lotus represents the Buddha and becomes an important symbol in Buddhism, and the lotus pattern has also become the most common decoration in Buddhist art. Buddhism's love for lotus flowers stems from the agreement between its character and Buddhist teachings, and the lotus pattern is regarded as the most common element in Buddhist decoration. In Buddhism, lotus is regarded as a state of rebirth from one shore to the other, and the state it enters is the lotus state, which is an ideal state of mind in Buddhism.

There is a saying in Avatamsaka Sutra: "Five color stems come out from the land of Buddha. The flower represents the world, the wood represents life, the grass represents heaven, the leaf symbolizes Buddha".

4.3 Study Cautionary

In the form of Lotus, Da Gang serves as a warning of the buddhist precepts in the practice of monks. The sturdy lotus stem on its back has become one of Da Gang's symbolic religious languages. The complicated lotus stem guides believers to aspire in reaching the lotus realm as their ultimate goal. They should strictly require themselves to be "clean in body and mind", and as quiet and beautiful as a lotus flower. Its significance coincides with the core theory of Buddhist scriptures. The Sutra says:" lotus is the only one who has many fruits. It can be said that the fruit is round and virtuous, because it contains many ways of spiritual practice." There are three forms of the lotus stem of Da Gang, and its religious ritual connotation is unified in all interpretative designs (Fig.9).

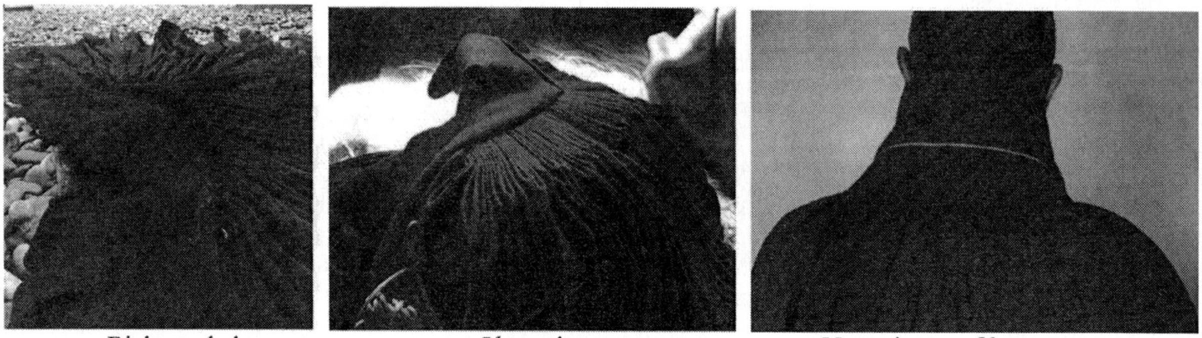

Right angle lotus stem Obtuse lotus stem Natural state of lotus stem
(From Qiangbarin Temple shooting in 2019.9)(From Bayon temple shooting in 2019.9) (From Gaden Temple shooting in 2017.10)
Fig.9 Da Gang lotus stem design

The lotus stem on the back of Da Gang and the lotus petal pattern shoulder piece have deep religious connotations related to women. In Hindu and Buddhist scriptures, the lotus flower is metaphorized as a sacred uterus or vagina. Lotus is a synonym with the signifier for "vagina" in Sanskrit [14]. Da Gang's frill of lotus stem and lotus petal pattern on shoulder piece form a cautionary deterrent, which is in line with Freud's "tabu" theory. Buddhism's tabu to women is general knowledge, which is also true in Tibetan Buddhism. In rules of council for tribal affairs in the Qing Dynasty, there were strict regulations on the contact between lamas and women, saying that lamas ``live in the families of unmarried women, no matter whether they commit adultery or not, they commit adultery or not, the Lama would be deprived of title of yellow, whip one husband times, and order to return to the secular." [15] The lotus stem frill on the back of Da Gang is the embodiment of that strict celibacy. The formation and implementation of this kind of religious rule is directly related to the restoration of Buddhist order by Gelug Sect of Tibetan Buddhism. Da Gang produces the" Tabu "effect through the shape of lotus flower, enabling the monks to obtain spiritual power from a thing that can be seen and tangible. The lotus meaning could strengthen their will and determination to practice.

4.3 Both Religion and Secularity of Da Gang

The monk's Da Gang of Tibetan Buddhist applied the secular "lotus flower" pattern to the religious costume modeling. The lotus figure with auspicious meaning is most acceptable to the Tibetan people. They consciously and purposefully apply the Buddhist concept to daily life, so as to exorcise evil spirits, place good wishes and achieve the goal of perfect transcendence. According to Feuerbach, only religion close to secular life is more convincing. Da Gang is unique clothing of Tibetan localization of Indian monk clothing. It is also the product of religionization of Tibetan cloak clothing. The traditional culture of "out of the mud but not stained" coincides with the "transcendence" in Buddhism. The social custom of religion and secularity in Tibet has laid a foundation for the prosperity of Tibetan religious culture. Once religion takes shape, the secular society will naturally depend on religion. They live together and depend on each other.

5. Conclusion

Lotus flower is a plant found in nature, but massive cultural impacts of the lotus flower pattern should start has had its roots as a symbol of Buddhism [16]. Compared with other symbols, lotus flower is the closest to daily life, the most aesthetic, and the easiest to localize. Lotus flowers have long been placed in a high renown during Buddhist classics. The natural attributes of lotus flower are similar to Buddhist doctrines, commandments and rules, forming the supreme worship of lotus. The application of Da Gang's "lotus flower" figure reflects the simplicity of Buddhist culture, enriching Da Gang's religious significance. In order to restore the strict law of Buddhism, this paper defines the differences between monks and their customs. The garments' design makes a positive contribution to regulating the behavior of monks. Just as Shan Jian Lv (The law of good opinion) says: "only by practicing the precepts, can we better protect the Dharma. And the monks who wear Da Gang express that they have the mind of keeping precepts, which undoubtedly has an impact on the secular world. Da Gang's "lotus flower" figure warns monks to practice forever in order to achieve perfection".

The presentation of Da Gang's "lotus flower" figure is not an intuitive pattern expression, but conveys religious ideas through clothing outline and internal structure. The transformation of lotus decoration is not only the result of the absorption and integration of foreign cultures, but also the result of in-depth understanding and development of Tibetan local religious and secular culture.

6. Acknowledgement

This work was supported by the doctoral project in Beijing Institute of Fashion Technology. The project name is "Rescue and inheritance of Chinese traditional costume culture and design innovation project". The project number is NHFZ20190110.

References

[1] Dao X, Yuan ZH, Hong Y: Si Fen Lv Xing Shi Chao Zi Chi Ji Jiao Shi9[M]. Beijing: Religious Culture Press: 2015; 2373.
[2] Chen SQ. General synopsis of the Tibetan Classics, Shanghai Ancient Books Publishing House: 2008; 194.
[3] Li YQ: Research on Tibetan costume culture; Bei Jing: People's Publishing House: 2010; 178-189.
[4] Ji CL: the theory of clothes and name of the traditional Tathagata and Monk image: Collected works of 2004 Longmen Grottoes international academic seminar; Zhengzhou: Henan people's Publishing House: 2006; 629.
[5] Luo WH, Gesangqupei. Mural in Gonggar Choide Monastery.Palace Museum press: 2015; P. 212.
[6] Luo WH,Gesangqupei. Mural in Gonggar Choide Monastery.Palace Museum press: 2015; P. 202.
[7] Luo WH,, Gesangqupei. Mural in Gonggar Choide Monastery.Palace Museum press: 2015; P. 212.
[8] Ding FB: essential dictionary of Buddhism; Beijing: Religious Culture Press: 2014; 144.
[9] Yi G. Notes on the Lotus Sutra; Beijing: Religious Culture Press: 2017; 7.
[10] Gong J. E Dun Zha Ba, Re Se. Gong Jue Jia Cuo, Gong Jue. Sang Dan: Collection of Tibetan Buddhist instruments and rituals; Lhasa: Tibet people's Publishing House: 2012; 29.
[11] Zhang PC. The flower of the sky in the cosmic schema-Persimmon pattern discussion: Decoration: 2002; 12: P5.
[12] Arnheim R. Art and visual perception; Bei Jing: China Social Science Press: 1994; 633.
[13] Zhang HC: Selfless and nirvana: the essence of Buddhist Ethics; Changsha: Hunan University Press: 1999; 75.
[14] Bill R. Symbols and utensils of Tibetan Buddhism;Bei Jing: China Tibetology Publishing House: 2014; 8.
[15] Li FY (Council for Tribal Affairs). Li-fan" system, Yang XD, Jin F annotated. Hailar: Inner Mongolia Culture Press: 1998; 412.
[16] Zhang YF. The symbol of Buddhism: The Voice of Dharma: 1992; 07: 22-23.

Cite as: J.W. Yi et al.,
TBIS 10.3993/tbis (2021).

Artistic Characteristics and Evolution Of Modern Fashion Vest

Jiang-Wan Yi, Li-Zhao Qing*

Xiamen Academy of Arts and Design, Fuzhou University, Jimei District, Xiamen, Fujian, 361000, China

*Corresponding author's email: 1304814327@qq.com

Abstract

Based on recent first-hand literature, this paper interprets the style changes and artistic characteristics of modern fashion vest, systematically combining its historical evolution characteristics, and summarizes the popular law of fashion vest from the late Qing Dynasty to the 1930s. The literature and iconology are further analyzed to evaluate the artistic characteristics and evolution law of modern fashion vest.

Keywords: Modern Times; Fashion Vest; Artistic Characteristics; Evolution

Introduction:

The vest is also called "half back", commonly known as "shoulder" or "vest". As a sleeveless garment, its production method is similar to "two crotches" (a vest style garment popular in the Jin and Southern and Northern Dynasties) [1]. It is a one piece that covers the chest and back area. "Two crotches" refers to two pieces of clothing, one in front of the chest and the other behind. Wang Xianqian of the Qing Dynasty noted: "The "half back" of the Tang and Song Dynasties is the same as the present vest." With a long history, the general body shape of vest remains unchanged in the process of development. As the vest can protect the body and is easy to wear, it became popular in different dynasties.

Due to the great changes in modern society, Chinese traditional costumes have changed significantly in just one hundred years. As one of the traditional clothing styles, vest not only inherits the characteristics of Chinese clothing, it has also boldly absorbed the characteristics of Western clothing, evolving into a functional and beautiful clothing style suitable for women. According to current relevant research, few people have studied the fashion vest's popular law and artistic characteristics. Based on the style changes of modern fashion vest, this paper analyzes and summarizes the artistic characteristics of fashion vest and expounds the popular characteristics of fashion vest in different periods, hoping to provide a theoretical reference for further research on modern fashion vest.

1. Analysis of Changes in Styles of Modern Fashion Vest

1.1 Late Qing Dynasty and Early Republic of China

In the late Qing Dynasty, Manchu and Han Cultures blended with each other, and people broke the boundary between flag clothing and Han clothing. As according to traditional fashion, Chinese fashion vest style have simultaneously maintained the key elements of broad outline and plane cutting. Due to the establishment of the Republic of China and the prosperity and development of Shanghai's economy, the fashion center gradually moved the south, and a new style of vest appeared [2].

In terms of shaping, there is little difference between the late Qing Dynasty vest and the traditional vest, and the structure is generally similar. That is to say, the vest is horizontal and vertical, with a trapezoidal or rectangular outline. The side seams on both sides of the body have no obvious arc change. The main ways to open the placket include a big placket, one word placket, pipa placket, antithesis placket and so on. Take Fig.1 as an example is given. The vest with standing collar has six buttons. As far as the way of placket opening is concerned, it belongs to pipa placket originating from the design of partial door without placket. Therefore, it is also known as the missing flap. The trend of the placket is similar to the placket, but it is not connected with the armpit. Instead, it swings down vertically, turns to the horizontal line, and finally aligns with the middle of the collar. Its overall structure is horizontal and vertical, the style is relatively wide, giving people a neat and generous, traditional classic image. The structural lines of shoulder, side seam and hem are straight. Although the cutting of armhole presents a curve shape, it still belongs to two-dimensional plane cutting, which does not highlight the beauty of human body lines.

Fig.1 Pipa-breasted vest in the late Qing Dynasty (From http://www.biftmuseum.com)

In the early years of the Republic of China, the country gradually opened, leading to the influence of Western style on Shanghai's clothing. It is during this time that the vest changed: the typical characteristics of the fashion vest in the early Republic of China included the excessive collar shape, the length of the buttocks and the slightly flared shape. In the early years of the Republic of China, brothel women became the leading fashion pictures at that time. Due to their occupation, they needed to dress well to attract customers, and they were bound less by traditions. In the film "The Shadow of the Prostitutes in the Country", many brothel women took pictures in long vests with a collar and knee length [3]. The collar type of this period is called Yuanbao collar. As the collar was extremely high, every woman has elongated her neck and their faces looked slim and more slender. The front part of the vest consists of two pieces, with an upper and lower part. The vest is cut horizontally from the collar, and several pairs of turnbuckles are nailed at the cutting part to connect the upper and lower parts. Since the buttons are arranged in a straight line, it is called a straight placket. With traditional Chinese plane cutting, the vest has an arc-shaped design in the hem, shoulder line and armhole, and the vest profile is A-shape, which changes the traditional H-shape, making the wearer more slender and straight visually.

Fig.2 Prostitute in long vests in 1910s [4]

In general, the vest style had very fewchanges in the late Qing Dynasty and the early Republic of China. Most of the vests are cut horizontally and vertically, the body is wide and out of body, and there is no prominent curve shape.

1.2 The Middle Period of the Republic of China

The biggest difference between the fashion vest in the middle of the Republic of China and the traditional fashion vest is the emergence of more diversified styles. Western clothing culture gradually influenced womeniggest differe in the Republic of China. During these times, women prefer more suitable fashion vest. With the characteristics of Chinese style and Western style, these small vests appeared in the wardrobe of women in the Republic of China. In addition, the innovative long vest was greatly welcomed.

As it was cold in spring in the early 1920s, it was popular to wear a fitting vest on the outside of the coat. This vest was called "half arm" at that time [5]. In terms of structure, although the pony vest is not different from the traditional Chinese style, its slim and concise design is particularly similar to western style clothing. The style is tight and thin, with a narrow stand collar. The length of the vest ends at the waist, and the bottom shows an arc angle, which can highlight the beauty of the body lines.

In the mid-1920s, long vests became popular. People usually put a long vest on the outside of a short coat or cheongsam. Because fashionable Shanghai women realized that the short clothes shelved in the wardrobe or covered in the cheongsam can be used as new clothes, they put the long vest on the outside of the old clothes, giving the vest a new fashion use [6]. In 1925, an article in Shanghai pictorial wrote: the new dress of Shanghai women has brought us into the era of big vest. People add a vest to their tulle, which has many colors. The form of vest varies in length. The short vest only reaches the position of the navel, and the long vest can be dragged to the ground. People used to like small vest and corsets, but now people are turning to big vest, which is one of the phenomena of decoration changes. In the 1920s-1930s, vest not only changed significantly in length, they were also influenced heavily by western style clothing, with an emphasis on highlighting the female body curve. The big and long vest is the right lapel of the big skirt, the body is loose, and the hem reaches the ankle or calf. However, in the chest and buttocks, there are lines in the long vest, which changes the traditional straight vest style. In the late 1920s, the long vest not only followed the Chinese plane structure, it also integrated western style modeling techniques. People use the bust dart and waist dart of Western clothing structure to make the vest fit better. At the same time, people sew shoulder seams to make the shoulders fit better. At this time, the long vest with S-shape appearance appeared [7].

Fig.3 Long Vest [8]

Through the pictures, we can see that in the middle of the Republic of China, the vest changed from short to long, which modified the beauty of women's body lines. The widespread popularity of long vests was in the mid-1920s. It was evolved from the back of the Chinese dress [9]. Later, this kind of vest evolved into a long vest style cheongsam, known to women as the cheongsam vest. As far as the traditional women's wear in modern China is concerned, Qipao vest breaks through the limitation of the skirt under the top and plays an innovative role in the traditional women's wear.

1.3 Late Republic of China

From 1930 to 1940, women in the Republic of China ushered in the peak stage of fashion women's wear development, and a large number of western fashion poured into China. In 1930, the length of the vest shortened again. With a simple and elegant overall design, it is full of traditional Chinese women's clothing charm and modern sense [10]. At that time, there were various types of vest, including cheongsam vest, knitted vest, western vest and so on.

In the 1930s, braided clothing entered China and was widely popular among fashionable women [11]. There are many kinds of knitted clothing in China, including gloves, scarves, socks and other accessories as well as vest, sweater, coat and other clothing styles. Because of its wide class applicability, practical heat preservation and air permeability as well as convenient matching, people have published the relevant knitting methods of vest in many newspapers and magazine, from which we can see that vest was greatly loved by women in this period. In 1934, ARK Magazine published the detailed knitting method and materials of a women's vest, and attached the matching display and plan. Simple and fashionable, this woven vest has no cardigan and a waist length design. It not only considers fashion and warmth, it also does not affect body movement. It fully demonstrates the aesthetics and practicability of the clothes popular in the 1930s.

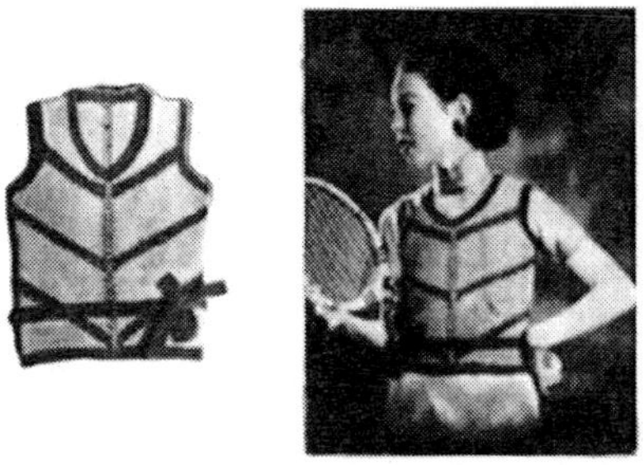

Fig.4 Upper body and plan of women's vest [12]

In the 1930s, female students preferred simple vests with pockets. Because they can put their belongings in the vest bag and wear the vest in a loose way, the vest is both practical and fashionable. At that time, it was still popular for women to wear men's suit vests. These vests are solid green fruit necked open vests. Because most of them are not buckled, women can show their long white arms. This kind of modeling is widely popular among fashionable women, and the short vest with narrow body shows the curve of women's body. This development symbolizes the change of traditional aesthetic value, and the public acceptance of western aesthetic values.

In the late period of the Republic of China, the popular vest was the western style knitted vest or the cardigan vest with simple design, which had strong practicability and showed the body lines of female Miaoman generously. After 1940, due to the popularity of Western women's wear and war, there was no new style change of vest, so it is not discussed here.

2. The Artistic Characteristics of Modern Fashion Vest

According to the changes of modern fashion vest styles, we further summarized and analyzed the patterns and compatible accessories of modern fashion vest.

2.1 Pattern

Table 1. characteristics of modern fashion vest pattern

Patterns	Common elements	Quantity
Auspicious patterns	Curved patterns, Huizi pattern	10
Traditional flowers and plants	Lotus, Biaoxiang flower	13
Western-style flowers, roses, vines	Rose, rattan	7
Geometric pattern	Horizontal and vertical stripes	19

We sorted out more than 100 documents and pictures. After eliminating the blurred image, we analyze and summarize the pattern of fashion vest. Modern fashion vest patterns mainly include auspicious patterns, traditional flowers, Western flowers and geometric patterns, as shown in Table 1.

Due to the social and political changes, economic turbulence and the changing aesthetic fashion, from the late Qing Dynasty to the late Republic of China, the overall pattern of vest presented changed from complex to simple, from dense to sparse, from existence to non-existence. In the late Qing Dynasty, there were many patterns of vest, including auspicious patterns and traditional flowers. Let us take the pan length as an example, as its knots are continuous in shape and have no beginning or end. This shows the circular implementation of the Dharma and conveys the meaning of long-term eternity [13]. People use it in clothing, expecting the prosperity of the family, the continuation of the descendantsas well as auspiciousness. This kind of design contains the beautiful vision of the people at that time. A design must have a unique moral, and its moral should express and bring good luck. These patterns reflect the traditional Chinese culture. In the early years of the Republic of China, complex decoration gradually reduced, but there are still many decorative details. Western clothing culture not only affects the style. Influenced by the "new art" style, Western plant and flower patterns appear in a new and natural form. In addition, in the middle and late Republic of China, geometric abstract patterns frequently appeared on the vest. For example, the rose pattern symbolizing love, the grass and vine pattern with warm vitality, the wave point of the living wave, the horizontal and vertical stripes, and the lattice pattern are not the patterns of traditional Chinese vest.

2.2 Accessories

Because people have long used a long history of vest with other clothing, so different matching accessories can also reflect the era and best fashion.

In the late Qing Dynasty, no matter how old people were, people used vests to match Manchu flags. In the period of the Republic of China, the collection of vest was more diversified. As the mainstream, people used small vests to match with simple plain color coats, short and wide sleeves or with floor length skirts, and embellished the same color lines on the edge of the trousers. With the overall unity and harmony, people chose similar colors of vest and skirt, walking with aswaying posture. At that time, wearing hairpin was very popular. People often wear decorative flowers on the shoulder and neck to match with vest. Based on the fabric of dress, people would collect different hairpins and sprayed perfume. In the 1930s, combined with the cheongsam, the vest presented a mixed effect of rich layers and the combination of Chinese and western styles, highlighting the curvilinear beauty of women. The historical records of this period are as follows: "People chose a light-colored long cheongsam with short sleeves, coupled with a woolen vest scattered on the chest, or chose a long cheongsam with short sleeves and silk edges and coupled with a one foot long vest with a tight belt around the waist." [14]. The overall modeling is fashionable and neat, and some women will match with a narrow scarf with printing, which fully shows the image reshaped by Chinese and western clothing culture.

Table 2. characteristics of modern vest accessories

Accessories	Quantity
Manchu outfit	23
Epididymis	24
Hairpin	10
cheongsam	33
scarf	4

3. Analysis on the Evolution Law of Modern Vest

Based on the above analysis, we can see that the overall shape of modern vest remains unchanged, and the overall length presents a curve change. Moreover, the decoration has change from extremely cumbersome to simple, and then to no decoration. The change of clothing style is from traditional Chinese aesthetics to western clothing elements, and finally a vest is formed that captures both the Chinese and western clothing culture. The evolutionary law is shown in Fig.5.

Fig.5 the evolution of modern fashion vest (drawn by the author)

In the late Qing Dynasty, the style of vest was relatively simple, except for the skirt design. The clothing structure was relatively simple, but the decoration was complex. In the early period of the Republic of China, there was a new development of vest, with a new version and collar. At the end of the Qing Dynasty and the beginning of the Republic of China, the vest had the characteristics of simple style and dense decoration; In the middle period of the Republic of China, compared with the previous period, vest had a more abundant development. The first is the appearance of small fitting vest, and then the popularity of long vest. With regard to compatibility, the hairpin flower matches the vest, but generally ornamentation reduced. Due to the variety of styles and collection of vests, vests were more diversified in this period. The characteristics of this period are "more styles" and "less decoration"; After 1930, the simple design of the knitted vest and suit vest became widely popular amd without decoration. In the late period of the Republic of China, the style of vest returned to "simple", and there was no accessory.

4. Conclusion

Due to the variety of styles, simple structure, functionality and fashion, modern vest became a popular clothing type for women. The research shows that we can divide the evolution from the late Qing Dynasty to the 1930s into three stages; In the late Qing Dynasty and the early Republic of China, the vest was inherited from the traditional Chinese vest, which was influenced by Manchu clothing, and the clothing was broad and complex; In the middle period of the Republic of China, the vest was influenced by the western culture, and began to highlight female body lines, with diverse styles and more simple decoration; In the late period of the Republic of China, the vest had no decoration and the style was more westernized. The graphic pattern of fashion vest is more concise, and the matching accessories are consistent with the fashion of the times. As a unique style in modern clothing, vest not only reflects the aesthetic elements of modern clothing development, it also shows the cultural integration of Western clothing style.

References

[1] Gao G. A detailed description of the graphic version of Chinese clothes. Beijing: Guangming Daily Publishing House: 2010.
[2] Bian XY. History of Shanghai style clothing in modern China. Shanghai: Donghua University Press: 2014.
[3] The famous prostitutes in Shanghai: Qixianglou. Yanling Huaying: 1911 Xiaoying of Famous Prostitutes in All Ports of the Country: p. 114.
[4] The famous prostitutes in Shanghai: Libaolan. Yanling Huaying: 1911 Xiaoying of Famous Prostitutes in All Ports of the Country: p. 98.
[5] Sun JZ. "Spring Cold". Beijing News Supplement: Issue 119: 1925.
[6] Shan S. Shanghai women's vest. Shanghai Pictorial, Issue 23: 1925.
[7] Li N. Origin of modern women's wear. Comparison of Chinese and Western women's wear in 1920s. Beijing: China Textile Publishing House: 2012.
[8] Celebrities. Liangyou Pictorial: 1928, Issue 33; p. 46.
[9] Introducing celebrities. Liangyou Pictorial: 1928, Issue 31; p. 22.
[10] Li ZQ. Research on fashion in old Shanghai (1910-1940s). Shanghai: Theatre Academy: 2015.
[11] Liu Y. Research on the "new clothes of civilization" in the Republic of China and its popular replacement with the improved cheongsam. Decoration: 2020; (1): 80-83.
[12] S HT. Women's vest (applicable for playing tennis). Fang Zhou: 1934: Issue 12; p. 57-58.
[13] Wang H. "Analysis of casual dress in the middle and late Qing Dynasty ". Beijing Institute of Fashion Technology: 2007.
[14] Ke DZ. Brilliance language. Shenbao Newspaper: April 7; 1933.

Cite as: J.F. Zheng et al.,
TBIS 10.3993/tbis (2021).

Research and Contemporary Application of Modern Shanghai Fashion Dance Clothes

Juan-Fen Zheng[1], Zhao-Qing Li[2*]

[1]*Xiamen Academy of Arts and Design, JiMei Street, Xiamen, FuJian, 361000, China*
[2]*Xiamen Academy of Arts and Design, JiMei Street, Xiamen, FuJian, 361000, China*

[*]Corresponding author's email: 961226603@qq.com

Abstract

By studying and combing the design elements and characteristics of modern Shanghai fashion dance clothes, it conveys the fashion trend and cultural connotation of that time. At the same time, by combining the current design its contexts, the traditional design element characteristics and contemporary women's clothing are merged to iterate on design choices found in female dance clothes in contemporary context that meets the Chinese aesthetics.

Keywords: Modern; Fashion Dance Clothes; Clothing Elements; Dance Clothes Design

Foreword: In the history of Chinese clothing development, modern times has been an important transitional period between tradition and modernity. The strong implantation of Western clothing cultures has brought a major impact to China's local clothing culture, but also injected new momentum in its development. Among Chinese cities, Shanghai's clothing has changed the most. Relying on its unique geographical advantage as a port at that time, Shanghai had continued to accept foreign fashions and became the largest fashion center in China at that time.

1. The Development Background of Modern Shanghai Fashion Dance Clothes

Modern Shanghai fashion dance clothing refers to fashionable and modernized dance clothing, which has a friendly, communicative and casual temperament, while also being appropriate for dancing and conducive to stage performance. At that time, the clothes were worn according to applicable occasions, resulting in slightly different designs. Categorically, dance clothes can be divided into tea dance clothes, banquet social clothes and stage costumes.

The specific factors that promoted the development of Shanghai's fashion dance clothes at that time were as follows:

1.1 Changes in the Way of Life, Entertainment and Social Interaction.

In modern times, Shanghai became the commercial center of China in the 1920s and 1930s due to its unique geographical location, resulting in a most prosperous development of entertainment and its subsequent consumption by its populace. Ballroom dancing, which is widely popular in the West, was introduced to China. At the same time, Shanghai's entertainment venues are becoming increasingly refined leading to major hotels, clubs, etc. toset up ballrooms, all of which made people's enthusiastic for dancing. Dancing has quickly become a social activity that people followed and pursued, and dance halls and dance schools have quickly emerged. The Dadong Ballroom opened by Yongan Company in 1927 became the first commercial ballroom in Shanghai. As of 1933, there were 39 ballrooms with business licenses in Shanghai. In addition to the rising popularity of dance studiosdance-related textbooks were also popular. At the same time, many singers and dance stars have received widespread attention. They wore flowing dance clothes and showed their moving dances under the spotlight, earning the peoples' yearning. At this point, ballroom dancing has gradually become an important activity amongst the upper class and has managed to attract the attention of general public during weekdays. Peoples' entertainment and social practices have gradually changed, which promotes the development of dance clothes.

1.2 The Promotion of Mass Media and Fashion Shows, and the Change of Aesthetic Concepts.

The development momentum of modern mass media has been flourishing, and many influential periodicals have had extensive discussions and dissemination of women's thoughts, women's fashion, and women's makeup. These newspapers display fashionable photos of celebrities, sketches made by fashion designers, and media coverage from Europe and the United States, etc, so that people can keep up with current beauty and fashion trends, and were thus popular. The development of the modern clothing industry has provided an important material basis for the changes in women's clothing. Through holding fashion shows and various beauty pageants, the invitation of celebrities as models, as well as the promotion of novel clothing styles. These factors are subtly changing people's aesthetic concepts and clothing choices, thereby promoting the development of clothing and accessories.

2. Fashion Dance Clothes Design Elements

If a piece of clothing is to be regarded as a whole, then the whole should be considered as acomposition of its constituent elements. The constituent elements of clothing are diverse, including styling, fabric, color, decoration, and so on. Through the analysis of theirstyle, fabric colors, decorative details and other elements of modern dance clothes, specifically in the context of Shanghai in the Republic of China, it is possible to summarize the characteristics of fashion dance clothes elements and discover its inherent artistic value, so as to better stimulate modern clothing design inspiration and innovation.

2.1 Styling and Structure

Fashion dance clothes present a trend of gradual westernization in the styling and structure. Through research and analysis, one can determine that most fashionable dance clothes in China in the 1920s and 1930s were skirts. Affected by the western trend for narrow clothing and accessories, its overall styling has changed from H-shaped to S-shaped. Specifically, by cutting the waist, hip, chest and abdomen curves, the natural presentation of the human body curve is more emphasized. With the strong collision of Chinese and Western clothing cultures, fashion dance clothes are gradually moving closer to the West, partially showing exaggerated and illusions resembling naked forms. These changes have also benefited from the awakening of women's free consciousness and the change in aesthetic concepts. This freed them from feudal ethics, to pursue freedom and equality between genders. This also prompted them to proactively participate in social activities beyond the domestic sphere. Women gradually allowed themselves to show off their independence, expressing themselves through adopting fashionable dance wear. As a result of this aesthetic development, women were able to achieve double liberation in both the body and mind.

For different situations, there are variants in the styling and design of different dance clothes

2.1.1 Tea Dance Clothes

"When the 'Entertainment of Tea Dance' enters the dance hall, people can not only enjoy the exquisite refreshments, but also talk freely. When the music sounds, people can dance together and enjoy themselves physically and mentally. This is really normal and noble communication [1]." Tea dance clothes are more appropriate for daily casual social interaction, resulting in more daily and practical designs, focusing on creating communicative and casual temperaments, while maintaining its mobility for dancing. It is generally an improvement of the upper Chinese style and the lower Western style. That is to say, assuming the natural waistline acts as its boundary, the top retains the traditional Chinese elements of the cheongsam, such as the original stand-up collar, placket, and raglan sleeves, while the waist resembles a western-style flared or asymmetrical skirt style that is convenient for dancing. As shown in Fig.1, this is the tea dance suit worn in the fashion show. Its overall shape is slender and fits the curve of the body. The raglan sleeves and half sleeves are decorated with ruffles, the neck and shoulders are embellished with colored diamonds, and the hem is naturally hanging down, creating an elegant feeling.

2.1.2 Social Banquet Clothing

"During the Republic of China, Western ballroom dancing was an important social event, so dancing clothes were necessary. At that time, Western evening dresses were also used for dancing clothes [2]." This kind of dance dress resembled contemporary dresses of its time, reflecting strong influences from European and American fashion styles: narrow upper bodies and exaggerated lower hems. The design prioritizes mobility over aesthetic qualities found in evening dresses. At the same time, compared to tea dance clothes, they are still more grandiose compared to tea dance clothes, often using elegant and light fabrics, including intricate designs such as lotus leaves and folded decorations to make the presentation of the dance more perfect. Such dance clothes are often unique in shape, as shown in Fig.2. This dance suit "has a wide chest, narrow shoulders into a ribbon styling, and buttons at both ends. When wearing two straps on the back, the chest will be scratched. The back is completely exposed from the waist up. The buttocks have a starting line to widen the hem, which allows for mobility. There is also a scarf, which is made of black yarn against small dazzling glass pieces. It is knotted from the back of the shoulders and is called the beauty knot [3]. "As shown in Fig.3, the upper part is narrow, the lower part is wide, and it is flared. There is a tight waistband around the waist, which is hung from the left waist, and both ends are decorated with black glass. A lotus leaf-shaped shawl is draped on the shoulders, and tongue-shaped black glass pieces are embellished on the front and back of the neck.

2.1.3 Stage Costumes

These dance clothes were mostly worn by singers and dancers in dance halls and during performance. Therefore, instead of focusing on setting up casual and friendly temperaments, these stage costumes pay more attention to creating dramatic and exaggerated expressions, influenced by European and American fashion styles. As shown in Fig.4, this dance suit adopts pleated elements, and the overall shape is exaggerated and unassuming, thoroughly drawing from Western styles. The upper body is shows a pleated cloak with a hedging, and the collar, hands and skirt are decorated with ruffles. The front and back slits of the lower body skirt and the wide skirts increase the elegance of the dancer, which is more beautiful and agile when looming. Due to the emphasis of Chinese clothing culture on human body wrapping and privacy, even as people are beginning accept Western cultural norms of curves and exposure, they retain the charm of oriental implicit beauty its their garments' overall design. Therefore, compared with the West, there are fewer exposed areas, and it pays more attention to proper decency.

Fig.1 Fig.2 Fig.3 Fig.4
"China (Shanghai)" 1935.31 [4] "Art Life" 1934.7 [3] "Art Life" 1934.7 [3] "Extravehicular Pictorial" 1936.3 [5]

2.2 Fabrics and Colors

Fashionable dance clothes are mostly light and revealing skirts with bright colors. As shown in Fig.5, "It is made of tulle, which is extremely light and cool, and is especially suitable for dancing [6]." Out of functional considerations for the convenience of dancing, dance clothes are mostly made of tulle, which is slightly transparent and has a hazy beauty. At the same time, it is suitable for stacking or pleating, which is considered vogue. The most fashionable new would be displayed at dance parties. These dancing yarns are not only transparent, but they are also often decorated with colorful beads, in colors such as light green and orange. As the record shows, "A new kind of women's evening dance gown is a successful product of a famous factory. It is made of fine transparent yarn, crafted by famous craftsmen, and tailored out. It is also decorated with colored diamonds, and the surroundings are in an excellent style and are extremely gorgeous. There are many colors, such as rose, blue, green, white, black and so on. This dance costume glitters with brilliance, and is reflected in the spotlight in the dance hall, which will surely amaze the audience." The cut material is usually silk fabric, using extremely gorgeous silk or Western-style floral silk, and the fine clothing is made of Western dancing yarn."[7]

Fig.5 "Ling Long" 1931.10 [6]

2.3 Decoration and Matching

The fashionableness of fashion dance clothes is mostly embodied in its decorative design. These decorative design details are mostly concentrated on the collar, sleeves, and skirt hem. Their overall shapes basically fit with the upper body, the hem is spread out, with a floor-length skirt. It pays much attention to the use of lotus leaf, on the shoulders of the collar, the waist of the sleeves, or the skirt; the lotus leaf decoration can be seen, mainly conveying a sense of elegance in dancing and reflecting the beauty of women. Different variations of bell sleeves and waistlines are divided horizontally in three sections. "The method of opening the collar is square or round, or triangle, or shoulder-length." In addition to the change of length, it attaches great importance to the change of sleeve shape, and often adds single or multi-layer pleated lotus leaf sleeves on the sleeves. To increase the dynamic feeling of dancing, the skirts are usually larger, and the common types are diagonally cut, flared, multi-layer wave, and so on. Occasionally, folds or layered decorations are used to bring out its gorgeousness. In the later period, to dance freely and conveniently, collarless, sleeveless, and even skirts longer than the thigh appeared (Fig.8), with lace and flared styles becoming quite popular. The decoration of its edges are different from the previous cumbersome and delicate patterns, the dance clothes put emphasis on the splicing decoration of lace, velvet and other laces (see Fig.9), or the expression of geometric abstract patterns. The expression of skin exposure is usually more subtle, mostly concentrated on the back and so on. [8]

In terms of costume collocation, to enrich the stage effect, stage costumes often wear exaggerated hats, hair accessories and jewelry. These accessories usually use the same fabric color as the dance clothes, paying attention to the coordination of the overall outfit. The banquet social clothes etc. pay more attention to daily practicability in its coordination, usually with exquisitely small handbags with high-heeled patent leather shoes. Western-style coats were also recommended for outings, to show off an exquisiteness and elegance from the inside and the outside. [9]

Fig. 6
"Qiong Bao" 1929.9[10]

Fig. 7
"Art Life" 1934.2[11]

Fig. 8
"The Times" 1934.2[12]

Fig.9
"Ling Long" 1931.5[13]

3. Application of Fashionable Dance Clothes Elements in Contemporary Dance Clothes Design

As Western influences spreadto the East, China's clothing styles gradually gravitated closer to the west. The origin of the western clothing system did not depend on the Chinese cultures enthusiasm for complex decorative totem patterns, and at present, the western clothing culture and fashion industry remain as the mainstream. Therefore, Chinese designers are actively exploring ways for traditional elements to be incorporated into modern design practice. Chinese-style fashion is not simply a fashion made of traditional Chinese and Western-style clothing elements, nor is it a fashion that refuses all foreign cultures in its adherence to tradition. It should be resonant: a contemporary fashion that fits the current Chinese aesthetics.

The fashion industry attaches great importance to the development of local clothing culture. Designers strive to push the unique Chinese aesthetic clothing into the world, with the influence of Western clothing culture. This important clothing node in modern China, including fashionable dance clothes with unique artistic charm, has gradually entered the vision of the public and designers. Through combining and researching its design elements such as its shape, fabric, color, decoration, and matching, it can discover its unique artistic value and portray the image of an exquisite and elegant modern girl. In addition, it can provide designers with more anachronistic inspirations, thereby better exploring potential applications of traditional elements in dance dress design with respect to contemporary Chinese aesthetics.

The design of this series (Fig.10) is inspired by the exquisite, elegant, and luxurious image of modern girls summarized in the research and combing process. At that time, women were deeply oppressed by the wide-clothing culture and did not dare to expose their body curves at all. Therefore, at that time, the emphasis on female body curves and the exploration of femininity became representative of modern fashion. Asethetic elements such as bow knots, ruffles, clothing tightly wrapped around the body curve, and colorful beads embossed on the tulle, show the charm of women. Furthermore, in today's society, fashion has become more inclusive. Therefore, the design of this series attempts to break the inherent stereotype of social dance clothes for banquets, and incorporate feminist imagery of both feminine and tough women. By combining yarn, leather, shoulder pads and other elements to highlight the image of rigidity and softness in opposition and coordination, it tries to create an expression of exquisiteness, luxury, and opulence in the context of contemporary design.

Fig.10 The author's self-drawn "Qi Dream" costume renderings

4. Conclusion

Modern Shanghai fashion dance clothes have gradually moved closer to Western costumes in style, with more emphasis on tailoring to fit the curves of the human body. Most of them are decorated with lotus leaf and folds to increase its elegance whilst dancing. Morever, according to different occasions, dance clothes can divided into three categories: tea dance clothes, banquet social clothes and stage costumes. They are slightly different in shape, and the fabrics are mostly decorated with gorgeous colors, with dazzling tulle and colored beads. The decorative details are mostly concentrated on the collar, cuffs, and skirt, and lotus leaf edges are applied to delicately portray delicate and elegant female images. By combing the design elements of fashionable dance clothes and combining them with contemporary design language, it can better represent current dance clothes designs in line with Chinese aesthetics. This is not only conducive to designers to better explore local clothing culture and establish self-confidence, but also help to build the right to speak in Chinese clothing culture.

5. Acknowledgment

Fund: The National Social Science Fund of China: History of Modern Fashion Design (1900-1949) in China, 18CG192.

References

[1] Lin L. The Entertainment of Tea Dance. Ling Long: 1931; 2: 62.
[2] Zhou QC, Zhao F, Bao MX. General History of Chinese Textiles. Shanghai: Donghua University Press: 2018.
[3] Anon. Fashion. Art Life: 1934; 7: 20-39.
[4] Wang K. The Fashion Show. China (Shanghai): 1935; 31: 15.
[5] Anon. Extravehicular Pictorial: 1936; 3. 6: 1.
[6] Anon. Going out with another boyfriend. Ling Long: 1931; 10: 335.
[7] Wu H. Chinese Women's Clothing and Body Revolution (1911-1935). Oriental Publishing Center: 2008.
[8] Zhang JQ, Li M. History of Chinese clothing. Shanghai: Donghua University Press: 2018.
[9] Li ZQ. Research on old Shanghai Fashion (1910-1940s). Shanghai Academy of drama: 2015.
[10] Anon. Qiong Bao: 1929; 9. 17: 1.
[11] Wang K. New Clothes. Art Life: 1934; 2: 33.
[12] Anon. New Dance Clothes. The Times: 1934; 2.13: 4.
[13] Anon. Ling Long: 1931; 5: 148.

Cite as: J.M. Zhang et al.,
TBIS 10.3993/tbis (2021).

The Design of Regional Elements in Cultural and Creative Products

Jie-Min Zhang

Liming Vocational University, fengzhe ,quanzhou, fujian , 362000,

*Corresponding auther's email: 417155565@qq.com

Abstract

With the development of the times, the new cultural creation and the national trend are gradually emerging. Moreover, regional elements are used by more and more men in the widespread. In the beginning, the article explains the concept of Dunhuang's cultural and creative product design. The Cultural and Creative Products are generated and then taken through individual or collective intellectual property rights, and they also can promote social and economic development to pass on historical and cultural heritage. Creative products rely on the designer's intelligence to enhance their creativity. Then we outline the classification of Dunhuang decorative patterns, and finally, we focus on the design application of Dunhuang cultural and creative products. We start from the application principles, then analyze the styles and methods of applications. We hope to analyze the colors, graphics, and materials of Dunhuang's cultural and creative products through actual cases, to promote Dunhuang's tourism culture.

Keywords: Regional Elements; Cultural Creation; Dunhuang Patterns

Dunhuang, located in ancient Chinese and Western communications, is a city that maintains the traditional national characteristics, Chinese and Western culture. It is a world cultural heritage site and has attracted foreign scholars to explore and study. There has been much research on Dunhuang culture at home and abroad, and numerous articles and books based on archaeology and history, most of which are academic research. For example, painting techniques, sculpture techniques, and arrangement of patterns, but there is almost no research on applying Dunhuang cultural elements to cultural and creative products.

1. Concept of Design in Dunhuang Cultural and Creative Product

Cultural and creative products, generated and accessed through individual or collective intellectual property, can create wealth and contribute to social and economic development. The Cultural products and resources can be enhanced through the wisdom and talent of people. Cultural and creative products have their origins in developed countries. Furthmore, because of their special and uniquely creative ways, they approach all over the world. The article talks about the design of Dunhuang patterns, which is designed by CG illustration, and the illustrated patterns are promoted and printed on silk scarves, books, ornaments, and other derivatives, making the products both culturally and practically valuable [1].

2. Design Techniques of Regional Cultural Elements

The connotation and cultural carrier of regional cultural elements are clearly defined. When designing creative cultural products, the following aspects can be started:

2.1 Pattern of Typical Things

After the precipitation of time, the baptism of time, each city has its cultural characteristics. "Pattern of typical things" is the most common design technique in the design of creative cultural products. It is to select the most representative sites and use abstract methods to carry out visual symbols of patterns, and at the same time, add color, techniques, and other design techniques to convey cultural connotations [2]. This design method has obvious cultural region identification, and consumers can intuitively interpret the product information. However, this kind of product design is simple, the product's added value is low, and product similarity will occur.

2.2 Transformation of Abstract Consciousness

Much cultural consciousness is the belief of people's values and spirits, which is an abstract existence. Such ideology exists in people's lifestyles and habits, and these ideologies accumulate around us. Therefore, such ideology is applied to the design of cultural and creative products to make them appealing [3]. This design method pays attention to the expression of "artistic conception," which is the extension of the product spirit level. Such products can let people savor. Such a design method has specific problems. Designers need to understand the regional culture in various ways and design elements in line with sea silk culture. Moreover, consumers' understanding of culture, only with a certain degree of recognition of regional culture can they understand the "artistic conception" expressed by-product theory and trigger consumers' purchasing behavior.

2.3 "Grafting" Design of Elements

"Grafting" design applies some cultural elements to daily necessities through design, so these elements exist in another form. These cultural elements can be patterns, colors, characters, buildings, and figures [4]. The biggest difference between such a design technique and the "pattern of typical things" is that it redesigns the extracted elements and applies such a design technique, which can popularize culture and bring economic benefits.

3. The Design Application of Dunhuang Cultural and Creative Products

3.1Common Patterns in Dunhuang

Dunhuang's decorative patterns are vibrant. They are divided into basic and geometric forms because of the terms of pattern expression. There are Dunhuang Mogao Caves, West Thousand Buddha Caves, Anxi Yulin Caves, with more than 500 grottoes in Dunhuang murals [5]. There is much diversity between different locations and dynasties of the pattern. Furthermore, there are the flat ceiling pattern, decorative patterns at the top of the grottoes, decorative caisson patterns, niche lintel patterns, and cross flower joint bead decorative patterns, and geometric patterns. Now, we will analyse a few common patterns.

3.1.1 Dunhuang Pattern - flying in the Sky

The flying pattern is a signature feature of Dunhuang. There is a difference from the usual flying pattern: no wings and feathers in the Dunhuang flying pattern, and no colourful clouds. We can feel the flying through the two hands in the air, by flowing the ribbons and long skirts of clothes in the sky. The flying pattern Feitian is the incarnation of qiankaipo and jinnaro, symbolizing good luck and happiness [6]. Flying in the sky for the Buddha flowers, accompanied by song and dance into one, presents the beauty of a thousand shapes and forms. Fig.1 shows the Dunhuang flying pattern.

Fig.1 Dunhuang flying pattern

3.1.2 Dunhuang Pattern - Caisson

The Dunhuang Caisson Patter is generally used on the dome, symbolizing sublime power. There are polygonal, concave round, geometric bodies in the Caisson, surrounded by many beautiful textures. The Caisson of the pattern is often decorated with lotus leaves, water lilies, and water ling. Well sustenance represents water [7]. The ancients made patterns in the grottoes and on the top of the palace for Caisson because they hope to suppress the fire demon through the water god, ensuring the practicality and safety of the building. [8]

The Dunhuang Caisson simplifies the traditional Han dynasty stacked wood structure by tilting it on all sides and assembling it toward the center; it looks like an inverted bucket. There are many layers of relief and painting beside the caisson, which gives a sense of longevity and sublimity. Fig.2, for example, the picture shows the flying lotus caisson in Cave 420 at Dunhuang during the Sui Dynasty, with blue as the dominant color and red and green as secondary colors, which looks bright and vivid. At the center , it is a finely carved lotus flower with a spiral pattern around it, creating a deep visual effect.

Fig.2 Flying Lotus caisson in Cave 420 at Dunhuang, Sui Dynasty

3.1.3 Dunhuang Pattern - Flat Ceiling Pattern

As a form of roof construction in traditional wooden buildings in China, the Flat ceiling pattern comprises branched wooden panels that resemble a chessboard square. Unlike the Caisson, the flat ceiling pattern is the only one in a cave, and has multiple rows of decoration at the top. During the Middle Tang Dynasty, the flat ceiling pattern was used to decorate the niche roofs of flat-topped tent-shaped Buddhist niches. There is a variety of pattern combinations and delicate colors. In addition to the niche roofs, the cave roofs are also decorated with flat ceiling pattern around the caissons. The pattern is characterized by a predominantly floral style with the colors of the earth red and yellowish-brown [9]

3.1.4 Dunhuang Pattern -Gable Ceiling Pattern

The roofs of lambdoidal shape form are common in traditional wooden architecture. There are imitations of such traditional wooden lambdoidal shape roofs on early caves and some Sui dynasty caves. It shows the emergence of nationalized features in the form and style of the caves. The lambdoidal shape roofs are composed of relief or painted ridge pillars, eaves pillars, rafters, and inter-rafter lookouts, all of which are painted with patterns, and because of their similarity of the form of the building structure, the patterns are distinctive in their compositions and decorations [10].

3.1.5 Dunhuang Pattern - border Decoration Pattern

The border decoration pattern in Dunhuang patterns is a continuous pattern of bands, used to separate the edged, and it is a bipartite continuity. The banded edge decoration is the dividing element around the figures of Caisson, pingqi, huizi, and costume, and the textures inside are single branch scrolling grass, twig scrolling grass, lotus, rhombus, geometric, and flower cluster patterns, and the patterns are colorful and varied. During the Tang Dynasty, this texture reached a level of power and splendor. The entwined scrolling grass pattern is used the most, including tea flower scrolling grass, pomegranate scrolling grass, and multi-stemmed scrolling grass. It is also more abundant than ever before. Furthermore, there are more styles such as side, front and back, as well as different levels of color.

3.2 Integration and Practice of Cultural and Creative Product Design of Regional Cultural Elements

There is a question raised on how we can integrate the regional cultural elements into the design of cultural and creative products so that the cultural and creative products have the attributes of practical products and bear the regional culture so that the products become an invisible publicity card and carry forward the regional culture. Another problem worth pondering is how to apply modern design methods to regional culture [11].

3.2.1 Design Integration of Regional Cultural and Creative Products

When designing creative products of regional cultural elements, it is necessary to decompose the regional cultural elements in structure first and then reshape them according to the characteristics of the decomposed elements. In order to make cultural and creative products diversified, multi-level and personalized, and to prevent similarity in the integration of sea silk elements, designers can use design techniques to integrate sea silk cultural elements into cultural and creative products and improve the fun and playability of cultural and creative products, mainly from the following aspects [12].

3.2.2 Break the Traditional Rigid Element Structure, Split, Extract and Design, to Make the Product Interesting

In this way, compared with ordinary products, interesting cultural creative products are not easy to make consumers feel boring visually, have no interest, and have poor interactivity. Fun products will bring pleasant psychological changes to consumers in the process of use, and such products will be more affinity and surprise consumers when reflecting the functional value of the products themselves [13]. Only in this way can such products stand out from the increasing number of products. In the fun design of products, designers can start from the aspects of graphics, color modeling, and materials, and then use modern design techniques to carry out fractal fusion so that the original regularity and geometric elements show a exciting side.

3.2.3 Break the Single Structure of Traditional Products, and Carry out Portable, Practical and People-oriented Design

As a cultural and creative product, it can be favored by consumers only by combining it with practical and beautiful. Therefore, in the design and the time can choose a part of the elements for the practical design of modeling design, form modeling to retain and increase the practical function [14]. The elements can also be selected for larger deformation in proportion. Humanize and otherwise impractical element. Such a form that breaks the product structure will bring a new visual effect to the finished product, and while showing the original elements, it will also give consumers a new experience.

3.2.4 Integrate with the Implied Meaning of Elements to Show the Inheritance of Regional Culture

Regional culture is handed down from generation to generation regarding a residents' long-term lifestyle, way of thinking, outlook on life, values, and other aspects. It is distinct from the culture of other regions. Cultural and creative products, a means of inheriting culture in the new era, take the essence and remove the dross of the culture that has been baptized by the years through designers as a bridge of communication to bring the culture back to the public view in an innovative form, promote the culture and enhance the national cohesion [15]. Therefore, we can take the implied meaning of the elements as the center, re-design, and combine from the graphics, continue the original font form in the text, break the original state in shape, and derivative the product. In the design, it is necessary to ensure the legibility of the elements, and not to blindly pursue beauty and lose the discrimination.3.2 Elements of Dunhuang cultural and creative product and style design.

3.3 Dunhuang Elements in the Application of Cultural and Creative Products Form

Dunhuang Flat ceiling pattern, caisson, gable ceiling pattern, flying pattern, and other patterns have a strong decorative effect, whether, in the form of color or pattern composition, they can be reasonably used in cultural and creative products to strengthen the visual impact, for highlighting the profundity of traditional culture. There are different characteristics in different times about the different elements. The contemporary designers will design them in two forms- the direct use of patterns and indirect design applications.

The direct use of Dunhuang patterns shows the patterns directly, in the overall or partial position of the cultural and creative design works. These patterns are copied by the artist and utilized on the cultural and creative products, and the patterns used directly basically do not change the shape and color, restore and retain the images of flat ceiling pattern, Caisson, gable ceiling pattern, flying pattern to the maximum extent, respecting the tradition [16]. Taking Fig.3 as an example shows a square scarf that directly utilizes the Caisson pattern. The square scarf is divided into two colors, blue and green, and the pattern is directly utilized from the Caisson and the flying pattern, but in more vivid color. The square border of the Caisson and the emitted composition perfectly fit the shape of the square scarf, which is beautiful and elegant.

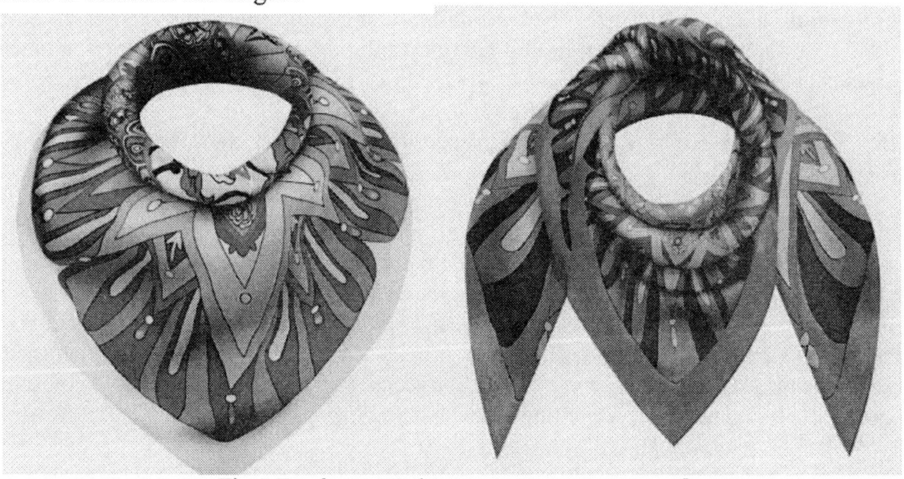

Fig.3 Dunhuang Caisson pattern square scarf

The indirect design of Dunhuang patterns is an evolutionary distillation of most elements rather than direct or mechanical repetition. Applying modern design concepts to traditional Dunhuang pattern elements shows a new artistic form, and designers will simplify and extract them based on Dunhuang elements, incorporating the characteristics of the times. With the development of the times, digital media technology is applied to the design field, gradually replacing traditional hand-drawn forms, drawing images through various software, and post-processing to design works that people better. Combining traditional patterns and contemporary fashion patterns can solve the problem of a single stereotype of traditional patterns. The most characteristic representative elements can be summarized in the extraction technique, such as the three rabbits with ears, the nine-colored deer, the bouncing lute, the treasure phase flower pattern, the tassel beaded trim, and the flying pattern. The basic shape is retained in the design, and the abstract pattern is used to break up and reconstruct, which follows the beauty of contrast, rhythm and rhyme. In modern design, the concept of "less is more" and "simple but not simple" is incorporated into it, and some overly complicated decorations are abandoned and expressed through more fashionable metaphors. For example, Fig.4 is a cultural and creative product designed by using the Dunhuang Buddha statue. Through large generalization and thick lines, the Dunhuang Buddha statue is expressed vividly with smooth lines, reasonable colors, simple and bright, and a very modern sense. These patterns are printed on black shirts, wallets, and wine bottles for a more stylish look.

Fig.4 Dunhuang Cultural Creations

4. Conclusion

Through field research, interviews, literature review, and other methods, this paper takes cultural and creative products as the entry point of the design, explores their market status, consumer demand, and development significance, and proposes design principles and design methods for Dunhuang cultural and creative products according to the research on the development process of cultural and creative products. This paper summarizes the social values such as cultural identity and confidence enhancement expressed by cultural and creative products and the economic values such as the embodiment of cultural connotation and the visualization of product publicity carried by them. Designers break the limitations of tradition and interpret what standards of the new era. The application of Dunhuang elements pays attention to the inheritance of traditional culture and attaches great importance to the development of modern fashion spirit. Theory and practice are combined to promote the later regional cultural and creative design.

References

[1] Jiang XO. The visual design and application of "Great Beauty Dunhuang" cultural and creative products [D]. Hubei. Hubei University of Technology: 2018; 5.
[2] Li WJ. Research on Protective Development of Dunhuang Cave Cultural Tourism Resources [F]. Xinjiang: Shihezi University College of Literature and Arts: 2016.
[3] Chen XY. Research on Quanzhou Higher Vocational Students' Inheritance of Quanzhou Excellent Traditional Culture -- Taking New Media Communication as the Breakthrough point [J]: [Academic Journal], Journal of Ningbo Vocational And Technical College: 2017.
[4] Yan XM. Analysis of cultural Creative Product Development and Design Ideas [J]: [Academic journal], Ge Hai: 2016.
[5] Zhang Y. Research on Regional Cultural And Creative Product Design Based on Context Construction [D]: [Master's Thesis], Zhejiang Gongshang University: 2018.
[6] Chen Q. Analysis of Quanzhou Traditional Residential Decoration "Red Brick Culture "[J]: [Academic journal], Journal of Nanjing Arts Institute (Fine Arts and Design edition): 2015.

[7] Zhou R. Research on the development strategy of cultural creative products based on wine culture tourism [J]. Food Research: April 2015.
[8] Dai JJ. Research and Development of Peking Opera Cultural Creative Products and Engineering Practice [J]. Beijing University of Technology: 2014; 06.
[9] Gao Y. Research on the Application of Yangzhou criticism Culture in cultural And Creative Product Design [J]. East China University of Science and Technology: 2016; 11.
[10] Qiao X. Research and Application of interactive design of cultural and creative products [J]. Beijing Institute of Graphic Technology: 2016; 12.
[11] Cheng ZP. Research on Cultural Creative Product Design Based on User Experience [J]. East China University of Science and Technology: 2016; 11.
[12] Zhou LL. Research on Cultural Creative Product Design Based on Context Integration [J]. Hunan University: 2016; 05.
[13] Yong JC. Regional Culture Research and Its Contemporary Value [J]. Journal of Ningxia University (Humanities and Social Sciences edition): 2008; 30 (3): 52-57.
[14] Xu BX. Reflections on the Development of Cultural and Creative Products in Taiwan [D]. [Master's Thesis] Beijing: Beijing Institute of Technology: 2015.
[15] Stuart C. From cultural Industry to creative Industry theory, industry and policy meaning (World Cultural Industry development [M]. Beijing: Social Sciences Academic Press: 2006.
[16] Countries. Research on cultural Creative Product Design of Shenyang Imperial Palace under symbolic Perspective [J]. Shenyang Aerospace University

Construction of Sunlight-resistant Color and Process Control of Meta-aramid Fiber

Hong-Hui Xia[1], Xin Chen[1], Dan Sheng[1,2], Yun-Li Wang[1], Rui-Qing Li[3], Gen-Yang Cao[1*]

[1]*State Key Laboratory of New Textile Materials and Advanced Processing Technologies, Wuhan Textile University, Wuhan, Hubei, 430200, China*
[2] *School of Textile and Clothing, Jiangnan University, Wuxi, Jiangsu, 214122, China*
[3] *CTES Research Institute, Suzhou, Jiangsu, 215000, China*

*Corresponding author's email: genyang.cao@wtu.edu.cn

Abstract

To increase the color fastness to sunlight of dyed meta-aramid, meta-aramid fabrics are dyed with vat dyes with excellent color fastness to sunlight. Optimization was achieved through conducting the suspension dyeing method and the leuco-acids dyeing method. The effects of dye concentration, thiourea dioxide concentration, reduction temperature, dyeing temperature and dyeing time on dyeing performance were studied. The dyeing process is as follows: vat dye 60 g/L, thiourea dioxide 5 g/L, reduction was carried out at 60°C, followed by dyeing at 130°C for 60 minutes. The results show that vat dyeing can deal with the issue of poor color fastness to sunlight of meta-aramid fiber, and leuco-acid dyeing with vat dyes can improve meta-aramid fiber color fastness to sunlight, up to 3-4.

Keywords: Vat Dyes; Meta-aramid; Suspensions; Leuco-acids; Color Fastness to Sunlight

1. Introduction

Meta-aramid is a high-performance fiber that is commonly used in various fields such as aerospace, military, industry, and civil due to its excellent performance. However, the disadvantage of weak solar resistance limits its further application [1-2]. The most serious damage inflicted on armored fiber during sun exposure is from ultraviolet light [3-4]. The wavelength range of ultraviolet light on a surface is 290nm to 400nm. Aramean macromolecules mainly absorb 290nm to 310nm and 320nm to 355nm. The spectrum of two bands are in nm [5]. The conjugated structure in the amide functional group is in an excited state after absorbing ultraviolet light, and chemically cracked under the auxiliary action of oxygen in the air to generate carboxylic acid ends and nitrogen-oxygen double bonds, leading to discoloration of agamid fibers [6-7].

It is reported in the literature that the poor colour fastness to sunlight of meta-aramid after dyeing is not only related to the aging and discoloration of the fiber itself, it is also related to the type of dye and the chemical structure of the dye itself [8]. The commonly used dyes of meta-aramid are disperse dyes and cationic dyes, etc [9-10], but the dyed meta-aramid has minimal colour fastness to sunlight, which still limits the application and promotion of meta-aramid.

Vat dyes are a class of dyes with exceptional performance. They have vivid shades, complete chromatograms, and excellent color fastness, especially high colour fastness to sunlight. It is an important outstanding variety in the dyeing of various textiles that require high dyeing quality [11]. Vat dyes are used in the dyeing of cellulose fibers and can be used in the dyeing of chemical fibers such as polyester [12], nylon [13], PLA [14]. The dyed fabric has good color fastness.

This paper discusses the dyeing effect of vat dye suspension dyeing method and leuco-acids dyeing method on meta-aramid fabrics, from the dyeing K/S value, color fastness, colour fastness to sunlight, dye particle size and dye penetration rate to determine the best dyeing method. After determining the best dyeing method, the dyeing process is optimized in terms of vat dye concentration, reducing agent thiourea dioxide concentration, vat dye reduction temperature, dyeing temperature and dyeing time.

2. Materials and Methods

2.1 Materials

Meta-aramid woven fabric (plain weave, 520 (J) × 300 (W) roots/10 cm) was acquired from Shandong Yantai Taihe New Material Co., Ltd. China. Vat Blue BC and Disperse Blue 2BLN were provided by DyStar Group, Germany. Thiourea dioxide was purchased from Bailingwei Technology Co., Ltd. China. Sodium hydroxide and Acetic Acid were all purchased from Sinopharm Chemical Reagent Co., Ltd. China.

2.2 Dyeing Process

2.2.1 Suspension Dyeing Method of Vat Dyes

Using 6 g/L of Vat Blue BC with a liquid ratio of 30:1, dyed in an infrared dyeing machine at a heating rate of 1°C /min to 130°C for 60 minutes and then cooled to 50°C at a cooling rate of 2°C/min. The sample was reduced with 6 g/L sodium hydroxide and 6 g/L thiourea dioxide at 30°C at a ratio of 30:1 for 10 minutes, and then the sample was subjected to a process of soaping and washing with running water. Finally, the samples were dried in an oven at 70°C.

2.2.2 Leuco-acid Dyeing Method of Vat Dyes

First, use 6 g/L sodium hydroxide and 6 g/L thiourea dioxide to reduce 6 g/L reduced blue BC at a ratio of 30:1 at 30°C for 10 minutes, and then add glacial acetic acid to make the pH of the dyeing solution reach After 4-5, dye for 60 minutes at a heating rate of 1°C/min to 130°C in an infrared dyeing machine, and then cool to 50°C at a cooling rate of 2°C/min, then perform oxidation, reduction, saponification and running water washing on the sample process. Finally, the samples were dried in an oven at 70°C.

2.2.3 Dyeing Method of Disperse Dyes

Using 3% owf Disperse Blue 2BLN with a liquid ratio of 20:1, the pH of the dyeing solution reached 4-5 by adding glacial acetic acid. After dyeing in an infrared dyeing machine at a heating rate of 1°C/min to 130°C for 60 minutes, the temperature is reduced to 50°C at a cooling rate of 2°C/min. The sample was decreased and washed with 2 g/L EF with a washing solution ratio of 30:1 at 75°C for 15 minutes, and then washed in running water. Finally, the samples were dried in an oven at 70°C.

2.3 Testing Methods

The K/S value of the sample was based on the X-Rite colori7 colorimeter (USA). Using D65 light source and 10 observation angles. Each sample was measured 5 time (valid data E≤0.5). And the average of the results for values. The color fastness to rubbing was set up according to GB/T 3920-2008 "Textile Color fastness test color fastness to rubbing". The color fastness to sunlight is set up according to GB/T 8427-2008 "Textile Color fastness test color fastness to artificial light: Xenon arc".

The vat dye suspension dye solution and the leuco-acid dye solution were respectively attached to the surface of the copper mesh by the hanging drop method, dried at 25°C for 24 hours, and placed in a Tecnai G2 20 S-TWIN transmission electron microscope (Czech Republic) to observe the dye The aggregation morphology and structure of particles in the dye solution.

A Malvern laser particle size analyzer was used to evaluate and analyze the particle size and distribution index of the dye in the suspension dye solution and the leuco-acid dye solution, and the average value was taken after 3 tests. The dyed fiber is wrapped and fixed with resin to make an ultra-thin section, which is scanned in the Chirascan-SF digital section scanner (USA), and then processed with the corresponding software to observe the dye entering under the same magnification The depth of the fiber.

EDX-7000 X-ray fluorescence spectrometer (Czech Republic) was used to test the crystalline structure of undyed meta-aramid fiber and meta-aramid fiber dyed with vat leuco-acid method, the test range is 5°-45 °, scanning speed is 7 (°)/min, Cu target Kα ray, wavelength is 0.154 nm.

According to GB/T 3923.1-2013 "Textile Fabric Tensile Properties Part 1: Determination of Breaking Strength and Breaking Elongation (Strip Method)" to test the breaking strength and breaking elongation of the fabric, the size of the cloth sample is determined as the length (250±2) mm, width (50±2) mm; standard distance is 200 mm; stretching speed is 100 mm/min; pre-tension is 200 cN. Each sample was tested 5 times and the average value was reached.

The vertical burning test is carried out in accordance with the standard of GB/T5455-1997 "Vertical Method of Textile Burning Performance Test". The sample size is (300±2)mm×(80±2)mm. Test Process: Fix the sample on the U-shaped clamp, hang it above the igniter, ignite the igniter, and adjust the gas flow to ensure the blue flame height is (40±2) mm. Place the igniter at 10mm from the lower end of the sample and aim it at the center of the lower end of the sample, and start timing. After the flame is ignited for 12 s, remove the flame and immediately record the subsequent burning time t1 and the overcast burning time t2 after the frame is removed. Rating Standard: B_1 Level: continuous burning time 5 s, smoldering time 5 s; B_2 level: continuous burning time 15 s, smoldering time 15 s.

3. Results and Discussion

3.1 Effect of Different Dyeing Methods of Meta-aramid Fibers

Table.1 K/S values and color fastness of samples dyed with different dyeing methods

Dyeing method	K/S value	Rubbing fastness		Colour fastness to sunlight
		dry	wet	
Suspension	3.67	4	4	3
Leuco-acid	4.53	4	3-4	3-4

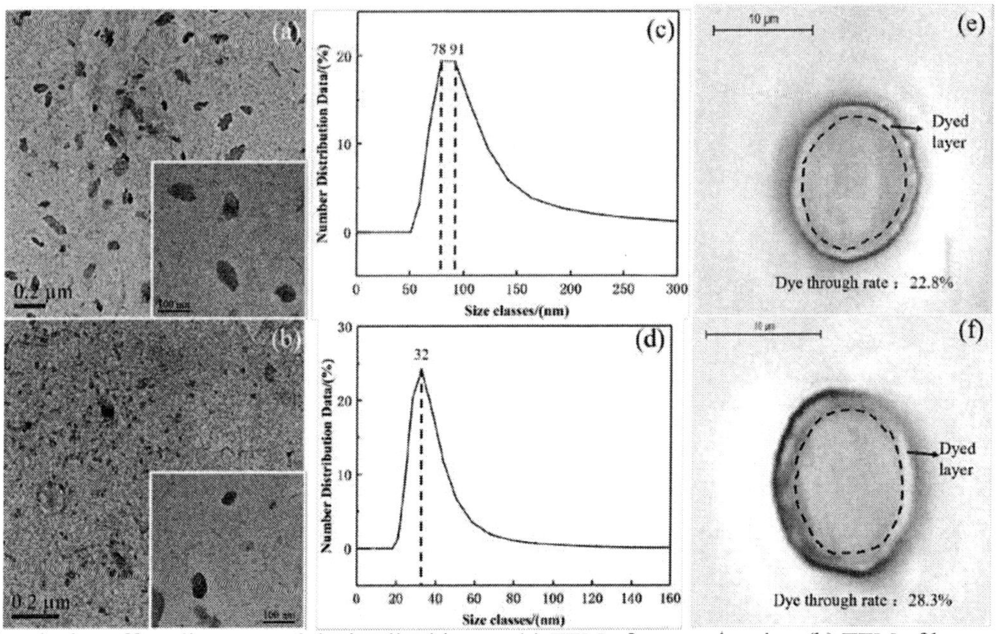

Fig.1 Fiber dyeing effect diagram and dyeing liquid state. (a) TEM of suspension dye; (b) TEM of leuco-acid dye; (c) Particle size distribution diagram of suspension dye; (d) Particle size distribution diagram of leuco-acid dye; (e) Fiber section diagram of suspension dyeing method; (f) Fiber section diagram of leuco-acid dyeing method

It can show in table 1 that the dyeing effect of the leuco-acid dip dyeing method is preferable to that of the suspension dip dyeing method, and the colour fastness to sunlight of the meta-aramid fabrics dyed by the leuco-acid dip dyeing method is higher. From Fig.4(a) and Fig.4(b), it can be observed that the dye is evenly distributed and the particles have no observable aggregation phenomenon. The particle diameter of the leuco-acid dye solution is about 40nm, which is significantly smaller than the particle diameter of the suspension dye solution, which is close to 100nm. Both are nanoparticles. It can be seen from Fig.4(c) and Fig.4(d) that the particle size distribution of the dye is consistent with the results of transmission electron microscopy. Because the particle diameter of the leuco-acid dye solution is much smaller than the particle diameter of the suspension dye solution, the leuco-acid particles are not difficult to enter the interior of the meta-aramid fiber and enter more particles. Therefore, the K/S value of fabrics dyed by leuco-acid dyeing method is greater than that of suspension dyeing method. The fiber dye penetration rate is the percentage of the dye penetration layer deep to the fiber radius, which is used to measure the dyeing effect. It can be seen from Fig.4(e) and Fig.4(f) that the dye penetration rate of the leuco-acid sample is 28.3%, and the suspension is dyed. The dye penetration rate of the method is 22.8%. The leuco-acid with smaller molecular volume is more conducive to the coloration of meta-aramid fiber, with greater color depth and better colour fastness to sunlight. In summary, the leuco-acid dyeing method is selected as the best dyeing method.

3.2 Effect of Different Dyeing Conditions on dyeing Properties of Meta-aramid Fibers

3.2.1 Effect of Dye Concentration

It can be seen from Fig.2 that as the dye concentration increases, the K/S value of meta-aramid fabrics after dyeing increases first and then decreases. The effect is best when the vat dye concentration is 6 g/L. Vat blue BC is an insoluble vat dye. It requires to reduce to a local body in an alkaline environment. And then define the pH to 5-6 to convert the leuco-acids body into leuco-acid and dye the fiber. Therefore, the dyeing temperature is compatible, and the swelling degree of meta-aramid fiber is constant. As the dye concentration increases, the leuco-acid concentration increases, which increase the leuco-acid particles entering the fiber, and dye concentration increases to 6g when /L, the internal space of the fiber absorbs leuco-acid particles to reach saturation. As the concentration continues to rise, the more noticeable the dye agglomeration, the less the amount entering the fiber, resulting in a decrease in the K/S value.

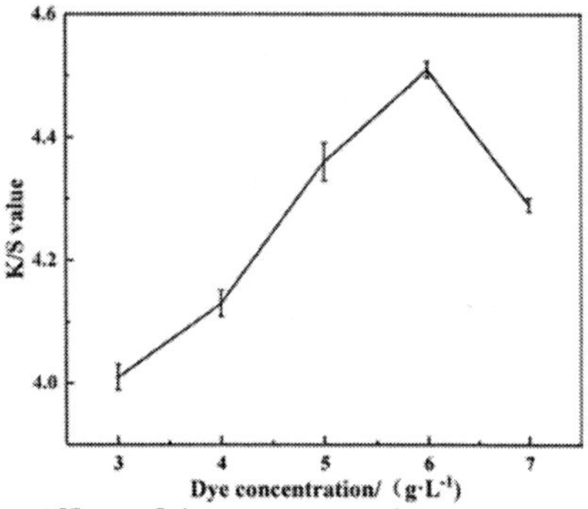

Fig.2 Effect of dye concentrations on K/S value

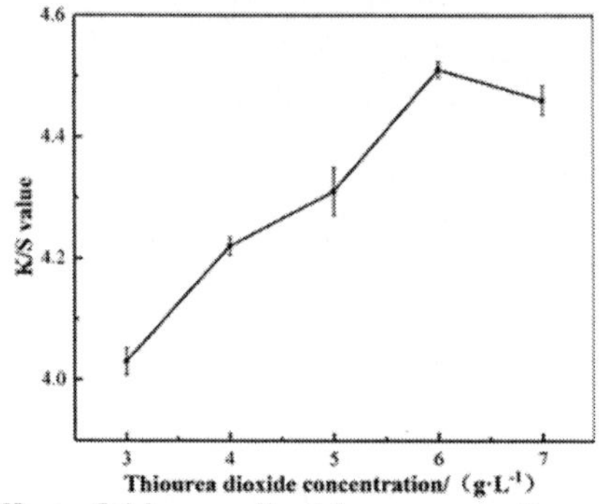

Fig3. Effect of thiourea dioxide concentrations on K/S value

3.2.2 Effect of Thiourea Dioxide Concentration

It can be observed in Fig.3 that the K/S value of meta-aramid fabrics after dyeing increases first and then decreases with the increase of thiourea dioxide concentration. The effect of thiourea dioxide concentration is 6 g/L. Thiourea dioxide transforms into active isomers under the action of an alkaline environment and temperature, and then decomposes into urea and strong reducing sulfuric acid. Strong reducing sulfuric acid reduces reduced blue BC into leuco-acids body. When the dye concentration is coherent, as the concentration of thiourea dioxide increases, the reduction degree of Vat Blue BC increases, the concentration of leuco-acid increases, and the number of leuco-acid particles entering the fiber increases. When the concentration of thiourea dioxide increases to 6 g/L When the dyeing K/S value reaches the maximum, continue to increase the concentration of thiourea dioxide, which may reduce the reduction of Vat Blue BC too much, because the molecular structure of Vat Blue BC dye contains halogen groups. When the reduction conditions are too severe, dehalogenation is prone to take place. Narrow the directness, vividness and color fastness of dyes.

3.2.3 Effect of Dyeing Temperature

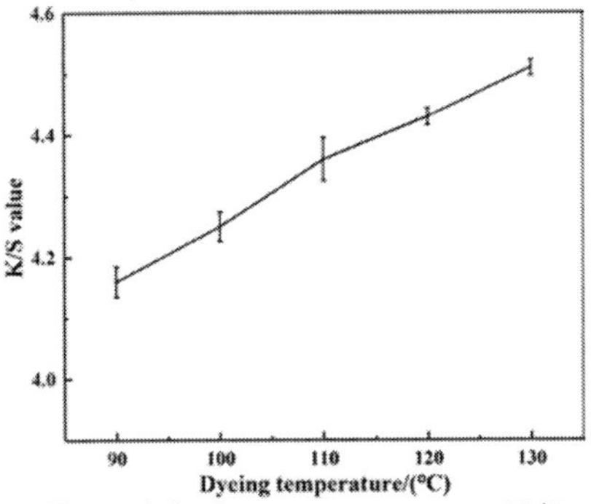

Fig.4 Effect of dyeing temperature on K/S value

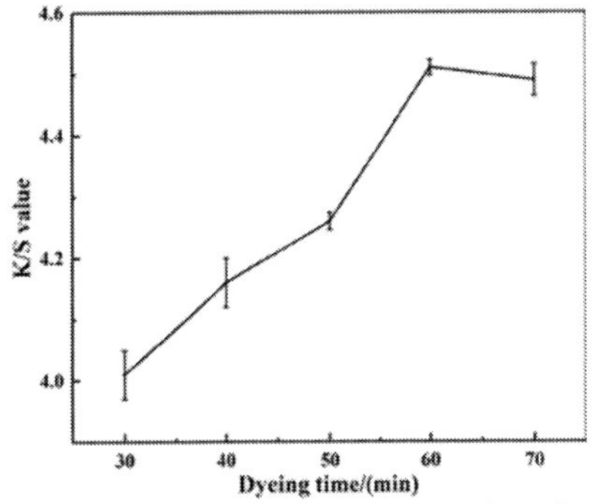

Fig.5 Effect of dyeing time on K/S value

It can be seen from Fig.4 that as the dyeing temperature increases, the K/S value of meta-aramid fabrics after dyeing continues to increase, but the rising rate tends to slow down. The effect is best when the dyeing temperature is 130°C. Because 130°C is the highest temperature widely used in experimental process production. No further research on higher temperatures has been carried out. As the dyeing temperature increases, the kinetic energy of the leuco-acid molecules increases, and the diffusion rate increases. At the same time, the increase in temperature makes the movement of the molecular segments of the fiber intensified, and the pores between the macromolecular chains increase or increase, which is beneficial to The leuco-acid adsorbs on the fiber surface and enters the interior, so the K/S value of meta-aramid fabric increases after dyeing.

3.2.4 Effect of Dyeing Time

It can be seen from Fig.5 that as the dyeing time increases, the K/S value of meta-aramid fabric first increases and then decreases after dyeing. The effect is best when the dyeing time is 60 min.

3.2.5 Effect of Reduction Temperature

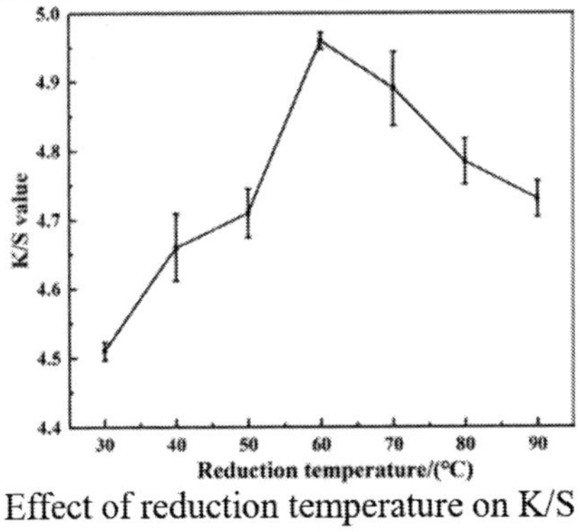

Fig.6 Effect of reduction temperature on K/S value

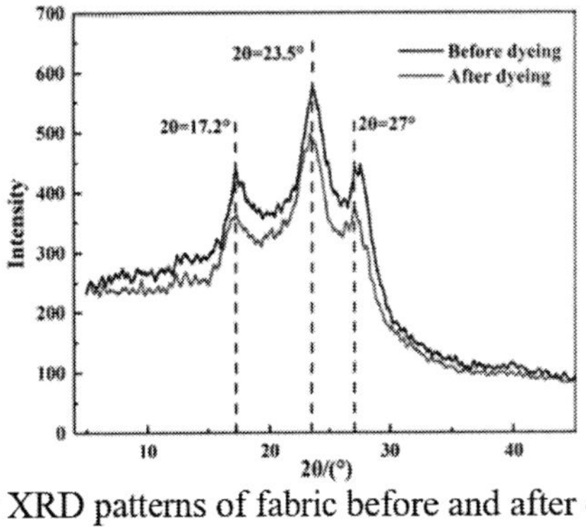

Fig.7 XRD patterns of fabric before and after dyed

It can be seen from Fig.6 that as the reduction temperature increases, the K/S value of meta-aramid fabrics after dyeing increases first and then decreases, and the effect is best when the reduction temperature is 60°C. This is because as the reduction temperature increases, the molecular motion accelerates and the reaction speed increases, making it easier for the reduction blue BC to be reduced to leuco-acids by sulfuric acid. Under the relentless reduction time, as the reduction temperature increases, the reduction reaction steps up. The higher the reduction degree of reduced blue BC, the higher the concentration of the local body and the concentration of leuco-acid. When the reduction temperature is raised to 60°C, the dyeing K/S value achieves the maximum. Continue to improve the reduction temperature, the reduction conditions are too severe, the reduction of the reduction Blue BC is excessive, and the K/S value of the meta-aramid fabric decreases.

3.3 Performance Changes of Meta-aramid Fiber before and after Dyeing

Table.2 Performance comparison before and after dyed with vat dye by leuco-acid dyeing method

Samples	Mechanical performance		Flame retardant performance	
	Breaking strength/N	Breaking elongation /%	Burn time (t_1)/s	Smoldering time(t_2)/s
Before dyeing	1503	32.4	1.6	2.2
After dyeing	1312	33.1	1.9	2.1

It can be seen from Fig.7 that the shape and position of the diffraction peaks before and after the leuco-acid dip dyeing method did not change significantly, indicating that the dyeing did not change the structure of the meta-aramid fiber macromolecules and had little effect on the aggregate structure of the main body. After getting the crystallinity of the sample before and after dyeing, the crystallinity of the fiber after dyeing decreased by 5.18%. The reason may be the case that the high temperature intensified the movement of the fiber molecular segments, and some macromolecules changed from crystalline regions to amorphous regions, resulting in a decrease in crystallinity. It can show in table 2 that the breaking strength of meta-aramid fabrics reduced after dyeing, while the elongation at break increased, and the flame retardant properties of meta-aramid fabrics did not change before and after dyeing. The flame retardant grades were all level B_1. This is because dyeing does not destroy the macromolecular structure of meta-aramid fiber, but the fiber crystallinity decreases, which makes the slip and drawing of the fiber easier, so the breaking strength of the fiber decreases and the breaking elongation increases.

3.4 Comparison of Dyeing Effects between Disperse Dyes and Vat Dyes

Table.3 Comparison of K/S value and color fastness of disperse dyes and vat dyes

Samples	K/S value	Structure	Rubbing fastness		Colour fastness to sunlight
			Dry	Wet	
Disperse Blue 2BLN	4.45	Anthraquinone	4	3-4	1
Vat Blue BC	4.53	Anthraquinone	4	3-4	3-4

It can be seen from Table 3 that the K/S value and rubbing fastness of meta-aramid fabrics dyed with Disperse Blue 2BLN and Vat Blue BC are not much different, indicating the appearance of meta-aramid fabrics dyed with disperse dyes and vat dyes The color depth is similar to the spatial distribution on the fiber. The disperse blue 2BLN and vat blue BC used are both blue dyes, the dye structure is anthraquinone structure, and their colour fastness to sunlight is 7 grades, but the colour fastness to sunlight of meta-aramid fabric dyed with disperse blue 2BLN The degree is only level 1, while the colour fastness to sunlight of meta-aramid fabric dyed with vat blue BC is 3-4 levels, which is related to the macromolecular structure of the agamid fiber and the molecular structure of the dye. Because anthraquinone dyes have long been considered to be a class of dyes with better colour fastness to sunlight, when anthraquinone is exposed to light, the carbonyl group on the molecule is easily excited to generate $1(n\pi^*)$ excited molecules, $1(n\pi^*)$ State further becomes $3(n\pi^*)$ state [15-16]. As the electron cloud density of oxygen atoms decreases, $3(n\pi^*)$ state has strong electrophilicity, so it is too difficult to be oxidized by oxygen. After the amide bond in the macromolecular structure of the agamid fiber absorbs ultraviolet light, the conjugated structure in the amide functional group is in an excited state, and chemically cracked under the assist of oxygen in the air to generate carboxylic acid ends and nitrogen-oxygen double bonds, the fiber has photo-aging phenomenon and causes the meta-aramid fabric to turn yellow [6-7]. With the same degree of dyeing and binding force, both nano-diameter disperse dyes and vat dyes can go into the interior of meta-aramid fibers, but the vat dyes are reduced and disbanded in water by a reducing agent. After dyeing the fiber, it is re-oxidized by air. The formation of insoluble vat dyes is established on the fiber. During this process, the vat dyes may form a protective film on the surface of the fiber, which absorbs most of the ultraviolet energy under the action of light but is difficult to be oxidized and faded by oxygen, making the amide in the macromolecular structure The functional group absorbs less or does not absorb the energy of ultraviolet rays, the probability of light aging of the fiber is reduced, and the protective layer plays a protective role, thereby improving the colour fastness to sunlight of the meta-aramid fabric.

4. Conclusion

The dyeing effect of vat leuco-acid dyeing method for dyeing meta-aramid is better than that of vat dye suspension dyeing. When the concentration of vat dye is 6 g/L, the concentration of sodium hydroxide is 6 g/L, the concentration of thiourea dioxide is 6 g/L, the reduction temperature is 60°C, the dyeing temperature is 130°C, and the dyeing time is 60 min. The dyeing method gets the best dyeing effect. And the colour fastness to sunlight of meta-aramid dyed with vat dye leuco-acid dyeing method is 3-4, indicating that dyeing with vat dye can deal with the issue of poor colour fastness to sunlight of meta-aramid.

5. Acknowledgement

This work was supported by National Natural Science Foundation of China (51325306).

References

[1] Cao G, Sheng D, Xu W, Wang X. Structural and dyeing properties of aramid treated with 2-phenoxyethanol. Coloration Technology: 2015; 131 (5): 384-388.1
[2] Li S, Gu A, Xue J, Liang G, Yuan L. The influence of the short-term ultraviolet radiation on the structure and properties of poly(p-phenylene terephthalaramide) fibers. Applied Surface Science: 2013; 265: 519-526.
[3] Xu F, Fan W, Zhang Y, Gao Y, Jia Z, Qiu Y, Hui D. Modification of tensile, wear and interfacial properties of Kevlar fibers under cryogenic treatment. Composites Part B: 2017; 116: 398-405.
[4] Sun J, Yao L, Sun S, Qiu Y. ESR study of atmospheric pressure plasma jet irradiated aramid fibers. Surface & Coatings Technology: 2011; 205 (23): 5312-5317.
[5] Zhang X, Jiang L. Overview of UV-resistant Textiles. China Science and Technology Information: 2006; 5: 70.
[6] Day M, Wiles DM. Effect of light on the flammability of Nomex fabrics. Textile Research Journal: 1974; 44: 888-891.
[7] Hargreaves G, Bowen JH, Combined effects of gamma and ultraviolet radiation plus heat on fibrous polyamide. Textile Research Journal: 1973; 43: 568-576.
[8] Zhang S, Tan Y, Huo Q, Li W. Analysis of factors exerting influence on sunlight fastness of aramid fabric. Printing & Ayeing: 2018; 44: 24-27.
[9] Zhao J, Cui S. Disperse dyeing of meta-aramid fiber with a novel carrier CZ-12. Printing & Ayeing: 2016; 42 (04): 26-29.
[10] Liu ZX, Zheng JX, Chen GJ. Carrier dyeing of aramid fiber. Printing & Ayeing: 2008; 13: 19-21+30.
[11] Ashraf U, Hussain T, Jabbar A. Effect of cotton fiber and yarn characteristics on color variation in woven fabric dyed with vat dyes. The Journal of the Textile Institute: 2014; 105 (12): 1287-1292.
[12] Seok Jeong D, Lee M, Nishi K, Tokuyama T, Wakida T, Doi C. Dyeing of Polyester Fabric with Vat Dyes by Thermosol Process. Fiber: 2004; 60 (8): 252-256.
[13] Burkinshaw SM, Son YA. The dyeing of supermicro fibre nylon with acid and vat dyes. Dyes & Pigments: 2010; 87 (2): 132-138.
[14] Jeong DS, Chun TI. Color Fastness of PLA Fiber Dyed with Vat Dyes. Journal of the Korean Society of Dyeing and Processing: 2013; 25 (1): 75-81.
[15] Zuang D, Chen C, Zhang L, He JX. The photo-induced reactions of dyes and their fading products. Dyestuffs and Coloration: 2007; 44 (5): 33-37.
[16] He L, Lu L, Tang B, Zhang S, Yang J. Advances in relationship between light fastness of anthraquinone dyes and it's molecular structures [J]. Dyes and Coloration: 2007; 44 (3): 10-13.

Cite as: M.R. Du et al.,
TBIS 10.3993/tbis (2021).

Dynamic Impact Model of Stab-resistant Panels

Mei-Ru Du, Bo-An Ying*, Jing Qi, Long Wu

Apparel & Art Design College, Xi'an Polytechnic University, Xi'an, Shaanxi, 710048, China

*Corresponding author's email: yingba2006@163.com

Abstract

Composite materials have been widely used in the field of stab-resistant prevention, but the cost of physical experiments is too high. Therefore, in order to save test funds, it is necessary to carry out research on numerical simulation experiments. In this paper, glass-fiber reinforced epoxy resin composites and graphite-fiber reinforced epoxy resin composites are used to design the layer structure of the stab-resistant layer, and the dynamic impact simulation model of the stab-resistant layer with different layer structures on the backing is established in Abaqus software. After the simulation calculation, the stress change nephogram and energy change curve of the four stab-resistant layer models are obtained. Then the simulation results are compared and analyzed. The results show that the stab resistance of graphite-fiber reinforced epoxy resin composites is better than glass fiber reinforced epoxy resin composites. The stab resistance of the alternate laying method of the two materials is also better than the former laying method and the laying method of single material superposition.

Keyword: Finite Eleent Simulation; Layer Structure; Dynamic Impact

1. Introduction

In recent years, composite materials are widely used in automobile, aerospace, sports, safety protection and other fields due to their specific strength, high specific stiffness and designability.

Kun Liu et al. [1] used the Abaqus software to model the carbon fiber laminate composites with different angles and two kinds of laminate structures, then simulated the mechanical properties and failure modes of the laminate under tensile load. It is proved that the stress-strain curve of the laminate obtained by simulation aligns with the sample results. Huamei Zou et al. [2] used finite element method to simulate the impact process of tool piercing UHMWPE fabric. Based on the analysis of this course, the energy change of impact system is obtained. Zhaowen Gu [3] investigated the principle of stab-resistance armor by establishing a finite element stab-resistant model of quarter structure, but the validity of the model was not verified by experiments. Fangfang Qiao [4] used Abaqus software to construct the equivalent stab-resistant model of the stab resistant armor, and the effectiveness of the model was verified. Xiangyang Zhu [5] created a hemispherical model of composite laminates with low-velocity impact by ABAQUS software. Based on the model analysis of this course, conclusions of locating the low-velocity impact upon the composite laminates are presented. Hanjian Sun [6] constructed a micro-model and monadic model of two different structures of textile composites that are three-dimensional orthogonal hybrid woven composites by ABAQUS software and carried out finite element simulation. It was proved that the damage results and load-displacement curves were consistent with the experimental results, verifying the validity of the model. However, the current research on composite materials for stab-resistant materials mainly relies on physical experiments, and most of the research on numerical experiments is on the simulation of a single stab-resistant material. There are few numerical experiments on the structural design of different stab-resistant composites.

In this paper, the impact simulation model of stab-resistant panel on the backing is established in Abaqus, and the laminated structures of glass fiber epoxy resin and carbon fiber epoxy resin composite stab-resistant materials is designed. Then the Hashin progressive failure criterion is introduced into the property setting of stab-resistant material. Finally, the stress nephogram and energy curve of the model are obtained by numerical simulation, and the simulation results are compared and analyzed.

2. Construction of Dynamic Impact Finite Element Model

2.1 Geometric Model Construction

Construction of puncture geometry model in Abaqus / CAE. Including impact tool, backing material and stab-resistant panel for test [4, 7].

1) Geometric model and meshing of impact tool for test. The tool geometric model is established according to the size required by the national police stab-resistant clothing standard GA 68-2019. The tool geometric model is constructed by solid stretching in Abaqus/CAE, and the geometric model is shown in Fig.1. Since the tool has irregular shape in the tip part and the tip impact is a critical position, the tip grid segmentation is necessary to ensure the accuracy of the results, and avoid the tip part from producing the tip grid, the division method of the structural grid can be used .The C3D4 element is used to divide the grid, and a total of 5439 elements are generated.

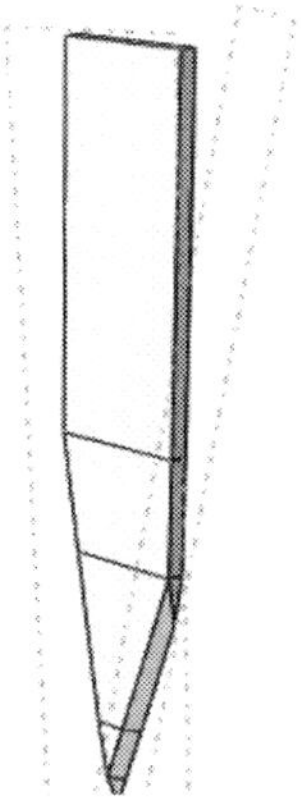

Fig.1 The impact tool geometric model

2) Backing material geometry model and mesh generation. The geometric model of backing material was established according to the standard size required in police stab-resistant clothing, and the geometric body with the size of 400 mm*400 mm *67 mm was created by solid stretching. From top to bottom, it was divided into four layers of 6 mm thick chloroprene rubber sponge, one layer of 30 mm thick foam plastic and two layers of 6.5 mm thick natural rubber. The model uses eight-node linear hexahedron element C3D8R to divide sweeping grids, generating a total of 44800 elements. The geometric model of the backing material is shown in Fig.2.

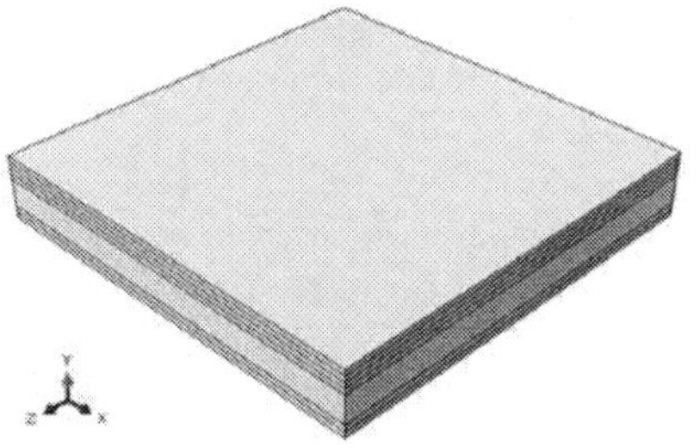

Fig.2 The backing material geometric model

3) Geometric model and mesh generation of stab-resistant panel. The geometric model size of stab proof material is set to 250 mm *250 mm*12 mm. The model uses eight-node linear hexahedron element SC8R to divide sweeping grids, generating a total of 11520 elements. The geometric model is shown in Fig.3.

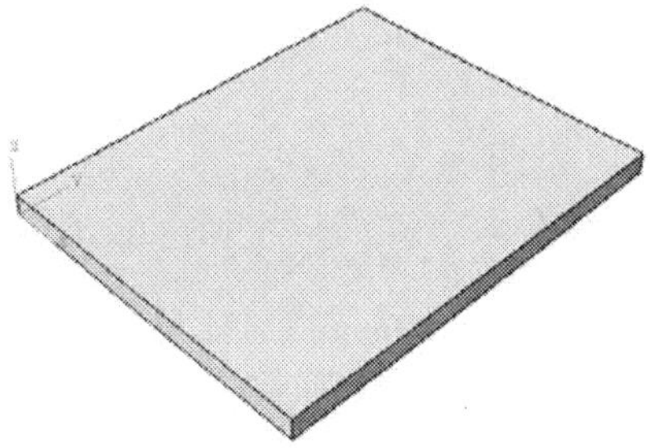

Fig.3 The stab-resistant panel geometric model

2.2 Material Property Settings

In the finite element model of stab-resistant impact, the property settings of impact tool, backing material and stab-resistant material are completed in the Property module of Abaqus.

1) Tool and backing material property settings. According to the requirements of the national police stab-resistant clothing standard GA 68-2019, the impact tool material used in the test is 9Cr18Mo, which is a stainless steel material. Three kinds of materials are used for the backing material. The above four layers of chloroprene rubber sponge are used to simulate human tissue, one layer of polyethylene closed-cell foam plastic and two layers of natural rubber to play a buffer role. The material parameters of tools and backing materials are obtained by consulting relevant information and literature as shown in Table 1 [4].

Table 1. the material parameters of impact tool and backing

Properties	Density/t·mm^{-3}	Elastic Modulus(E)/Mpa	Poisson's ratio(v)
Impact tool	7.83E-9	2.14E5	0.3
neoprene sponge	1.085E-9	500	0.3
natural rubber	9.13E-10	2000	0.45
polyethylene closed-cell foam	1.05E-9	1360	0.3

The tool and backing material properties are constructed by traditional modeling methods. Firstly, the corresponding materials are created according to the mechanical parameters of materials, and then the cross-section properties of each material are created, and then the material assignment area is assigned by assigning the cross-section.

2) The property setting of stab-resistant materials. In this paper, two kinds of composite materials are selected for the design of the layer structure of the stab-resistant panel, and the material mechanical parameters of the two stab-resistant materials are obtained by referring to the relevant literature [8]. Material 1 is glass-fiber epoxy resin composites, and material 2 is graphite-fiber reinforced epoxy resin composites. The mechanical parameters of stab-resistant materials are shown in Table 2.

Table 2. The mechanical parameters of stab-resistant materials

Materials	Density /t·mm^{-3}	E_1 /MPa	E_2 /MPa	Nu	G_{12} /MPa	G_{13} /MPa	G_{23} /MPa	X_T,X_C /MPa	Y_T /MPa	Y_C /MPa	S_L,S_T /MPa
Material 1	1.678E-9	30500	6900	0.344	4650	4650	1600	700	100	237	64
Material 2	1.58E-9	139000	9400	0.309	4500	4500	2980	2070	74	237	64

The property setting of stab-resistant material and the direction of each layer of stab-resistant panel are defined by the Composite Layup tool in the ABAQUS property module. The ply angles involved in the paper were all 0 degrees to reduce the complexity of the model analysis. A total of 24 layers of stab-resistant materials are laid in the stab-resistant layer, and the layer structure of stab-resistant panel are shown in Table 3. The layups and ply angles of the stab-resistant material were shown in Fig.4.

Table 3. The layer structure of stab-resistant panel

stab-resistant panel	layer numbers	layer structure
1	24 layers	24 * material 1
2	24 layers	24 *Material 2
3	24 layers	12*Material 1+12*Material 2
4	24 layers	12*(Material 1+Material 2)

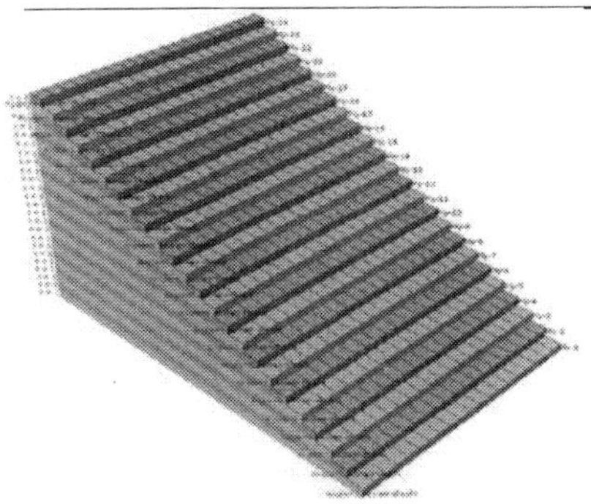

a. Stab-resistant panel 1

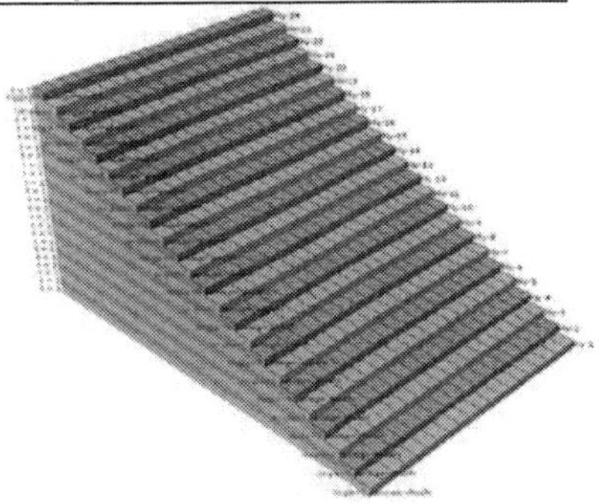

b. Stab-resistant panel 2

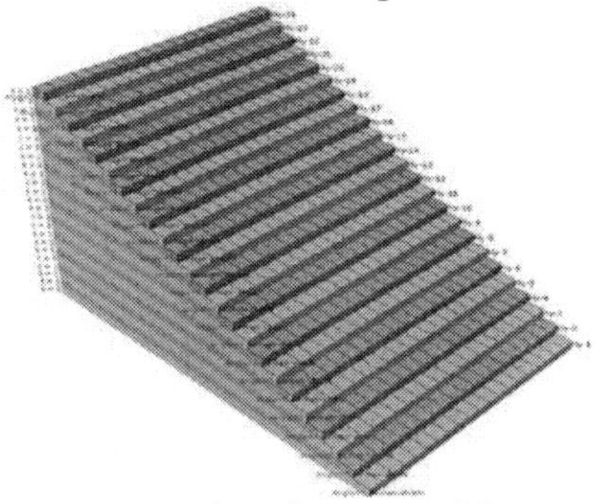

c. Stab-resistant panel 3

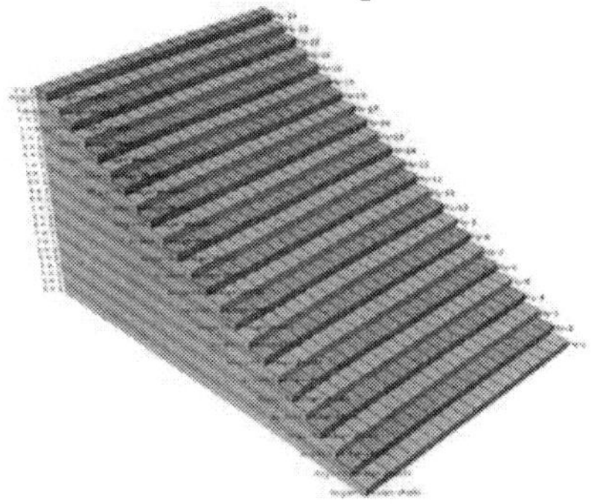

d. stab-resistant panel 4

Fig.4 The layups and ply angles of the stab-resistant material

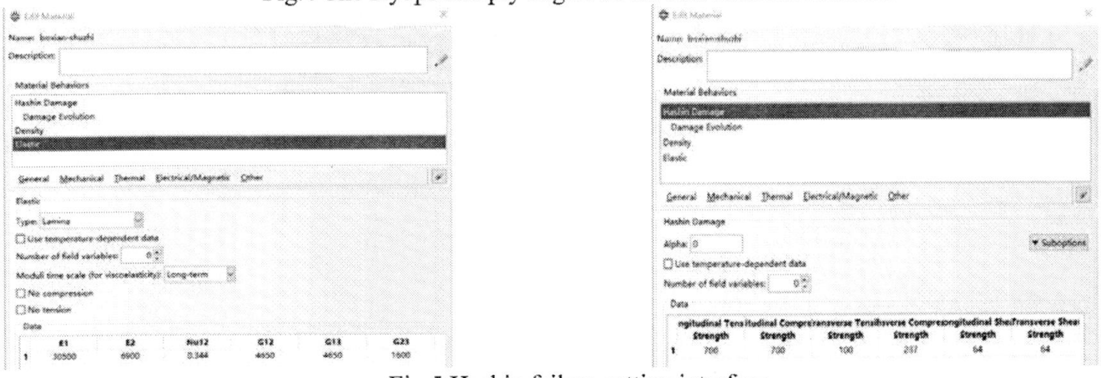

Fig.5 Hashin failure setting interface

A variety of failure criteria for 2D composites are embedded in ABAQUS, such as Hashin criterion, Puck criterion, Tsai-Wu criterion, Tsai-Hill criterion, maximum stress criterion and maximum strain criterion. Hashin criterion divides the material failure into two types: matrix control and fiber control. It contains four different failure modes: fiber tensile failure, fiber compression failure, matrix tensile failure and matrix compression failure. Therefore, Hashin progressive failure criterion is added to the property setting module of stab-resistant materials to set parameters related to damage initiation and damage evolution [9]. The Hashin failure setting interface is shown in Fig.5.

2.3 Interaction and Load Settings

Model contact is defined in the Abaqus. Firstly, the contact surface between stab-resistant panel and backing materials is defined by surface contact method. According to the general friction state between human skin and clothing, the tangential friction coefficient is set to 0.3.Secondly, the contact between the impact tool and the stab-resistant panel is defined by the general contact method. According to the national police stab-resistant clothing standard GA 68-2019 and the free fall motion formula, when the impact tool and counterweight impact the stab-resistant armor with the impact energy of 24 J, the instantaneous velocity of the tool and counterweight falling to the surface of the stab-resistant clothing according to the free fall state is -4430mm/s. Therefore, a 4430mm/s initial velocity constraint is applied to the reference point of the tool by predefined field. ABAQUS / Explicit solver is used to calculate the motion under the influence of gravity. In addition, the constraints in X and Z directions are applied to the tool, and the complete constraints in three directions at the bottom of the backing material. The load and boundary conditions of the model are shown in Fig.6.

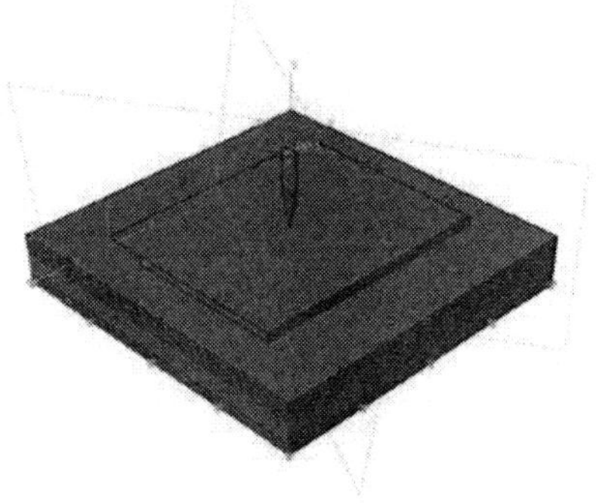

Fig.6 Load and boundary condition

3. Results Output

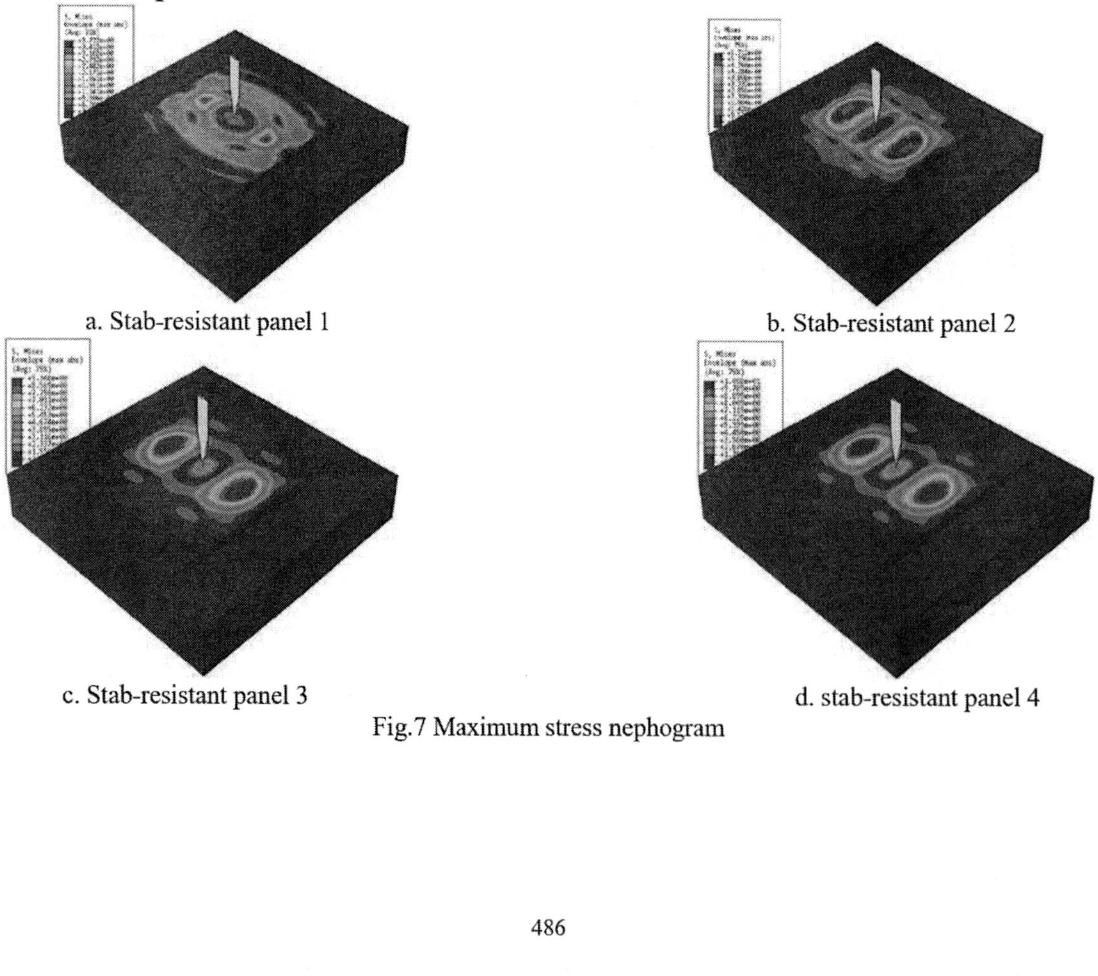

a. Stab-resistant panel 1

b. Stab-resistant panel 2

c. Stab-resistant panel 3

d. stab-resistant panel 4

Fig.7 Maximum stress nephogram

In Abaqus software, the energy change of the overall model can be output by defining the historical output. The parameter change of the X-Y curve of the model can be seen in the visualization function module [10]. Set the whole impact process time of the model to 0.003 s. In the process of the impact tool penetrating into the anti-stab material from the surface of the anti-stab material, the force of the stab-resisyant area contacting with the tool tip increases gradually, and the maximum impact force is reached when the tool tip penetrates. The maximum stress contours of four models are shown in Fig.7.

When the stab-resistant panel is a single stab-resistant material superimposed by 24 layers, the maximum stress of stab-resistant layer 1 is 3.722 MPa when the tool penetrates, and the stress distribution gradually decreases from the middle to the edge. The maximum stress of model 2 is 5.712 MPa. Compared with model 1, the stress area of is larger.

When the stab-resistant panel is composed of two kinds of stab-resistant materials by 24 layers, the maximum stress of the anti-burring layer 3 is 9.348 MPa, and the maximum stress of the model 4 is 10.68 MPa at this moment. The stress distribution of the two layers is similar.

4. Results Analysis

According to the comparative analysis of the stress nephogram in Fig.7, it can be known that the stress of different stab-resistant materials is different when subjected to impact. The greater the stress the layer, the better the performance.

When the stab-resistant panel is laid 24 layers for a stabproof material, the maximum impact stress of anti-stab layer 2 is 1.99 MPa higher than that of anti-stab layer 1, indicating that the anti-stab performance of carbon fiber / resin composite is better than that of glass fiber / resin. When the stab-resistant panel is laid by the combination of two materials, the maximum stress of the stab-resistant panel 3 and 4 is higher than that of the stab-resistant layer composed of one material. Among them, the stress value of the stab-resistant layer 4 is the largest when the prop is impacted, reaching 10.68 MPa.

It can be seen from the energy change curve in Fig.8 that the energy change trends of the four models are basically the same. When the impact tool penetrates into the Stab-resistant panel, the internal energy of the whole model begins to increase, and the kinetic energy begins to decrease. Then the tool stops impacting the material and pops up.

When the impact time is 6.037E-5s,the internal energy of stab-resistant layer 1 reaches the maximum of 146.849mJ, and the kinetic energy reaches the minimum of 45.122mJ; at 4.539 E-5s, the maximum internal energy of stab-resistant layer 2 is 145.246 mJ, and the minimum kinetic energy is 33.932 mJ; the maximum internal energy of stab-resistant layer 3 is 157.29 mJ, and the minimum kinetic energy is 38.065 mJ at 4.526E-5s; At 4.525E-5s, the internal energy of stab-resistant layer 4 is 159.029 mJ, and the kinetic energy is 38.382 mJ.

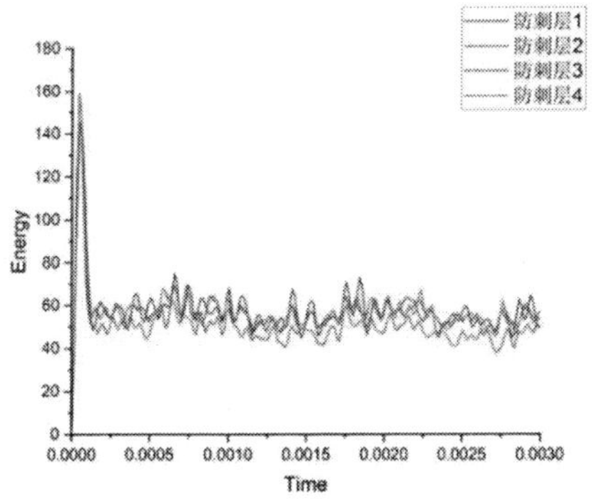

a. Internal energy curve

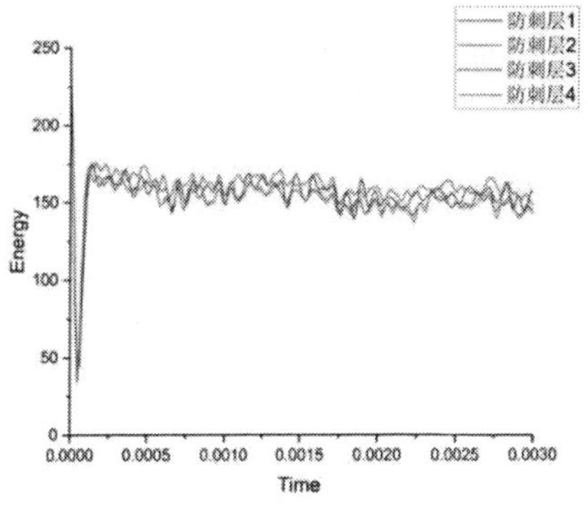

b. Kinetic energy curve

Fig.8 Energy variation curve

5. Conclusion

1) In this paper, the dynamic impact finite element models of stab - resistant pannel with different layer structures

are constructed.. The establishment of the model can provide modeling ideas and technical methods for the impact test of stab-resistant materials, and has certain practicability.

2) By analyzing and comparing the stress changes between different models and the energy changes of the overall model, it is found that the stresses of the stab-resistant layers with different laminated structures are different under impact. When the stab-resistant panel is laid 24 layers for one material, the maximum stress of stab-resistant panel 1 is 3.722 MPa, and the stab-resistant panel 2 is 5.712 MPa, indicating that the stab-resistant performance of carbon fiber resin is better than that of glass fiber resin. When the anti-piercing layer is composed of two materials, the maximum stress of the stab-resistant panel 3 is 9.348 MPa, and the stab-resistant panel 4 is 10.68 MPa, indicating that the anti-piercing performance of the alternate laying of the two materials is better than that of the former laying method.

3) The simulation results show that different anti-piercing materials and layer arrangement will affect the anti-piercing effect of the anti-piercing layer, and compared with the superposition of single anti-piercing material, the anti-piercing performance of the layer structure with alternate laying of the two materials is better.

6. Acknowledgement

Thanks to the Shaanxi Engineering Research Center Of Apparel Intelligent Design and Manufacturing for the warm help of the author.

References

[1] Liu K, Zhang F, Yu LJ. Finite Element Analysis of Tensile Properties of Carbon Fiber Laminate Composites [J]. Materials for Mechanical Engineering: 2014; 38 (09): 109-112+116.

[2] Zou ZM, Liu XJ, Xie CP, Su XZ, X u BJ. Simulation of puncture impact and response analysis of ultra high molecular weight polyethylene plain fabric [J]. Silk: 2019; 56 (06).

[3] Gu ZW. Study on Mechanism of Flexible Compound Anti-stab Clothing [J]. Journal of Textile Research: 2006 (08); 80-84.

[4] Qiao FF. Research on Protective Performance of Anti-stab Clothing Based on Finite Element Simulation Analysis [D]. Xi 'an Engineering University: 2019.

[5] Zhu XY, Guo Xu, He Liang. Research on Low Speed Impact Load Localization of Composite Laminates [J]. Aeronautical Computing Technology: 2018; 48 (06): 29-31+39.

[6] Sun HJ. Quasi-static Penetration Experimental Analysis and Finite Element Simulation of Textile Composites with Different Structures [J]. Fiber Composites: 2008 (03); 45-49.

[7] Shi YP, Zhou YR. Detailed Analysis of ABAQUS Finite Element Analysis [M], Beijing: China Machine Press: 2006.

[8] Li CF, Cheng JG, Hu N, Yao ZH. Numerical method for simulating impact damage of long fiber reinforced resin laminates [J]. Journal of Mechanics: 2002 (06); 989-996.

[9] Hashin Z. Failure Criteria for Unidirectional Fiber Composites. Journal of Applied Mechanics: 1980; 47 (2): 329-334.

[10] Zhuang Z, Zhang FN, Qin S, et al. ABAQUS nonlinear finite element analysis and examples [M]. Beijing: Science Press: 2004; 08.

Prediction of Fabric Pilling Propensity of Woven Worsted Fabrics Based on Regression Analysis and Artificial Neural Networks

Eman Mustafa[1], Abdel Salam Malek[2], Sherien El kateb[1], Sherwet El Gholmy[1], Adel El Geiheini[1]

[1] *Faculty of Engineering, Alexandria University, Egypt*
[2] *Faculty of Technology and Education, Suez University, Egypt*

*Corresponding author's email: mustafa.eman188@gmail.com

Abstract

This paper compared between two different models to predict the degree of pilling of blended worsted fabrics. The first model was based on regression analysis and the second on artificial neural networks. Two different materials with five counts of weft yarn (Ne 25/1, Ne 18/1, Ne 50/2, Ne 40/2, and Ne 32/2) along with different ranges of picks per inch and three structures of fabric weaves were used. The results of the two models indicated a good estimation on the degree of pilling, however, the artificial-neural-network-based model was better than the model that used regression analysis in predicting the degree of pilling.

Keywords: Prediction; Pilling Degree; Woven Worsted Fabrics; Regression Analysis; Artificial Neural Network.

1. Introduction

Pilling formation is a fabric defect that affects the quality of cloths and consumer needs satisfaction. When pilling starts to take place, the fabric starts to deteriorate, until a spot with a lot of pills turns into a hole. Pills do not interfere with the functionality of textile, until the formation of the hole [1]. Pill formation is a dynamic process, represented by pills forming on the surface of the fabric and then wearing off. If the pill formation rate is higher than the breaking-off rate, pills will accumulate on the fabric's surface. The formation rate is influenced by fiber, yarn, and fabric parameters. Together they affect pilling tendency and the quality of the end-product in a complex manner [2]. Thus, many research have been conducted to quantify pilling formation. These studies tried to isolate pill pixels from images of fabric features, by attenuating these features, through the adjustment of the contrast or brightness of fabric images [3-9]. Pill characteristics, such as count or size, were extracted from images of fabric features, using an optical triangulation system. This was done to predict fabric pilling [10, 11]. Most of the studies in this area focused on the estimation of pilling degree for knitwear and tested various factors used in the construction of fabric [12-15], compared to the different factors used in this research. Despite the many existent methods to predict fabric pilling tendency, the subjective method is still used to evaluate pilling formation. Fabric appearance is compared to a set of photographic standards from grade 1 (severe pilling) to grade 5 (no pilling). Despite the simplicity of this method, it remains uncertain due to its subjectivity [10]. Therefore, this work aims to find a simple objective method to predict fabric pilling grades. This paper aims to find a simple statistical tool to predict the pilling degree of blended worsted woven fabrics, based on the small number of factors used in fabric construction.

2. Experimental Work

Two different materials of weft yarn of woven worsted fabrics were used, a 60/40 blend of wool/polyester fibers and a 35/65 blend of viscose/polyester fibers of different counts (Ne 25/1, Ne 18/1, Ne 50/2, Ne 40/2, and Ne 32/2). Warp yarns were made from a 60/40 blend of wool/polyester fibers for yarn count Nwr 32/2. The range of picks per inch was selected between 44 and 50 and the twist coefficients were 3.8 and 4.2 for single thread weft and plied yarns, respectively. The fabrics were woven on Rapier and Picanol machines, with different weave structures.

Pilling tests for fabric samples were carried out on an ICI box tester, according to the standard British method (BS 5811:1986). To evaluate the degree of pilling, samples were rated from 1 to 5; whereas grade 5 means no pills on the surface of the fabric and grade 1 means a severe change in fabric surface due to pilling.

Regression analysis was applied to predict the grade of fabric pilling, based on the type of raw material, weft yarn count, weft doubling (single/ply), picks per inch, and fabric weave structure, in the first model.

Artificial neural networks (ANN) are mathematical models, inspired by biological neural networks, which process information using a connectionist approach to computation. ANN was used in the second model to improve the performance of a statistical tool used to predict the degree of pilling. In this study, the Levenberg-Marquardt algorithm was used with the "trainlm" training function. The model consisted of one input layer, one hidden layer with seven neurons, and one output layer. Mean square error (MSE) was used as an indication of the performance of the model.

Table (1) shows the experimental work design, which was executed to predict the grades of fabric pilling, based on fabric construction.

Table 1. Experimental work design

Factor	Type
Material type (X1)	Wool / Polyester 60/40
	Viscose / Polyester 35/65
Weft yarn count (X2)	Ne 25/1
	Ne 50/2
	Ne 40/2
	Ne 18/1
	Ne 32/2
Doubling (X3)	Single
	Ply
Picks/inch (X4)	44
	45
	47
	50
Fabric structure (X5)	Plain
	Twill
	Satin

3. Results and Discussion

Two different models were employed to predict the level of pilling, as per the given fabric characteristics. Model (1) was based on regression analysis and model (2) was based on ANN. Both tried to improve the efficiency of prediction.

3.1 Model (1): Regression Analysis

The equations below (Eq.1 and Eq.2) were concluded to predict the degree of fabric pilling, by using regression analysis. The first equation contains the type of fabric material, weft count, doubling of weft yarn, picks per inch, and fabric weave structure. The second equation was concluded after neglecting non-significant parameters, such as weft yarn count and picks per inch.

$$\text{Pilling grade} = 2.987 - (0.524*\text{Material type}) + (0.175*\text{Weft count}) + (0.399*\text{Doubling}) + (0.059*\text{Picks/inch}) - (1.265*\text{Structure}) \quad 1)$$

Pilling grade = 3.024 - (0.477*Material type) + (0.322*Doubling) - (1.265*Fabric weave structure)   2)

Table (2) shows a comparison between Equations (1) and (2). Despite that R^2 is higher and MSE is lower in Equation (1) than in Equation (2), the later equation is the best formula to predict the grade of fabric pilling, due to its higher significance. After neglecting non-significant factors, i.e., weft yarn count and picks per inch (p-values more than 0.05), the significance of model (1) seems to have increased.

According to the coefficients of regression, it can be concluded that the weave structure according to the type of fabric has the most significant effect of all factors on the pilling tendency of fabric, followed by the type of weft of the yarn material (single or plied).

Table 2. Coefficients of regression of Equations (1) and (2)

	Factors	P-Value	R-Square	Significance	MSE
	X1	6.18E-06			
	X2	0.214			
Equation (1)	X3	8.26E-04	0.885	3.23E-13	0.216
	X4	0.569			
	X5	1.05E-14			
	X1	3.78E-06			
Equation (2)	X3	6.80E-04	0.879	9.708E-15	0.229
	X5	2.98E-15			

3.2 Model (2): Artificial Neural Network

As can be seen from Fig.1, the applied neural network consisted of five inputs, which are type of weft material, weft yarn count, number of doubling of weft yarn, picks per inch, and fabric weave structure; one hidden layer with seven neurons; and the degree of pilling as the output of the network.

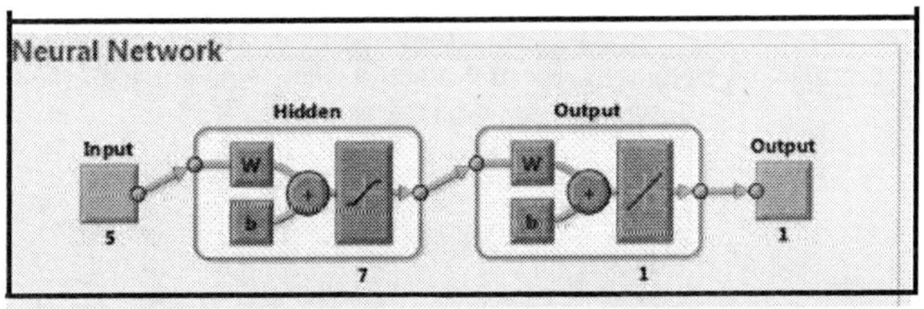

Fig.1 Artificial neural network model for the prediction of the degree of pilling

Fig.2 shows that the correlation and regression coefficients between the real and predicted data were rather high. For the training data, the regression coefficient between the real and predicted values was 99.9%, whereas for testing data, this value was found to be 98.5%. The overall regression coefficient of model (2) was found to be equal to 98.4%, whereas the performance of this model (comparable to the MSE) was 0.003. It can be, thus, concluded that in this study, the model of the artificial neural network is a more effective tool in predicting pilling tendency compared to the regression analysis model.

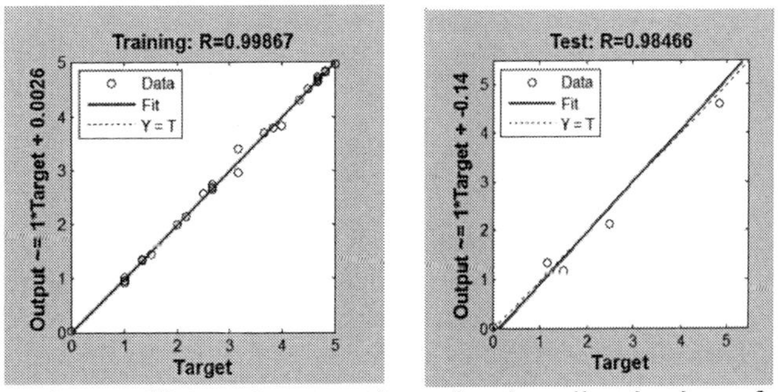

Fig.2 Regression coefficients between the real and predicted values of model (2)

Fig.3 compares between actual and predicted values of the pilling grade using the ANN and the multiple regression analysis models. It can be seen that the predicted values of pilling grades using ANN are closer to the real values of pilling grades than those predicted using multiple regression analysis.

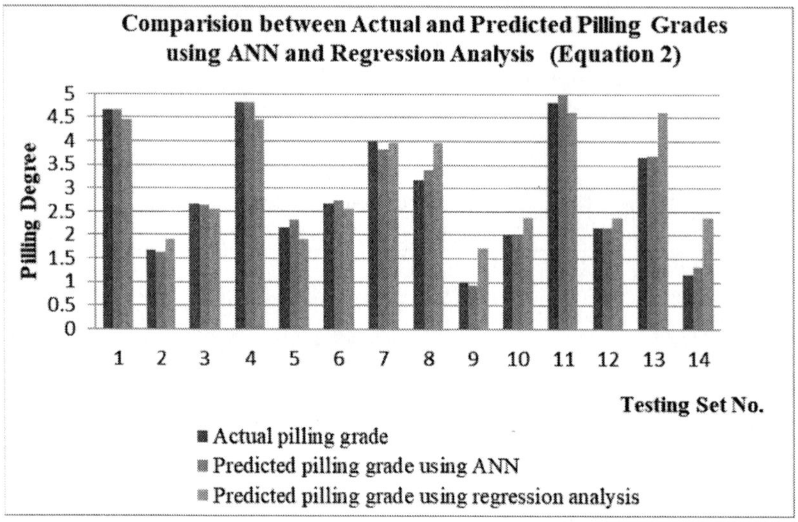

Fig.3 Comparison between actual and predicted values of models (1) and (2)

Fig.4 compares between the errors of the actual and predicted values of the pilling grade, using ANN and regression analysis. It can be seen that the predicted values of pilling grades using regression analysis have higher values of error compared to those predicted using ANN.

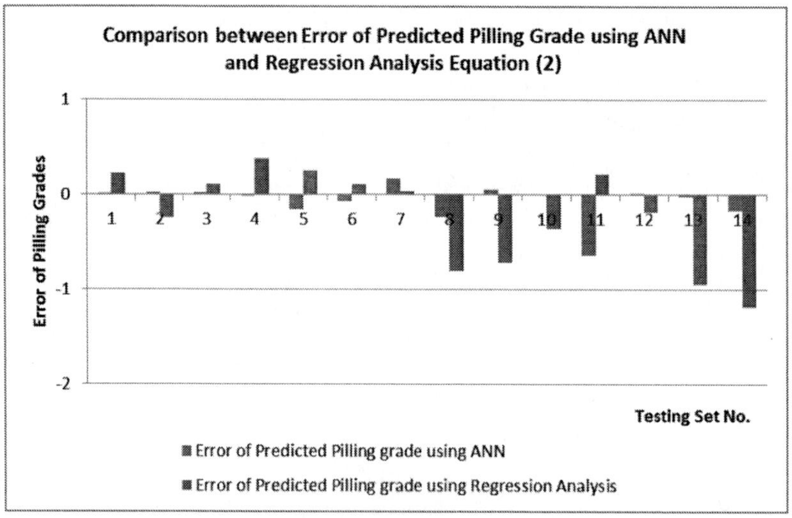

Fig.4 Comparison between Error of Predicted pilling Grades of Models (1) and (2)

4. Conclusion

Due to the high competition to produce high quality fabrics at low cost, this study tried to find a simplified technique to evaluate the pilling tendency of woven worsted fabrics. Two different models were used to make this prediction, where model (1) used regression analysis and model (2) used the Levenberg-Marquardt algorithm of artificial neural networks. Model (2) achieved a higher performance and resulted in lower MSE in predicting the pilling tendency of blended worsted fabrics compared to model (1).

References

[1] Hassan, Eldeen NN. An Examination of the Impact of Interfacing Fabric on Woolen Clothes Pilling Resistance. IDJ: 2016; 6.4: P. 101-110.

[2] Ukponmwan JO, Mukhopadhyay A, Chatterjee KN. Pilling. TP: 1998; 28.3: P. 1-57.

[3] Deng Z, Wang L, Wang X. An integrated method of feature extraction and objective evaluation of fabric pilling. JTI: 2011; 102.1: P. 1-13.

[4] Jung MH, Rhodes PA, Clark M. Objective evaluation of fabric pilling using digital image processing. 12th AIC: 2013; P. 1033-1036.

[5] Mayekar, Vijay M, Nachane RP. Noncontact capture of pilling profile on fabric surface-objective assessment method. SRP: 2016; 6.5: P. 86-90.

[6] Techniková L, Tunák M, Janáček J. Pilling evaluation of patterned fabrics based on a gradient field method. IJFTR: 2016; 41.1: P. 97-101.

[7] Techniková L, Tunák M, Janáček J. New objective system of pilling evaluation for various types of fabrics. TJTI: 2017; 108.1: P. 123-131.

[8] Guan S, Shi H, Qi Y. Objective evaluation of fabric pilling based on bottom-up visual attention model. TJTI: 2017; 108.4: P. 597-604.

[9] Guan S, Li W, Wang J, Lei M. Objective evaluation of fabric pilling based on data-driven visual attention model. IJCST: 2018; 30 (2): P- 210-221.

[10] De Oliveira Mendes A, Torrão Fiadeiro P, Lopes Miguel RA, Lucas JM. Optical estimation of a set of pilling coefficients for textile fabrics. TRJ: 2009; 79.5: P. 410-417.

[11] Saharkhiz S, Abdorazaghi M. The Performance of Different Clustering Methods in the Objective Assessment of Fabric Pilling. JEFF's: 2012; 7 (4): P. 35-41.

[12] Beltran R, Wang L, Wang X. Measuring the Influence of Fiber-To-Fabric Properties on the Pilling Of Wool Fabrics. TJTI: 2006; 97 (3): P. 197-204.

[13] Yap PH, Wang X, Wang L, Ong KL. Prediction of wool knitwear pilling propensity using support vector machines. TRJ: 2010; 80 (1): P. 77-83.

[14] Rejali M, Hasani H, Ajeli S, Shanbeh M. Optimization and Prediction of the Pilling Performance of Weft Knitted Fabrics Produced from Wool/Acrylic Blended Yarns. IJFTR: 2014; 39 (1): P. 83-88.

[15] Kayseri GÖ, Kırtay E. Part II. Predicting the Pilling Tendency of the Cotton Interlock Knitted Fabrics by Artificial Neural Network. JEFF: 2015; 10 (4).

Preparation and Properties of Mulberry Fiber/ Sodium Alginate Composite Hydrogel

Gui-Cui Chen[1,2], Li-Feng Zhang[1,3*], Lei Zhao[1,2] Ling Liu[1,2]

[1.] School of Textile and clothing, Yancheng Polytechnic College, Yancheng Jiangsu, 224000, P.R. China;
[2.] Jiangsu Province Engineering Research Center of Biomass Functional Textile Fiber Development and Application, Yancheng Jiangsu, 224000,P.R. China;
[3.] Logistics Infrastructure Management Division, Yancheng Jiangsu, 224000, P.R. China

*Corresponding author's email: ycfycgc@126.com

Abstract

In order to prepare hydrogel, mulberry fiber was used as the framework structure, and sodium alginate was selected as the functional component. Double-network porous structure hydrogel bead of mulberry fiber/sodium alginate were prepared by blending method. The morphology, rheological properties of hydrogel spheres were analyzed, and the preparation technology of mulberry fiber/sodium alginate hydrogel was optimized in the paper.

Keywords: Double-network; Mulberry Fibre; Sodium Alginate; Hydrogel; Rheological

1. Introduction

Hydrogel, as a kind of cross-linked polymer with polyelectrolyte properties and three-dimensional network structure. It is a type of functional polymer material that has properties of integrated water absorption, water retention and sustained-release [1-3]. Hydrogels have high water content, excellent biocompatibility, flexibility, and can be used in biomedical fields, such as tissue engineering scaffolds, wound dressings, contact lenses, and controlled-release of drugs [4-6].

Mulberry fiber are widely available and biodegradable. Sodium alginate is a kind of anionic polyelectrolyte polysaccharide, that can form gel microspheres by cross-linking its spindle group with divalent cation [7]. To prepare porous aerogel balls with a core-shell type. Mulberry bark fiber was used as the skeleton structure, and sodium alginate was used as flexible lamellar functional component. Mulberry fiber/sodium alginate composite hydrogel was successfully prepared by blending method. The mechanical properties of hydrogel could be improved by cross-linking polymers [8]. The morphology of hydrogels was characterized by TM3030 table electron microscope. The rheological properties of hydrogels were evaluated by a rheometer. The preparation process of mulberry fiber/sodium alginate hydrogels was optimized by taking advantage of the porous network structure of hydrogels, to provide reference for the application of mulberry fiber/sodium alginate hydrogels.

2. Experiments

2.1 Experiment Materials

Mulberry bark (Dongtai, Yancheng, Jiangsu province). Sodium hydroxide (analysis of pure, Aladdin reagent (Shanghai) co., LTD.). Sodium tripolyphosphate (analysis of pure, Aladdin reagent (Shanghai) co., LTD.). Wool oil (analysis of pure, Aladdin reagent (Shanghai) co., LTD.). Sodium alginate (chemical pure CP500g, Shanghai national medicine group chemical reagent co., LTD.). Deionized water (laboratory)

2.2 Experiment Instruments

J-HH-6A Precision digital display constant temperature water bath (Shengwei Electronic Technology Co., Ltd., Shanghai, China). FA2004 Analytical Balance (Tongjun Scientific Instrument Co., Ltd., Jiangsu, China). YB-4500A medicinal material crusher (Sufeng Industry & Trade Co., Ltd., Yongkang, China). FD-1A-80 Vacuum Freezed Dryer (Tianling Instrument Co., Ltd., Jiangsu, China). PH400HD101A-2 Electric Blowing Oven (Hongda experimental instrument Co., Ltd., Nantong, China). ITT-1100s Vacuum Defoaming Machine (Intaite Laser Co., Ltd., Shenzhen, China). LIAEBO Mixer (LIAEBO Machinery Equipment Co., Ltd, Changzhou, China). TM3030 Desktop Electron Microscope (Shanghai Shuangxu Electronic Co., Ltd.). AR2000 rheometer (TA company, USA)

2.3 Preparation of Mulberry Fiber/Sodium Alginate Composite Hydrogel

1) Preparation of mulberry fiber: The mulberry bark, sodium hydroxide, sodium tripolyphosphate, crude oil and water were mixed at the dosage ratio of 2g: 3g: 3g: 2mL: 100mL. Microwave treatment was performed for 2h each time, and deionized water was used to rinse repeatedly until the mixed solution was neutral. The mulberry fiber was obtained after three times of treatment. The degummed mulberry fiber was freezed and dried in a vacuum machine, then the mulberry fiber were crushed o prepare the mulberry fibers with a diameter of micron (5 min each time, 3times, 10-100μm length), and 2g fibers were weighed (6 kinds of samples).

2) Preparation of Mulberry Fiber/Sodium Alginate (SA) hydrogels. Mulberry fibers (0 wt%, 0.5 wt%, 1 wt%, 1.5 wt%, 2 wt%, 2.5 wt%) were put respectively in a 2 wt% SA solution (50 mL) and were evenly stirred and dispersed in SA solution (10 min) at room temperature. Mixed solution after defoaming were divided into six beakers (MBSH0, MBSH1, MBSH2, MBSH3, MBSH4, MBSH5) and sealed for cross-linking for 12h (see Fig.1).

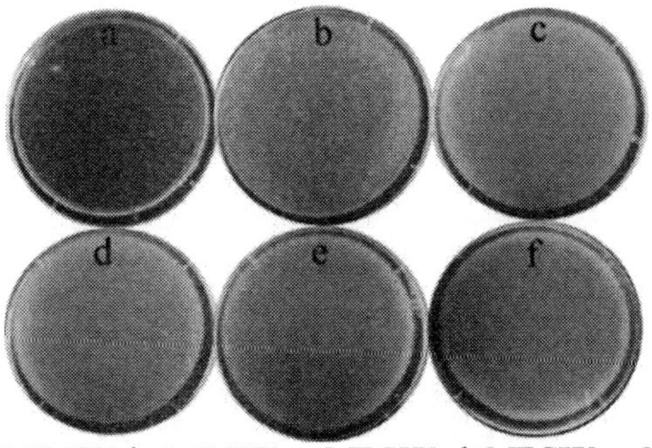

Fig.1 Hydrogel, a, MBSH0 b, MBSH1 c, MBSH2 d, MBSH3 e, MBSH4 f, MBSH5

3. Characterization of Mulberry Fiber/SA Composite Hydrogel

3.1 Morphology of MBSH

TM3030 electron microscope can intuitively capture the micro-morphology of the hydrogel material after freeze-drying. The hydrogel was cut with scissors and pasted for a while to restore the original appearance on the conductive adhesive. The internal morphology of the hydrogel was observed after spraying gold.

3.2 Rheology of MBSH

3.2.1 Rheological Properties of MBSH

The pressure-controlled rheometer AR2000 (TA, USA) (Diameter, 40 mm. Angle of conical plate , 1°. Shear rate, 0.01 to 500s-1) was used to test the rheological properties of mulberry fiber/SA hydrogel (MBSH0, MBSH1, MBSH2, MBSH3, MBSH4, MBSH5). Hydrogels (0.8 mL-0.9mL) was extracted by syringes and placed them on the center of the bottom plate to obtain the steady viscosity.

3.2.2 MBSH Viscoelasticity

The viscoelastic region was determined by amplitude scanning before frequency scanning, and the linear viscoelastic region of hydrogel was determined by dynamic strain scanning at a fixed frequency (ω = 1 rad/s).

4. Results and Conclusions

4.1 MBSH Morphology

This double-network porous hydrogel network structure and lamellar structure could be controlled by the addition amount of mulberry bark fiber. With the increase of cellulose content, the network structure of mulberry bark fiber alginate composite hydrogel tended to be uniform. Sodium alginate was attached to the cellulose chain of mulberry bark, and the main body showed a network interpenetrating three-dimensional network structure. When the fiber content was low, the hydrogel morphology contained soft SA lamellae. When the cellulose content reached the critical value (Fig.2(c) and Fig.2(d)), the pore and lamellar structure of the hydrogel would become disorganized. This is because the excessive mulberry was unable to satisfy the crosslinking of sodium alginate, and the excess fibers were attached to the surface of the lamellar and intersperse in the interior of the pores, forming the structural morphology as shown in Fig.2(e).

There were many interconnected porous three-dimensional holes in the hydrogel, as well as some smaller laminated structure were formed by the cross-linking of mulberry fiber and SA, which double-network porous hydrogel network structure and lamellar structure were regulated by the mulberry fiber content (see Fig.2). The double-network porous structure of hydrogel tended to be uniform with the increase of fiber content. Sodium alginate was attached to the cellulose chain of mulberry fiber, and the main structure showed a network interpenetrating three-dimensional network structure. When the fiber content was low, the hydrogel morphology contained soft SA laminated structure. The porous and lamellar structure of the hydrogel became disorganized after the fiber content reached the critical value (Fig.2(c) and Fig.2(d)). It was the excessive fiber was unable to satisfy the crosslinking of sodium alginate, and the excessive fibers were attached to the surface of the lamellar and interspersed in the interior of the pores (Fig.2(e)).

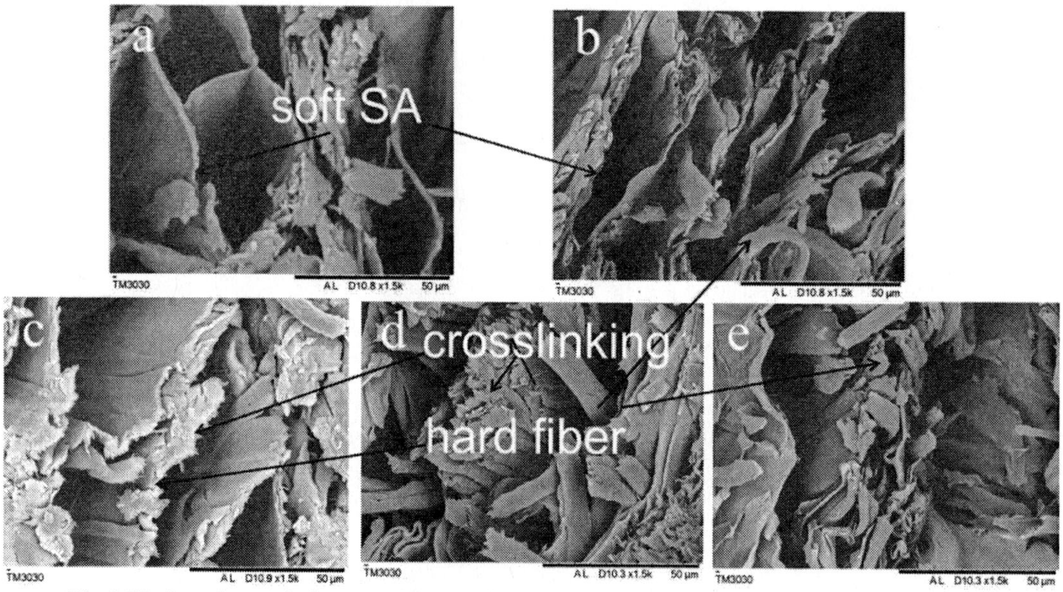

Fig.2 Hydrogel morphology a, MBSA1, b, MBSA2, c, MBSA3, d, MBSA4, e, MBSA45

4.2 Rheological Property Analysis of MBSH

4.2.1 Static Rheological Properties Analysis of MBSH

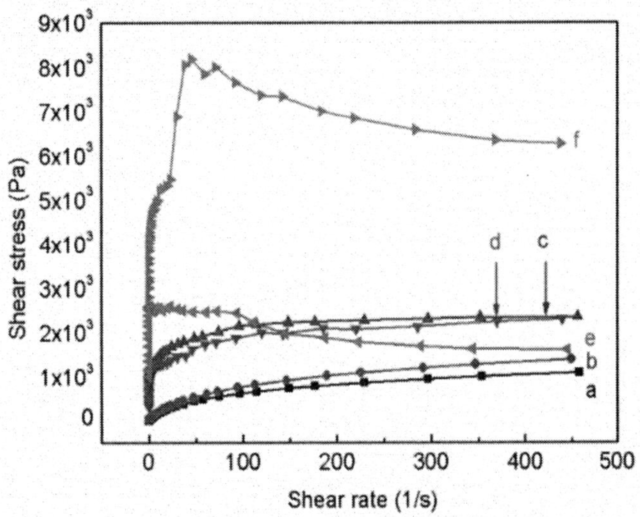

Fig.3 Static rheological curve at full shear rate
a, MBSH0, b, MBSH1, c, MBSH3, d, MBSH2, e, MBSH5, f, MBSH4

The static rheological property trends of MBSH were basically consistent at the same shear rate. The shear stress of hydrogel increased significantly at medium and low shear rates of 0.01 ~ 10 s-1. The shear stress of hydrogel tended to be stable at the high shear rates of 10 ~ 500 s-1. When the shear stress of hydrogel reached maximum, it decreased and tended to be stable. The general trend that the shear stress of curve (a ~ f) increased gradually with the increase of mulberry fiber content. The maximum shear stress of MBSH4 (8×103 Pa) was higher than that of SA hydrogel. However, especially when the shear rate increased, the shear stress of MBSH5 was smaller than that of MBSH2 and MBSH3. when the content of mulberry fiber was between 1.0g and 3.0g, the experimental results showed that the molecular crosslinking points between the fiber and SA increased, however the intermolecular entangling points in the composite hydrogel system decreased, among which MBSH4 had the most crosslinking points and the highest viscosity. The crosslinking point of MBSH5 began to decrease after reaching the saturation, when the mulberry fiber content was 3.0g, the main reason was that there was excess mulberry fiber in the hydrogel, and there was no cross-linking, which led to the decrease of hydrogel viscosity.

4.2.2 Viscoelastic Properties Analysis of Hydrogels

G ' represented the elastic storage modulus of the solution, and G "represented the viscous loss modulus of the solution, and G 'intersected with G' ' at a certain shear frequency as shown in Fig.4. According to the definition of gel forming point of hydrogel, the intersection point was the gel forming point of hydrogel. Six solutions had different relationships between G'and G"at different frequency range as shown Fig.4. The viscous loss modulus G " was higher than elastic storage modulus G 'of MBSH0 solution indicated that the solution presented fluid state at the 0~200rad·s-1 amplitude frequency range as shown in Fig.4(a)~ 4(b). The elastic storage modulus G ' was higher than viscous loss modulus G " indicated that the solution exhibited hydrogel properties at the 0~200rad·s-1 amplitude frequency range as shown in Fig.4(c)~ 4(e). The gelling point of MBSH5 solution indicated that MBSH5 solution presented different fluid states at the 0~200rad·s-1 amplitude frequency range as shown in Fig.4(f).

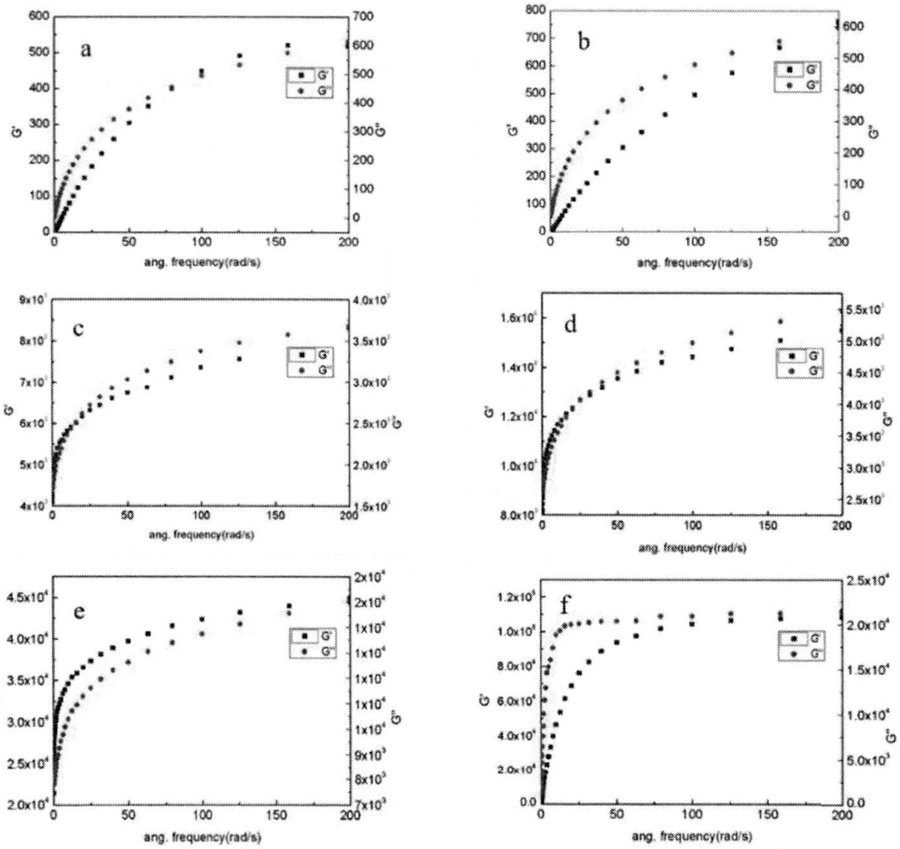

Fig.4 Correlation curve between dynamic viscoelastic parameters and frequency, a, MBSH0, b, MBSH1, c, MBSH2, d, MBSH3, e, MBSH4, f, MBSH5

5. Conclusion

The composite hydrogel spheres containing different mulberry fiber were prepared by blending method with sodium alginate as functional component and droplet suspension gel pellet-forming method, and the mulberry fiber/SA composite hydrogel with core-shell structure were prepared by freeze-drying method. The experimental results showed as follows.

1) Mixing mulberry fiber with SA improved the stability of hydrogel to some extent. The network structure of the composite hydrogel tended to be uniform with the increase of fiber content. The SA was attached to the cellulose chain of mulberry fiber, and the main body showed network interpenetrating three-dimensional structure.

2) The rheological properties of the hydrogels showed that the fluid types of the six hydrogels wre pseudoplastic fluid. The number of intermolecular crosslinking points between cellulose and SA increased in the hydrogel, and the number of intermolecular entangles decreased in the cellulose solution system, among which MBSH4 had the most crosslinking points and the highest viscosity.

6. Acknowledgement

This work was funded by QingLan Project of Educational Department of Jiangsu Province, China (Education Department of Jiangsu Province ((2019)03), the project of Yancheng polytechnical college (ycgy201901), the Integration Platform of Industry and Education of Jiangsu Higher Vocational Education, Grant number: Jiangsu Vocational Education 2019. No 26, and Jiangsu Province Higher Vocational Education High-level Major Group

Construction Project-Modern Textile Technology Major Group. Science and Technology Guidance Project of Chinese Textile and Apparel Association, No.2021055/ No.2021022.

References

[1] Xu Y, Xie J, Gao H, et. al, Interpenetration Enterpenetration Enhancing of Chitosan-Peglm Double Network (DN) Hydrogel and its Properties[J]. Macromolecular Research: 2015; 23 (1): 2-12.

[2] Zhou HT, Fang DW, Wang B, et al. Preparation and Characterization of CS/AMPS/PEGDA Photocrosslinked Porous Hydrogel via Freeze-drying [J]. Journal of Materials Science & Engineering: 2015; 33 (3): 432-437.

[3] Sun JX, Tan PX, Feng XL, et al. Preparation and characterization of hightoughness ALG-Ca/PAM double network hydrogel [J]. Journal of Sichuan University (Natural Science Edition): 2015; 3: 645-650.

[4] CalóE, Khutoryanskiy VV. Biomedical Applocations of Hydrogels: a Review of Patents and Commercial Products[J]. European Polymer Journal: 2015; 65: 252-267.

[5] Sergeeva AS, Gorin DA, Volodkin DV. In-situ Assembly of Ca- alginate Gels with Controlled Pore Loading/ release Capability [J]. Langmuir: 2015; 31 (39): 10813-10821.

[6] Zhong M, Liu XY, Shi FK, et al. Self-healable, Tough and Highly Stretchable Ionic Nanocomposite P hysical Hydrogels [J]. Soft Matter: 2015; 11 (21): 4235-4241.

[7] Morch YA, Donati I, Strand BL, et al. Effect of Ca2+,Ba2+, and Sr2+ of Alginate Microbeads [J]. Biomacromolecules: 2006; 7: 1471-1480.

[8] Shao Y, Wu CY, Cheng Ch, et al. Preparation and and Controlled Release Behavior of Sodium Alginate- g-Polyacrylamide/Graphene Oxide Composite Hydrogel [J]. Polymer Materials Science and Engineering: 2006; 7: 1471-1480.

Research on Properties of Aramid Filament/Stainless Steel Microfilament Woven Fabric

Yong-Gui Li[*1, 3], Ming-Ming Kang[1], Dan Li[1], Zong-Yao Bao[2], Xiao-Han Shi[1], Chao-Qiang Huang[4]

[1] *College of Clothing and Design, Minjiang University, Fuzhou 350108, Fujian, China*
[2] *School of Textile Science and Engineering, Xi'an Polytechnic University, Xi'an 710048, Shanxi, China*
[3] *Fujian Key Laboratory of Novel Functional Textile Fibers and Materials, Fuzhou 350108, Fujian, China*
[4] *Fujian Qianglun New Materials Co., Ltd., Longyan 366012, Fujian, China*

[*] Corresponding author's email: lygwxjd@sina.com

Abstract

In order to obtain a new type of electromagnetic shielding fabric, aramid filament and stainless-steel microfilament is used to prepare composite yarns, and then the fabrics are made by sizing, weaving, and finishing processes. The properties of electromagnetic shielding and flame retardancy are studied. The results show that the twist of yarn should be fixed quickly to keep it stable when the composite yarn is made. Among woven fabrics, the electromagnetic shielding efficiency of the plain fabric is the best, while that of the satin fabric is the worst. The highest shielding efficiency is obtained when the layering angle of the fabric is 0°/45°. The structure of fabric has a significant effect on the shielding efficiency. Meanwhile, the fabric has good flame retardant performance.

Keywords: Aramid Filament; Stainless Steel Microfilament; Woven Fabric; Electromagnetic Shielding

1. Introduction

The electromagnetic waves are everywhere and without a trace, which is called "electronic radiation pollution" by scientists. The harm brings to people cannot be underestimated. Excessive electromagnetic radiation will affect the normal operation of electronic equipment, and cause significant harm to the human body. At the same time, if not appropriately handled, the explosion of explosives or explosive mixtures may be triggered to cause a fire. Electromagnetic wave pollution in the environment has become a serious social hazard [1-3].

At present, the market's electromagnetic shielding/flame retardant fabrics mainly include polymer conductive fabric, metal fiber fabric, and filled composite shielding fabric [4-6]. Although the cost of polymer conductive fabric is relatively low, it is not resistant to washing and is easily oxidized. The ferromagnetic material in metal fiber fabric has a good shielding effect in the low hertz frequency region, but its poor mechanical properties make it difficult to be widely used in the market. The filled composite shielded fabric can obtain shielding effectiveness by adding conductive coatings or additives through extrusion or injection, but the weaving process is complicated [7-8]. To obtain electromagnetic shielding/flame retardant fabric with excellent performance, there is a method of blending the metal fiber with ordinary fiber to weave. For example, Song Yuling [9] et al. studied the electromagnetic shielding efficiency of stainless steel/polyester blended fabrics; Zhang Zuoping [10] prepared wool/polyimide fiber blended functional fabrics and studied its flame-retardant and anti-ultraviolet properties. Huang Huayou [11] et al. weaved electromagnetic shielding fabric with a good performance by using aramid /stainless steel fiber blended yarn, but it is challenging to blend aramid staple fiber with stainless steel fiber since the yarn is thick; hence the woven fabric feels stiff and heavy.

To obtain a lighter and more comfortable electromagnetic shielding fabric with flame-retardant property, the aramid 1313 filament and stainless steel microwire (filament) are merged and twisted so as to weave woven fabric with both flame-retardant and electromagnetic shielding properties, and the properties of the fabric are studied. It is hoped that this paper can provide a new idea for the development of novel electromagnetic shielding fabric with flame retardant property.

2. Experiment

2.1 Materials and Equipments

Aramid 1313 (denier of fiber: 200D, breaking strength: 4.01cN/dtex, oil content: 1.5%, elongation at break: 20%) is produced by Yantai Tayho Advanced Materials Co., Ltd.

Type 316L stainless steel microfilament (diameter: 0.035mm, fineness: 7.83tex, breaking strength: 1.2cN/dtex, elongation at break: 21.3%) is produced by Qianglun New Materials Co., Ltd.

Digital sample doubler winder DSDw-01-2, digital sample doubler twister DSDt-01, semi-automatic sample loom SLo-01, fully automatic single-yarn sizing equipment DSSi-01-2 (Tianjin Jiacheng Electromechanical Equipment Co., Ltd.), yarn twist counter Y331C (Jiangsu Hengxing Machinery Manufacturing Co., Ltd.), YG (B) 815D-III fabric flame retardant tester, DR-913G fabric electromagnetic shielding radiation tester (Wenzhou Darong Textile Instrument Co., Ltd.), YG063 automatic single yarn intensity tester (Quanzhou Meibon Instrument Co., Ltd.), and TM4000Plus desktop electron scanning microscope (Hitachi Co., Ltd.).

2.2 Preparation of Composite Yarn

First, an aramid filament and a stainless-steel filament is combined by double winder and then twisted on doubler twister. It is processed into strands with a twist of 30 T/10cm, a winding tension compensation number of 1.0 N and a spindle speed of 6000 r/min, and these strands are finally sized by sizing equipment and wound for later use.

2.3 Preparation of Aramid/Stainless Steel Microfilament Fabrics

Electromagnetic shielding fabrics with different weaves, such as plain weave, 1/2 twill weave and 5/3 satin weave, were woven by sample loom. The fabric specifications are presented in Table 1.

Table 1. Fabric parameters

Fabric weave	Plain weave	Twill weave	Satin weave
Warp density/ (tuft/10cm)	210	210	210
Weft density/ (tuft/10 cm)	205	205	205
Thickness of fabric /mm	0.51	0.52	0.55

2.4 Test Method

2.4.1 Twist Stability of Composite Yarn

According to GB/T 2543.2-2001 Textiles-Determination of twist in yarns, the yarn twist counter is used for testing. The stability of the twist can be expressed by the twisting stability. When the twisting stability is 40%-60%, the weaving requirements can be met, and the twisting stability can be expressed via Formula (1).

$$P=\left(1-\frac{b}{a}\right)\times 100\% \quad 1)$$

In Formula (1): a is the measured length of twisted yarn, which is generally 50 mm, b is the distance between one end of the tested yarn and the other end moving towards the fixed end until the yarn begins to kink.

2.4.2 Tensile Property of Composite Yarn

In accordance with GB/T 3916-2013 Textiles-Yarns from packages-Determination of single-end breaking force and elongation at break using constant rate of extension (CRE) tester, the breaking strength and elongation are calculated through a fully automatic single yarn intensity meter.

2.4.3 Electromagnetic Shielding Efficiency of Fabrics

In accordance with GB/T30142-2013 Measuring method for shielding effectiveness of planar electromagnetic shielding materials, the electromagnetic shielding efficiency of the samples is tested through an anti-electromagnetic radiation performance tester. The electromagnetic shielding effect is represented by shielding efficiency SE, and the shielding efficiency of electromagnetic waves passing through shielding materials can be calculated by the formula (2) [12].

$$SE=20\lg\left(\frac{E_0}{E_1}\right)=20\lg\left(\frac{H_0}{H_1}\right) \quad 2)$$

In Formula (2):

- SE is shielding efficiency in dB.
- E_0 is the electric field intensity when shielding materials are not present.
- E_1 is the electric field intensity when shielding materials are available.
- H_0 is the magnetic field intensity without shielding materials.
- H_1 is the magnetic field intensity where shielding materials are available.

According to 2.7.1 of GB/T 26667-2011 Terminology for electromagnetic shielding materials: electromagnetic shielding efficiency SE is generally negative, but its absolute value is customarily used. SE value is 0 when there is no shielding, and the more significant the SE value, the better the anti-electromagnetic radiation effect. The shielding efficiency grade is obtained according to the existing standards, as shown in the table below.

Table 2. Shielding efficiency level

	Shielding efficiency (SE/dB)	Shielding effect	Remark
1	<0	Poor	/
2	10~30	Inferior	≥20 dB may be commercialized
3	30~60	medium	≥35 dB meets civil requirements
4	60~90	good	≥75 dB meets military requirements
5	>90	superior	/

2.4.4 Flame Retardant Performance Test

In accordance with GB/T5455-2014 Textiles- Burning Behavior-Determination of damaged length, afterglow time and afterflame time of vertically oriented specimens, the flame retardancy of the samples is tested. After the ignition time, igniter is removed, then the after flame time and afterglow time are recorded, and the damaged length is measured.

3. Results and Discussion

3.1 Preparation Technology of Composite Yarn and its Influence on Yarn Properties

3.1.1 Preparation Technology of Composite Yarn

The following precautions should be paid attention when twisting twice because of the high modulus, high hardness, and challenging combination of aramid and stainless steel microfilament. In the first place, to ensure the simultaneous unwinding of stainless-steel microfilaments and aramid filaments, and to improve the mechanical properties of composite yarns, the quantitative filament tension disc should be added. Besides, during twisting, it is requisite to control the spindle speed properly so that the machine speed is stable, and the leather roller is not easy to fall off. After twisting, the yarn's initial modulus is larger than before, which causes severe twisting between fibers, and the internal stress cannot be neutralized promptly. Therefore, sizing treatment is needed to stabilize its shape and reduce fiber scuffing.

3.1.2 Specifications of Composite Yarn

Based on the method of 1.4.1, the fineness and the actual twist of the yarn are 30.2 tex and 29.118 twists /10 cm. It is different from the design twist of 30 twists /10 cm by 1 unit. It is chiefly because after sizing treatment, the vibration of the molecular links of aramid fiber and stainless-steel microfilament is intensified, and the kinetic energy between the molecules is increased so that the amorphous region molecules are rearranged. The interaction force between the macromolecules is weakened. The fast elastic deformation of fibers is transformed into slow elastic deformation, which expedited the stress relaxation process between fibers and finished the set twist in a short period [13]. The P value of twist stability is 58.1%, in line with the yarn requirements of basic woven fabrics. Thus, composite yarn can be used to prepare electromagnetic shielding fabrics.

3.1.3 Mechanical Properties of Composite Yarn

The primary performance results of composite yarns tested by the method mentioned in Article 1.4.2 are presented in Table 2. In combination with Table 2 and Fig.1, it can be learned that the surface of composite yarn is smooth and hairless. The yarn structure is similar to wrapped yarn, with stainless steel microfilaments and aramid filaments alternately wound on the yarn surface. With the addition of stainless-steel microfilament, the breaking strength and elongation at the break of the yarn distinctly increased, owing to its high intensity, high rigidity, and corresponding tensile force. The mechanical properties of the composite yarn were improved as well.

Table 3. Mechanical properties of composite yarns

Samples	Breaking strength (cN/dtex)	Elongation at break (%)
Aramid 1313	4.01	25.93
Stainless steel microfilament	1.21	21.31
Composite yarn	4.45	29.24

(a) Aramid filament (b) Stainless steel microfilament (c) Composite yarn

Fig.1 Longitudinal surface of samples

3.2 Flame Retardant Properties of Fabrics

The flame-retardant properties of different fabrics are presented in Table 3. Three types of fabrics are subjected to varying degrees of carbonization reaction, but there is no dripping and smoldering phenomenon. After leaving the fire, twill and satin fabrics underwent a comparatively short afterglow time, while plain fabric keeps burning for 1s. The length of damage of the three types of fabrics in the burning direction is more than 10mm, but all of them meet the requirements of the B1 level of GB/T 17591-2006 Flame retardant fabrics.

Table 4. Flame retardant properties of different fabrics

Fabric structure	Test direction	afterflame time (s)	afterglow time (s)	length of damage (mm)
Plain weave	warp	0	1.2	11
	weft	0	1.5	26
Twill weave	warp	0	0.8	10
	weft	0	0.9	13
Satin weave	warp	0	0.7	16
	weft	0	0.3	18

Note: combustion characteristics: carbonization, no dripping and afterglow.

3.3 Electromagnetic Shielding Efficiency of Fabrics

3.3.1 Influence of Structure on Electromagnetic Shielding Efficiency of Fabrics

The shielding efficiency curves of fabrics with different structures in the range of 30-3000 MHz are presented in Fig.2.

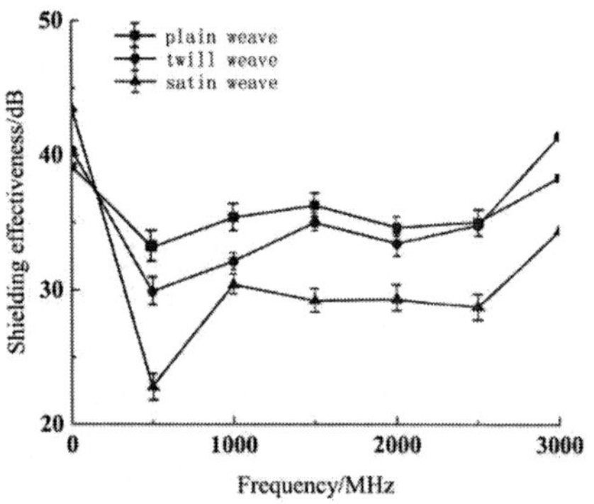

Fig.2 The electromagnetic shielding efficiency curves of fabrics with different structures

As observed in the figure, in the frequency range of 30-3000MHz, the shielding efficiency of plain fabric is the best while that of the satin fabric is the worst, and the peak value of electromagnetic shielding efficiency of plain weave and twill weave is above 30dB. When the fabric is considered a shield with meshes, the pores will influence the shielding efficiency of the fabrics. The pore or slit formed by the fabric is equivalent to a rectangular waveguide, and the cutoff frequency of the waveguide is chiefly associated with the wire diameter of the aperture. The larger the wire diameter, the lower the cutoff frequency of the waveguide. There is no shielding frequency when the electromagnetic wave frequency is up to the cutoff frequency [14]. In the three elementary weaves, the average floating length of plain weaves is comparatively short. There are numerous interweaving times of warp and weft to increase the tightness of fabric, and the gaps between yarns decreased. Hence, the one with the best shielding efficiency is plain fabric. Furthermore, twill and satin fabrics are of longer floating length, which is easy to cause snagging in the dressing process. Therefore, the transmission rate of electromagnetic wave in the fabric will be uneven, affecting the fabric's electromagnetic shielding efficiency.

3.3.2 Influence of Layers on Electromagnetic Shielding Efficiency of Fabrics

The curves of shielding efficiency of fabrics with different layers in the range of 30-3000MHz are presented in Fig.3.

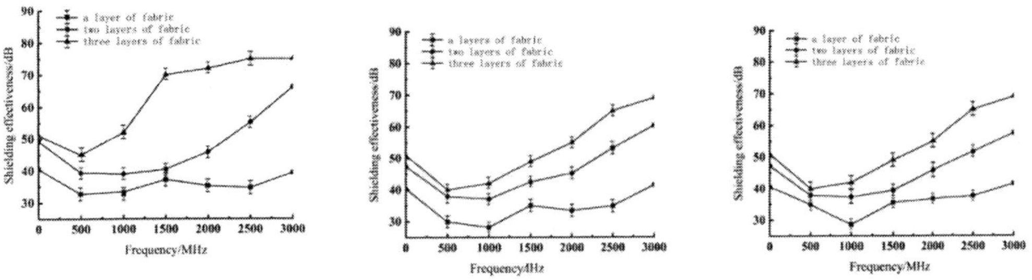

(a) Plain fabric (b) Twill fabric (c) Satin fabric

Fig.3 Electromagnetic shielding efficiency curves of fabrics with different layers

It can be known from Fig.3 that in the frequency range of 30-3000MHz, the shielding efficiency of three-layer fabrics is the best, while that of the one-layer fabric is the worst. The electromagnetic shielding efficiency of three-layer fabrics with three types of weaves is more than 60dB. According to Schelkunoff's shielding theory, the shielding performance of shielding materials is determined by shielding efficiency (SE), that is, SE = R+A+B. As indicated in Fig.4, when electromagnetic waves propagate to the fabric surface, there are generally three different reflections of electromagnetic waves: (1) attenuation of reflection of electromagnetic wave incurred by sudden changes in impedance of different structures on the incident fabric surface(R); (2) attenuation that is not reflected but enters the interior of the fabric and is absorbed by the specific material of the fabric(A); (3) multiple reflection attenuation into the interior of the fabric(B). As the number of fabric layers was on the rise, the thickness of the fabric increased as well, resulting in the increased number of times and loss of internal reflection of electromagnetic wave (B) and the accelerated loss of electromagnetic wave so that the shielding efficiency increases.

3.3.3 Influence of Laminated Angle on Electromagnetic Shielding Efficiency of Fabrics

Double-layer fabrics are used in most cases as overalls and protective clothing in the market should possess the characteristics of breathability, softness, and portability. Therefore, this experiment probed into the influence of the lamination angle on the electromagnetic shielding efficiency of fabrics based on two-layer fabrics. In this research, the lamination angles of fabrics are classified into 0°、45° and 90°. By arrangement and combination, three types of fabrics are obtained: 0°/45°, 0°/90° and 45°/90° fabrics. As the 0°/45° and 45°/90° fabrics only change positions in the warp and weft directions, the electromagnetic shielding efficiency is not enhanced. Hence, the shielding efficiency of the lamination angles with 0°/45° and 0°/90° is taken [15], as exhibited in Fig.4.

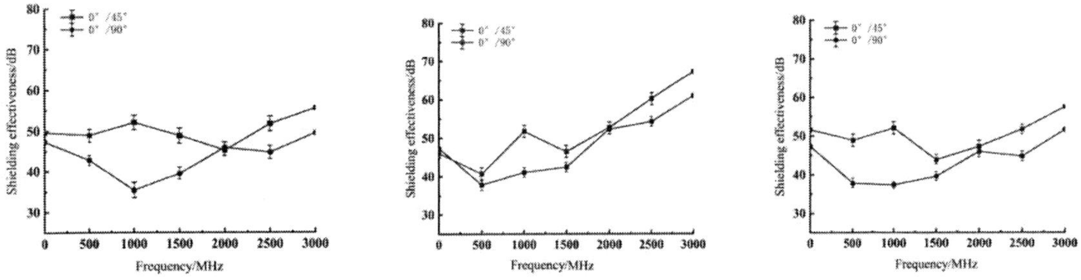

(a) Plain fabric (b)Twill fabric (c) Satin fabric

Fig.4 Electromagnetic shielding efficiency curves of fabrics with different laminating angles

It can be learned from Fig.4 that when the fabric's lamination angle is arranged at 0°/45°, the shielding efficiency is the best, and that of the 0°/90° fabric is the second. On the one hand, when the lamination angle is arranged at 45°, the interlacing superposition between yarns increased, and the gap decreased, resulting in the decrease of electromagnetic wave transmittance. In contrast, when the fabric is arranged with a lamination angle of 0°/90°, there exists no distinct difference in shielding efficiency; only the warp and weft directions are changed. On the other hand, when the fabric with a lamination angle of 0°/90°, the vertical magnetic field presented vertically in the warp and weft directions can be offset and shielded, so the shielding effect is comparatively poor [16, 17].

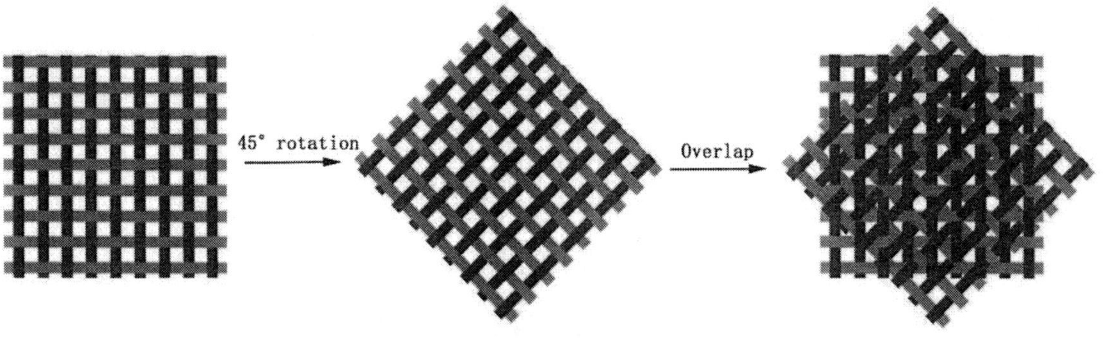

Fig.5 0°/45° superimposed renderings

3.4 Two-factor Analysis of Variance

Matlab's two-factor analysis of variance is applied to explore the significant factors that influence the electromagnetic shielding efficiency of fabrics: fabric structure and laminated angle. Most functional garments are made in two layers; the characteristics of permeability, softness, and weight of fabrics should be considered. Therefore, the influence of layers as a factor on dependent variables will not be elaborated. The mean value of shielding efficiency of fabrics with plain weave, twill weave, satin weave, 0°/45° lamination angle, and 0°/90° lamination angle is taken within the range of 30-3000MHz.

A two-factor analysis of variance is launched on the strength of SPSS [18]. Within the range of 30-3000MHz, α is set at 0.05; the results of two-factor analysis of variance are presented in Table 4. The original assumption is that the structure and laminated angle in the model exert no effect on shielding efficiency. However, it can be learned from the table that the P value of this test is far less than 0.05, which means a variance model provided with statistical significance, suggesting that at least one factor, i.e., structure and laminated angle, imposes a significant impact on shielding efficiency.

Table 5. Intersubjective effect test
Dependent variable: shielding efficiency

Source	Square sum of a type	Degree of freedom	Mean square	F	Significance
Modified model	220.495[a]	20	1	911.7	.001
Intercept	188857.924	20	1.2	897.0	.000
Weave type	216.321	30	1.2	935.6	.000
Lamination angle	4.174	30	1.5	915	.201
Error	3.395	40	1.5	823.4	
Total	19081.814	40	1	870.7	
Revised total	223.891	904.35	891.2		

a. R square = .985 (adjusted R square = .970)

The third and fourth rows in the table check the structure and laminated angle in the model. The original assumptions are that the weave structure exerts no effect on the shielding efficiency, and all its coefficients α_i are equal to 0. Laminated angle imposes no effect on shielding efficiency, and all its coefficients β_j are equal to 0. Concerning the data in the table, the P value of the third row is far less than the set α value, that is, 0.05. The P value of the fourth row is 0.201, which is greater than 0.05. Thus, different weave structures impose a comparatively significant impact on shielding efficiency, but the impact of laminated angle is not significant.

4. Conclusion

When preparing composite yarns of aramid filament/stainless steel microfilament, it is requisite to make sure that stainless steel microfilament and aramid filament are unwound simultaneously when the bobbin is unwound in a bid to decrease the twist unevenness of composite yarns. Meanwhile, a heat setting process or sizing process should be launched to make the yarn quickly set to facilitate subsequent weaving after the yarns are prepared. The woven fabrics of aramid filament/stainless steel microfilament are provided with favorable flame retardant/ electromagnetic shielding efficiency. The electromagnetic shielding efficiency of plain fabric is the best, while satin fabric is comparatively poor. The increase in the number of fabric layers will cause the rising shielding efficiency. When the fabrics are arranged at a laminated angle of 0°/45°, the shielding efficiency of the fabric is the best. In combination with the two-factor analysis of variance, it is concluded that fabric structure imposes a relatively significant influence on shielding efficiency, but the influence of laminated angle is not significant.

5. Acknowledgement

This word was supported by Project of Central Leading Local Science and Technology Development (No. 2018L3012), the Science and technology projects of Fujian Province (No. 2019H6019), and the Research Project of Minjiang University (No. YSZ20007) and Innovation and Entrepreneurship Project for University Students in Fujian Province (202010395003).

References

[1] Yang XX, Li SL, Yu P et al. Electromagnetic wave pollution in the environment and its hazards. Frontier Science: 2014; 8 (1): 13-26.
[2] Zhang ZW. Analysis of fire hazards in reprocessing plants. Industry and Technology Forum: 2019; 18 (15): 78-79.
[3] Song B, Huang YW, Zu WH, et-al. Research progress of electromagnetic shielding materials. Guangzhou Chemistry: 2021; 46 (01): 1-7.
[4] Wanasinghe D, Aslani F, Ma G. Electromagnetic shielding properties of cementitious composites containing carbon nanofibers, zinc oxide, and activated carbon powder. Construction and Building Materials: 2021; 285.
[5] Liu PZ, Zhang L, Wang WF, et al. Synthesis and electromagnetic shielding properties of graphene-Fe_3O_4-$BaTiO_3$/ silicone rubber nanocomposites. Materials Science Forum: 2019; 950: 97-102.
[6] Sun T, Zhao X. Research progress of electromagnetic shielding materials. Journal of Textile Science and Engineering: 2018; 35 (2): 118-122.
[7] Zeranska CK, Wróblewska A, Kowalczyk S, et al. Graphene infused ecological polymer composites for electromagnetic interference shielding and heat management applications. Materials: 2021; 14 (11): 1-11.
[8] Wang Z, Cheng Z, Fang CQ, et al. Recent advances in MXenes composites for electromagnetic interference shielding and microwave absorption. Composites Part A: 2020; 136: 1-17.
[9] Song Y, Dong C, Cai X, et al. Electromagnetic shielding effectiveness of stainless steel/polyester blended fabrics. Synthetic Fibers: 2019; 48 (11): 39-42.
[10] Zhang ZP, Yang L, Shao L, et al. Production practice of wool/polyimide fiber blended functional fabrics. Wool Spinning Technology: 2020; 48 (8): 42-45.
[11] Huang HY, Wu YL, Ni HY, et al. Study on electromagnetic shielding performance of aramid/stainless steel fiber blended woven fabrics. Chemical Fiber and Textile Technology: 2020; 49 (1): 17-22.
[12] Liu L, Zhang D. Research progress of electromagnetic shielding materials. Functional Materials: 2015; 46 (3): 3016-3022.
[13] Zhu Z, Liu Z. Monophosphorylation of acid-thinned starch to enhance the quality of viscose yarns sized at reduced temperature. Starch - Starke: 2010; 61 (3-4): 139-144.
[14] Shi FW, Wang Q. Electromagnetic functional textile materials. Beijing: Science Press: 2015; 153-159.

[15] Mondal S, Nayak L, Rahaman M, et al. An effective strategy to enhance mechanical, electrical, and electromagnetic shielding effectiveness of chlorinated polyethylene-carbon nanofiber nanocomposites. Composites Part B: 2017; 109 (2): 155-169.
[16] Maiti S, Shrivastava NK, Suin S, et al. Polystyrene/MWCNT/graphite nanoplate nanocomposites: Efficient electromagnetic interference shielding material through graphite nanoplate-MWCNT-graphite nanoplate networking. ACS Applied Materials & Interfaces: 2013; 5 (11): 4712-4724.
[17] Ortlek HG, Saracoglu OG, Bilgin SS. Electromagnetic shielding characteristics of woven fabrics made of hybrid yarns containing metal wire. Fibers & Polymers: 2012 (13): 63-66.
[18] Saiyidi M, Roni, Hadrian, Geri D. Data Analysis with SPSS for Survey-based Research: 2021-07-15.

Cite as: H.L. Xu et al.,
TBIS 10.3993/tbis (2021).

The Effects of Needle Configurations on the Electrospinning Triangle

Hui-Lin Xu[1], Zhi-Yong Liang[2], Xiao-Hong Qin[1*]

[1] *Key Laboratory of Textile Science & Technology, Ministry of Education, College of Textiles, Donghua University, Shanghai, 201620, P. R. China*
[2] *College of Science, Donghua University, Shanghai, 201620, P. R. China*

*Corresponding author's email: xhqin@dhu.edu.cn

Abstract

The geometry and stability of the electrospinning triangle has an affect on nanofiber yarn formation. However, it is difficult to describe the formation process of the electrospinning triangle due to its complex features. This paper divides the state of the electrospinning triangle into a static electrospinning triangle and a dynamic electrospinning triangle for analyzing its formation process. The electrospinning jets are simplified as a certain number of charged particles to track the jet trajectory and to investigate the effect of the needle configurations on the electrospinning triangle. Experiments are performed to verify the theoretical predictions. The range of the dynamic electrospinning triangle for preparing continuous yarn is explored. This analytical method combining theory with experiment provides an idea for predicting the optimal configuration conditions for the stable E-triangle and continuous nanofiber yarns.

Keywords: Electrospinning Triangle; Charged Particle Tracking Method; Needle Configurations; Nanofiber Yarn

1. Introduction

Electrospinning is a simple method to produce micro and nano sized fibers. The continuous electrospinning yarn is a type of nanofiber that can be easily converted into fabrics, making them potentially useful in electronic textiles, moisture transmission, tissue scaffold, antibacterial and other fields [1-5]. To continuously to meet the high demand of nanofiber yarns, researchers have introduced various methods into the electrospinning system. For example, rotating drum, rotating sharpy disk or liquid bath is applied to collect the fibers into bundles [6-8]. There are also some studies that uses the self-organization of electrospinning jets to form continuous yarn. Double needle conjugate electrospinning method seems to be one of the most successful approaches. The method makes use of two opposing placed needles applied with positive and negative voltages respectively and twist the self-organized nanofibers into yarns via a rotating collector. The collector started off as a rotating shaft, which was later modified into hollow funnel with a collector of conducting tines. The collection system is composed of a neutral metal disc and an opposite arranged hollow metal rod [9-12].

Many published papers on these methods focuses on the properties of obtaining nanofiber yarns and their applications. The factors of yarn formation have also been studied. The main concern is the effect of process parameters on the properties of the final product, such as voltage, humidity, twist, and solubility, conductivity, and viscosity of the solution. While little attention has been paid to the electrospinning triangle zone before a yarn is formed. Hajiani defined this zone as E-triangle, which is the area between the collector surface and yarn convergence point. Hajiani investigated the effect of twist on the geometry of E-triangle [13]. He founded out that the geometry of E-triangle has a great influence on the performance of obtaining yarn, similar to the twist triangle in traditional ring spinning. Maleki also adopted the concept of E-triangle and experimentally studied the influence of the shape change of E-triangle on the performance of pure nanofiber yarns and core-shell yarns [14, 15]. But these researches only focused on the plane E-triangle without considering the hollow cone-shaped E-triangle when the collector surface is circular. Liu analyzed the spinning triangle of modified ring spinning system, such as compact spinning, and believed that the fibers in the spinning triangle at the front nip line were not arranged in the same plane but in a concentric circular cone [16]. Like spinning triangles of the traditional and modified ring spinning mentioned above, the E-triangles are plane E-triangle and hollow cone-shaped E-triangle when the collector of electrospinning yarn forming setup is rotating or not. Moreover, Due to the complicated whipping behavior of the electrospinning jet, it is difficult to see and control the jet trajectory which determined the nanofiber deposition. While the nanofiber deposition will seriously affect the formation of the E-triangle. Therefore, it is necessary to discuss the forming of E-triangle by predicting the jet trajectory and nanofiber deposition. The electric filed plays an important role in determining the jet trajectory and nanofiber deposition [17]. Gupta used COMSOL software to study the effect of electric field distribution on resultant nanofibers, but he only considered a unipolar jet and did not consider the interaction between jet and electric field [18].

In this study, we will use the modified double needle conjugate method developed by Wu to investigate the forming process of E-triangle and the effects of needle configurations on E-triangle [11]. Firstly, we simulate the effect of needle configurations on electric field distribution without considering the jets. Then the movement of jets is considered in following simulations, because the jet movement and deposition are very important to the forming of the E-triangle. During electrospinning yarn forming process, the complex relationships of electrostatic fields, jets-electrostatic field interaction, nanofiber winding and the rotation of E-triangle are all involved. It is difficult to simulate the E-triangle forming process directly since it involves the rotation of E-triangle which is complicated to added to simulations. We define the plane E-triangle with a fixed disc collector as static E-triangle, and hollow cone-shaped E-triangle as dynamic E-triangle when the disc collector is rotating. In simulations, we only consider the static E-triangle, the dynamic E-triangle will be investigated experimentally. And a certain number of charged particles are used to represent electrospinning jets in simulations. Through the charged particle tracking model of simulation software COMSOL, the trajectory of the charged particles can be simulated to identify the fiber deposition position and predict the approximate range of the static E-triangle. When all electrospinning parameters are fixed, even only changing the configuration of the needles may affect the collection efficiency of nanofibers, that will affect the formation of the E-triangle. In order to reduce the variables and simplify the simulations, we only focus on the effect of needle configuration in this paper. First, we set the angles between the two needles and the axis of the neutral disc and the hollow metal rod at 60°, 90°, 120° to predict the fiber collection efficiency and the geometry of the E-triangle. Then, the distance between the two needles and the neutral disc is increased from 20mm to 50mm at an interval of 5mm in the highest collection efficiency angle to predict the results and determine the distance range for favorable E-triangle. In addition, experiments are also carried out to verify the theoretical predictions. Based on the analysis of the static E-triangle, the dynamic E-triangle is changed by controlling the convergence point. And the distance range of convergence point in a dynamic E-triangle that is conducive to continuous yarn formation is discussed.

2. Materials and Method

2.1 Materials

Polyacrylonitrile (PAN) is widely used as nanofiber templates and carbon precursors to obtain carbon nanofibers, and is also commonly used in electrospinning. Therefore, we choose PAN as our polymer material [19]. PAN with a molecular weight of 75,000 was purchased from the Shanghai Chemical Fibers Institute, and N, N-Dimethylformamide (DMF) as a solvent was purchased from Shanghai Lingfeng Chemical Reagent Co. Ltd. PAN was dissolved in DMF with concentrations of 10 wt% and stirred magnetically at room temperature until becoming homogeneous. A CCD digital camera (Canon EOS7D) was used to capture the geometry of the E-triangle.

2.2 Electrospinning Apparatus

The electrospinning yarn apparatus, shown in Fig.1(a), used here consists of two oppositely placed metal needles, a neutral metal disc (hereinafter referred to as disc or NMD, with 60 mm of diameter),a neutral hollow metal rod (hereinafter referred to as rod or NHMR, with non-insulated part length of 10 mm, inner diameter of 8 mm and outer diameter of 12 mm) and a roller. The disc and the rod are placed opposite with a distance of 70mm. The two needles are placed perpendicular to the axis of the disc and rod, and applied with a positive and negative voltage of 10kV respectively. The distance between two needles was 20mm, and the solution supply velocity is 0.8ml/h.

When electrospinning starts, positive and negative charged jets meet in the middle plane to form a fiber web, which will be guided through the rod to the roller to form yarns. When the disc is fixed and the roller is rotating, a planar E-triangle which defined as static E-triangle is formed as shown in Fig.1(b). And Fig.1(c) shows the diagram of dynamic E-triangle which is hollow cone-shaped when the disc and the roller are rotating at the same time.

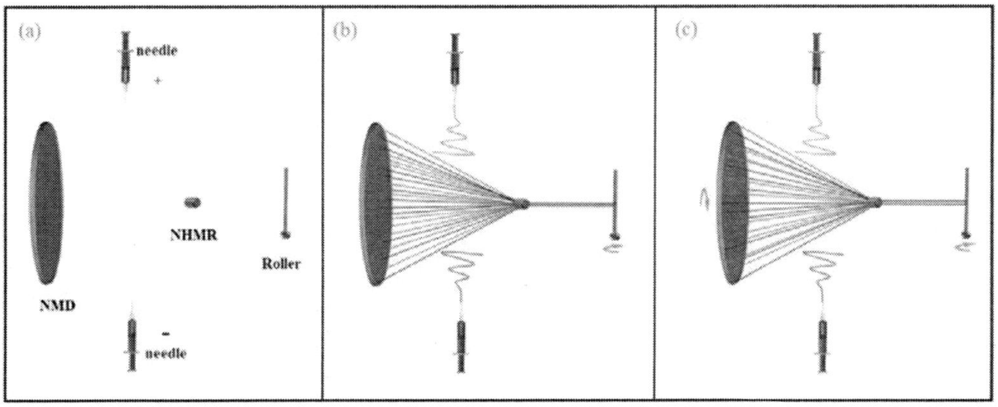

Fig.1 Diagram of electrospinning apparatus (a), the forming process of static E-triangle (b) and dynamic E-triangle (c) between the disc (NMD) and rod (NHMR).

2.3 The Simulation Method

COMSOL is a well-known Multiphysics coupling simulation software based on finite element method. And the particle tracing module can be used for computing the trajectory of particles in a fluid or electromagnetic field, including particle-particle, fluid-particle, and particle-field interactions. The physical simulation model adopted in this paper is shown in Fig.2. There are two needles, the disc and rod, and a cylindrical area is used to define the research area filled with air. We first simulate the electric field distribution of the model without considering the jets. Then, electrospinning jets are simplified as amounts of charged particles emitted from the needles at a given initial speed, and their trajectory under the action of fluid and electrostatic field is studied to indicate the approximate deposition position of the jets. In this model, the size of the needle is slightly enlarged in order to simplify the calculation, which will not affect the simulation results. The positive and negative jets are represented by red and blue particles, respectively. And the velocity of these particles is considered equivalent. The simulation time ends when the charged particles reach the middle surface of two needles, because this means that the positively and negatively charged jets meet in the middle surface and self-organize into a film to form static E-triangle.

3. Results and Discussion

3.1 Simulation of Electric Field Distribution

First of all, we simulate the electric field distribution of the electrospinning setup before taking the jets into consideration. The physical model is explained above and the research area contains only the front end of the needles because the other parts of needles have little effect on the electric field distribution. In addition, subsequent simulations need to set the back ends as the inlets of particles, which needs to be arranged outside the research area. In the premise of not changing the results, such a model is conducive to accelerating simulation speed and maintaining coherence with subsequent models.

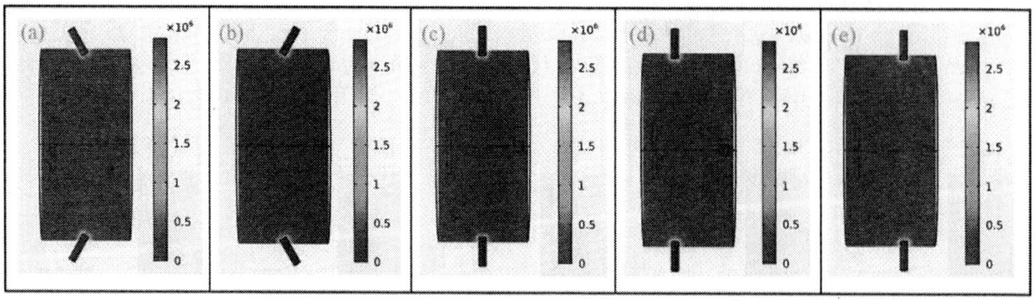

Fig.2 Simulation results of electric field distribution when the two needles placed at the middle of disc and rod with an angle of 60° (a), 90° (b), and 120° (c) and placed closer to disc (d) or rod (e) with an angle of 90°, respectively.

Fig.2 simulates the electric field distribution without considering the influence of electrospinning jets when positive and negative voltages are applied to two needles with different needle configurations. It can be seen that the highest electric field strength is concentrated at the tips of the two needles, and the distribution of the needles has little effect on their field strength. This may be due to the fact that the voltage difference between the two needles and the overall structure of the spinning equipment have not changed much. Therefore, it can be concluded that the electric field distribution is not sufficient to characterize the difference of electrospinning yarn forming process with different needle configurations without considering the movement of jets. However, in the actual experiments, the needle configurations do make a difference. So the movement of the jet should be considered in the following analysis.

3.2 The Effect of the Needle Angle

Wu [20] placed two needles perpendicular to the axis of the disc and the rod, while others indicated that it is beneficial to yarn forming when the relative angle of the needles and drum collector is 60° [21]. This means that the relative angle between the needles and the collector might have different effects on the fiber collection efficiency and yarn forming when different collectors are used. Therefore, we simulated the trajectory of the charged particles when the angle between the two needles and the axis of the disc and the rod is 60° (I), 90° (II), and 120° (III), respectively. Fig.3(a) and Fig.3(b) shows the physical model and simulation results under the configurations of above three angles. The red particles represent the positively charged jet, and the green particles represent negatively charged one. From Fig.3 can see that the particles are distributed at the middle surface of the two needles. And the particle distribution range can roughly represent the jet deposition which shows the approximate shape of the static E-triangle. Experiments were also carried out to verify the simulation results in Fig. 3(c).

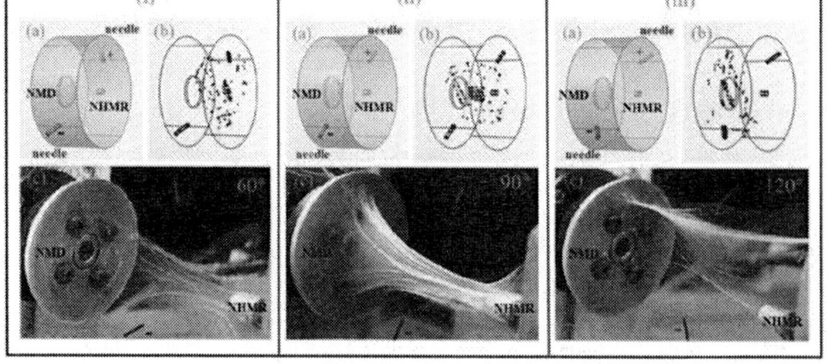

Fig.3 The physical model (a), charged particle tracking simulation results (b) and experimental photos (c) when the angle between the two needles and the axis of the disc (NMD) and the rod (NHMR)are 60° (I), 90° (II), and 120° (III), respectively.

As can be seen from Fig.3(b): the charged particles are mainly concentrated around the rod, while few fibers are deposited on the disc when the angle is 60° (I); the charged particles are distributed between the area of the rod and the disc when the angle is 90° (II); and the charged particles are mainly concentrated around the disc with fewer deposited on the rod when the angle is 120° (III). As static E-triangle is formed when the nanofibers are connected to the rod and the disc and will collect more nanofibers and get a thicker E-triangle when the angle is 90°. The experiment shown in Fig.3(c) verified the above simulation results. It can be seen from the experiment that when the angle is 90°, the collected fiber is significantly more than the other two, and the nanofibers are firmly adhered to the disc and the rod. Therefore, 90° is considered as the optimal angle in this electrospinning yarn apparatus.

3.3 The Effect of the Needle Position

From the above analysis, it is obvious that the fiber needs to be connected between the disc and the rod to form a thick and stable static E-triangle. Therefore, the distance between the needles and the disc may also be an important factor in addition to the angle. As follow, we investigate the formation of the static E-triangle under different distances when the angle is fixed at 90° and the rotating of the disc is not involved. Fig.4 shows the particle tracking simulation results and experimental photos of the static E-triangle when the distances between the two needles and the disc is changed from 20mm to 50mm at an interval of 5mm.

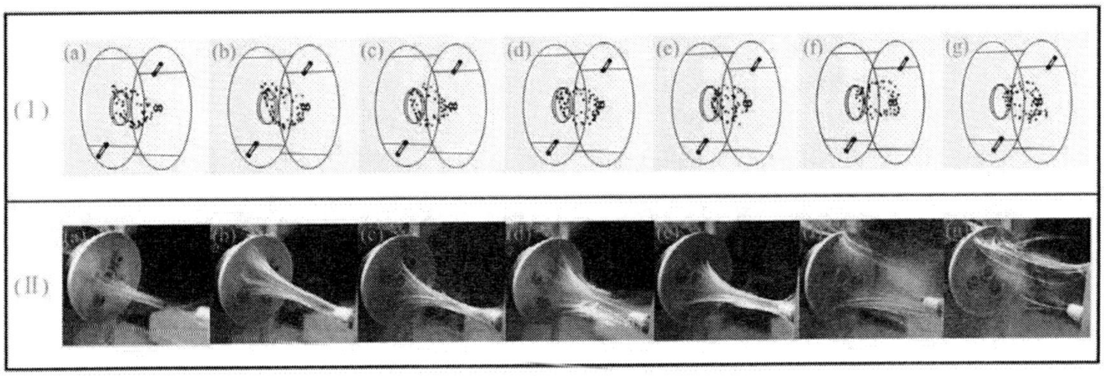

Fig.4 The particle tracking simulation results (I) and experimental photos of the static E-triangle (II) when the distances between the two needles and the disc is 20mm (a), 25mm (b), 30mm (c), 35mm (d), 40mm (e), 45mm (f), and 50mm (g), respectively.

Carefully observing the simulation results in Fig.4(I), it is found that when the distance between the needle and the disc gradually increases from 20mm to 50mm, the charged particles gradually tend to deposit to the rod direction from the beginning when most of the charged particles were deposited on the disc and only a few particles deposited in front of the rod (Fig 4(I .a)). The experimental results in Fig.4(II) also show similar trends with the simulation results. In Fig.4(II .a), most of the nanofibers are deposited on the disc and only a thin layer of fibers attached to the rod. As the distance increases, more nanofibers are connected to the disc and the rod to form an E-triangle that is more suitable for continuous yarn forming. To further increase, the fibers tend to deposit to the rod direction and connect to the surrounding objects, which is not conducive to the nanofiber collection and the static E-triangle forming. Therefore, it can be concluded that the distance ranges from 25mm to 35mm, when charged particles are well connected to the disc and rod and mainly concentrate between them, is more adequate for the forming of static E-triangle and continuous yarn.

3.4 The Effect of Convergence Point

After analysis the effect of the angle and the position of the needles that will lead to a stable static E-triangle, it is necessary to take the rotation of the disc into account. Since it is a key step to form a dynamic E-triangle and continuous nanofiber yarns. After introducing the rotation of the disc, the nanofibers attached to the disc are swinged to the disc edge, and the dynamic E-triangle become as a uniform hollow cone. When the diameter of the disc is fixed, the location of the convergence point determines the geometry of the dynamic E-triangle. Fig.5 shows several forms of dynamic E-triangle when the location of the converged point is changed with varied collecting speed. The angle and position of two needles are fixed at 90° and 35mm. When the distance between the convergence point with the disc is slightly larger than fibers' deposition range which is predicted by Fig.4(d), the dynamic E-triangle is evenly formed as shown in Fig.5(a,b). While the dynamic E-triangle will not be able to involve all nanofibers when the distance is smaller as shown in Fig.5(c,d). It may cause hairiness on the obtained yarn or even yarn breakage. Therefore, it is feasible to determine the stable dynamic electronic triangle forming conditions by using the charged particle tracking method. Therefore, it is conducive to produce continuous yarn by using the charged particle tracking method to determine the fiber deposition, and controlling the convergence point of the dynamic electronic triangle to be slightly away from the deposition area.

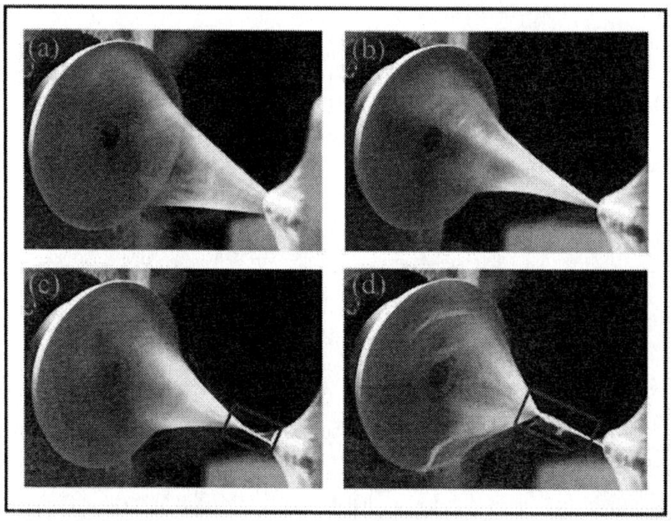

Fig.5 Photograph of the dynamic E-triangle when the distance between the converged point with the disc is slightly larger than fibers' deposition range (a, b), and when it is smaller (c, d).

4. Conclusion

In the process of electrospinning yarn formation, the E-triangle will affect the performance of the obtained yarns. The strength and distribution of electric field and the movement of jets have great influences on the geometry of the E-triangle. Firstly, we simulate the electric field distribution and find that the needle configurations have little influence on the strength and distribution of electric field, which means that the electric field distribution is not the main reason for the different geometries of the E-triangle. Then we define the E-triangle as static E-triangle or dynamic E-triangle when the rotation of E-triangle is considered or not. Therefore, we begin with the simulation of static E-triangle by applying the charged particle tracking method to simulate the trajectory of the electrospinning jets. The simulation results of charged particles on the middle surface of two needles are used to represent the deposition range of the jets and predict the approximate range of the static E-triangle. Through this method, the effect of the angle between the needles and the axis of the disc and rod on the geometry of the static E-triangle is studied, and the effect of the distance between the needles and the disc is also investigated. The results indicate that the 90° and 20mm-35mm configuration parameters are more adequate. Corresponding experiments are performed, and the experimental results prove the feasibility of this theoretical predictive method. In addition, we experimentally study the dynamic E-triangle by changing the convergence point of the E-triangle with a varied collecting speed. And found that it is beneficial for yarn formation when the length of the E-triangle is slightly larger than predicted static E-triangle. In summary, this paper provides an analytical method combining theory and experiment to predict the optimal configuration conditions for a stable E-triangle of continuous nanofiber yarns.

5. Acknowledgement

This work was partly supported by the Fundamental Research Funds for the Central Universities (2232020D-15, 2232020A-08, 2232020G-01, 2232020D-14 and 2232019D3-11) and grants (51773037, 51973027, 51803023, 52003044 and 61771123) from the National Natural Science Foundation of China. This work has also been supported by the Chang Jiang Scholars Program and the Innovation Program of Shanghai Municipal Education Commission (2019-01-07-00-03-E00023) to Prof. Xiaohong Qin, and the Fundamental Research Funds for the Central Universities and Graduate Student Innovation Fund of Donghua University (CUSF-DH-D-2019035) to Ms. Huilin Xu.

References

[1] Krumme A, Vassiljeva V, Plamus T, Savest N, Viirsalu M. Novel method for producing electrospun composite nanofibre yarns. Proceedings of the Estonian Academy of Sciences: 2018; 67 (2): p. 169.

[2] Barani H. Antibacterial continuous nanofibrous hybrid yarn through in situ synthesis of silver nanoparticles: Preparation and characterization. Materials Science and Engineering: C: 2014; 43: p. 50-57.

[3] Hosseini Ravandi SA, Mehrara S, Sadrjahani M, Khodaparast Haghi A. Tunable wicking behavior via titanium oxide embedded in polyacrylonitrile nanofiber strings of yarn. Polymer Bulletin: 2019; 77 (1): p. 307-322.

[4] Wu S, Duan B, Liu P, Zhang C, Qin X, Butcher JT. Fabrication of Aligned Nanofiber Polymer Yarn Networks for Anisotropic Soft Tissue Scaffolds. ACS Applied Materials & Interfaces: 2016; 8 (26): p. 16950-16960.

[5] Gao Y, Guo F, Cao P, Liu J, Li D, Wu J, Wang N, Su Y, Zhao Y. Winding-Locked Carbon Nanotubes/Polymer Nanofibers Helical Yarn for Ultrastretchable Conductor and Strain Sensor. ACS Nano: 2020; 14 (3): p. 3442-3450.

[6] Frank Ko YG, Ashraf A, Nevin N, Ye HH, Yang GL, Christopher L, Peter W. Electrospinning of Continuous Carbon Nanotube-Filled Nanofiber Yarns. Advanced Materials: 2003; 15 (14): p. 1161-1165.

[7] A Theron EZa ALY. Electrostatic field-assisted alignment of electrospun nanofibres. Nanotechnology: 2001; 12 (3): p. 384.

[8] Teo WE, Gopal R, Ramaseshan R, Fujihara K, Ramakrishna S. A dynamic liquid support system for continuous electrospun yarn fabrication. Polymer: 2007; 48 (12): p. 3400-3405.

[9] Pan H, Li L, Hu L,Cui X. Continuous aligned polymer fibers produced by a modified electrospinning method. Polymer: 2006; 47 (14): p. 4901-4904.

[10] Ali U, Zhou Y, Wang X, Lin T. Direct electrospinning of highly twisted, continuous nanofiber yarns. Journal of the Textile Institute: 2012; 103 (1): p. 80-88.

[11] Wu SH, Qin XH. Uniaxially aligned polyacrylonitrile nanofiber yarns prepared by a novel modified electrospinning method. Materials Letters: 2013; 106: p. 204-207.

[12] Joseph J, Nair SV, Menon D. Integrating Substrateless Electrospinning with Textile Technology for Creating Biodegradable Three-Dimensional Structures. Nano Letters: 2015; 15 (8): p. 5420-5426.

[13] Hajiani F, Jeddi AAA, Gharehaghaji AA. An investigation on the effects of twist on geometry of the electrospinning triangle and polyamide 66 nanofiber yarn strength. Fibers and Polymers: 2012; 13 (2): p. 244-252.

[14] Maleki H, Gharehaghaji AA, Dijkstra PJ. Electrospinning of continuous poly (L-lactide) yarns: Effect of twist on the morphology, thermal properties and mechanical behavior. Journal of the Mechanical Behavior of Biomedical Materials: 2017; 71: p. 231-237.

[15] Maleki H, Rahbar RS. Processing and Tensile Properties of Twisted Core-Shell Yarns Fabricated by Double Nozzle Electrospinning Device. Fibers and Polymers: 2021; 22 (5): p. 1256-1265.

[16] Liu X, Su X, Zhang H. Theoretical study of spinning triangle with fiber concentric circular cones arrangement at front nip line. The Journal of The Textile Institute: 2015; 106 (4): p. 377-382.

[17] Zheng Y, Zeng Y, Dong X. Simulation of Jet Motion during Electrospinning Process through Coupled Multiphysics Method. Fibers and Polymers: 2019; 20 (1): p. 113-119.

[18] Gupta A, Ayithapu P, Singhal R. Study of the electric field distribution of various electrospinning geometries and its effect on the resultant nanofibers using finite element simulation. Chemical Engineering Science: 2021; 235: p. 116463.
[19] Zhai Y, Liu H, Li L, Yu J, Ding B. Electrospun Nanofibers for Lithium-Ion Batteries. Electrospinning: Nanofabrication and Applications: 2019.
[20] Wu S, Zhang Y, Liu P, Qin X. Polyacrylonitrile nanofiber yarns and fabrics produced using a novel electrospinning method combined with traditional textile techniques. Textile Research Journal: 2016; 86 (16): p. 1716-1727.
[21] Wang XE, Gao WJ, Wang K, Liu Y, Fan J. Effect of Process Parameters on Continuous Nanofiber Yarn Using Two Oppositely Charged Spinnerets. Advanced Materials Research: 2013; 821-822: p. 32-35.

Cite as: A.M. Xiao et al.,
TBIS 10.3993/tbis (2021).

Weaving Technology Innovation of Kirgiz Traditional "Kajiare" Wool Fabric

Ai-Min Xiao [1], Han Yu[1]

[1] *Textile and Clothing College, Xinjiang University, Urumqi, Xinjiang 830046, China*

*Corresponding author's email: 495178065@qq.com

Abstract

"Kajiare" is the most distinctive fabric in the traditional "Yuermaike" wool fabric of the Kirgiz. It is woven on the original triangular ground loom using the principle of "hand-made jacquard". The weaving process is complicated, requiring intensive labor and long weaving time. The efficiency is low which affects the inheritance and development of this traditional craft. In this paper, a small sampling loom is used to test the "Kajiare" fabric. Through the analysis of the pattern characteristics and formation principles of the traditional "Kajiare" wool fabric, related technical parameters are tested. Based on the sampling loom, the paper determines the number of warp and weft yarns, conduct fabric design with on-machine drawing and CAD simulation diagram to realize the trial weaving sample on the sampling loom, with an aim to provide solutions for the inheritance and innovation of Kirgiz traditional webbing.

Keywords: Kirgiz; Kajiare; Weaving Technology; Innovation

1. Foreword

The emergence and development of intangible cultural heritage as a traditional cultural form is rooted in the practice of manual labor and its social significance. The survival of its characteristics also depends on the technical nature of the differences in manual production methods [1]. As an intangible cultural heritage, Kirgiz textile "Yueernaike" has its own characteristics due to the different fabric mechanisms and pattern styles presented by the different weaving processes. "Kajiare" is a kind of fabric in "Yueernaike" wool fabrics. It is unique among handmade woolen fabrics with due to its specialized weaving techniques, bright colors and rich patterns [2].

The traditional Kajiare weaving method requires high accuracy in the arrangement of colored yarns and the number of warps. However, due to the acceleration of modern society and the transformation of the Kirgiz lifestyle, there are fewer people to inherit the handicraft skills. Therefore, the traditional weaving technology needs continuous improvement and breakthroughs in order to produce more exquisite woolen products [3]. Through literature studies on Kirgiz's traditional wool fabric weaving technology, this paper studies the Kirgiz's "Qiepukan" technique [3] as well as the Kirgiz's traditional wool fabric weaving technique [4]. and analyzes the weaving techniques and clothing structure of the Kirgiz ground loom [5]. Currently, there are no relevant literature on the innovative research on the trial weaving of the traditional "Yueermaike" wool fabric on the small sampling loom. There are some research on the modern technology based on other traditional fabrics, such as the development of traditional styles of summer cloth such as modern crafts [6]. Based on the formation principle of Kirgiz's traditional wool fabric "Kajiare" and the weaving of Kirgiz's traditional wool fabric by sampling loom, this paper uses modern looms instead of manual warp picking, and uses the characterization method of fabric structure to illustrate the weaving technique. It also records webbing patterns through pattern design and weaving practice, which is conducive to the inheritance and development of Kirgiz's traditional webbing weaving skills.

2. Characteristics of Traditional Kajiare

The traditional "Kajiare" of the Kirgiz is a complex warp double weave fabric with warp threads woven on the original triangle loom using the warp thread layered arrangement and three shuttles to control the layering method.

The long floating lines on the reverse side are the same as those on the front side, which are usually formed by handmade jacquard of female workers based on experience and preset patterns [7]. The width of the traditional "Kajiare" wool fabric is 20-35cm. The wool fabric is stitched together to form a 1.8m*2.2m jacquard blanket, which is stitched together with the felt "Xierdake" into a tapestry, which serves as a dowry for yurt decoration [8, 9]. The patterns of Kajiare are different and the most common is the claw pattern. The Kirgiz people who are engaged in nomadic and hunting have a special affection for animals and regard them as their guardian gods. They put the graphic patterns of these animals on the body or embroidered them on the fabric and hung them in the yurt, which formed the totem worship of these animals [10]. Kajiare fabrics are closely related to the natural environment in which the Kirgiz people live in terms of color matching, artistic expression and decorative effects, presenting a distinctive pattern style and fabric mechanism [11].

Fig.1 "Kajiare" and "Xierdake" fabric stitched tapestry

Through field research in three counties and one city in Kezhou, the physical picture of the traditional Kajiare tapestry is shown in Fig.1. The middle part is a 20cm wide Kajiare fabric spliced woven carpet, surrounded by "Xierdake" floral felt. The main color of the woven carpet is mainly red and black, with blue and yellow patterns. It is a cycle of A, B, C, and D, which are composed of comb pattern (harpoon pattern), petal pattern, and horn pattern, impressing people with a happy and enthusiastic sense [12,13].

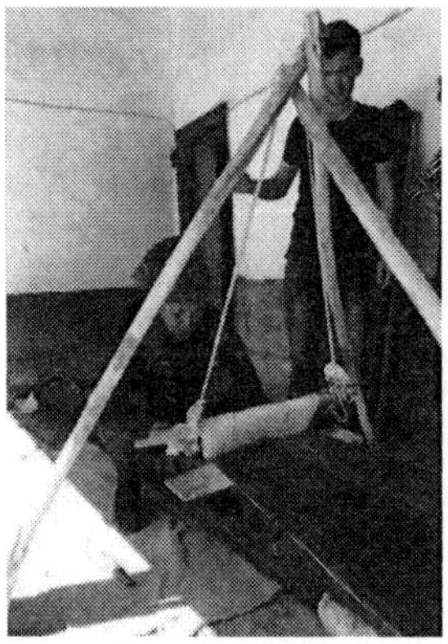

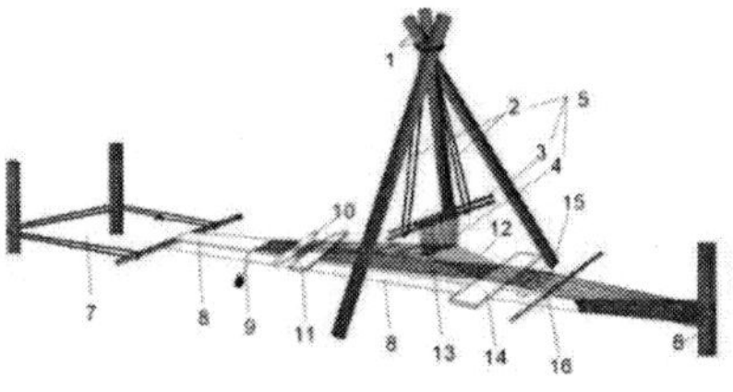

1	Triangle support	9	weft
2	Lanyard	10	Cross stitch knife
3	Stick heald	11	Odd layer longitude
4	Meridians	12	Even layer meridian
5	Lifting heald system	13	Meridional plate
6	warp beam	14	Meridians clue
7	Cloth axis	15	Propping up stick
8	Textile	16	

Fig.2 Traditional Kajiare weaving equipment and hand-made jacquard

According to the technological process of Kirgiz's traditional handmade woolen fabrics, the spinning process is as follows: raw wool, wool selection, impurity removal, hand-made wool top, hand-spinning, finishing into a tuft, and dyeing; rough finishing and sorting of raw materials is an important step before weaving. The weaving process is: sizing, warping, drawing in, dividing warp, and weaving on the machine. The multi-color jacquard fabric Kajiare uses the original triangle loom, the method of handmade jacquard, where the drawing-in and jacquard are based on the weaving experience of the Kirgiz women [14]. As shown in Fig.2.

(a) Front side of fabric (b) Back side of fabric

Fig.3 The physical picture of the traditional Kajiare sample

3. Analysis of Traditional Kajiare Thermal Fabric

3.1 Yarn

Through field investigations, samples of Kajiare were collected in areas where the Kirgiz ethnic group lived. The uneven warp yarn is an important factor affecting the quality and appearance of Kajiare fabrics. Hand-spinning causes serious unevenness of fineness and twist, which reduces the overall strength of the yarn and causes more hairiness. The resulting fabric has a rough surface and a rough style. As shown in Fig.3, due to the greater stiffness of the hand-twisted wool, Kajiare with more wool but less velvet will have a more obvious thorny feeling.

Determine the linear density of the yarn according to the test methods and requirements of GB/T 14343-2008 Chemical Fiber Filament Linear Density Test Method, GB /T 4743-2009 Textile Package Yarn, Hank Method Linear Density Determination .

Use Y (B) 331C digital yarn twist counter to measure the twist of hand-made wool yarn according to the test methods and requirements of GB/T 2543.2-2001 Determination of Twist of Textile Yarn Part 2: Untwisting and Twisting Method. The basic parameters of Kajiare are shown in Table 1.

Use YG (B) 063 automatic single yarn strength machine to test the strength of hand-made wool yarn according to the test method and requirements of GB/T 3916-1997 Determination of Breaking Strength and Breaking Elongation of Single Yarn of Textile Package Yarn. The basic parameters of Kajiare fabric are shown in Table 1.

Table 1. Basic parameters of the traditional Kajiare fabric

Warp and weft direction	Yarn type	yarn Density/tex	Density/tex (root•10cm)	Fabric shrinkage
Warp	wool	616.5	115.5	5.13
weft	chemical fiber	51.2×4	75.75	17

It can be seen from Table 1 that the warp yarn of the fabric is thicker than the weft yarn. The difference between the longitude density and the weft density of Kajiare is extremely large, with a ratio of 4:1. Through analysis, we get that: the warp and weft yarn wave diameter index HD>1, so the fabric structure phase formed is the warp support surface structure phase [4]. See Table 2 for warp twist and breaking strength.

Table 2. Twist and breaking strength of warp

project	twist		Breaking strength/cN	Elongation at break/mm
	Z	S		
average value	1726.9	1850.0	1729.6	22.1
standard deviation	257.1	331.8	322.5	7.1
CV%	14.9	17.9	18.6	31.8

It can be seen from Table 2 that the warp yarns of Kajiare fabrics are spun purely by hand, and are used alternately with Z twist and S twist. The warp yarn twist and strength show uneven characteristics, and the quality of hand-spinning depends on the experience and technology of the operator.

3.2 Fabric Organization Chart

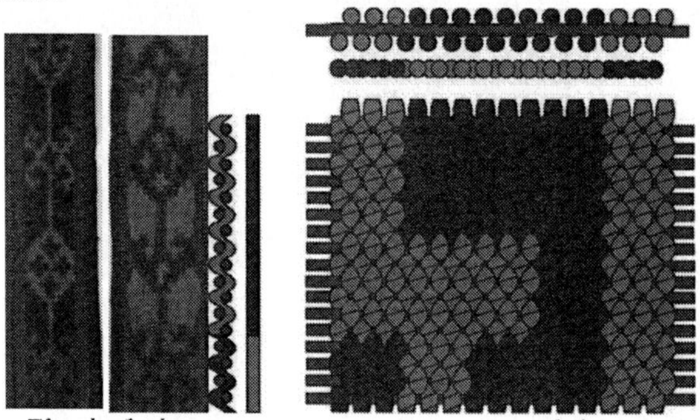

Physical picture a-a b-b

Fig.4 Physical picture and organization chart of traditional Kajiare

The author of this paper draws an organization chart as shown in Fig.4 based on the physical part and the physical pattern as an example. Through the restoration of the weaving process and the analysis of the fabric structure, it can be seen that the fabric is actually a plain weave structure. The diagonal line on the warp in the figure indicates that the twist direction of the yarn is Z twist. The colors of the warp and weft yarns are: surface warp: red, inner warp: pink, weft: red; where two warps and one weft are interwoven. The warp yarns are layered according to color or pattern unit. Select the warp yarns at the jacquard position. The surface warp yarns form a pattern on the front side of the fabric, and the warp yarns that are not woven form floating lines on the back side of the fabric. Since the reverse side of the fabric is composed of floating long threads, in order to ensure the durability and stability of the webbing, it is usually woven by changing the pattern or increasing the joints. Through the a-a cross section, we can see the floating law of the warp weave point and the weft weave point on the surface. The left side of the longitudinal section taken from b-b shows the interlacing of the warp and weft yarns on the surface, and the rectangular color bars indicate the long floating lines on the back. The bending strength of the warp yarn is obvious, and the bending amplitude of the weft yarn is extremely small. This characteristic makes Kachare "change the warps between the surface and the inside, and show the pattern with the warp." The dense and thick warp yarns are closely arranged, and the thinner weft yarns are covered inside the fabric, making almost no weft yarns appear on the cloth surface.

4. Kajiare Design Plan and Trial Weaving Sample on the Sampling Loom

4.1 Experimental Materials and Instruments

Material: Tools and materials: pure wool; combs with high and low densities, scissor, heddle hook, reed knife.

Instrument: Y200S small sampling loom (semi-automatic sampling loom)

4.2 Process Flow

Analyze the sample- draw the organizational chart, draw the machine drawing- draw the CAD simulation diagram warping, drafting, denting, and put on the loom.

4.3 Fabric Analysis

1) Sampling: Select fabric (10cm×40cm).

2) Analyze the organization of the fabric and the coordination of the colored yarns, and get the pattern A, B, C, D fabric on-loom pictures through the method of unraveling:

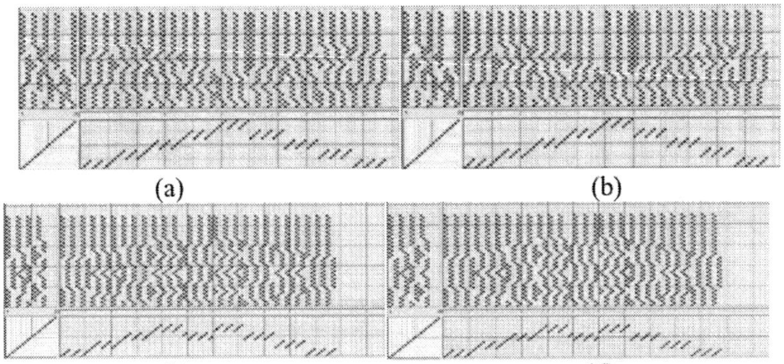

(a) (b) (c) (d)

Fig.5 Fabric on-loom pictures

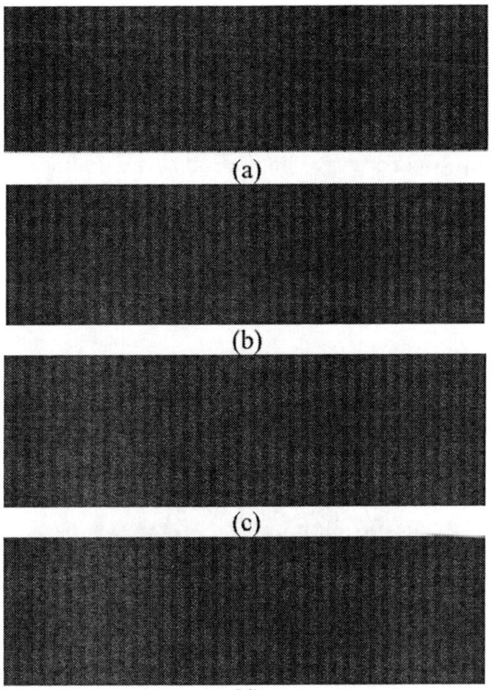

(a) (b) (c) (d)

Fig.6 CAD simulation diagram: (a)(b)(c)(d)

3) The drafting method used according to the organization chart: when weaving complex weave fabrics with two or more warps, for the convenience of craftsmanship and embossing work, the partition drafting method [15] is used, (a total of 20 drafts). According to the on-loom pictures, make the pattern board drawing on the loom. According to the drafting, the penetration is made in sections, and the denting is four-entry (two red surface warps and two pink inner warps penetrate into a reed tooth).

Where the formula

$$R_j = \left(\frac{\text{The least common multiple of Rm and m}}{\text{m}} \text{ and } \frac{\text{The least common multiple of Rn and n}}{\text{n}}\right) \text{ Least common multiple} \times (m+n)$$

(R_j is the number of cyclic yarns in the double weave, m、n is the arrangement ratio of the front and back warp yarns, R_m is the number of cyclic yarns in the surface weave, and R_n is the number of cyclic yarns in the inner weave) The number of yarns in the organization cycle can be calculated as $R_j = 1 \times 4 = 4$.

4) Warping process

Perform weaving on Y200S electronic sampling loom. The pattern is woven horizontally in 2 cycles, with 108 front and back warps in one cycle. Warping is arranged in the order of 2 white and 2 red colored yarns. The drafting is carried out according to the automatic drafting sequence of the textile CAD software, a total of 20 draft frames: Drafting order 2 red 2 white (1,2,3,4) × 3, 2 red 2 white (5, 6, 7, 8) × 3, 2 red 2 white (9, 10, 11, 12) × 3 , 2 red 2 white (13,14,15,16)×3, 2 red 2 white×3, (17,18,19,20) ×3, 2 red 2 white (13,14,15,16)× 3, 2 red 2 white ((9,10,11,12)×3, 2 red 2 white (5,6,7,8)×3, 2 red 2 white (1,2,3,4)×3. Reed No. 120, 4 per reed, upper loom width 6.6cm, lower loom width 5.8cm, total warp roots 108, fabric warp density 552 roots/10cm, weft density 570 roots/10cm.

5) Loom operation

Adjust the warp tension and perform a beating-up trial weaving. To use the small sampling loom, only the machine drawing is required, and the correct pattern of the pattern must be ensured, and only repeated beating work is required. Pay attention to the uniform tension to ensure the neatness of the fabric, and the beating force needs to be uniform. The choice of wool is a yarn with greater strength, which can improve the efficiency of use. The cut yarn has good elasticity, and the fabric feels firm and soft. However, the friction is greater when weaving and beating up the weft, so the hairiness produced will affect the clarity of the opening. The Kajiare double-layer fabric woven by a sample loom has clear lines, a good sense of richness, unique style and strong three-dimensionality. The sample of the finished product weaving is shown in Fig.7.

6) Conclusions

By adopting the principle of warp backed weave on the loom, the traditional Kirgiz nationality webbing can be fully automated through the loom. And the appearance style of the woven fabric conforms to the characteristics of Kirgiz traditional webbings, which greatly improves the weaving efficiency.

Due to the limited number of loom heald frames, in the actual design and weaving process, there are often problems such as a single weft pattern cycle and the inability to weave cloth edges. The pure wool is selected, and the friction rate is higher when weaving on the loom, and the hairiness produced will affect the clarity of the opening.

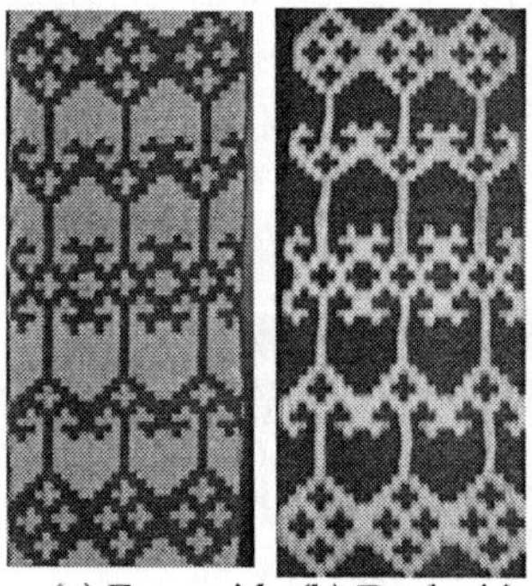

(a) Front side (b) Back side

Fig.7 The Actual Kajiare Weaving on a Sampling Loom

5. Conclusion

Based on the weaving principle formed by the traditional Kajiare weaving on the original triangle loom, this paper first analyzes the fabric, then uses the computer graphics and the CAD simulation diagram to preview the expected effect, and finally get the trial weaving product. On the one hand, the method of improving the weaving skills of Kajiare solves the uneven yarn evenness caused by manual twisting, and gets rid of the labor intensity of manual beating-up. On the other hand, in textile CAD, the organization chart is determined to generate the pattern chart, which simplifies the process. The Kaijare woven on the sampling loom features bright color, softness and fineness, and strong practicability. It has modern characteristics, while retaining the characteristics of traditional patterns, realizing the possibility of weaving with modern technology. Compared with the traditional triangle loom, the small sampling loom can get rid of manual beating-up when weaving Kajiare, solve the shortcomings of uneven yarn evenness caused by manual twisting, and greatly shorten the weaving time. Using the small sampling loom, the operator can more easily understand the related operations of the dyed yarn arrangement and drawing-in, which is conducive to inheriting and innovating the traditional Kirgiz woolen.

References

[1] Lu PT. Protection and development in production -- on the "productive protection" of traditional handicrafts. Art observation: 2009; 7: 5-7.
[2] Zhang JD. A preliminary study of Kirgiz history [D]. Lanzhou University: 2013.
[3] Qian J, et al. Study on weaving technology of Kirgiz traditional wool fabric. Decoration: 2017; 12: 116-117.
[4] Qian J, et al. Research on Inheritance and innovation of "Chepukan" of Kirgiz camel hair fabric. Wool textile technology: 2018; 8: 72-75.
[5] Yan WQ, Ge MJ, et al. A study on weaving technique and clothing structure of Kirgiz loom in Xinjiang. Silk: 2019; 009: 103-109.
[6] Yu HM, Yu CW. Development of traditional style summer cloth with modern technology. Acta textile Sinica: 2016; 37: 37-41.
[7] Xiao AM, Ge MJ, et al. An analysis of the weaving process of Kirgiz traditional "Kajiare" tapestry. Silk: 2019; 9: 35-39.
[8] Wan XY. Kirgiz traditional needlework technique based on field investigation. Journal of Xinjiang University (PHILOSOPHY, HUMANITIES AND SOCIAL SCIENCES EDITION): 2014; 5: 72-76.
[9] Wan XY. Kirgiz traditional handicrafts in the period of social transformation [J]. Journal of Guizhou Normal University: 2016; 4: 83-90.
[10] Xue JL. Kirgiz primitive religious totem and epic Manas. Youth and Society I: 2013; 7: 337.
[11] Liu TT, Xiao AM. Analysis of traditional ribbon types and color design of Kirgiz people. Daguan: 2019; 12: 176-177.
[12] Wan XY. Analysis of Kirgiz yurt culture. Journal of Xinjiang University (PHILOSOPHY, HUMANITIES AND SOCIAL SCIENCES EDITION): 2013; 2: 90-95.
[13] Zhang AJ. Kirgiz nationality, Yili Kazak Autonomous Prefecture Study on traditional webbing Technology. DAGUAN: 2019; 11: 132-133.
[14] Wan XY. Kirgiz traditional needlework skill based on field.National Forum: 2018; 03: 108-112.
[15] Yu WD. Textile materials [M]. Beijing: China Textile Press: 2006; 276.

Cite as: L.F. Zhang et al.,
TBIS 10.3993/tbis (2021).

Development of Miyabit/ Viscose/Wool Interwoven Leisure Fabric for Work Uniforms

Li-Feng Zhang[1,2], Gui-Cui Chen[1,3*]

[1.] *School of Textile and clothing, Yancheng Polytechnic College, Yancheng Jiangsu, 224000, P.R. China;*
[2.] *Logistics Infrastructure Management Division, Yancheng Polytechnic College, Yancheng Jiangsu, 224000, P.R. China;*
[3.] *Jiangsu Province Engineering Research Center of Biomass Functional Textile Fiber Development and Application, Yancheng Jiangsu, 224000, P.R. China*

[*]Corresponding author's email: ycfycgc@126.com

Abstract

To research the production process and fabric performance of Miyabit/Viscose/Wool interwoven leisure fabric for work uniforms, miyabi fiber was replaced with acrylic fiber and blended with viscose fiber. Reasonable textile production measures were adopted. The process parameters were optimized. The key technology of textile process was solved to produce Miyabit/Viscose/Wool interwoven fabric. The content change of fabric miyabit composition was adjusted, then the general rule of miyabit influence on the fabric performance was explored. The results demonstrate that the tensile strength, wrinkle resistance, anti-fuzzing ball resistance of the fabric improved. Moreover it has excellent warmth retention performance with the increase of the weft throw ratio of Miyabit/Viscose blended yarn。 The fabric is a high-quality leisure fabric for work uniforms.

Keywords: Miyabi; Wool; Interwoven Fabric; Work Uniforms; Production Technology

1. Introduction

The word tooling fabric first appeared in the early 20th century. With the gradual expansion factories, fabric manufacture was widely popular. Workwear fabrics were developed from the original gauka, polyester plain cloth, Oxford cloth, Gongshang cloth and antistatic silk, etc. Workwear fabrics were also developed from the original work clothes to be fashionable and made with popular leisure clothing fabrics, to meet people's demands.

Miyabit is a functional acrylic fiber produced by Mitsubishi Riyang Co., Ltd., Japan, with superior thermal insulation and easy maintenance [1]. The viscose is composed of regenerated cellulose fiber with high wet modulus and had good hygroscopic and spinnable properties [2].Wool fiber will not melt and drip when burning, and would not stick to the skin [3, 4]. As an ideal intrinsic flame retardant raw material, wool fiber was suitable for making flame retardant tooling fabrics [5]. Adding meiabi fiber and viscose fiber into wool fabric not only improves aesthetic through enahcning color and lustre, it also help alleviate functional defects of the original fabric. The mixed fabric is not easily deformed, pill and fade.

This design adopts reasonable spinning process measures to spin 15tex viscose fiber/Miyabit 70/30 compact spinning yarn, which endured the blended yarn with good hand feel and mechanical properties. The blended yarns, wool yarns and blended yarns were interwoven to develop the Miyabit/Viscose/Wool interwoven leisure workwear fabrics, which could improve the poor anti-pilling performance of wool fabrics and effectively reduced the production cost, thus providing a novel design idea for the preparation of high-grade popular fabrics.

2. Fiber Properties

2.1 Miyabit Fiber

Miyabit fiber was a super-fine acrylic fiber, which had a large specific surface area and could absorb more static air to provide better thermal effect. This product selects the beauty fiber bright color, high strength, fluffy, soft feel, hygroscopic than ordinary acrylic fiber double. Unlike natural fibers, which wrinkle easily and were not as airtight as man-made fibers, Miyabit fibers are light, waterproof, mildew free and insect-free.

2.2 Viscose Fiber

Viscose fiber had good hygroscopic property, viscose fiber fabric was not easy to recover after elongation, poor elastic recovery property, bright color and good color fastness after dyeing. The main characteristics of the viscose fiber used in this product were uniformity of fineness, good length uniformity, high cleanliness of appearance, and had not stain, dirty and colorful spot. Selective performance index of cotton and viscose. Performance indicators of raw materials were as shown Table 1.

Table 1. Performance indicators of raw materials

Fiber	Fineness /dtex	L /mm	Dry strength /cN/dtex	Wet strength /cN/dtex	Dry elongation /%	Wet elongation /%	Moisture regain /%
Miyabit	0.89	38	3.5	3.38	35	40	2
Viscose	1.33	38	2.6	2.0	17.5	22	13

3. Textile Process Flow and Process Design

3.1 Textile Process Design

Miyabit and Viscose: FA002 automatic cotton grabbing machine (FA121 metal removal device) →FA104 six drum cotton opener →ZFA026 automatic cotton mixing machine →FA106 porcine cotton opener →A092AST vibrating double-box cotton feeder →FA141 single hand rolling machine →FA201B card →FA306 drawing machine (pre-drawing can get Miyabit strips and adhesive strips respectively)

Miyabit sliver + Viscose sliver: FA306 drawing frame →FA458 roving frame →FA507B spinning frame Autoconer338 automatic winding machine →GA126P warping machine →G177-180 automatic reed machine

Wool/viscose blended yarn: ZA205I-190 air-jet loom → singing → drop water → plain washing and boiling → double boiling → roll cloth → baking → medium inspection → ripe repair → shearing → soft → steaming.

3.2 Spinning Process Design

The process of opening and cleaning cotton did not require high effect of opening and removing impurity, and the number of opening and counting could be small. In order to ensure the uniform structure of the coil, the ration supply and the stability of the cotton flow density, the designed speed, spacing and the installation Angle of the dust rod can be small. The specific process parameters were as follows: dry weight 369.1g/m, porcupine beater speed 800r/min, comprehensive beater speed 960r/min, roll roller speed 14.364r/min, feeding roller and beater spacing 7mm, beater and dust rod spacing 7mm×18mm, dust rod and dust rod spacing 8mm.

In order to realize the smooth transfer of the fiber, the fiber was fully carded without damaging the fiber. The low-speed, small spacing between the feeding board and the cylinder, the spacing between the cylinder and the doffer, and the larger spacing between the dust cutter and the roller were adopted. The specific process parameters were as follows: the strip production quantity was 22 g/5m, the mechanical draft multiple was 91.455, the actual draft multiple was 96.27, the roller rotation speed was 860r/min, the cylinder rotation speed was 330r/min, the doffer rotation speed was 30.4r/min, the five point spacing between the cylinder and the moving cover plate was 8×7×7×7×8 ", and the strip stripping speed was 60m/min.

In order to realize the drawing process, the strips were mixed with the viscose fiber, and the pre-drawing was carried out separately, which could make the parallel straightening degree of the two fibers consistent and reduce the tension difference between them. Appropriately increase the roller spacing and cots pressure, running speed selection was three pieces Miyabit fibers and three pieces viscose fibers were combined, and the specific process parameters were shown in Table 2.

Table 2. Performance indicators of drawing

process sequen ce	Dry weight g/5m	waste percentage /%	Draft multiple allocation					
			mechani cal	practical	front tensi on	front sec tion	back zo ne	back ten sion
Miyabit pre-drawing	21.3	0.63	6.118	5.813	1.0175	5.55	1.08	1.08
Viscose pre-drawing	22	0.33	6.294	5.980	1.0175	5.56	1.08	1.08
First drawing	22	0.5	7.158	6.801	1.0175	5.58	1.08	1.08
Second drawing	21	0.5	7.275	6.912	1.0175	5.82	1.06	1.06
Third drawing	19	0.5	7.275	6.912	1.0175	6.04	1.09	1.06

For the roving process, the four roller double short leather ring was chosen because of the puffiness of Miyabit fiber. Heavy pressure and large spacing were adopted, small back draft, small running speed could be selected, the purpose was to improve the fiber straightness and the evenness of yarn. The production of this product was the use of compact spinning yarn, roving quantitative light. In spun yarn was the key to the "rainbow" cases, appropriate decrease roving winding density and lifting dynamic range, reasonable adjustment of roving forming Angle, ensure good roving forming, adjust good roving tension, to reduce the front and the back and the difference between big and small yarn roving elongation, avoid excessive produced in the process of winding and unwinding of stretch. This product was fine yarn, the bell mouth should be small. The specific process parameters are as follows: the roving quantity was 4.49g /10m, the actual draft ratio of roving was 29.97, the mechanical draft ratio was 31.547, and the roving twist coefficient was 96.

The spinning process adopts tight spinning method, which could effectively reduce hairiness and improve the quality of yarn. The speed and tension should be selected slightly to reduce the breakage rate of fine count blends. The specific process parameters were as follows: spinning quantity was 1.47g /100m, spinning actual draft ratio was 8.417, mechanical draft ratio was 8.86, spinning twist coefficient was 380.

3.3 Weaving Process Design

Using 60 tex ×2 wool as the warp, 15tex×4 viscose/Miyabit blended yarn, 60 tex ×2 wool as the weft material, the fabric structure was plain weave, the warp density was 450 /10 cm, the weft density was 360 /10 cm, through the change of weft filling ratio to change the content distribution of the fabric warp and weft.

In the warping and weaving process, in order to avoid yarn breakage affecting production efficiency and minimize knitting defects, the speed of warping was 550m/min, the number of batches of warping was 18, and the number of warping was 380. The loom adopts the technology of reducing primary and auxiliary air pressure and weft humidification. The specific process parameters were speed 500, machine tension 350, opening time 280, main injection pressure 3.2, auxiliary injection pressure 3.7. After finishing process to ensure that the head and tail of the straight alignment, to avoid weft inclined, crease defects .Boil the cloth for 40 min, and the temperature was not higher than 85°C. The speed of drying setting was 25 ~ 30 m/min, and the heat setting was160°C.

4. Fabric Performance Test and Analysis

4.1 Experimental Instruments and Test Standards

Under standard temperature and humidity conditions, HD026+ electronic fabric strength meter was used. The weft stretch is 20cm apart and the stretching speed was 100mm/min.Y522N disc type fabric wear-resistant meter was used. The diameter of disc was Φ90 mm, the rotation speed was 70 r /min, and the number of pilling of 25cm2 fabric was calculated after 200 turns. The fabric crease recovery was tested by YG541E digital fabric crease elastometer. The thermal conductivity of the fabric was tested by KES-F7 contact temperature tester.

4.2 Experimental Results Analysis

The fabric performance test results were shown in Table 3.

Table 3. Test results of fabric properties

Samples	Weft ratio (Blended yarn: wool)	Strength /N	The count / the quantity /25cm^2	Angle of rapid elastic recovery /°	Angel of slow elastic recovery / °	Instantaneous maximum heat transfer / J·cm^{-2}
1#	5:1	700.4	5	215.24	224.24	1.06
2#	5:2	683.8	7	172.34	192.74	1.03
3#	5:3	674.6	10	167.71	192.98	1.02
4#	5:4	665.6	13	156.13	156.13	0.97
5#	1:1	656.8	17	150.12	179.46	0.92

With the same fabric structure, it was the yarn raw material that plays a decisive role in the tensile strength of the fabric. As can be seen from the experimental results in Table 3, the tensile strength of the fabric increases with the increase of the content of Miyabit/Viscose blended yarn in the filling ratio, indicating that the addition of the blended yarn was conducive to improving the tensile mechanical properties of wool fabric. The pilling resistance of the fabric was improved with the increase of the filling ratio of Miyabit/Viscose blended yarn. As a super fine denier fiber, the rigidity of the fiber was less than that of wool, so the crease recovery Angle increases with the increase of weft casting ratio, and the crease resistance was better. At the green fiber as the superfine fiber had strong adsorption capacity, air will be beautiful elegant garden superfine fiber filling between blended yarn, maked the fabric of the original porosity change changeable, increased the air flow resistance, effectively reduced the air layer of liquidity, reduced the air convection, thermal properties of the fabrics without decline because of the weft insertion rate increase.

5. conclusion

Using reasonable textile technology measures, the fabric specification parameters and different weft filling ratio of Miyabit/Viscose/Wool interwoven leisure tooling series fabrics were designed to adjust the content changes of Amby components in the fabrics, and then the general law of the influence of Miyabit fiber raw materials on the properties of interwoven leisure tooling fabrics was studied. With the increase of the filling ratio of Miyabit /viscose blended yarn, the tensile strength of the fabric increased, the anti-wrinkle and anti-pilling properties were improved, and the thermal properties of the leisure fabric did not decrease with the increase of filling ratio.

6. Acknowledgement

This work was funded by QingLan Project of Educational Department of Jiangsu Province, China (Education Department of Jiangsu Province ((2019)03), the project of Yancheng polytechnical college (ycgy201901), the Integration Platform of Industry and Education of Jiangsu Higher Vocational Education, Grant number: Jiangsu Vocational Education 2019. No 26, and Jiangsu Province Higher Vocational Education High-level Major Group

Construction Project-Modern Textile Technology Major Group. Science and Technology Guidance Project of Chinese Textile and Apparel Association, No.2021055/ No.2021022.

References

[1] Lei BY, Liu WP. Development of Jacquard Piquette Thermal Knitted Fabric Using Miyabi Fiber [J]. The knitting technology: 2014; (5): 1-3.
[2] Sheng LY, Chun KW, Li J. Spinnability Analysis of Dralon Acrylic Viscose Cotton Blended Yarn [J]. Cotton Textile Technology: 2019; 47 (4): 53-555.
[3] Zhang LL. Research and development technology of worsted fabric with the properties of flame retardant and comfortable [D]. Xi' an Polytechnic University: 2013.
[4] MA JZ, Hong GE, Wang D. Preparation and Property Study of Flame-retardant Viscose Fiber [J]. Wool Textile Journal: 2017; 45 (9): 18-21.
[5] Zhang XJ, Fu AF, Zhan YJ, et al. Development of wool /viscose /polyamide flame retardant fabrics for work uniforms [J]. Wool Textile Journal: 2020; 48 (2): 11-13.

The Basic Pattern Design Model of Men's Top Clothing Oriented for Personalized Customization

Bo-An Ying*, Chao-Jun He, Jing Qi, Long Wu, Xiao-Feng Wang

Apparel & Art Design College, Xi'an Polytechnic University, Xi'an, Shaanxi, 710048, China

*Corresponding author's email: yingba2006@163.com

Abstract

In order to improve the lack of fit and flexibility in the current personalized clothing customization technology, a basic pattern design model for men's top clothing for personalized customization is proposed to improve the efficiency of clothing personalized pattern designs. Based on geometric principles, the human body is unfolded and breathing looseness is added along with the completed design of the basic pattern of male tops. The parametric relationship model of the pattern is constructed to sort out the knowledge of the basic pattern design. The design model of the basic pattern is constructed based on the Auto Pattern system. The gap amount evaluation method is used to evaluate the fit of the basic pattern, and the directed graph model is used to analyze the fit of the pattern and modify the pattern design model. The experimental results show that the revised basic pattern has a good fit, and the effectiveness of the design method of the basic pattern design model has been verified. The intelligent design of the basic pattern can be realized by adjusting different human body parameters and interactive parameters through the pattern design model. This research can improve the fit, flexibility, and design efficiency of personalized clothing customization, and reduce the company's dependence on the experience of pattern makers.

Keywords: Personalized Customization; Basic Model for Men's Tops; Clothing Pattern Design Model

1. Introduction

With the continuous improvement of living standards, people not only require comfortable and aesthetic clothing, they also require clothing to be more fit and distinctive. In this context, herd mentality of consumer behaviors are declining, and the value of individualization is becoming more prominent. The ability to quickly and accurately design and manufacture products that consumers need has become key for companies to gain market competitive advantages [1].

The clothing prototype method is simple to operate. It combines planes and is three-dimensional. It can adapt to changes in various clothing styles and body shapes. It has the characteristics of accurate drawing and strong application changes [2], which can meet the personalized customization demand to improve fit and the need for flexibility. With the continuous exploration of experts and researchers in the field of clothing, a variety of clothing prototypes have been launched. Taking the prototypes of men's tops as an example, the more common ones are "Japanese cultural menswear prototype", "Men's Donghua prototype" and "Men's standard basic pattern". Despite these development, none of these three prototypes designed waist darts. Even if men's clothing will not be as tight as women's waists, but for menswear with a fit style, part of the chest-waist difference must be eliminated. The general way to eliminate the chest-waist difference is to make darts. The number and location of darts and the size of each dart have a greater impact on the shape of clothing [3]. Moreover, the current prototypes of men's tops serve a group of similar body types, while the prototypes needed for personalized customization serve a single individual. Some scholars such as Hyo Kim C [4], Mesuda Yuko [5] and others also use the method of converting three-dimensional clothing to two-dimensional planes. This method requires operators to have high technical and artistic accomplishments. Only by relying on the experience of the pattern maker can we design a good clothing pattern, and the dependence on the experience of the pattern maker is still significant.

In order to improve the lack of fit and flexibility in personalized clothing customization [6], the study of the basic pattern of men's clothing for personalized customization is proposed, and a basic model design model for men's clothing is built to realize the automation of model design knowledge. This method can liberate the pattern maker from the complicated design work, improve the efficiency of clothing pattern design, and improve the response of clothing companies to consumer needs.

2. The Method of Basic Pattern design for Men's Tops

Everything in the world can be approximated by geometry, and the human body is no exception [7]. The body surface of the human body is unfolded according to the principle of geometry. The basic pattern design of the normal body shape of an adult male is taken as an example. The design of the basic pattern is as follows:

(1) Pattern basic structure line design

According to the circumference of the human body, the length of the body surface determines the horizontal and vertical basic frame lines of the pattern. For example, determine the horizontal length of the pattern according to the front bust width, the back width, and the arm roots width. The principle of the pattern base line design is shown in Fig.1. Because clothing needs to meet the basic physiological needs of the human body, a basic breathing loose volume of 2 ~ 4cm is set in the girth direction of the model.

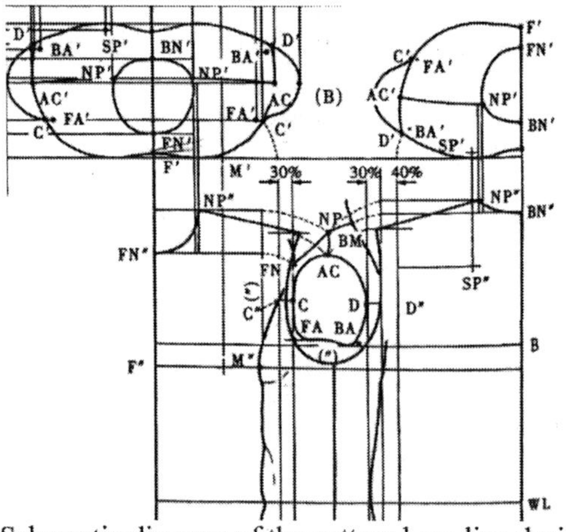

Fig.1 Schematic diagram of the pattern base line design [8]

(2) Design of the basic pattern shoulder line

1) Determination of the front shoulder line: Measure the length of the arc from the shoulder point to the anterior axillary point as L1. The position of the anterior axillary point B is determined through the position relationship between the position of the chest width line, the position of the neck fossa point FNP, and the value of the front chest width. The position of the SNP at the side neck point is determined by the front neck width and front neck depth. Draw an arc with the point SNP as the center and the small shoulder width as the radius. Draw an arc with point B as the center and L1 as the radius. The intersection of the two arcs is shoulder point A, and the connection between SNP and A is the front shoulder line.

2) Determination of the back shoulder line: Draw an arc with back neck point BNP as the circle point, the total shoulder width as the radius. Draw an arc with side neck point SNP as the circle point, and the small shoulder width as the radius. The intersection of the two arcs is the back shoulder point. Connecting this intersection and the SNP is the back shoulder line.

(3) The basic pattern armhole dart design

The male chest is approximated as a semicircular platform, the radius of the small circle of the circular platform is the distance between the bust line and the axillary line. Fig.2 shows the unfolding of the human breast. Two concentric circles were used to flatten the chest, and the arc length needed to be removed for the larger circle was obtained. The center angle of the circle with radius r2 corresponding to the arc length was the armhole dart angle $\alpha 1$. The dart cusp point is set at the foot drop point where the bust point BP is on the bottom line of the axilla. The location of the dart is at the line connecting the foot drop point and the anterior axillary point. The calculation steps of the province angle are as follows:

$$r_3 = r_1 + h * \tan\beta_1 , r_2 = \frac{h}{\cos\beta_1}$$

$$\Delta = \pi * [(r_1 + r_2) - r_3] = \pi * \left(\frac{h}{\cos\beta_1} - h * \tan\beta_1\right)$$

$$\alpha_1 = \frac{\Delta}{2\pi * r_2} * 360^\circ = 180^\circ * (1 - \sin\beta_1)$$

(4) The basic pattern shoulder dart design

The back convex part of the human body can be approximated as a 1/4 cone with the convex point as the apex as shown in Fig.3. The apex is the most prominent point of the scapula, and it is also the apex of the shoulder dart after the curved surface is unfolded [9]. The unfolding schematic diagram of the human back convex is shown in Fig.3. Moreover, the arc length of the sector is half the circumference of the bottom circle of the cone, so the calculation formula for the shoulder angle α_2 can be calculated as follows:

$$\pi * l_2 = \pi * l_1 * \left(1 - \frac{4 * \alpha_2}{180}\right), l_2 = l_1 * \sin\beta_2$$

$$\alpha_2 = \frac{180 * (1 - \sin\beta_2)}{4}$$

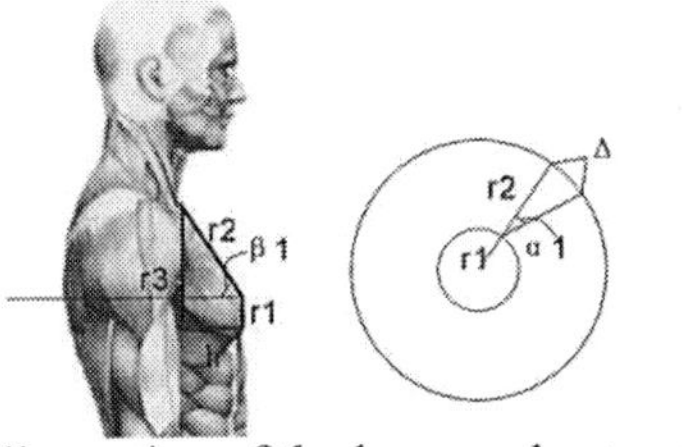

Fig.2 Illustration of the human chest unfolded

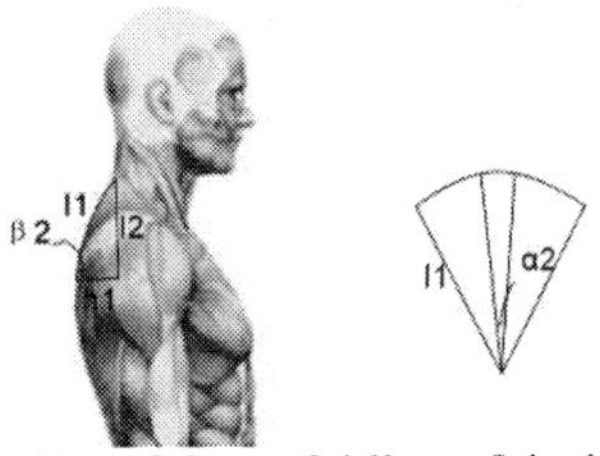

Fig.3 Illustration of the unfolding of the human back convex

(5) The basic pattern waist dart design

The body surface of the human body is concave and convex, and the pattern should be formed into a structure consistent with the human body. The position of the dart should be set on the prominent protruding parts of the torso such as the chest point and the shoulder blade. The front abdomen of the male body is almost vertical. The back shoulder blades are prominent, and the back slope from the shoulder blade to the waist is very obvious. Therefore, the darts of the basic pattern are set at the chest point, the back center line, the back axillary point, the shoulder blade point and the side seam. Three draping experiments of the tight-fitting prototype of ALVANON's M-size mannequin were done. The average distribution ratio of the darts of each part was measured by flattening the draperies, chest point: side seam: back axillary point: scapula point: back center =16%: 16%: 36%: 24%: 8%.

(6) Garment loose volume distribution design

In order to facilitate the activities of the human body, the looseness of the clothing should be distributed at the front chest width, back width, and sleeve cage parts, and the distribution ratios are 30%, 40%, and 30% [10].

3. The Construction of the Parametric Relation Model of the Basic Pattern

The parametric relationship model of the clothing pattern is essentially to transform the topological relationship between the points and lines in the pattern into a set of geometric data constrained by parameters and at the same time construct the corresponding relationship of the graphic data of the structural parameters. When assigning different values to the parameters, the information of the geometric data in the driving pattern directly forms a new clothing pattern [11].

3.1 Human Body Measurement Items and Results

The normal body shape of adult males was used as the research object. A concrete example of this body shape, ALVANON's M men's mannequin was used to design the basic pattern design model of men's tops. According to the basic pattern design method, the items and results of the mannequin measurement required by the basic pattern are shown in Table 1.

Table 1. The measurement data of the mannequin required for the basic pattern

No.	Data name	Variable name	Data
1	Front neck point to waist	L1	39.9cm
2	Front neck width	N1	7.5cm
3	Front neck depth	N2	7.8cm
4	From front neck point to chest width line height	L2	9.0cm
5	Front bust width	B1	36.8cm
6	Shoulder length	B2	14.8cm
7	From anterior axillary point to shoulder point along the root of the arm	L3	12.1cm
8	The distance between waist line and bust line	L4	20.2cm
9	Arm root width	B3	13.1cm
10	Back width	B4	40.0cm
11	Back neck point to waist	L5	46.2cm
12	Back neck width	N3	7.5cm
13	Back neck depth	N4	2.4cm
14	Back shoulder width	B5	43.6cm
15	From scapula point to back neck point	L6	12.6cm
16	Scapula point width	B6	20.4cm
17	Scapula inclination angle	$\alpha 1$	44.6°
18	From the bottom of the armpit to waist line	L7	24.8cm
19	Bust point width	B7	19.4cm
20	The inclination angle of the chest above the axillary line	$\alpha 2$	69.3°
21	Back neck point to back width line.	L8	15.0cm
22	Bust girth	B8	95.2cm
23	Waist girth	B9	81.0cm

3.2 Interactive Parameter design in the Parametric Relational Model of the Basic Pattern

The interactive input parameters are designed in the parametric relational model of the clothing pattern to better extract the experience of the pattern designer and to facilitate the use of the customer. The selection of parameter values allows different pattern designers to modify and adjust the pattern according to their own design experience and customer requirements. This is also a concrete manifestation of human-oriented design ideas and designer experience. Table 2 shows the interaction parameters designed in the parametric relationship model of the basic pattern.

Table 2. Interactive parameter design in the relational model Unit: cm Unit: cm

Parameter	Variable name	Default	Range	Explanation
Breathing ease	R1	4	2-4	The clothing should meet the basic physiological needs of the human body, and the ease of the bust should be set to the basic breathing ease of 2-4cm.
Waist ease	R2	2	1-2	In order to ensure the comfort of the clothing, the waist should be set with a certain amount of ease, and the body-fitting clothing is generally 1-2cm

3.3 Construction of the Parametric Relationship Model of the Basic Pattern

The following rules should be followed when constructing the pattern parameterized relationship model: All points and lines appearing in the pattern design process must be all naming if there are constraints (numerical value, formula, geometry, etc.) in the process flow of generating each point and line, all must be noted. Part of the parameterized relational model of the basic pattern based on this rule is shown in Table 3. Fig.4 is a basic pattern structure diagram based on this relational model.

Table 3. Part of the parameterized relationship model of the basic pattern

No.	Structural line	Build process (Unit: centimeter or degree)
1	Front center line	It takes point P0 as the starting point and vertically upward L1 to point FNP. After connecting point FNP and point P0, the front center line L_FCL is completed.
2	Front neck line	It takes point FNP as the starting point and horizontally moved to the right N1 to point P1, and then vertically upward N2 to point F_SNP. After connecting point FNP and point F_SNP, the front neck line L_C1 is completed.
3	Front bust width line	It takes point F_SNP as the starting point and vertically downed L2 to point P2. Point P2 is moved horizontally to the right B1/2 to point F_AP, and then starting from point F_AP, horizontally moved to the right R1/2*30% to point P3. After connecting point P2 and point P3, the front bust width line LS_1 is completed.
4	Front shoulder line	Draw an arc Arc_1 with point FNP as the center and the shoulder length B2 as the radius. Draw an arc Arc_2 with point F_AP as the center and L3 as the radius. Arc_1 and Arc_2 two arcs intersect at point F_SP. After connecting point F_SNP and point F_SP, the front shoulder line L_SL is completed.
5	Waist line	Suppose the width of the half-length pattern is B10, B10=B1/2+B3+B4/2+R1/2. It takes point P0 as the starting point and horizontally moved to the right B10 to point P4. After connecting point P0 and point P4, the waist line L_WL is completed.

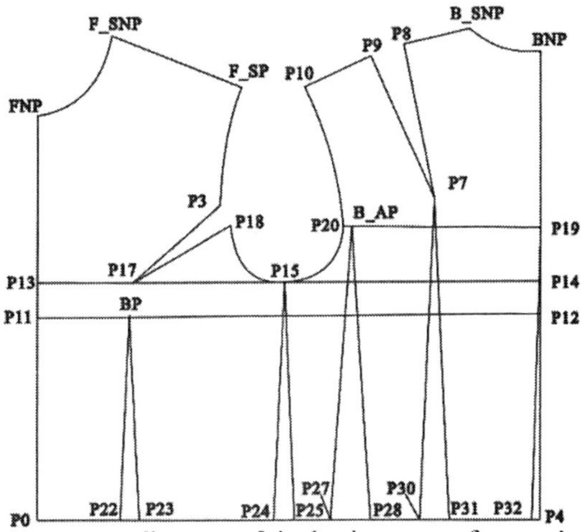

Fig.4 Structure diagram of the basic pattern for men's tops

4. Construction and Verification of Basic Pattern design Model for Male Tops

4.1 The Basic Pattern design Model Construction Process

The clothing pattern design model is to transform the clothing pattern design process into a clothing pattern design principle and method that is digitally expressed in the computer, that is, to establish a data model in the computer. The design of the basic pattern for men's tops is used as an example to illustrate the construction method of the pattern design model.

Combining the parametric relationship model of the basic pattern, 40 points, 19 auxiliary lines, 28 structural lines, and 26 parameters can be extracted during the pattern design process. According to the combed clothing pattern design knowledge, build a basic pattern design model based on the Auto Pattern system. The construction process of the basic pattern design model of the male top is shown in Fig.5.

The construction of the basic pattern design model is essentially to establish a database of the basic pattern The specific operations are as follows: Corresponding to the parameterized relationship model of the basic pattern, enter the pattern making parameters under the category of C plate and basic pattern respectively. In this step, all the points and lines that constitute the basic pattern are defined, and the generation sequence of the points and lines that constitute the pattern, the generation rules (formula, inter-interaction, etc.) and constraint relationships (line segments, Curve, etc.), and finally complete the drawing of the basic pattern.

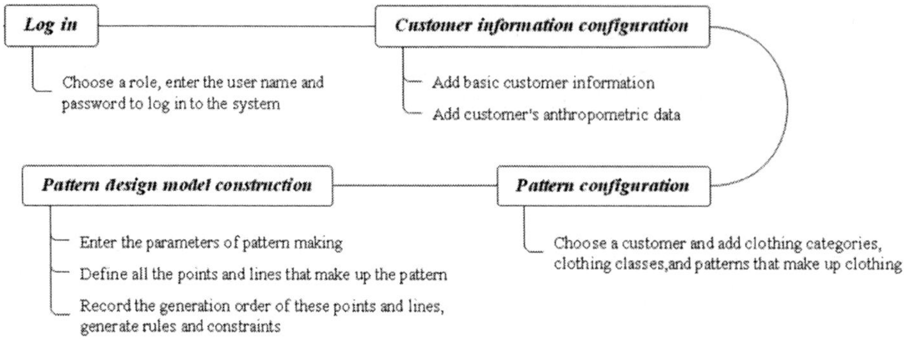

Fig.5 The flow chart of building a pattern design model in the Auto Pattern system

4.2 Intelligent Pattern Making of Basic Pattern for Men's Tops

According to ALVANON's M size mannequin and the default data of interactive parameters, the design model of the basic pattern can automatically complete the calculation, drawing and other design work according to the sequence of point and line generation in the model construction and the constraint conditions according to the drive of the data. Output the basic pattern graphic that meets the design requirements, as shown in Fig.4.

When the customer needs to modify the design parameter value, in the pattern design model, it can automatically locate and modify the points and lines that need to be modified according to the modified parameters, and it can quickly complete the calculation and automatically generate the adjusted basic pattern. In the system, you only need to modify the design parameter input box. For example, to adjust the adjustment parameter of breathing ease, just change the data 4cm to 2cm in the breathing ease adjustment box, confirm the modification of the parameters, and the system will automatically complete the calculation and drawing of the basic pattern according to the modified parameters. After the design work of the model, the modified basic pattern is shown in Fig.6

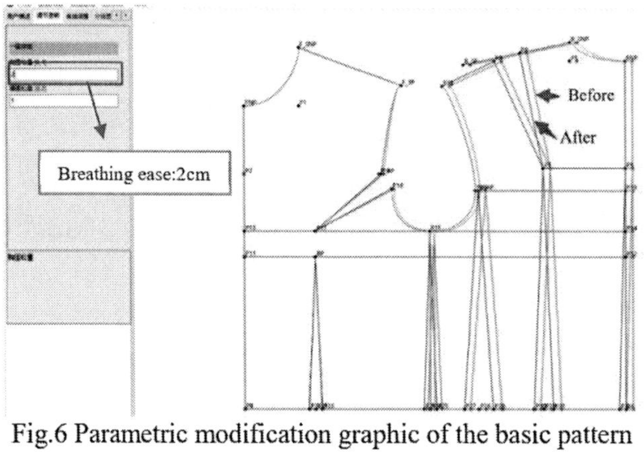

Fig.6 Parametric modification graphic of the basic pattern

This method makes it very convenient to design a clothing pattern. Using this method to build a clothing pattern design model, when the designer applies the model, he only needs to select the corresponding human body data and the clothing category to be designed. The clothing pattern can be automatically generated and changed with the input parameters of the pattern. This method of using a design model to generate a pattern provides an automatic calculation mechanism and method for the intelligent design of patterns.

5. Model Verification of Basic Pattern design for Male Tops

The basic pattern of the M size mannequin output based on the pattern design model, and the sample clothing of the basic pattern was sewn with grey cloth. The physical experiment of clothing gap evaluation was used to evaluate the fit of the sample clothing. Through the size and distribution of the gap between the clothing and the mannequin, it can be judged whether the basic pattern generated by the basic pattern design model fits.

The clothing ease evaluation system of Shaanxi Engineering Research Center of Apparel Intelligent Design and Manufacture was used to carry out the trial-wear evaluation experiment. First, the camera was used to take pictures of the undressed mannequin and dressed mannequin at 8 different angles (0°, 45°, 90°, 135°, 180°, 225°, 270°, 315°) photos, superimpose and see-through the images of 8 corresponding angles in these two states, and get the perspective of the dressing effect of the sample clothes at eight angles Figure, as shown in Fig.7.

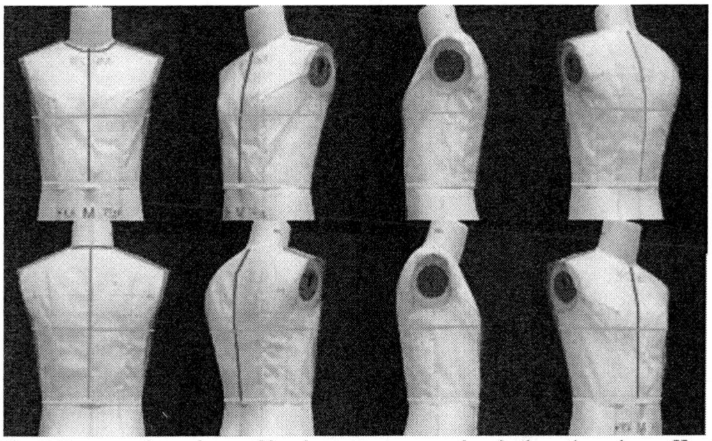

Fig.7 Perspective view of basic pattern sample clothes dressing effect

Photos taken with the same camera and at the same distance have the same number of pixels. Place a horizontal ruler next to the photographed stage, and use image processing software to determine how many pixels are contained in a length of 1 cm. The measurement of the gap between the clothing and the mannequin includes several pixels so that the size of the gap can be estimated, and the gap between the clothing and the mannequin can be digitized. According to the feature line of the human body, a total of 30 points are collected as shown in Fig.8, and they are numbered in sequence. Measure the gap between the position of the sample and the mannequin at these points, and the measurement results are shown in Table 4.

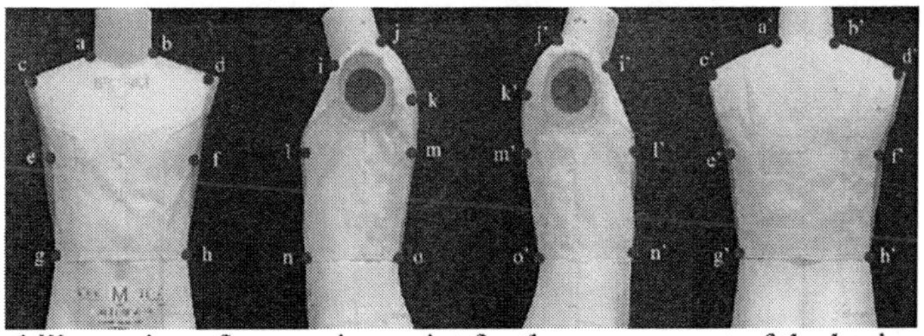

Fig.8 Graphical illustration of measuring point for the gap amount of the basic pattern sample

In Fig.7, from the horizontal view, the neck line (the purple line) of the sample clothes is in good agreement with the base line of the mannequin pedestal neck and is relatively smooth. The position of the bust line (the blue line), bust width line, back width line, and waist line (the yellow line) of the sample clothes are in good agreement with the corresponding characteristic lines of the mannequin, and in a horizontal state. The front shoulder line (the orange line) of the sample suit coincides with the shoulder line of the mannequin, but it does not fit well with the shoulder arc. The end of the shoulder line on the back of the sample is slightly raised, and analysis from other angles shows that the amount of space between the back of the sample and the mannequin is a bit large. Viewed longitudinally, the front centerline (the red line) and back centerline (the green line) of the sample are in good agreement with the corresponding characteristic lines of the mannequin, and they are in a vertical state. The side sutures are consistent with the side sutures of the mannequin. The arc of the sleeve cage of the sample clothes coincides with the arc of the arm base of the mannequin, only the shoulder blades are slightly raised.

Table 4. The measurement result of the gap between the sample and the mannequin Unit: mm

Point	Gap	Point	Gap
a	1.9	a'	0.6
b	1.9	b'	1.3
c	6.9	c'	10.6
d	6.9	d'	10.6
e	21.3	e'	10.0
f	20.0	f'	10.6
g	1.3	g'	1.3
h	1.9	h'	1.9
i	0.6	i'	0.6
j	1.9	j'	1.9
k	13.8	k'	10.0
l	8.8	l'	8.1
m	10.0	m'	2.5
n	0.6	n'	1.9
o	6.9	o'	2.5

It can be analyzed from Table 4: From the gap amount distribution at points a, b, i, j, a', b', i', j', it can be seen that the gap amount distribution between the sample garment and the neck of the mannequin is relatively uniform. The gaps at points e, f and e', f' are similar; the gaps at points l and m are similar. However, the gap amount at points l' and m' is quite different. It should be because the sample clothes were not arranged and leveled when taking pictures, and the amount of voids on the bust of the sample clothes is evenly distributed. The gaps at points g, h, g', and h' are similar; the gaps at points n' and o' are similar, taken together, the amount of voids at the waist of the sample is evenly distributed. From the points c, d, k, c', d', k', the gaps are more evenly distributed, but the gaps are too large, so the gaps on the back of the shoulders of the sample clothing can be larger. It does not fit well with the back of the mannequin.

In summary, the sample clothes fit on the upper body of the mannequin as a whole, only the back of the sample clothes does not fit well with the back of the mannequin, and the gap is too large, which needs to be corrected. Generally speaking, the design of the basic pattern design model is reasonable, but some details need to be adjusted.

6. Modification of the Basic Pattern design Model for Men's Tops

6.1 Deviation Analysis of the Basic Pattern design Model for Male Tops

The amount of space between the shoulders and back of the basic model sample and the mannequin is relatively large, but the shoulder points of the sample clothes coincide with the shoulder points of the mannequin. It can be seen that the amount of space on the back of the shoulder should be affected by the structural line of the shoulder dart. The directed graph model generated by the two structural lines that constitute the shoulder dart is extracted, as shown in Fig.9. Using the directed graph model, the root cause of unfit can be traced back to the source [12].

It can be seen from Fig.9 that the generation of L_L2 depends on the generation of line L_L1, which is formed by connecting points P7 and P8. Point P7 is the shoulder blade point, which is determined by human body parameters, so the determination of L_L1 is mainly determined by point P8. The determination of point P8 is mainly affected by point B_SP. Point B_SP is the intersection of Arc_3 and Arc_4. The positions of Arc_3 and Arc_4 are determined by point B_SNP and shoulder length B2, point BNP and back shoulder width B5/2 respectively. The positions of BNP and B_SNP are in good agreement with the positions of the back neck point and the side neck point of the mannequin. Therefore, the deviation of the position of L_L1 should be due to the insufficient accuracy of the method for determining the point B_SP.

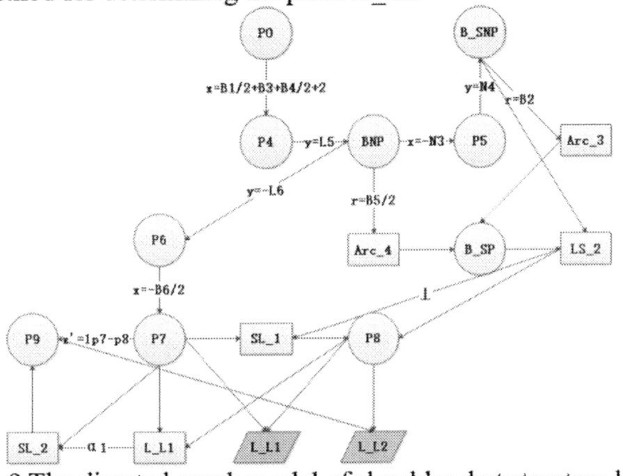

Fig.9 The directed graph model of shoulder dart structure line

6.2 The Revision Process of the Basic Pattern design Model of the Male Top

Analyzing the inconsistency of the shoulder and back of the basic pattern from the directed graph is because the method of determining the point B_SP is not accurate enough. Therefore, the method of determining the point B_SP, which is the shoulder point, is changed to be determined by the shoulder angle and the shoulder length Supplementary measurement of the shoulder angle α3 of ALVANON's M size male mannequin is 21°. The parameterized model of the structure line related to the revised basic pattern is shown in Table 5.

Table 5. Parameterized Model of Modified Structure Line of the basic pattern

No.	Structural line	Build process (Unit: centimeter or degree)
1	Back shoulder dart	It takes point B_SNP as the starting point and horizontally moved to the left B2 to point t1. After connecting point t1 and point B_SNP, line LS_4 is completed. Pass point B_SNP and make the angle line SL_9 with the angle α3 from line LS_4. Pass point P7 to make the vertical line SL_9 of line SL_10, and intersect line SL_7 at point t2. After connecting point t2 and point P7, shoulder dart line 1' L_L1_1 is completed. Measure the length of L_L1_1 as l1', pass point P7 and make the angle line SL_11 with L_L1_1 with an angle β1=(180°*(1-sinα1))/4. Taking point P7 as the starting point, go up l1' along SL_11 to point t3. After connecting point t3 and point P7, shoulder dart line 2' L_L2_1 is completed.
2	Back shoulder line	After connecting point t2 and point B_SNP, back shoulder line 1' L_L3_1 is completed. from line LS_4. Pass point t3 to make the vertical line SL_12 of line SL_9. Measure the length of L_L1_3 as l3'. Taking point t3 as the starting point, go left B2-l3' along SL_10 to point t4. After connecting point t3 and point t4, back shoulder line 2' L_L4_1 is completed.
3	Back armhole curve	After connecting point t4, P20 and P15, back armhole curve L_C5_1 is completed.

According to the revised parametric relationship model of the basic pattern, the directed graph model of the basic model is revised, and then the design model of the basic pattern is revised.

6.3 Verification of the Revised design Model of the Basic Pattern

Use the same steps as 5. to make a modified sample clothing of the basic modification pattern. According to the experimental procedure in 5., a perspective view of the dressing effect of the basic modified pattern sample clothes is obtained, as shown in Fig.10. The measured gap data of each point is shown in Table 6.

In Fig.10, from the horizontal view, the neck line (the purple line) of the sample clothes is in good agreement with the base line of the mannequin pedestal neck and is relatively smooth. The position of the bust line (the blue line), bust width line, back width line, and waist line (the yellow line) of the sample clothes are in good agreement with the corresponding characteristic lines of the mannequin, and in a horizontal state. The front shoulder line (the orange line) of the sample suit coincides with the shoulder line of the mannequin, the end of the shoulder line of the sample is slightly raised but not obvious, which is a normal state. Viewed longitudinally, the front centerline (the red line) and back centerline (the green line) of the sample are in good agreement with the corresponding characteristic lines of the mannequin, and they are in a vertical state. The side sutures are consistent with the side sutures of the mannequin. The arc of the sleeve cage of the sample clothes coincides with the arc of the arm base of the mannequin, and the sample clothing is smooth on the shoulder blades.

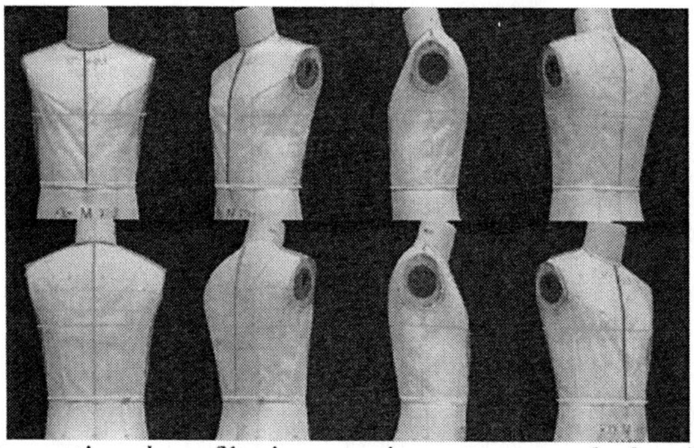

Fig.10 Perspective view of basic correction pattern sample dressing effect

Table 6. Comparison of the measurement results of the sample gap before and after the correction

Point	Gap 1	Gap 2	Difference	Point	Gap 1	Gap 2	Difference
a	1.9	1.9	0.0	a'	0.6	2.5	1.9
b	1.9	1.9	0.0	b'	1.3	2.5	1.3
c	6.9	5.0	-1.9	c'	10.6	4.4	-6.3
d	6.9	5.0	-1.9	d'	10.6	4.4	-6.3
e	21.3	17.5	-3.8	e'	10.0	11.9	1.9
f	20.0	18.1	-1.9	f'	10.6	10.0	-0.6
g	1.3	0.6	-0.6	g'	1.3	0.6	-0.6
h	1.9	1.3	-0.6	h'	1.9	0.6	-1.3
i	0.6	1.3	0.6	i'	0.6	6.9	6.3
j	1.9	1.9	0.0	j'	1.9	0.6	-1.3
k	13.8	5.0	-8.8	k'	10.0	4.4	-5.6
l	8.8	8.1	-0.6	l'	8.1	8.8	0.6
m	10.0	2.5	-7.5	m'	2.5	4.4	1.9
n	0.6	0.6	0.0	n'	1.9	0.6	-1.3
o	6.9	3.1	-3.8	o'	2.5	2.5	0.0

In Table 6, gap 1 is the gap amount measured at each point before correction, gap 2 is the gap amount measured after the correction, and the difference is the difference between gap 2 and gap 1. The gap amount difference in Tab.7 is mostly negative. It can be seen that the corrected gap between the basic pattern sample and the mannequin. is reduced, and the sample is more fitted. There are some abnormal values in the table, which should be due to the fact that the samples were not completely arranged and leveled when taking pictures. These abnormal values can be skipped during analysis. It can be seen from the table that the gaps at points c, d, k, c', d', and k' are significantly reduced, and the gaps on the back and shoulders of the corrected basic pattern sample are evenly distributed and fit well with the mannequin. From the gap amount distribution at points a, b, i, j, a', b', i', j', it can be seen that the gap amount distribution between the correction sample and the neck of the mannequin is relatively uniform. From the gap amount distribution at points e, f, l, m, e', f', l', m', it can be seen that the gap amount distribution between the correction sample and the bust of the mannequin is relatively uniform. From the gap amount distribution at points g, h, n, o, g', h', n', o', it can be seen that the gap amount distribution between the correction sample and the waist of the mannequin is relatively uniform.

In summary, the basic modified pattern model suits the upper body of the mannequin as a whole, the structural line of the pattern is in good agreement with the characteristic line of the male upper body, the space between the sample clothing and the mannequin is more evenly distributed, and the sample clothing fits well. It can be seen that the design method of the basic modified pattern design model is reasonable and meets the requirements of the basic pattern to fit the human body.

7. Conclusion

The normal body shape of adult males is used as the research object. Based on geometric principles, the human body is approximately unfolded and breathing ease is added, the design of the basic pattern of male tops is completed, and then the design model of the basic pattern is constructed. Based on the model, a basic pattern is generated according to the data of ALVANON's M-size mannequin. The distribution and size of the gaps are used to evaluate the fit of the basic pattern. The directed graph model is used to analyze the reasons why the pattern does not fit, and based on this, the basic pattern design model is revised. The experimental results show that the revised basic pattern has a good fit, verifying the effectiveness of the design method of the basic pattern design model. The design model can realize the intelligent design of the basic pattern by adjusting different human body parameters and interactive parameters. This research provides a new method for the extraction, collection, processing, sharing, and reuse of model design knowledge, provides a new idea for the realization of the intelligent design of clothing patterns, and improves the efficiency of the generation, the fit, and the design flexibility of personalized customized clothing.

8. Acknowledgment

Thanks to Shaanxi Engineering Research Center of Apparel Intelligent Design and Manufacture for supporting the research work.

References

[1] Wei N, Bai LC. Current situation and research direction of apparel customization. Melliand China: 2019; v. 47; No. 342 (08): 63-66.
[2] Guan XF, Deng YM, Jiang RF. Study on the new prototype plate making method based on the individual data. Progress in Textile Science & Technology: 2017 (11); 33-37.
[3] Wang WF. Prototype classification of men's tops. West Leather: 2015; 37 (015): 37.
[4] Hyo KC, Hwan SI, Kyu PC, et al. Automatic basic garment pattern generation using three-dimensional measurements. International Journal of Clothing Science and Technology: 2010; 22 (2/3): 101-113.
[5] Mesuda Y, Inui S, Horiba Y. Virtual draping by mapping. Computers in Industry: 2018; 95: 93-101.
[6] Li XF. Domestic personalized 3D garment customization technology and its development trend. Progress in Textile Science & Technology: 2018; 214 (11): 21-24.
[7] Ma XP. Interpretation of garment structure with "Approximate Body Principle". Art science and technology: 2016; 029 (012): 132-133.
[8] Nakazawa Y. Human body and apparel. China Textile & Apparel Press: 2005.
[9] Wang XX, Dai JG, Duan D, et al. Study on dart size distribution of prototype of women's dress from body surface angle. Journal of Zhejiang Sci-Tech University: 2013; 30 (003): 316-320.
[10] Wang X. Analysis study of the ease allowance needed for garment. Journal of Textile Research: 2005 (04); 126-128.
[11] Duan J. Research on Relational Model for Pattern of Menswear based on Digitizing 3D Human Model. Xi'an Polytechnic University: 2017.
[12] He CJ, Ying BA, Wang XF, et al. Application of the directed graph method in the knowledge analysis of clothing pattern design, Virtual, Online: Textile Bioengineering and Informatics Society (TBIS): 2020.

Cite as: C. Chi et al.,
TBIS 10.3993/tbis (2021).

Application of Human Feature Curves in 3D Human Model Hole Repair

Cheng Chi[1,2*], Pascal Bruniaux[2], Xian-Yi Zeng[2], Guillaume Tartare[2]

[1] *Wuhan Textile University, No.1, Fangzhi Road, 430073, Wuhan, China*
[2] *ENSAIT, GEMTEX Laboratory, 59100, Roubaix, France*

[*]Corresponding author's email: cheng.chi@ensait.fr

Abstract

In this study, we propose a hole repair method based on real human feature curves, taking the arm root as an example, to solve the problem wherein large and complex holes are difficult to be repaired accurately in human scan models. Different types of holes can be repaired by the human feature lines. Firstly, the real body feature points are extracted from the three-view of the human body which are mapped onto the 3D human model. Then the body feature curves are drawn by set parameters. The parameters are applied directly to the 3D human model to precisely manage its shape and volume according to its different dimensions. The experimental results show that the repair method based on human feature curves can accurately repair large and complex holes in scanned human models, with good results for holes caused by human manipulation or lack of information, providing an effective method for repairing 3D human models.

Keywords: 3D Human Model; Hole Repair; Real Human Feature Curve; Arm Root

1. Introduction

3D human models are significant for ergonomic product design, especially in digital fashion design. With the development of technology, more designers are using digital tools such as computer-aided design and 3D scanning to design a garment. These tools help designers to pre-demonstrate the effect of the final garment with a computer without actually making the garment. As such, 3D human model are becoming an essential conceptual tool in the garment design and production process. Currently, 3D human models are widely used in garment mass customization, 3D garment design [1], and virtual try-on [2]. However, due to the scanning environment and body self-occlusion, the human models obtained by 3D scanning are missing information [3-4]. For example, the shape of hidden areas such as the arm root, groin, and crotch are difficult to observe from the human body itself or 3D human models [5, 6]. The missing information have a significant impact on the accuracy of the model [7]. Most 3D human models are currently rigid geometric surface models in a standing posture, without considering body composition (bone, fat, and muscle) and body deformation related to skin elasticity [8]. If these factors are not integrated into the 3D human model, the results of garment fitting will not be very accurate. Complete personalized human body model data are vital for personalizing garments [9]. Therefore, the accurate repair of large and complex human body holes is an important task in the post-processing of 3D human models.

Two general methods to repair holes were mentioned in the existing literature: one is to repair the holes directly on the scattered point cloud model [10-12]. The other is to reconstruct the point cloud to obtain a triangular mesh model, identify the mesh topological relationship's hole boundaries, and then repair the hole on mesh models [13-17]. However, these methods are all applied to goods, and there are few researches on hole repair of 3D human body models at present.

In order to repair the large area holes in 3D human models due to self-occlusion in the natural standing posture, this paper proposes a hole shape repair method based on the feature curves of the human body, taking arm roots as an example. The method's effectiveness can be demonstrated based on a comparison of accuracy rates by classifying the arm root body shapes of models before and after the repair, respectively. This research is significant in improving the accuracy of 3D human body models, and fashion design and virtual try-on techniques based on

3D body models. The following section introduces the method of repairing the arm root shape based on the arm root feature curve.

2. Methodology

The human body cannot be considered as a simple object. According to the analysis results of human body data, the feature points and curves are selected to capture the body shape features and generate similar shapes. Basic rules are applied to the human body to adjust its different parameters. There are three operations in the process of hole repair of the human model. The first operation is to determine the location of feature points to create a feature curve that is similar to the real human body. The second operation is to select and extract feature curves to create human models. The idea is to respect totally the morphology (more or less muscles or fat). The third operation is to adjust the parameters of the human model. According to the different size and shape of the feature curve to adjust.

The human model in this study is a surface model defined by the shape curve network extracted from the scanned human body. The CAD model is generated from the scanning data of the main body by using reverse engineering and modelling software. To achieve this, we use DC3D software edited by Lectra. The process of hole repair in a human model can be described in 5 steps, from human data acquisition to the final result.

2.1 3D Data Post-treatment

3D scan data are generally noisy and present defects due to the inaccessibility of the hidden areas. The hidden areas are generally located in the tangential areas of the laser beam, i.e., the top of the head and the shoulders, or in the hidden areas of the body itself, such as the armpits and the crotch. Various data pre-treatment operations such as filtering and hole filling are performed on the mesh of the point cloud to be treated as a surface model represented as a set of faces, as shown in Fig.1.

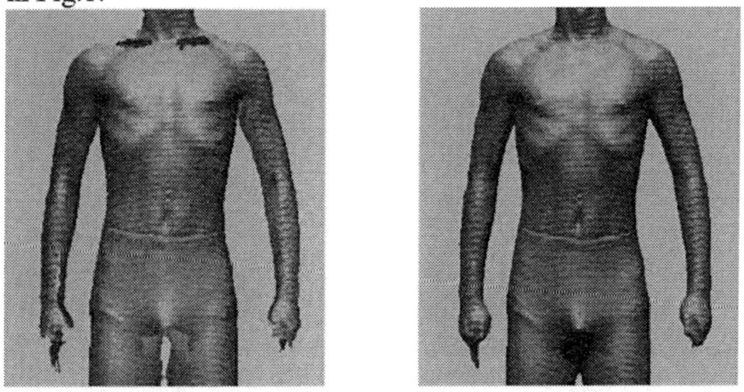

(a) (b)

Fig.1 Comparison of 3D human models with pre-treatment and post-treatment.

2.2 Feature Points Position Determination

In the previous study [18], four feature points affecting the shape of male arm roots were identified, including anterior axillary point, posterior axillary point, armpit point, and acromion. These four points greatly influence drawing the shape of the arm root and determining the position of the arm root. Due to the large area of holes in the arm root of the human model, in order to prevent the recognition of feature points is not accurate, auxiliary points can be marked on the real human body before scanning. Combined with the three-views of the human body, the position of the four feature points is determined by the relationship between the auxiliary points and the above four feature points. The operations are as follows:

1) Combining anthropometric definitions and expert experience, as shown in Table 1, marks six auxiliary points on the human body, i.e., lateral neck root point, waist point, fossa jugularis point, cervical, front central point, and back central point. Black marking stickers that cannot be recognized by 3D body scanners are used, as shown in Fig.2. The purpose is to facilitate the fast and accurate finding of these points on the 3D human model.

Table 1. List of anthropometric landmarks used in this study

NO.	Body Feature Point	Abbreviation	Definition
1	Cervical	BNP	The point of the tip of the seventh cervical spine.
2	Fossa Jugularis Point	FNP	The intersection of the line connecting the superior border of the right and left lateral clavicular terminal with the median sagittal plane.
3	Lateral Neck Root Point	SNP	At the lateral cervical triangle, the intersection of the anterior border of the trapezius muscle and the curve connecting the cervical fossa to the cervical point on the lateral part of the neck.
4	Acromion		The most lateral point of the outer edge of the scapula, usually equal to the shoulder height.
5	Anterior Axillary Point	AAP	The point superior to the anterior axillary fissure.
6	Posterior Axillary Point	PAP	The point at the upper end of the posterior axillary fissure.
7	Armpit point	AP	The point at the lower edge of the axillary fissure, with a wooden stick inserted to identify this point.
8	Waist Point	WP	The maximum point of lumbar concavity on the left side of the body.
9	Front Central Point	FCP	The intersection point of the waistline with the front central line.
10	Back Central Point	BCP	The intersection point of the waistline with the back central line.

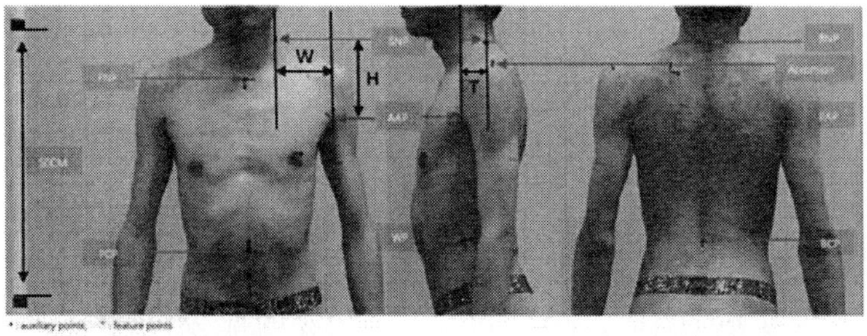

Fig.2 A three-view of the human body with 10 landmark points.

2) A three-view photograph of the human body was obtained based on 2D photogrammetry. The position of the feature points was measured and calculated according to the scale and the position of the auxiliary points marked in the figure. As shown in Fig.2, the position relationship between the lateral neck root point and the anterior axillary point, i.e., width(W), height(H), and thickness (T), is measured starting from the lateral neck point.

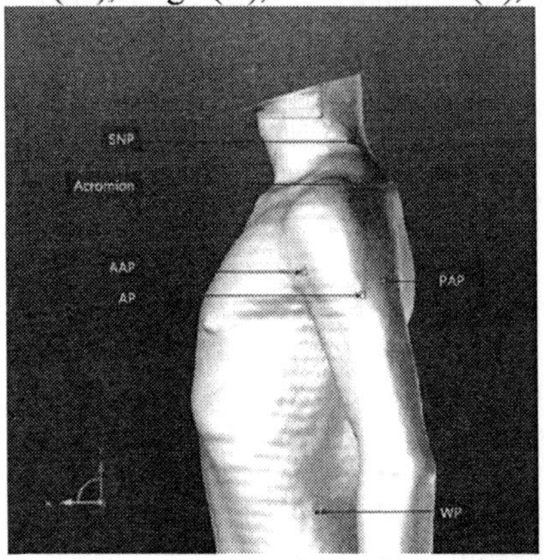

Fig.3 3D human model side view with landmark points.

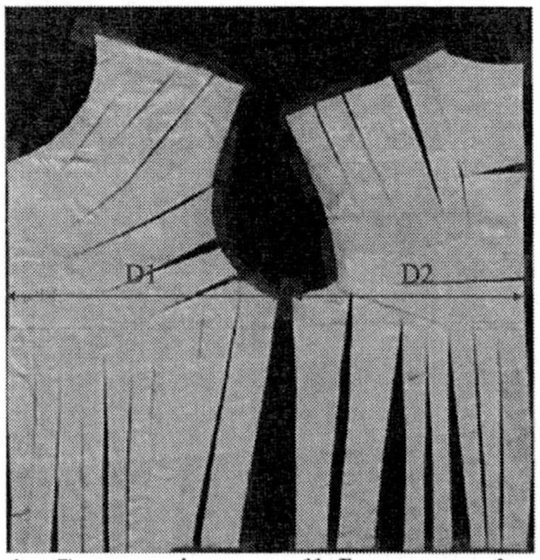

Fig.4 Gypsum inner wall flatten graph

3) With the auxiliary point as the origin, the coordinate system is established. By adjusting the value of the coordinate system, the position relationships between the auxiliary points and the feature points are mapped onto

the 3D human body model to obtain the position of the feature point. The process is repeated twice with the lateral neck root and waist points as the coordinate origin. The final coordinate value of the feature point is an average value, as shown in Fig.3.

4) However, the position of the armpit point cannot be determined directly from the three-view photograph because it is hidden. According to the definition and scaling relationship, the actual height of the armpit point and the intersection of the armpit horizontal line and the front and back central lines can be determined. Then the distance from the armpit point to the front and back central lines, D1 and D2(excluding the volume of dart), can be measured according to the paper film flatten graph of the human body, as shown in Fig.4. The position of the armpit point is determined by combining the cross-section of the armpit horizontal line and the D1, D2.

2.3 Creating Arm Root Curve

According to the anthropometric definition of the arm root circumference: the acromion's circumference through the anterior and posterior points of the axilla to the starting point. The arm root circumference curve is an irregular closed curve that does not lie in the same plane. Therefore, different planes are created based on the arm root feature points related to the arm root, as shown in Fig.5. Firstly, we divide the arm root circumference curve into front and back arm root curves. The front arm root curve is constructed on the plane with the acromion point, the armpit point, and the anterior axillary point. The back arm root curve is constructed on the plane of the acromion, armpit, and posterior axillary points, as shown in Fig.5. The principle is that the curve must pass through all three feature points in the same plane. There is still a certain difference between the completed arm root curve (black curve) and the real arm root curve of the human body. In order to obtain a more accurate arm root curve, it is necessary to set the parameter values for the black curve separately. The parameter values are represented as equal lines vertical to the black curve. Connect the other end of all lines to obtain the parametric curve (yellow curve and green curve), as shown in Fig.5. The goal is to adjust the parameters to obtain an arm root curve that is the same as the real human body.

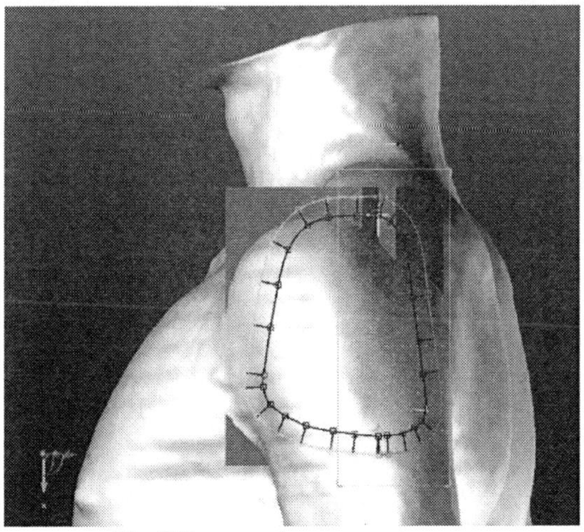

Fig.5 Parametric arm root curve

2.4 Create a Surface Model

Based on the auxiliary and feature points, draw feature lines on the surface of the 3D human body model, including the front and back central line, shoulder line, side suture line, waistline, and left neck circumference line, as shown in Fig.6. The drawing method follows anthropometric definitions, as shown in Table 2. In order to create a surface model of the upper left side of the body that more closely the real human body, add some auxiliary lines along the surface of the model. Then, generate a surface model, and the new surface model is meshed and processed from 3D to 2D by the software DC3D, as shown in Fig.7.

Table 2. Upper body feature curves

NO.	Body Feature curve	Abbreviation	Definition
1	Neck Base Girth	NBG	The circumference of the cervical root measured through the seventh cervical point, Lateral Neck Root Point , and the Fossa Jugularis Point.
2	Shoulder Line	SL	Measure the straight line distance from the lateral neck root point to the shoulder crest point.
3	Upper Chest Girth	AG	The horizontal circumference length through the posterior point of the left and right axilla.
4	Chest Girth	CG	The horizontal circumference measured through the scapula, axilla, and nipple.
5	Minimum Waist Girth	WG	The horizontal circumference of the smallest part of the waist between the inferior border of the ribs and the superior border of the iliac crest.
6	Front central Line	FCL	The length of the curve from the anterior neck point to the minimum waist circumference on the anterior centerline of the torso.
7	Back central Line	BCL	The length of the curve from the posterior cervical point to the minimum waist circumference on the posterior centerline of the torso.
8	Lateral Line	LL	Solid length from the lateral point of the horizontal waistline to the armpit point.

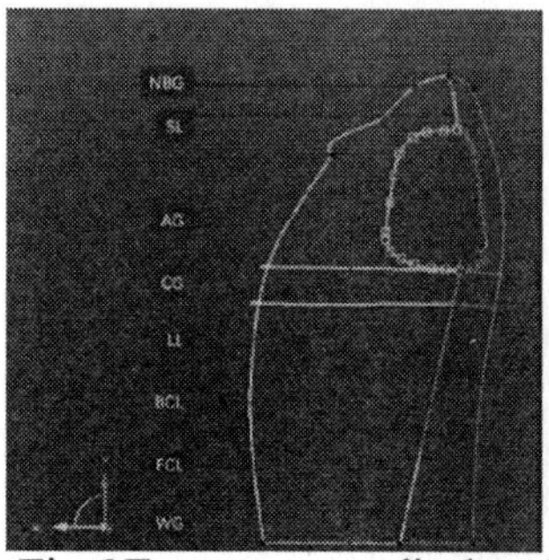

Fig.6 Feature curve display chart

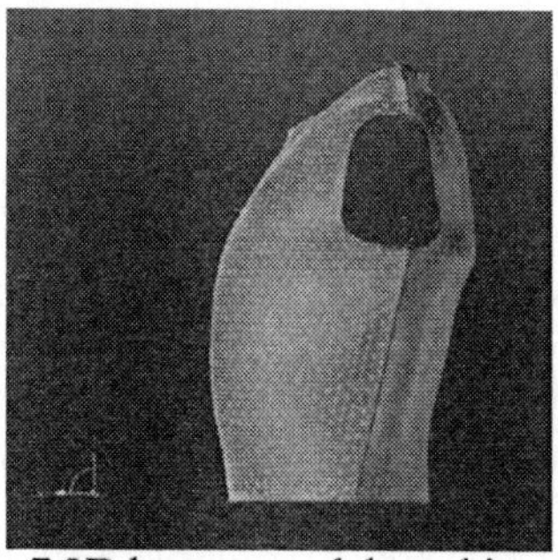
Fig.7 3D human model meshing

2.5 Adjusting the Shape of the Arm Root Curve

The surface model was treated from 3D to 2D by the software DC3D to obtain a 2D morphology of the surface model. The 2D morphology of the surface was compared with the shape of the arm root curve (red curve) based on the human paper film flatten graph, as shown in Fig.8(a). The 2D morphology is similar to the red curve, but there are still some differences. In order to match the two curves more closely, it can be realized by adjusting the values of the previously set parameters, as shown in Fig.8(b). Superposition results of 2D morphology and real arm root curve show the morphological similarity of this method.

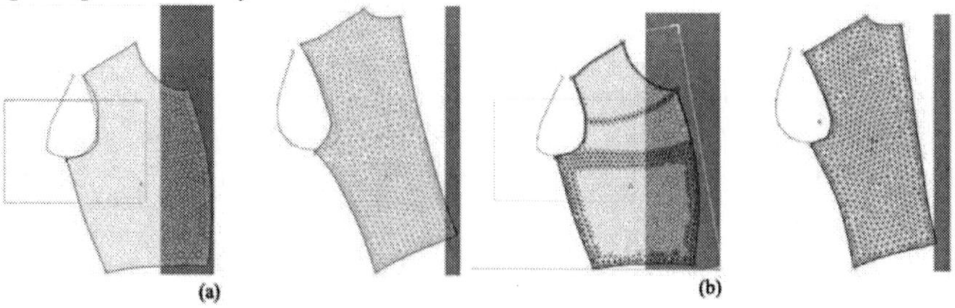

Fig.8 3D human model flatten graphs

3. Results and Discussion

With the adjustment of the parameters, the shape of the 3D arm root also changes accordingly. Finally, the shape of the arm root is closer to that of the real human body, as shown in Fig.9. Moreover, this method is not too much affected by the size of the hole. Once the position of the feature point is determined, we can get a satisfactory result. We can obtain the same shape of arm root shape with different sizes.

Arm root curves have their special design, and it is not easy to control and design the arm root curve by just the position of the points to match the real human body. It is difficult for designers to grasp the radian and shape of the curve without reference. Only obtaining an accurate arm root shape, designers can have a reference when designing the ease of armhole curve. The reasonable classification of arm root shape [18] can provide a basis for the reconstruction of arm root curve and improve the accuracy of the 3D human model.

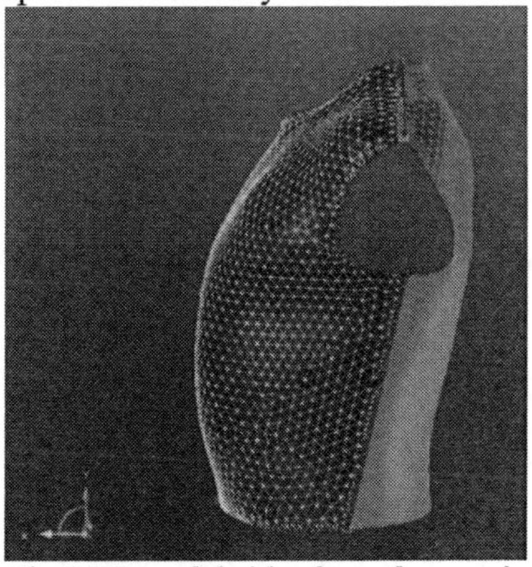

Fig.9 3D model side view after repair

In order to demonstrate that feature curve can be effectively corrected for the 3D human model. In this paper, feature items were measured on the 3D human model before and after correction, and the body shape classification results were compared. Using the arm root classification results from the previous study [18] as the standard (as shown in Table 3), the accuracy of the arm root classification before the correction was 17% and the accuracy of the arm root classification after the correction was 91.1% (as shown in Table 4). The comparison of the classification results concluded that the 3D human model's accuracy could be significantly improved by correcting the 3D human model based on feature curves.

Table 3. Final cluster center and capacity of arm root shape

Items	Final Cluster Centers						
	Cluster						
	1	2	3	4	5	6	7
ARD	16.1	13.7	16.2	15.5	17.6	17.1	16.2
ARW	10.7	9.9	11.0	9.1	12.2	8.3	10.9
RACD	2.8	4.8	4.2	4.0	8.6	6.2	10.4
FACD	6.9	5.4	13.1	10.4	15.2	5.8	10.8
DFR	4.1	0.6	8.9	6.3	6.6	-0.4	0.4
RDW	1.7	1.8	2.0	1.9	1.9	1.3	1.6
DCB	-0.9	-4.1	-2.8	0.9	0.1	0.2	-2.5
Sample capacity	6	5	2	7	4	3	6

Table 4. Classification accuracy of arm root shape

Existing groups	Classification based on 3D model before correction								Classification based on 3D model after correction							
	AR1	AR2	AR3	AR4	AR5	AR6	AR7	Classification accuracy (%)	AR1	AR2	AR3	AR4	AR5	AR6	AR7	Classification accuracy (%)
AR1	1	0	0	0	0	0	1	16.7	5	0	0	0	0	0	0	83.3
AR2	0	3	0	0	0	1	1	60	0	5	0	0	0	0	0	100
AR3	0	0	0	0	0	1	1	0	0	0	2	0	0	0	0	100
AR4	1	1	0	0	1	2	0	0	0	0	2	5	0	0	0	71.4
AR5	0	1	0	0	0	1	0	25	0	0	0	0	4	0	0	100
AR6	0	1	0	0	0	1	0	33.3	0	0	0	0	0	3	0	100
AR7	0	0	0	0	0	0	0	0	0	0	0	0	1	0	5	83.3
Total	-	-	-	-	-	-	-	17	-	-	-	-	-	-	-	91.1

Note: Unit, number of people

4. Conclusion

Nowadays, in the apparel industry, about 3D virtual try-on simulation and 3D garment design, the fitting and design of garments are done based on an idealized virtual human body without considering the feature of customers. This paper proposes a virtual 3D human body model reconstruction method based on feature curve, taking the arm root shape as an example. We attribute the reconstruction problem to an optimization problem. All reconstructed human models need to be based on human feature points and feature curves. The results show that the accuracy of restructured human models are much higher than before. However, the method proposed in this paper still has some limitations. For example, the reconstruction method is based on professional anthropometric knowledge. The reconstructed human model cannot be obtained quickly.

Reconstruction and design based on 3D human models is still an open issue with enough room for innovation. Our future work will present findings in garment modeling and numerical simulation. The strategic point of the study will be the rapid and accurate reconstruction of the new 3D human model obtained from the scans.

References

[1] Liu K, Zhu C, Tao X, et al. Parametric design of garment pattern based on body dimensions[J]. International Journal of Industrial Ergonomics: 2019; 72: 212-221.

[2] Hong Y, Zeng X, Bruniaux P, et al. Interactive virtual try-on based three-dimensional garment block design for disabled people of scoliosis type [J]. Textile Research Journal: 2017; 87 (10): 1261-1274.

[3] Ju T. Fixing geometric errors on polygonal models: a survey[J]. Journal of Computer Science and Technology: 2009; 24 (1): 19-29.

[4] Bakirman T, Gumusay MU, Reis HC, et al. Comparison of low cost 3D structured light scanners for face modeling [J]. Applied optics: 2017; 56 (4): 985-992.

[5] Chiu CY, Pease DL, Sanders RH. Effect of different standing poses on whole body volume acquisition by three-dimensional photonic scanning [J]. IET Science, Measurement & Technology: 2016; 10 (6): 553-556.

[6] Pickup D, Sun X, Rosin PL, et al. SHREC'14 track: Shape retrieval of non-rigid 3D human models[C]//Proceedings of the 7th Eurographics workshop on 3D Object Retrieval. Eurographics Association: 2014; 1 (2): 6.

[7] Attene M, Campen M, Kobbelt L. Polygon mesh repairing: An application perspective[J]. ACM Computing Surveys (CSUR): 2013; 45 (2): 1-33.

[8] Pickup D, Sun X, Rosin PL, et al. SHREC'14 track: Shape retrieval of non-rigid 3D human models[C]// Proceedings of the 7th Eurographics workshop on 3D Object Retrieval. Eurographics Association: 2014; 1 (2): 6.

[9] Lanman D, Taubin G. Build your own 3D scanner: 3D photography for beginners[M]//ACM SIGGRAPH 2009 Courses: 2009; 1-94.

[10] Sharf A, Alexa M, Cohen-Or D. Context-based surface completion[M]//ACM SIGGRAPH 2004 Papers: 2004;

878-887.
[11] Kumar A, Shih A, Ito Y, et al. A hole-filling algorithm using non-uniform rational b-splines [C]//Proceedings of the 16th International Meshing Roundtable. Springer, Berlin, Heidelberg: 2008; 169-182.
[12] Gai S, Da F, Zeng L, et al. Research on a hole filling algorithm of a point cloud based on structure from motion [J]. JOSA A: 2019; 36 (2): A39-A46.
[13] Barequet G, Sharir M. Filling gaps in the boundary of a polyhedron [J]. Computer Aided Geometric Design: 1995; 12 (2): 207-229.
[14] Pernot JP, Moraru G, Véron P. Filling holes in meshes using a mechanical model to simulate the curvature variation minimization [J]. Computers & Graphics: 2006; 30 (6): 892-902.
[15] Zhao W, Gao S, Lin H. A robust hole-filling algorithm for triangular mesh [J]. The Visual Computer: 2007; 23 (12): 987-997.
[16] Gong Y, Wang L, Guo R, et al. Multi-scale orderless pooling of deep convolutional activation features [C] //European conference on computer vision. Springer, Cham: 2014; 392-407.
[17] Girshick R, Donahue J, Darrell T, et al. Rich feature hierarchies for accurate object detection and semantic segmentation [C] //Proceedings of the IEEE conference on computer vision and pattern recognition: 2014; 580-587.
[18] Chi C, Zeng XY, Pascal B, et al. Arm root shape classification of young male based on artificial anthropometry [C]// Proceedings of the 14th International FLINS Conference: 2020; 622-630. https://doi.org/10.1142/9789811223334_0075

Cite as: S.Q. Wang et al.,
TBIS 10.3993/tbis (2021).

Classification and Identification of Young Males' Shoulder Shapes using Extreme Learning Machine based on 3D Body Measurement

Shu-Qing Wang[1], Juan-Feng Jin[1,2,3,*]

[1]Faculty of Clothing and Design, Minjiang University, Fuzhou, Fujian, 350108, China
[2]Engineering Research Center of Textile and Clothing (Minjiang University), Fujian Province University, Fuzhou, Fujian, 350108, China
[3]Fujian Clothing Industry Technology Development Base, Fuzhou, Fujian, 350108, China

*Corresponding author's email: jjf6859@163.com

Abstract

In order to meet the requirements of clothing shoulders, the young male shoulder shape ELM identification model based on three-dimensional body measurement was designed and developed. Based on the measurement of three-dimensional human body, shoulder data of 213 young male students aged from 18 to 26 years old were obtained. Taking the characteristic point curvature radius of the cross-sectional curve and the ratio of sagittal to frontal diameter as body type classification indices, 4 types of shoulder types were subdivided and verified by variance analysis. According to the clustering result, the Extreme Learning Machine (ELM) method was introduced, and the ELM identification model constructed by MATLAB software was used to identify the shoulder shape types. The model had high recognition rate and excellent identification performance, which could provide a new method for human body type identification, and can also broaden the application field of Extreme Learning Machine method.

Keywords: 3D Human Body Measurement; Shoulder Shape Classification; Extreme Learning Machine; Identification Model

1. Introduction

For consumers, clothing fit is an important factor for clothing aesthetics [1]. The same human body size with the same clothing can lead to different wearing comfort if the body shape is different. Body shape analysis is therefore an important research hotspot [2], and the accurate recognition of human body shape is an important premise to improve clothing fit.

In recent years, 3D anthropometer is becoming an important tool to obtain human data [2]. In the research of human body shape classification, Yong LC proposed to use body side angle to represent body side shape characteristics [3]; Yang YC introduced trunk volume index (TVI) to divide the human body trunk into four types: X, H, A and V [4]; Zhang JH extracted two shoulder typical indexes by three-dimensional measurement to subdivide the shoulder shape of young women into four types [5]; Huang CY extracted 4 neck typical indexes and applied K-means clustering method to subdivide the neck morphology of male college students [6]. However, the change of human body shape can be divided into two directions: length (shoulder width, chest, waist, hip circumference, etc.) and girth (shape change of cross-section curve). Targeting the difference of human body cross-section curve shape, Gu L proposed the method of body shape classification based on parts (chest, waist, hip) by using the ratio index of feature point data [7].

The curvature radius can be used to describe the degree of curve bending, and it can be used as an index of body shape classification. Lee HY used reverse engineering software to extract the curvature radius of the feature points of the breast bottom contour line of women and subdivided the breast shape [8].

In the research of human body shape recognition, more emphasis is put on the recognition of the overall body shape, and the application of mathematical statistics is more, such as graphic classification [2, 9], clustering [10, 11]; artificial intelligence methods such as neural network have the unique advantages of good operability, fast and efficient processing of batch data, and become the main research direction of current body shape recognition. Ding XJ [12] used genetic neural network method, Li XJ [13] used Fisher discriminant analysis method, Zhang SY [14] proposed body type classification model based on SVM, Chen MZ [15] designed and developed female body type recognition expert system, and Jin JF [16] designed and developed probabilistic neural network body type recognition model. However, the boundary of human body shape has strong fuzziness, so we need to constantly seek better classification methods to improve the recognition rate.

Shoulder is the key part of male upper body. This paper focused on the changes of shoulder girth section curve, and studied the male shoulder shape characteristics. With the help of 3D measurement, the shoulder section curve and the curvature radius of feature points were extracted; through cluster analysis and variance analysis, the classification of body shape was subdivided, and the Extreme Learning Machine (ELM) method was used to recognize the shoulder body shape. The results showed that the performance of the recognition model was excellent, which provided a new method for human body shape recognition, and could also be extended to other age and gender groups.

2. Experiment

2.1 Subjects and Instrument

Subjects, 213 male university students with ages ranging from 18 to 26 years, were scanned with the $[TC]^2$ 3D body scanner ($[TC]^2$ Corporate, USA).

2.2 Extraction of Section Curve

With the help of three-dimensional anthropometer, the point cloud data of human shoulder was obtained, and according to the height of human shoulder endpoint, the point cloud data of shoulder section curve which could represent the shoulder morphological characteristics was extracted, as shown in Fig.1; using MATLAB software, the minimum enclosing rectangle (MER) [17] was used to adjust the axis of the shoulder section, and the symmetry was processed; the least square method [18] was used to fit the curve of the shoulder section, and a curve mathematical model with high goodness of fit was established to effectively characterize the morphological characteristics of the human shoulder, as shown in Fig.2.

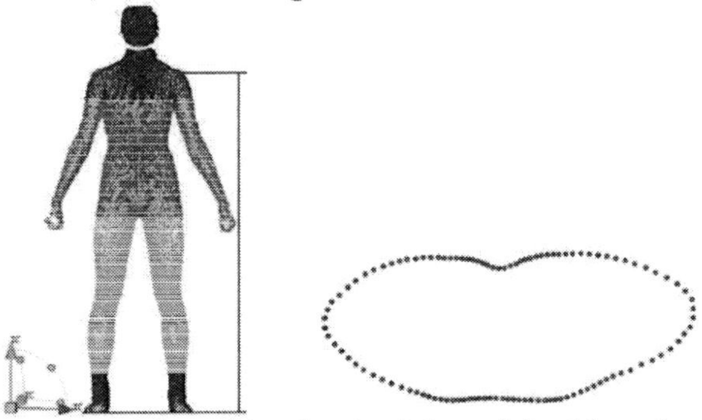

(a) shoulder point height (b) point cloud data of shoulder point cross-section

Fig.1 Point cloud data

Fig.2 Fitting curve of shoulder cross section

2.3 Experimental Data Parameters

2.3.1 Ratio of Sagittal Diameter to Frontal Diameter of Shoulder Section

In human body orientation, the ratio of sagittal diameter to frontal diameter is the ratio of short axis to long axis. The ratio of rise to forehead to diameter of cross section R is the ratio of thickness a to width b of cross section, that is, $R = a/b$, which represents the degree of oblateness of cross section. For the same section girth, if the ratio of sagittal to frontal diameter is larger, the ratio of sagittal to radial direction is larger, and the ratio of frontal to radial direction is smaller, which shows that the shoulder section is thicker and rounder.

2.3.2 Curvature Radius of Shoulder Section Curve

The curvature of a plane curve is the rotation rate of the tangent direction angle of a point on the curve to the arc length. The larger the curvature is, the greater the curvature of the curve is, and the greater the deviation of the curve from the straight line is. The curvature formula of the curve [19] is as follows:

$$K = \frac{|y''|}{(1+y'^2)^{3/2}} \tag{1}$$

where y' is the first derivative of the fitting curve, y'' is the second derivative of the fitting curve.

The radius of curvature is equal to the reciprocal of the curvature, that is, the radius of the circle formed when the arc is a part of a circle. Because the curvature value of shoulder section fitting curve is relatively small, it is not convenient for subsequent analysis, so the curvature radius is selected to evaluate the degree of curve bending. The larger the radius of curvature is, the smoother the arc is. The smaller the radius of curvature is, the steeper the arc is. Therefore, the radius of curvature can accurately reflect the shape of shoulder section.

The human body is approximately left-right symmetrical, so the left half of the male shoulder section curve was analyzed, and the curvature radius was extracted with 5° interval angle [20, 21]. A total of 36 curvature radii were extracted to reflect the morphological characteristics of the section curve, as shown in Fig.3.

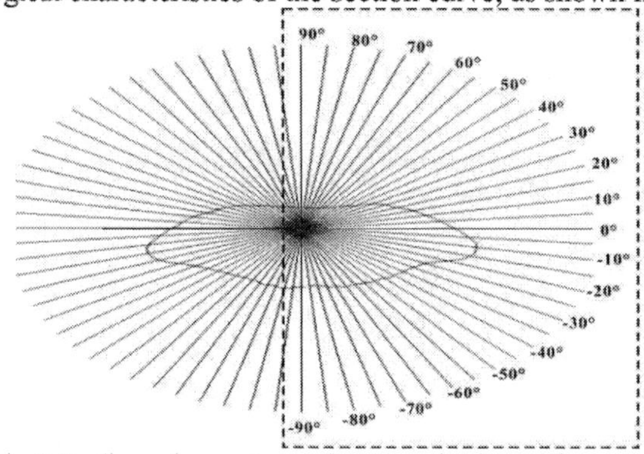

Fig.3 Radius of curvature of each point in section curve

3. Data Processing and Analysis

3.1 Feature Points Extraction of Shoulder Section Curve

In order to ensure the accuracy and reliability of the curvature radius values of each feature point, it was necessary to preprocess the data before data analysis, so as to eliminate the influence of singular values and wrong data on the experimental results. Spss19.0 was used to preprocess the obtained data through box graph, and the useful information was extracted by eliminating the existing "abnormal data". Descriptive analysis was conducted on the retained 200 experimental sample data, and the mean change of curvature radius was obtained, as shown in Fig.4.

It could be seen from Fig.4 that the curvature radius fluctuated continuously from -90° to 90° and the curvature radii at -50° , -35° , 0° , 50° and 90° were all local minimum values. The curve bending degree in the local range was the largest, which could accurately reflect the local shape of the section curve and the overall change of the section shape. Therefore, the curvature radii at these five angles, namely RC_{-50°, RC_{-35°, RC_{0°, RC_{50°, RC_{90°, were selected as the follow-up points. The corresponding position of the shoulder point section curve was shown in Fig.5.

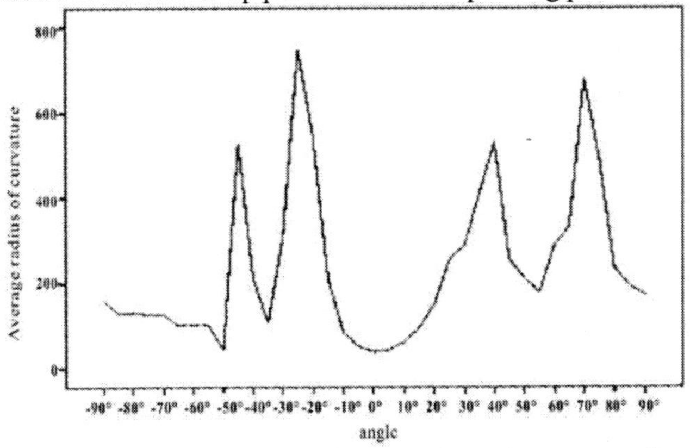

Fig.4 Distribution of mean radius of curvature of angles

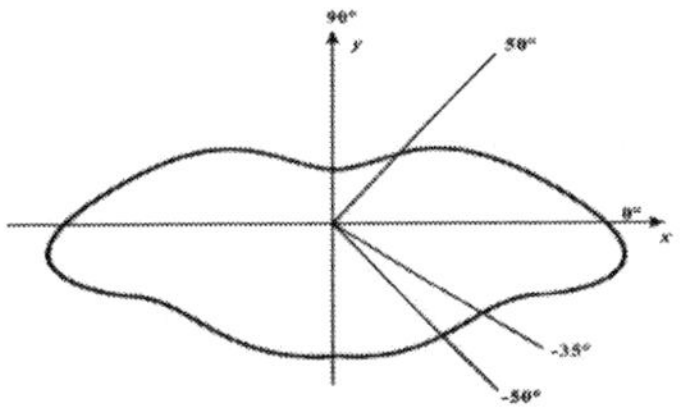

Fig.5 Feature points of section curve

3.2 Shoulder Shape Subdivision

In order to find the characteristics of shoulder data, the k-means method was used to cluster the shoulder data dynamically according to the curvature radius of five feature points and the ratio of sagittal diameter to frontal diameter (R), which could quickly distribute the experimental samples to various types of shoulder data.

Table 1. Analysis of variance about shape types of shoulder

Clusters	Indices	Mean square Between groups	Mean square Within group	F	Sig.
3	R	0.093	0.103	3.124	0.001
	RC_{50°	135827.660	3525.145	49.852	0.000
	RC_{-50°	1105.480	421.218	8.526	0.000
	RC_{-35°	110376.109	258.187	601.938	0.410
	RC_{0°	175.311	119.275	2.133	0.008
	RC_{90°	201831.257	5088.021	112.038	0.000
4	R	0.198	0.015	3.395	0.037
	RC_{50°	143156.257	2375.692	53.818	0.000
	RC_{-50°	2651.428	380.726	6.284	0.000
	RC_{-35°	125347.166	221.462	535.372	0.000
	RC_{0°	287.163	74.420	3.426	0.005
	RC_{90°	263963.845	5007.271	52.893	0.000
5	R	0.248	0.031	7.407	0.025
	RC_{50°	131050.837	2964.435	51.342	0.000
	RC_{-50°	3274.242	371.161	5.598	0.000
	RC_{-35°	163591.510	200.344	580.225	0.081
	RC_{0°	386.528	54.484	2.753	0.000
	RC_{90°	419124.721	5011.225	150.148	0.000

Table 1 showed the variance analysis results of the curvature radius and the ratio of sagittal diameter to frontal diameter of the feature points when the shoulder data were clustered into 3, 4 and 5 categories according to the six section curve morphological indexes. It could be seen from Table 1 that the probability of F-test was less than 0.05 when the shoulder shapes were divided into four categories, so the four categories were the best choice for shoulder shape classification. When the shoulder shapes were divided into four categories, the first category contained 49 samples, the second category contained 25 samples, the third category contained 106 samples, and the fourth category contained 20 samples.

Fig.6 was the cross-sectional curve of each type of shoulder, which clearly and intuitively showed that there were significant differences in the cross-sectional shape of the four types of shoulder. Type 1: The shoulder thickness was moderate, the curve of shoulder ends and the middle part were obvious. There was a sharp corner on the side of the shoulder, and the back shoulder arc was smoother. Type 2: The thickness was thick, the shoulder curve was relatively smooth, the total was unique, the shoulders were rounded, and the front and rear shoulder arc was round; after the unevenness was obvious, the front shoulder was more round. Type 3: The shoulder curve had a clear contour, the front and rear shoulders were uneven, and the two sides were obvious. Type 4: The back had a thick thickness, the front and rear shoulders were more gentle, and the shoulders were slightly forward.

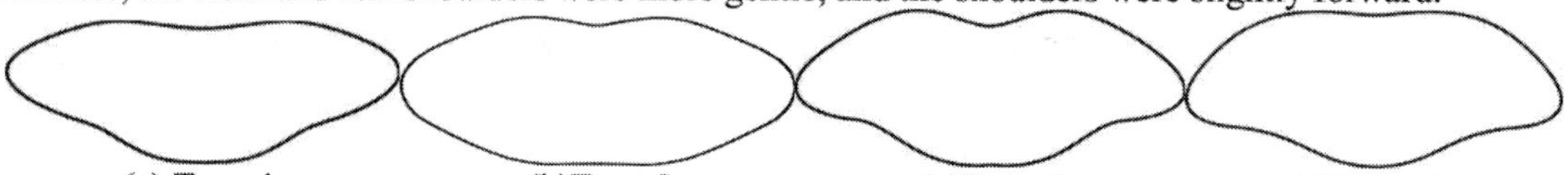

(a) Type 1 (b)Type 2 (c)Type 3 (d)Type 4

Fig.6 Curves of all types of shoulder point sections.

4. Shape Identification Model based on Extreme Learning Machine (ELM)

4.1 Reason for Constructing ELM Identification Model

At present, neural network is widely used in the classification and recognition of human body, but the traditional neural network has a long training time, the number of nodes and threshold are difficult to determine, the optimal solution needs multiple iterations, which is not easy to determine (for example, the iterative steps of BP network, SVM and so on often make the parameter training process take up a lot of time, Thus, the efficiency of network training process can not be guaranteed); it is easy to fall into local minimum, resulting in "false saturation", "oversaturation" and other phenomena.

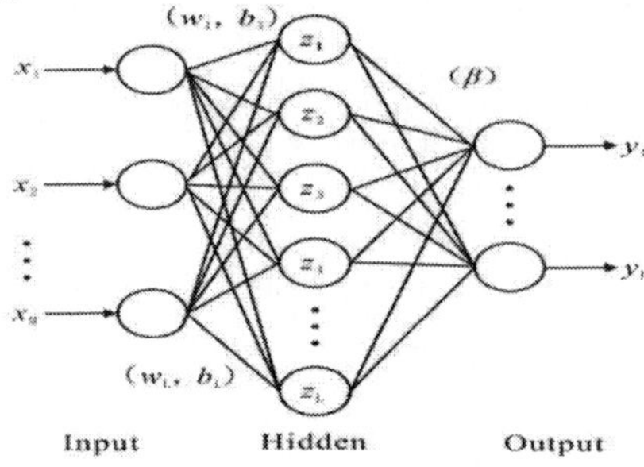

Fig.7 ELM network structure

Different from the defects of the traditional learning methods mentioned above, the input weights and thresholds of extreme learning machine (ELM) are randomly obtained by the system, which makes the network constructed by elm approach any continuous system. This method is a new feedforward neural network training method proposed by Professor Huang GB of Nanyang University of technology in 2006. The network structure is shown in Fig.7.

Given a data set $(x_i, y_i), i = 1, \ldots, N$, with sample size N, where $x \in R^M$ is the input eigenvector, its dimension is M, and y is the sequence of tag values. The structure of ELM neural network model can be expressed as following:

$$\sum_{j=1}^{L} \beta_j g(w_j \cdot x_i + b_j) = y_i,\ (i = 1, \ldots, N) \qquad 2)$$

where, L is the number of hidden layer nodes, $g(x)$ is the activation function, $w_j = [w_{j,1}, w_{j,2}, \ldots, w_{j,m}]^T \in R^M$, is the input weight between the input layer neuron and the j th hidden layer neuron, b_j is the bias value of the j th hidden layer neuron, and β_j is the output weight between the j th hidden layer neuron and the output layer neuron.

The above formula can be expressed as $H\beta = Y$, where

$$H = \begin{bmatrix} g(w_1, x_1 + b_1) & \cdots & g(w_L, x_1 + b_L) \\ \vdots & \ddots & \vdots \\ g(w_1, x_N + b_1) & \cdots & g(w_L, x_N + b_L) \end{bmatrix} \qquad 3)$$

$$\beta = [\beta_1^T \cdots \beta_L^T],\ Y = [y_1^T \cdots y_N^T] \qquad 4)$$

Where H is the output matrix of hidden layer of ELM neural network. Since the weights w and bias value b of the input layer are determined randomly, the training process of ELM neural network is essentially to find the least square solution $\hat{\beta}$ of the linear system $H\beta = Y$.

The training steps of ELM neural network are as follows: (1) set the input layer weights w_j and bias values b_j, $j = 1, \ldots, L$; (2) calculate the hidden layer output matrix H and its pseudo inverse H^+; (3) calculate the hidden layer output weights β: $\hat{\beta} = H^+Y$.

It can be seen that ELM network has a good recognition function, and has the following advantages that traditional neural network does not have: (1) the training speed of the model is high, the learning process is simple; (2) the parameter selection is easy, only need to select the number of hidden layer nodes, no need to adjust the network weight and the bias of hidden elements, and the resulting solution is the only solution; (3) realize nonlinear approximation, will not fall into the trap; (4) good generalization performance.

4.2 Construction of ELM Shoulder Shape Recognition Model

4.2.1 Establish ELM Model and Training

The constructed ELM shoulder shape recognition model framework was shown in Fig.8, with six characteristic indexes, namely $RC_{-50°}$, $RC_{-35°}$, $RC_{0°}$, $RC_{50°}$, $RC_{90°}$ and R, as the input variables of the model, and the type of shoulder shape as the output variable.

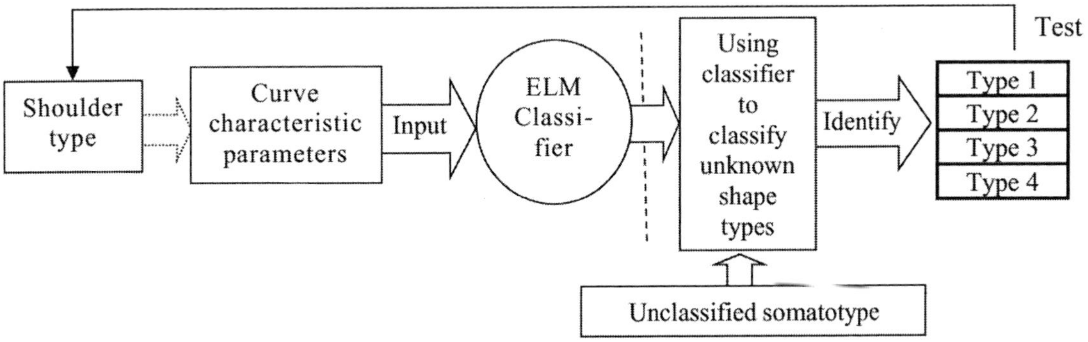

Fig.8 Shoulder shape recognition model framework

When training the body recognition model, 3/4 samples of each category were selected as training samples, and the remaining 1/4 samples were selected as test samples. The 200 samples were numbered, and the number of training samples for each category was shown in Table 2.

Table 2. The capacity of samples to train in each type

Items	Type				Total
	1	2	3	4	
Sample size	49	25	106	20	200
Training sample size	37	19	79	15	150

4.2.2 Excitation Function and Determination of the Number of Hidden Layer Nodes

In order to make the recognition model of extreme learning machine have higher recognition rate, it was necessary to select the excitation function and the number of hidden layer nodes. Firstly, four kinds of common excitation functions (including Sine function, Radial Basis function, Hardlim function, Triangular Basis function, Sigmoidal function and so on) were selected for comparison. The number of hidden layer nodes of each excitation function was initialized to 10, and the number of hidden layer nodes was increased by 20. The influence of different excitation functions and the number of hidden layer nodes on the classification accuracy of ELM model was analyzed. The training accuracy of each excitation function with different number of nodes was shown in Fig.9.

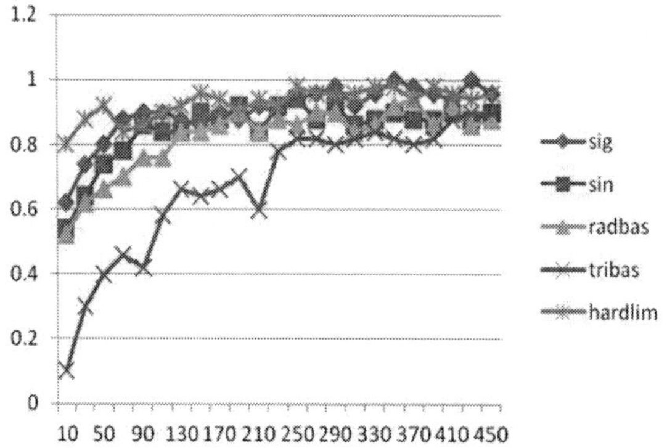

Fig.9 the influence of different excitation function and node number on shoulder shape classification accuracy

It was not difficult to see from the Fig.9 that in the process of model training, when the number of hidden layer nodes exceeded 250, the classification accuracy of each incentive function tended to be stable. When the number of nodes was 350, the accuracy of Sigmoidal incentive function was the highest, reaching 100%; the accuracy of Hardlim incentive function was the second, and the accuracy of Tribas incentive function was the lowest. Therefore, the Sigmoidal function was selected as the incentive function of the model, and the number of hidden layer nodes was determined to be 350 to construct the identification model.

4.2.3 Model Testing

After the model training, the remaining 1/4 sample size was used as the test sample to test the model, so as to explore the recognition rate of shoulder shape. The recognition rate was calculated according to the following formula:

Recognition rate = Recognition feature value/Total number of feature values 5)

Using MATLAB software, the test samples were input into the trained Extreme Learning Machine model for testing, so as to obtain the body recognition rate. Table 3 was shoulder shape ELM model performance test result, Table 4 was shoulder shape recognition rate table.

Through comprehensive analysis of Table 3 and table 4, it could be seen that the overall recognition rate of the shoulder shape ELM recognition model was 94.00%. Among them, in the automatic recognition of the model, the type 2 and type 4 were fully recognized, and both of their recognition rates were 100%; in the type 1 with 12 samples, one sample was misjudged, and the 30th sample was misjudged as the type 4, so the recognition rate of the type 1 was 91.67%; in the type 3 with 27 samples, two samples were misjudged, namely the 6th sample and the 13th sample, were misjudged as the type 1 and the type 4 separately, thus, the recognition rate of type 3 was 92.60%. The sample size was not large enough, and some local human body of samples were similar, so that the model had misjudgment.

Table 3. Prediction results of ELM model

Expected value \ Estimate value	1	2	3	4
1	11	0	0	1(30)
2	0	6	0	0
3	1(6)	0	25	1 (13)
4	0	0	0	5

Note: "( )" indicated the test sample number.

Table 4. Each type recognition ratio

Type	Total categories	Number of misjudgments	Recognition rate (%)
1	12	1	91.67
2	6	0	100.00
3	27	2	92.60
4	5	0	100.00
Total	50	3	94.00

5. Conclusion

In this paper, we quantitatively studied the data of male shoulder body shape and obtained the following conclusions:

1) The classification indexes which could be used to quantify the shape characteristics of shoulder cross-section curve were extracted: curve curvature radius and vector frontal diameter ratio; by analyzing the change of mean curvature radius, the angles of five feature points, namely -50°, -35°, -0°, -50°and 90°, of shoulder cross-section curve were determined, and the corresponding curvature radius were extracted, that is, RC_{-50°, RC_{-35°, RC_{0°, RC_{50°, RC_{90°.

2) In this paper, we established a method to subdivide the male shoulder shape by using the curvature radius and the ratio of vector to frontal diameter of the characteristic points of the cross-section curve; by using the K-means clustering method and variance analysis, we divided the shoulder shape into four categories, quantitatively distinguished the morphological differences of the cross-section curve, and obtained the number of different types of people;

3) The ELM recognition model of young male's shoulder shape was constructed by using MATLAB software, and the recognition rate was 94.00%; the recognition rate of the type 1 was 91.67%, the type 2 was 100%, the type 3 was 92.60%, and the type 4 was 100%; the model had high recognition rate and excellent performance, which provided a new method for human body shape recognition.

6. Acknowledgement

This research was financially supported by the Program of Fujian Natural Science Foundation (No. 2020J01848), the Open Project Program of Fujian Clothing Industry Technology Development Base, Minjiang University, China (No. MJXY-KF-201907), the Open Project Program of Fujian Province University Engineering Research Center of Textile and Clothing, Minjiang University, China (No. MJXY-KF-201902), the Science and Technology Pre-research Program of Minjiang University (No. MJY17005), and the Program of Fuzhou Science and Technology

Bureau (No. 2018-G-97).

References

[1] Fan J, Yu W. Clothing appearance and fit: science and technology. Woodhead, Cambridge: 2004.
[2] Arzu V, Ender B. Identification of female body shapes based on numerical evaluations. International Journal of Clothing Science and Technology: 2011; 23 (1): 46-60.
[3] Young L, Yun J. Classification of upper lateral body shapes for the apparel industry. Human Factors and Ergonomics in Manufacturing & Service Industries: 2010; 20 (5): 378-390.
[4] Yang YC, Chen MZ, Zou FY. Female body shape parameters analysis based on 3D scan data. Journal of Textile Research: 2009; 30 (8): 117-122.
[5] Zhang JH, Wang HF. Study of womens' shoulder based on 3D body measurement. Journal of Textile Research: 2011; 32 (5): 95 -97.
[6] Huang CY, Cui LN. The subdivision of men's neck shape based on 3-dimensional anthropometry. China Textile Leader: 2008; 2: 94-95.
[7] Gu L, Zhang X. Curve form and classification method of youth female based on three dimension. Journal of Beijing Institute of Clothing Technology: 2008; 28 (1): 22-25.
[8] Lee H, Kyunghi H. Measurement protocol of women's nude breasts using 3D scanning technique. Applied Ergonomics: 2004; 35: 353-359.
[9] Kada S, Cynthia LI. Female Figure Identification Technique (FFIT) for Apparel part 2: development of shape sorting software. Journal of Textile and Apparel Technology and Management: 2004; 4 (1): 1-14.
[10] Wang YX, Li XJ, Liu H. Fuzzy clustering analysis of body type in 2D non-contacted body measurement system. Journal of Textile Research: 2007; 28 (2): 100-103.
[11] Qian XN, Yin B. Research on subdividing of human body recognition based on 3D body measurement. Journal of Textile Research: 2011; 32 (2): 107-111.
[12] Zou FY, Ding XJ, Zhang SY, et al. Application of neural network to identification of young females' body type [C]//LEE Tsutian, ZHOU Mengchu. 2006 IEEE International Conference on Systems, Man and Cybernetics. Taipei: IEEE Press: 2006; 2257-2261.
[13] Li XJ, Wang YX, Liu H. Classification and identification of body type in non-contacted body measurement system. JOURNAL TIANJIN POLYTECHNIC UNIVERSITY: 2007; 26 (5): 33-35.
[14] Zhang SY, Zou FY, Ding XJ, et al. Research on the young women's body classification based on SVM. Journal of Zhejiang Sci-Tech University: 2008; 25 (1): 41-45.
[15] Chen MZ, Zou FY, Zhang WY. Design and implementation of an expert system for female figure identification. Journal of Textile Research: 2009; 30 (11): 141-144.
[16] Jin JF, Yang YC, Zou FY. Developing an intelligent model for construction a hip shape recognition system based on 3D body measurement. Fibres & Textiles in Eastern Europe: 2016; 24 (5): 110-118.
[17] Wu XG, Wang DQ, Sheng H. An algorithm and implementation for obtaining minimum extrerior rectangle of image region. Computer Engineering: 2004; 30 (12): 123-125.
[18] Shen JH. Numerical calculation basis. 2nd ed. Shanghai: Tongji University Press, 2006:37-45.
[19] Department of Applied Mathematics, Tongji University. Advanced Mathematics. 5th ed. Beijing: Higher Education Press: 2002; 163-171.
[20] Deng WY. Research on method of 3D personalized human body modeling based on photos and neural network. Zhejiang: Zhejiang University: 2008; 61-62.
[21] Liu H, Zhang M, Chen DS. Study on bust curve of standard female figure. Journal of Textile Research: 2011; 32 (2): 117-120.

Construction and Application of Parameterized Model of Cheongsam Clothing Model

Xiao-Yan Zhuang, Long Wu*, Bo-an Ying, Jing Qi, Xin Zhang

Apparel & Art Design College, Xi'an Polytechnic University, Xi'an, Shaanxi,710048, China

*Corresponding author's email: wulong7398@163.com

Abstract

In order to better realize the application of customized digital design in cheongsam model design and solve the problems of long periodicity and poor fit of traditional cheongsam model generation methods, we propose a knowledge-driven digitalization based on digital technology and means. The method of cheongsam model design realizes the custom design of "one person, one version" cheongsam. First, analysis and research was conducted on the knowledge of cheongsam garment model design; then, a parametric model of cheongsam garment model was constructed according to the principles, processes, methods, and experience of fashion model design. Finally, static analysis was performed on the constructed cheongsam garment model parameterized model Dynamic virtual try-on evaluation and practical application. The verification results show that the design of the parametric relational model of the cheongsam garment model is reasonable, and it can provide a relational model for the automatic generation of the customized model of the cheongsam.

Keywords: Cheongsam; Clothing Model; Parametric Model; Relational Model

1. Introduction

With the continuous improvement of people's living standards, consumers increasingly pursue personalized clothing customization with good quality, strong fit, style fit. Mass production of ready-to-wear garments has been unable to meet the growing individual needs of consumers. Therefore, the research on clothing technology that combines personalized clothing products with intelligent manufacturing technology has become more meaningful.

At present, there are two main ways to realize the automatic generation of digital templates: the first is to use clothing CAD software to complete plate making by using the relationship between human body data, clothing styles, and model structure. There are mainly three methods: semi-automatic traditional release Code technology, parameterized pattern, and intelligent pattern design. Semi-automatic traditional grading technology [1] By combining traditional plate making methods with digital technology, the semi-automatic generation of templates is realized; parametric pattern design realizes automatic generation of clothing templates through parameters or other parameterized expressions [2, 3], Bao Chen et al. [4] proposed an interactive 2D-3D clothing parametric model generation method and established a clothing design system; intelligent pattern generation technology is the use of intelligent technology to build a clothing pattern knowledge base. According to logical reasoning, the process of garment plate making [5, 6] is obtained, to realize the automatic generation of flat garment model. It is an application of artificial intelligence in clothing. Hu et al. [7] proposed an interactive, collaborative Garment pattern design system, Inui [8] used an iterative genetic algorithm to develop a CAD system to create users' favorite garment designs. The automatic generation technology of garment patterns based on two-dimensional CAD software relies on the rich experience and knowledge of the pattern maker, but it is simple, effective, and widely used; however, it is very difficult to develop and implement intelligent pattern design, and there is still a certain distance to realize practicality.

The second is the curved surface development technology, which expands the 3D model to form a 2D template. There are two main ways: build a 3D model of the human body based on a 3D scanner and then expand it into a 2D pattern. Hyo Kim C et al. [9] use the free deformation method to deform the predefined clothing model and flatten the model to generate a two-dimensional model of the clothing; The other is like a virtual cutting method, first construct the clothing model in the three-dimensional virtual space, and then Directly derive the two-dimensional clothing model from the clothing model [10]. The clothing model development technology based on the three-dimensional human body model relies solely on computer simulation technology. There is inevitably the problem of curve deformation in the mapping process, so it cannot be used well in the generation of clothing models [11].

In order to improve the accuracy and efficiency of the clothing model, this paper proposes a way to provide patternmakers with knowledge-driven intelligent design of digital clothing model based on digital technology and means and construct the parametric relationship model of cheongsam clothing model to realize cheongsam. The automation of prototype design knowledge meets the needs of different customers in a high-quality and low-cost manner to achieve a rapid response to the rapid changes in the market.

2. Knowledge Analysis of Model design of Cheongsam Clothing

Cheongsam is one of the cultural products of historical development, which can represent traditional Chinese clothing. The changes in cheongsam styles mainly focus on collar type, sleeve type, lining, dress length, skirt hem, gusset position, craftsmanship, etc. [12, 13]. Table 1 shows a collection of cheongsam style elements.

Table 1. Cheongsam style elements collection

Location	Style
Sleeve Type	Sleeveless, wiping sleeves, lantern sleeves...
Collar	No collar, stand-up collar, flat collar, square collar, lapel...
Lining	Single, double, straight, curved, pipa, big round, double round...
Length	Super short, short, medium long, long, super long...
Skirt	Wide pendulum, straight pendulum, fishtail pendulum, A-shaped pendulum...
Slot	No splits, single-sided splits, double-sided splits, high splits, low splits...
Craft	Selvage, piping, welt, buckle, embroidery...

Cheongsam's structural design mainly has the following elements: neck circumference, shoulder width, bust circumference, waist circumference, hip circumference, dress length, sleeve length, waist height, and hem. Through the analysis and comparison of the style and structure of cheongsam, it can be found that the cheongsam body structure has some common characteristics: there are front and back bodies, which are connected by shoulder seams; there is a provincial design on the garment, and in general, the front piece is mostly used. The design of "waist province + armpit province", while the back panel is only a waist province design; tightening the waist and highlighting the chest is the biggest feature of the cheongsam structure; the sleeve line design is related to the sleeve shape of the cheongsam; the slit design makes the cheongsam and women. The perfect combination of body shape features fully reflects the beauty of female body [14].

Cheongsam structure plane drawing methods are divided into two types: indirect drawing method and direct drawing method. The indirect drawing method also includes the prototype method and the basic method. The direct drawing method can also be divided into the proportional and actual sizes [15]. The prototype method is simple and highly applicable but requires rich experience. The basic method is to select the garment model closest to the needs, and make local adjustments on this basis. The actual measurement method is to make a plane drawing of multiple parts of the actual measurement, with more data and a high degree of integration. Proportional cartography is based on the bust circumference of the human body and uses proportions to calculate the dimensions of other parts for cartography. The formula is large in calculations, and changes in looseness are not easy to grasp. The qipao mapping method in this article also uses the proportional method.

3. Establishment of the Parametric Relationship Model of the Cheongsam Model

3.1 Principles and Methods of Parametric Modeling of Cheongsam Version

The clothing model design model is to transform the clothing model design process into a clothing model design principle and method that is digitally expressed in the computer, that is, to establish a data model in the computer. This article uses this technical method to design the parametric relationship model of the cheongsam model. The construction platform of the cheongsam design model is the Auto Pattern plate-making system researched by the "Shaanxi Apparel Intelligent Design and Manufacturing Engineering Technology Research Center."

In the clothing model modeling process, there are essentially two aspects. One is the construction of clothing model design models, which is also a critical step; the other is the use of clothing model design models, applying the built model, and inputting different for human body data, the system generates corresponding clothing templates, and the design parameters can also be modified according to needs.

The specific steps include category definition, classification definition, clothing component definition, global variable setting, and clothing model process modeling. As shown in Fig.1 below, in the design interface of the "model building" part, a database of women's cheongsam is established, and a tree-shaped component list is generated, including C piece, back piece, front piece, collar, and a model relationship model of the cheongsam is established on the relevant component design interface. Select the model relationship model to be designed in "Smart Plate Making," generate the model of the required style by inputting the selected parameters and interactive size modification, and output the model design result.

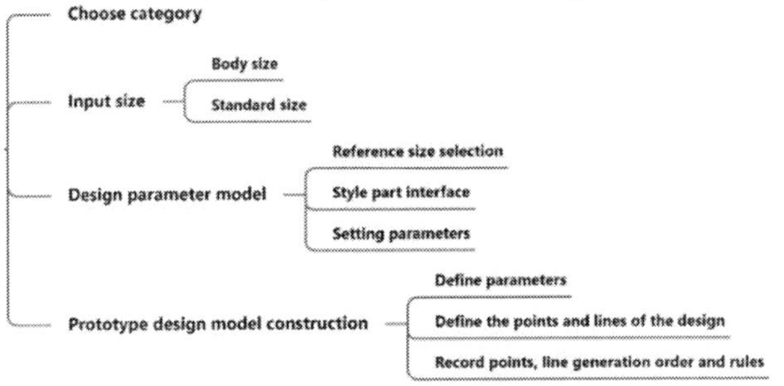

Fig.1 Auto Pattern modeling process

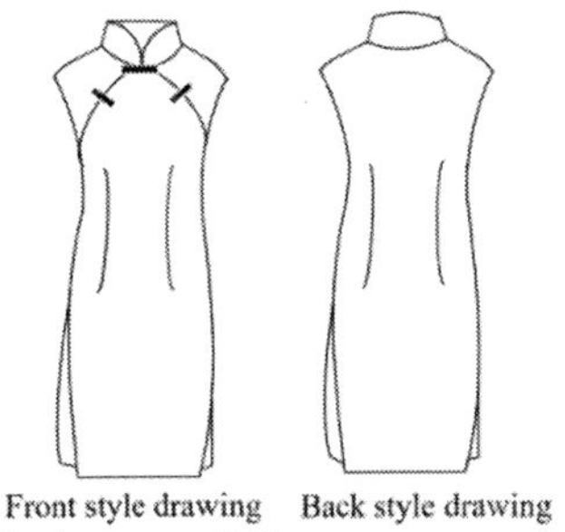

Fig.2 Style map of cheongsam version

When using the constructed clothing model, the user only needs to select the customer, select the clothing category, select the style part, and enter the body shape. The system automatically completes the calculation, drawing, and other design work according to the selected input data and outputs the clothing that meets the design requirements. Sample graphics.

3.2 Construction of the Parametric Relationship Model of Cheongsam

The proportional method is currently the most used clothing pattern-making method. In this study, the proportional method is used to design and revise the cheongsam model, and the empirical values and rules of cheongsam pattern-making are obtained. Based on this, the parametric model of the cheongsam model is constructed - Design method for analysis.

Fig.2 is a more common style of cheongsam selected in this study, a sleeveless short cheongsam. The size data used in cheongsam plate making is based on the design experience to design the 160/84A platform.

To realize the parametric design of a cheongsam model, we must first establish a cheongsam sample template. The design of the model is based on the parameterized template. To design of different sizes of the same board, we only need to design according to any specifications or make local adjustments. The prerequisite for the establishment of the model parameterized relationship model is the determination of the model parameters. The parameters in the model can be divided into key parameters, secondary parameters, and composite parameters. Tables 2, 3and 4 are the design parameters used in the cheongsam model design system.

Table 2. Key parameters of human body in cheongsam design

Number	Human Body Parameters	Variable Name
1	Bust chest circumference	AB2
2	Shoulder width	AB10
3	Neck circumference	AN1
4	Length of neck and waist section	AL12
5	Waist circumference	AW1
6	Hip height	AH4
7	Hip circumference	AH2

In the design of the cheongsam pattern, adding input parameters interactively can better extract the experience of the model, and the size selection range of the interactive value can enable different model designers to modify and adjust the model according to their design experience and customer requirements.

Table 3. Description of interactive input size information in cheongsam model design

Number	Name	Defaults (cm)	Size Range (cm)	The main factor of Influence
1	V Change in length	33	[30,35]	Style
2	V Bust loose volume	12	[10,14]	Clothing fit inside the fabric
3	V Shoulder looseness	2	[1,3]	Shoulder movement range
4	V Neck loose volume	5	[3,7]	Collar shape
5	V Waist loose volume	18	[8,20]	Style and beauty
6	V Loose hips	2	[1,6]	Hip fit

Table 4. Compound parameters in the design of cheongsam model

Number	Parameter Name	Variable Name	Calculation Formula
1	V dress length	VL	AL12+AH4+V Change in length
2	V clothing bust	VAB	AB2+V Bust loose volume
3	V shoulder width	VAB10	AB10+V Shoulder looseness
4	V waist	VAW1	AW1+V Waist loose volume
5	V neckline	VAN1	AN1+V Neck loose volume

Because of the differences in the plate-making methods and language expressions adopted by different pattern designers, before establishing the parameterized relationship model of the cheongsam garment plate-making process, the following aspects of pre-processing should be carried out on the template design process.

First, all points and lines required in the plate making process must be named; second, if there are constraints (numerical values, geometry, formulas) in the process of generating each point and line, they must all be noted; third, the formation of a cheongsam All the structural lines of the must be named one by one according to the process description. Take the preprocessing of generating the description sentence of the back shoulder slash of a cheongsam as an example: take P2 as the starting point and measure AN1/5-0.5 fixed point P3 horizontally to the right, connect P2 and P3 as L and then open the collar line horizontally, and then use P2 as the starting point is right AB10/2 and then down 3.5 fixed point P4, connect P3, P4 to determine the L back shoulder diagonal.

Table 5. Parameterized relationship model of partial cheongsam back model

Number	Structure Line	Generation Process (unit: cm)
1	Back length	Take the point P1 as the starting point and vertically upward the length of V to point P2, and connect P1P2 to the rear centerline of LV.
2	Back neckline Back shoulder slash	Take P2 as the starting point and measure AN1/5-0.5cm fixed point P3 horizontally to the right, connect P2 and P3 as L and then open the neckline horizontally, then take P2 as the starting point to the right AB10/2 and then down 3.5cm fixed point P4, connect P3, P4 Determine the back shoulder oblique line of L, and then use P2 to go down 2.5cm to determine the depth of the back collar of P5L. P5 stretches one-third of the length of the collar to the right. Set the point Pa. Connect the three points of P5, Pa, and P3 to draw the arc of the L back neckline.
3	Back wide line Back bust line	Take P2 as the starting point and go down 24.2cm to the fixed point P6, and then horizontally to the right AB2/4 to the fixed point P7, connecting P6 and P7 to the rear bust line, P6 horizontally to the right 0.15*AB2+4.5cm fixed point P8, P8 upwards 18cm fixed point P16, Connecting P8 and P16 is the wide rear line of L back.

Due to the complex relationship between dots and lines in the process of cheongsam plate making, in order to make the relationship model tidier, it is divided into three modules, the back piece, the front piece, and the collar, to construct the parameterized model. The points, lines, and directions in the plate making process are the three basic elements in the process of template formation, and the data (numerical value, size formula, plate making formula), parallel, vertical, and the sequential relationship of the process direction are the relationships formed by connecting the three elements - bridge. Therefore, the plate-making knowledge involved in the plate-making process includes the data between dots and lines, the relationship between parallel and vertical, and the process sequence of plate-making.

The parametric relationship model of the back piece of the cheongsam is composed of 25 points, 11 straight lines, and six curves. It is confirmed that there are 24 size data of the back piece. As shown in Table 5, it is the generation process of the parametric relationship model of part of the model of the cheongsam back piece. The generation process of the parametric model in the table is drawn using the proportional method, and the style and empirical values determine some specific parameters.

The establishment of the parametric relationship model between the front piece and the collar of a cheongsam is the same as the back piece.

3.3 Realization of the Parametric Relationship Model of the Cheongsam Model

The parameterized relational model constructed above is established in the Auto Pattern plate-making system. First, define the category, classification, and clothing components; then, set the global variable parameters of the cheongsam; finally, model the cheongsam model process and input the parameterized relationship model of the cheongsam model.

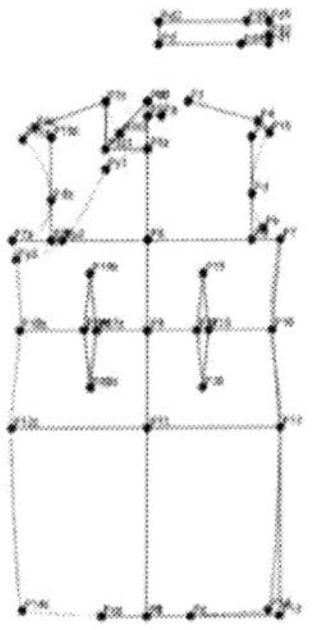

Fig.3 Cheongsam model automatically generated in Auto pattern

In the cheongsam process modeling, the pixels of each component are defined, and the generation order, generation rules, and constraint relationships of the pixels are recorded. The modeling process is based on the knowledge of the clothing model. According to the geometric topology of the clothing model parts, the model design and editing are carried out to meet the actual assembly relationship of the model, including the use of different rules (formula, interaction, internal division, etc.) to determine the key point location, and use different rules (line segments, curves) to connect key points to complete the drawing of model structure lines and components. Fig. 3 shows the cheongsam model automatically generated in Auto Pattern.

4. Verification and Application of the Parametric Relationship Model of Cheongsam Version

4.1 Verification of the Parameterized Relationship Model of Cheongsam

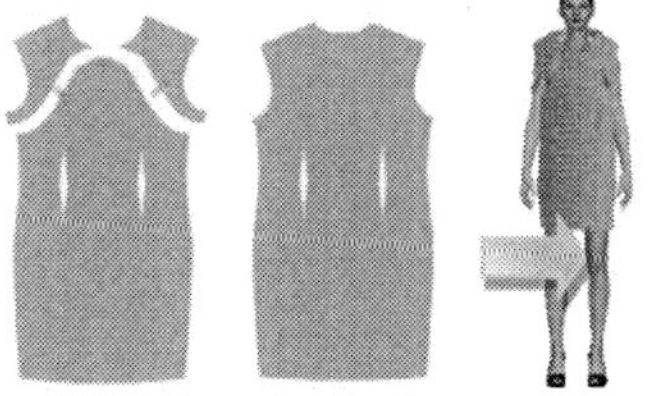

Fig.4 Virtual stitching process

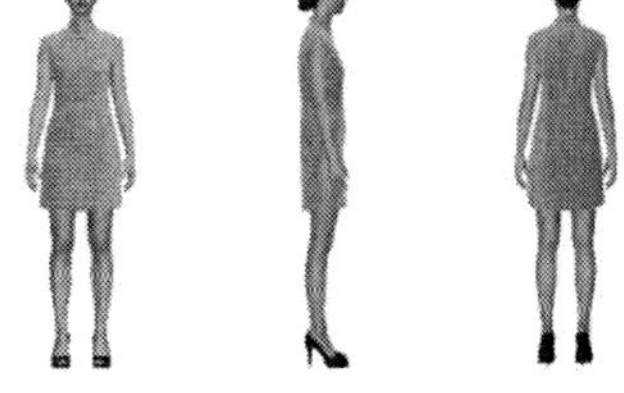

Fig.5 Virtual try-on process

Import the generated cheongsam model into the three-dimensional virtual design software CLO 3D for virtual try-on, as shown in Fig.4. In order to ensure the rationality and scientific of the verification process, the virtual model is transferred into the three-dimensional data size of the characteristic human body 160/84A female human table, and the virtual fabric is set to mulberry silk satin in the physical properties window to simulate the fabric and the human body. During the test, observe the accuracy of the position of each suture, mainly check the shoulder line, side suture, and back centerline, and secondly check the looseness of the chest, waist, hip, neckline, and armholes. It is necessary to ensure the functionality of the human body. It is also necessary to fit and be comfortable, and finally observe whether the clothing is smooth and beautiful, as shown in Fig.5.

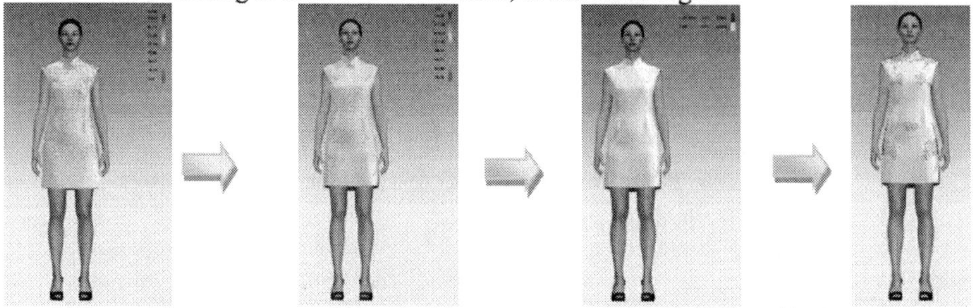

(a)Stress Map (b)Strain Map (c)Fit Map (d)Show Pressure Points

Fig.6 Garment Fit Maps

Cheongsam is a more considerate garment. When performing a virtual try-on, we can verify the fit of the clothing through four aspects: Stress Map, Strain Map, Fit Map, and Show Pressure Points display. The three-dimensional virtual fitting software used in this article comes with the detection functions of these four aspects. Fig.6 shows the static virtual detection map of these four aspects. Taking virtual pressure detection as an example, the pressure can intuitively show the size of the clothing pressure and test the fit of the clothing. The gradual change of pressure from blue to red indicates that the pressure changes from weak to strong. The color distribution from dark yellow to red appears when the clothing is worn on the human body, which indicates that the pressure of the clothing is too high. The appearance of a yellow area indicates that the pressure comfort zone of the clothing is exceeded, and a blue area appears. It shows that there is no apparent pressure on the human body wearing clothing. We can also view the pressure range value from the garment fit properties in Fig.7. In addition to static virtual pressure detection, we can also observe clothing fit through dynamic virtual pressure detection. As shown in Fig. 8, the dynamic catwalk virtual pressure detection, the cheongsam, and the human body contact surface are blue during the dynamic catwalk; the result shows that the virtual model wears the cheongsam in a comfortable and stress-free state, and the cheongsam model fits well.

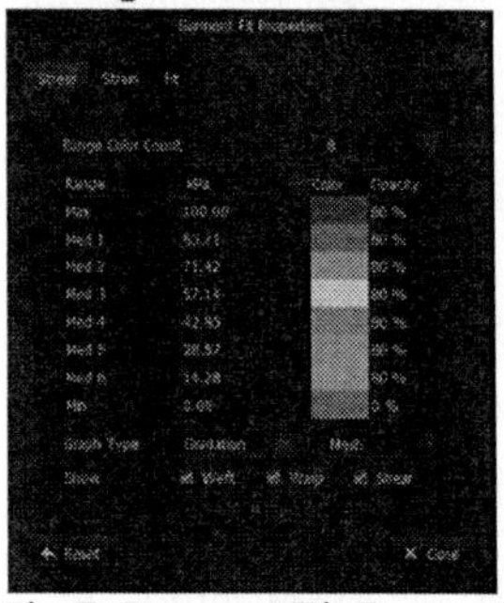

Fig.7 Garment Fit Properties

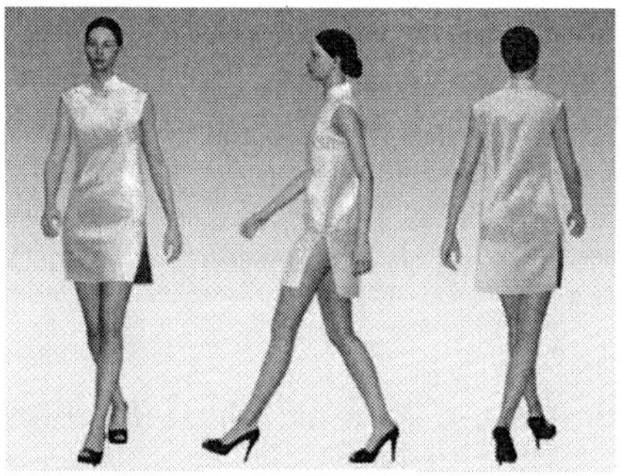

Fig.8 Dynamic catwalk pressure detection

Check the fit of the parametric model of the cheongsam clothing model through the virtual fitting process, trace the roots, and find out where the cheongsam model does not fit. After that, the parametric model of the cheongsam garment model needs to be revised. The revised cheongsam clothing model has a good fit, which verifies that the design method of the cheongsam clothing model is reasonable.

4.2 Application of Cheongsam Parametric Relation Model

When using the cheongsam model design model constructed in this article, the user only needs to select the user who has entered the size data, select the clothing category, select the style part, enter the body shape, and the system automatically generates calculations, drawings and other designs based on the selected input data. Work, output the cheongsam model graphics that meet the design requirements. For example, select the customer "Liu Chang" in the interface, select the clothing category "Women's Cheongsam," and select the cheongsam model automatically generated by the style component "Short Cheongsam."

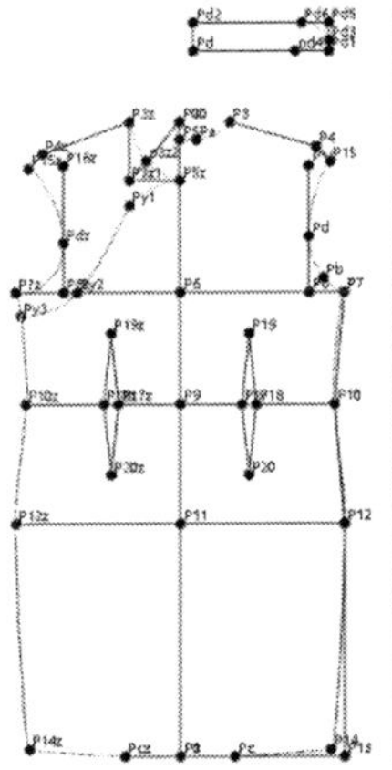

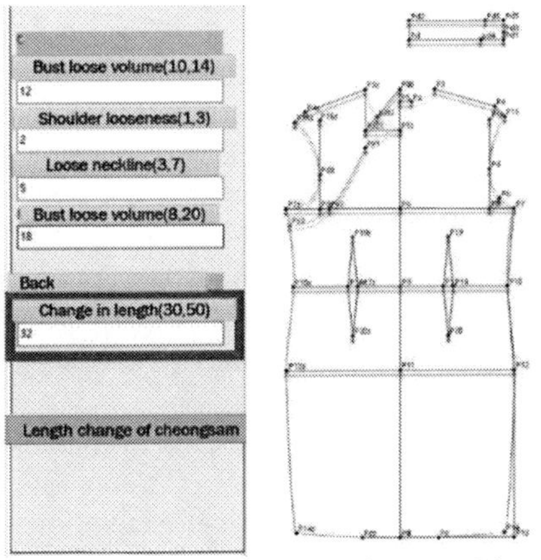

Fig.8 Cheongsam model graphics Fig.9 Parameterized modification graphics of cheongsam model

When the designer needs to modify the design parameter value according to consumer demand, he only needs to modify the interactive design parameter box in the system. After confirming the modified parameter, the system automatically completes the calculation, drawing, and other template design work according to the modified parameter. For example, to modify the adjustment parameter of cheongsam dress length, we only need to change the data from 33 to 32 in the dress length adjustment box, and press Enter to confirm the modification of the parameter. Figure 3-6 shows the automatic change after shortening the dress length parameter - the generated cheongsam models.

5. Conclusion

This paper proposes a knowledge-driven digital cheongsam model design method using digital technology and means. This method combines clothing model design knowledge with digital technology to construct a parametric model of the cheongsam model. The static and dynamic virtual fitting pressure detection shows that the design method of the parametric relationship model construction of the cheongsam garment pattern is scientific and effective. From the perspective of the entire design process, this method can adjust the pattern in real-time, quickly conduct comparative experiments, and confirm the wearing effect of the clothing. The construction of the parametric model of cheongsam in this article takes the standard body as the research object and does not consider the body shape of the unique human body. Further research can be done later.

6. Acknowledgement

This research work is supported by the (National Treasury) Personalized Silk Cheongsam Virtual Dynamic Simulation Research (Project No. 11310031252003) and the Shaanxi Apparel Intelligent Design and Manufacturing Engineering Technology Research Center.

References

[1] Zou QZ, Xie H, Li J. Tailor-made technology for women's suits based on different body types. Shanghai Textile Technology: 2010; 38 (4): 8-10.
[2] Zhao XM. Research on sleeve structure technology based on parameterized features, Shanghai Donghua University: 2007.
[3] Kim S. Simulation of bespoke garments using parametrically designed patterns. IJCST: 2012; 24 (5): 350-362.
[4] Bao C, Miao YW, et al. 3D interactive garment parametric pattern-making and linkage editing based on constrained contour lines. IJCST: 2021; 2.
[5] Noor A, Saeed MA, et al. A review of artificial intelligence applications in apparel industry. The Journal of The Textile Institute: 2021; 24 (5): 350-362.
[6] Kang Y, Oh J, Kim S. Development of parametric garment pattern design system. IJCST: 2021.
[7] Hu ZH, Ding YS, Zhang WB, Yan Q. An interactive co-evolutionary CAD system for garment pattern design. Computer-Aided Design: 2008; 40 (12): 1094-1104.
[8] Inui S. A Combined System of Computer Aided Design and Genetic Algorithm for Apparel Designing. Sen' I Gakkaishi: 1996; 52 (11): 605-611.
[9] Kim CH, Sul IH, et al. Automatic basic garment pattern generation using three - dimensional measurements. IJCST: 2010; 22 (2/3): 101-113.
[10] Liu K, Zeng X, et al. 3D interactive garment pattern-making technology. Computer-aided Design: 2018; 104: 113-124.
[11] Xu Q. Research on parametric prototype model generation system in virtual environment, Zhejiang Sci-Tech University: 2019.
[12] Chen RF, Chen WR. Research on the evolution of cheongsam and structural design changes. Journal of Zhejiang Sci-Tech University: 2007 (02); 155-159.
[13] Cui M. Research on the structure design of cheongsam based on neural network, Qingdao University: 2020.
[14] Ma Q. Research on women's clothing style design method based on personalized three-dimensional human body model, Xi'an Polytechnic University: 2017.
[15] Zhang WB. Clothing structure design. China Textile Publishing House: 2013.

Hunchback Body Type Classification

Ru-Yi Yang[1], Yi-Fan Chen[1], Ming-Hai Cui[1*]

[1]*Beijing Institute Of Fashion Technology, Sakura East Street, Beijing, 100029, China*

[*]Corresponding author's email: fzycmh@bift.edu.cn

Abstract

The main purpose of this research was to develop a method for the body type classification of Hunchbacks, using a three-dimensional body scanner and anthropometric measurement data. A sample contains 102 female test subjects, aged 18 to 25 years. Anthropometric body measurement was conducted. The body samples were divided into three categories according to the size of the upper back angle. The three categories were chest-out, normal, and hunchbacked. Professionals in the apparel industry were invited to conduct a visual assessment to examine classification result, the result shows that body shape classification based on the upper back angle is precise, with an accuracy percentage of 93.3%. By using the K-means cluster analysis method, hunchbacked bodies were divided into five categories according to five body surface angles.

Keywords: Hunchbacked Body Type; Body Surface Angles; Body Type Classification; K-means Cluster Analysis

1. Introduction

Due to improper postures, many people's front and back areas of their spine accumulates fat resulting in a condition known as hunchback. There is a certain difference between a hunchbacked body type and the standard body type. Therefore, there will be a certain discomfort when wearing existing clothing, as such there is a necessity to subdivide the existing body type classification system.

The problems of people's daily postures are the focus of researchers and experts from various fields such as medical sciences, kinesiology and anthropology, given the importance of body posture for every individual and the serious consequences of improper body posture [1-4]. The concept of the hunchbacked body type are varied in literature. According to Walk G [5], the correct posture of the human body is an upright position with relaxed arms at the side and palms pointed forward. Characteristics of correct body posture can be seen from the anteroposterior and sagittal views. Viewed from the front side, all individual body parts should be in perfect symmetry, and in the sagittal view all segments should be aligned with the gravity line positioned through the center of the body. Barrow and Mec Gi [6] developed a method for the examination of school-aged children to conduct body posture assessment in the sagittal plane. The method is based on a comparison of body parts in the lateral and in the anteroposterior position with a particular card representing each of 13 body parts.

Fengqin X [7] used a mathematical method to convert the lateral curves into an angle function curves on 60 longitudinal sections of the scanned human body models. Fengqin X used Cluster analysis to divide the human bodies into 4 categories which is described by the differences in the back form, the center line of the back and the lateral thickness of abdomen. Mi Kyung Yoon [8] used the 3D data from 317 men to defined and measure six space vector angles of the lateral forms. According to the upper lateral somatotype classification method and using a prediction model to classify 317 Korean men. As a result, 67.82 percent of samples were found to be of the straight type, 9.15 percent were of the swayback type, 10.73 percent were of the bend-forward type, and 12.30 percent were of the back-forward type. Young Lim Choi [9] and other experts visually classified the lateral body shapes of 246 women aged between 18 and 49. the 246 three-dimensional scan data were classified into the straight type (28.7%), swayback type (24.3%), lean-back type (22.8%), and bend-forward type(24.3%). According to Müller [10], Armstrong [11], Rasband [12], Nam [13], and Jung [14], the upper lateral somatotype can be categorized into three to five types. Maja Mahnic Naglic [15] developed a method for the classification of body posture types. According to the defined posture indicators, the developed method differentiates three different types of upper posture and three types of lower body posture. Classification of the whole body posture can be achieved by combining the two proposed methods. In fact, three-dimensional properties of a human body inherently relate to both lateral body shapes and frontal shapes. It means that three-dimensional properties of the frontal body shapes may contaminate the classification of lateral body type. Accordingly, it is necessary to employ only the lateral properties of a human body for the classification of lateral body shape[8].But, little attention was paid for the classification of the lateral body-type among most previous studies when using the statistical analysis methods.

2. Problem Formulation

2.1 Subjects and Instruments

200 non-pregnant young women aged form 18 to 25 years old and height between 160±2.5cm were selected to be the subjects of this experiment. Body scanning experiment performed at a laboratory with an ambient temperature of 26±2°Cand relative humidity is (60±10)% . According to National Standard [16] requirement, subjects stand naturally and look straight ahead, wear light colored, tight underwear, and a swimming cap, without any accessories. The specific posture is shown in Fig.1.

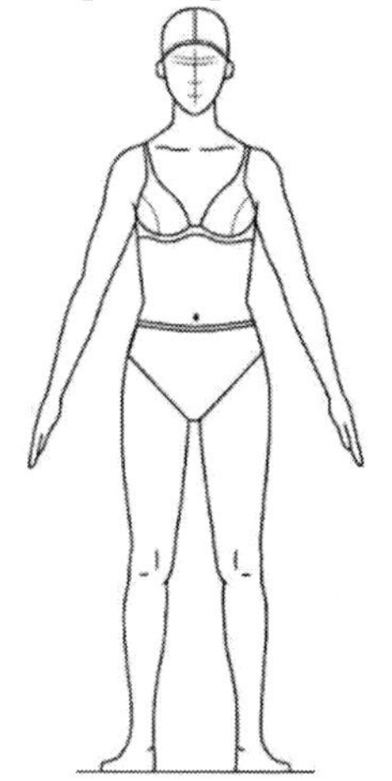

Fig.1 Scanning gesture.

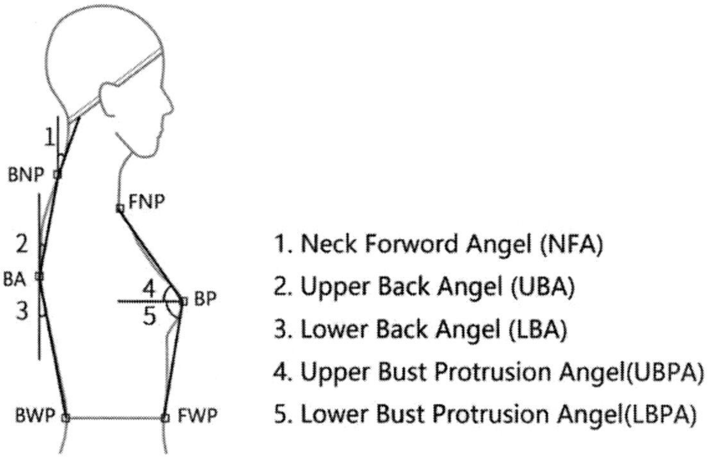

Fig.2 Lateral back angle measurement.

2.2 Lateral Back-Angle Measurement

At this step, data on a body shape is acquired by using a 3D scanner named VitusSmart XXL. Then, the data from the scanned body shape is imported to another software, called RapidForm, to edit and correct the defects of the 3D meshed object fundamentally.

The five lateral angles were measured from the lateral silhouette of the three-dimensional scan data. For angle measurement, the front neck point(FNP), back neck point(BNP), back protrusion(BA), bust point(BP), front waist point(FWP), and back waist point(BWP) were set as basic measurement points. The definition of each lateral surface body angle can be seen in table 1. The lateral angles were measured on the basis of the measurement points as shown in Fig.2.

Table 1. Definition of lateral angle

Name	Measurement Points	Definition
NFA	BNP	The angle between the tangent through BNP and the plumb line
UBA	BNP, BA	The angle between the line through BNP and BA and the plumb line
LBA	BA, BWP	The angle between the line through BA and BWP and the plumb line
UBPA	FNP, BP	The angle between the line through FNP and BP and the horizontal line
LBPA	BP, FWP	The angle between the line through BP and FWP and the horizontal line

2.3 Data Preprocessing

In order to ensure the data reliability, before analyzing the relationships of lateral body angles, the data is pre-processed, and four wrong or unusual data were deleted. The information of 200 samples are shown in Fig.3. Using the Q-Q graph to do a normal test on remaining 196 samples, and the data conform to the Normal distribution, which can be seen in Fig.4.

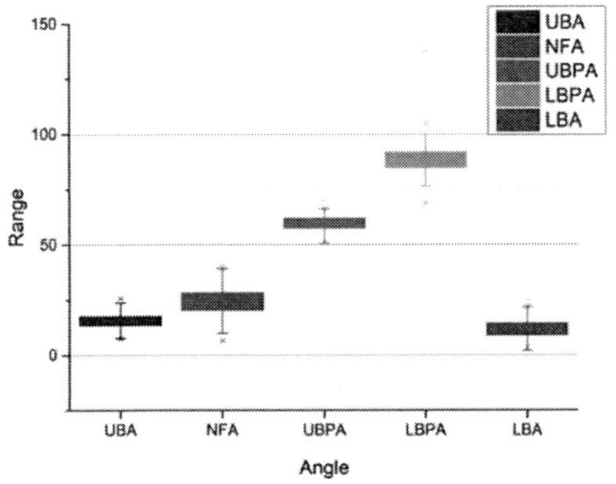

Fig.3 Singular value.

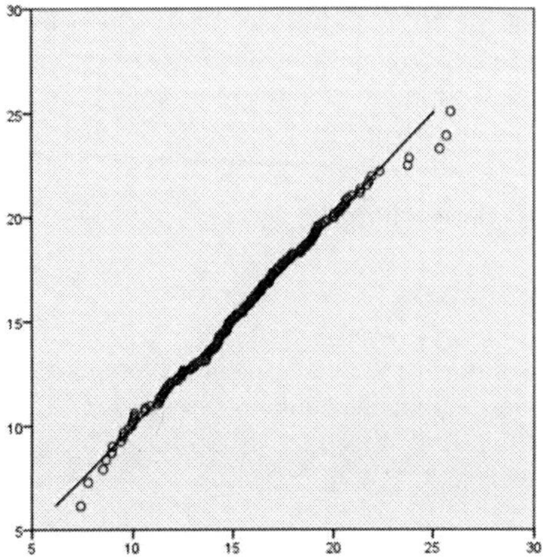

Fig.4 Q-Q graph.

2.4 Body Type Division and Analysis

The upper back angle (UBA) can accurately reflect the characteristics of the back [17], so choose the UBA as the criterion to distinguish whether it belongs to hunchbacked body type. Descriptive statistics of the UBA is shown in table 2. The mean and standard deviation of UBA are used to limit the range of the normal body type. If it is greater than this range, it belongs to the hunchbacked body type, and it belongs to the up-chested body type when it is less than or equal to this range. The range of each body type are shown in Table 3. In order to verify the accuracy of this discriminant method, a team includes five experts conducted a body-shape classification by visual assessment. The result shows that the similarity of both classification methods reaches 93.3% which proves using UBA to classify body type is feasible and accurate.

Table 2. Descriptive statistics

	Min	Max	Mean	SD
UBA	7.38	25.88	15.61	3.47

Table 3 Definitions and proportions

Type	Range of UBA	Sample Size	Proportion(%)
up-chested	[7.38, 12]	31	15.82
normal	[12, 19]	134	68.36
hunchbacked	[19, 25.88]	31	15.82

Table 4. The distribution of samples in "National standard size + new standard size"

Type	Y		A		B		C	
	PCT.1(%)	PCT.2(%)	PCT.1(%)	PCT.2(%)	PCT.1(%)	PCT.2(%)	PCT.1(%)	PCT.2(%)
up-chested	22.58	17.5	61.29	15.97	16.13	14.29	0	0
normal	20.15	67.5	59.70	67.22	18.66	71.42	1.49	100
hump-backed	19.35	15.00	64.52	16.81	16.13	14.29	0	0

Note. PCT.1 indicates the proportion of each national standard body type in the new body type. PCT.2 indicates the proportion of each new body type in the national standard body shape.

In order to analyze the distribution of all samples, the three new body types were combined with the current national standard body types. Table 4 shows the distribution of "National standard size + new standard size" samples.

We can see from table 4 that in the new body type, the national standard size A has the largest proportion in every new standard size reached about 60%. The second proportion is the national standard size Y reached about 20%. Among every national standard body types, the normal body type accounts for the largest proportion, reaching about 70%, and about 15% people belong to hunchbacked body type. that means the clothes made according to existing national standard size are unsuitable for the 30% people. So it is necessary to subdivide the body shapes and improve garment prototype for the those people. The next part of this paper is to subdivide the humpbacked body type.

3. Problem Solution

3.1 Data Analysis

In order to improve the subsequent prototype of the hunchbacked body shape, the current 31 hunchbacked body samples were subdivided again. Correlation Analysis between body surface angles is the basis of body type classification. SPSSWIN 12.0 (SPSS Inc., Chicago, IL) was used for this statistical analysis method. the result can be seen in Table 5.

Table 5. Correlation analysis

	NFA	UBA	LBA	UBPA	LBPA
NFA	1	-	-	-	-
UBA	-0.217	1	-	-	-
LBA	0.299	-0.414*	1	-	-
UBPA	0.071	-0.256	-0.166	1	-
LBPA	0.412*	-0.284	0.298	0.188	1

Note. * represent significant correlation i.e. p value <0.05

According to table 5, some conclusions can be list as following.

The UBPA has no obvious relationship with any other angles, indicating that it is an independent characteristic angle which can represent the characteristics of the upper front part of the body.

There is a significant correlation between The NFA and the LBPA(0.412),the UBA and the LBA(-0.414). The relationship between this angles needs to be examined by Partial Correlation Analysis.

Correlation Analysis can represent linear relationship between two variables, but factors would be influenced by the third variable easily by using this method. Partial Correlation Analysis is mainly to analyze the correlation of more than two variables after controlling other variables. This is a good way for main factors to avoid being affected by others. In the case of controlling other angles, the correlation coefficient between the NFA and the LBPA is -0.029, and the correlation coefficient between the UBA and the LBA is -0.086, indicate that there is almost no relationship between them. Partial Correlation Analysis results are shown in Table 6.

Table 6. Partial correlation analysis

Controlled Angle			NFA	LBPA
UBPA,UBA,LBA	NFA	R	1	-0.029
	LBPA		-0.029	1
			UBA	LBA
UBPA,NFA,LBPA	UBA	R	1	-0.086
	LBA		-0.086	1

According to both analysis results, the size of NFA can explain the lean forward extend of neck. the size of UBA can representative the extend of back protrusion, and the size of LBA relate to the lean back extend of the center of gravity. The the size of UBPA can describe the extent of chest protrusion accurately and the LBPA can explain the extent of abdomen protrusion. So chose these five body surface angles to characterize the lateral body shape.

The structure of the subjects was examined by using the k-means cluster analysis method in order to determine posture types, that is, define the number of most various groups of test subjects. Since the method need predefining the desired number of groups, and the number of body type classifications is unknown, so the Sum of Squares (SSE) is used to determine the optimal number of clusters. The formula of SSE is as follows:

$$SSE = \sum_{i=1}^{n} w_i \left(y_i - \bar{y} \right)^2 \quad 1)$$

As the number of clusters increasing, each category contains fewer samples, and the distance within the group is getting closer, so the change in the slope of the inflection point graph is used to define the number of clusters. The result of SSE changes with the number of classes as shown in Fig.5. It can be seen that the number 5 is an inflection point, so the optimal number of clusters is 5. So, 31 hunchbacked body samples were subdivided into 5 categories.

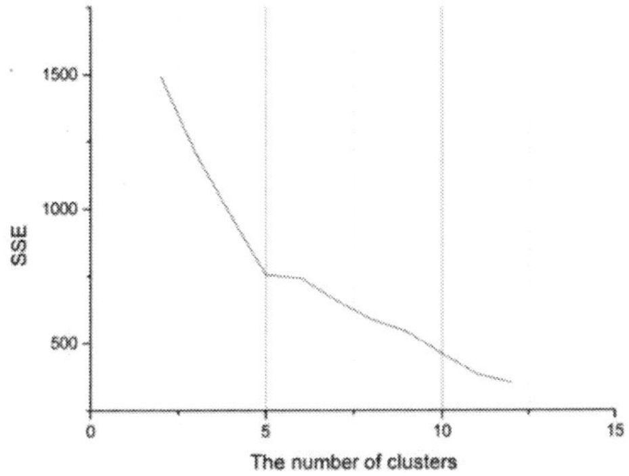

Fig.5 SSE.

3.2 Subdivide Hunchbacked Body Type

Table 6 shows the subdivision result and sample sizes of various body types. According to this result, The right side view of representative samples of each subdivided body type are extracted from 31 hunchbacked body types which is shown in Fig.6.

Table 6. Partial Correlation Analysis

Project (°)	Category				
	1	2	3	4	5
NFA	34.83	22.99	30.52	29.2	19.39
UBA	19.78	21.72	20.40	21.62	21.83
LBA	10.23	8.73	11.34	12.93	10.26
UBPA	63.55	63.35	63.14	55.59	57.06
LBPA	85.14	83.96	93.03	87.63	80.75

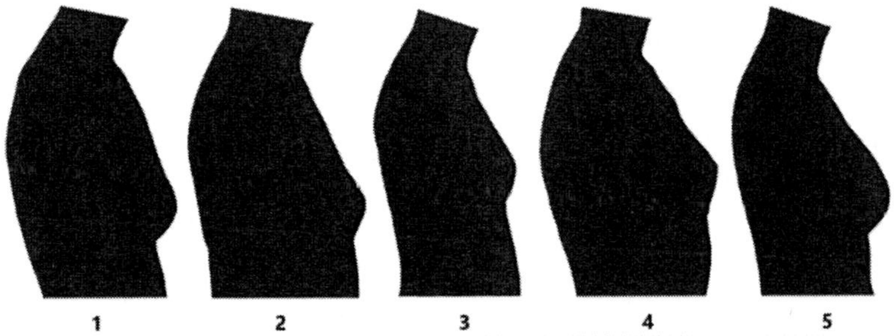

Fig.6 representative Samples of each body type.

From table 6 and Fig.6, we can get some conclusions as follows.

1) During the hunchbacked body type, samples in the 1st subdivided body type have smaller upper back angel (UBA), and have bigger neck foreword angle (NFA), upper bust protrusion angle (UBPA) and lower back angle (LBA) than people in other types. Visually, their neck is tilted forward and the center of gravity is backward. When drawing the paper pattern, it is necessary to heighten the height of the back shoulder line and deepen the depth of the front collar line appropriately.

2) During the hunchbacked body type, people in the 2nd subdivided body type have smaller neck foreword angle (NFA) and lower back angle (LBA), and have bigger upper back angel (UBA) than people in other types. Visually, their back is protruding. When drawing the paper pattern, it is necessary to widen the length of back width line.

3) During the hunchbacked body type, people in the 3rd subdivided body type have bigger neck foreword angle (NFA) and upper bust protrusion angle (UBPA) than people in other types. Visually, their back and abdomen is protruding. When drawing the paper pattern, it is necessary to widen the length of back width line and waist line.

4) During the hunchbacked body type, people in the 4th subdivided body type have bigger neck foreword angle (NFA), upper back angel (UBA) and lower back angle (LBA) than people in other types. Visually, their neck is tilted forward, the center of gravity is backward and their back is protruding. When drawing the paper pattern, it is necessary to widen the length of back width line and increase the size of the waist darts.

5) During the hunchbacked body type, people in the 4th subdivided body type have smaller neck foreword angle (NFA) and lower bust protrusion angle (LBPA), and have bigger upper back angel (UBA) than people in other types. Visually, their back is protruding but not obvious relative to others. When drawing the paper pattern, just widen the length of back width line appropriately is enough.

4. Conclusion

Five lateral body surface angles of young women was extracted, and all of body samples were divided into three categories according to their size of the upper back angle, they are chest-out body type, normal body type, and hunchbacked body type. Professionals in the apparel industry were invited to conduct a visual assessment, the result shows that accuracy of body shape classification based on the range of upper back angle reached 93.3%. Aside form this, we have also counted the proportion of hunchbacks in the national standard body types, it was found that the hunchbacks accounted for a large proportion, and the coverage of existing clothing patterns was incomplete.

After conducting Correlation Analysis and Partial Correlation Analysis between five body surface angles, there is a conclusion that the five lateral body surface angle of the hunchback is less correlated. According to the range of five body surface angles, we used K-means clustering analysis method to subdivide the 31 hunchbacked body samples, and finally the hunchbacked body samples is divided into five types. The problems when drawing patterns to the five representative types were also summarized.

These conclusions supplement the existing national standard body type classification system, and can also provide data for the following hunchbacked body patterns study.

References

[1] Canhadas Belli JF, Chaves TC, Siriani de Oliveira A, et al. Analysis of body posture in children with mild to moderate asthma. Eur J Pediatr: 2009; 168: 1207-1216.

[2] Dunk NM, Lalonde J, Callaghan JP. Implications for the use of postural analysis as a clinical diagnostic tool: reliability of quantifying upright standing spinal postures from photographic images. J Manip Physiol Ther: 2005; 28: 386-392.

[3] Fedorak C, Ashworth N, Marshall J, et al. Reliability of the visual assessment of cervical and lumbar lordosis: how good are we? Spine: 2003; 28: 1857-1859.

[4] Watson AWS, Mac Donncha C. A reliable technique for the assessment of posture: assessment criteria for aspects of posture. J Sport Med Phys Fit: 2000; 40: 260-270.

[5] Walk G, Lindsey R, Corbin CB. Concepts of physical fitness. New York: McGraw-Hill Companies: 2000.

[6] Barrow HM, Mec Gi R. Measuring in physical education. Belgrade: Vuk Karadzic: 1975.

[7] Fengqin X, Ge W. Research of Somatotype Based on the Sagittal Plane Silhouette of Human Body. Journal of Textile Research: 2017; 38 (06): 86-91.
[8] Mi KY, Yun JN, Woong K. Classifying male upper lateral somatotypes using space vectors [J]. International Journal of Clothing Science and Technology: 2016; 28 (1).
[9] Young LC, Yun JN. Classification of upper lateral body shapes for the apparel industry [J]. Human Factors and Ergonomics in Manufacturing & Service Industries: 2010; 20 (5).
[10] Müller M. Der Zuschnitt für die Herren-Schneiderei, System M. Müller & Sohn, München Deutsche Bekleidungs-Akademie: 1951
[11] Armstrong HJ. Patternmaking and Fashion Design, Hopper Collins, New York, NY: 1995.
[12] Rasband J. Fabulous Fit, Fairchild Publications, New York, NY: 1994.
[13] Nam YJ. "A study on classification of somatotype based on the lateral view of women's upper body", doctoral dissertation, Seoul National University, Seoul: 1991.
[14] Jung MS. "Classification of side somatotype of the trunk by analysing photographic data", Korean Association of Human Ecology: Vol. 12; No. 5: pp. 767-776: 2003.
[15] Maja MN, Slavenka P. A method for body posture classification of three-dimensional body models in the sagittal plane [J]. Textile Research Journal: 2019; 89 (2).
[16] GB/T 23698-2009.General requirements for 3D scanning anthropometric methods[S].Beijing:China Standard Publishing: 2010.
[17] Jiayan X, Caiyuan K. Influence of characteristic parameters of body surface on long-range body measurement precision [J]. Jouranl of Silk: 2019; 56 (08): 36-45.

Cite as: S.Y. Yang et al.,
TBIS 10.3993/tbis (2021).

Clothing Recommendation with Style Recognition

Si-Yu Yang[1], Yue-Qi Zhong[1,2*], Xin Wang[1]

[1] *College of Textiles, Donghua University, Shanghai, 201620, China*
[2] *Key Laboratory of Textile Science & Technology, Ministry of Education, Donghua University, Shanghai, China*

*Corresponding author's Email: zhyq@dhu.edu.cn

Abstract

People tend to purchase the same style clothing during online shopping. However, current clothing recommendation systems often suffer from style diversity. To recommend clothing with the same style, we proposed a system with a recognition module (RM1) and a recommendation module (RM2). RM1 used EfficientNet-b3 network to identify user input and filter clothing dataset. RM2 then acquires user input clothing feature vector (ICFV) and clothing dataset feature vectors (CDFVs). CDFVs were extracted from the filtered dataset through the Resnet-50 neural network and ICFV was extracted from user input though Resnet-50 neural network. RM2 encoded feature vectors with Locality Sensitive Hashing Function and calculated the similarity between ICFV and CDFVs. Finally, RM2 recommended the most similar clothing to customers. RM1 was fine-tuned on a dataset with 5 common clothing style preferences, achieving an accuracy of 92.88%. The recommendation accuracy of the system with RM1 and RM2 was 30.00% higher than before.

Keywords: Clothing Recommendation; Style Recognition; Clothing Dataset; Convolutional Neural Network

1. Introduction

When customers click or collect clothing during online shopping, they want the system to recommend same style clothing [1-3]. According to the excellent performance of convolutional neural network, many works reported the usage of neural network to extract high level clothing features or merge Multi-features for retrieval or recommendation [4-6]. Wu [7] contacted a series of multiple features to recognize different clothing. However, these methods could not identify the clothing style. Xu [8] used DenseNet to extract the feature vectors, and Multi-Similarity Loss for the training. This method used local information to reduce noise interference. However, since the network features were difficult to control, the recommendation results still demonstrated different styles.

To recommend clothing with the same style, we proposed a system with two modules: the style recognition module (RM1) and recommendation module (RM2). A new dataset with 5 common clothing styles preferences was built for training RM1, since the division of styles in prior datasets cannot satisfy the experiments due to the lack of universality. [9-10].

2. Method

The recommendation system was built with two components: the style recognition module (RM1) and recommendation module (RM2). RM1 was based on the EfficientNet-b3 [11] network, and it worked for recognizing user style preference and filtering noisy style clothing. RM2 used the Resnet-50 [12] pre-trained module features to extract feature vectors. The Locality Sensitive Hashing technology (LSH) [13] was used to encode feature vectors. The Cosine Similarity was used to calculate the similarity between user input clothing feature vector and filtered dataset feature vectors. The system recommended the most similar clothing which was in the same style to user input for users. The architecture of the proposed system was shown in Fig.1.

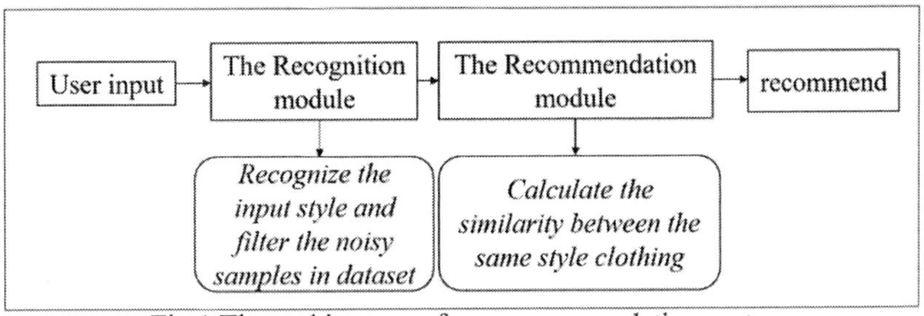

Fig.1 The architecture of our recommendation system

2.1 The Style Recognition Module

For style recognition purpose, we use EfficientNet-b3 network, which can achieve state-of-the-art performance on ImageNet dataset and the Top-1 accuracy is 81.7%. The EfficientNet was introduced by Google in 2019[11], which achieved joint expansion of network depth, image size, and resolution. This method improved model accuracy while reducing the number of parameters. EfficientNet are based on MBconvblock in the MobileNetV2 [14], and they merge the Squeeze-and-Excitation block in the SENet [15]. The architecture of EfficientNet-b3 is shown in Fig.2, and it can be represented as:

$$N = F_k \odot \ldots \odot F_2 \odot F_1(X_1) = \odot_{j=1\ldots k} F_j(X_1) \quad 1)$$

Where N indicates the entire network; F_k is the k-th layer of the net, and $\odot$ represents the network operations.

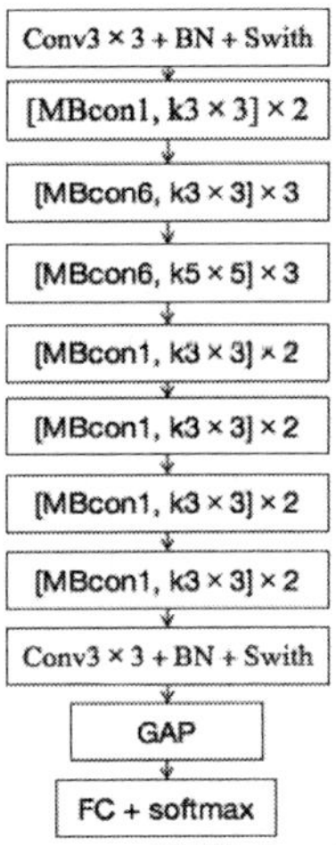

Fig.2 The architecture of EfficientNet-b3 Network

In this paper, the Entropy Loss was used for training the EffcientNet-b3, which could be expressed as:

$$L(\hat{y}, y) = -[y log\hat{y} + (1-y)log(1-\hat{y})] \quad 2)$$

Where $\hat{y}$ represents the prediction; y represents the ground truth, and $L(\hat{y}, y)$ represents the Cross Entropy Loss for our classification task.

2.2 The Recommendation Module

After filtering noisy styles using RM1, we used RM2 to calculate the most similar clothing to user input. RM2 was mainly based on a pre-trained Resnet-50 network [12]. In addition, the Locality Sensitive Hashing and Cosine Similarity were used for encoding and searching.

2.2.1 Resnet-50

Deep learning combined into clothing similarity calculation shows an excellent performance. Thus, we selected the commonly used pre-trained Resnet-50 as feature extractor.

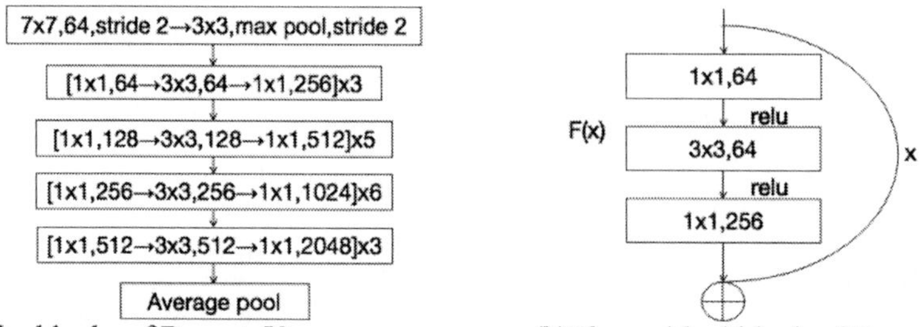

(a)The blocks of Resnet-50 (b)The residual block of Resnet-50

Fig.3 The architecture of Resnet-50

As shown in Fig.3(a), there are blocks of the Resnet-50. In the module $[1 \times 1{,}64 \rightarrow 3 \times 3{,}64 \rightarrow 1 \times 1{,}256] \times 3$, 1×1 or 3×3 represents the kernel size; 64 or 128 represents the number of kernels, and $\times 3$ implies the module runs 3 times. Each module crossly uses 1×1 and 3×3 kernels to decrease the number of parameters, and the residual block are applied in each module, which can be expressed as:

$$y = F(x) + x \tag{3}$$

Where $F(x)$ represents the residual value, x means the origin input value. The residual blocks can resolve the problems of vanishing gradient in the deep neural net. Such with the deep layers trained net can extract high level features.

2.2.2 The Locality Sensitive Hashing.

Locality Sensitive Hashing (LSH) [13] was used to encode the feature vectors to accelerate the computation. It assumed that the similarity among vectors will not be changed after those vectors being mapped by hash functions. Hence after extracting the feature vector with the size of 1*2048 via Resnet-50, LSH was employed to encode it. With eq.(4), the hash functions $F(x)$ changed the element values of the feature vector into 0 or 1 to simplify the computation and hence to accelerate the recommendation.

$$\begin{bmatrix} -0.01764908 \\ 0.06506907 \\ \cdots \\ 0.01393 \\ 0.01036 \end{bmatrix} \overset{F(x)}{\rightarrow} \begin{bmatrix} 1 \\ 0 \\ \cdots \\ 1 \\ 1 \end{bmatrix} \tag{4}$$

2.2.3 The Cosine Similarity.

In this paper, we used the Cosine Similarity to calculate cosine of the angle between clothing feature encoding (vectors).

$$\cos\theta = \frac{A \cdot B}{\|A\| \cdot \|B\|} = \frac{\sum_{i=1}^{n} A_i \times B_i}{\sqrt{\sum_{i=1}^{n} A_i^2} \times \sqrt{\sum_{i=1}^{n} B_i^2}} \tag{5}$$

Where A_i and B_i respectively represent the sub-vectors of the clothing style feature codes A and B.

2.3 Dataset

A new dataset which set manually with 5 common clothing styles was used to fine-tune the EfficientNet-b3 as the style recognition module. The dataset with complicated backgrounds was collected from sales websites, including Taobao (https://www.taobao.com/) and JD (https://www.jd.com/). Data augmentation (random horizontal flip and random vertical flip) was employed to prevent overfitting. The dataset was split into train: test: validation at the ratio 8: 1: 1. The details were shown in the Table1.

Table1. The details of the dataset

Category	Characteristics	Representative clothing	Number
Classical	wide sleeves and gauze	Hanfu, embroidered skirt, etc	603
Neural	single tone	Hard collar suit, professional outfits, etc	571
Sports	casual and comfortable	Sportswear, baseball jersey, tights, etc	567
Hip-Hop	graffiti and personality	Loose T-shirt, loose hoodie, etc	512
Celebrity	noble, elegant, formal	Evening dress, small fragrance suit, etc	603

3. Experiment

To verify the proposed method, the EfficientNet-b3 was trained with our dataset to recognize the user input clothing style. We set different learning rates and weight decay to improve the accuracy.

For the recommendation module, Resnet-50 was pre-trained on the retrieval-SfM-120K [16], a public dataset, to extract clothing feature vectors. The Resnet-50 pre-trained module has been exceeding many feature extraction models. We directly used it as the feature extractor. Cosine Similarity was used to calculate the similarity between user input and filtered dataset. The recommendation module would propose the most similar clothing to users.

3.1 The Training Details

Our experiment was performed on a PC equipped with a NVIDIA 1080 GPU. The processor was Inter(R) Core (TM) i7-3770K CPU@3.50GHz, and the experiment environment was python 3.7 and pytorch1.2.0. For the recognition module, we trained 25 epochs with Adam optimization. The image size was set as 300×300.

3.2 Recognition Module Evaluation

We used accuracy to evaluate the performance of recognition module. There are four types of outputs in classification. Positive samples are predicted as positive class (TP); positive samples are predicted as negative class (TN); negative samples are predicted as positive class (FP), and positive samples are predicted as negative class (FN). Hence the accuracy can be expressed as

$$Accuracy = \frac{TP+TN}{TP+TN+FP+FN} \quad 6)$$

3.3 Recommendation Module Evaluation

3.3.1 The Number of Wrong Style Clothing

To verify the recommendation performance of our system, we input query images to the recommendation module. During two recommendation methods with a recognition module (ours) or without, we counted the number of wrong style clothing in Top10, Top20, and Top50 results in each images, respectively. We randomly selected 10 query images. The results were the average number of 10 images results.

3.3.2 User Study

We selected 30 experienced female online shoppers at the age ranging from 28 to 35, to participate in the user study. In the questionnaires, we showed 10 query images. Each query image had top-3 results returned from two recommendation module with a recognition module (ours) or without. The participator chose the best recommendation method from two methods. According to the 10 choices determined the best method.

4. Experiment Results and Discussion

4.1 Recognition Accuracy

Table 2. The accuracy of different learning rates

Learning rate	0.001	0.0005	0.0001	0.00005
Accuracy	87.26%	88.76%	90.26%	92.88%

To find the appropriate learning rate, the weight decay was set to 0.01. The different learning rates were set as shown in Table2. When the learning rate was set to 0.001, the test accuracy was 87.26%. When the learning rate was set to 0.0005, the test accuracy was 88.76%.When the learning rate was 0.0001, the test accuracy was 90.26%. And when the learning rate was 0.00005, the accuracy was 92.88%. From our observation, we found the accuracy gradually increased as the learning rate decreased.

Table 3. The accuracy of different weight decay

Weight decay	0.01	0.005	0.001	0.0001
Accuracy	92.88%	88.01%	90.00%	89.13%

To find the appropriate weight decay, the learning rate was set to 0.00005. The different weight decay was set as shown in Table3. When the learning rate was set to 0.01, the test accuracy was 92.88%%. When the learning rate was set to 0.005, the test accuracy was 88.01%. And when the weight decay was 0.001, the accuracy was 90.00%.

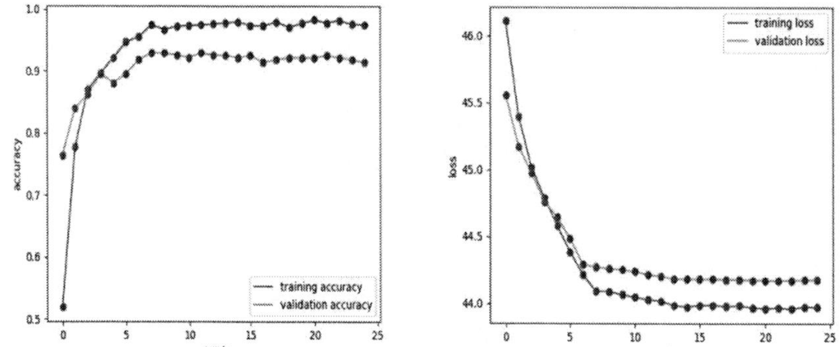

Fig.4 The accuracy and loss of the appropriate weight decay.

As shown in Fig.4, by setting weight decay and learning rate to 0.01 and 0.00005, the highest accuracy could be reached. The best accuracy during test was 92.88%. However, the training accuracy was 99.80%, which indicated the recognition module somewhat produced overfitting on the proposed clothing dataset.

4.2 The Style Recognition Module Performance

To verify the capability of the style recognition module in identifying the complicated images, we randomly selected several clothing images similar in the dataset, as shown in Fig.5.

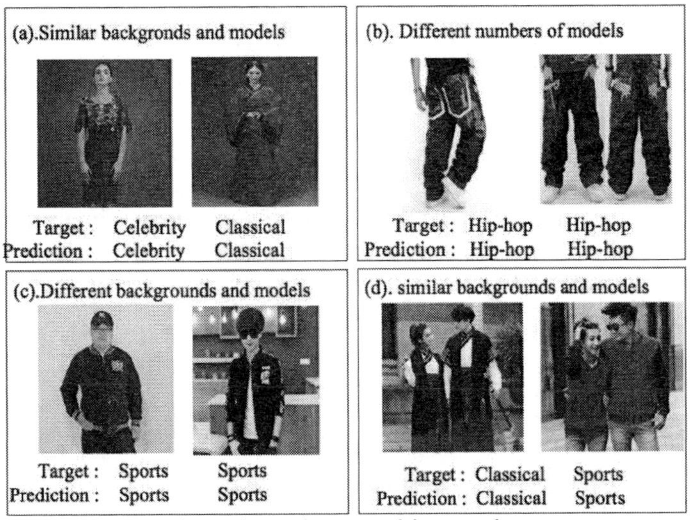

Fig.5 The style recognition results

In Fig.5, Target represents the true style of clothing in Fig.5. Prediction represents the prediction style of clothing. As shown in Fig.5, (a) or (d) both had two images. The backgrounds of images are similar. The postures of models were also consistent. From the prediction results under the images, we knew that the recognition module could identify they were different clothing styles. Though they were similar in the backgrounds and models, they were different in clothing styles. As show in Fig.5(c), we also selected the same style clothing. But the two images are different in backgrounds. The recognition module still could identify they were both sports. The number of models was different in Fig.5 (b), but the style of pants could also be recognized.

4.3 The Recommendation Module Performance

To compare the recommendation results with the recognition module (WRM) and without the recognition module (WORM), we demonstrated two randomly selected query images, and compared the top-3 results returned by both methods. In addition, we counted the number of wrong styles. To further investigate the performance of our proposed method, we conducted a user study, and the results were depicted in Table5.

The query examples were shown in Fig.6, the WRM method filtered the wrong style clothing (the red block on the first row returned by WORM), and proposed clothing in the same style (on the second row).

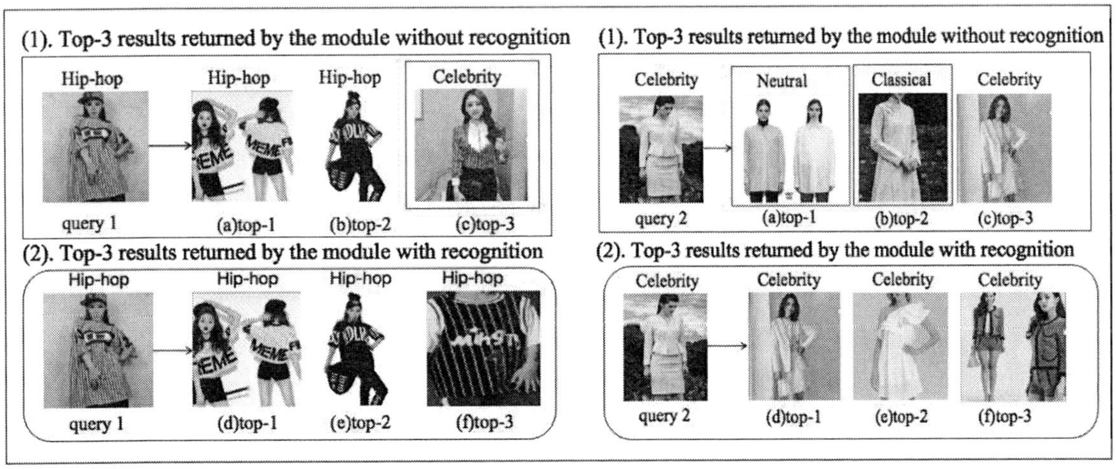

Fig.6 The recommendation performance with and without the recognition

Table 4. The number of wrong style clothing returned by two methods (WRM and WORM)

Number of wrong style clothing	Top-10	Top-20	Top-50
Without the recognition module(WRM)	4	10	27
With the recognition module(WORM)	0	1	3

From Table 4, the number of wrong style clothing was significantly reduced by the proposed recognition module (WRM). In the top-10 results, there were 4 wrong style clothing returned by WORM, but 0 returned by WRM. In the Top-20 results, the number of wrong style clothing had reached 50% of the total returned by the WORM. The number of wrong style clothing was just 1 returned by WRM. When counting in the top-50 results, the number of wrong style clothing proposed by WORM was 27. The number of wrong style clothing recommended by WRM was just 3.

Table 5. The results of user study

Situations	WRM=WORM	WRM>WORM	WRM<WORM
Votes	1	19	10
Percentage	3.33%	63.33%	33.33%

The results of user study were shown in Table 5 , where WRM=WORM meant the participator thought two recommendation methods were same; WRM>WORM meant the participator thought WRM was better than WORM, and WRM<WORM meant the participator thought WRM was bad than WORM. From the results, the method with the recognition module was preferred by 63.33% customers, which was 30% higher than the other.

5. Conclusions

To recommend clothing in the same style, we proposed a system with a recognition module followed by a recommendation module. We fine-tuned the EfficientNet-b3 on our dataset as recognition module to filter noisy style clothing. And used the Resnet-50 pre-trained module features combined with Locality Sensitive Hashing for recommendation. According to the experiment results, the accuracy of our recognition module could achieve 92.88%. The satisfaction rate of the recommendation system with recognition module was 30.00% higher than the one without recognition module by the user study.

6. Funding

This work was supported by the National Natural Science Foundation of China (Grant No. 61572124).

References

[1] Ren JH. China's niche apparel brand development and niche design style analysis [D]. Beijing Institute of Fashion Technology: 2016.
[2] Hou DY, Li JH. Discussion on the artistic style of fashion design and brand cultural connotation [J]. Tomorrow's Fashion: 2020 (20); 14-15.
[3] Zhang YD. Analysis on the marketing development model of e-commerce clothing brand [J]. Textile Report: 2021; 40 (01): 35-36+49.
[4] Chen YJ. Online clothing retrieval and recommendation based on deep network [D]. Beijing University of Posts and Telecommunications: 2018.
[5] Zhang Q. Application of Deep Learning Algorithm in Clothing Recommendation System [J]. Digital World: 2019 (08); 125.
[6] Xu H, Bai ML, Wan TR, Xue T, Tang W. Semantic Analysis and Retrieval Recommendation of Clothing Image Based on Deep Learning [J]. Journal of Basic Science of Textile Universities: 2020; 33 (03): 64-72.
[7] Wu F, Deng ZJ, Shang SF.A method of multi-feature clothing image retrieval based on convolutional neural network[J].Journal of Hunan Institute of Technology (Natural Science Edition): 2020; 30 (03): 48-53.
[8] Xu FF, Tian Y, Zheng S. A Clothing Retrieval Model Based on DenseNet and Multi-Similarity Loss [J].

Journal of Physics: Conference Series: 2020; 1684 (1).
[9] Hu MY, Zhong YQ. Analysis and Evaluation of Women's Wear Style Based on Deep Learning[D]. Donghua University: 2020.
[10] Yin GC. Research on Clothing Style Recognition and Recommendation Based on Convolutional Neural Network [D]. Zhejiang Sci-Tech University: 2020.
[11] Tan M, Quoc V L. Efficient Net: rethinking model scaling for convolutional neural networks [J]. arXiv preprint: 2019: arXiv: 1905.11946v3.
[12] He K, Zhang X, Ren S, et al. Deep residual learning for image recognition [J]. arXiv preprint: 2015: arXiv: 1512.03385v1.
[13] Filip R, Giorgos T, Ondřej C. CNN Image Retrieval Learns from BoW: Unsupervised Fine-Tuning with Hard Examples, arXiv preprint: 2016: arXiv: 1604.02426.
[14] Mark S, Andrew H, Zhu ML, Andrey Z, Chen LC. MobileNetV2: Inverted Residuals and Linear Bottlenecks [J]. arXiv preprint: 2017: arXiv: 1801.04381.
[15] Hu J, Shen L, Albanie S, Sun G, Wu E. Squeeze-and-Excitation Networks [J]. arXiv preprint: 2017: arXiv: 1709.010507.
[16] Filip R, Giorgos T, Ondřej C. Fine-tuning CNN Image Retrieval with No Human Annotation [J]. arXiv preprint: 2017: arXiv: 1711.02512.

Human Clothing Model for Tailor-Made VR Experience of Clothing Wear

Jing Qi, Chao Huang, Bo-An Ying*, Xiao-Feng Wang

Apparel & Art Design College, Xi'an Polytechnic University, Xi'an, Shaanxi, 710048, China

*Corresponding author's email: yingbo2006@163.com

Abstract

Clothing fitting and fabrics are the key factors in the process of personalized customization. For the tailor-made virtual reality experience of clothing wear, the visual expression of fitness and fabric of models are important. We compared and analyzed several common modeling methods of both the human body and garment, then constructed a clothed 3D Mannequin Model and tested its visual effect in a virtual reality space. This paper aims to provide a reliable way to create a human clothing model for VR tailor-made experience.

Keywords: Garment Modeling; Personalized Garment Customization; Virtual Reality

1. Introduction

Personalized clothing customization means that customers combine personal preferences and requirements with the production process of clothing design according to their needs [1]. It is a more advanced, more diversified, and more individual service model. With personalized customization becoming a new profit model for apparel companies, the user's customized experience has become increasingly vital.

The fit and fabric of the clothing are the key factors in the personalized customization process. Traditional online customization generally only involves the selection process of clothing fabric styles. For more professional users, the acquisition of measurement information still needs to be completed offline. Therefore, the complete customization experience is split; for offline customization stores, users participating in the customization process, cannot see the actual effect of the product in real-time, they can only imagine the effect of the sample. Clothing, the human body, and the environment are inseparable. The current online and offline customization experience is difficult to intuitively integrate the element of the environment into the customization process. More novel technology proposed to improve these experiences include Immersive Virtual Reality.

Immersive Virtual Reality (VR) refers to technology that blurs the boundaries between the physical, virtual, and simulated worlds. It creates a deeper immersive experience for users through more complex tracking systems and interactive methods [2]. Users usually need to wear a head-mounted display and be surrounded by a virtual environment. In recent years, domestic and foreign research and application of immersive virtual reality technology in the field of clothing have mainly focused on clothing product design, clothing virtual display, clothing knowledge teaching, and clothing retail. Hong, Y [3] and others proposed a custom clothing design method based on a collaborative design for people with scoliosis combined with immersive VR technology. Jung, J [4] and others have conducted theoretical and empirical research on how consumers can obtain meaning from VR experience in their research on luxury brands combined with virtual reality technology marketing. Park, M [5] and others developed the VR clothing retail store application, tested and evaluated the value of VR technology in the clothing retail store design process. Despite its wide adoption in fashion retail, VR are seldom used for tailor made clothing.

The clothing model is an important element of experience during the VR clothing customization experience. On the one hand, the clothing model needs to give users a visual experience and provide a reference for customized effects; on the other hand, the realism of clothing is also a necessary factor that affects the immersive experience.

This paper aims to provide a reliable way to create a human clothing model for an immersive VR tailor-made experience. Firstly, we analyzed the methods of modeling for the human body and garment. Next, we build a human body model based on the anthropometric data and use interactive tools to complete clothing modeling. After that, we use a fabric scanner to obtain the texture maps and finished texture processing in unity3D. Finally, we tested the visual effect of the 3D model in virtual reality space.

2. Analysis of the Construction Method of Human Body and Clothing Geometric Model

The construction of the human body model is the basis of 3D clothing shape design and virtual display. In the subsequent construction process of the clothing geometric model, it is necessary to perform clothing modeling based on the constructed human body model.

The methods of human body and clothing geometric models are mainly divided into the following four categories: scanner-based, image-based, interaction-based and parameter-based.

2.1 Scanner-based Modeling

The scanner-based human-body modeling refers to a method for geometrically reconstructing the real human body using professional scanning equipment [6]. This method mainly uses three-dimensional human body data obtained from a 3D scanning instrument in multiple directions and uses techniques such as keypoint extraction and template matching to enlarge and process the acquired data to obtain a complete set of three-dimensional human body models. Liu Hongyan [7] obtained a complete clothing model by reverse processing the three-dimensional scan data.

Because there are blind spots in clothing or human body scanning, obtaining clothing models in this way requires a series of reverse engineering processing, and the overall steps are cumbersome.

2.2 Image-based Modeling

By collecting and processing the image information of the clothing, clothing or human body modeling can be completed. Zheng Z et al. [8] proposed an image-guided volume-to-volume translation CNN restructured 3D human from a single RGB image. Decaudin et al. [9] directly mapped the two-dimensional sketch design to the construction process of the three-dimensional clothing model based on the clothing sketch information; Zhou [10] et al. realized clothing modeling from a single photo by designing a general outline.

Limited to the two-dimensional attributes of the original image information, the model constructed based on the two-dimensional image information is only suitable for simple-style clothing, and due to the difference in algorithm processing, the ability to express the details of the model is insufficient, and the realism of generated model is weak.

2.3 Interaction-based Modeling

Interactive 3D modeling technology is mainly based on software to complete the construction of geometric models. Softwares such as 3DS Max and Maya can be used to complete the model construction. In addition, some clothing simulation software, such as CLO 3D, Marvelous Designer, combined with the discrete model calculation ideas of clothing, has further improved the quality of clothing model construction.

Interactive 3D modeling technology requires users to be dependent on professional knowledge, and the modeling process is more complicated, but the final geometric model effect is closest to reality.

2.4 Parameter-based Modeling

The construction of the geometric model by Parameter-based modeling, expresses the shape characteristics of the human body model through geometric constraints [11]. By using the existing model, through modifying the abstracted characteristic parameters and affecting the association relationship between the model parts, a new human body or clothing model can be obtained, thereby realizing parametric modeling.

Parametric modeling technology is efficient and is a quick way to obtain models, but the morphological effect of the model depends on the accuracy of the parameters provided by the user and the program algorithm.

2.5 Summary

Among the immersive design elements, the representative fidelity of the model is an important factor. Therefore, this article has two requirements for the human body model: the size of the model should resemble reality and has a certain aesthetics. The size of the human body model needs to refer to the real human body and virtual human table data and has a high degree of fidelity in terms of the scale and size of the model. Secondly, the human body model, as the basis of the virtual display of clothing, needs to have a certain aesthetics, and the model surface should be smooth. The proportions of the lines are balanced. In summary, this article chooses parameter-based human-body modeling technology to carry out research work on the construction of human body models.

Personalization experience needs to support the customization of clothing styles, so it is required to constitute the overall model of clothing in the form of traditional clothing structure, for example, the clothing model is specifically composed of sub-models of clothing parts such as the front piece model, the back piece model, and the button model. The garment models obtained based on interactive technology are more in line with the requirements of the personalization experience.

3. Construction of 3D Mannequin Model with Clothing

The workflow of the 3D mannequin model with clothing obtained is shown in Fig.1. First, this paper completes the construction of the human body geometric model based on parametric modeling. Then the clothing is modelled based on the human body model. Finally, the fabric material of the obtained clothing model is processed.

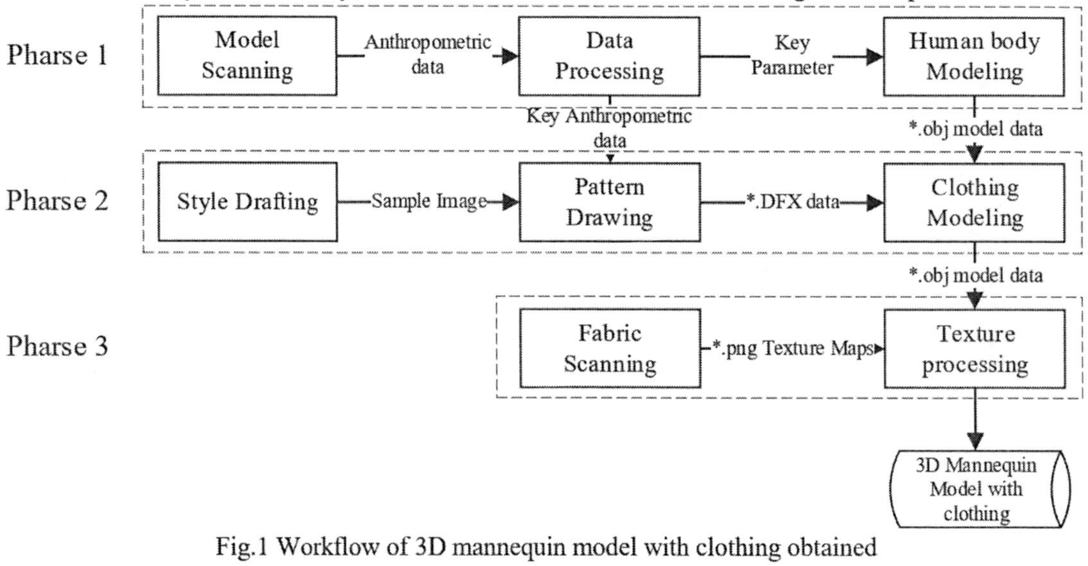

Fig.1 Workflow of 3D mannequin model with clothing obtained

3.1 The Process of Parameter-based Human-body Modeling

To obtain real human body size data, the Human Solutions body scanner is used to obtain 40 sets of anthropometric data. Some of the data is used to guide the production of clothing patterns, and the key parameter for human-body construction need further processing.

This paper use make-human-community 1.2.0 for parameter-based modeling, which provides the function of inputting real measurement data to generate model data [12]. According to the parameter types, the MakeHuman tool can support, the data parameters obtained in the previous section. Some parameters require calculation to obtain 20 measurement data as shown in Table 1.

Table 1. List of parameters supported by the parameter-based modeling tool MakeHuman

Category	Measurement Items	Length(cm)	Category	Measurement Items	Length(cm)
Neck	Neck circ	44.3	Upper leg	Height	42.7
	Neck height	11.2		Thigh circ	55.2
	Front chest distance	41	Knee	Knee circ	36.6
	Bust circ	107.9	Lower leg	Heigh	44.5
Torso	Underbust circ	98.4		Calf circ	36.7
	Waist circum	83.1	Ankle	Ankle circ	28.1
	Nape to waist	43.8	Upper arm	Upper arm circ	34.1
Torso	Waist to hip	24		Upper arm length	37.1
	Shoulder distance	13.9	Lower arm	Lower arm length	30.5
Hips	Hips circ	96.8		Waist circ	16.6

*circ=circumference

Finally the measurement option in makehuman tool was selected and the above table data was entered into the parameter edit box by category. The tool will generate a human body model of the corresponding size, then export the data file in *.obj format which assists garment modeling. Fig.2 shows the 3D preview of the mannequin.

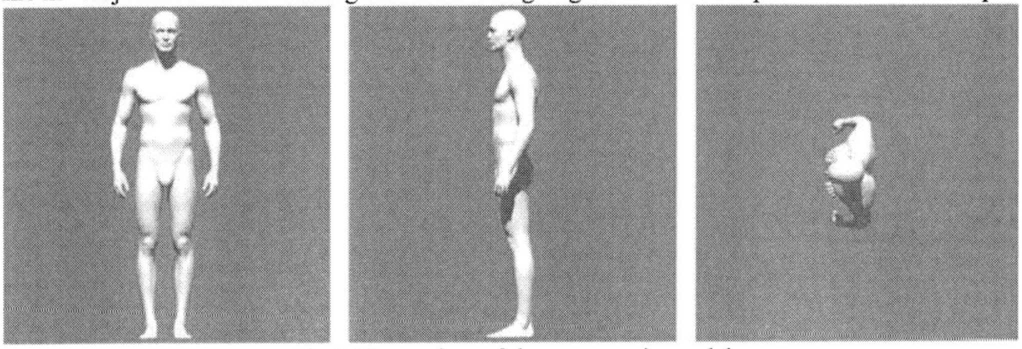

Fig.2 3D preview of the mannequin model

3.2 The Process of Interaction-Based Garment Modeling

This paper chooses CLO 3D interactive tool for garment modeling. CLO 3D is a 3D CAD tool used in the clothing field. It realizes the shape simulation of clothing fabrics through simulation calculations and supports the reading of clothing panels in DXF format [13]. The process of garment modeling by interactive technology is shown in Fig.3.

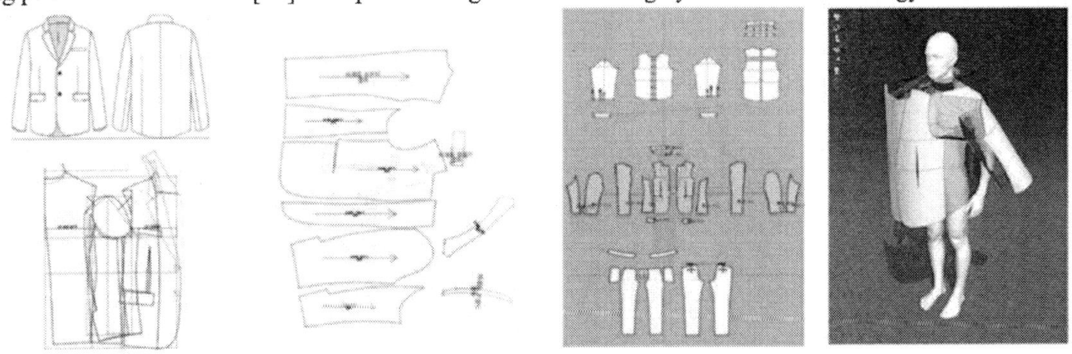

(a) Style drafting **(b) Pattern drawing** **(c) Virtual sewing** **(d) Simulating**

Fig.3 The process of garment modeling by interactive technology

First, the clothing style to be made is clarified, and the corresponding clothing structure diagram is drawn according to the style diagram. As shown in Fig.3(a), the style of clothing selected for this topic is a six-open body insert pocket men's suit, a single row with two buttons, the body shape is a close-fitting body, the collar is a flat turn-over collar, and the fit is two-piece bent sleeves.

Secondly, according to the size information of the wearing model, the clothing CAD pattern of the corresponding size are drawn in the clothing CAD software, and the pattern is exported to a file in DXF format. The pattern includes front, side, back, sleeve, collar, and placket.

After, a new CLO 3D project is careated and the DXF file as well as the human-body model file is imported. A 3D pattern for the imported model is arranged, and the virtual sewing thread for each pattern is set. After the above work is completed, fabric attributes such as thickness, fabric type, particle spacing, etc. are set for the clothing model, and then simulated try-on is started. Different fabrics and apparel fabrics will simulate different appearance styles. At the same time, during the virtual try-on process, it is necessary to constantly debug the try-on results. When multiple virtual garments are tried-on simulated, it is also necessary to set up levels for simulation. After the simulation result is appropriate, an ironing tool is used to iron the folds of the garment evenly. Preview of the human clothing model shown in Fig.4.

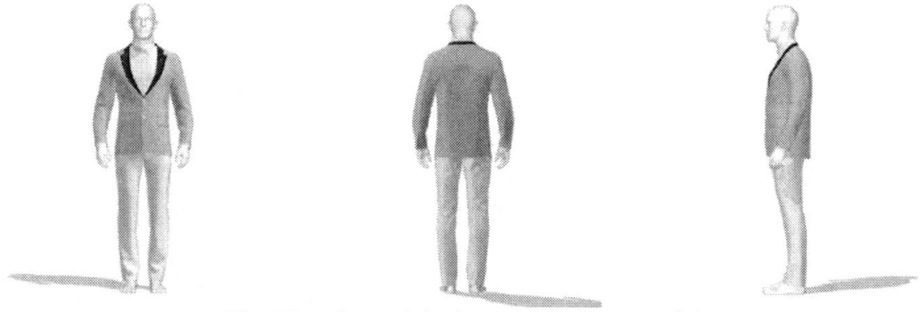

Fig.4 Preview of the human clothing model

3.3 Texture Processing for Garment Model

In computer graphics, to reflect the visual details of an object under different light sources and angles, it is necessary to use multiple texture maps to form a material, and then form a visual effect through a rendering engine.

Physically Based Rendering (PBR) is a texturing workflow in the 3D modeling process, designed to simulate the way light reacts with the model, to simulate materials in real life. Its main advantages are as follows: (1) It is easier for PBR to create realistic-looking model materials; (2) Under all lighting conditions, digital assets appear to be accurate; (3) PBR provides a workflow for creating consistent works. Based on the principle of PBR, this paper adopts a metal/roughness workflow for the construction of digital clothing fabrics and uses a fabric scanner to obtain the texture map of the fabric. Finally, the cross-platform 3d content production tool unity3d is used to make PBR materials to express the visual information of clothing [14]. Fig.5 shown the workflow of fabric material obtained.

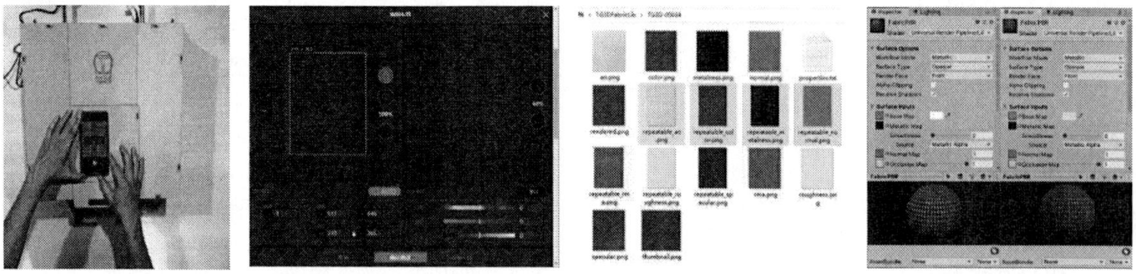

(a) Scanatic™ Nuno Fabric Scanner for Textures | (b) Adjust the texture details of the cloth | (c) Export texture maps | (d) Create PBR material

Fig.5 Workflow of PBR material obtained

At first, we used the Scanatic™ Nuno Fabric Scanner for fabric texture maps. Scanatic™ Nuno Fabric Scanner is a portable fabric scanning device. It only needs a smart phone and a special carton to complete digital fabric production with cloud applications [15]. Different maps correspond to different models that have been created to describe light behavior. For example, the Base Color map describes the color of diffuse light, and the Metalness map represents the proportion of light that has specular reflection and diffuse reflection [16]. Then the hue, saturation, brightness, and other color information of the scanned fabric are adjusted. After, the texture maps are imported into Unity and the relevant maps are assigned to the material's inspector correctly. Fig.6 shows the the effect comparison before and after the texture is shown. The geometric shape and appearance of the fabric can make relatively realistic simulation effects.

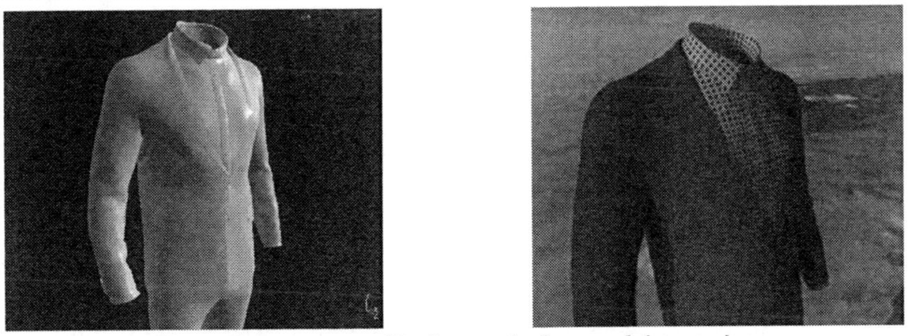

Fig.6 The contrast of before and after model texturing

4. Result and Discussion

The model is imported into the "VR Customization Store" application developed by the laboratory, providing VR experiences for suit customization and scene exploration for users. The HTC VIVE Pro VR headset is used to test the display effect of the model under the VR perspective [17]. Fig.7 shows the clothing display effect in virtual reality.

Fig.7 Custom clothing display effect under the VR perspective

To evaluate the fit of virtual clothing and human body, we observe and judge its visual effects from different directions such as front and back, left and right, up and down by adjusting the VR viewing angle. Since the human body model is constructed based on anthropometry data, and the clothing model is also made based on the human body model, the human body model matches the clothing. By moving the headset position, it can be observed that as the observation angle is different, the surface fabric of the clothing model also presents a three-dimensional display effect with the angle change. the silhouette and fabric details of clothing have a better display effect.

5. Conclusion

This paper provides a reliable way to create a human clothing model for an immersive VR tailor-made experience. In the first part, we discussed the construction method of the human clothing model and analyzed its applicability. Then we build a human body model based on the anthropometric data and use interactive tools to complete the clothing modeling. The fabric scanner was obtained for the texture maps and the unity 3D is used for texture processing. Finally, we tested the visual effect of the 3D model in virtual reality space. The result indicated that the 3D mannequin model with Clothing constructed by the method proposed in this paper satisfies the two factors of fit and fabric material in the VR tailor-made experience.

6. Acknowledgment

We would like to thank the support of Shaanxi Engineering Research Center of Apparel Intelligent Design and Manufacture.

References

[1] Li H. Research on Problems of Smart Garment Ecosystem Oriented to Personalized Customization. Beijing Jiaotong University: 2017.
[2] Lee HG, Chung S, Lee WH. Presence in virtual golf simulators: The effects of presence on perceived enjoyment, perceived value, and behavioral intention. New media society: 2013; 15: 930-946.
[3] Hong Y, Bruniaux P, Zeng XY, et al. Virtual reality-based collaborative design method for designing customized garment for disabled people with scoliosis. Int J Cloth Sci Tech: 2017; 29: 226-237.
[4] Jung J, Yu J, Seo Y, et al. Consumer experiences of virtual reality: Insights from VR luxury brand fashion shows. Journal of Business Research: 2019.
[5] Park M, Im H, Kim DY. Feasibility and user experience of virtual reality fashion stores. Fashion and Textiles: 2018; 5.
[6] Wang QL, Xu ZB, Tian BQ. Research Progress of Personalized 3D Human Body Modeling in Virtual Fitting. China Textile Leader: 2020; 07: 79-82
[7] Liu HY. Reusable Techniques in Constructing Physically-based 3D Scanned Garment Model. Dong Hua University: 2011
[8] Zheng Z, Yu T, Wei Y, et al. Deephuman: 3d human reconstruction from a single image. Proceedings of the IEEE/CVF International Conference on Computer Vision: 2019; 7739-7749.
[9] Decaudin P, Julius D, Wither J, Boissieux L, Sheffer A, Cani MP. Virtual garments: A fully geometric approach for clothing design. Comput Graph Forum: 2006; 25: 625-634.
[10] Zhou B, Chen X, Fu Q, Guo K, Tan P. Garment modeling from a single image. Wiley Online Library: 2013.
[11] Seo H, Magnenat-Thalmann N. An automatic modeling of human bodies from sizing parameters. Proceedings of the 2003 symposium on Interactive 3D graphics: 2003; 19-26.
[12] Briceno L, Paul G. MakeHuman: a review of the modelling framework. Congress of the International Ergonomics Association: 2018; 224-232.
[13] Kang TS, Lee DY, Kim J. Production of Digital Fashion Contents based on Augmented Reality Using CLO 3D and Vuforia. Journal of the Korea Computer Graphics Society: 2020; 26: 21-29.
[14] Information on https://unity.com.
[15] Information on https://tc.tg3ds.com/fabric-digitization.
[16] Burley B, Studios WDA. Physically-based shading at Disney: 2012; 1-7.
[17] Information on https://www.vive.com/eu/product/vive-pro/.

Jacket Strain Measurement for Different Amounts of Ease

Mizuki Yamakoshi[1], KyoungOk Kim[2*], Masayuki Takatera[2]

[1]*Shinshu University, Graduate School of Science and Technology, 3-15-1 Tokida, Ueda, Nagano, 386-8567, Japan*
[2]*Shinshu University, Faculty of Textile Science and Technology, 3-15-1 Tokida, Ueda, Nagano, 386-8567, Japan*

[*]Corresponding author's email: kimko@shinshu-u.ac.jp

Abstract

We measured the maximum principal strain of a ladies' tailored jacket when worn on two sizes of dress forms during arm movement, using a non-contact optical three-dimensional deformation measurement system. The maximum principal strains around the armhole were larger than the centre back for both body sizes. However, the strain distribution differed according to dress form size. More deformation occurred when the jacket is worn on a larger body. The deformation directions were also affected by the difference in the dress form sizes. It could be due to the different contact area between dress forms and the jacket. The amount of ease around the armhole affected the jacket's deformation on different-sized dress forms during arm movement.

Keywords: Strain Measurement; Tailored Ladies' Jacket; Ease; Different-sized Dress Form

1. Introduction

Clothing deformation is related to wearing comfort. Clothing deformation is caused by "fabric stretch" and "garment slip" [1]. Clothes are designed to have some ease for garment slip during movement. However, there is a limitation on garment slip. When garment slip exceeds the limitation, the fabric stretches, and this causes clothing pressure. Thus, the tensile properties of fabrics are usually considered when an article of clothing is made. However, the tensile properties of the fabric are not the only factor in clothing deformation. In addition to the characteristics of the fabric, garment shape and body movement also affect clothing deformation. To investigate garment deformation, it is necessary to quantitatively measure the strain of clothing caused by body movement.

Some researchers have investigated the measurement of clothing deformation. Kirk and Ibrahim [1] described the importance of fabric stretch for clothing comfort by measuring skin deformation and the tensile properties of fabric. However, not all clothing can be made out of stretchable fabric. Seo and Tamura [2] used a circular stamp to measure the local strains on long flared skirts after they were hung for two weeks. They also investigated the influence of the mechanical properties of fabric on local deformation using multiple regression analysis [3]. However, they measured the deformation of flared skirts, which is not strongly related to clothing comfort.

A jacket, which is made of woven fabric, is sometimes evaluated as uncomfortable to move in because of the low elongation of woven fabric. Kim et al. [4] investigated the comfort of a ladies' tailored jacket using arm movement. They clarified the places where there was a feeling of clothing pressure. However, the strain of the jacket that resulted from arm movement was not measured quantitatively. Moreover, a jacket is made to include some ease of movement and appearance [5]. Jacket deformation differs according to the amount of ease. However, the effect of movement ease on the jacket's deformation distribution has not yet been investigated.

In this study, we measured the strain of a ladies' tailored jacket on two sizes of dress form during arm movement. By comparing the strain differences between two dress forms, we investigated the effect of movement ease on jacket deformation during arm movement. By clarifying the deformation distribution of the jacket, we can provide information to set the appropriate amount of movement ease for a jacket.

2. Experiment

We prepared a ladies' tailored jacket (wool 100%, plain, Fig.1) and two ladies' dress forms (85 and 88 sizes) with arms. Table 1 shows the sizes of the jacket and dress forms. Table 2 shows the mechanical properties of the wool fabric. Assuming movement in daily life, we placed the jacket on each dress form, raised both arms forward. We measured the maximum principal strain on the shoulder of the jacket using a non-contact optical three-dimensional deformation measurement system (ARAMIS, GOM GmbH, Germany). We operated the movement of the dress form's arms automatically using an electric head tripod at a constant speed to reproduce the same movement. The initial posture of the arms is set at the height of the initial position of the electric tripod. Then, the arms moved to the position that the wrist and the shoulder of the dress foam were horizontal. The arms moved through an angle of 23.7 degrees. The measuring points were the left shoulder (s1-s21), as shown in Fig.2. We marked the measurement points with black thread. We randomly patterned the jacket using a black spray to enable image recognition using ARAMIS. We took measurements seven times for each dress form, and used the average value of five of the measurements, excluding the maximum and minimum values for each measurement point, as the result.

Table 1. Jacket and dress form sizes (cm)

Jacket and dress forms / Body parts	Jacket	Dress form 85 size	Dress form 88 size
Bust	92.8	85.0	88.0
Waist	81.1	64.0	67.0
Hip	99.8	89.0	92.0
Shoulder width	39.5	38.5	39.0
Back length	37.5	39.0	39.0
Upper arm circumference	32.4	24.0	26.0

Table 2. Mecanical prorerties of wool fabric

Mechanical properties	Wool fabric
Elongation at 500 gf/cm (%) in warp direction	4.73
Elongation at 500 gf/cm (%) in weft direction	6.04
Shear stiffness (gf/cm・degree) in warp direction	0.67
Shear stiffness (gf/cm・degree) in weft direction	0.66
Bending rigidity (gf・cm^2/cm) in warp direction	0.06
Bending rigidity (gf・cm^2/cm) in weft direction	0.05

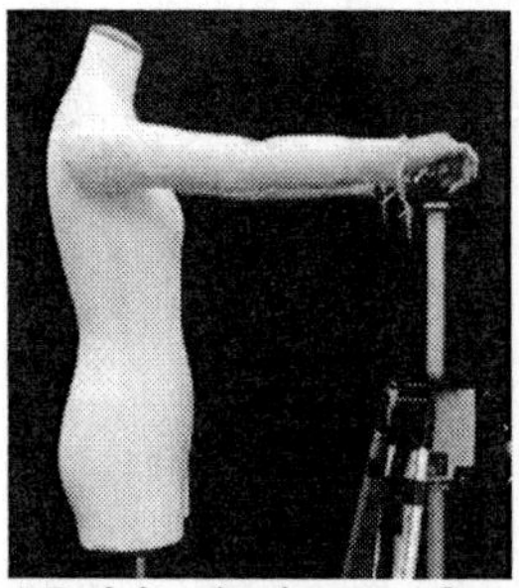

Fig.1 Raising both arms forward

Fig.2 Photo of the jacket

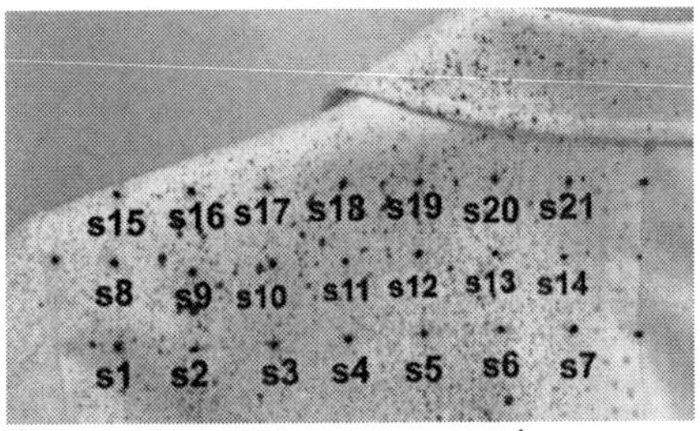

Fig.3 Measurement points

3. Results and Discussion

Figs.4 and 5 show the average values of the maximum principal strains of each measurement point for each dress form size. For the 85 size dress form, the measurement points for which the measured maximum principal strains were over 0.9% were s1, s2, s4, s8, s9, s11, and s15. These points were around the armhole. The largest strain was 2.5% at s15. By contrast, for the 88 size dress form, the measurement points for which the measured maximum principal strains were over 1% were s3, s4, s7, s10, s15 and s16. The largest strain was 3.4% at s16. Thus, it was found that the strain distribution differed according to the amount of ease.

Figs.6 and 7 show the directions of the maximum principal strains at the shoulder. For both dress forms, we confirmed that the directions of the maximum principal strains were toward the armpit around the armhole. For the 85 size dress form, we measured the elongation from the collar to the centre of the back. On the other hand, for the 88 size dress form, we measured the elongation from the collar to the armpit at the centre back. We found that the deformation directions were also affected by the difference in the dress form sizes.

This could be due to the different contact areas between dress forms and the jacket. It will be necessary to confirm the contact areas with cross-sectional data. More deformation occurred on the large dress form, which allowed less ease than the other body. This means that the jacket was more deformed when there was a small amount of ease.

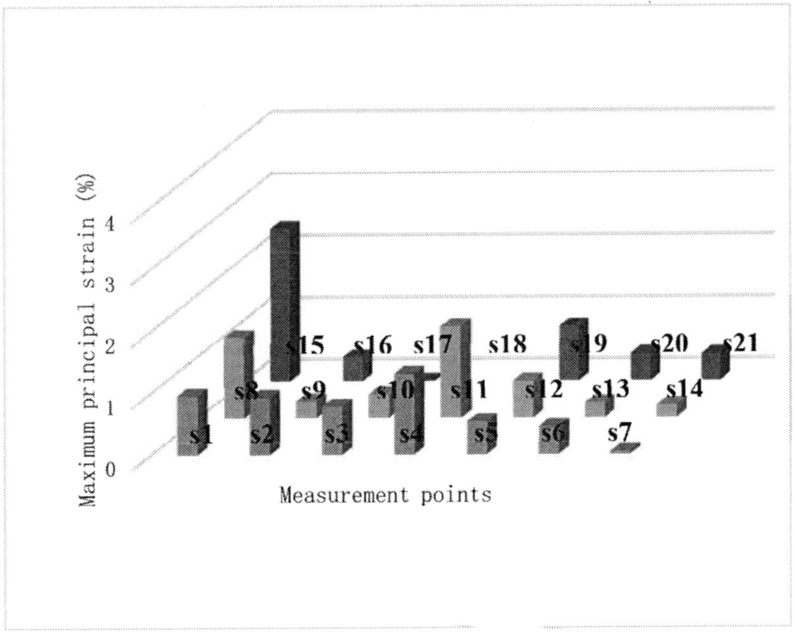

Fig.4 Maximum principal strains of the jacket on the 85 size dress form

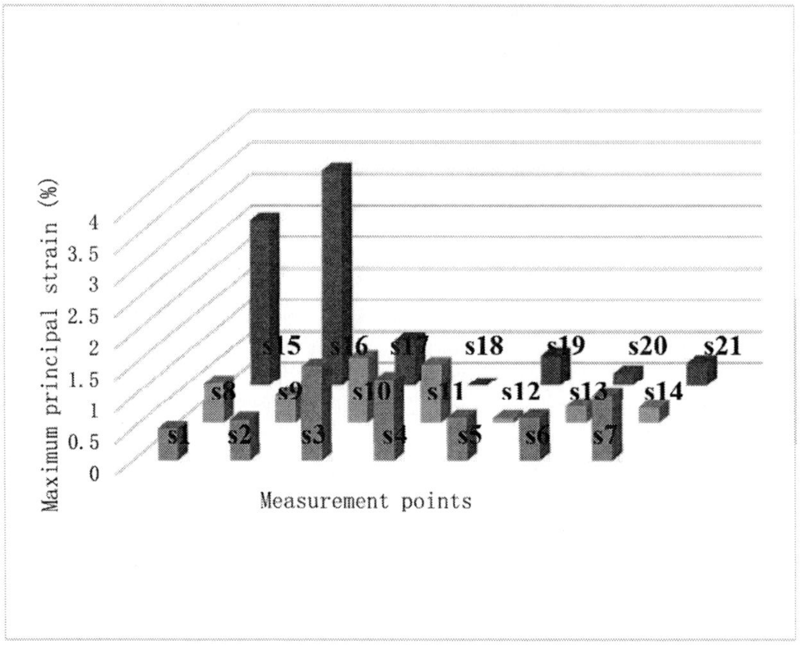

Fig.5 Maximum principal strains of the jacket on the 88 size dress form

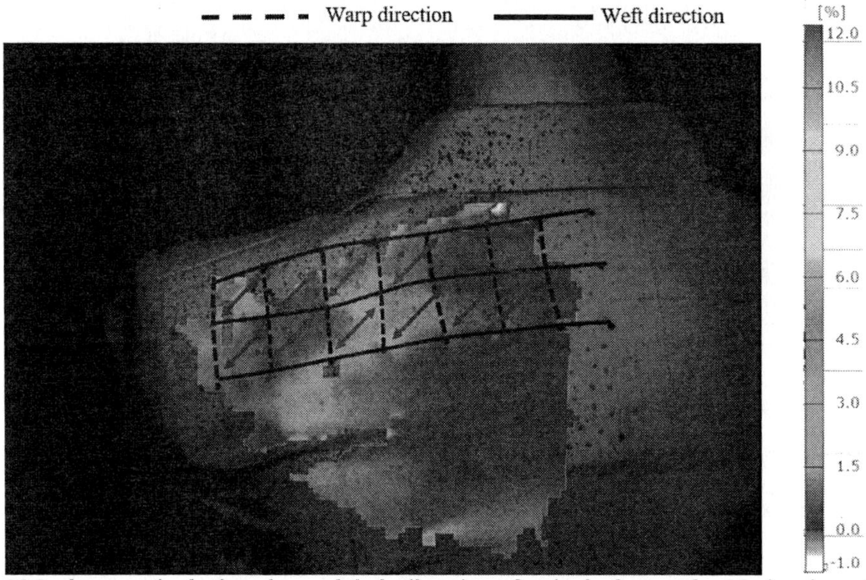

Fig.6 Maximum principal strains and their directions for the jacket on the 85 size dress form

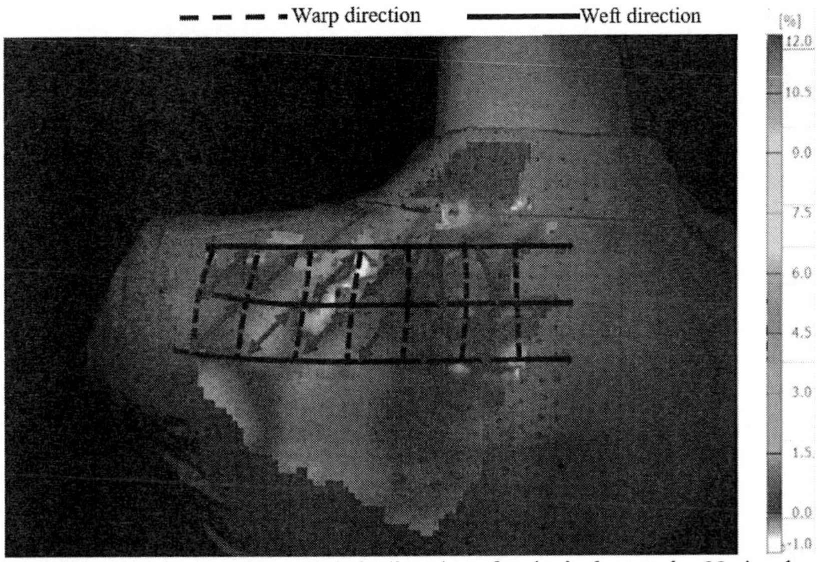

Fig.7 Maximum principal strains and their directions for the jacket on the 88 size dress form

4. Conclusion

We successfully measured the maximum principal strains of a tailored ladies' jacket worn on two sizes of dress forms during arm movement. We found that the strain distribution differed according to the amount of ease. The maximum principal strains around the armhole were larger than that at the centre back. More deformation occurred on the large dress form, which allowed less ease than the other body. This means that the jacket was more deformed when there was a small amount of ease. We also found that the amount of ease around the armhole greatly affected jacket deformation during arm movement.

In the future, we will take measurements of other parts of the jacket, such as the lower and side parts of the back. Moreover, further research of changes in the contact areas is necessary to understand clothing deformation more specifically.

References

[1] Kirk W, Ibrahim SM. Fundamental relationship of fabric extensibility to anthropometric requirements and garment performance. Text Res J: 1966; 36 (1): 37-47.

[2] Seo K, Tamura T, Local deformation on flared skirts under clothing's own weight and the relationship to the mechanical characteristics of fabric. J Jap Res Ass Text End-Uses: 2016; 57 (10): 763-774.

[3] Seo K, Tamura T, Influence of the mechanical characteristics of fabrics on local deformation of flared skirts under the clothing's own weight-using multiple regression analysis. Journal of the Japan Research Association for Textile End-Uses: 2017; 58 (8): 686-696.

[4] Kim K, Sugiyama C, Takatera M. Relationship between jacket comfort and stiffness of adhesive interlining. Int J Affect Eng: 2015; 14 (3): 137-142.

[5] Monobe A, Kim K, Takatera M. Effect of the differences between body dimensions and jacket measurements on the appearance of ready-made tailored jacket. Int J Cloth Sci Tech: 2017; 29 (5): 627-645.

MaxScript-based 3D Modeling of the Kirgiz "Tierman"

Han Yu, Ai-Min Xiao*, Shen-Ling Li

College of Textiles and Clothing, Xinjiang University, Urumqi, 830000, China

Corresponding author's email: 495178065@qq.com

Abstract

With the rapid development in society and changing lifestyles, many Chinese traditional weaving techniques are being practiced less and less. It is important to preserve Kirgiz traditional fabrics and integrate them in modern society. We propose to use three-dimensional modeling to analyze the weaving process and fabric organization of the Kirgiz traditional fabric Tierman. The weaving process of Tierman fabric, the 3D model of spinning, warping and weaving process are established, and the Tierman fabric is analyzed in terms of organization. The sample curves of the warp and weft yarns of the fabric are constructed by using MaxScript script to establish the 3D structure model of the fabric organization. The analysis of traditional weaving techniques through 3D modeling software promotes the dissemination of traditional weaving techniques, and to a certain extent solves the many problems faced by traditional weaving techniques in modern inheritance, and also provides a basis for the establishment of a digital display platform and the development of an interactive system for Tierman fabrics.

Keywords: Tierman; 3D Modeling; Weaving Process; Tissue Analysis

Foreword

The long history of handicrafts created by Kirgiz people have strong cultural significance, which has profoundly influenced the development process of the entire ethnic culture of the Kirgiz people [1]. The handcrafted fabric of the Kirgiz ethnic minority, Yueermaike, has a long history, strong cultural characteristics, ethnic personality, and plays an irreplaceable role in the cultural life of the Kirgiz ethnic minority. Tierman fabric is one of the hand-made fabrics of the Yueermaike non-heritage of the Kirgiz people, which has historical, cultural, spiritual, and social values. Many scholars have studied and investigated the Yueermaike, weaving techniques, and Ge Mengjia [2] studied the weaving techniques of the Kirgiz ground loom in many, describing and examining the characteristics of the ground loom fabric, fabric style, weaving principles, and the weaving process in detail. Yan Wenqi [3] examined Xinjiang Kirgiz weaving techniques as well as garment structure, mainly describing the ground loom fabric features fabric specifications in a textual manner, the principle of the ground loom and the weaving process in a graphic form, and also explains in detail the influence of fabric width on the structure of traditional Kirgiz clothing. This paper presents a systematic analysis of the Tierman weaving process and fabric organization using three-dimensional modeling. The weaving process is a key step in learning traditional weaving techniques. In the analysis of the weaving process, fabric spinning, warping, and weaving processes are introduced and the tools used are modeled in 3D structure. In the three-dimensional modeling of Tierman fabrics, the fabric tissue is the main object of study, and the fabric tissue is analyzed by the structural phase, and the MaxScript script is used to model the fabric tissue in a three dimensional perspective. The 3D fabric tissue analysis reflects the spatial relationship of warp and weft yarns, including the interweaving pattern, yarn cross-sectional shape, and yarn axial flexion pattern [4].

1. Introduction

Tierman fabric, as one of the traditional fabrics of the Kirgiz people is strongly represented through weaving techniques. The raw material of Tierman fabric consists of mainly wool camel hair and it is has been used for an extremely long time [5]. In the Kirgiz tradition Tierman fabrics are used most often for binding fences in felted houses, and in different regions also for decorating felted house enclosures or for making horse clothes to protect horses. The weaving process of the fabric is carried out after the process of, washing, loosening, felting strips, twisting yarns, dyeing, combining yarns, warping the spun yarns. [6].

2. Fabric Structure Characteristics Analysis

Tierman fabric for the warp line has flower warp floating pattern in the fabric organization. It is divided into plain weave and jacquard, according to the preset pattern color arrangement warp line, using 1 up 1 down plain warp and weft organization point alternation of the basic law, according to the pattern requirements change the organization point law, in the fabric form warp floating pattern.

By analyzing the front and back of the Tierman fabric as shown in Fig.1 and 2, it is concluded that both sides of the fabric contains plain tissue. The left warp is 6 green, 4 red, 1 green 1purple, 4 red arrangement, the right is 4 red, 1 purple, 1 green, 4 red, 6 green arrangement of warp threads to form a longitudinal stripe between the red and green. The middle section consists of two warp threads at the top and bottom of the warp floating pattern, 6 red and 6 green, 14 red and 14 green warp threads arranged and two warp threads of the same color warp thread jacquard formed pattern, Fig.3(a) for the plain fabric in Tierman fabric organization.

Tierman fabric in plain jacquard 1 as shown in Fig.3(b) the fabric organization structure follows certain procedure: 1 of green and red warp forms a group, the adjacent two groups of green warp line are picked out at the same time. Three consecutive green warp tissue point in the surface of the fabric is used to form a short warp floating line, then the back of the second tissue point weft is exposed; Similarly, the adjacent two groups of red warp line is picked out at the same time, three consecutive red warp tissue point in the surface of the fabric is used to form a short warp floating line, then the back of the second tissue point weft is exposed; Similarly, the adjacent two groups of red warp line at the same time is picked out, three consecutive red warp t1ssue ro1ntin the surface of the fabric to form a short warp floating line, then the back of the second tissue point weft exposed, the two colors of the short warp floating in the fabric surface are used to form a pattern. Tierman fabric jacquardis 2 as shown in Fig.3(c) in the jacquard tissue pattern is the color of Almuz to form the pattern of the number of warp yarn is 14 + 14. The green warp threads form a natural opening at the bottom and the red warp threads form a thread heald opening at the top.

Fig.1 Front Tierman fabric

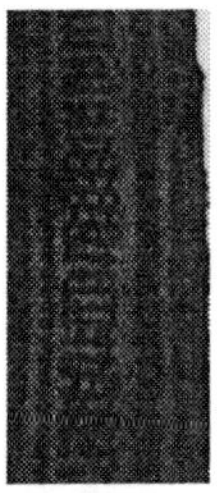

Fig.2 Reverse side of Ti1lman fabric

Fig.3(a) Plain tissue Fig.3(b) Plain jacquard 1 Fig.3(c) Plain jacquard 2
Fig.3 Tierman fabric physical organization diagram

3. 3D Simulation of the Weaving Process

3.1 Spinning

The traditional spinning tool of Tierman fabric is the Kirgiz. homemade spinning pendant and it was found in the research that the Kingiz have been using this traditional spinning tool and method to weave various "Yueermaike" woolen fabrics. The spinning pendant consists of around a spinning wheel and a twisting rod, and there is a horizontal wood at the top of the twisting rod to fix the pendant. The 3D modeling of the spinning pendant includes the modeling of each structure and the simulation of the appearance of the material as shown in Fig.4(a) and (b), and the detailed structural features of the spinning pendant can be observed through the 3Dstructural model.

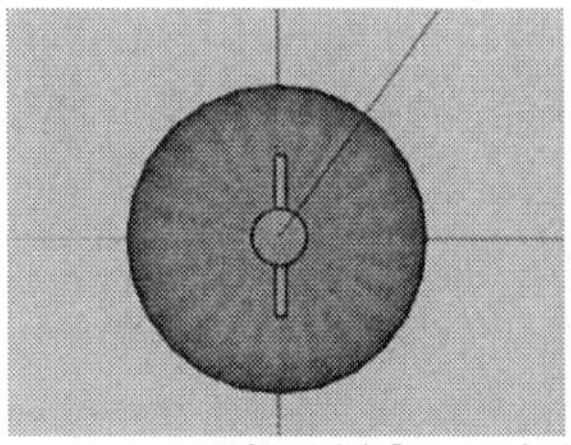
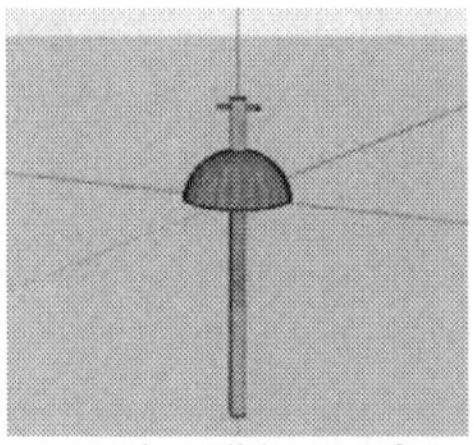

Fig.4(a)front view Fig.4(b) top view

Fig.4 Spinning pendant 3D model

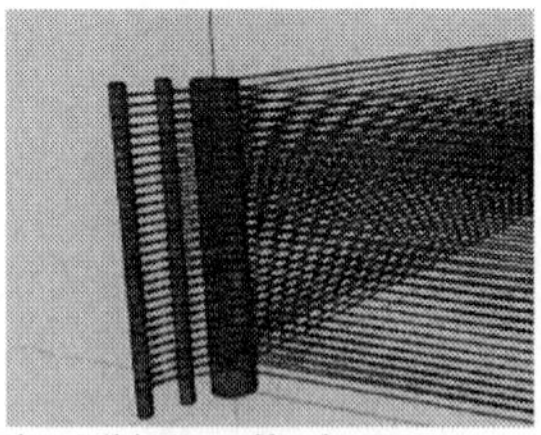

Fig.5(a)Full view Fig.5(b)Detail view 1

Fig.5 3D model of the whole warp

3.2 Warping

The finishing of the Tierman fabric warp is based on the jacquard pattern and color variation of the fabric, and the warping process is carried out by traction of the warp threads on triangular warping piles [7]. The 3D structural model is created based on the overall structural components of the warp by first setting up the required stakes and the brown thread device and then creating a 3D model based on the odd and even warp threads as shown in Fig.5. The creation of the 3Dstructure model allows the overall structure of the warp to be observed and reveals how the odd-numbered warp threads are fixed in the heald to be observed, facilitating the mastery and learning of traditional warping methods.

3.3 Weaving

The weaving of the Tierman fabric requires the original ground loom of the Kirgiz people. The assembly and structure of the original floor loom are simple. The main structure is a simple loom transformed from three wooden piles for warping [8], which is then turned into a complete 3D loom according to the order of the components, as shown in Fig.6(a) and (b) below. The floor loom structure is constructed by adding weft knives and warp dividers to the whole warp pile and the portable thread pallet, as well as the three wooden piles that support the whole floor loom.

By creating a 3D model of the fabric weaving tools and simulating the materials in each structure, we were able to present the specific size and shape of each tool and learn about the organization of each tool and the relationship between the operation of each part of the Tierman fabric weaving process.

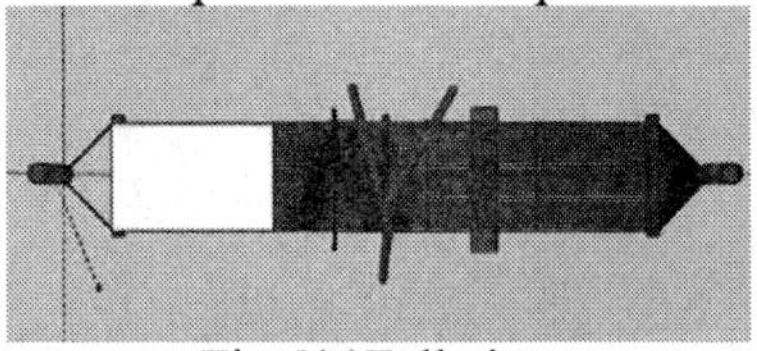

Fig.6(a)Full view Fig.6(b) Top view

Fig.6 Ground loom 3D model

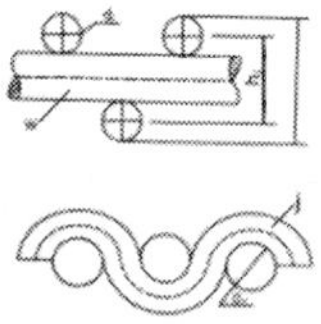

Fig.7 schematic diagram of the ninth structural phase of the fabric

4. Fabric Structure 3D Modeling

4.1 Fabric Structure Phase

The structural phase theory of fabrics can quantitatively describe the spatial relationship between warp and weft yarns, and the changes in the spatial geometry of warp and weft yarns in the fabric tissue are more complex, which are described by the flexural state of the yarns [9]. The fabric structure phase is divided by a flexural wave, and the flexural wave height is divided into eight equal parts, and nine equal points are obtained in the whole flexural height， each equal point corresponds to a structure phase, and the relationship between flexural wave and diameter of yarn can be analyzed by using the division of structure phase. The yarn is bent in the fabric to form a flexural wave, and the vertical distance between the wave crest and the wave valley is called the flexural wave height of the yarn in the fabric [10]. The flexural wave height is divided into the flexural wave height of warp yarn, expressed by hj, and the flexural wave height of weft yarn, expressed by hw. The relationship between warp and weft yarn flexural wave height and diameter: hi+hw-di+dw (hi warp flexural wave height, hw weft flexural wave height; dj warp yarn diameter, dw weft yarn diameter). Warp yarn is not bent, the weft yarn is bent then hj=0, hw=dj+dw; weft yarn is not bent, warp yarn bent then hw=0, hj=dj+dw [12]. The structural phase corresponding to the Tierman fabric tissue is the ninth structural phase, where the weft yarn is not bent and the warp yarn is bent, in a limit state, as shown in Fig.7 the center point of each weft yarn is on a straight line and the flexural wave height of the weft yarn is zero (hw = 0) the flexural wave height of the warp yarn is the sum of the warp and weft yarn diameters. In this limit state structural phase, the Tierman fabric tissue is modeled with the center of the fabric located on the weft yarn center line and the center point of the weft yarn on the reference line [11].

4.2 Geometric Relationship of Fabric Weave

Assuming that the plain weave section of warp and weft yarn is the warp and weft yarn interleaving model along the yarn axis, the dotted line in the figure is the yarn axis of warp and weft yarn, LJ (LW) is the distance between the yarn axes of two adjacent warp (weft yarn), and the length and width of a plain weave are 2lj and 2lw respectively.

As shown in the figure, it is the cross-sectional diagram of meridional plain weave θ J is the included angle between the warp axis and the center line of the cloth, which is called the weaving angle; L is the length of the yarn in the unit structure; LJ is the center distance between two yarns (i.e. the reciprocal of the warp and weft density of the fabric); HW is the buckling wave height of weft yarn; HJ is the buckling angle of the warp (i.e. the angle corresponding to half of the warp arc); H1 represents the height of weft in plain fabric; H2 represents the height of warp yarn in plain fabric; B J represents the included angle between the connecting line lo1o2 of two adjacent weft center points and the vertical line; MJ is the length of the warp cutting line; RJ is the warp radius; RW is the weft radius;DJ is the warp diameter; DW is the weft diameter; D is the sum of warp and weft yarn diameters; H is the thickness of the fabric [12].

Correlation from meridional diagram:

1) H1=2rw+hw;
2) Tanβj=Lj/hw;
3) Lj=(L-Dθ)cosθj+Dsinθj;
4) HJ= （Lj-Dθj） sinθj+D(1-cosθj)=1/2dj*cosθ;
5) D=dw+dj=hj+hw.

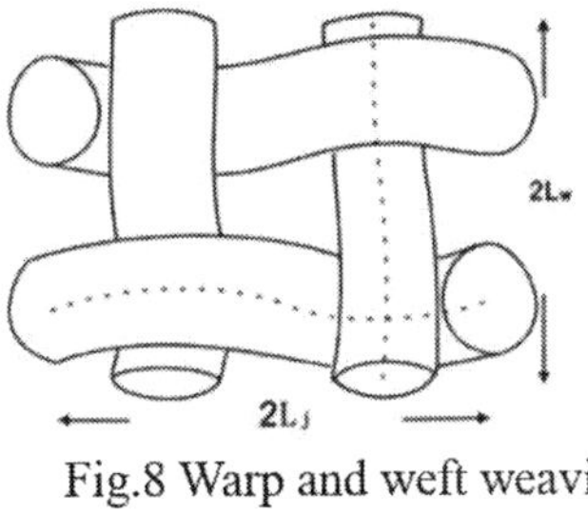

Fig.8 Warp and weft weaving model

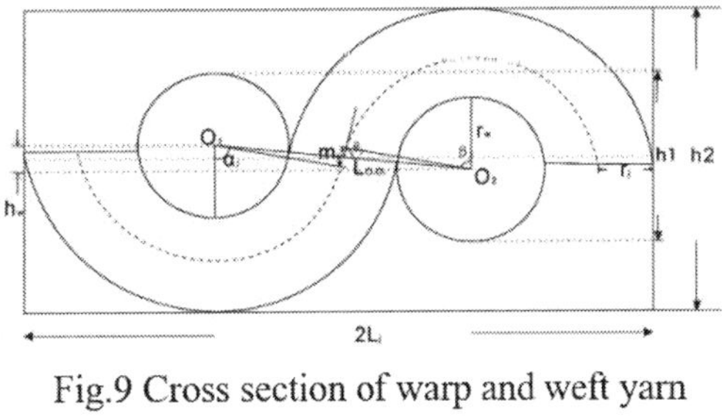

Fig.9 Cross section of warp and weft yarn

4.3 Basic Tissue 3D Structure Model

Structural modeling of fabric tissues starts by determining the type value points within that yarn using the complete fabric organization relationship and the flexural state of the yarn. The type value points are few coordinate points on the curve that describe the geometry obtained by measurement or calculation [13]. Tierman fabric is the ninth structural phase, and the type value points are selected to describe the yarn at the tissue points where the warp and weft yarns are interwoven and at the intermediate points of the tissue points. The yarn axis is generated by the type value point coordinates, and the fabric is formed according to the yarn organization cycle rule. MaxScript is the built-in scripting language of Max, which is quite powerful. It not only has the features that a programming language has, but can also be implemented on almost all the interactive operations within the program under the 3dmax interface. MaxScript is good at creating 3D vectors, matrices, quaternion algebra, and other fields, and can be used to solve complex problems in real-world production scenes. In the 3D modeling of fabric tissues especially complex fabric tissues, the use of MaxScript can present the 3D structure of the fabric more easily [14].

4.3.1 Plain 3D Structure Model

The Tierman fabric tissue contains a plain weave as well as two jacquard sections. The three-dimensional modeling of the plain tissue only needs to establish a complete tissue cycle and determine the coordinates of the type-value points of each warp and weft yarn, which can be sorted in an array along the overall axial direction to obtain a complete fabric tissue model. From the analysis of the fabric organization, it can be seen that the weft yarn of Tierman plain fabric does not bend, so the flexural wave of the weft yarn is zero, and the flexural wave height of the warp yarn is the sum of the warp and weft yarn diameters. In the plain tissue yarn axial diagram can be obtained from the number of warp and weft yarns in each complete tissue and the yarn correlation between the type value point. The following equation is obtained: number of warp yarn type value points = number of weft yarns *2+1; the number of weft yarn type value points = number of warp yarns *2+1 [15].

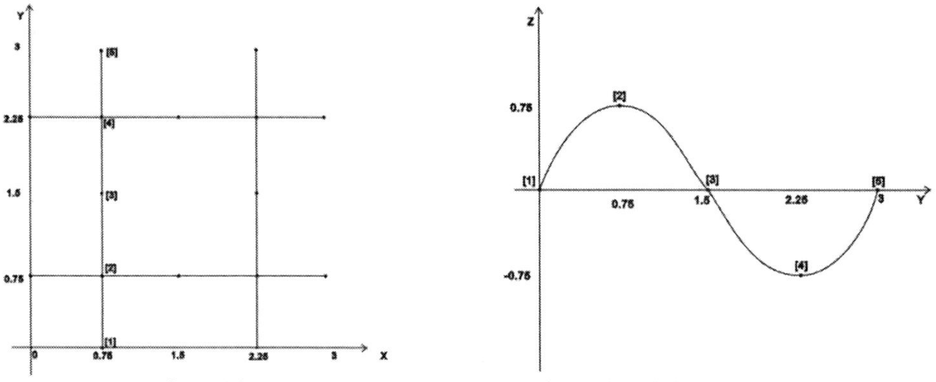

Fig.10(a)Top Fig.10(b) Right view

Fig.10 Schematic diagram of plain tissue yarn

In the plain tissue of Tierman fabric, the warp yarn flexural wave height is 1.5 when the warp and weft yarn diameter is 0.75. The control points of warp yarn are 5 and weft yarn is 5 in the complete tissue cycle according to the interweaving law plain. From Fig.10(a)(b), it can be seen that the X-axis coordinates are all 0.75 in the first warp yarn, 2.25 in the second warp yarn, and the Y-axis coordinates are also all 0.75 and 2.25 in the weft yarn, and a correlation between the type value points on the same yarn and different yarns can be found by the graphical representation of this complete tissue. When the diameter of yarn is 0.75 and the distance between warp and weft yarns is 1.5, the coordinates of each type value point are shown in Table.1 based on the organization points between yarns and the midpoint of the organization points.

Model the fabric organization in the MaxScript scripting language. The specific steps are as follows.

1) Curve first to create a Splineshape, and give an empty line, add a Spline in the empty line (AddNewSpline) at this time the line is still in the empty state.

2) Add the type value point coordinates to the empty line spline [addKnot. ss1#corner#line] 1 means add points to the first spline, corner means corner point, smooth, corner, bezier, bezierCorner are available. The line means a straight line, a curve and line are available.

3) Wirecolor to adjust the color of warp and weft yarn, color plus RGB color theory expression, the RGB value of 0-255, according to the fabric jacquard design for the color setting.

4) Array sorting of complete organizations using For loops

Finally, the three-dimensional structure model of the fabric under the constraints is obtained.

This plain tissue through the above steps to get the complete plain tissue three-dimensional model is shown in Fig.11, you can observe the characteristics of the fabric tissue through four views, the use of For cycle to the entire three-dimensional complete tissue array sorting is shown in Fig.12.

Table 1. Coordinates of type value points

Yarn	Type value point1	Type value point2	Type value point3	Type value point4	Type value point5
Chaine1	[0.75,0,0]	[0.75,0.75,0.75]	[0.75,1.5,0]	[0.75,2.25,-0,75]	[0.75,3,0]
Chaine2	[2.25,0,0]	[2.25,0.75,-0.75]	[2.25,1.5,0]	[2.25,2.25,0.75]	[2.25,3,0]
Weft yarn1	[0,0.75,0]	[0.75,0.75,0]	[1.5,0.75,0]	[2.25,0.75,0]	[3,0.75,0]
Weft yarn2	[0,2.25,0]	[0.75,2.25,0]	[1.5,2.25,0]	[2.25,2.25,0]	[3,2.25,0]

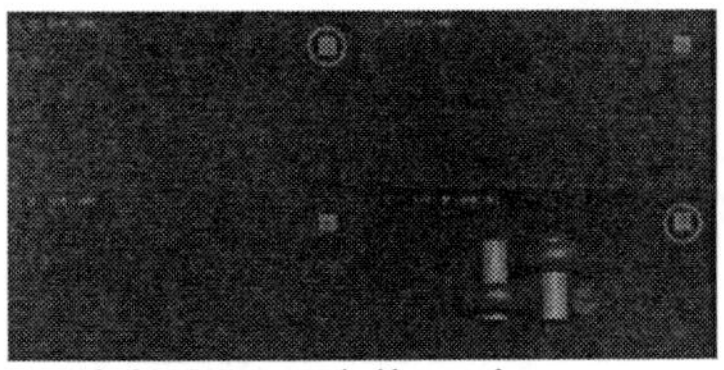

Fig.11 Plain 3D modeling view

Fig.12 Plain 3D modeling fabric

4.3.2 Plain Jacquard tThree-dimensional Structure Model

Tierman fabric jacquard organization is divided into two parts, weaving characteristics for plain warp and weft tissue point alternation based on the pattern requirements to change the tissue point pattern.

Jacquard fabric interweave law has periodicity. A complete tissue cycle contains two warp yarns and six weft yarns, according to the warp and weft yarn type value point calculation formula derived from the jacquard organization in the weft yarn type value point is 5, the warp yarn type value point is 13, in the jacquard fabric three-dimensional coordinate system X-axis direction for the weft direction, Y-axis direction for the warp direction, Z-axis direction for the plane normal as shown in Fig.13(a). The fabric is the ninth structural phase of the fabric structure phase, the warp yarn flexural wave height value is equal to the sum of warp and weft yarn diameter, so in the warp and weft yarn is interwoven tissue point part of the Z-axis type value point coordinate absolute value is 0.75, in other type value point Z-axis coordinate is 0, in Fig.13(b) can be seen in the distribution of the warp and weft yarn type value point in the jacquard fabric, the warp yarn of 13 The relationship between the yarn son the X, Y and Z axes can be seen in the coordinates of the yarns and the pattern between them. When the yarn diameter is 0.75 and the distance between the warp and weft yarns is 1.5, each type value point is derived from the relationship between the yarns (the type value point coordinates are omitted). In the jacquard fabric, the same sample curve is used for the 3D modeling of the complete organization, and the coordinates of the type value points of each yarn are determined, and the 3D model of the jacquard fabric is constructed through the script program to form the 3D model of the complete organization of the warp pattern as shown in Fig.14. The command of For loop is executed for the complete tissue loop, and the 3Dmodel of the jacquard one structure of the Tierman fabric is produced as shown in Fig.15.

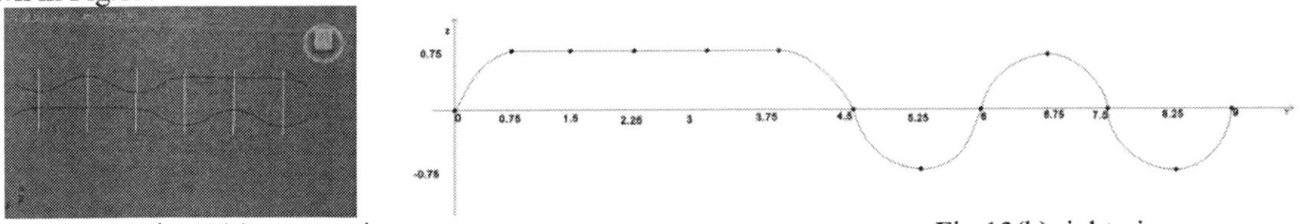

Fig.13(a) perspective Fig.13(b) right view

Fig.13 Schematic diagram of plain jacquard tissue yarn

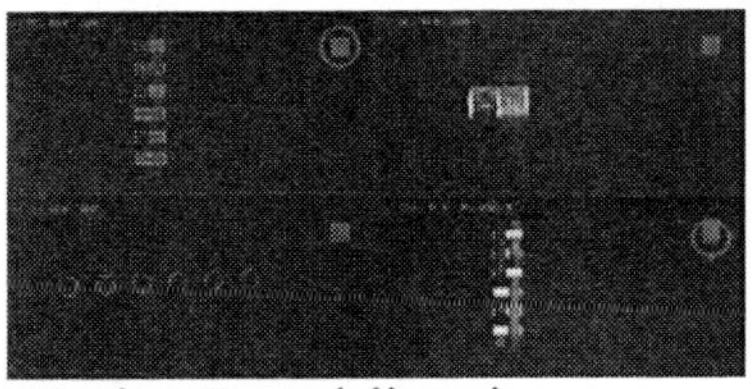

Fig.14 Jacquard 1 3D modeling view

Fig.15 Jacquard 1 3D modeling fabric

The jacquard two part of the Tierman fabric, in which the cross-section of each warp and weft yarn is obtained by type-valued point coordinates, enables the analysis of this jacquard fabric tissue. The 3D model of the fabric is constructed by SplineShape and For loop computer program. The control points of warp and weft yarns are used to model the complete tissue loop of Jacquard ll as shown in Fig.16, and the warp and weft yarns are rendered separately to form a 3D model of the complete tissue of a warp showing a plain pattern as shown in Fig.17.

The multi-perspective view of a complete tissue loop of Jacquard ll can be seen in Fig.18, through which the tissue structure can be observed. The cross-sectional view of each warp and weft yarn within the complete tissue cycle can be observed for each warp yarn in this Jacquard complete tissue as shown in Fig.19.

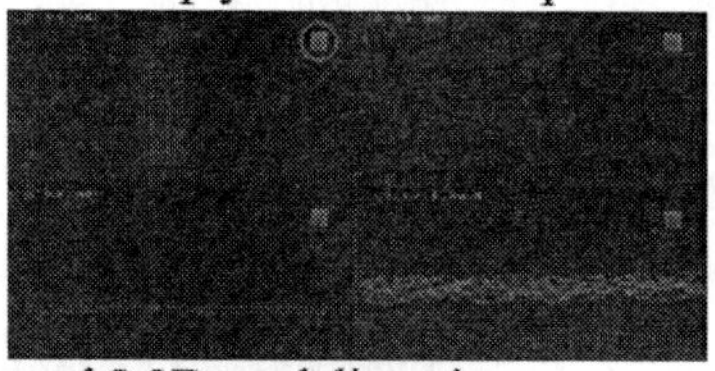

Fig.16 jacquard 2 3D modeling view

Fig.17 jacquard 2 3D modeling fabric

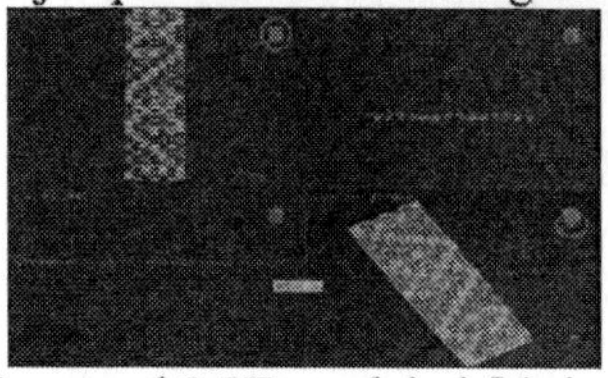

Fig.18 Jacquard 2 3D modeled fabric view

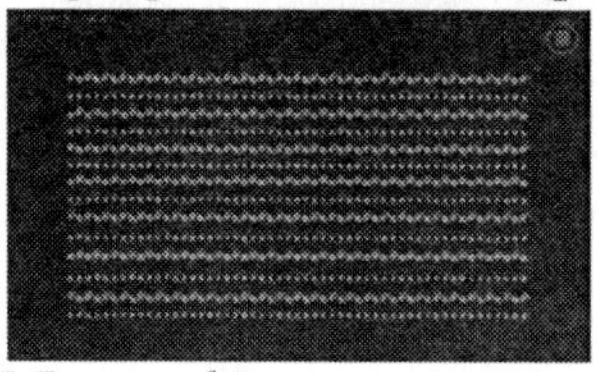

Fig.19 Jacquard 2 warp yarn cross-section

4.4 3D Structure Modeling of Tierman Fabric

Through three-dimensional modeling and finishing of the plain and jacquard tissues, a complete three-dimensional model of the entire Tierman fabric has been obtained as shown in Fig.20, which is the front view of one tissue cycle of the fabric. Fig.21 is the fabric three-dimensional tissue cycle on top view, front view, left view, and orthogonal view. The established three-dimensional model of the fabric tissue provides for a more realistic image, closer to the fabric itself. The three-dimensional organization of the plain tissue jacquard tissue is aggregated to form a complete Tierman three-dimensional model.

Fig.20 Tierman fabric 3D modeling

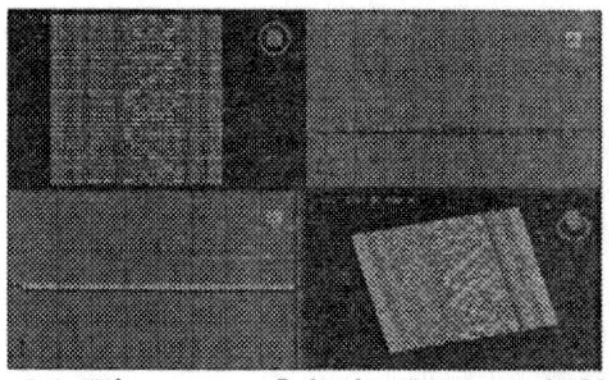

Fig.21 Tierman fabric 3Dmodeling view

5. Summary

In this paper, a three-dimensional analysis of fabric is proposed by analyzing the traditional Kirgiz. Tierman fabric. The structural phase of the fabric is analyzed with the physical fabric, and the three-dimensional modeling of the weaving process and fabric organization of the Tierman fabric is realized using modeling software and MaxScript. The realization of three-dimensional modeling of fabric can help cognize the weaving process of Tierman fabric thus is an innovative way to inherit and spread Tierman fabric. The use of MaxScript script for modeling purposes can also increase the efficiency compared to manual modeling while avoiding poor scalability of the model in 3D,, As a result, the fabric can be simulated quickly and thoroughly thus creating a digital representation of the Kirgiz traditional weaving skills.

References

[1] Zhang JD. A preliminary study of Kirgiz history. Lanzhou University: 2013.
[2] Ge MJ, Xiao AM, et al. Analysis of weaving skills of Xinjiang local loom. Silk: 2020; 57: 81-86.
[3] Yan WQ, Ge MJ. Analysis on weaving skills and clothing structure of Xinjiang Kirghiz local loom. Silk: 2019; 56: 103-109.
[4] Wang X. Research on 3D modeling method of woven fabric structure. Journal of Henan University of Engineering (NATURAL SCIENCE EDITION): 2013; 25: 6-10.
[5] Wan XY. Kirgiz traditional handicrafts in the period of social transformation. Journal of Guizhou Normal University: 2016; 4: 83-90.
[6] Qian J, Xiao AM, etal. Study on the weaving process of Kirgiz traditional wool fabric. Decoration: 2016; 12: 116-117.
[7] Xiao AM, Ge MJ, etal. Analysis on the traditional "Kaqiare" flower blanket weaving process of Kirgiz nationality. Silk: 2019; 56: 35-39.
[8] Qian J, Xiao AM. Research on Inheritance and innovation of Kirgiz camel hair fabric "Qiepukan" .Wool textile technology: 2018; 46: 72-75.
[9] Liu TT, Xiao AM. Analysis of traditional ribbon types and color design of Kirgiz people. Daguan: 2019; 12: 176-177.
[10] Yang H, Tong XT. Theoretical design and implementation of 3D woven fabric simulation based on 3DS MAX. Silk: 2011; 04: 24-27.
[11] Zhang SJ. Construction of geometric structure phase of woven fabric in virtual reality technology. Cotton textile technology: 2019; 12: 21-25.
[12] Gu DP. Space parametric model and simulation of plain woven fabrics. Journal of Textile Research: 2012; 10: 37-42.
[13] Wang X. 3D modeling of single layer woven fabric structure using MAXscript language. Acta textile Sinica: 2019; 40: 159-165.
[14] Wang H. Complete learning manual of 3dsmaxscript script language. Ordnance Industry Press: 2006.
[15] Yu WD. Textile materials. China Textile Press: 2018.

The Pattern Structure of Elastic Movement Armguard Based on the Pressure Changes

Shu-Tian Duan[1], Yu-Ting Fan[1], Ming-Hai Cui[1*]

[1]*Beijing Institute of Fashion Technology, Sakura East Street, Beijing, 100029, China*

[*]Corresponding author's email: fzycmh@bift.edu.cn

Abstract

To explore the relationship between the stretch rate, stretch length and pressure of the elastic fabric on the sports arm. The main research content of this paper is to conduct fabric experiments and pressure experiments on nylon/spandex and polyester/spandex fabrics with different blending ratios, compare the structure pattern of inelastic fabrics and analyze the application performance of elastic fabrics on the upper limbs of the human body. The aim of the research is to obtain the pattern prototypes of different fabrics under constant pressure to acquire the regression relationship between stretch rate, stretch length and corresponding pressure value. The results reversely predict the pressure value of the known fabric under different stretching conditions and can provide a reference for the structural design of elastic fabric sports armguards in the future.

Keywords: Elastic Fabric; Tensile Properties; Pattern Structure; Clothing Pressure; Sports and Protection

1. Introduction

With the vigorous development of national fitness activities, physical fitness activities have become one of the lifestyles pursued by the younger generation. However, if fitness exercises are practiced without professional guidance, it is very likely to result in sports injuries.[1] Therefore, there is an increasing market consciousness on wear sports protective gear to avoid injury during exercise. In fitness activities, training involving upper limbs such as lifting barbells, planks, rock climbing, tennis, and badminton smashing movements are common. Due to the repeated use of upper limbs, research on upper limb sports protective gear has practical significance. [2]

Since the development of sports protective gear, there is no corresponding industry standard for its protective effectiveness, so it is difficult to judge the true effect of such products. The development of sports protection products involves many interdisciplinary subjects such as biomechanics, sports medicine, and material mechanics. Therefore, it is necessary to study the protective performance of sports armbands under different fabrics.

1.1 Research Status of Sports Arm Guards

The current research on sports protective gear is currently limited. From the point of view of sports injury mechanism and biomechanics, Li Yuan changed the way of fabric compounding and the way that the protective gear wraps the upper limbs. It is concluded that the elasticity, thickness, composite method of the fabric and the wrapping method of the protective gear influences the change of the upper limb motion parameters.[3] Moreover, most of the current stress studies on sports protective gear are studies on the lower limbs and torso of the human body. For example, cycling pants, compression stockings, abdominal belts, and tights. [4] There are are lack of related research on human upper limbs.

The arm guard has two main functions, to maintain the temperature of the arm and to secure the tightness of the muscles. Therefore, this research combines the various physical performance parameters of elastic fabrics, starting from the stress of elastic fabrics on the upper limbs of the human body due to stretching and deformation. This study aims to discover the relationship between the plane model structure of the sports armband, the elastic elongation of the fabric and the pressure change.[5~7]

2. Fabric Experiment

Nowadays, there are sports armguards of different prices and materials on the market.[8] Through the previous market research and questionnaire survey, the experiment selected 11 common elastic woven fabrics from 106 valid questionnaires and 69 existing products in the market. They are fabrics made of nylon and spandex blends and polyester and spandex blends with different proportions. The basic parameters of each fabric are shown in Table 1. In Table 1, N stands for nylon, SP stands for spandex, and T stands for polyester.

Table 1. Basic parameters of fabric

Fabric number	fabric composition[%]	Fabric thickness [mm]	Fabric weight [g/m^2]
1#	N/SP 87/13	0.83	286.38
2#	N/SP 81/19	0.59	226.82
3#	N/SP 80/20	0.47	153.35
4#	N/SP 76/24	0.70	218.54
5#	N/SP 70/30	0.74	242.50
6#	T/SP 92/8	0.50	191.00
7#	T/SP 87/13	0.70	247.20
8#	T/SP 82/18	0.54	216.26
9#	T/SP 80/20	0.72	253.85
10#	T/SP 77/23	0.84	254.39
11#	T/SP 73/27	0.69	246.64

2.1 Fabric Comfort Performance Test

Refer to the standard GB/T3820 "Determination of Thickness of Textiles and Textile Products"[9], GB/T3923.1-2013 "Tensile Properties of Textile Fabrics Part 1: Determination of Breaking Strength and Breaking Elongation (Strip Method)"[10], GB/T3923.1-2013 T5453.1977 "Determination of Air Permeability of Textile Fabrics"[11], GB/T12704.1991 "Method for Measuring the Moisture Permeability of Fabrics" Moisture Permeable Cup Method[12]. The thickness, dry weight per square meter, elastic elongation and recovery rate, fabric air permeability, and moisture permeability of the above 11 elastic knitted fabric samples were tested. Three samples of each fabric were selected for fabric experiment. The experimental results are shown in Table 2.

Table 2. Fabric comfort performance parameters

Fabric number	Elastic elongation[%]		Elastic recovery rate[%]		Air permeability[mm/s]		Moisture permeability[g/m^2·h]	
	warp	weft	warp	weft	face	rear	face	rear
1#	34.13	41.25	86.83	88.22	87.33	75.22	117.00	261.61
2#	41.27	43.95	90.86	97.72	105.13	110.95	201.52	174.04
3#	37.82	66.37	100.00	89.34	156.22	156.78	316.51	114.84
4#	25.98	30.55	93.48	96.07	92.65	90.79	131.33	139.25
5#	35.48	37.82	97.93	97.82	157.18	152.01	311.91	102.98
6#	44.13	57.78	90.52	92.94	217.74	204.69	176.71	164.12
7#	12.93	47.23	97.84	100.00	92.57	89.10	414.46	359.98
8#	24.43	38.65	97.61	86.13	96.72	88.43	108.55	104.51
9#	39.32	49.98	94.48	97.73	61.68	61.90	140.66	323.50
10#	47.68	63.95	97.73	99.84	64.42	64.81	96.20	167.99
11#	33.52	45.98	99.76	99.48	140.08	173.80	115.05	228.12

2.2 Fabric Experiment Results and Analysis

2.2.1 Fabric Elongation and Elastic Recovery Rate

In the experiment, the YG065C electronic fabric strength meter was selected as the experimental instrument, and the operation was carried out in accordance with the standard GB/T3923.1-2013. Set the pre-tension to 5N. The elastic recovery rate experiment adopts the method of constant force and repeated stretching, and the test result is expressed as the average value of the experimental data of 3 samples. In order to better analyze the various properties of the experimental fabrics, the elastic elongation results of the above 11 fabrics are plotted as a bar graph, as shown in Fig.1.

On the whole, the elongation in the weft direction of the 11 fabrics is significantly better than that of the fabric in the warp direction. The difference in the elastic recovery rate in the warp and weft directions is not obvious, and both remain above 85%. Among them, 3# fabric has excellent weft elongation and elastic recovery. The warp elongation and elastic recovery rate of 10# fabric are second.

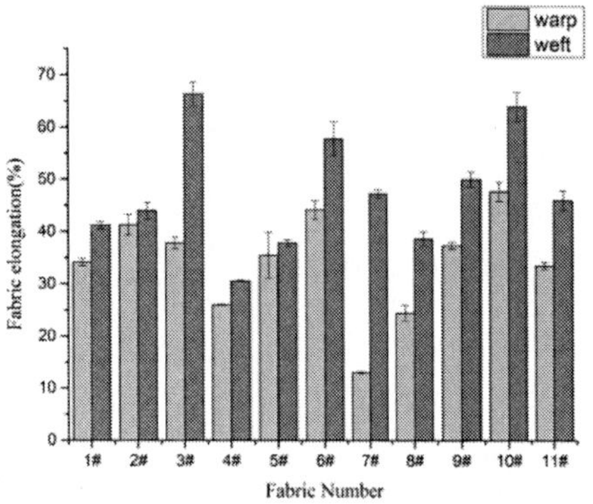

Fig.1 Fabric elongation histogram

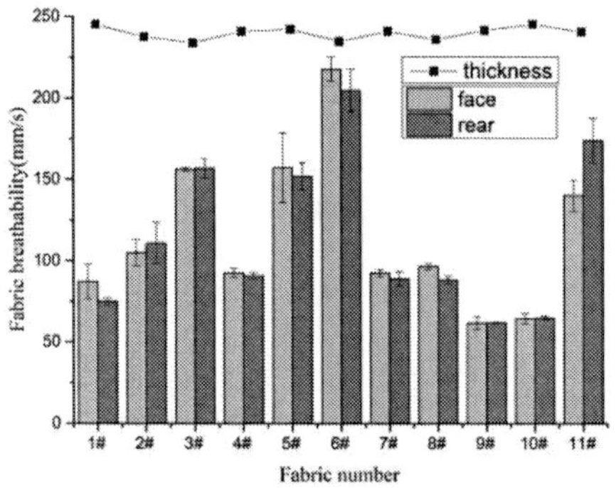

Fig.2 Fabric breathability and thickness histogram

2.2.2 Fabric Breathability and Moisture Permeability

The air permeability test results of 11 elastic knitted fabric samples are shown in Fig.2. It can be seen from the analysis in Fig.2 that the air permeability of the experimental fabric varies with the thickness of the fabric. Among them, 6# fabric (92% polyester/8% spandex) has the best air permeability, and 9# fabric (80% polyester/20% spandex) has the worst air permeability. This may be due to the thickness of the knitted samples and the different weaving methods.

From the analysis of Table 2, it can be seen that the moisture permeability of the 11 experimental fabrics is different, and the moisture permeability of the face and rear sides of some fabrics is very different. Among them, 7# fabric (87% polyester/13% spandex) has good moisture permeability on both sides, and 8# fabric has the worst moisture permeability.

2.2.3 Fabric Experiment Data Analysis

In summary, the performance parameters of the 11 elastic knitted fabric samples are different. In order to further simplify the selection of fabrics, the factor correlation analysis was carried out through IBM SPSS Statistics 21 software. Thereby, the main influencing factors of various performance parameters of the fabric are obtained.

Taking fabric elongation as an example, the KMO test coefficient is 0.523 (greater than the threshold 0.5), and the significance level of Bartlet is 0.009 (less than 0.05). It can be seen that there is a significant correlation between the variables, indicating that this result is suitable for factor analysis. Using principal component analysis to extract the main factors, it can be seen from Table 3 that the initial eigenvalues describe the initial factors. The characteristic value of the first factor is 2.163, which explains 54.071% of the original 4 fabric performance. The characteristic value of the second factor is 1.327, which explains 33.184% of the original 4 fabric performance. The cumulative contribution rate of these two factors reached 87.255%. And its factor characteristic values are all greater than 1, these two factors can better explain most of the content of the fabric performance.

Next, according to the two common factors extracted by factor analysis, multiply the variance contribution rate after the corresponding factor rotation as the weight value. Calculate the comprehensive score of the factor, and process it in descending order. In the same way, the elastic recovery rate, air permeability, and moisture permeability of the fabric were analyzed by factor analysis. Comprehensive consideration of fabric experiment results and factor comprehensive score ranking. In the end, the top four fabrics are 10#, 11#, 7# and 9#. Choose these four outstanding fabrics for follow-up pressure experiments.

Table 3. Explained total variance

Ingredients	Initial eigenvalue			Extract the sum of squares			Rotated sum of squares		
	Total	Variance [%]	Accumulation [%]	Total	Variance [%]	Accumulation [%]	Total	Variance [%]	Accumulation [%]
1	2.163	54.071	54.071	2.163	54.071	57.071	1.941	48.526	48.526
2	1.327	33.184	87.255	1.327	33.184	87.255	1.549	38.729	87.255
3	.414	10.348	97.604						
4	.096	2.396	100.000						

Extraction method: principal component analysis.

3. Elastic Fabric Pressure Test

3.1 Pressure Test Program

The pressure experiment uses the German Novel Pliance multifunctional dynamic pressure distribution measurement system (Fig.3), a standard rigid cylinder with a radius of 4 cm.

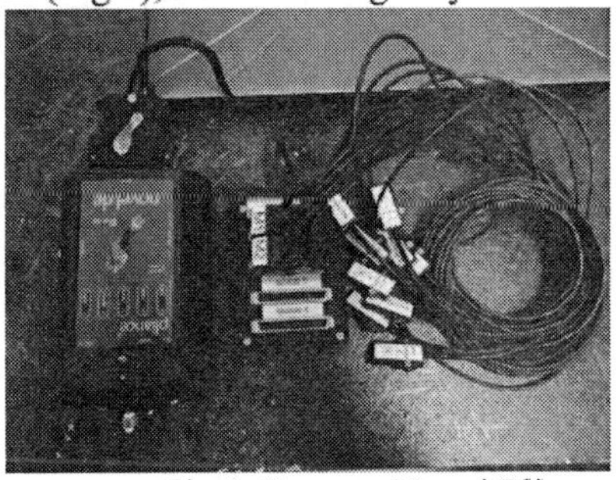

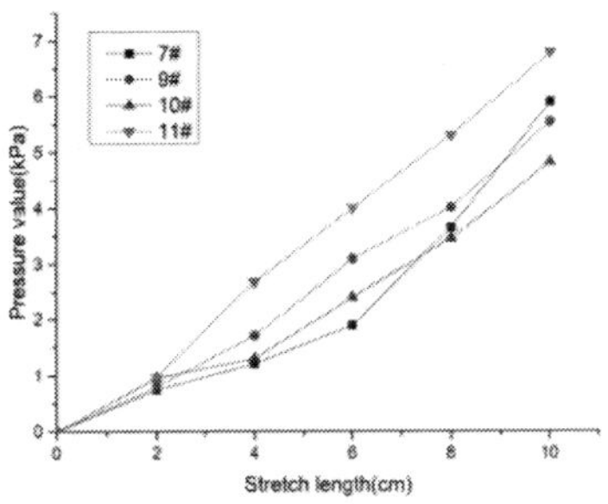

Fig.3 German Novel Pliance pressure test instrument(left)
Fig.4 Line chart of stretched length and pressure of fabric(right)

Refer to the excerpt from "Chinese Adult Human Body Dimensions (GB/T 10000-1988)", it is known that the average full arm length of women aged 18-25 is 50cm, and the arm circumference is 24cm. Therefore, a cylinder with a radius of 4cm, which is close to the arm circumference of a female human body, is selected for the pressure experiment. Put the 4 fabrics selected in the fabric experiment on a standard cylinder in an unstretched manner, record the length of the fabric, and the pressure is 0 at this time. Put the 4 fabrics on the standard cylinder in the manner of stretching 4cm, 6cm, and 8cm in turn, and record the length of the fabric and the pressure on the standard cylinder at this time. Obtain the pressure values of the 4 fabrics under different stretch lengths, see Table 4.

3.2 Analysis of Pressure Test Results

The pressure experiment uses a rigid cylindrical tube to simulate the actual wearing pressure of the human body, assuming an ideal state.[13~15] Therefore, the influence of variables such as human tissue structure and body surface curvature is not considered. The results of the pressure experiment are shown in Fig.4, the pressure values of the four elastic fabrics all increase with the increase of the stretched length. Perform correlation analysis and regression analysis on the stretch rate, stretch length and the corresponding pressure value, and establish a regression equation:

$$Z=1.628+0.567X-0.037P \quad 1)$$

In formula (1): Z is the total pressure, kPa; X is the stretched length of the fabric, cm; P is the stretch rate of the fabric, %. The correlation coefficient of the equation is 0.960 (greater than 0.95), indicating that the regression equation has a good fit. Therefore, it can be seen that the stretch rate, stretch length and pressure value of different fabrics are highly positively correlated.

Table 4. Fabric stretch length and pressure test results

Fabric number	Stretching length [cm]	Pressure[kPa]	Fabric number	Stretching length [cm]	Pressure[kPa]
7#	0	0.000	10#	0	0.000
	2	0.740		2	0.957
	4	1.205		4	1.299
	6	1.896		6	2.400
	8	3.650		8	3.460
	10	5.900		10	4.822
9#	0	0.000	11#	0	0.000
	2	0.801		2	0.966
	4	1.706		4	2.679
	6	3.091		6	3.999
	8	4.010		8	5.300
	10	5.540		10	6.792

4. Pattern Structure of Sports Arm Guard

In order to analyze the influence of different fabrics on the structure of sports armbands. In this research, the plane prototypes of different fabrics on the upper limbs were obtained through the methods of fixed-point positioning, drawing grid processing, and three-dimensional cutting. [16] And compare the human upper limb structure obtained by the non-elastic fabric with the flat prototype under the elastic fabric.

4.1 Prototype Structure Acquisition of Elastic Fabric

4.1.1 Experiment Preparation

Experimental materials: 9# elastic fabric, marking line, gel pen, pins, scissors.

The subjects selected women whose BMI index was in the range of 18.5-23.9 and arm circumference was within 26cm. Refer to the excerpt from "Chinese Adult Human Body Dimensions (GB10000-88)", it is known that the average full arm length of women aged 18-25 is 50cm, and the arm circumference is 24cm. And stipulate that the subject's arm is extended horizontally, and the palm of the hand is upward as the basic action. Based on the data of the human upper limbs and most of the sports protective products on the market, this study set the sample length to be 38cm and the girth to be 16-24cm. Measure 48cm in the longitudinal direction of the fabric, draw a vertical horizontal line as the length, and then take the vertical bisector of the length 34cm as the width.

4.1.2 Picture Frame Processing

The drawing grid processing of the fabric refers to the longitudinal and horizontal drawing lines on the experimental fabric to form a grid. [17] In this study, grids were drawn with a spacing of 5cm × 5cm (length × width) to observe the elastic elongation changes in the warp and weft directions of the fabric during the three-dimensional cutting process.

Secondly, draw a standard line in the measured area of the subject's left limb. Before drawing, please stretch your arms horizontally with your palms facing upwards. It is stipulated that the straight line starting from the inner midpoint of the wrist, passing through the inner midpoint of the elbow through the flexor group along the elbow, to the axillary position in front of the shoulder is the meridian standard line. And the latitude standard line requires a circle around the elbow horizontally. The purpose of drawing the standard line is to unify the alignment standard for subsequent experiments.

4.1.3 Three-dimensional Cutting

The three-dimensional cutting of the one-piece elastic sports armband needs to reach a state where the upper limbs are wrapped tightly and without wrinkles. Therefore, when performing three-dimensional cutting of elastic fabrics, the seams of the fabrics should be flat and aligned. According to the structure of the upper limbs of the human body, the circumference is marked with 5cm in each length direction. The purpose is for the subsequent stripping and rubbing of the paper pattern. At the same time, pay attention to observe the changes in the stretched length of the fabric in the warp and weft directions.

4.1.4 Three-dimensional Cutting into a Flat Pattern

Take off the three-dimensionally cut fabric and iron it to make sure that the deformation of the fabric is fully restored. Use a ruler to measure and calibrate to obtain a three-dimensional cut sample of the elastic fabric. The obtained vertical cut sample is connected to the dots, the sulfuric acid paper is rubbed, and the curve is corrected, and the net sample is finally obtained. Organize the adjusted net sample data on key parts, such as the length of the sleeve, the width of the upper arm circumference, elbow circumference, and lower arm circumference. Fig.5 shows the plane structure of the elastic fabric sports arm guard drawn in the rich CAD software.

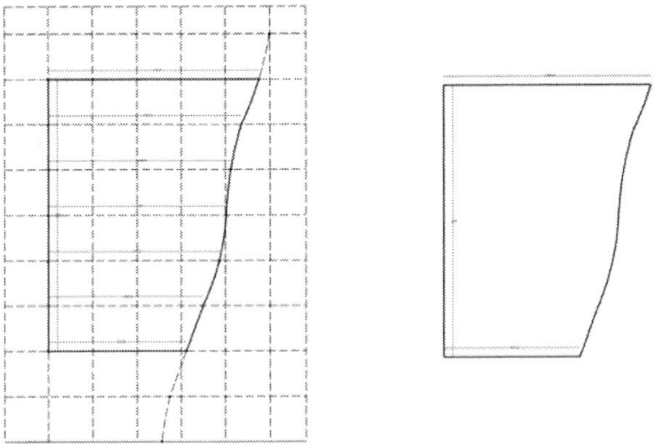

Fig.5 Plane structure of elastic fabric sports arm guard

4.2 Prototype Structure Acquisition of Non-elastic Fabric

4.2.1 Experiment Content

The experimental materials are non-elastic non-woven tape, marking line, gel pen and scissors. The requirements of the experimental object and the prototype structure of the elastic fabric are the same.

Similar to the method of obtaining the prototype structure of elastic fabrics, the prototype structure of non-elastic fabrics first requires three-dimensional extension. The non-woven tape is fixed on the test area of the subject's arm according to the position of the standard line of the human body in the order of longitudinal first and then transverse to obtain a completely body-fitting paper pattern, which can well show the structural characteristics of the upper limbs of the human body. In the specific operation process, it is also necessary to mark the circumference of every 5 cm in the length direction according to the structure of the upper limbs of the human body to facilitate the subsequent stripping and rubbing of the paper pattern. Then, cut from the side close to the body (that is, along the armpit, through the upper end of the elbow and the ulnar styloid of the wrist) to obtain the three-dimensional structure of the upper limb of the human body, as shown in Fig.6.

4.2.2 Convert Three-dimensional Rubbings into Flat Patterns

Different from the process of three-dimensional cutting of elastic fabrics into flat patterns, since non-woven tapes have no elastic stretch recovery performance, the three-dimensional structure of human upper limbs obtained by vertical cutting of non-elastic fabrics. Therefore, it is first necessary to transform the three-dimensional volume into a two-dimensional plane. The protrusions and depressions of the three-dimensional structure of the upper limbs of the human body are respectively kneaded and cut, so that the fabric can be spread on a horizontal tabletop. Secondly, the three-dimensional rubbing is scanned into the computer through the scanning software (Fig.6), and the provincial roads and cuts are converted and corrected to make the overall outline structure clearer. Then, organize the data of the key parts of the adjusted plane pattern, and draw the plane structure diagram of the non-elastic fabric through Fuyi CAD software (Fig.6).

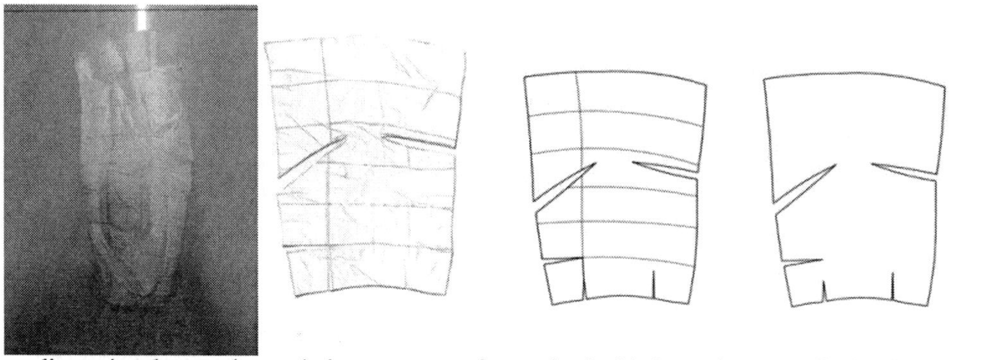

Fig.6 Three-dimensional extension and plane patterns of non-elastic fabric on the upper limbs of the human body

4.3 Comparative Analysis of Prototype Patterns of Elastic Fabrics and Non-elastic Fabrics

Measure the girth data of each file of the prototype under the elastic fabric and the non-elastic case, and record it, analyze the difference between the prototype pattern of the elastic fabric and the non-elastic fabric, and find the law of change. The comparison of the circumference data of the upper limbs of the elastic fabric and the non-elastic fabric is shown in Table 5, and the prototype pattern comparison of the elastic fabric and the non-elastic fabric is shown in Fig.5 and Fig.6.

Table 5. Comparison of girth data of elastic fabric and non-elastic fabric prototype pattern[cm]

Prototype pattern	0cm	5cm	10cm	15cm	20cm	25cm	30cm	length
Stretch fabric	15.7	18.7	20.4	20.2	20.6	22.7	23.8	29.9
Non-elastic fabric	18.4	20.4	21.6	21.3	22.3	23.4	24.4	30.6
Difference	-2.7	-1.7	-1.2	-1.1	-1.7	-0.7	-0.6	-0.7

By comparing the girth data of the elastic fabric and the non-elastic fabric prototype, we can find that compared with the non-elastic fabric prototype, the dimensions of the elastic fabric prototype in the vertical and horizontal directions show a shrinking trend. And there will be differences according to the changes in the structure of the upper limbs of the human body.

By comparing the prototype patterns of elastic fabrics and non-elastic fabrics, we can find that in terms of structural treatment, the prototype of elastic fabrics mainly eliminates excess float through the elastic stretch recovery performance of the fabric itself. However, the prototype of non-elastic fabrics needs to deal with the floating margin through the way of saving and saving. In terms of graphics, the elastic fabric prototype takes the savings in the non-elastic fabric prototype at the seams by offsetting the elastic stretching. Therefore, compared with the prototype obtained from the non-elastic fabric, the prototype drawing of the elastic fabric does not appear to be a province. Therefore, the size of each girth is relatively reduced.

5. Results and Discussion

This study combines the linear relationship between fabric stretch rate, stretch length and pressure, and obtains the influence of elastic fabric performance parameters on the change of pattern structure. Under the same pressure, the stretch of different elastic fabrics is different, and the structure of the sports armband is different. In contrast, in the case of the same fabric properties, different stretching conditions will produce different pressure values.

Compared with the structural pattern of inelastic fabrics, the stretch properties of elastic fabrics can simplify the dart and cut in the pattern of inelastic fabrics. When designing paper patterns with complex human body structures, the design can be optimized by using the stretch recovery performance of elastic fabrics. It is hoped that this research result can provide a reference for the structural design of elastic fabric sports armguards in the future. The design of the tight-fitting sports arm guard can make full use of the stretchability of the fabric under the condition of certain control pressure.

References

[1] Li YN, Lu H. Development of smart clothing for waist sports injury protection. Journal of Textile Research: 2020; 41(02): 119-124.
[2] Duffield R, Portus M. Comparison of three types of full-body compression garments on throwing and repeat-sprint performance in cricket players. [J]. British journal of sports medicine: 2007; 41 (7): 409-14; discussion 414.
[3] Li Y, Xie H, Deng HQ, Wu XB. The influence of sports protective gear on upper limb protection in tennis . Wool Spinning Technology: 2018; 46(06): 71-77.
[4] Wang XB Yao M. Discussion on the pressure comfort and sports function of sports protective equipment. Journal of Northwest Institute of Textile Technology: 2001 (02); 56-59+65.
[5] Hogan MC, Kurdak SS, Arthur PG. Effect of gradual reduction in O2 delivery on intracellular homeostasis in contracting skeletal muscle. Journal of applied physiology (Bethesda,MD.: 1985); 1996: 80 (4): 1313-21.
[6] Jia XR, Xue WL, Wei JB. Research status of clothing pressure and sports comfort. International Textile Herald: 2020; 48 (09): 48-53.
[7] Zhang M, et al. Finite element simulation on clothing pressure and body deformation of the top part of men's socks using curve fitting equations [J]. International Journal of Clothing Science and Technology: 2015; 27 (2): 207-220.
[8] Rui D, et al. Numerical simulation of the relationship between pressure and material properties of the top part of socks[J]. Journal of The Textile Institute: 2013; 104 (8): 844-851.
[9] GB/T 3820-1997, Determination of thickness of textiles and textile products [S].
[10] GB/T3923.1-2013, Textiles—Tensile properties of fabrics—Part 1:Determination of maximum force and elongation at maximum force using the strip method [S].
[11] GB/T5453.1977, Textiles—Determination of the permeability of fabrics to air [S].
[12] GB/T 12704.1-2009, Textiles—Test method for water-vapour transmission of fabrics—Part 1:Desiccant method [S].
[13] Samuel B, Michel C, Timothée M, François C, Fiona E, François B. Compression Garments and Exercise: No Influence of Pressure Applied [J]. Journal of Sports Science and Medicine: 2015; 14 (1): 75-83.
[14] Zeng XY, Li Y, Ruan D, Koehl L (2007). Studies in Computational Intelligence Computational Textile Volume 55 || The Simulation of Elastic Human Body Deformation and Garment Pressure with Moving Mesh Method: 289-300.
[15] Shi JL, Xie ZQ. The Cause of Sports Injury in Colleges and Universities and Its Safety Protection Technology [J]. International Journal of New Developments in Education: 2020; 2 (4).
[16] Liu YM, Yang YF, Zhao WS. The structure design of knitted one-piece tights based on specific elongation[J]. Knitting Industry: 2019 (09); 57-62.
[17] GB/T 10000-1988, Human dimensions of Chinese adults [S].

Research on Automatic Generation of Kirgiz Felt Pattern based on Template Reconstruction

Man Zhou, Xiao-Yu Xin*, Hui Zhang, Juan Qian

College of textile and clothing, Xinjiang University, Urumqi, 830046, Xinjiang

*Corresponding author's email: 479790302@qq.com

Abstract

Kirgiz pattern is an important part of the traditional Chinese culture. The unique patterns and the neat structure with rich decorative themes contain very high research value. To seek the way of integration art-technology-industry to explore new methods in pattern design, this paper utilizes computer technology, combined with the construction principle of Kirgiz felt, to analyze and explore the construction law of Kirgiz felt. The personalized development of the Web interface based on HTML technology is proposed. SVG primitives are applied in the innovative design of patterns with JavaScript, allowing the rapid generation of patterns based on templates.

Keywords: Kirgiz; Pattern; Reconstruction; Vector; Structure; Meta-library

1. Introduction

Kirgiz folk textile pattern art is an important component of the ethnic minority culture and art in China. However, this cultural treasure is slowly dying out. It is our duty and obligation to protect national treasures, especially detailed in "The 13th Five-Year Plan for Promoting the Development of Ethnic Minority Areas and Ethnic Minority Areas with Small Populations". The plan emphasizes that it should be a priority to accelerate the development of ethnic minorities and ethnic minority areas, and ethnic cultural resources should be fully explored and brought into play. The rapid development of computer technology plays a supporting role in the digital protection and inheritance of these folk arts to a certain extent [1]. Many scholars have established a digital protection platform integrating learning, innovative design and application for traditional handicraft patterns of different nationalities. In addition, there are also digital platforms and museums, based on immersive experience of virtual tour developed with web technology. These information platforms featuring computer networks, on the one hand, enable the public to develop a basic understanding of the culture; on the other hand, they also have a positive influence on cultural protection and are conducive to cultural output. The purpose of this paper is to put forward a new idea for the innovative design of the Kirkzi pattern, and to provide some reference for the research of subsequent scholars. It also provides case support for the protection and inheritance of other ethnic patterns.

2. Research Status of Ethnic Patterns

Minority patterns are the essential elements of folk textiles, but also the core expression carrier of ethnic cultural characteristics and aesthetic taste. In recent years, a large of number of scholars have devoted themselves to the study of ethnic patterns, for example, Xue-Yu Wan explored Kirgiz textiles and techniques, including textile culture, style, color schemes, pattern composition, origin of patterns, development and changes of uses. Thus, Xue-Yu Wan's study provided a basis for further research by subsequent scholars [2-5]. Miao-Miao Yi summarized the patterns, structures, colors and techniques of Kirgiz textile patterns, and analyzed the cultural origin of Kirgiz textile patterns from the perspective of various cultures such as religious, totem, regional and nomadic [6, 7]. Deng Teng extracted elements of Shao yang blue calico flower patterns, and designed typical cases of different works by segmenting, merging, and reconstructing them; her work provided a theoretical basis for the transformation and development of traditional patterns [8]. Wang Hui proposed a pattern element extraction method based on contrast and the GrabCut algorithm, in which users can extract the apparel pattern elements with contour, color and structure, and store them quickly and effectively, making the design of patterns more intelligent [9]. Wen-Qing Li developed the Xinjiang Uyghur pattern recombination design system, studied key technologies of image local vectorization, pixel recombination and self-stitching texture, and finally realized the prototype system [10].

The current study of ethnic patterns can be divided into three categories. The first category is the study of historical and cultural origin of ethnic patterns, through religion, history, geography and other aspects of exploration. The second is the research on extraction of national designs and the respective methodologies, which is then used for the manual extraction feature and computer algorithms of the drawing software. The extraction often contains more than one unit in textile design patterns elements, and each element has its own meaning and is key to designing information extraction. When precision for extraction is high, the software is also slower and less efficient when processing large amounts of images. However, the accuracy of software extracted by computer algorithm is low, and the existing automatic extraction methods are limited to contour tracing and extraction, ignoring the internal structure of textile patterns. The third category is the study of innovative design and application of ethnic patterns. The same method of innovative design is drawing software and computer algorithms; the former is not as efficient as the latter, while the latter is not as flexible as the former. To sum up, all three types of studies included the structural study of the pattern. To sum up, none of the three types of studies left the structural study of the pattern. Xiao-Yu Xin et al. analyzed several existing structures of Kirgiz patched felt 'Hildak' in their research and pointed out that its layout could be divided into two parts: central main flower area and surrounding floral area [11, 12], which is shown in Fig.1. Through the analysis of these structures, it is straightforward to seek that the floral felt structure is mainly composed of two parts, which are the motif decoration in the middle and the auxiliary pattern decoration in the corners, as shown in Fig.2, confirming the above statement. The material of the main pattern is mostly the pattern with a complete unit, and the auxiliary decorative pattern is more jagged, with plant vines and undulating veins.

Fig.1 Felt

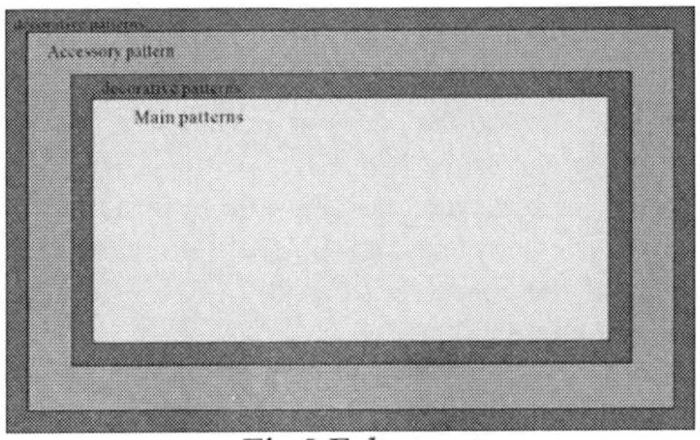

Fig.2 Felt structure

3. Research Methods

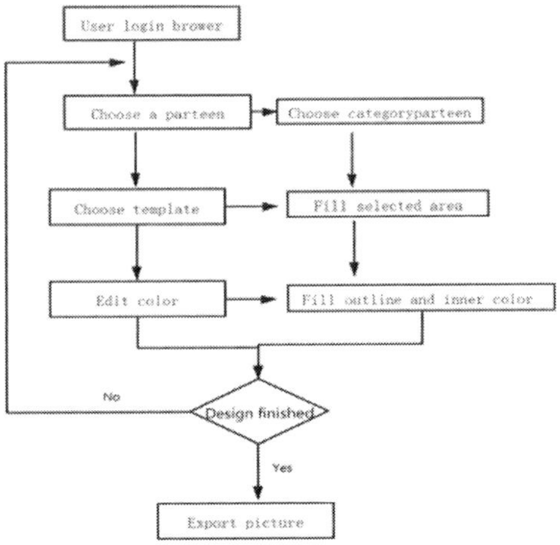

Fig.3 System operation structure

This paper utilizes HTM l5 information processing technology and JavaScript in the effective fusion of web page production and the establishment of the digital platform for pattern independent innovation design. We also use JavaScript to add the SVG pattern fill command, allowing one to move quickly to execute the filling and removal of the template; and realize the rapid generation method based on templates, meeting the demand of the user. Operating the system is simple; the user can directly open the web page in the browser, and realize the pattern of independent innovation design. The structure of the web interface system is shown in Fig.3

3.1 Production of the Web Page

HTML is a markup language and a necessary tool for web page creation. HTML5's adaptive web design allows designers to create a design once, and other porting changes can be adjusted automatically according to the new environment. The standard HTML5 is applied for specification, which is convenient for later testing, modification and maintenance [13]. It also updates automatically and efficiently. The drawing function is also included in HTML5, in which JavaScript can be used in addition to draw pictures and create animation. In HTML5, drawing can be done via Canvas, and SVG.HTML5 can be used across platforms. We chose to develop with HTML5 for template generation of pattern design due to the language's simplicity and flexibility. In the future, an app and a program can also be developed to provide convenience for the audience.

3.2 Selection of Primitive Format

Vector graphics have the feature of infinite zoom, which is related to the complexity of the graphics and is independent of the file size. Vector graphics editor SVG format is the most application friendly of all file formats. SVG was developed by the W3C organization with an open standard of vector graphics expression language. It is used to describe two-dimensional vector graphics, irrespective of the platform, scripts and the amount of data; the characteristics of external programs and scripts help improve the DOM and SAX interface to deal with each of these elements and operation [14, 15], and they can be produced in any resolution for high quality printing. In SVG, every graph drawn is treated as an object. If the properties of the SVG object change, the browser automatically reproduce the graphic. The current mainstream vector graphics editing design softwares such as Adobe Illustrator and CorelDraw directly edit and save the SVG files. Professional SVG editor software is also available, as shown in Fig.4. In this paper, the Kirgiz ethnic visual effect of the flower felt design in SVG format provides an effective theoretical basis and methods for the rapid generation of patterns.

Fig.4 SVG editor (taken from https://c.runoob.com/more/svgeditor/)

3.3 The use of the Scripting Language

JavaScript is a scripting language that is widely used in client-side web development to add dynamic functionality to HTML web pages, such as responding to various user actions. This article uses JavaScript, and its source code can be run directly by the browser. It facilitates the preview of the pattern template reconstruction design for timely adjustments. The cross-platform advantage provides convenience for follow-up research and development of a mobile app.

4. The Experimental Process

The experimental process is divided into two parts. First, establish a stable environment to prepare experimental materials, followed by the experimental part of pattern design.

4.1 Data Construction

4.1.1 Metadata Library Construction

Information of all patterns collected will be classified according to different patterns category. First, the pattern is imported from CorelDraw. Primitive segmentation and extraction are carried out by the use of dot-dash line tracing simulations of pattern. With Bessel curve primitives adjustments, the complete primitive is extracted and exported locally or onto the server in SVG format. The process is shown in Fig.5. In this experiment, 12 typical patterns were used to establish the SVG primitive database, including 5 corner patterns and 7 parent decorative patterns, which played an auxiliary decorative role.

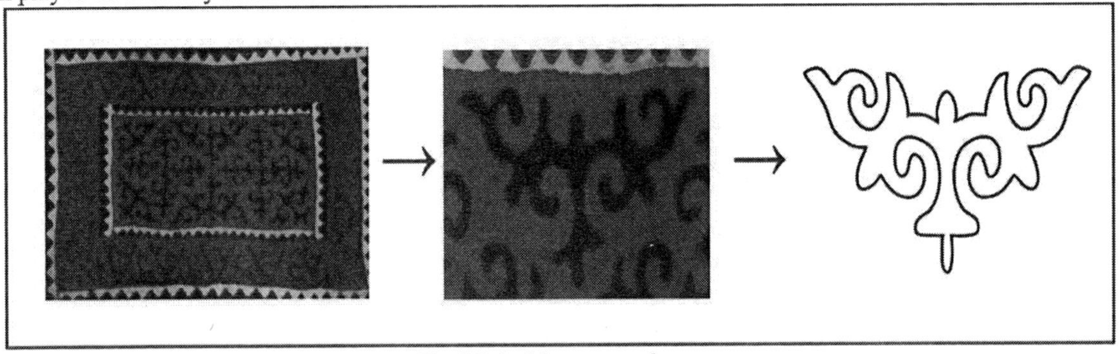

Fig.5 Primitive extraction

4.1.2. Structure Library Construction

During the experiment, for the structure selection, a typical structure inherent to the Kirgiz floral felt was selected. On the one hand, it retains the unique national flavor of Kirgiz. On the other hand, the user can play an innovative role in the design and adjust the pattern according to their own preferences. They can use CorelDraw to draw the structure and export it in SVG format. This is shown in Fig.6.

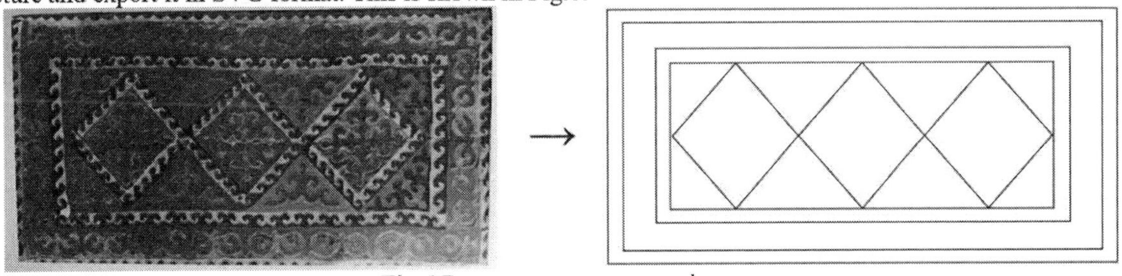

Fig.6 Pattern structure extraction

4.1.3 Color Library Construction

Color is one of the essential elements of pattern design, the colors of Kirgiz textiles are bold, demonstrating its strong national characteristics. On the webpage, users can design their own color library, making color matching more convenient and flexible, as shown in Fig.7.

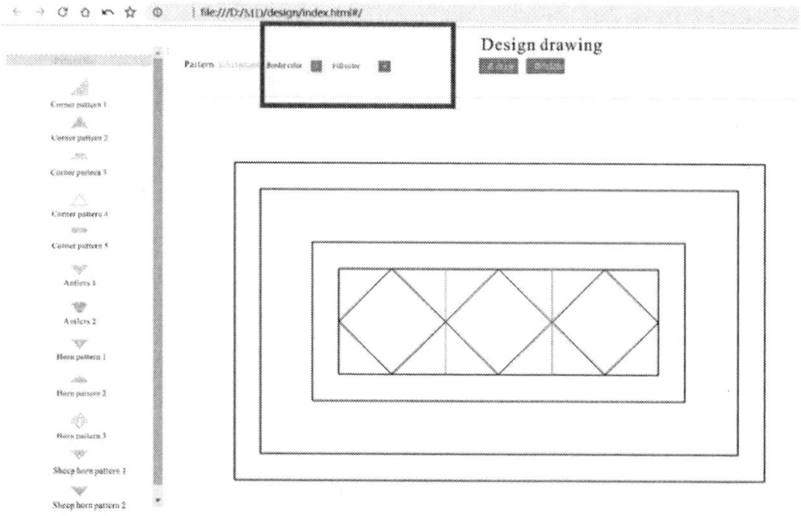

Fig.7 Color library construction

4.2 Pattern Design

4.2.1 Fill the Template

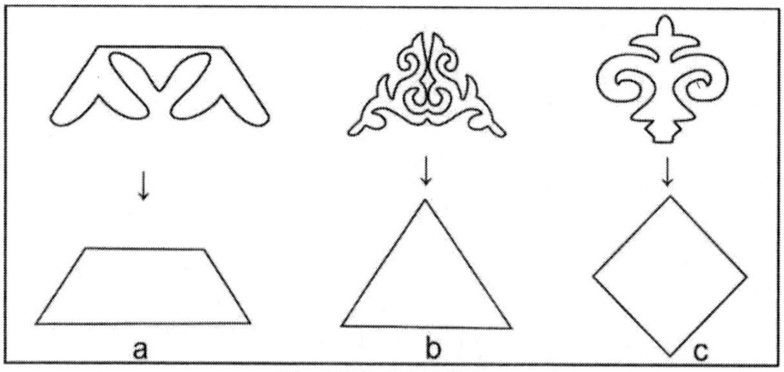

Fig.8 Primitive shape

The structure of each element is different, making the form of the template that is arranged accordingly also different. The author uses different structures of the graphic elements to define whether the graphic elements are regular or not. According to the shape of the Kirgiz pattern, the structure is geometric, making it easier for the computer to perform the calculations then fill the template. The demarcation elements of an equilateral triangle, a square and a rectangle are set as regular patterns, while the rest are irregular patterns, as shown in Fig.8. Fig.8a shows irregular patterns, while Fig.8b and c show regular patterns. The observation of the felt revealed that the irregular pattern are mostly the edges and corners of the auxiliary decorative patterns, while the regular patterns are mostly the motif decoration of the central patterns.

For regular patterns, conventional arrangement methods include displacement, mirror image and flip based on typical mathematical function model. For irregular patterns, the arrangement mode is combined with its regional location to design algorithm for recognition and filling. Adaptive filling for regular patterns is shown in Fig.9, 9a and b, and adaptive filling for irregular patterns in Fig.9c. The arrangement mode is an action description of the above distribution rules of various primitives and provides a basis for the automatic generation of SVG vector graphics, which automatically distributes specific primitives in a certain plane according to the rules. The users can select the pattern on the left side of the webpage by clicking the fill area, followed by draw. One can automatically fill the element and complete the pattern design efficiently.

Fig.9 Primitive fill

4.2.2 Color Editing

In the web interface, you can edit the pattern color after importing the color library,. The clear key enables users to select pixel filling independently in the pixel library according to their preferences. As shown in Fig.10, the filled pixel can be switched freely during color editing, and the contour and interior of the pattern can be colored respectively, which is efficient and convenient.

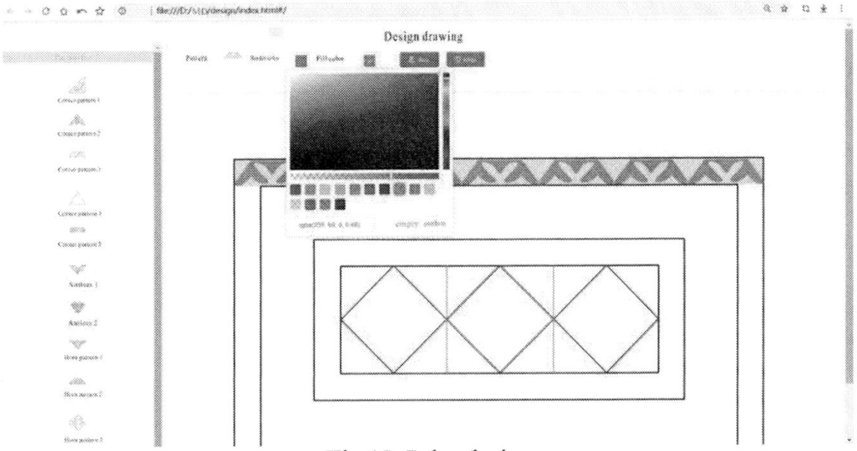

Fig.10 Color design

4.2.3 Pattern Export

The experimental renderings are shown in Fig.11. The user-designed patterns can be exported directly and saved either locally or onto the server. The patterns can be exported in many formats, such as PSD, PNG and SVG, which is convenient for users to create and apply again.

Fig.11 Design renderings preview

5. Conclusion

In this paper, an interactive pattern generation interface is developed by analyzing the structure of Kirgiz floral felt and to improve efficiency of the current rapid pattern generation method. Using the method of template matching and on the basis of human-machine collaborative design, pattern design is personalized. This is done to promote the inheritance and digital protection of Kirgiz textile pattern art, and also to provide reference cases for Kirgiz researchers. The method and source code described in this paper can also be used in the innovative design of other ethnic patterns, to broaden the design scope of ethnic patterns. The code should also slay a foundation for the development of mobile applications and small program on WeChat, providing a new idea for the sustainable development of ethnic textile patterns.

References

[1] Li PF. Research on the innovation of high-quality development path of cultural industry in border ethnic areas --Take Xinjiang as an example. Journal of Yunnan National University (Philosophical and Social Sciences Edition): 2020; 37 (01): 38-45.
[2] Wan XY. Kirkiz Felt Analysis, Journal of Xinjiang Academy of Art: 2010; 8 (03): 14-17.
[3] Wan XY. Analysis of the Culture of the Kirkzi Felt Room, Xinjiang University Journal (Philosophy, Humanities and Social Sciences): 2013; 4102: 90-95.
[4] Wan XY. Field-based Traditional Female Red Skills of the Kirkiz (J. Journal of Xinjiang University (Philosophy, Humanities and Social Sciences): 2014; 4205: 72-76.
[5] Wan XY. traditional crafts of the Kirkiz ethnic group during the period of social transformation. Journal of Guizhou Normal University (Social Sciences Edition): 2016; 04: 83-90.
[6] Yi MM. Xinjiang Kirkiz ethnic decorative pattern of the art form of Art and Design (Theory): 2016; 2 (8): 104-106.
[7] Yi MM. Cultural Analysis of China's Kirkzi Ethnic Decorative Pattern Art Dyeing technology: 2018; 40 (6): 48-51.
[8] Deng T. Study on the extraction and reconstruction of the artistic elements of the flower pattern of The Blue Printed Cloth of Fuyang. Hunan Normal University: 2017.
[9] Wang H. Costume Pattern Element Extraction based on Contrast and GrabCut Algorithms. Zhejiang University of Technology: 2020.
[10] Li WQ. Key Technical Research on Uighur Pattern Reorganization Design. Hangzhou University of Electronic Science and Technology: 2011.
[11] Xin XY, Qian J, Xiao AM. Study on the pattern and technology of Kirgiz flower felt "Hildak". Minyi: 2019 (06); 147-152.
[12] Qian J, Xin XY, Xiao AM, Repukaiti, YLTZ Chen YS. Analysis and application of Kirgiz traditional textile pattern primitives [J]. Silk: 2021 (03); 127-133.
[13] Zhang YW. HTML5-based online dressing room APP design and implementation of software engineering: 2019; 22 (07): 31-33.
[14] Song SD, Xiong ZZ, Li WG, Tang XF. Design and implementation of the SVG-based vector graphics editor. Computer Engineering and Science: 2003; 02: 91 -94.
[15] Zhao L, Yang Q. Vector graphics auto-generation method based on meta-libraries and distribution rules. Modern Machinery: 2015 (03); 90-94.

Functional Bra Cup Inner Surface Parametric Modeling for Personalized Customization

Jia-Jian Ma, Bo-An Ying*, Long Wu, Jing Qi, Xin Zhang

Apparel & Art Design College, Xi'an Polytechnic University, Xi'an, Shaanxi, 710048, P.R. China

*Corresponding author's email: yingba2006@163.com

Abstract

This paper is based on the idea of personalized customization and proposes a forward hierarchical design method to solve the problems in the production process of the bra mold cup and meet the demands of consumers. This paper analyzes the feature points of the human body and the mold cup to find the relationship between the human body characteristics and the mold cup control parameters and establishes a quantitative formula for the constraint relationship between the basic mold cup and the functional mold cup based on the human body, and passes UG OPEN GRIP programs the generation process of the inner surface of the mold cup in UG, and realizes the automatic generation of the functional mold cup by controlling the design parameters in the UG environment. Furthermore, this paper discusses the application of the parameterized design model of the inner surface of the personalized custom middle mold cup, including the application for individuals with different human body data and the application of selecting different functional parameter values for the same human body data. This method provides a method reference for the digital design method of mold cups in the future.

Keywords: Bra mold Cup; Personalized Customization; Parameterization; Hierarchical Design

1. Introduction

With the advancement of society and science and technology, people are pursuing a higher quality of life, and the demand for "individual customization" is increasing day by day. The underwear industry is also the same. Due to the difference in individual details, consumers have more demand for underwear fit. As strong. Therefore, applying the concept of "individual customization" in the underwear industry is essential.

In the design of bra mold cups, Pan Sichen [1] uses three-dimensional anthropometric technology to extract the detailed dimensions of the human body. By analyzing and extracting the characteristic points, curves, and surfaces required for the mold cup design, he proposes establishing a complete set in the UG modeling environment. Model cup models at all levels based on individual consumer data by analyzing the characteristics of women's breasts and bra cups, Cui Xue [2] divides them into three relationship levels in the modeling design process of the formwork cups, obtained the relationship model among the geometric attributes of the bra cups, and developed the user-defined mold cup modeling design module by using UG secondary development tool. By analyzing the morphological characteristics of the chest, Lv Feifei [3] carries out parameterized representation of the feature points and curves related to the chest, establishes the parameterized constraint relationship between the bra mold cup model and the digital human body, and proposes to design the design plate of the bra mold cup through UG Open GRIP secondary development technology. Li Mingju [4] combines the Bezier curve and parameterization concept to analyze the difference of human body shape in different regions, establishes a mathematical model suitable for underwear design on the Visual C++6.0 and OpenGL platform, and applies it to the tight-fitting prototype and the basic version of the bra cup – interactive design method. Based on the female breast shape, Han Jing [5] studies the design method of the digital mold cup, establishes the digital bra mold cup model, and summarizes the modeling technology of the digital mold cup. Xing Chao [6] uses 3D scanning technology to transform the bra cup into a digital model, extract the features of the cup, classify and analyze the inner and outer surface surfaces according to the similarity classification of the cup, extract the basic model of the outer surface, realize the parametric design of the curve of the inner surface, and build the model library of the cup. Wang

Shuanghu [7] and Ding Yaru analyze the breast shape of middle-aged women, use several main parameters describing breast shape, and adopt the method of forwarding design to construct the characteristic curve required by the mold cup prototype in the UG environment and establish the mold cup model. Peng Hui [8] studied the four design factors of the unrimmed molded cup bra, including the thickness of the cup, the angle of the center of the heart, the height of the side ratio, and the material of the bottom, and explore the relationship and influence between the design factors of the unrimmed molded cup bra and the bra shaping effect, the size of the pressure and the fit of the wearing. Xu Yaoyao [9] obtains the point cloud data of the mold cup through a 3D scanner, and obtains the function relationship between the arc length of the characteristic curve and the depth of the mold cup through analysis. The morphological structure of the curved surface could be controlled by changing the geometric parameters on the morphological characteristic curve. Heh Soon Jung and Mi Hyang Na [10] aim to develop no-wire mold cups to commercialize bras that make the breast appear beautiful by keeping the shape without wire supporting them. Okkyung Lee, Kyunghi Hong and Yejin Lee [11] analyze the breast's shape change according to the replica combination method of the lower cup and suggest a pattern construction for the end use purpose. Lee C W [12] Independent variables of breast size, underband length, cup material modulus, shoulder strap, and underband material modulus, underwire, and gait speed are regarded as important variables in predicting vertical breast displacement. The soft manikin shows a satisfactory result validated by a subject wear trial, thus providing a reliable and objective approach for future bra studies. Tian Y, Ball R. [13] develops a parametric personalized bra design algorithm that can generate personalized bra by using self-measurements of women users. Hyun-Young Lee [14] proposes the suitable sensation for cup volume and the natural wearing silhouette could be accomplished by removing the surplus ease amounts from the upper edge of the mold cup to reduce the gaps between brassieres and the breasts, which could be accomplished through an approach reducing the volume near the upper edge of mold brassiere cup and make the volume concentrated at the lower cup. Sun Y, Yick K, Cai Y [15] propose a subject-specific model based on numerical simulation to investigate the influence of different bra cup materials on the shape of the breasts and the amount of pressure on the breasts. Through the study and research of the literature related to the design of digital bra cup in recent years, it is found that the research on the parameterized relationship between women's breasts and the cup is gradually enriched. In contrast, the content of the physical realization process of the bra cup has not been explained in detail through examples.

Therefore, in order to meet the consumer type of mold cup bra product customization urgent demand trend of the development of industry, this paper based on the personalization of digital functional underwear cup mold design problem, with its design constraints model as the research object, put forward "customization oriented functional bra cup inner surface parameterized modeling."

2. Analysis of the Relationship between Human Body Characteristics and the Control Parameters of the Mold Cup

2.1 Relationship between Human Body Feature Points and Mold Cup Feature Points

In anthropometrics, human body feature points are an obvious boundary point used to mark the position. In general, human body feature points are located at the prominent bone or joint joints of the human body. Based on the digital design requirements of mold cup modeling, this paper defines the points that can reflect the curved shape of the female chest as feature points, as shown in Fig.1.

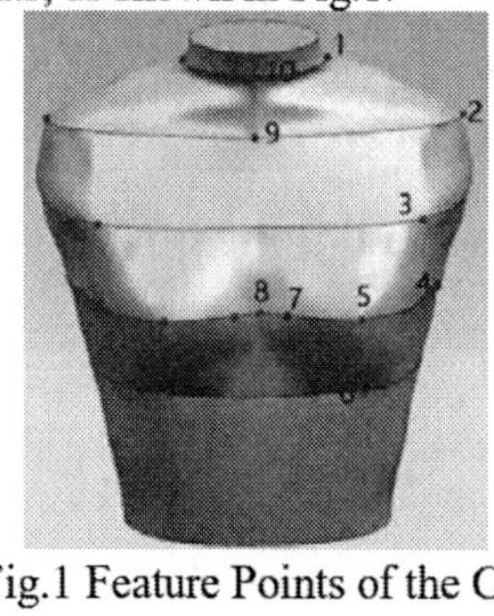

1.side neck point
2.Shoulder point
3.Upper edge of breast
4.Lateral margin of the breast
5.breast point
6.Lower edge of breast
7.Inner edge of breast
8.Front center point
9.Anterior cervical fossa reference point
10.Anterior neck

Fig.1 Feature Points of the Curved Shape of Female Human Chest

In the design of the bra mold cup, the parts from the upper bust to the lower bust are mainly referred to, that is, Upper edge of breast, lateral margin of the breast, breast point, lower edge of breast, inner edge of breast, and the front center point.

2.2 Relationship between Human Body Characteristic Line and Mold Cup Characteristic Line

The characteristic line of the curved surface of the female body's chest is a free curve formed by connecting two or more characteristic points on the surface of the human body. These characteristic curves, which identify the characteristic size and position of the human body, are the key geometric elements in the bra mold cup shape, and they provide an important reference standard for the mold cup shape design. In the digital design of the bra mold cup shape, it is essential to make the designed mold cup meet the individual needs of consumers, and to grasp the characteristic curve of the consumer's breast shape. In this article, the bust line, the princess line, the front centerline, the shoulder line, the BP spacing line, the lower bust line, and the upper bust line are defined as characteristic curves that characterize the curved surface of the female human body, as shown in Fig.2.

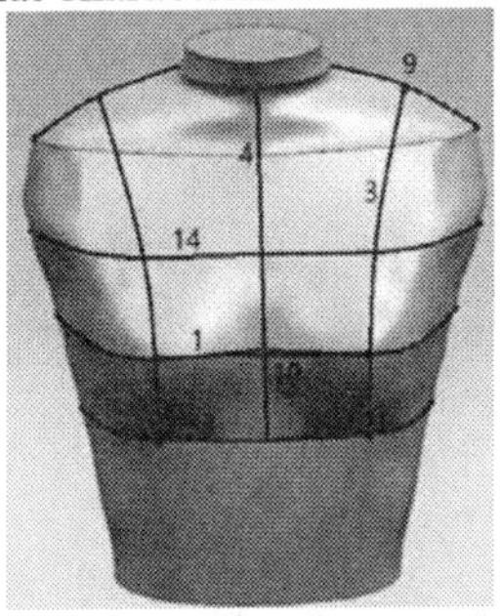

1.breast line
3.Princess line
4.Front center line
9.Shoulder line
10.Breast spacing line
13.Under breast line
14.Upper breast line

Fig.2 The characteristic curve of the curved shape of the female breast

2.3 Design and Analysis of Functional Control Parameters of the Inner Surface of the Mold Cup

According to the theory of human physiology, the left and right breasts of the human body are not entirely symmetrical according to the front midline. Therefore, the left and right breast's feature points directly extracted from the human body are not entirely symmetrical. In the process of digital design of bra mold cups, the human body is regarded as asymmetrical body, and a complete pair of mold cups are created by first designing a molded cup and then symmetrically generated according to the front centerline of the human body. In order to achieve such a design process, it is necessary to scan asymmetry. The feature points of the left and right breasts are symmetrically corrected. In order to correct the asymmetry of the left and right halves of the human body, it is necessary to offset the coordinate of the feature points from the feature points directly extracted from the human body to construct the human feature points which are symmetrical in the front centerline of the human body. First measure the left and right side of each feature point distance from the centerline of the body before the chest. Then, according to the difference of the two distance on the left side of the chest feature point to point offset. In order to achieve the feature points around each symmetrical about the center line of the former, according to the feature points generated model known as the base type. In this research, the human body model after symmetry correction is used as the basic model of the functional mold cup; that is, the feature points of the functional mold cup are coordinate offset according to the feature points of the basic mold cup through the symmetry correction.

Based on the basic model cups generated according to the human body feature points, the functional model cups adjust the position of the functional feature points according to the individual characteristics of the consumer's breast shape so that the functional model cups drive the corresponding changes in the characteristic curves and curved surfaces, to realize the partial lifting and gathering function of the bra model cups. The feature points of functional cups are offset by the coordinates of the feature points based on the feature points of the basic cups. The offset of each feature point is realized by influencing the parameters of the modeling design of the cups. The parameters are allocated to each functional feature point purposely, and the quantization formula expresses the quantization relationship. Therefore, the quantification formula of the offset between the coordinates of the feature points of the functional cups and those of the basic cups is the most direct embodiment of the designer's design intention.

3. Construction of the Parametric design Model for the Inner Surface of the Mold Cup

3.1 Displacement Analysis of Feature Points on the Inner Surface of the Mold Cup based on Human Body Feature Points

As for the function of functional bra cups, the function of functional bra cups on the breast is mainly reflected in the two aspects of upper lifting and gathering and side collecting. Among them, the lifting function is mainly manifested in the upward movement of the breast in the vertical direction, and the improvement of the chest to body ratio by controlling the vertical position of the BP point, to achieve a relatively perfect proportion visually. The gathering function is mainly shown in the horizontal direction of the breast to the centerline of the human body, reduce the distance between BP points so that the breast presents a fuller state.

3.2 Exposition of Feature Points on the Inner Surface of the Mold Cup based on Human Body Feature Points

Based on the characteristic points of the basic mold cup, the coordinate points of the functional mold cup can be expressed as the following formula (2-1):

$$Cup2\_i\_j=Cup1\_i\_j+\Delta P \qquad 2\text{-}1)$$

The formula indicates that the offset of production quantity generates the feature points of functional cups based on the feature points of basic cups. Cup2_i_j represents the feature points of the functional mold cup, and Cup1_i_j represents the feature points of the basic mold cup. ΔP denotes the coordinate offset of characteristic points of the basic mold cup.

The offset formula assigned to each point is realized by the parameter value and the empirical characteristic coefficient. Taking the gathering function as an example, its quantitative relationship can be expressed by the formula as follows (2-2):

$$\Delta Pij=Ki\_jji \qquad 2\text{-}2)$$

Since the main influencing parameter for the realization of the gathering function is the distance between the breasts, the characteristic points in the three-dimensional coordinates are mainly offset in the horizontal direction. ΔPij is the offset in the x-axis direction of each point of the functional mold cup, Ki_j is the offset coefficient, and D is the design parameter value showing the degree of gathering. This example represents the difference between the target inter-milk spacing value after gathering and the size of the human inter-milk spacing. The offset coefficient is the empirical value assigned by the designer to the offset of each point according to the total parameter offset. Therefore, the Ki_j value is set differently for different mold cup designers.

Based on the above formula, the X-axis offset of each point of the functional mold cup can be expressed as the following matrix expression (2-3):

$$Cup2\_i\_j=\begin{bmatrix} Cup1\_i\_j & \cdots & Cup1\_i\_j \\ \vdots & \cdots & \vdots \\ Cup1\_i\_j & \cdots & Cup1\_i\_j \end{bmatrix}+\begin{bmatrix} Ki\_j & \cdots & Ki\_j \\ \vdots & \cdots & \vdots \\ Ki\_j & \cdots & Ki\_j \end{bmatrix}\times D \qquad 2\text{-}3)$$

3.3 Establish a Parameterized Design Model for the Inner Surface of the Mold Cup

In order to realize the personalized customization of "one cup for one person," the realization of the bra mold cup surface in this paper adopts the programming and operation of the program. By quantifying the offset formula of the basic and functional mold cups at their respective base points, the offset of the characteristic points of each level of the mold cups and the process of the surface generation of the mold cups at all levels are reflected through programming, and the automatic generation of the surface model of the mold cups in the UG environment is realized.

By designing interactive Windows, generating points, generating mesh curves, and generating mesh surfaces in the program, designers can achieve the effect that functional design parameters can generate functional surfaces of different functional degrees when inputting different values. Fig.3 shows the design of the interactive window in UG. Take the gathering function as an example, the designer edits the offset formula of each point into the program and inputs the specific parameter total offset in the interactive window to control the feature points of the surface to drive the curve and then generate the inner surface of the mold cup.

Fig.3 Interactive window (amount of aggregation)

In the program design of the inner surface of the functional cup, it is necessary to assign the offset value of the feature points of the basic mold cup in the direction of x-axis to generate new feature points of the functional cup. To achieve the gathering function as an example, through the interactive window for the input of D value and then control the basic mold cup inner surface of the offset of each point, Cup2_i_1, I ={1,2,.. 13} Functional bra cup feature points By the method of quantitative formula assignment, the point coordinate change of the offset in the direction of X-axis and the definition of the point are changed. Then mThennd the definition of the point are cmed. In this paper, 13 hennd the definition of the point are cmed. In this paper, quantitative formula aIn this example, the first four feature point positions determined are the BP point, the inner point of the mold cup, the upper point of the mold cup, and the outer point of the mold cup. Therefore, to ensure the operability in the UG software and the accuracy of the generated surface, this example selects 13he first four feature point podensity. In the UG interface, as shown in the figure below (Fig.4):

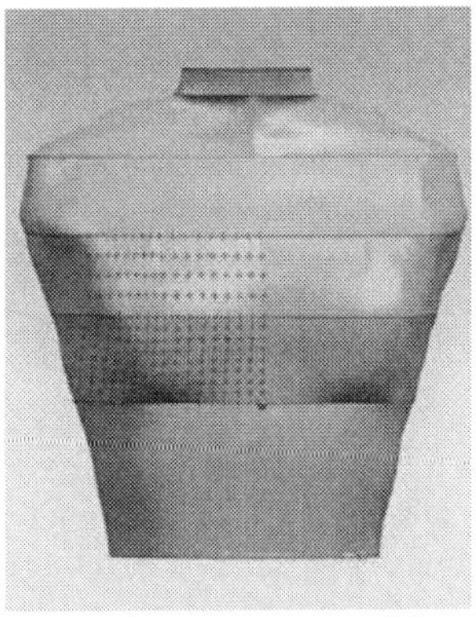
Fig.4 Establishment of feature points of functional mold cups

In the program, 17 curves in the vertical direction and 13 curves in the horizontal direction in the 13×17 curve grid are defined and generated, and then the grid surface is generated through the grid curve. The grid surface is the inner surface of the mold cup, with the gathering function controlled by the input value of the interactive window. In the UG interface, as shown in the following figure (Fig.5, Fig.6):

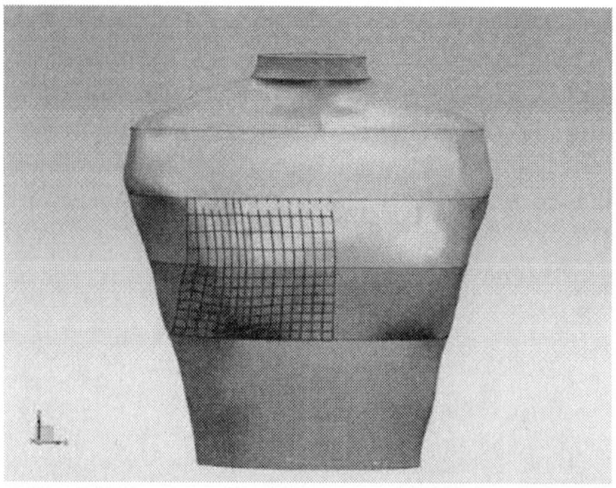
Fig.5 Establishment of grid curve of functional mold cup

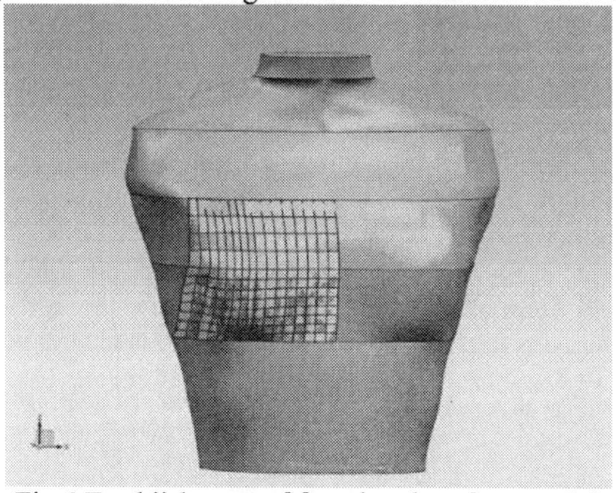
Fig.6 Establishment of functional surface model

4. Discussion and Analysis

The parameterized modeling method of the inner surface of the functional bra cup is customized discussed in this paper can be applied to the following two situations: the application for individuals with different human body data, and the application of selecting different functional parameter values for the same human body data.

When this method is applied to modeling functional bras with different human body data, the selection rule of basic feature points is still grid feature points, but the 3D coordinate values of corresponding basic feature points will be different due to different human body shapes. If the requirements for the functional degree of different human body data are similar, then the offset formulas based on the basic feature points are still similar, and the model construction methods are the same.

The method is applied to the same human body data by selecting different functional parameters values to build functional bra mold cup according to the human body to establish the basis of the appropriate type grid feature points. As the demand for functional level is different, different parameter values can be inputted in the interactive window, and in the generated module cup programming to modify the migration coefficient. The inner surface of a functional bra cup shows different functional effects.

As shown in Fig.7, the different surfaces generated by the gathering parameter values are 1, 5, and 10. The larger the value of the gathering parameter, the better the gathering effect, that is, the smaller the distance between the BP points after the bra is adjusted. At the same time, the curvature of the curve corresponding to the inner surface will also change accordingly. (see Fig.8).

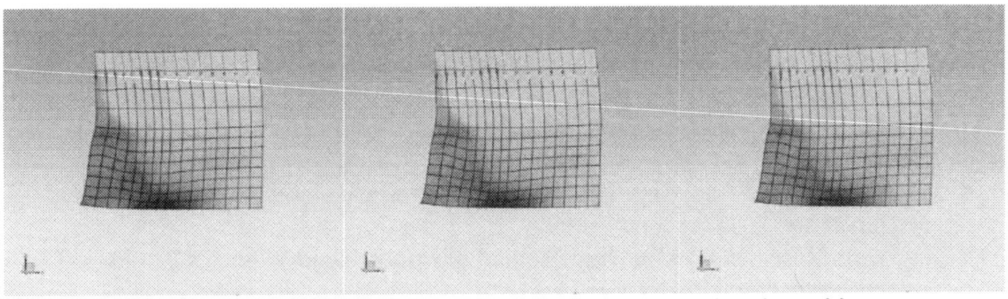

Fig.7 Comparison of different curved surfaces of functional mould cups

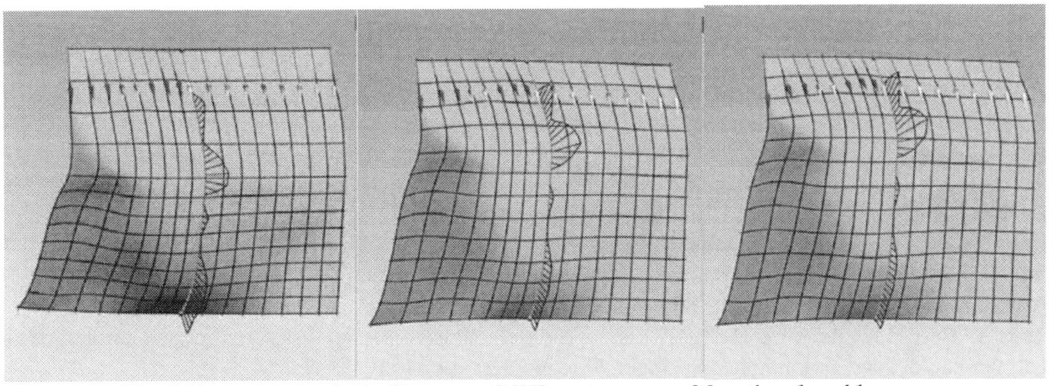

Fig. 8 Curvature analysis diagram of different curves of functional mold cups

5. Conclusion

In this paper, based on individualized customization, the design method of the functional bra mold cup is studied by looking up domestic and foreign documents. The main conclusions of this study are as follows: (1) Analyze the relationship between the human body and the characteristic points of the mold cup. (2) According to the relationship between the human body and the feature points of the mold cup, the definition of the basic and functional mold cup relationship models in the modeling design is introduced. (3) Combining the functional bra mold cup with the basic bra mold cup from the human body to clarify the generation relationship between the mold cups at all levels. (4) The parametric design method of program design is proposed, and the design feature points of the functional bra mold cup are realized through the feature point offset of the basic bra. the offset of the coordinates drives the basic mold cup based on the feature points of the human body. The curve change then generates a curved surface, which is reflected in the program design through a parameterized formula. Different design parameter values can be entered in the interactive window to generate mold cup surfaces with different functional levels in the program. This method provides a more efficient and convenient method for constructing a personalized functional bra cup design model.

6. Acknowledgement

This research is supported by Shaanxi Engineering Research Center of Apparel Intelligent Design And Manufacture Platform.

References

[1] Pan SC. Design and Research of Mold Cup Bra Based on 3D Human Body Measurement and Surface Modeling Technology [D]. Xi 'an Polytechnic University: 2017.
[2] Cui X. Study on Relational Model of Bra Mold Cup Modeling Design [D]. Master's Thesis of Xi 'an Polytechnic University: 2013.
[3] Lv FF. Research on the Design of Bra Mold Cup Based on Human Body [D]. Master Dissertation of Xi 'an

Polytechnic University: 2012.
[4] LI MJ. Research on the Structure Composition and Digital Design of Underwear Based on the Analysis of Female Body Type [D]. Donghua University: 2001.
[5] Han J. Research on Digital Mold Cup Design Method Based on Breast Morphology [D]. Master Dissertation of Xi 'an Polytechnic University: 2005.
[6] Xing C. Research on Feature Analysis of Bra Mold Cup Surface and Construction of Model Library [D]. Xi 'an Polytechnic University: 2015.
[7] Wang SH, Ding Yaru. Research on Modeling Method of Middle-aged Women's Bra Mold Cup [J]. Progress in Textile Science and Technology: 2016 (08); 24-30.
[8] Peng H. Study on the Influence Factors of the Shaping Effect and Pressure Comfort of the Unrimmed Mold Cup Bra [D]. Suzhou University: 2017.
[9] Xu YY, Zhu L, Du L Zou FY. Extraction and Analysis of Morphological Characteristic Curve of Bra Mold Cup Surface [J]. Journal of Textile Research: 2016; 37 (11): 103-108.
[10] Heh SJ; Mi HN. Development of a Water-Droplet-Shaped Bra Mold Cup Design [J]. Indian Journal of Science and TechnologyVolume 9, Issue 35: 2016.
[11] Lee O, Hong K, Lee Y. (2021) "Development of the Fundamental Methodology of Lower Cup Pattern depending on 3D Shape Analysis of Breast," Journal of the Korean Society of Clothing and Textiles. The Korean Society of Clothing and Textiles: 45 (1); pp. 94-105.
[12] Lee CW. Ergonomic design of supportive moulded bras [J]. Hong Kong Polytechnic University: 2020.
[13] Sun Y, Yick K, Cai Y, et al. Finite Element Analysis on Contact Pressure and 3D Breast Deformation for Application in Women's Bras [J]. Fibers and Polymers: 2021; 1-12.
[14] Tian Y, Ball R. Parametric Design Method for Personalized Bras[C]//International Conference on Applied Human Factors and Ergonomics. Springer, Cham: 2020; 145-151.
[15] Lee HY. Shape design method of mold brassiere cup for small-breasted women in their twenties [J]. Fashion & Textile Research Journal: 2015; 17 (6): 988-995.

Acknowledgement to Reviewers - TBIS 2021

A	Ling	An
	Majid	Ali
	Muhammad	Asif
	Seung Kook	An
	Sheraz	Ahmad
	Simon	Annaheim
B	Soamshine	Boonyananta
C	Aizheng	Chen
	Yan	Chen
	Youfang	Chen
	Ying	Chen
D	Bo	Deng
	Emiel	DenHartog
	Xiaoqun	Dai
	Yongmei	Deng
	Zhaoqun	Du
E	Ibrahim	Elhawary
	Mahmoud	Elshakankery
G	Jing	Guo
	Yuanyuan	Gu
H	Aris	Huang
	Haiqiao	Huang
	Jianhan	Hong
	Yan	Hong
	Yosuke	Horiba
I	Shigeru	Inui
J	Xiaowen	Jiang
K	Andrey	Koptyug
	Fandong	Kong
	Hiroyuki	Kanai
	KyounghHou	Kim
	Nawar	Kadi
L	Chenxi	Li
	Cui	Lina
	Fengzhi	Li
	Gang	Li
	Jiashen	Li
	Raechel	Laing
	Long	Lin
	Xuqing	Liu
	Yingchun	Liang
	Zhe	Li
	Zhi	Li
	ZhongHua	Liu
M	Jiri	Militky
	Haiwen	Mao
	Marjo	Määttänen
N	Maryam	Naebe
	Siti Hana	Nasir
P	Rajiv	Padhye
Q	Jing	Qi
	Wenbao	Qin
R	Abher	Rasheed
	Shaghaeygh	Rezaei Arangdad
	Tan	Rong
	Eugenija	Strazdiene
	Guowen	Song
	Sachiko	Sukigara
	Tetsuya	Sato
	Xiaping	Shu
S	Xiaolin	Shen
T	David James	Tyler
	Masayuki	Takatera
	Miao	Tian
	Zhihong	Tian
W	Chunxia	Wang
	Long	Wu
	Qufu	Wei
	Xin	Wang
	Yuanfeng	Wang
	Zhaohui	Wang
	Zhujun	Wang
	Zhenzhen	Wei
X	Aimin	Xiao
	Achang	Xu
	Fujun	Xu
	Kehan	Xu
	Lan	Xu
	Maobin	Xie
	Tieling	Xing
	Xiaoman	Xiong
	Xiaoyu	Xin
Y	Gong	Yan
	Guoxiang	Yuan
	Jie	Yang
	Kai	Yang
	Nannan	Yang
	Tao	Yang
	Xiaoli	Yu
	Zhicai	Yu
Z	Chang-Lan	Zhou
	Chi	Zhang
	Chunhong	Zhu
	Jun	Zhang
	Ru	Zhang
	Shangyong	Zhang
	Xianyi	Zeng
	Yongkai	Zhou
	Yuxiao	Zhao
	Yuyang	Zhou
	Zhaozhu	Zheng
	Zheng	Zhao

We sincerely thank every reviewers for providing valuable comments to the papers submitted to TBIS 2021.
We are looking forward to your continuing support in TBIS 2022 (Czech Republic)

Textile Bioengineering and Informatics Society (TBIS)
Unit 503, 5/F. Silvercord
Tower 2, 30 Canton Roa
Tsim Sha Tsui
Hong Kong

ISBN 978-1-7138-3559-2

9 781713 835592

CURRAN ASSOCIATES INC.
proceedings.com